KU-166-226

HANDBOOK OF OFFSHORE ENGINEERING

Volume I

Elsevier Internet Homepage – http://www.elsevier.com
Consult the Elsevier homepage for full catalogue information on all books, major reference works, journals, electronic products and services.

Elsevier Titles of Related Interest in this series

Ocean Engineering Series
The Elsevier *Ocean Engineering Book Series* edited by Rameswar Bhattacharyya and Michael McCormick (both at the U.S. Naval Academy) addresses the need for a comprehensive and applied source of literature relevant to both researchers and practitioners alike. For a complete listing of books in this series visit:
http://www.elsevier.com/locate/OEBook

L.K. KOBYLINSKI, S. KASTNER, V.L. BELENKY, N.B. SEVASTIANOV
Stability and Safety of Ships (2 Volume Set)
2003, ISBN: 0-08-044355-9

M. OCHI
Hurricane-Generated Seas
2003, ISBN: 0-08-044312-5

A. PILLAY, J. WANG
Technology & Safety of Marine Systems
2003, ISBN: 0-08-044148-3

J. BROOKE, N. BOSE
Wave Energy Conversion
2003, ISBN: 0-08-044212-9

Related Journals:
Elsevier publishes a wide-ranging portfolio of high quality research journals, encompassing the materials science field. A sample journal issue is available online by visiting the Elsevier web site (details at the top of this page). Leading titles include:

Ocean Engineering
Coastal Engineering
Applied Ocean Research
Marine Structures

All journals are available online via ScienceDirect: www.sciencedirect.com

To Contact the Publisher
Elsevier welcomes enquiries concerning publishing proposals: books, journal special issues, conference proceedings, etc. All formats and media can be considered. Should you have a publishing proposal you wish to discuss, please contact, without obligation, the publisher responsible for Elsevier's Composites and Ceramics programme:

Nick Pinfield
Publisher
Elsevier Ltd
The Boulevard, Langford Lane — Phone: +44 1865 84 3352
Kidlington, Oxford — Fax: +44 1865 84 3700
OX5 1GB, UK — E-mail: nick.pinfield@elsevier.com

General enquiries, including placing orders, should be directed to Elsevier's Regional Sales Offices – please access the Elsevier homepage for full contact details (homepage details at the top of this page).

HANDBOOK OF OFFSHORE ENGINEERING

SUBRATA K. CHAKRABARTI
Offshore Structure Analysis, Inc.
Plainfield, Illinois, USA

Volume I

2005

Amsterdam – Boston – Heidelberg – London – New York – Oxford
Paris – San Diego – San Francisco – Singapore – Sydney – Tokyo

Elsevier
Linacre House, Jordan Hill, Oxford, OX2 8DP, UK
Radarweg 29, PO Box 211, 1000 AE Amsterdam, The Netherlands

First edition 2005
Reprinted 2005, 2006 (twice), 2008

British Library Cataloguing in Publication Data
A catalogue record for this book is available from the British Library

Library of Congress Cataloging-in-Publication Data
A catalog record for this book is available from the Library of Congress

ISBN: 978-0-08-044568-7 (vol 1)
ISBN: 978-0-08-044569-4 (vol 2)
ISBN: 978-0-08-044381-2 (set comprising of vols 1 & 2)

For information on all Elsevier publications
visit our website at books.elsevier.com

Printed and bound in *Great Britain*

08 09 10 12 11 10 9 8 7 6 5

PREFACE

Due to the rapid growth of the offshore field, particularly in the exploration and development of offshore oil and gas fields in deep waters of the oceans, the science and engineering in this area is seeing a phenomenal advancement. This advanced knowledge is not readily available for use by the practitioners in the field in a single reference.

Tremendous strides have been made in the last decades in the advancement of offshore exploration and production of minerals. This has given rise to developments of new concepts and structures and material for application in the deep oceans. This has generated an obvious need of a reference book providing the state-of-the art in offshore engineering.

This handbook is an attempt to fill this gap. It covers the important aspects of offshore structure design, installation and operation. The book covers the basic background material and its application in offshore engineering. Particular emphasis is placed in the application of the theory to practical problems. It includes the practical aspects of the offshore structures with handy design guides, simple description of the various components of the offshore engineering and their functions.

One of the unique strengths of the book is the impressive and encompassing presentation of current functional and operational offshore development for all those involved with offshore structures. It is tailored as a reference book for the practicing engineers, and should serve as a handy reference book for the design engineers and consultant involved with offshore engineering and the design of offshore structures. This book emphasizes the practical aspects rather than the theoretical treatments needed in the research in the field of offshore engineering. In particular, it describes the dos and don'ts of all aspects of offshore structures. Much hands-on experience has been incorporated in the write up and contents of the book. Simple formulas and guidelines are provided throughout the book. Detailed design calculations, discussion of software development, and the background mathematics has been purposely left out. The book is not intended to provide detailed design methods, which should be used in conjunction with the knowledge and guidelines included in the book. This does not mean that they are not necessary for the design of offshore structures. Typically, the advanced formulations are handled by specialized software. The primary purpose of the book is to provide the important practical aspects of offshore engineering without going into the nitty gritty of the actual detailed design. Long derivations or mathematical treatments are avoided. Where necessary, formulas are stated in simple terms for easy calculations. Illustrations are provided in these cases. Information is provided in handy reference tables and design charts. Examples are provided to show how the theory outlined in the book is applied in the design of structures. Many examples are borrowed from the deep-water offshore structures of interest today including their components, and material that completes the system.

Contents of the handbook include the following chapters:

Historical Development of Offshore Structures
Novel and Marginal Field Offshore Structures
Ocean Environment
Loads and Responses
Probabilistic Design of Offshore Structure
Fixed Offshore Platform Design
Floating Offshore Platform Design
Mooring Systems
Drilling and Production Risers
Topside Facilities Layout Development
Design and Construction of Offshore Pipelines
Design for Reliability: Human and Organisational Factors
Physical Modelling of Offshore Structures
Offshore Installation
Materials for Offshore Applications
Geophysical and Geotechnical Design

The book is a collective effort of many technical specialists. Each chapter is written by one or more invited world-renowned experts on the basis of their long-time practical experience in the offshore field. The sixteen chapters, contributed by internationally recognized offshore experts provide invaluable insights on the recent advances and present state-of-knowledge on offshore developments. Attempts were made to choose the people, who have been in the trenches, to write these chapters. They know what it takes to get a structure from the drawing board to the site doing its job for which it is designed. They work everyday on these structures with the design engineers, operations engineers and construction people and make sure that the job is done right.

Chapter 1 introduces the historical development of offshore structures in the exploration and production of petroleum reservoirs below the seafloor. It covers both the earlier offshore structures that have been installed in shallow and intermediate water depths as well as those for deep-water development and proposed as ultra-deep water structures. A short description of these structures and their applications are discussed.

Chapter 2 describes novel structures and their process of development to meet certain requirements of an offshore field. Several examples given for these structures are operating in offshore fields today. A few others are concepts in various stages of their developments. The main purpose of this chapter is to lay down a logical step that one should follow in developing a structural concept for a particular need and a set of prescribed requirements.

The ocean environment is the subject of chapter 3. It describes the environment that may be expected in various parts of the world and their properties. Formulas in describing their magnitudes are provided where appropriate so that the effect of these environments on the structure may be evaluated. The magnitudes of environment in various parts of the world are discussed. They should help the designer in choosing the appropriate metocean conditions that should be used for the structure development.

Chapter 4 provides a generic description of how to compute loads on an offshore structure and how the structure responds to these loads. Basic formulas have been stated for easy references whenever specific needs arise throughout this handbook. Therefore, this chapter may be consulted during the review of specific structures covered in the handbook. References are made regarding the design guidelines of various certifying agencies.

Chapter 5 deals with a statistical design approach incorporating the random nature of environment. Three design approaches are described that include the design wave, design storm and long-term design. Several examples have been given to explain these approaches.

The design of fixed offshore structures is described in Chapter 6. The procedure follows a design cycle for the fixed structure and include different types of structure design including tubular joints and fatigue design.

Chapter 7 discusses the design of floating structures, in particular those used in offshore oil drilling and production. Both permanent and mobile platforms have been discussed. The design areas of floaters include weight control and stability and dynamic loads on as well as fatigue for equipment, risers, mooring and the hull itself. The effect of large currents in the deepwater Gulf of Mexico, high seas and strong currents in the North Atlantic, and long period swells in West Africa are considered in the design development. Installation of the platforms, mooring and decks in deep water present new challenges.

Floating offshore vessels have fit-for-purpose mooring systems. The mooring system selection, and design are the subject of Chapter 8. The mooring system consists of freely hanging lines connecting the surface platform to anchors, or piles, on the seabed, positioned some distance from the platform.

Chapter 9 provides a description of the analysis procedures used to support the operation of drilling and production risers in floating vessels. The offshore industry depends on these procedures to assure the integrity of drilling and production risers. The description, selection and design of these risers are described in the chapter.

The specific considerations that should be given in the design of a deck structure is described in Chapter 10. The areas and equipment required for deck and the spacing are discussed. The effect of the environment on the deck design is addressed. The control and safety requirements, including fuel and ignition sources, firewall and fire equipment are given.

The objective of chapter 11 is to guide the offshore pipeline engineer during the design process. The aspects of offshore pipeline design that are discussed include a design basis, route selection, sizing the pipe diameter, and wall thickness, on-bottom pipeline stability, bottom roughness analysis, external corrosion protection, crossing design and construction feasibility.

Chapter 12 is focused on people and their organizations and how to design offshore structures to achieve desirable reliability in these aspects. The objective of this chapter is to provide engineers design-oriented guidelines to help develop success in design of offshore structures. Application of these guidelines are illustrated with a couple of practical examples.

The scale model testing is the subject of Chapter 13. This chapter describes the need, the modeling background and the method of physical testing of offshore structures in a small-scale model. The physical modeling involves design and construction of scale model, generation of environment in an appropriate facility, measuring responses of the model subjected to the scaled environment and scaling up of the measured responses to the design values. These aspects are discussed here.

Installation, foundation, load-out and transportation are covered in Chapter 14. Installation methods of the following substructures are covered: Jackets; Jack-ups; Compliant towers and Gravity base structures. Different types of foundations and their unique methods of installation are discussed. The phase of transferring the completed structure onto the deck of a cargo vessel and its journey to the site, referred to as the load-out and transportation operation, and their types are described.

Chapter 15 reviews the important materials for offshore application and their corrosion issues. It discusses the key factors that affect materials selection and design. The chapter includes performance data and specifications for materials commonly used for offshore developments. These materials include carbon steel, corrosion resistant alloys, elastomers and composites. In addition the chapter discusses key design issues such as fracture, fatigue, corrosion control and welding.

Chapter 16 provides an overview of the geophysical and geotechnical techniques and solutions available for investigating the soils and rocks that lay beneath the seabed. A project's successful outcome depends on securing the services of highly competent contractors and technical advisors. What is achievable is governed by a combination of factors, such as geology, water depth, environment and vessel capabilities. The discussions are transcribed without recourse to complex science, mathematics or lengthy descriptions of complicated procedures.

Because of the practical nature of the examples used in the handbook, many of which came from past experiences in different offshore locations of the world, it was not possible to use a consistent set of engineering units. Therefore, the English and metric units are interchangeably used throughout the book. Dual units are included as far as practical, especially in the beginning chapters. A conversion table is included in the handbook for those who are more familiar with and prefer to use one or the other unit system.

This handbook should have wide applications in offshore engineering. People in the following disciplines will be benefited from this book: Offshore Structure designers and fabricators; Offshore Field Engineers; Operators of rigs and offshore structures; Consulting Engineers; Undergraduate & Graduate Students; Faculty Members in Ocean/Offshore Eng. & Naval Architectural Depts.; University libraries; Offshore industry personnel; Design firm personnel.

Subrata Chakrabarti
Technical Editor

ABBREVIATIONS

List of Acronyms

ABS	American Bureau of Shipping
ABL	Above Base Line
API	American Petroleum Institute
BOP	Blowout Preventor
CFD	Computational Fluid Dynamics
CG	Center of Gravity
CRA	Corrosion Resistant Alloys
CVAR	Compliant Vertical Access Riser
DDCV	Deep Draft Caisson Vessel
DNV	Det Norske Veritas
DTU	Dry Tree Unit
EP	Equivalent Pipe
FE	Finite Element
FEA	Finite Element Analysis
FPDSO	Floating Production, Drilling, Storage and Offloading System
FPS	Floating Production System
FPSO	Floating Production Storage and Offloading
FSO	Floating Storage and Offloading
Fr	Froude number
GOM	Gulf of Mexico
HLV	Heavy Lift Vessel
IACS	International Association of Classification Societies
IMO	International Maritime Organization
IRM	Inspection, Repair and Maintenance
ISSC	International Ship Structures Congress
JIP	Joint Industry Project
JONSWAP	Joint North Sea Wave Project
KC	Keulegan–Carpenter Number
ksi	Kips per Square Inch
LF	Low Frequency
LRFD	Load and Resistance Factor Design
MODU	Mobile Offshore Drilling Unit
MPa	MegaPascals (N/mm^2)
NDP	Norwegian Deepwater Programme
NDT	Non-Destructive Testing
PDF	Probability Density Function

PIP	Pipe-In-Pipe
PM	Pierson–Moskowitz
psi	Pounds per Square Inch
QA/QC	Quality Assurance/Quality Control
Re	Reynolds Number
RFC	Rainflow Counting
SCF	Stress Concentration Factor
SCR	Steel Catenary Riser
SLC	Sustained Load Cracking
SPM	Single Point Mooring
SSCV	Semi-Submersible Crane Vessel
St	Strouhal Number
SWL	Still Water Level
TDP	Touch-Down Point
TDZ	Touch-Down Zone
TLP	Tension-Leg Platform
TSJ	Tapered Stress Joint
TTR	Top Tensioned Riser
UKCS	United Kingdom Continental Shelf
UOE	Pipe formed from plate, via a U-shape, then an O-shape, then Expanded
UTG	Upstream Technology Group
VIV	Vortex-Induced Vibration

CONVERSION FACTORS

	English	Metric
Length	1 in	25.4 mm
	1 ft	0.3048 m
	1 lbf	4.448 N
	1 lbm	0.4536 kg
	1 knot	0.5144 m/s
	1 mile	1.609 km
Area	1 ft^2	0.0929 m^2
Volume	1 ft^3	0.0283 m^3
	1 gallon	0.003785 m^3
Velocity	1 ft/s	30.48 cm/s
	1 mile/hr	1.609 km/hr
Acceleration	1 ft/s^2	30.48 cm/s^2
Mass/Force	1 ton (long)	1.016 m. ton
Density	1 lb/ft^3	16.0185 kg/m^3
Pressure	1 psi	6894.76 pascals
Moment	1 ft-lb	1.3558 N-m
Mass Moment of Inertia	1 lbm-ft	20.0421 $kg\text{-}m^2$

LIST OF CONTRIBUTORS

Chapter 1
Historical Development of Offshore Structures
Subrata Chakrabarti, Offshore Structure Analysis, Inc., Plainfield, IL, USA,
Cuneyt Capanoglu, I.D.E.A.S., Inc., San Francisco, CA, USA, and
John Halkyard, Technip Offshore, Inc., Houston, TX, USA

Chapter 2
Novel and Marginal Offshore Structures
Cuneyt Capanoglu, I.D.E.A.S., Inc., San Francisco, CA, USA

Chapter 3
Ocean Environment
Subrata Chakrabarti, Offshore Structure Analysis, Inc., Plainfield, IL, USA

Chapter 4
Loads and Responses
Subrata Chakrabarti, Offshore Structure Analysis, Inc., Plainfield, IL, USA

Chapter 5
Probabilistic Design of Offshore Structure
Arvid Naess and Torgeir Moan, Norwegian University of Science and Technology, Trondheim, NORWAY

Chapter 6
Fixed Offshore Platform Design
Demir Karsan, Paragon Engineering Services Inc., Houston, TX, USA
Vissa Rammohan, Stress Offshore, Inc., Houston, TX, USA (Contributed to the Jackup section)

Chapter 7
Floating Offshore Platform Design
John Halkyard, Technip Offshore, Inc., Houston, TX, USA
John Filson, Consultant, Gig Harbor, WA, USA (Contributed to the Semi, TLP and Hull Structure sections)
(assisted by Krish Thiagarajan, The University of Western Australia, Perth, Australia on Static Stability)

Chapter 8
Mooring, Cables and Anchoring
David T. Brown, BPP Technical Services Ltd., London, UK, and Houston, TX, USA

Chapter 9
Drilling and Production Risers
James Brekke, GlobalSantaFe Corporation, Houston, TX, USA (Drilling section)
Subrata Chakrabarti, Offshore Structure Analysis, Inc., Plainfield, IL, USA (Production section)
John Halkyard, Technip Offshore, Inc., Houston, TX, USA (Contributed to the Top Tension Risers)
Thanos Moros, Howard Cook, BP America, Houston, TX, USA (Contributed to the Steel Catenary Risers) and David Rypien, Technip Offshore, Inc., Houston, TX, USA (Contributed to the Materials Selection)

Chapter 10
Topside Facilities Layout
Kenneth E. Arnold and Demir Karsan, Paragon Engineering Services Inc., Houston, TX, USA
Subrata Chakrabarti, Offshore Structure Analysis, Inc., Plainfield, IL, USA

Chapter 11
Pipeline Design
Andre C. Nogueira and Dave S. McKeehan, INTEC Engineering, Houston, TX, USA

Chapter 12
Design for Reliability: Human and Organizational Factors
Robert G. Bea, University of California, Berkeley, CA, USA

Chapter 13
Physical Modeling of Offshore Structures
Subrata Chakrabarti, Offshore Structure Analysis, Inc., Plainfield, IL, USA

Chapter 14
Offshore Installation
Bader Diab and Naji Tahan, Noble Denton Consultants, Inc., Houston, TX, USA

Chapter 15
Materials for Offshore Applications
Mamdouh M. Salama, ConocoPhillips Inc., Houston, TX, USA

Chapter 16
Geophysical and Geotechnical Design
Jean M.E. Audibert and J. Huang, Fugro-McClelland Marine Geosciences, Inc., Houston, TX, USA

TABLE OF CONTENTS

Handbook of Offshore Engineering
S. Chakrabarti (Ed.)

Chapter 1

Historical Development of Offshore Structures

Subrata Chakrabarti
Offshore Structure Analysis, Inc., Plainfield, IL, USA

John Halkyard
Technip, Houston, TX, USA

Cuneyt Capanoglu
I.D.E.A.S., Inc., San Francisco, CA, USA

1.1 Introduction

The offshore industry requires continued development of new technologies in order to produce oil in regions, which are inaccessible to exploit with the existing technologies. Sometimes, the cost of production with the existing know-how makes it unattractive. With the depletion of onshore and offshore shallow water reserves, the exploration and production of oil in deep water has become a challenge to the offshore industry. Offshore exploration and production of minerals is advancing into deeper waters at a fast pace. Many deepwater structures have already been installed worldwide. New oil/gas fields are being discovered in ultra-deep water. Many of these fields are small and their economic development is a challenge today to the offshore engineers. This has initiated the development of new structures and concepts. Many of these structures are unique in many respects and their efficient and economic design and installation are a challenge to the offshore community. This will be discussed in more detail in Chapter 2. In order to meet the need for offshore exploration and production of oil/gas, a new generation of bottom-supported and floating structures is being developed.

The purpose of this chapter is to introduce the historical development of offshore structures in the exploration of petroleum reservoirs below the seafloor.

The chapter covers both the earlier offshore structures that have been installed in shallow and intermediate water depths and the various concepts suitable for deep-water development as well as those proposed as ultra-deep water structures. A short description of these structures is given and their applications are discussed.

1.1.1 Definition of Offshore Structures

An offshore structure has no fixed access to dry land and may be required to stay in position in all weather conditions. Offshore structures may be fixed to the seabed or may be floating. Floating structures may be moored to the seabed, dynamically positioned by thrusters or may be allowed to drift freely. The engineering of structures that are mainly used for the transportation of goods and people, or for construction, such as marine and commercial ships, multi-service vessels (MSVs) and heavy-lift crane vessels (HLCVs) used to support field development operations as well as barges and tugs are not discussed in detail in this book. While the majority of offshore structures support the exploration and production of oil and gas, other major structures, e.g. for harnessing power from the sea, offshore bases, offshore airports are also coming into existence. The design of these structures uses the same principles as covered in this book, however they are not explicitly included herein.

We focus primarily on the structures used for the production, storage and offloading of hydrocarbons and to a lesser extent on those used for exploration.

1.1.2 Historical Development

The offshore exploration of oil and gas dates back to the nineteenth century. The first offshore oil wells were drilled from extended piers into the waters of Pacific Ocean, offshore Summerlands, California in the 1890s (and offshore Baku, Azerbaijan in the Caspian Sea). However, the birth of the offshore industry is commonly considered as in 1947 when Kerr-McGee completed the first successful offshore well in the Gulf of Mexico in 15 ft (4.6 m) of water off Louisiana [Burleson, 1999]. The drilling derrick and draw works were supported on a 38 ft by 71 ft (11.6 m by 21.6 m) wooden decked platform built on 16 24-in. (61-cm) pilings driven to a depth of 104 ft (31.7 m). Since the installation of this first platform in the Gulf of Mexico over 50 years ago, the offshore industry has seen many innovative structures, fixed and floating, placed in progressively deeper waters and in more challenging and hostile environments. By 1975, the water depth extended to 475 ft (144 m). Within the next three years the water depth dramatically leapt twofold with the installation of COGNAC platform that was made up of three separate structures, one set on top of another, in 1025 ft (312 m). COGNAC held the world record for water depth for a fixed structure from 1978 until 1991. Five fixed structures were built in water depths greater than 1000 ft (328 m) in the 1990s. The deepest one of these is the Shell Bullwinkle platform in 1353 ft (412 m) installed in 1991. The progression of fixed structures into deeper waters upto 1988 is shown in fig. 1.1.

Since 1947, more than 10,000 offshore platforms of various types and sizes have been constructed and installed worldwide. As of 1995, 30% of the world's production of crude came from offshore. Recently, new discoveries have been made in increasingly deeper waters. In 2003, 3% of the world's oil and gas supply came from deepwater (>1000 ft or 305 m) offshore production [Westwood, 2003]. This is projected to grow to 10% in the next fifteen years [*Ibid.*] The bulk of the new oil will come from deep and ultra-deepwater production from three offshore areas, known as the "Golden Triangle": the Gulf of Mexico, West Africa and Brazil. Figure 1.2 illustrates the recent growth in ultra-deepwater drilling in the Gulf of Mexico. Drilling activity is indicative of future production.

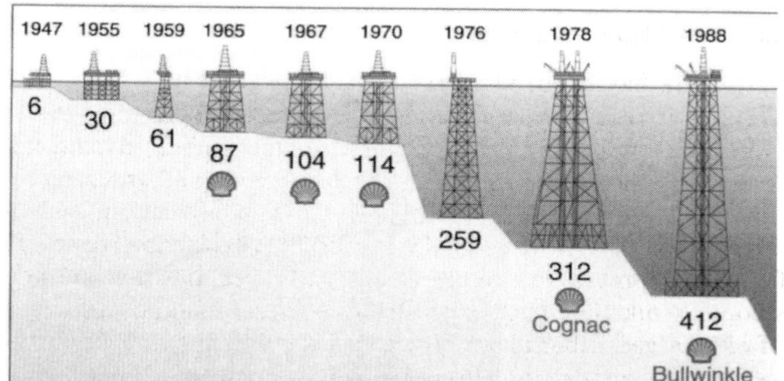

Figure 1.1 Progression of fixed platforms in the GOM – depths in meters (Courtesy Shell)

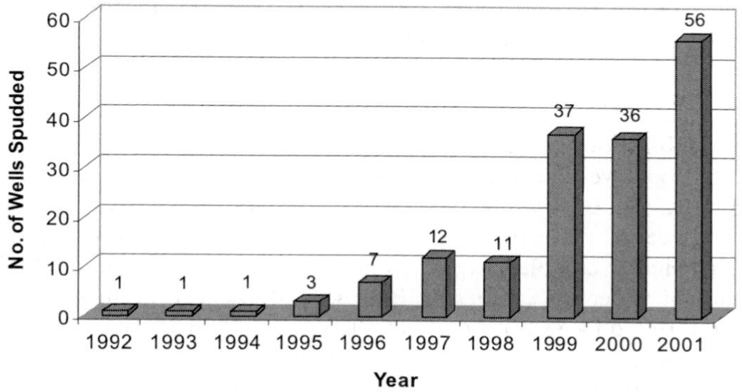

Figure 1.2 Ultra-deepwater (> 5000 ft or 1524 m) wells drilled in the Gulf of Mexico [adopted from MMS, 2002]

The importance of deepwater production to the US is illustrated in fig. 1.3. US oil production is on the decline, dropping from about 7.5 MM BPD in 1989 to 5.9 MM BPD in 2001. The current US oil consumption is about 20 MM BPD. Experts do not believe there are significant new resources onshore in the US. Deepwater production has grown from 9.5% of US production in 1989 to 26.4% in 2001 (from 750,000 to 1,500,000 BPD). The drilling activity shown in fig. 1.2 suggests that this percentage will continue to grow.

Fixed structures became increasingly expensive and difficult to install as the water depths increased. An innovative and cheaper alternative to the fixed structure, namely, the Lena guyed tower was introduced in 1983. The platform was built in such a way that the upper truss structure could deflect with the wave and wind forces. Piles extending above the sea floor could bend, and horizontal mooring lines attached midway up the platform could resist the largest hurricane loads. The Lena platform was installed in 1000 ft (305 m) of water. Two more "compliant" towers were installed in the Gulf of Mexico in 1998: Amerada Hess Baldpate in 1648 ft (502 m) and ChevronTexaco Petronius in 1754 ft (535 m). Petronius is the world's tallest free-standing structure.

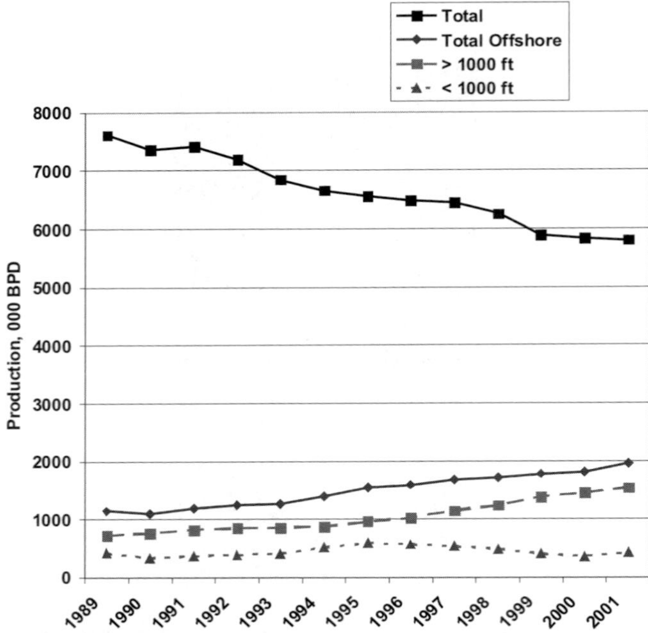

Figure 1.3 US crude oil production trends: importance of deepwater (Source: Westwood (www.dw-1.com) and OGJ Database (www.ogj.com))

Although nearly all of these platforms are of steel construction, around two dozen large concrete structures have been installed in the very hostile waters of the North Sea in the 1980s and early 1990s and several others offshore Brazil, Canada and the Philippines. Among these, the Troll A (fig. 1.4) gas platform is the tallest concrete structure in existence. It was installed offshore Norway in 1996. Its total height is 1210 ft (369 m), and it contains 245,000 m^3 of concrete, equivalent to 215,000 home foundations.

Gravity structures differ from other fixed structures in that they are held in place strictly by the weight contained in their base structures. The Troll platform, for example, penetrates 118 ft (36 m) into the seabed under its own weight.

The first floating production system, a converted semi-submersible, was installed on the Argyle field by Hamilton in the UK North Sea in 1975. The first ship-shaped floating production and storage system was installed in 1977 by Shell International for the Castellon field, offshore Spain. There were 40 semi-submersible floating production systems (FPSs) and 91 ship-shaped floating production and storage systems (FPSOs) in operation or under construction for deepwaters as of 2002 [*Offshore*, 2002]. Petrobras has been a pioneer in pushing floating production to increasingly deeper waters in their Campos Basin fields, offshore Brazil. Table 1.1 lists the progression of field development offshore Brazil in ever-increasing water depths. Some of the unique features of innovation and records are included in the last column.

Figure 1.4 Troll A gas platform, world's tallest concrete structure

Table 1.1 Field development in offshore Brazil

Field	Well	Water Depth ft (m)	Year	Remarks
Marimba'	RJS-284D	1355 (413)	1987	Wet Christmas tree
Marlim	MRL-3	2365 (721)	1991	Monobuoy & FPS
Marlim	MRL-4	3369 (1027)	1994	Subsea completion
Marlin Sul	MLS-3	5607 (1709)	1997	Deepest moored production unit
Roncador	RJS-436	6079 (1853)	1998	FPSO depth record
2000 BC 200 Block	RJS-543	9111 (2778)	2000	Drilling depth record at that time

1.1.3 Selection of Deepwater Production Concepts

The types of production concepts available for deepwater production are illustrated in fig. 1.5.

Most floating production systems, and virtually all of the semi-submersible, FPSs and FPSOs, produce oil and gas from wells on the seabed, called "subsea wells". Unlike wells

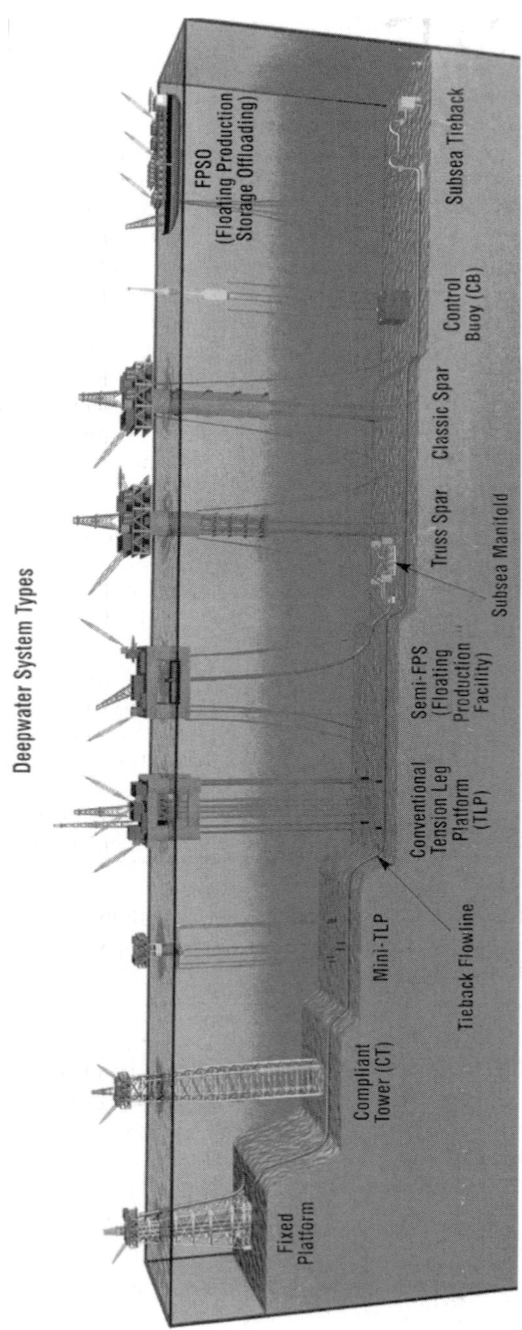

Figure 1.5 Deepwater system types [*Offshore Magazine*, 2002]

on fixed platforms and on land, subsea wells do not allow operators to have direct access to the wells for maintenance, or for re-completion (drilling into new reservoirs from an existing well).

The well consists of a wellhead, which supports the well casing in the ground, and a pod, which contains valves to control the flow and to shutoff the flow in the case of an emergency or a leak in the riser. This pod is called a "submerged Christmas tree", or simply a "wet tree". Subsea wells are expensive, but not as expensive in deepwater as placing a platform at the site. If a subsea well ceases to produce, or if its rate of production falls below economic limits, it is necessary to bring in a mobile drilling unit to remove the tree and perform the workover. This can be an extremely expensive operation and if the outcome of the workover is in doubt, the operator may choose to abandon the well instead. Because of this, much of the oil and gas in reservoirs produced through subsea trees may be left behind. Subsea wells may also result in lower reservoir recovery simply because of the physics of their operation. The chokes and valves placed in a subsea tree result in a pressure drop in the flow of oil or gas. When the well formation drops below a certain threshold, production ceases to flow. The difference in cut-off pressure between a subsea well and a surface well can be as much as 1000 psi vs. 100 psi [OTRC, 2002].

These facts motivated operators to seek floating platforms, which could support Christmas trees at the surface, "dry trees". Fixed and compliant platforms were safe for this kind of production because they could protect the well casings from the environment. Floating platforms generally had too much motion to protect the wells during extreme storms. A group of engineers in California invented a floating system in the early 1970s, which could be tethered to the sea floor, effectively making it a tethered compliant platform [Horton, et al 1976]. This gave rise to what is called the Tension Leg Platform (TLP) [Horton, 1987]. The first commercial application of this technology, and the first dry tree completion from a floating platform, was the Conoco Hutton TLP installed in the UK sector of the North Sea in 1984 [Mercier, et al 1980]. Dry trees are possible on a TLP because the platform is heave-restrained by vertical tendons, or tethers. This restraint limits the relative motion between the risers and the hull, which allows for flowlines to remain connected in extreme weather conditions.

The deep draft Spar platform is not heave-restrained, but its motions are sufficiently benign that risers can be supported by independent buoyancy cans, which are guided in the centerwell of the Spar. Both the Spar and the TLP designs are discussed in more detail in Chapter 7.

Today, many deepwater fields in the Gulf of Mexico are being developed by a combination of surface and subsea wells. Operators are able to develop a number of smaller marginal fields by combining subsea production with hub facilities [Schneider, 2000; Thibodeaux, et al 2002]. There is a growing trend towards third party ownership of the floating facilities, which opens the possibilities of several operators sharing production through one facility [Anonymous, 2003]. A consequence of this is that floaters may be designed with excess capacity for a given reservoir, in effect adding an "option cost" into the facility investment banking on future tiebacks from additional reservoirs.

Deepwater floating production systems are generally concentrated in the "Golden Triangle" of the Gulf of Mexico, offshore West Africa and Brazil (fig. 1.6). As of this writing,

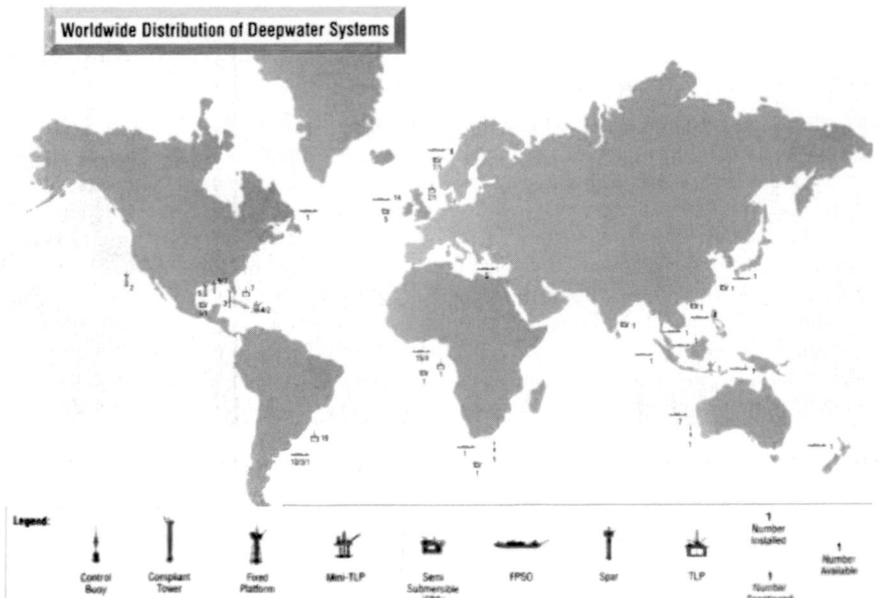

Figure 1.6 Worldwide distribution of floating production platforms [*Offshore*, 2002]

production Spars have only been installed in the Gulf of Mexico. TLPs have been installed in the Gulf of Mexico, West Africa, the North Sea and in Indonesia. FPSOs have been installed in virtually all of the offshore oil producing areas of the world with the exception of the Gulf of Mexico. Semi-submersible FPSs are prolific in the North Sea and Brazil.

According to industry sources (Westwood), the floating production systems will be growing at a rate of almost 30 per year through 2006, mostly in deepwater.

There is no simple answer to the question of which concept is "right" for a particular project. Selection of a concept for deepwater production is often a multi-year effort involving numerous studies and analyses. The primary drivers are reservoir characteristics and infrastructure, which will dictate the facility size, number of wells, their location, and whether wet or dry trees are called for. Drilling often represents over 50% of the value of deepwater projects, so that the method of drilling often dictates the type of surface facility required, e.g. whether the facility needs to support a drilling rig or whether a leased Mobile Offshore Drilling Unit (MODU) will be used.

Further discussion on concept selection is included in Chapter 7.

1.1.4 Offshore Disasters

Although most of the offshore structures constructed to date have withstood the test of time, there have been several catastrophic failures of offshore structures as well. Weather,

Figure 1.7 Accident of P-36 converted semi-submersible after flooding in one column [Barusco, 2002]

blowout, capsizing and human errors have resulted in the loss of a substantial number of fixed and floating structures. Between 1955 and 1968, nearly two dozen mobile drilling units have been destroyed. Within the two-year period between 1957 and 1959 alone, hurricanes Hilda and Betsy inflicted losses of hundreds of millions of dollars to drilling, production and pipeline facilities. Two semi-submersibles capsized and sank in the 1980s: Alexander Keilland, an accommodation vessel in the Norwegian north sea (1980), and Ocean Ranger offshore Hibernia, Canada (1982), resulted in the loss of hundreds of lives. The worst offshore disaster occurred when the Piper Alpha oil and gas platform caught fire in 1988. One hundred and sixty-seven lives were lost. In March, 2001, the world's largest floating production system, the Petrobras P-36, sank in Campos basin (fig. 1.7) costing 10 lives [Barusco, 2002].

1.2 Deepwater Challenges

The progression of platforms placed in deeper waters worldwide through the years is illustrated in fig. 1.8.

This figure also shows the progression of drilling and subsea completions. It is interesting to note from this figure, the gap between drilling and production. For example, the first drilling in 2000 ft of water took place in 1975. However, the first production from this water depth did not occur until 1993, i.e. 18 years later. This gap appears to be narrowing as recent advances in floating production systems and moorings have allowed rapid extension of this technology to deeper and deeper water depths.

In the sixties, production platforms designed for installation in less than a hundred meters of water were considered deep-water structures. In the seventies, platforms were installed

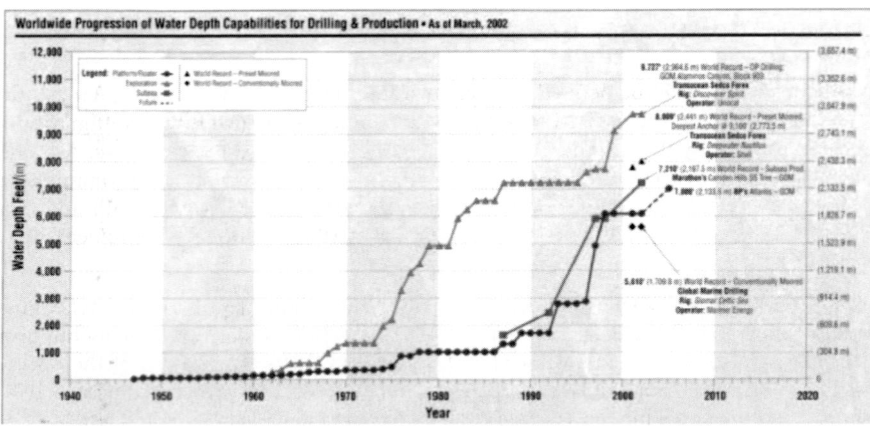

Figure 1.8 Progression of water depths [*Offshore Magazine*, 2002]

and the pipelines were laid in nearly 300-meter water depths. FPSs and FPSO systems were designed for similar water depths in the late seventies. Early FPSs and FPSOs took advantage of a surplus supply and low cost of semi-submersible drilling units and tankers to reduce the cost of deepwater development. Currently, FPSs and FPSOs are in demand all over the world in record water depths. FPSOs are yet to operate in the Gulf of Mexico perhaps due to both regulatory requirements and the availability of infrastructure for production export, i.e. pipelines make storage unnecessary. As of this writing, the deepest floating production system is Shell's Nakika semi-submersible in 6300 ft (1920 m) water in the Gulf of Mexico. This record will be extended to 7000 ft (2133 m) in 2005 with BPs Atlantis project and there will undoubtedly be a progression into ever-deeper waters in the future.

At present, deep water is typically defined to cover the water depth greater than 1000 ft (305 m). For water depths exceeding 5000 ft (1524 m), a general term "ultra-deep water" is often used. Bottom-supported steel jackets and concrete platforms are impractical in deep water from a technical and economic point of view giving way to floating moored structures. In deep and especially ultra-deep water, risers and mooring systems provide considerable challenge. These water depths are demanding new materials and innovative concepts. Synthetic fibre ropes, which are lighter, stronger, and more cost-effective are beginning to replace wire ropes and chains. Taut synthetic polyester mooring lines produce less vertical load on the floating platform. Several deepwater floating production systems using polyester moorings are now operating in Brazil, and two Spar platforms will use polyester moorings in the Gulf of Mexico in the near future [Petruska, 2003].

Flexible risers used for subsea tiebacks to floating structures are currently limited to about 5900 ft (1800 m) water depths. Steel catenary risers are becoming more common in deep and ultra-deep waters. New risers are being designed with titanium steel with high strength to weight ratio and favourable fatigue characteristics. Titanium and composite materials are also being developed for top tensioned risers.

1.3 Functions of Offshore Structures

Offshore structures may be defined by their two interdependent parameters, namely their function and configuration. A Mobile Offshore Drilling Unit (MODU) configuration is largely determined by the variable deck payload and transit speed requirements. A production unit can have several functions, e.g. processing, drilling, workover, accommodation, oil storage and riser support. Reservoir and fluid characteristics, water depth and ocean environment are the variables that primarily determine the functional requirements for an offshore facility.

Although the function of the structure, together with the water depth and the environment primarily influences its size and configuration, other factors that are just as important are the site infrastructure, management philosophy and financial strength of the operator as well as the rules, regulations and the national law. The structural design of the offshore structure is distinct based on the type of structure, rather than its function. These two categories will be addressed separately. First, we discuss the structure based on its function.

1.3.1 Exploratory Drilling Structures

Some of the desirable characteristics applicable to exploratory drilling units, such as limited structure motions and good station-keeping characteristics in relatively severe environment, are equally applicable to production units. MODUs must accommodate highly variable deck loads due to the different drilling requirements they will encounter, and they are usually designed for relatively high transit speeds to minimise mobilisation costs. Three of the most common forms of drilling structures are drillships, jack-up barges and semi-submersibles. Submersible gravity structures are also used for drilling in shallow water. These structures with buoyant legs and pontoons are set on seafloor by ballasting, thus allowing the structure to be deballasted, and moved to another location. Drillships are ship-shaped and self-propelled, which can accommodate the drilling equipment on board. They have the advantage of rapid transit between stations and can take up and leave stations quickly, especially if they are dynamically positioned instead of being moored in place. However, the large motions and thruster (or anchor) capacity limit the weather conditions in which they can drill.

The mobile semi-submersible drilling unit hulls typically consist of four or six columns connected with horizontal pontoons and support a large deck on top. Most of these structures do not have thrusters or dynamic positioning and are usually towed like the barges or transported on large purpose-built transport vessels. The semi-submersibles have good motion characteristics in severe environment and thus have the advantages of being able to stay in the drilling modes longer than a typical drillship.

The jack-up barges are usually buoyant during transit and are towed from station to station. Once they reach the drilling site, the legs of the jack-up (usually three in number) are set on the ocean bottom and the deck is jacked up above water level on these legs. During drilling the jack-ups act like fixed platforms. However, the water depth of about 150 m limits their operations to shallow-to-moderate water depths. Jack-up units have other constraints, such as being largely affected by the seafloor terrain and material characteristics and the time and environmental constraints associated with jacking operations.

The mooring systems for MODUs do not have to meet the same severe environmental requirements of production vessels. If severe weather is forecast, MODUs can disconnect the drilling riser and leave station, or slacken mooring lines to avoid damage. Permanent production facilities cannot afford this luxury and are required to remain within a safe watch circle under the most extreme weather conditions.

Each of the three configurations discussed as exploratory structures in this section, with suitable modifications, is suitable for use as production structures. Many of the FPSs are converted drilling units with the drilling equipment being replaced by production equipment.

1.3.2 Production Structures

Production platforms are required to stay on station during its lifetime, which is usually from 20 to 30 years. In shallow waters, the most common type of production platforms is the fixed piled structures, commonly known as jackets in the offshore industry. These are tubular structures fixed to the seafloor by means of driven or drilled and grouted piles. The economic water depth limit for fixed platforms varies by environment. In the North Sea, the deepest fixed jacket platform, the BP Magnus platform, is in 610 ft (186 m) of water. The deepest concrete structure in the North Sea, the Shell Troll Gravity Base Structure, extends the fixed structure limit there to nearly 1000 ft (305 m) water depth. In the Gulf of Mexico, the Shell Bullwinkle platform holds the water depth record at 1352 ft (412 m) of water.

When the water depth exceeds these limits, compliant towers or floating production platforms become more attractive. Three compliant towers have been installed in the Gulf of Mexico. The deepest is the ChevronTexaco Petronius platform in 1754 ft (535 m) water depth. This is probably the economic limit for these types of structures.

Fixed and compliant platforms support conductor pipes, which are essentially extensions of the well casing from the seafloor. The conductors are supported along their length and are not free to move with the dynamics of the waves. The wellhead is at the deck of the platform and well operations are similar to land-based operations.

One of the most important requirements for floating production systems is their interface with risers. Production may originate from wellheads on the sea floor (wet trees), or from wellheads located on the structure (dry trees). The selection is driven by reservoir characteristics and has a significant impact on the selection of the structure. Dry tree risers are nearly vertical steel pipes, which must be designed to contain well pressure in all operational conditions. This places limits on the motions of the production platform. To date, only tension leg platforms and Spars have been used with these types of risers. Subsea production risers are typically composite flexible risers, which are more tolerant to vessel motions. These risers have been used with all types of floating platforms. Steel catenary risers (SCRs) have been employed from TLPs, Spars and semi-submersibles.

In the same manner in which production may originate from wet or dry trees, drilling may be performed with a subsea blowout preventor (BOP) or a surface BOP. The BOP provides the safety shutoff capability in the case of an unexpected release of well pressure during drilling. Exploratory drilling units employ a subsea BOP and a low pressure drilling riser.

The drilling riser can be disconnected from the BOP in an emergency. Most floating production systems with drilling capability use a surface BOP. The drilling riser in this case must be designed to take the full well pressure, and the vessel and mooring must be designed to support this riser in the harshest environmental conditions. Again, TLPs and Spars are the only floating production systems, which are currently performing drilling with a surface BOP.

There is an important distinction between the requirements of Gulf of Mexico and North Sea platform, which impact platform design. Extreme sea states in the Gulf of Mexico are associated with hurricanes. Platforms in the Gulf of Mexico shut down operations and are abandoned when a hurricane threat arises. North Sea platforms are subject to many fast moving weather fronts, which can create extreme sea states. They cannot be abandoned and operations continue through the weather conditions, which are as bad as or worse than the hurricanes in the Gulf of Mexico This distinction leads to differences in safety factors and design criteria for these locations. For example, North Sea floating production systems are currently designed to survive with two mooring lines missing. Gulf of Mexico standards only require designing for one mooring line missing. The environments in the other major offshore regions, i.e. West Africa, Brazil and Southeast Asia are generally more benign than those in either the North Sea or the Gulf of Mexico.

1.3.3 Storage Structures

During the production of offshore oil, it may be desirable to store the crude temporarily at the offshore site before its transportation to the shore for processing. Storage capacity is dictated by the size of shuttle tankers and frequency of their trips. Historically, storage capacities have typically been between 15 and 25 days at peak production. These values are appropriate for FPSOs employed on remote, marginal fields. This was the original province of FPSOs. During the 1990s, FPSOs became more popular for large fields in more developed areas. Storage requirements and shuttle tanker specifications could be optimised. North Sea shuttle tankers, for example, are usually purpose-built vessels and the storage requirements can be optimised for a particular project. West African FPSOs are generally sized to store and load VLCCs for long voyages. Southeast Asian FPSOs typically offload to tankers of opportunity. Storage capacities for recent FPSO projects range from as low as three days production (BP Foinhaven in the North Atlantic) to as much as eleven days production (CNOC Liuhua in the South China Sea). The ship-shaped production platforms (i.e. FPSOs) possess large enclosed volume, and are ideally suited for the combination of production and storage. Floating Storage and Offloading vessels, i.e. without processing, (FSOs) may also be used in conjunction with floating or fixed production platforms. The Shell Expro Brent Spar was used for this purpose for 20 years in the North Sea. FPSOs are the most prolific floating production systems. As of 2002, there were 91 installations [*Offshore*, 2002]. There were also 63 fields developed with FSOs as of 1993.

Oil storage tanks are usually maintained at atmospheric pressure with an inert gas blanket. According to international regulations, water ballast tanks are required to be segregated from cargo tanks [IMO, 1978]. Cargo and ballast management is an important aspect of FPSO operations.

Figure 1.9 500,000 barrel oil storage structure

Oil storage may also be accommodated at ambient pressure in tanks, which are open to the sea at the tank bottom. Oil, being lighter than water, may be pumped into the compartment displacing seawater out the bottom. The tank wall pressure is only a function of the height of the oil column and the difference in density between the water and the oil. A photograph of such a structure using this principle, called Khazzan storage tank, is shown in fig. 1.9 being towed to site. Three of these structures built in the sixties in the Persian Gulf in a water depth of about 150 ft (46 m) are still operating by Conoco. They are open at the bottom, having a storage capacity of 500,000 barrels and situated on the ocean floor by driven piles. These platforms do not perform any oil processing.

The same storage principle was also used on the floating structure Brent Spar, and on most of the large concrete gravity-base structures built in Norway in the 1980s and 90s (Condeeps). The Gullfaks C platform shown in fig. 1.10 is capable of storing almost 2 million barrels of oil in 712 ft (217 m) of water. The oil-over-water storage principle has been proposed for a drilling and production Spar [Bugno and Horton, 2000].

1.3.4 Export Systems

The oil produced offshore requires transportation from the site to the shore. If the structure is located close to the shore or there is an existing infrastructure to tie-in, then underwater pipeline may be used for this purpose. The sequential development of fields from shallow-to-moderate-to-deep water in the Gulf of Mexico has resulted in the development of a network of these subsea pipelines. These pipelines are supported on the floor of the ocean, connecting the production wells to the shore or other platforms. The design of subsea pipelines has been covered in Chapter 11.

However, for remote offshore locations, this method of transport is not economically feasible and other means of transportation is required. This transportation is usually

Figure 1.10 Gullfaks C production platform with 2 million barrel wet oil storage (Statoil)

accomplished by shuttle tankers. Sometimes, these tankers are moored directly to the storage or production structure. Often, however, special structures are required to moor the tankers. These structures take the form of floating buoys or articulated structures. The tankers are moored with the help of a floating hawser, which is retrieved from the ocean surface. The transfer of oil from the structure to the tanker is accomplished by a loading hose. Since a single line is used for the mooring, these special structures are called Single Point Mooring (SPM) or Single Buoy Mooring (SBM).

1.4 Offshore Structure Configurations

Offshore structures may be defined as being either bottom-supported or floating. Bottom supported structures are either "fixed" such as jackets and gravity base structures, or "compliant" such as the guyed tower and the compliant tower. Floating structures are compliant by nature. They can be viewed either as "neutrally buoyant", such as the semi-submersible-based FPSs, ship-shaped FPSOs and monocolumn Spars, or "positively buoyant", such as the Tension Leg Platforms.

1.4.1 Bottom-Supported Structures

Bottom-founded structures, with the notable exception of the Gravity Base Structures (GBS e.g. Condeeps), are typically constructed from welded steel tubular members. These members act as a truss supporting the weight of the processing equipment, and the environmental forces from waves, wind and current. Bottom-founded structures are called "fixed" when their lowest natural frequency of flexural motion is above the highest frequency of significant wave excitation. They behave as a rigid body and must resist the full dynamic forces of the environment. "Compliant" bottom-founded structures are usually designed so that their lowest natural frequency is below the energy in the waves. Waves, wind and current cause these structures to deflect, but the magnitude of the dynamic loads is greatly reduced. This allows economical bottom-founded structures to be designed for water depths, which would not be practical for fixed structures.

Another type of bottom-supported structure behaves like a fixed structure in a mild environment. Such a structure is designed with the means to behave both as a fixed and as a compliant structure. Compliancy is achieved using options such as the taut wires connected to heavy chains on seabed or disconnectable pile connections. Thus when the applied lateral wind, wave and current forces exceed the design limit, chains are lifted off the seabed or the pile connections released, to turn the fixed structure into a rotationally compliant structure (i.e. from zero degrees of freedom to two degrees of freedom about the seabed).

1.4.2 Floating Offshore Structures

Floating structures have various degrees of compliancy. Neutrally buoyant structures, such as semi-submersibles, Spars and Drillships are dynamically unrestrained and are allowed to have six degrees of freedom (heave, surge, sway, pitch, roll and yaw). Positively buoyant structures, such as the Tension Leg Platforms (TLPs) and Tethered Buoyant Towers (TBTs) or Buoyant Leg Structures (BLS) are tethered to the seabed and are heave-restrained. All of these structures with global compliancy are structurally rigid. Compliancy is achieved with the mooring system. The sizing of floating structures is dominated by considerations of buoyancy and stability. Topside weight for these structures is more critical than it is for a bottom-founded structure. Semi-submersibles and ship-shaped hulls rely on waterplane area for stability. The centre of gravity is typically above the centre of buoyancy. The Spar platform is designed so that its centre of gravity is lower than its centre of buoyancy, hence it is intrinsically stable. Positively buoyant structures depend on a combination of waterplane area and tether stiffness to achieve stability.

Floating structures are typically constructed from stiffened plate panels, which make up a displacement body. This method of construction involves different processes than those used in tubular construction for bottom-founded structures.

Neutrally buoyant floating structure motions can be accurately determined as a single six-degrees of freedom system subjected to excitation forces. Positively buoyant floating structures in deep water will have restraining systems with substantial mass, and the restraining systems are subjected to excitation forces as well. The motions of the platform are coupled with the dynamics of the mooring system. The coupling of motions between the platform, risers and mooring systems becomes increasingly more important as water depth increases. The discussion of the dynamic analysis of these platforms subjected to environment is included in Chapter 4.

1.4.3 Floating vs. Fixed Offshore Structures

Table 1.2 summarises the main differences between bottom-founded and floating structure design. Fixed and floating platforms are very different not only in their appearance but also in their structural members. They are unique in how they are constructed, transported and installed, what kind of excitation forces they are subjected to, how they respond to these excitation forces and how they are decommissioned and reused/recycled at the end of their design lives. The major common characteristic of each type of structure is that they provide deck space and payload capacity (i.e. real estate) to support equipment and variable weights used to support drilling and production operations.

The fixed platform deck loads are directly transmitted to the foundation material beneath the seabed. Thus, fixed platform jackets supporting the deck are typically long, slender steel structures extending from seabed to 20–25 m above the sea surface. Floating structure deck loads are supported by the buoyancy forces of the hull supporting the deck.

Conductor pipes on bottom-founded structures effectively extend well casings to the deck of the structure. Drilling and well operations are identical to those on land. Floating

Table 1.2 Bottom-founded vs. floating structures

Function	Bottom-Supported	Floating
Payload support	Foundation-bearing capacity	Buoyancy
Well access	"rigid" conduits (conductors) surface wellheads and controls	"dynamic" risers subsea wellheads subsea or surface controls
Environmental loads	Resisted by strength of structure and foundation, compliant structure inertia	Resisted by vessel inertia and stability, mooring strength
Construction	Tubular space frame: fabrication yards	Plate and frame displacement hull: ship yards
Installation	Barge (dry) transport and launch, upend, piled foundations	Wet or dry transport, towing to site and attachment to pre-installed moorings
Regulatory and design practices	Oil industry practices and government petroleum regulations	Oil industry practices, government petroleum regulations and Coast Guard & International Maritime regulations

structures require dynamic risers to connect with wellheads on the seafloor. Drilling and production require a tieback at the mudline to the subsurface casing. Well control can require expensive subsea control systems (wet trees), or special low-motion vessels, which can support vertical risers in all weather conditions with well controls at the surface (dry trees).

Fixed platform jackets are constructed on their side, loaded out on to a barge (except for jackets with flotation legs), transported to the installation site, launched and upended (or lifted and lowered) and secured to seabed with driven or drilled and grouted piles. Floating structures, except for Spars, TBTs and BLSs, are constructed upright, either dry or wet towed to installation site and connected to the mooring system or secured to the seabed with tethers.

Fixed platform jackets need to have adequate buoyancy (i.e. more than their own self-weight) to stay afloat during installation. Thus, they are typically constructed of small diameter tubulars that form a space frame. Floating structure hulls need to have adequate buoyancy to support the deck and various other systems. Thus, they are typically constructed of orthogonally stiffened large-diameter cylindrical shells or flat plates. Small-diameter tubulars are susceptible to local instability and column buckling, while orthogonally stiffened systems are designed to meet hierarchical order of local, bay and general instability failure modes.

Fixed platform design is typically controlled by their functional gravity loads and the lateral forces and overturning moments due to wind, wave and current. For a preliminary design, wind, wave and current forces can be applied quasi-statically to a structure along with the dead loads from the deck and structural self-weight. A single load case defined by a "design wave" can, in most cases, be adequate to determine the required strength of a fixed structure.

A unique aspect of floating structures is that, in addition to the applied functional deck gravity loads and environmental forces acting on the body, it is necessary to determine the inertial loads due to acceleration of the body in motion. A floating structure responds dynamically to wave, wind and current forces in a complicated way involving translation and rotation of the floater.

Thus, while fixed structures in shallow and moderate water may be designed by applying the laws of static equilibrium to the structure, most fixed structures in deep water and all floating structures require the application of the laws of dynamics.

Weight control is more important to the design of floating structures than to the fixed structures.

Mooring and station-keeping are unique requirements of floating structures. "Mooring" refers to the means for providing a connection between the structure and the seafloor for the purposes of securing the structure against environmental loads. "Station-keeping" is a term used to define a system for keeping the facility within a specified distance from a desired location. This is typically a requirement of drilling or riser connections to the seafloor, or for running equipment to the sea floor. The station-keeping requirement may be achieved by means of mooring lines, which may be adjustable, or by means of a dynamic positioning system using thrusters, or a combination of the two.

Another unique characteristic of the floating structure is that typically it can be decommissioned readily and moved to another site for reuse. A decommissioned fixed platform has to be removed in whole or in part, requiring the use of heavy lift equipment and the reverse of the installation procedure. Typically, such a structure has to be taken to shore for use as scrap steel or possibly modified and given a second life. Thus, the capital expenditures (CAPEX) for fixed platforms need to allocate substantial sums to cover future decommissioning costs.

1.5 Bottom-Supported Fixed Structures

1.5.1 Minimal Platforms

For the marginal field development in shallow water, fixed production platforms with a small deck are often used. At a minimum these structures may support the following: (1) a few wells typically less than 10; (2) a small deck with enough space to handle a coil tubing or wireline unit; (3) a test separator and a well header; (4) a small crane; (5) a boat landing; and (6) a minimum helideck. Chevron [Botelho, et al (2000)] carried out a study to identify and select, among existing production platform concepts, the ones that would optimise the development of fields in 150 ft (46 m) and 200 ft (61 m) of water, and for three different design return periods (25, 50 and 100 years). Figure 1.11 depicts three of these concepts.

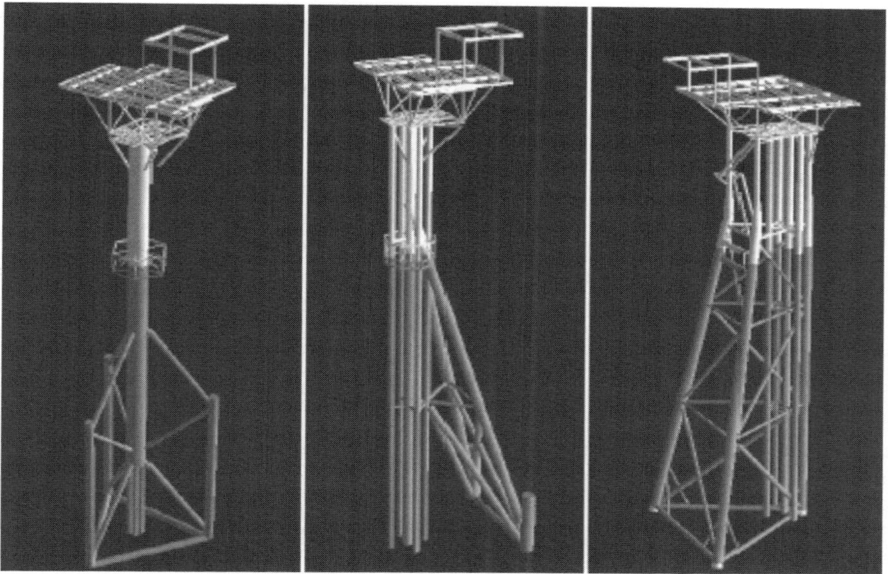

Figure 1.11 Example minimal platform concepts (LINX, MANTIS II and TRIPOD) for marginal field

1.5.2 Jacket Structures

The jacket, or template, structures are still the most common offshore structures used for drilling and production. Some structures contain enlarged legs, which are suitable for self-buoyancy during its installation at the site. Fixed jacket structures consist of tubular members interconnected to form a three-dimensional space frame (fig. 1.12). These structures usually have four to eight legs battered to achieve stability against toppling in waves. Main piles, which are tubular, are usually carried with the jackets and driven through the jacket legs into the seafloor. The term jacket structure has evolved from the concept of providing an enclosure ("jacket") for the well conductors. These platforms generally support a superstructure having 2 or 3 decks with drilling and production equipment, and workover rigs. The use of these platforms has generally been limited to a water depth of about 500–600 ft (150–180 m) in the harsh North Sea environment (typical design wave of 100 ft/30 m). In the more intermediate Gulf of Mexico environment (typical design wave of 75 ft/23 m) half a dozen jackets have been installed in deeper water. A large three-part jacket weighing 34,300 tons was installed in 1979 in the

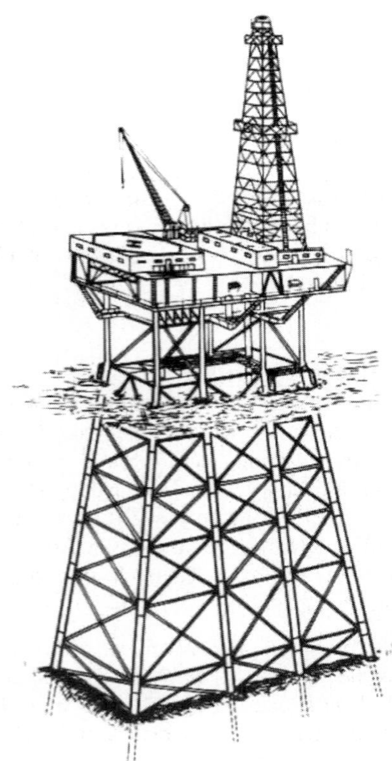

Figure 1.12 Fulmar jacket platform

Cognac field in the Gulf of Mexico off the Louisiana coast in a water depth of 1000 ft (300 m). This record-breaking structure was followed by other platforms including Cerveza and Cerveza Liguera (935 ft/285 m), Pompano (1290 ft/393 m) and the current record holder Bullwinkle Platform. The single piece Bullwinkle jacket weighs 49,375 tons and was installed at a site with 1350 ft (412 m) water depth in 1988. The initial deck weight for Bullwinkle was 2033 tons. Four thousand tons were added in 1996 to accommodate more production. A floating platform to accommodate this payload would weigh in the order of 6000–10,000 tons.

1.5.3 Gravity Base Structures

Offshore structures that are placed on the seafloor and held in place by their weight are called gravity structures. Thus these structures do not require additional help from piles or anchors. These structures are quite suited for production and storage of oil. They are built near-shore location or sheltered water, such as, fjords. Upon construction they are towed in the upright position to the final destination and submerged in place. It is often possible to carry the topside deck with the structure. Because of the nature of these platforms, they are often susceptible to scour of their foundation and sinkage.

The largest steel gravity structure was placed in 1984 in the UK Maureen field and operated by Phillips Petroleum. It had served its usefulness after about 20 years of operation and recently, has been floated up from site for its removal. Since the gravity structures require large volume and high weight, concrete has been a common material for gravity structures. In 1975, the first concrete structure called "Condeep B" in deepwater was built at the Stavanger, Norway fjord and placed in the Beryl field. Phillips Petroleum installed a one-million barrel storage capacity platform at the Norwegian Ekofisk field. It has a unique perforated outer wall, which was installed to dissipate wave energy from the structure. The platform underwent major renovation due to subsidence from depleted oil in the field and recently stopped its operation. The Statoil Gullfaks C platform, with 2 million barrel oil storage, is shown in fig. 1.10.

1.5.4 Jack-ups

The jack-up barges are typically three-legged structures having a deck supported on their legs. The legs are made of tubular truss members. The deck is typically buoyant. The jack-ups are used for the exploratory drilling operation and, therefore, are designed to move from site to site. The jack-up barges are towed while supported by the buoyancy of their own hull. Sometimes, they are transported on top of transport barges. They are called jack-ups because once at the drilling site, the legs are set on the ocean bottom and the deck is jacked up on these legs above the waterline. The jack-up barges behave like the stationary platform during the drilling operation.

Two self-installing jack-up production platforms have been installed in the North Sea: Elf Elgin Franklin and BP Harding. Both are based on the Technip TPG 500 design, which consists of a self-floating deck with jack-up legs. The Harding TPG 500, shown in fig. 1.13, was installed in 1995 on top of a gravity base, which includes 550,000 barrels of oil storage. These platforms are in 305 and 361 ft (93 and 110 m) water depths respectively. A similar structure is planned for installation in the Caspian Sea.

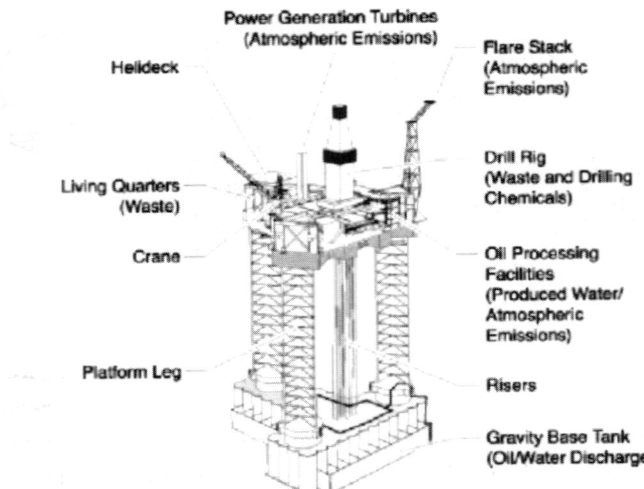

Figure 1.13 BP Harding TPG 500 platform with gravity base (Courtesy BP)

1.5.5 Subsea Templates

Subsea technology covers a wide range of offshore activities. Examples are subsea Xmas trees, manifolds, templates, flowlines and risers, control systems, well fluid boosters, multiphase pumping and metering, water separation, water injection, remote and diverless connections, guideline-free installations, seabed electrical power distribution systems, interventions, etc.

Subsea production is often the lowest cost alternative for marginal fields in deepwater when it is possible to tieback production from a few wells to an existing structure. Subsea systems consist of mechanical, hydraulic, electrical and structural components; usually packaged by a subsea vendor. The cost of a subsea development is partly due to the cost of this complex equipment, but in very deep waters the cost of installation and commissioning can be overriding. Large subsea systems can weigh up to several hundred tons, and they can only be lowered to the seafloor using dynamically positioned derrick vessels. There are only a few of these vessels in the world, and they can cost hundreds of thousands of dollars per day. Several deepwater fields offshore Brazil have seen successful installation of Xmas trees. In the Norwegian water at Njord field, a horizontal Xmas tree was installed in 1080 ft (330 m) water depth.

1.5.6 Subsea Pipelines

Subsea pipelines are used to transfer oil from the production platforms to storage facilities or to the shore. Installation of subsea pipeline is a common occurrence in moderately deep water up to a few hundred meters. As of April 1998, there were 26,600 miles of pipeline in the Gulf of Mexico. Nearly 50% of this is in deepwater (> 1000 ft or 300 m) and

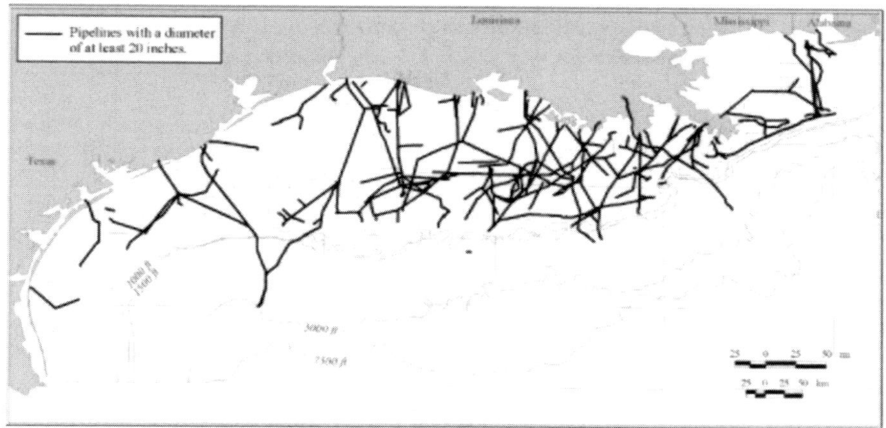

Figure 1.14 Pipelines in the Gulf of Mexico, 20 in. (508 mm) or greater

between 300 and 500 miles (480–800 km) of deepwater pipeline is being laid annually [MMS 2000].

Figure 1.14 shows the network of large pipelines in the Gulf of Mexico as of 2002. As of 2004, the Mardi Gras oil (30 in. or 760 mm and 28 in. or 711 mm) and Okeanus gas (24 in. or 610 mm) pipelines extend the Gulf of Mexico pipeline network to 6350 ft (1936 m). Another first was achieved, when the Blue Stream project pipeline system was completed by Saipem, permitting the flow of gas from Russia to Turkey. In addition to the water depths approaching 4920 ft (1500 m), the corrosive waters deep under the surface layer of Black Sea required the use of special titanium alloy pipeline.

Several pipeline installation methods have been used in the past, including:

- J-lay with an inclined tower
- J-lay with a vertical tower
- S-lay
- Reel method
- Piggyback
- Multiple lay

The selection process for the suitable method is governed by the laying speed, but may also be affected by the initiation and termination phases and by the need to install ancillary items, such as anodes, tees, in-line structures and risers. The essential variables to consider are geometrical steel pipe characteristics, diameter and wall thickness; chemical and mechanical steel properties; thermal insulation; pipe-in-pipe and internal and external coating requirements. The choice of the technique will be contingent on the quantity of work involved, total length to be laid and average length of each individual line [Borelli and Perinet, 1998].

The details of the design, construction and installation of subsea pipelines have been included in Chapter 11.

1.6 Compliant Structures

The definition of a compliant structure includes those structures that extend to the ocean bottom and directly anchored to the seafloor by piles and/or guidelines. These structures are typically designed to have their lowest modal frequency to be below the wave energy, as opposed to the fixed structures, which have a first modal frequency greater than the frequency of wave energy.

1.6.1 Articulated Platforms

One of the earliest compliant structures that started in relatively shallow waters and slowly moved into deep water is the articulated tower. An articulated tower is an upright tower, which is hinged at its base with a cardan joint and is free to oscillate about this joint due to the environment. The base below the universal joint on the seabed may be a gravity base or may be piled. The tower is ballasted near the universal joint and has a large enough buoyancy tank at the free surface to provide large restoring force (moment). The tower extends above the free surface and accommodates a deck and a fluid swivel. In deeper waters, it is often advantageous to introduce double articulation, the second one being at a mid-depth.

The articulated tower is used as a single-point mooring system (SPM) to permanently moor storage and production tankers or is utilised as a mooring and offloading medium for a shuttle tanker. The tower must survive its lifetime storm as well as the operating sea when attached to the tanker. Fatigue is an important criterion for this type of system. In intermediate water depths, the structure may need to be treated as a flexible structure for the fatigue stress evaluation. In fact, the earlier SPMs built for Petrobras offshore Brazil failed in fatigue near the J-tube entrance and had to be de-commissioned after only a few years of operation.

Examples of SPMs built so far are shown in table 1.3. It is recognised that the practical application and economics of SPMs are generally limited to a few hundred meters.

1.6.2 Compliant Tower

A compliant tower is similar to a traditional platform and extends from surface to the sea bottom, and it is fairly transparent to waves. However, unlike its predecessor, a compliant tower is designed to flex with the forces of waves, wind and current. It uses less steel than a conventional platform for the same water depth. Table 1.4 compares the structural weights for Bullwinkle, the world's deepest fixed platform, with the tallest compliant structures in the Gulf of Mexico.

As of this writing Petronius is the world's tallest freestanding structure (it is about 300 ft or 91 m taller than the CN Tower in Toronto) [Wikipedia encyclopedia].

Table 1.3 Examples of articulated platforms installed

Platform	Field/Location	Water Depth	Uniqueness	Year
2 SPMs	Campos Basin, Brazil	122 m	Permanently moored & Loading	1975
SALM	Thistle field, North Sea	174 m	Multi-hinged	1977
STATFJORD SPM	North Sea	145 m		1980
SALM	Hondo Field, Santa Barbara, Calif.	150 m	Permanently moored to 50,000 dwt	
ALT	Fulmar Field, North Sea	82 m	Permanently moored; shuttle tanker in tandem	1982
ALT	Maureen Field, North Sea	93 m	Concrete column	1982

Table 1.4 Structural weights of compliant towers

Platform	Water Depth ft (m)	Topside weight tons	Structure weight tons
Bullwinkle (fixed)	1353 (413)	2033 (originally)	49,375
Baldpate	1650 (503)	2400	28,900
Petronius	1754 (535)	7500	43,000

1.6.3 Guyed Tower

A guyed tower is a slender structure made up of truss members, which rests on the ocean floor and is held in place by a symmetric array of catenary guylines. A guyed tower may be applicable in deep hostile waters where the loads on the gravity base or jacket-type structures from the environment are prohibitively high. The guylines typically have several segments. The upper part is a lead cable, which acts as a stiff spring in moderate seas. The lower portion is a heavy chain with clump weights, which are lifted off the bottom during heavy seas and behaves as a soft spring making the tower more compliant.

Exxon in 1983 installed the first guyed tower named Lena Guyed tower in the Mississippi Canyon Block in 1000 ft (300 m) water depth. It resembles a jacket structure, but is compliant and is moored over 360° by catenary anchor lines.

1.7 Floating Structures

The floating structures have already been introduced. Here the various types of floating structures and their types are discussed.

1.7.1 Floating Platform Types

The floating structures may be grouped as Neutrally Buoyant and Positively Buoyant. The neutrally buoyant structures include Spars, Semi-submersible MODUs and FPSs, Ship-shaped FPSOs and Drillships. Examples of positively buoyant structures are TLPs, TLWPs and Buoyant Towers.

Floating platform functions may be grouped by their use as mobile drilling-type or production type. The number of units in these categories installed worldwide is shown in table 1.5 as of 2003.

There is little standardisation of floater units. Shell offshore and their partners achieved significant cost savings when they designed multiple TLPs following similar design practices (i.e. Auger, Ram-Powell, Mars, Ursa and Brutus). Kerr-McGee achieved some saving by designing the Nansen and Boomvang Spars identically. However, for the most part, each deepwater field has been developed with a "fit for purpose" design.

1.7.2 Drilling Units

Exploratory drilling precedes development drilling and production begins when development wells are drilled and completed. Thus, it is natural that there were many drilling units

Table 1.5 Floating systems as of 2002 [*Offshore*, 2002]

Class	Type	Structure	Unit
Drilling	Mobile Offshore Drilling Units (MODUs)	Semi-submersibles Ship-shaped Vessels Barges	112 25 12
Production	*Neutrally Buoyant*		
	Floating Production, Storage and Offloading Systems (FPSO)	Ship-shaped Vessels	85
	Floating Storage and Offloading (FSO)	Barges	67
	Floating Production Systems (FPS)	Semi-submersibles	41
		Spars	13
		Wellhead control buoys	2
	Positively Buoyant		
	Conventional TLPs		19
	Mini-TLPS	TLPs and TLWPs	7

before one of those units received a production module to initiate limited production. Although MODUs discussed in this book can be jack-up units that are bottom-supported in a drilling mode and relocated in a floating mode, the emphasis is placed on units that can drill in moderate as well as deep water. Such MODUs are by necessity floating units. Drilling barges are generally suitable for operation in mild environments. They are favoured for tender drilling in southeast Asia, for example.

The most versatile MODUs are either ship-shaped or semi-submersibles. These units are also ideally suited not only to develop the field but also to produce from it. The following two sections further discuss these structures.

1.7.3 Production Units (FPSO and FPS)

Most floating production units are neutrally buoyant structures (which allows six-degrees of freedom) which are intended to cost-effectively produce and export oil and gas. Since these structures have appreciable motions, the wells are typically subsea-completed and connected to the floating unit with flexible risers that are either a composite material or a rigid steel with flexible configuration (i.e. Compliant Vertical Access Risers). While the production unit can be provided with a drilling unit, typically the wells are pre-drilled with a MODU and the production unit brought in to carry only a workover drilling system.

The FPSO generally refers to ship-shaped structures with several different mooring systems. Early FPSOs in shallow waters and in mild environment had spread mooring systems. As more FPSOs were designed and constructed or converted (from a tanker) for deepwater and harsh environments, new more effective mooring systems were developed including internal and external turrets. Some turrets were also designed to be disconnect-able so that the FPSO could be moved to a protective environment in the event of a hurricane or typhoon.

The use of FPS in offshore oil and gas development is proliferating around the world. FPS technology has been in commercial use since the early 1970s when Hamilton Bros. utilized a converted MODU to produce from the Argyll Field in the UK sector of the North Sea. However, Petrobras gets the credit for widespread application of the FPS concept beginning in the late 1970s. The combination of depressed oil prices and advances in subsea production technology made the FPS concept more attractive. Another important reason for its popularity was that Petrobras had the insight on the cost and schedule advantages of MODU conversions and arranged the MODU lease/charter contracts to ensure the ownership transfer of the MODUs to Petrobras at the end of their contracts (typically a two- or three-year contract). FPS technology has become an effective solution for both the marginal and the deepwater field development. Although the advantage of converting semi-submersibles and other mobile offshore drilling units (MODUs) into FPS existed in the 1980s, with the surplus of such MODUs most FPSs put into service in the 1990s were based on newly constructed semi-submersible and Spar units. These structures have the advantages of versatility, mobility (in re-location, adverse weather or politics), relative low cost and self-containment.

Among the nations that are involved in the development and installation of FPS, Brazil has aggressively pushed into deepwater frontier. They first set the goal to produce from 1000 m depth and established a multi-faceted research and development programme to achieve this objective. Once this objective was achieved, they raised the bar and established a new goal of producing from 2000 m water depth. To achieve this target, Petrobras has created Procap 2000, Program for Technological Capability for Deepwater Production, to develop deep and ultra-deep waters of Campos Basin.

Floating Production System units were also installed in the US in the Gulf of Mexico. Unfortunately, the first three units to be installed (Placid Oil's Green Canyon Block 29, Enserch's Garden Banks Block 387/388, Tatham Oil's Ewing Bank Block 958/959) were less than successful due to poor reservoir conditions. The Gulf of Mexico has seen discoveries of more than 50 oil and gas fields with recoverable reserves of more than 40 million BOE in water depths greater than 1968 ft (600 m). It is likely that most of these fields will be developed utilising FPS and perhaps FPSO systems.

1.7.4 Drilling and Production Units

Typically floating units are designed to function as either drilling units or production units to minimise the deck payload and the overall size/displacement of the unit. The basic exceptions to this rule are the Tension Leg Platforms (TLPs) and Spars. These units have limited motions and provide a suitable facility for surface-completed wells.

For a ship-shaped FPSO with very large displacement, an increase in deck payload due to the introduction of a drilling system is not an issue. However, vessel motions have been the primary reason for the hesitation to develop Floating Production, Drilling, Storage and Offloading (FPDSO) units. Advances in technology and the potential for developing deepwater sites offshore Western Africa in a relatively mild environment may result in FPDSOs with mid-ship moonpools to be in service in the next few years.

1.7.5 Platform Configurations

The common floating production units already introduced are briefly described in the following subsections.

1.7.5.1 Semi-Submersible Platform

Semi-submersibles are multi-legged floating structures with a large deck. These legs are interconnected at the bottom underwater with horizontal buoyant members called pontoons. Some of the earlier semi-submersibles resemble the ship form with twin pontoons having a bow and a stern. This configuration was considered desirable for relocating the unit from drilling one well to another either under its own power or being towed by tugs. Early semi-submersibles also included significant diagonal cross bracing to resist the prying and racking loads induced by waves.

The introduction of heavy transport vessels that permit dry tow of MODUs, the need for much larger units to operate in deep water, and the need to have permanently stationed units to produce from an oil and a gas field resulted in the further development of the semi-submersible concept. The next generation semi-submersibles typically appear to be a square with four columns and the box- or cylinder-shaped pontoons connecting the columns. The box-shaped pontoons are often streamlined eliminating sharp corners for better station-keeping. Diagonal bracing is often eliminated to simplify construction.

1.7.5.2 Spar

The Spar concept is a large deep draft, cylindrical floating Caisson designed to support drilling and production operations. Its buoyancy is used to support facilities above the water surface [Glanville, et al 1991; Halkyard 1996]. It is, generally, anchored to the seafloor with multiple taut mooring lines.

In the mid-seventies, Shell installed an oil storage and offloading Spar at Brent Field, in the North Sea. The hull is 95 ft (29 m) in diameter, necks down to 55 ft (17 m) at the water plane, and the operating draft is 357 ft (109 m). Agip installed a flare Spar off West Africa in 1992. The Spar is 233 ft (71 m) long, with a diameter of 7.5 ft (2.3 m), which tapers to 5.5 ft (1.7 m) through the water plane. In 1993 Shell installed a loading Spar at Draugen. The hull diameter is 28 ft (8.5 m) and the operating draft about 250 ft (76 m).

The world's first production Spar was the Neptune Spar installed in 1996 by Oryx Energy Company (now Kerr-McGee) and CNG (fig. 1.15). The Neptune Spar has a hull 705 ft (215 m) long with a 32×32 ft^2 (10×10 m^2) centrewell and a diameter of 72 ft (22 m). The mooring system consists of six lines consisting of wire rope and chain (fig. 1.15).

As of this writing there are 13 Spars in production or under construction. Figure 1.16 shows the progression of Spars built by Technip Offshore, Inc. Three additional Spars have been built by J. Ray McDermott.

The first three production Spars consisted of a long cylindrical outer shell with "hard tanks" near the top to provide buoyancy. The middle section was void, free flooding and

Figure 1.15 Oryx/CNG Neptune Spar (Kerr-McGee)

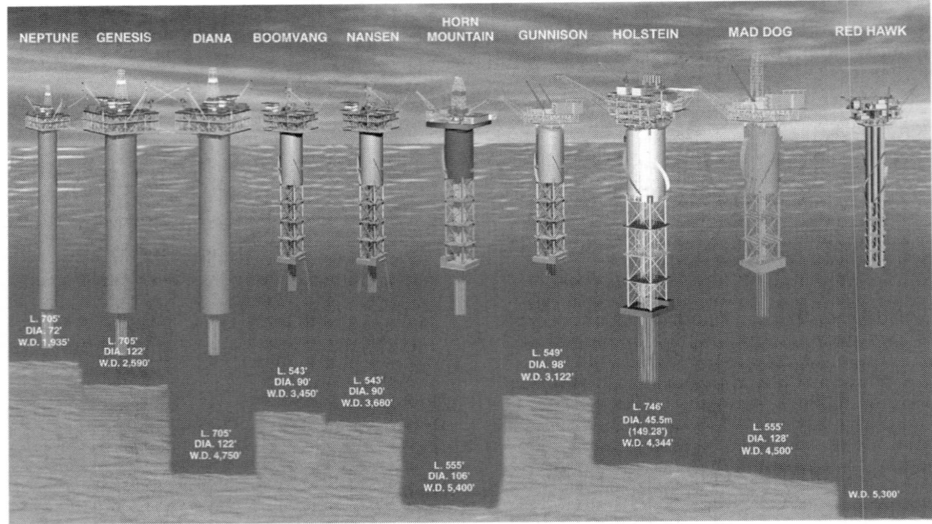

Figure 1.16 Progression of Spars (Technip Offshore)

the lower section consisted of "soft tanks" which were only used to allow horizontal flotation of the Spar during installation, and for holding fixed ballast, if necessary. Subsequent Spars replaced the middle section with a truss structure to reduce weight and cost, and to reduce current drag. Horizontal plates were included between the truss bays to trap mass in the vertical direction to minimise heave motions. Figure 1.17 shows these two types of Spars, the "classic" and the "truss" Spars.

A third generation "cell" Spar was introduced in 2004. It performs similar to the other Spars, but it is constructed differently. The hull consists of multiple ring-stiffened tubes, or "cells", which are connected by horizontal and vertical plates. This method of construction is cheaper than the traditional plate and frame methods.

Because of the length of a Spar, the Spar hull cannot be towed upright. Therefore, it is towed offshore on its side, ballasted to a vertical attitude and then anchored in place. The topside is not taken with the hull and is mated offshore once the Spar is in place at its site. The mooring cables are connected with pre-deployed moorings.

1.7.5.3 Tension Leg Platform

A Tension Leg Platform (TLP) is a vertically moored compliant platform. The floating platform with its excess buoyancy is vertically moored by taut mooring lines called tendons (or tethers). The structure is vertically restrained precluding motions vertically (heave) and rotationally (pitch and roll). It is compliant in the horizontal direction permitting lateral motions (surge and sway).

Several TLPs have been installed in several parts of the oceans of the world (table 1.6). The first TLP was installed in Hutton Field in about 148 m water depth in the UK sector of

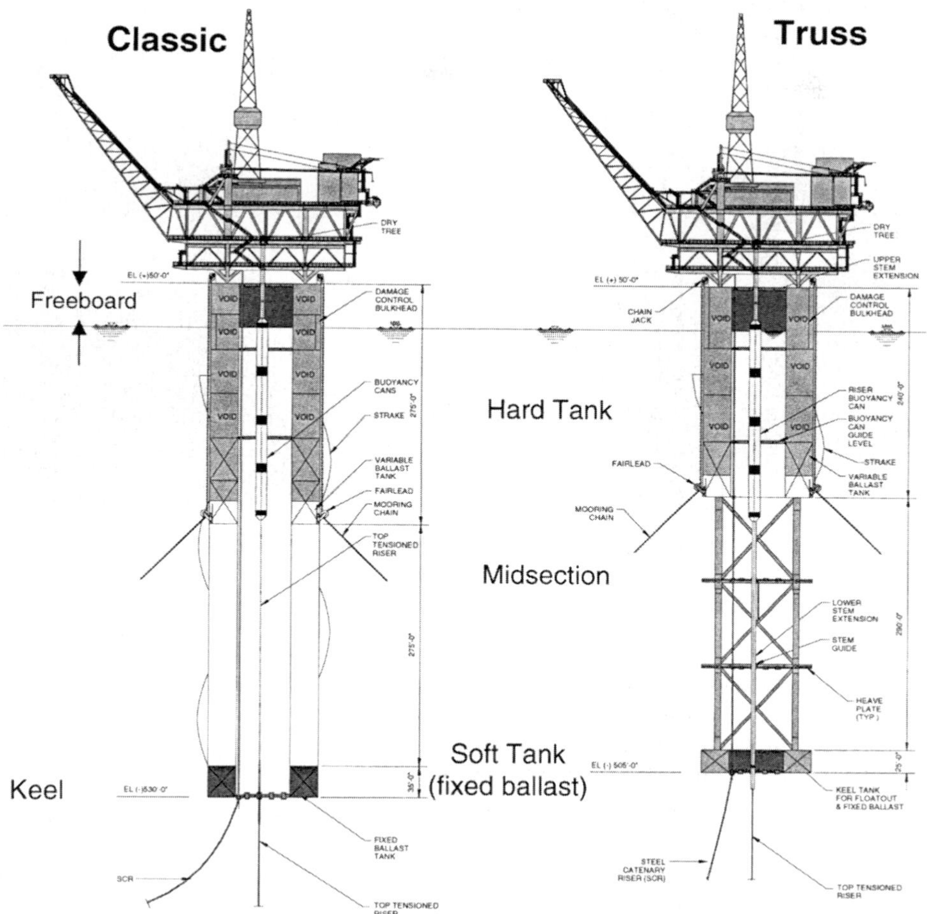

Figure 1.17 Classic and truss Spar (Technip Offshore)

North Sea in 1984. The Operator Conoco could have developed this field far cheaper with a fixed platform but utilised the TLP concept where diver intervention was feasible so that the experience would lead to the use of a TLP in deepwater. Conoco authorised the design of a small wellhead platform (TLWP) in 1986 and the unit was installed in the Jolliet field (1755 ft or 535 m water depth) in 1988. Other units to follow Jolliet in the Gulf of Mexico are Shell's deepwater TLPs. Auger was installed in 1994 in 2867 ft or 874 m water depth, followed by Mars in 1996 in 2930 ft or 893 m, Ram/Powell in 1997 in 3215 ft or 980 m, Ursa in 1999 in 3950 ft or 1204 m and Brutus in 2001 in 2998 ft or 914 m water depth (see fig. 1.18).

The Norwegian sector of North Sea has seen two TLPs: Snorre in 1992 (1017 ft or 310 m water depth) and Heidrun TLP in 1148 ft or 350 m water depth. Heidrun TLP is the first TLP to have a concrete hull.

Table 1.6 Tension leg platforms in-service or decommissioned

TLP	Field/Location	Uniqueness	Water Depth	Year
Hutton	North Sea, UK	First TLP removed in 2001	148 m	1984
Jolliet	Green Canyon, GOM	First deep water well head TLP	335 m	1989
Snorre	Norwegian North Sea		310 m	1992
Auger	Garden Banks, GOM	Has Tethers and Conv. Mooring	872 m	1994
Heidrun	Norwegian North Sea	First TLP with concrete hull	350 m	1995
Mars	Mississippi Canyon, GOM		894 m	1996
Ram/ Powell	Viosca Knoll, GOM	Copy of Mars	980 m	1997
Ursa	Mississippi Canyon, GOM	Largest TLP in GOM	1204 m	1999
Brutus	Green Canyon GOM	Korean Construction	914 m	2001

NOTE: MiniTLPs are not shown on this table. They are discussed in the following subsections

All of the "classic" TLPs consisted of four columns and ring pontoons without diagonal bracing.

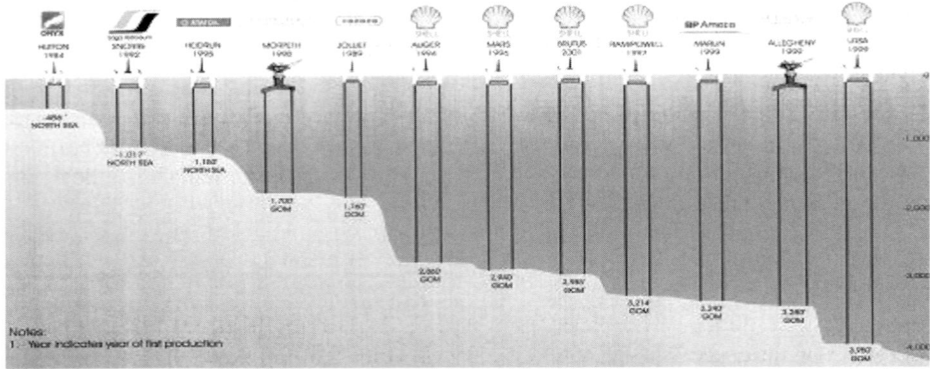

Figure 1.18 Tension leg platforms installed as of 2002 (Courtesy of Deepstar)

Figure 1.19 Extended leg TLP (Courtesy ABB)

A challenge for TLP designers is to keep the natural periods in heave and pitch below the range of significant wave energy. Heave period may be controlled by increasing the pipe wall thickness of the tendons. Pitch period may be reduced by placing the tendons on a wide spacing to increase stiffness. However, it makes the support of the deck with large spans expensive. The Extended Leg TLP, or ETLP (fig. 1.19) was introduced by ExxonMobil on its Kizomba A field in 2003. This concept has four columns on a closer spacing than normal, ring pontoons and pontoon extensions cantilevered to support the tendons on a wide moment arm [Huang, et al, 2000].

Tension Leg Platform technology preserves many of the operational advantages of a fixed platform while reducing the cost of production in water depths up to about 4900 ft or 1500 m. Its production and maintenance operations are similar to those of fixed platforms. However, TLPs are weight sensitive and may have limitations on accommodating heavy payloads.

A conventional TLP is towed to an offshore well site upright at its displacement draft, and then ballasted down so that the tendons may be attached to the TLP at its four corners. The mode of transportation of TLP allows the deck to be joined to the TLP at dockside before the hull is taken offshore.

1.7.5.4 MiniTLPs: SeaStar and Moses

SeaStar is a deepwater production and utility mini-platform [Kibbee, 1996, 1999]. It borrows from the concept of the tension leg platform and provides a cost-effective marginal field application. SeaStar is a small TLP with a single surface-piercing column (fig. 1.20). The column is necked down near the sea surface to reduce surface loads on the structure.

Figure 1.20 SeaStar Mini TLP (Courtesy Atlantia Corp.)

The submerged hull spreads into three structural members at the bottom in a triangular fashion, which are used to support and separate taut tubular steel tendons. The hull provides sufficient buoyancy to support the deck, facilities and flexible risers. The excess buoyancy provides tendon pretension.

SeaStar is generally towed or barged to site in a vertical position. But due to small waterplane area of its single column hull and low centre of buoyancy, it cannot carry the deck with it. Generally, the deck is mated on site similar to Spar once the tendons are connected and tensioned.

Moses MiniTLP (see fig. 2.13) appears to be a miniaturised TLP as the deck structure is supported by four columns and the columns are connected by pontoons. Motion characteristics of Moses is similar to that of SeaStar and, unlike the standard TLPs, miniTLPs need to dedicate a large percentage of their displacement (35–45%) for pretension.

1.8 Classification Societies and Industry Standard Groups

The first "Classification Society", Lloyd's Register of British and Foreign Shipping, was formed in 1834. Its purpose and the purpose of classification societies today is to provide standardised "rules" for designing, building and maintaining ships. This provides owners and insurance carriers with assurances that certain minimum standards have been met. To provide proof of this, an owner can elect to have a vessel "classed" by one of these societies in which case the society will certify that the vessel has been built and maintained according

Table 1.7 Members of IACS

Class	E-mail	Website
American Bureau of Shipping (USA)	abs-worldhq@eagle.org	http://www.eagle.org
Bureau Veritas (France)	veristarinfo@ bureauveritas.com	http://www.veristar.com
China Classification Society (China)	ccs@ccs.org.cn	http://www.ccs.org.cn
Det norske Veritas (Norway)	iacs@dnv.com	http://www.dnv.com
Germanischer Lloyd (Germany)	headoffice@ germanlloyd.org	http://www.GermanLloyd. org
Korean Register of Shipping (South Korea)	krsiacs@krs.co.kr	http://www.krs.co.kr
Lloyds Register of Shipping (UK)	Lloydsreg@lr.org	http://www.lr.org
Nippon Kaiji Kyokai (Japan)	mpd@classnk.or.jp	http://www.classnk.or.jp
Registro Italiano Navale (Italy)	info@rina.org	http://www.rina.org
Russian Maritime Register of Shipping (Russia)	004@rs-head.spb.ru	http://www.rs-head.spb.ru

to their rules. There is no requirement that offshore floater be classed, nor that the class rules be followed. However, there is at least no precedence where a governmental agency has deferred to a classification society for the enforcement of offshore regulations. In the US, an ABS classed floating offshore facility is deemed to meet the US Coast Guard regulations with regard to stability and mooring.

The major classification societies belong to the International Association of Classification Societies (IACS) and are listed in table 1.7, along with contact information. Certain minimum standards have been unified for similar requirements for all the IACS member societies. This includes standards on stability and strength of the structure in question.

Classification can be limited to components of a project. For example, the hull may be classed while facilities are not. Classification involves a chain of requirements, which extend to suppliers of equipment. The owner needs to decide if the benefits of class will outweigh the costs related to classification.

In any event the class rules provide a convenient and readily available tool for designers to use. They cover almost all aspects of design, especially stability, structure, materials, etc.

The rules are available for sale. Many are available free as downloads from the respective web sites listed in table 1.7.

Other documents of equivalent or more use to the designer are the industry standards and regulations, which have been developed by government and industry associations. Some of the most prevalent of these include those published by the organisations listed below:

- American Petroleum Institute (API),
- American Society of Mechanical Engineers (ASME),
- American Society of Civil Engineers (ASCE),
- US Minerals Management Service (MMS),
- United States Coast Guard (USCG),
- American Society for Testing and Materials (ASTM),
- American Welding Society (AWS),
- American Institute of Steel Construction (AISC),
- American National Standards Institute (ANSI),
- Norwegian Petroleum Directorate (NDP),
- UK Department of Energy (UK DOE).

These and other standards are discussed more extensively in various chapters of this book in describing specific offshore structures or requirements.

References

Anon (1992). "Floating production systems proliferating around the world". *Oil and Gas Journal*, Dec. 7, pp. 18–24. Anon. (1992). "FPS systems gain acceptance off Australia, Northwest Europe". *Oil and Gas Journal*, Dec. 14, pp. 26–29. Anon. (2003). "Marco Polo on course to chart virgin territory". *Upstream Magazine*, July 11.

Barusco, Pedron (2002). "The accident of P-36 FPS", *Proceedings of Offshore Technology Conference*, OTC 14159, May, Houston, Texas.

Borelli, A. J., and Perinet, D. (August, 1998). "Laybarge converted for service in deepwater West African fields Polaris to have flexibility to employ reel, S-lay and J-lay methods and two positioning capabilities", *Deepwater Technology*.

Botelho, D., Petrauskas, C., Vannan, M., Mackey, V., and Buddy Lang, B. (May 2000). "Life-cycle-cost-based design criteria for gulf of mexico minimum structures", *Proceedings of Offshore Technology Conference*, OTC 11889.

Bugno, Walter T., and Edward E. Horton (October, 2000). "Storage Spar for deepwater Brazil", *Proceedings, Rio Oil & Gas Conference, Rio de Janeiro, Brazil*, paper IBP-SE-013/00.

Burleson, Clyde. W. (1999). "Deep challenge: the true epic story or our quest for energy beneath the sea", Gulf Publishing Company, Houston, Texas.

Glanville, R., Paulling, J. R., Halkyard, J., and Lehtinen, T. (May 1991). "Analysis of the Spar floating drilling, production and storage structure", *Proceedings of Offshore Technology Conference*, OTC 6701, Houston, Texas.

Halkyard, J. E. (1996). "Status of Spar platforms for deepwater production systems", *Proceedings of ISOPE*, Los Angeles, Vol. I, pp. 262–272.

Halkyard, J. E. (2002). "Evolution of the Spar floating drilling, production and storage system", *Proceedings, Society of Naval Architects and Marine Engineers*, Texas Section Annual Meeting, Houston, Texas.

Horton, E. E. (1987). "Drilling, production and oil storage caisson for deep water", United States Patent 4,702,321, October 27.

Horton, E. E., Brewer, J. H., Silcox, W. H., and Hudson, T. A. (1976). "Means and methods for anchoring an offshore tension leg platform", US Patent 3,934,528, July 27.

Huang, E., Bhat, S., Luo, Y., and Zou, J. (May, 2000). "Evaluation of dry tree platform concepts", *Proceedings of Offshore Technology Conference*, OTC 11899, Houston, Texas.

International Maritime Organization, "The International Convention for the Prevention of Marine Pollution from Ships, 1973 as modified by the Protocol of 1978 relating thereto (MARPOL 73/78)", www.imo.org.

Kibbee, S. (June, 1996). "SeaStar minimal platform for small deepwater reserves", Offshore, pp. 46–79.

Kibbee, S. E., Leverette, S. J., Davies, K. B., and Matten, R. B. (May 1999). "Morpeth SeaStar Mini-TLP", *Proceedings of Offshore Technology Conference*, OTC 10855, Houston, Texas.

Mercier, J. A., et al (1980). "The Hutton TLP: a preliminary design", *Proceedings European Offshore Petroleum Conference and Exhibition*, London, England.

Minerals Management Service (2002). "Deepwater Gulf of Mexico 2002: America's expanding frontier", U. S. Minerals Management Service, Report MMS-2002–021.

Minerals Management Service, (May, 2000). "Gulf of Mexico Deepwater Operations and Activities Environmental Assessment", OCS EIS/EA MMS 2000–001 MMS Gulf of Mexico OCS Region.

OTRC, (2002). "2001–2003 OTRC Project: Investigation of Factors Affecting Ultimate Recovery from Subsea Wells", Offshore Technology and Research Center, Texas A&M University.

Petruska, D. (May, 2003). "Facilites & construction – BP brings polyester moorings to the Gulf", Society of Petroleum Engineers, Gulf Coast Section meeting.

Schneider, D. R. (May, 2000). "A history of a Gulf of Mexico deepwater subsea production system development strategy", *Proceedings of Offshore Technology Conference*, OTC 13116, Houston, Texas.

Thibodeaux, J. C., Vardeman, D. R., and Kindel, C. E. (May, 2002). "Nansen/boomvang projects: overview and project management", *Proceedings of Offshore Technology Conference*, OTC 14089, Houston, Texas.

Westwood, J. (2003). "Deepwater markets and gamechanger technologies", Presented at Deep Offshore Technology 2003 Conference, Nov., available at www.dw-1.com.

Handbook of Offshore Engineering
S. Chakrabarti (Ed.)

Chapter 2

Novel and Marginal Field Offshore Structures

Cuneyt Capanoglu
I.D.E.A.S., Inc., San Francisco, CA, USA

2.1 Introduction

Offshore structures differ from onshore structures both in terms of their pre-service and in-service characteristics. Offshore structure components are fabricated and assembled onshore, transported to an offshore site and then installed. Thus, construction methods, material specifications and the acceptable construction tolerances applicable to offshore structures substantially differ from those applicable to onshore structures.

Both offshore and onshore structures are designed to resist functional gravity loads and the site-specific wind and seismic forces. Offshore structures are subjected to additional forces associated with pre-service construction, transportation and installation and the in-service wave, wave drift, current and ice. Since functional requirements and the site characteristics define a set of given parameters that need to be incorporated into design, these parameters are often identified as "independent variables".

Offshore structures should be designed to minimise both the excitation forces and the response of the structure to these excitation forces. A structure can be designed to optimise functional buoyancy forces and hydrostatic pressure and to minimise the excitation forces associated with wind, wave, current, seismic event and ice. Since the component member-sizes of an offshore structure, their configuration and arrangement directly influence the magnitude of these forces, they are often defined as "dependent variables".

The study of dependent variables in developing an offshore structure configuration is not limited to minimising the excitation forces alone. Additional requirements often dictate trade-offs to meet several conflicting requirements. One such requirement is to have natural periods that would preclude resonant response of the structure to excitation forces. Another requirement for floating structures is to maintain positive metacentric height to ensure desirable stability characteristics.

The term "Innovation" is often defined either as "developing a unique solution to overcome a problem" or "developing a unique answer to a specific need". Many engineers and scientists are tasked to develop novel offshore structures or component systems, which are well suited for cost-effective development of marginal oil and gas fields.

The process of developing a novel structure that meets various offshore field requirements is presented in this chapter together with several examples. Many of these structures are already operating in offshore fields today and several novel concepts are in various stages of development. The main purpose of this chapter is to discuss the logical approach in developing a concept for a particular need and a set of prescribed requirements.

Different types of offshore structures and the history of their development have already been discussed in detail in Chapter 1. In this chapter, the logistics of a field development and the innovation of a structure that suits this development are addressed. Firstly, the offshore oil and gas field developments are reviewed. Then the field development parameters, structure types and the basis for determining a desirable field development option are discussed. The development of novel structures and novel field development systems, and the potential for future advances and innovations are also discussed.

2.2 Overview of Oil and Gas Field Developments

A preliminary decision to develop an oil or a gas field can be made after the completion of studies that define the reservoir and determine its functional requirements, evaluate the site and environmental characteristics and identify the technically feasible development concepts. Then, the capital and life cycle operating expenditures are estimated and economic studies are performed to determine the net return on investment. Since the operator will have limited capital and personnel resources and several commercial fields, only those fields with the highest return on investment are likely to be developed first. Jeffris and Waters (2001) provide an interesting overview of lessons learned from and strategies implemented on Shell-developed deepwater fields.

2.2.1 Field Development Parameters

The parameters that influence the development of a new field are the characteristics of a reservoir, the requirements for drilling, production and export of oil and gas, site and environment as well as design rules and regulations. These topics are discussed in the following sections.

2.2.1.1 Reservoir Characteristics and Modelling

It is possible to approximately define the reservoir size, its configuration and recoverable reserves by utilising all the available seismic data (in 2D, 3D and 4D) and a sophisticated reservoir model. Exploratory wells are drilled to confirm reservoir characteristics and the wells are allowed to flow for a limited period so that:

- the gathered data and the reservoir model can be used to determine the flow characteristics, to study the required number of wells and their arrangement, and to predict the production profile.

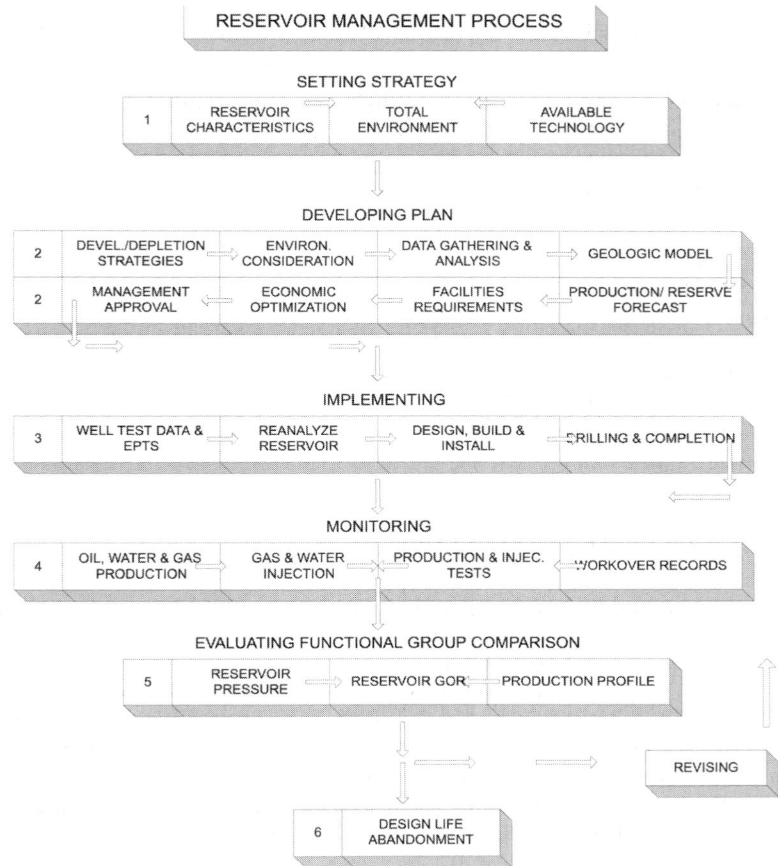

Figure 2.1 Flow chart for reservoir management process

- laboratory tests can be performed to determine the crude characteristics and define the functional requirements that directly affect the drilling, production and export systems for the field.

Figure 2.1 presents the sequence of activities associated with field development and reservoir management.

2.2.1.2 Drilling, Production and Export Requirements

The reservoir and crude oil characteristics permit definition of minimum requirements for field development. Typically, the following conditions are met:

- The number of production and injection wells defines the scope of development drilling.

- Production requirements are defined in terms of system capacities for processing fluids as defined by barrels of oil per day (BOPD), standard cubic feet of gas per day (MMscfd), barrels of produced water (bwpd) and the gas and water injection.
- Production specifics such as hydrates directly affect the deck area and load requirements and the crude oil characteristics, such as the pour and cloud points, directly affect the selection of the export system.

The minimum requirements may have to be adjusted upwards to ensure that the field development concept has excess capacity for the following two reasons:

- A decision to develop an oil or gas field is based on a conservative economic model and the actual recoverable reserves will be most often upgraded following initial production.
- Cost-effective production from a newly discovered adjacent field is possible only if excess capacity is available.

It should be noted that the most cost-effective field development option is to use an existing facility to process production from a newly discovered field. Figure 2.2 illustrates the key variables affecting the field development options.

2.2.1.3 Site and Environmental Characteristics

The site characteristics can readily make the field development not feasible due to technical and/or economic reasons. The key parameters defining the site characteristics, namely the water depth, foundation material, seismicity, ice, wind, wave and current, directly influence the selection of field development option and the magnitude of required investment.

2.2.1.4 Design Philosophy, Rules and Regulations

The other key variables affecting the field development option and the cost are the management philosophy and rules and regulations applicable to the site. Management philosophy may be too conservative to select a novel field development concept and the rules and regulations may substantially add to the cost of some field development concepts, making them financially unattractive.

2.2.2 Structure Types

Offshore structures discussed and illustrated in this book differ from each other and also exhibit common characteristics. All of the structures can be identified to be either "bottom-supported" or "floating". These in turn can be defined as "fixed" or "compliant" and "neutrally buoyant" or "positively buoyant", respectively.

2.2.2.1 Bottom-Supported Structures

There are a wide variety of bottom-supported structures that have been built and are operational today. Historical background on the development of bottom-supported structures is covered in Chapter 1.

Jacket- and tower-type fixed platforms, jackups, gravity base structures and the subsea production system are the typical fixed structures without appreciable compliancy. Guyed Tower, Delta Tower and other bottom-supported structures that rotate about their base, whenever the lateral excitation forces exceed the predetermined design limit, are defined as

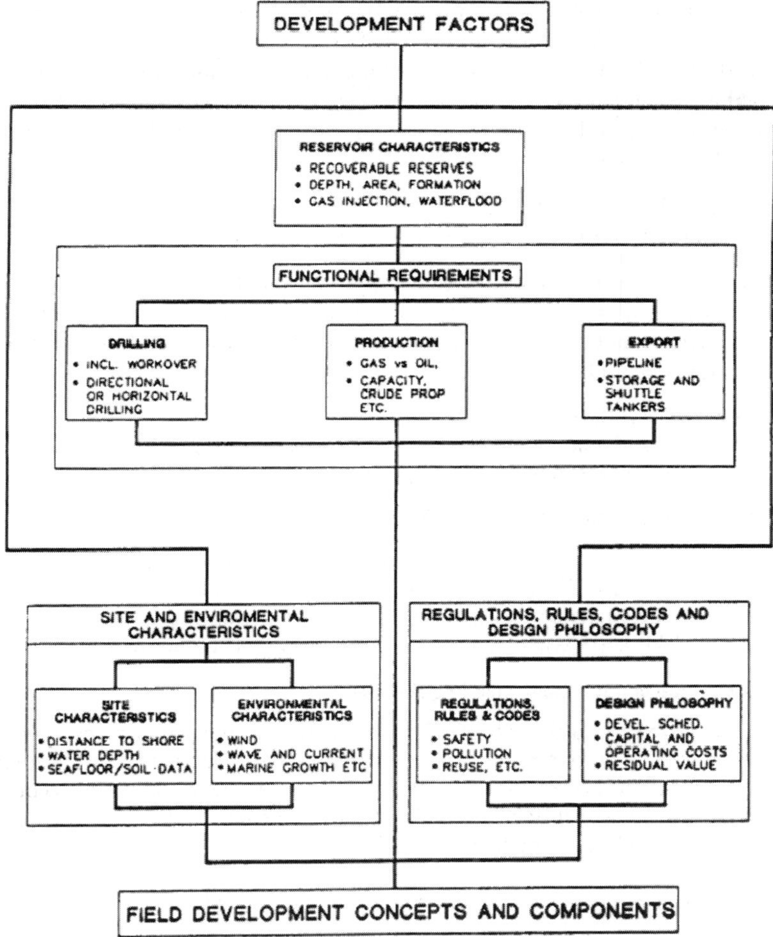

Figure 2.2 Factors affecting field development concepts and components

compliant structures. These compliant structures generate inertia forces due to their motions. Generated inertia forces act against the excitation forces, thereby reducing the net applied loads on the structure.

A discussion on the detailed design of bottom-supported structures is presented in Chapter 6.

2.2.2.2 Floating Structures

There are many in-service floating structures with a wide range of geometry. The history of the development applicable to floating structures is presented in Chapter 1.

Floating Production, Storage and Offloading (FPSO), Floating Production Facility (FPF), Surface Piercing Articulated Caissons (SPARs) and other structures, that freely float, are identified as neutrally buoyant structures. These structures are totally compliant and have six degrees of freedom, namely three displacement (i.e. heave, surge and sway) and three rotational (i.e. pitch, roll and yaw) freedoms. A discussion on the detailed design of these structures is presented in Chapter 7.

Two of the positively buoyant structures are the Tethered Buoyant Tower (TBT) and the Buoyant Leg Structure (BLS). Both TBT and BLS structures are single water-piercing columns tethered to the seafloor to preclude heave motions and therefore have only five degrees of freedom. Other positively buoyant structures have multiple tethers attached to the structure columns or pontoons far apart from each other to effectively preclude the heave, pitch and roll motions. These structures are commonly called Tension Leg Platforms (TLPs), Extended Tension Leg Platforms (ETLPs) and Tension Leg Wellhead Platforms (TLWPs) and may be considered to have essentially three degrees of freedom.

2.2.3 Selection of Field Development Option

The factors affecting field development concepts and components are illustrated in fig. 2.2. The successful development of an oil or a gas field also requires careful consideration of other parameters affecting project economics. These include the number of qualified construction yards and their backlog of work, number of suitable installation contractors and their backlog of work, concept patent rights, government requirements on local content, and taxes (i.e. royalty, value added tax, import duties, etc.).

A simplified field development selection process is discussed in Section 2.6. Further discussions of field development options are available on various publications including de Naurois, et al (2001), Johnston (1988), Clarkston, et al (2001) and Dorgant, et al (2001).

2.3 Technical Basis for Developing Novel Offshore Structures

There are numerous needs or problems that demand the development of innovative solutions. Most of these needs/problems can be grouped into three categories, namely, technical, financial and regulatory. Technically-driven developments and the basic principles dictating these developments are discussed in this section.

2.3.1 Overview of Historical Innovations

The historical development of offshore structures has been already introduced in Chapter 1. Here a brief introduction of the innovative nature of fixed and floating structures is given.

2.3.1.1 Bottom-Supported Structures

The offshore platforms installed in the Gulf of Mexico and offshore Southern California in the 1960s were simple four or eight leg platforms forming three-dimensional space frames and were supported by pin piles within the legs. Some of the platforms had additional skirt piles to accommodate the weak foundation material. These platforms were also constructed

Figure 2.3 Tower type cook inlet platform

of small-diameter tubular members with limited lengths, forming 40–50 ft high bays. Although the framing systems for these platforms were very similar to the framing systems for onshore buildings, an innovation was introduced to accommodate construction of the platform jacket horizontally and then the transportation and launching of the jacket from a barge. This first innovative component of the jacket, the launch trusses, becomes redundant once the jacket is upended, piled and put in service.

Earl and Wright introduced a truly effective innovation that overcame two site-specific problems during the design of offshore platforms for Cook Inlet, Alaska. Platforms in the Cook Inlet are typically subjected to large tidal changes, high current and ice loads. The site is also far removed from areas with fabrication yards and installation equipment. The well-known Cook Inlet-type tower, consisting of a deck supported by four large diameter legs, was introduced to overcome these problems (see fig. 2.3). The large-diameter legs provided adequate buoyancy to wet tow the tower (i.e. self-floater) and on reaching the installation site to permit upending of the tower without utilising heavy lift cranes. The wells were drilled through the drilled and grouted piles within each leg and the legs acted as shields against ice loading.

As the water depth increases, the platform base has to get larger to effectively resist the overturning moment generated by the wind, wave and current forces on the platform. Two considerations leading to innovation are the need to: (1) minimise projected surface area and reduce excitation forces, and (2) ensure that the platform natural periods remain small to preclude its dynamic response to excitation forces. The Extended Base Platform (see fig. 2.4) is a four-leg platform with an eight-leg base meeting both considerations listed above.

The Guyed Tower illustrates an innovation applicable to even deeper water, where the fixed Extended Base Platform is unsuitable. The Guyed Tower is a simple square space frame with a small projected area to minimise applied excitation loads. The principal function of the framing system is to transmit the functional deck loads to the foundation material. The guy wires attached to the tower resist the excitation forces due to wind, wave

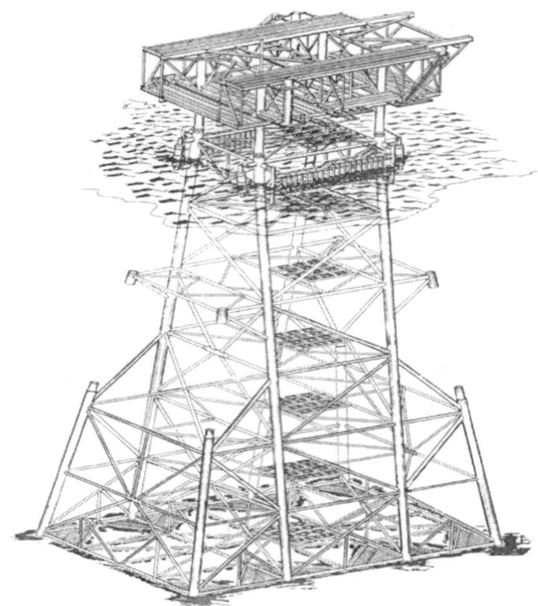

Figure 2.4 Extended base platform

and current as the unit behaves like a fixed platform. When the lateral excitation forces exceed the predetermined limit, the clump weights sitting at the mudline are lifted by the guy wires. In this orientation the guy wire and chain combine to provide the tower with a conventional mooring and turn the fixed tower into a compliant unit.

2.3.1.2 Floating Structures

The original neutrally buoyant floating structures are ships and barges and these vessels are subjected to substantial heave, pitch and roll motions detrimental to offshore operations in intermediate to harsh environment. Considering the green water on the deck and the undesirable motions, an innovation was introduced to separate the deck from the vessel and keep the major portion of the vessel buoyancy away from the water surface. This gave rise to the innovation of semi-submersible. The three-column Sedco 135 semi-submersible is a good example of not only precluding green water and minimising motions but also providing adequate positive stability through the use of large-diameter legs far enough apart.

The eight-column Sedco 700 semi-submersible illustrates additional innovations. The two-pontoons and the thrusters turn the unit into a Mobile Offshore Drilling Unit (MODU) that can move from one well to another under its own power at a reasonable speed. The dynamic positioning system added to some of the units allow the operation of these units at any water depth, unimpeded by the mooring system limitations.

The positively buoyant Tension Leg Platforms (TLPs) have multiple pretensioned tethers that restrict heave, pitch and roll motions. The first TLP installed to produce oil is the Hutton TLP in the North Sea. However, the tethered Screed Barge used to place gravel cover on top of the Bay Area Rapid Transit (BART) tube in San Francisco Bay, predates Hutton by some 15 yr. Heavy concrete weights lowered to the seafloor at the four corners of the barge pretensioned the unit and restricted the heave, pitch and roll motions, facilitating accurate placement of gravel for each lay of about 200 ft length.

2.3.2 Basic Technical Principles

This section describes the basic principles used in the design of these structures. The details of the technical designs for these and other offshore structures are addressed in a greater detail in Chapters 4, 6 and 7.

2.3.2.1 Stability

All free-floating (i.e. neutrally buoyant) structures should have adequate stability to preclude capsizing. Some of the positively buoyant TLPs are stable only in the tethered mode. Considering the potential need to relocate the unit to another field without removing the topside, a positive stability is desirable even for TLPs. The formula for the metacentric height GM (i.e. definition of static stability) is given as:

$$GM = CB + I/V - CG \tag{2.1}$$

where CB = centre of buoyancy, I = area moment of inertia, V = volume and CG = centre of gravity.

A platform consisting of a simple square deck weighing 2000 tons and a 40 ft diameter cylindrical shell column is selected to illustrate the stability characteristics. It is assumed that the keel is 200 ft below the water surface and the column is divided into four compartments. Since inertia of the column about its own centreline is of limited value, to obtain a positive metacentric height, the centre of gravity must be below the centre of buoyancy. When the ballast water alone cannot yield a positive metacentre, the lowest compartment needs to be filled with heavy metals. This is illustrated with an example in table 2.1.

2.3.2.2 Excitation Forces

The drag forces due to wind and current will cause the column to move laterally (giving an offset) while the inertia-dominated forces on the column due to the wave particle accelerations will cause column dynamic excursions. Since water particle accelerations and velocities rapidly decline with distance from the water surface, these drag and inertia forces can be reduced by shifting the structure displacement away from the water surface. The changes in the heave excitation forces affecting column response to excitation forces for a base case and two alternatives are presented in fig. 2.5.

Table 2.1 Illustration of parameters affecting the metacentre

Description	Variable	Comment
Deck		
1. Weight (tons)	2000.0	
2. CG elevation (ft)	235.0	above keel
Column		from + 30′ to −200′
3. Diameter (ft)	40.0	3-diaphragms @ −150′, −100′, −50′ = 4 tanks
4. Length (ft)	230.0	
5. Steel wt (tons)	2000.0	includes compartmentation
6. CG elevation (ft)	115.0	above keel
7. Ballast wt (tons)	4040.0	sea water or (heavy ballast)
8. CG elevation (ft)	50.2 (25.0)	50.2 vs. (25.0)
Displacement (tons)		
9. V = (1)+(5)+(7)	8040.0	$251{,}327^3$ ft of water
Stability		
10. CB (ft)	100.0	above keel
11. I (ft^4)	125,664.0	
12. I/V (ft)	0.5	
13. CG(ft)	112.3 (99.6)	water ballast alone is not adequate
14. GM = (10)+(12)−(13)	−11.8 vs. (+0.9)	heavy ballast is required

It is first observed that the water particle acceleration reduces with the distance from water surface. Alternative 1 appears to be the best option as the water particle accelerations and velocities contributing to the excitation force would be the least. Indeed, some structures are designed to keep the keel as far away from the water surface as feasible. However, two other important parameters should be considered:

Although the logic of keeping the displacement away from the water surface is reasonable for lateral excitation forces, the logic is valid for vertical excitation forces only if the inertia force is the pre-dominant force. The force that always acts in opposition to the inertia force is the variable buoyancy force associated with the wave profile being above or below the mean water surface. The base case configuration provides a reasonable balance between the hull heave (i.e. inertia) and the column float heave (i.e. variable buoyancy) forces for a wide range of wave periods (see fig. 2.5a). Alternative 1 configuration yields smaller hull heave

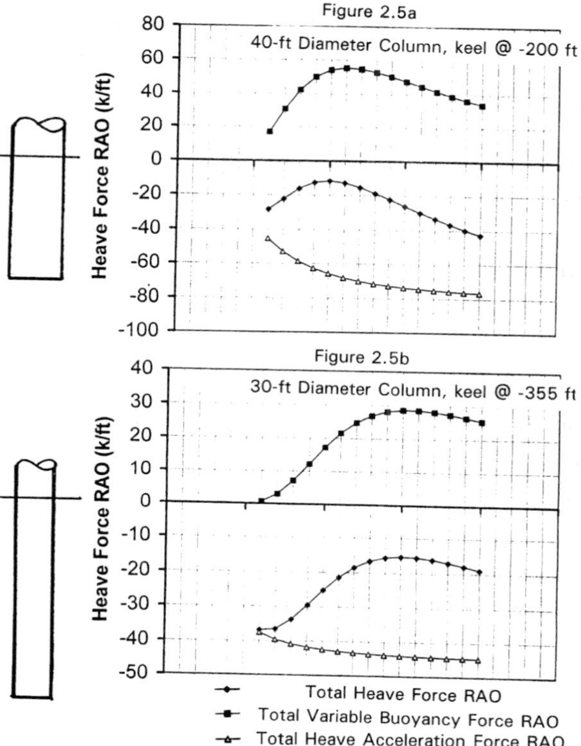

Figure 2.5 Heave force RAOs (kips/ft)

and variable buoyancy forces, yielding still acceptable net heave forces (see fig. 2.5b). Note that:

- Base case net heave forces, expressed as RAOs (force amplitude/wave height), range from 12 kips/ft for a 10 s wave to 28 kips/ft for a 15 s wave period and these forces are dominated by the variable buoyancy force.
- The Alternative 1 net heave forces range from 25 kips/ft for a 10 s wave to 15 kips/ft for a 15 s wave period and are also dominated by the variable buoyancy force.

Alternative 1 has the deepest keel and the structure has to be designed to resist much greater hydrostatic pressure. The extended length may complicate the construction and transportation of the unit.

2.3.2.3 Response of the Unit to Excitation Forces

The physical properties that affect the response of a linearly elastic structural system subjected to an external source of excitation or loading are its mass, elastic properties, and energy-loss mechanism, or damping. The general equation of motion for a structure having six degrees of freedom can be expressed as:

$$m_{ij}\ddot{x}_j + b_{ij}\dot{x}_j + c_{ij}x_j = f_i(t) \tag{2.2}$$

where m_{ij} = mass, moment of inertia and product of inertia of structure (a 6×6 matrix), $\ddot{x}_j$ = acceleration of structure, b_{ij} = linearised damping force coefficients, $\dot{x}_j$ = velocity of structure, c_{ij} = restoring force coefficients, x_j = displacement of structure, $f_i(t)$ = external excitation force.

Heave Characteristics

The vertical motion of the unit is considered first, which will have a similar equation of motion in the z direction

$$m\ddot{z} + b\dot{z} + cz = f(t) \tag{2.3}$$

The forces due to buoyancy, weight and ballast for a free-floating body are not shown on the right-hand side of the equation as they are in equilibrium

$$\rho g V - mg = 0 \tag{2.4}$$

Let the force function $f(t) = F \sin \omega t$ and on neglecting the damping term, equation (2.3) becomes

$$\ddot{z} + (c/m)z = (F/m)\sin \omega t \tag{2.5}$$

Using the operator D for the derivative, and solving for the complimentary solution (i.e. zero external force) only

$$(D^2 + c/m)z = 0 \tag{2.6}$$

and

$$\omega = D = (+/-)[c/m]^{0.5} \tag{2.7}$$

the restoring force coefficient c may be considered to be composed of (a) hydrostatic spring constant, $k_h = \rho g A$, and (b) tethering spring constant, $k_t = EA/L$ for pre-tensioned positively buoyant units, where A is the cross section and E is the modulus of elasticity.

The mass includes the added mass $\rho c' V$, where c' is the added mass coefficient. Thus, the definition of natural frequency becomes:

$$\omega = [\{(EA/L) + \rho g A\}/(m + \rho c' V)]^{0.5} \tag{2.8}$$

Free-Floating Column

Consider the example of the base case configuration in table 2.1 for a free-floating column. In this case

$$\text{column cross-sectional area, } A = 3.1418(20)^2 = 1257\,\text{ft}^2$$
$$\text{column mass, } m = 251{,}327\,\text{ft}^3(64\ \text{lb/ft}^3)/32.2 = 499{,}532\,\text{slugs}$$
$$\text{column added mass, } \rho c' V = \rho c'(D^3/6) = 1.99(40^3/6) = 21{,}227\,\text{slugs}$$

it follows that

$$\omega = [\{(0) + 1.99(32.2)1257\}/\{499{,}532 + 21{,}227\}]^{0.5} = 0.393$$

and the heave natural period

$$T = (2\pi/\omega) = (2 \times 3.1428/0.393) = 16.0\,\text{s}$$

If the base case column with a 40 ft diameter and a keel at 230 ft below water surface is replaced with Alternative 1 having a 30 ft diameter and a keel at 385 ft below water surface for a constant displacement, the smaller water plane area substantially reduces the stiffness term and results in a decrease in frequency and an increase in heave natural period:

$$\omega = \left[\{(0) + 1.99(32.2)707\}/\{499{,}532 + 8{,}955\}\right]^{0.5} = 0.298$$
$$T = (2\pi/\omega) = 21.1\,\text{s.}$$

If Alternative 2 were selected with the same smaller stiffness but larger added mass (i.e. 8955 vs. 75,544), the heave natural period would have been 22.4 s.

Tethered Column

Now consider the base case configuration on table 2.1 for a tethered column, where 2000 of the 4040 tons of ballast was allocated for pretension in water depth of 3000 ft with restraining tether having $A = 60$ of in.^{2}:

$$\text{column cross-sectional area, } A = 3.1418(20)^2 = 1257\,\text{ft}^2$$
$$\text{column mass, } m = \left\{251{,}327\,\text{ft}^3(64\,\text{lb/ft}^3) - 2 \times 2{,}000{,}000\right\}/32.2 = 375{,}308\,\text{slugs}$$
$$\text{column added mass, } \rho c' V = \rho c'(D^3/6) = 1.99(40^3/6) = 21{,}227\,\text{slugs}$$

Then

$$\omega = [\{(29{,}000{,}000 \times 60/3000) + 1.99(32.2)1257\}/\{375{,}308 + 21{,}227\}]^{0.5} = 1.463$$

and due to the high stiffness the heave natural period, $T = (2\pi/\omega) = (2 \times 3.1428/1.463) = 4.30$ s.

Lateral Offset Characteristics

Consider next the equation of motion in the horizontal direction (surge)

$$m\ddot{x} + b\dot{x} + cx = f(t) \tag{2.9}$$

The restoring (i.e. equivalent spring) force at a time, t, can be expressed as $P \sin \Theta$ or $P(x/L)$ for small angles. Using the operator, D, the complimentary solution for the equation becomes

$$(D^2 + P/Lm)x = 0 \tag{2.10}$$

$$\omega = D = (+/-)[P/Lm]^{0.5} \tag{2.11}$$

Thus, for the base case configuration, the mass and the added mass sum up to:

$$m = [\{251{,}327(64.0) - 2 \times 2{,}000{,}000\} + 251{,}327(64.0)]/32.2 = 874{,}840 \text{ slugs}$$

$$\omega = [4{,}000{,}000/(3000 \times 874{,}340)]^{0.5} = 0.0391$$

and the surge natural period, $T = 2 \times 3.1428/0.0391 = 161$ s.

This exercise allows one to select the desired geometry for the environment that is expected for a particular site. It considers the resonant period of the selected geometry of the novel structure and determines any problem with high responses that may be experienced by the chosen structure geometry.

2.4 Other Considerations for Developing Novel Offshore Structures

2.4.1 Financially-Driven Developments

Technical innovations are essential in overcoming site- and reservoir-specific problems. However, most technical innovations also need to be cost-effective to ensure their application to oil and gas field developments. The development of some novel offshore structures applicable to marginal fields may be identified as financially-driven. The following three examples illustrate the basis for such innovations.

2.4.1.1 Multi-Purpose Vessel

In the late 1980s, a vessel was constructed to perform extended production testing from marginal fields with complex reservoir configurations. The vessel would produce from a single well, process and store the oil, sail to a shore facility, transfer the oil and return to the well to repeat the production cycle again. The ability to use the single vessel to perform all of these functions and move from one field to another was attractive enough to have the operator make the financial commitment to construct the unit.

2.4.1.2 Risk and Incentive-Based Contract and a Novel Structure

A small independent operator acquired a marginal field lease from a major oil company not interested in developing the field in the late 1990s. The independent operator chose an innovative but unproven concept, which allowed the field to be developed cost effectively. The contracting arrangement permitted sharing of both the risks and the benefits. The project was successful and since the structure was proven to perform well, three other structures were constructed and the fifth will be constructed shortly.

2.4.1.3 Build, Operate and Transfer (BOT) Contracts

An operator with limited capital is able to develop their fields sequentially by using cashflow from one field to develop another. The offshore industry has adopted the Build, Operate and Transfer (BOT) concept, an innovation first implemented to construct onshore infrastructure projects in developing countries, to permit parallel development of two or more fields. The contractor will build a unit, such as an FPSO, assume operational responsibility for the field and transfer the unit to the oil company at the end of the contract period. Such contracts are based on present worth of the investment, including the cost of financing, insurance, and of course, the profit. The benefits to the oil company from this arrangement is evident as no capital expenditure (CAPEX) is recorded in their books and the total cost of the contract is treated as an operating expenditure (OPEX) and paid off from cashflow.

2.4.2 Regulatory-Driven Developments

Regulatory requirements do influence field development scenarios and sometimes lead to innovations. Offshore personnel are typically evacuated when the field is in the path of an oncoming hurricane or typhoon. An earlier requirement that would force the FPSO to be moved away from the path of a typhoon resulted in the development of disconnectable internal turret system.

Another example of a regulatory-driven development in the Gulf of Mexico is the use of pipelines rather than tankers for export purposes. The U.S. Minerals Management Service (MMS) will not allow flaring of gas to preclude wasting of resources and environmental damage. This requirement contributed the development of pipeline networks in the Gulf of Mexico with the oil companies sharing the cost investment.

2.5 Novel Field Development Systems

2.5.1 Bottom-Supported Systems

Many bottom-supported structure concepts that utilise existing technology in such a way that facilitate development of marginal oil and gas fields cost effectively can be defined as novel structures. Such structures, intended to support relatively small deck payloads and several wells, are often identified as minimal structures. These structures are designed, constructed and installed in months, rather than years, and their total installation costs are usually defined in terms of several million dollars.

2.5.1.1 Fixed Structures

Minimal structures that can be constructed quickly and cost effectively often lack redundancy and they are somewhat more susceptible to failure than other structures. There are more than 100 minimal structure designs and most of these were intended to support deck payloads of 400–1000 tons and transmit the functional and environmental loads to the seafloor through driven or drilled and grouted piles. Some of the other structures carry larger deck payloads and/or rely on gravity base structures, rather than piles, to transmit the loads to the seafloor. Although each minimal structure design is unique, these designs can be grouped into structure types defined as Tripods, Braced Caissons, Braced Monopods and Monotowers. Further details on minimal structures can be obtained from Craig (1995), Beims (1995) and the Offshore Magazine (January 2001).

(1) *Tripods*: A typical Tripod is a tubular space frame consisting of three legs and the bracing system that connects the legs. It is secured to the seafloor with three piles. Some of the more distinctive tripods are Atlantia's SeaHorse III, Enercon's EMOP-3S, Mustang's Skirt Pile Tripod and Pinnacle Engineering's Tripod (see fig. 2.6). Some of the basic characteristics of these structures are summarised in table 2.2 and more comprehensive data appear in Craig (1995).

(2) *Caissons and Braced Caissons*: A Caisson is a relatively large-diameter cylindrical shell that supports a small deck and this type of a structure is applicable to relatively shallow water depth sites. The Caisson structures installed in deeper water are provided with a bracing system to resist lateral loading. A Caisson that may be

Figure 2.6 Typical Tripod (Reproduced courtesy of Pinnacle Engineering)

Table 2.2 Novel Fixed Structures (from Offshore Magazine, January 2001)

Type	Name Unit	Company Name	Production MMscfd BOPD	Steel Weight (tons)	Water Depth m (ft)
Tripod	SeaHorse III	Atlantia	60	1200	92
			5000		(300)
	EMOP-3S	Enercon	100	2300	122
			8000		(400)
	Skirt pile	Mustang	80	2800	132
	Tripod		2000		(433)
	Tripod	Pinnacle	60	1500	114
			1000		(375)
Caisson and Braced Caisson	Caisson	Atlantia	25	300	27.4
					(90)
	Caisson	Petro-Marine	35	300	49
			3000		(161)
	Sea pony	Atlantia	25	520	61
	Braced Caisson				(200)
	Braced Caisson	Worley	50	620	73
			20,000		(240)
Monotower	Varg (NS)	Aker	55	4600	36.5
			57,000		(120)
	AMOSS	BPAmoco	180	1380	41
					(135)
	Monopod (NS)	Brown &	800	5100	33.0
	{modified AMOSS}	Root	80,000 300	1310	(108) 54.9
	SASP	Saipem			(180)

subjected to hurricane loading is typically limited to water depth sites of about 50 m (165 ft) while the Braced Caisson makes it cost-effective to utilise these Caissons to sites with water depths of 80–100 m (260–330 ft). Four different Caisson structures are listed in table 2.2.

(3) *Monotowers*: A typical monotower is a large-diameter cylindrical shell supporting a deck structure and it transfers the functional and environmental loads to the foundation through the framing system and the piles. Typically, a monotower is supported by four piles at four corners of the framing system. The size of the monotower and the restraining system (i.e. framing system and piles) depend on the deck payload and the environmental condition. Thus, as illustrated in table 2.2, the structures identified to be in the harsh North Sea environment would require substantial steel to resist wind, wave and current loads.

It is not possible to state that one structure type is superior to others. Whether an oil company selects a Tripod-, Caisson- or a Monotower-type structure depends on many factors including site water depth, foundation material and environmental characteristics, construction and installation considerations, decommissioning and removal cost, and most importantly, the management philosophy on field development option.

2.5.1.2 Compliant Structures

Following the installation of Exxon's Lena Guyed Tower, a number of other bottom-supported compliant tower designs were developed including Articulated Tower and Flex-Leg Tower. Amerada Hess' Baldpate Compliant Tower (CT) utilises axial tubes as flex elements (a.k.a., Articulated Tower) and Texaco's Petronius Compliant Tower utilises flex legs. These compliant tower designs are evolved from earlier CT technologies and are considered to be advances in technology, rather than technological innovations.

2.5.1.3 Hybrid Structures

A unique floating MODU, Kulluk, discussed in Section 2.5.2 was converted from a drilling unit to a production unit and installed at a shallow water site offshore Sakhalin Island. To match the concave hull profile with the water surface or a site-specific structural framing system was designed, constructed and installed prior to the arrival of Kulluk. The Kulluk unit (fig. 2.7) was ballasted and set on top of the framing system, making it a submersible sit-on-bottom unit.

Another novel hybrid structure is an expandable/retractable framing system with multiple fuel bladders. The retracted structure is expanded and fuelled upon arriving at the designated area and then ballasted to set it on the seafloor. When needed, the unit is deballasted and brought to the water surface and Fuel Cache becomes available for those in need.

2.5.2 Neutrally-Buoyant Floating Systems

Neutrally-buoyant floating systems are equally effective in both shallow and deepwater, and are designed to have six degrees of freedom (i.e. completely compliant). Most systems are conventionally moored and the water depth primarily affects the mooring system. The effect of mooring system weight on the floating unit displacement is relatively small on large units but very large on smaller units. In deep and ultra-deep water, dynamic positioning, initially introduced for effective station-keeping of MODUs, may be more effective than conventional mooring for the stationkeeping of floating production facilities.

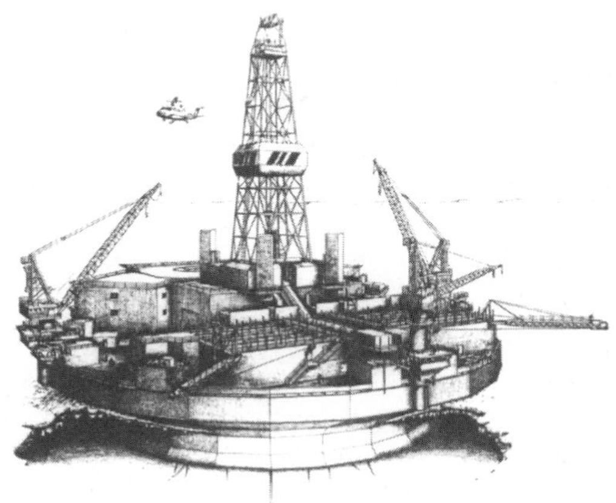

Figure 2.7 Kulluk Mobile Offshore Drilling Unit (MODU)

Some floating structure innovations are primarily technically driven. However, most floating structure innovations are driven by both technical and economic considerations.

2.5.2.1 Technically-Driven Innovations

Three technically-driven innovations are briefly discussed here to show that sometimes technical requirements alone are adequate for the development of novel structures.

Kulluk MODU, a novel floating unit, was designed and constructed to effectively work in ice-infested waters. The circular structure with the concave hull (fig. 2.7) effectively uses its mass and geometry to break the sheet ice.

Many floating systems are shown and their characteristics summarised in the Offshore Magazine (2000). One of the units, the Deep Draft Submersible (DDS) shown in fig. 2.8 [Halkyard, et al 2002] is designed not necessarily to minimise the excitation forces acting on it. Rather, the plate structure in DDS provides the unit with substantial added mass in the vertical direction so that the natural heave period of DDS is increased. Another DDS unit [Johannessen, et al 2002] has extended legs and a deep draft hull. This configuration was developed to reduce both the vertical and lateral excitation forces on DDS so that heave motions of the unit were reduced and mooring line forces due to wave drift and current were minimised.

A standard Spar unit developed for Oryx's (now Kerr-McGee's) Neptune field has its keel at about 213 m (700 ft) below the water surface. By keeping the keel and the centre of displacement far away from the water surface, both heave and surge excitation forces are reduced (fig. 2.9).

The next innovation in Spar concept was introduced to reduce both the lateral forces that dictate the mooring system design and the vertical forces that determine the heave motions. The ballasted compartments near the keel were replaced with a framing system that reduced both the displaced volume (i.e. reduction in inertial forces) and the projected

Figure 2.8 Deep Draft Submersible (DDS) (Reproduced courtesy of Technip)

Figure 2.9 Truss Spar in dry tow (Reproduced courtesy of Technip)

surface area (i.e. reduction in drag forces). These Truss Spars may be considered to be the Second Generation Spars.

The Third Generation Spars are made up of multiple cylindrical shells with the core cylindrical shells extending further than the secondary cylindrical shells to provide a lower keel and a smoother transition from one stiffness to another. This Cell Spar configuration has several advantages over standard Spars including reduction in steel requirements (i.e. less weight), elimination of complex and fatigue sensitive structural details, construction simplicity and the ability to construct the individual cells in parallel.

However, another major advantage of the Cell Spar may be its response characteristic to external excitation forces. A typical Spar is susceptible to vortex-induced vibrations (VIV) when subjected to high currents such as the Loop Current in the Gulf of Mexico (GOM). Initial data on the first two Spars installed in the GOM indicate that even with the helical strakes, VIV is not eliminated effectively. Back-calculated equivalent drag coefficients on these Spar units may be somewhere between 1.3 and 1.8. The Cell Spar's single advantage over the standard Spar is its geometry. Multiple cylinders are a passive and an effective method to minimise the potential for VIV. By providing a variable spacing between the cylindrical shells (i.e. both in a cross section and along the vertical axis) it is more likely that the VIV phenomena may occur but it will be limited to one region of the Cell Spar. The development of SPAR units is shown in fig. 1.16.

2.5.2.2 Technically and Economically-Driven Innovations

Since most innovations are driven by both technical and economic requirements, a substantial number of structures can be examined here.

The Truss-Spar shown in fig. 2.9 [Bangs, et al (2002); Beattie, et al 2002] was developed to overcome some of the disadvantages of a Standard Spar. The length of a Standard Spar (e.g. Neptune and Genesis Spars) necessitated the construction and transportation of these units in two parts across the Atlantic and joining of the two parts in the Gulf of Mexico. The Truss Spar is about 3/4th the length of the Standard Spar and was constructed and towed as a single unit.

The keel and the centre of displacement for the Truss Spar is closer to the water surface than the keel and the centre of displacement for the Standard Spar and therefore, the Truss Spar is subjected to higher average water particle velocities and accelerations. However, this disadvantage of the Truss Spar is offset as its displacement is less than that of a Standard Spar and the Truss Spar incorporates plate elements to increase the added mass and move the natural heave period further away from the energy-intensive wave spectra. The smaller size of the Truss Spar also requires less capital expenditure due to cost savings in construction and transportation.

The Cell Spar units will cost less than the Truss Spar due to the facts that (1) they require less steel, (2) are easier to construct because of minimisation of complex joints and details, and (3) construction schedule is shortened with parallel construction of cylindrical shells. Another major benefit of the Cell Spar over other Spars is in the reduction in operational expenditures (i.e. OPEX). The characteristics listed also indicate that the Cell Spar will have lower inspection, repair and maintenance costs than either the Truss or the Standard Spar.

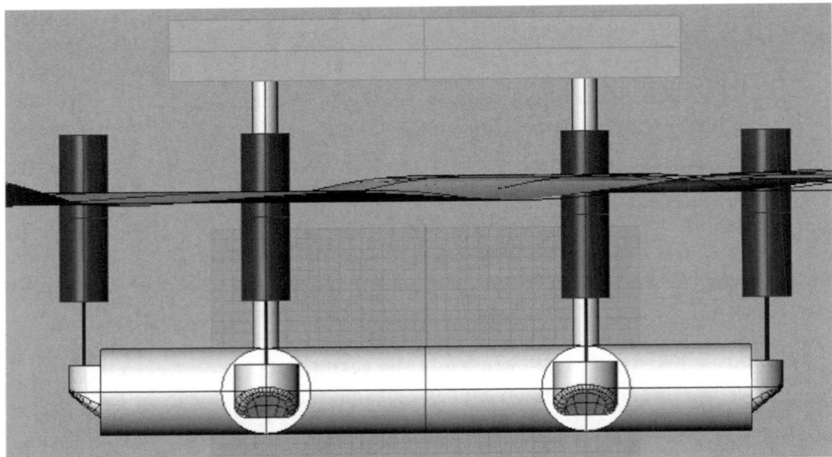

Figure 2.10 Articulated Stable Offshore Platform (ASOP) (Reproduced courtesy of Offshore Model Basin-OMB)

A concept known as Articulated Stable Offshore Platform (ASOP) is well suited for cost-effective development of marginal oil fields in remote areas without adequate infrastructure (see fig. 2.10). ASOP consists of a submerged hull, a multitude of tethered buoyant floats and a deck structure. ASOP hull is large enough to provide adequate buoyancy to support the deck structure and provide storage space for 5–7 days of crude oil production. ASOP's articulated floats are designed to minimise heave, pitch and roll motions by developing float forces opposing the forces acting on the hull. This innovation also limits the magnitude of the dynamic surge forces on ASOP as each float is compliant. Another ASOP concept utilises four cylindrical columns to support the deck, rather than the tubular framing system. Substantial cost advantages may be envisioned with ASOP as the components will be constructed in parallel and the deck structure will be fully commissioned at the yard quayside, eliminating the costly offshore installation, hookup and commissioning.

2.5.3 Positively-Buoyant Floating Systems

Positively-buoyant floating systems are pretensioned platforms tethered to the seafloor. Due to this tethering system all positively-buoyant floating systems are heave-restrained and can be categorised as Tension Leg Platforms (TLPs). These structures differ from standard TLPs that require huge investments. The positively-buoyant floating systems discussed in this chapter are unique in terms of their configuration, small size, performance characteristics and low cost that make these structures well suited for the development of marginal fields. While some of these structures may be called MiniTLPs, they can all be secured to the seafloor foundation system based on suction piles, template and drilled and grouted piles or gravity base structures. These positively-buoyant systems will also differ from each other in terms of their configuration, installation characteristics, in-service motions and post-service removal/relocation as indicated in table 2.3. Some of these structures are briefly discussed here.

Table 2.3 Positively-buoyant floating system characteristics

Characteristic	Options
Configuration	One, three or four columns supporting deck with or without pontoons
Transportation	Dry or wet tow
Installation	Ballasted hull upright or upended floated over or lift-installed deck
Well systems	Surface- or subsea-completed wells with or without lateral supports
Motions	Heave-restrained (five-degrees of freedom) Heave, pitch/roll restrained (three-degrees of freedom)
Relocation	Unstable or stable in free-floating mode

2.5.3.1 SeaStar MiniTLP

The deck structure of SeaStar is supported by a single column with three pontoons converging at the keel of the column. At the end of each pontoon, two symmetrical porches are built-in to attach the six tethers, two at each pontoon. The hull is dry towed to the installation site, ballasted and connected to the tethers. Then, the deck is lift installed on a stable platform.

By 2004 there were four of these units installed in the Gulf of Mexico, namely Agip's Morpeth and Allegheny, Chevron's Typhoon and TotalFinaElf's Matterhorn. Chevron's Typhoon MiniTLP [Young and Matten (2002)] is illustrated in fig. 2.11.

2.5.3.2 Moses MiniTLP

The deck structure of Moses is supported by four closely spaced columns connected with pontoons at the keel. Tethers are connected to pontoon extensions to increase the lever arm and reduce tether pretension requirements. Eight tethers, two at each pontoon extensions, connect the unit to the seafloor [Koon, et al 2002]. Moses TLP is shown in fig. 2.12.

By 2004 there were two Moses MiniTLPs installed in the Gulf of Mexico, namely, the Prince and the Marco Polo.

2.5.3.3 Buoyant Leg Structure (BLS)

The Buoyant Leg Structure (BLS) has four basic component structures: (1) deck, (2) buoyant column supporting the deck, (3) restraining leg connecting the BLS to the seafloor and (4) seafloor system composed of either a suction pile or a gravity base structure. The BLS unit, like SeaStar and Moses, is heave-restrained. However, unlike

Figure 2.11 SeaStar MiniTLP for Typhoon Field (Reproduced courtesy of Atlantia Corporation)

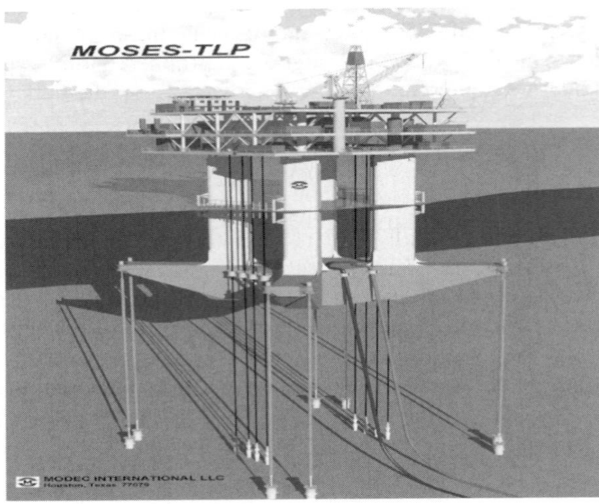

Figure 2.12 Moses MiniTLP for the Prince Field (Reproduced courtesy of Sea Engineering Associates)

SeaStar and Moses, it is not totally pitch- and roll-restrained. The restraining leg provides some rotational stiffness to minimise the pitch and roll motions. Thus, while the TLP is defined to have three-degrees of freedom and the Spar unit is a six-degree of freedom system, a BLS unit has five-degrees of freedom.

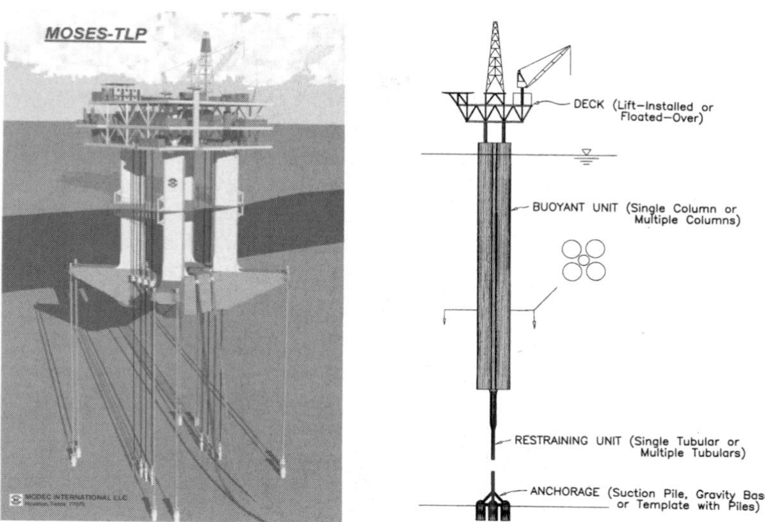

Figure 2.13 Buoyant Leg Structure (BLS) (Reproduced Courtesy of I.D.E.A.S., Inc.)

Work completed to date on BLS includes a two-phase Joint Industry Project (JIP), several preliminary designs and the model testing of a single column BLS and a multi-column BLS [Shaver et al (2001) and Capanoglu et al (2002)]. A typical BLS unit is illustrated in fig. 2.13.

2.6 Discussion of Selected Innovative Field Development Concept

It is technically and commercially feasible to develop many of the marginal oil and gas fields scattered around the world. The application of innovative concepts and technologies together with appropriate contracting strategies will most likely result in further reduction of life cycle costs. Estimated life cycle costs will remain valid only if the risks and uncertainties are correctly accounted for. Many neutrally- and positively-buoyant structures have been constructed and installed, while others have been proposed to develop marginal oil and gas fields.

An overview of field development components and functions is presented to first document the interacting elements with functional relationships. Then, an innovative field development concept is proposed for further scrutiny. Several innovative structures can then be selected for a comparative assessment. Selected structures are discussed in Section 2.7.

2.6.1 Overview

Based on the reservoir study results, site water depth, environmental characteristics and existing infrastructure, various screening studies are performed to rule out impractical field development options. After identifying the technically and economically feasible field development options, it is desirable to study them in detail to determine the most desirable option.

It is a standard practice to estimate the (1) capital expenditures (CAPEX), life cycle operating expenditures (OPEX), (2) cashflow based on recoverable reserves, production profile and the projected market price of barrel of oil equivalent (boe), and (3) the net return on investment. However, a very important parameter affecting not only the cost but potential damage to the environment is the "risk" factor. This parameter is especially important if the innovative concepts utilise less-than-mature technologies. Thus, it is necessary to assess the life cycle risk costs associated with the reliability characteristics of each concept/structure/component.

The risk costs include those due to inherent or natural causes (e.g. hurricane) and those due to external causes (e.g. operations and maintenance). The combined life cycle risk costs are intended to provide insight into the reliability characteristics of each system component. Four key considerations in determining the life cycle reliability characteristics are safety, durability, serviceability and practicality. These considerations can be defined as:

- safety: the requirement to prevent failure from natural hazard is met
- durability: the overall requirement for maintenance is met
- serviceability: the requirements for drilling and production are met
- practicality: the requirements for schedule and expenditures are met

Reliability can be defined as the likelihood or probability that the system components have the ability to meet the four considerations applicable to each system.

2.6.2 Field Development Concept

To facilitate comparative assessment of alternative innovative facilities, reservoir, crude and site characteristics are assumed to be:

(1) recoverable reserves of 100 million barrels of oil equivalent,
(2) reservoiur formation requiring frequent workovers,
(3) undesirable cloud and pour point, and
(4) 1000 m (3280 ft) water depth site susceptible to frequent hurricanes.

These assumptions indicate that the wellhead platform:

(1) should have a small deck payload to minimise displacement and cost,
(2) be used for development drilling to minimise CAPEX,
(3) will require substantial workover/interventions, and
(4) cannot utilise a pipeline to export oil due to undesirable cloud point and pour point characteristics.

A feasible field development concept for this hypothetical site requires:

(1) a drilling, wellhead and partial processing (DWPP) platform,
(2) a floating, production, storage and offloading (FPSO) vessel,

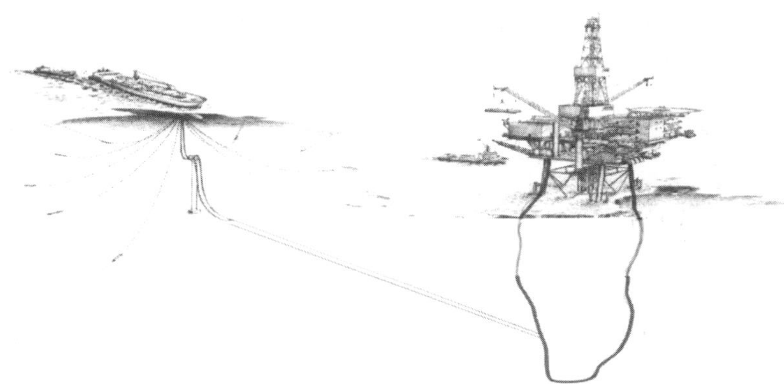

Figure 2.14 Hypothetical field development concept

(3) a dual pipeline between the DWPP platform and the FPSO vessel, and
(4) a gas to be re-injected into the reservoir for future recovery.

The partially processed fluids will be transferred from DWPP platform to the FPSO through the pipeline with a 30–40% water cut. The separated water is heated and returned to the DWPP platform. The dual pipeline permits emergency shutdown of the system and evacuation of personnel in the event of an imminent hurricane and still ensure that the pipeline will not be permanently clogged. The components suitable for this hypothetical field are illustrated in fig. 2.14 and several innovative DWPP structures are discussed in Section 2.7.

2.7 Discussion of Selected Innovative Structures

Many innovative concepts/structures have been designed and some of them constructed and installed. The objectives of this section include a discussion of the background for innovation, overview of innovative concepts/structures and a qualitative and quantitative comparison of three representative designs to provide the reader with a comprehensive understanding of innovative concepts/structures.

To ensure that both qualitative and quantitative comparisons remain valid, a topside design compatible with a marginal field was assumed. As indicated in table 2.4, a modular drilling rig is used for development drilling and replaced with a workover rig following the completion of development drilling. Equipment on DWPP platform, following the drilling programme, consist of minimal processing equipment, heating systems, pumps and generators. The total estimated deck payload during both the development drilling and production phases should remain close to 3000 tons.

Note that the deck structure steel requirements can differ from about 1000 tons for a deck structure supported by four columns to 1100 tons for a deck supported by a single column.

Table 2.4 Assumed topside loads for DWPP platform

Condition	Weight (tons)	Comments
Drilling		Modular drilling system to be removed
Modular drilling system	2400	
Living quarters/utilities	600	
Subtotal	3000	
Production		Workover drilling system and processing module added
Workover drilling system	700	
Living quarters/utilities	600	
Partial processing system	1500	
Subtotal	2800	

2.7.1 Structures Selected for In-Depth Discussion

The three concepts/structures selected for an in-depth discussion substantially differ in their configuration and their response to excitation forces. These characteristics also permit further discussion of key parameters affecting their motion characteristics and how these characteristics can be adjusted. The three concepts/structures selected are:

- Six-degrees of freedom system: Spar and Trussed Spar
- Five-degrees of freedom system: BLS
- Three-degrees of freedom system: MiniTLP

The basic configuration chosen for each system is defined in table 2.5.

The total displacements shown for each configuration, estimated steel weights, ballast and pretension requirements are based on sizing program algorithms and are considered to be approximate. The objective is to produce approximate data for each configuration and facilitate their comparative assessment.

It should be noted that the Buoyant Leg Structure – BLS [Shaver, et al 2001; Capanoglu, et al 2002] and the Tethered Buoyant Tower – TBT [Halkyard, et al 1991] are somewhat similar as they are both heave-restricted buoyant columns. BLS differs from TBT not only because its restraining leg offers partial rotational stiffness but also because larger BLS units are constructed of multiple smaller cylinders forming a single water-piercing column with a centrewell.

2.7.2 Construction and Construction Schedule

Construction simplicity translates into fewer defects during fabrication and assembly, faster construction and reduced cost. An unstiffened or ring-stiffened cylindrical shell is the simplest to construct. The next simplest construction forms are the orthotropically stiffened flat plates (i.e. typical pontoons) and large-diameter cylindrical shells (i.e. ring- and stringer-stiffened columns). Radially and transversely stiffened columns, such as the early Spars with large centrewells, are the most complex to construct. The deck structure is not

Table 2.5 Developed configurations for each selected system

Items and Selected Units	Spar	Truss Spar	BLS	MiniTLP
Water depth (ft)	3000	3000	3000	3000
Hull centrewell (ft × ft)	30×30	30×30		
Column dia. (ft)	60	60	44	55
Bot. of hard Tank (ft)	−400	−400	−400	100
Keel elev. (ft)	−700	−550	−400	100
Restraining leg dia. (ft)			5	
Restraining leg L (ft)			2600	
No. of pontoons				3
Pontoon length (ft)				80
Pontoon A×B (ft × ft)				25×25
No. of mooring lines	12	12		
No. of tendons				6
A. Deck payload (tons)	3000	3000	3000	3000
B. Deck steel (tons)	1100	1100	1100	1100
C. Riser force (tons)	–	–	500	500
Sum A+B+C (tons)	4100	4100	4600	4600
E. Column steel (tons)	1500	18,700	3600	1500
F. Truss steel (tons)	–	–	–	–
G. Restr. leg steel (tons)	–	–	1700	–
H. Pontoon steel (tons)	–	–	–	1400
I. Tendon steel (tons)*	–	–	–	–*
J. Mooring steel (tons)	2200	2200	–	–
Sum E to J (tons)	21,900	20,900	5300	2900
K. Appurtenances (tons)	1000	1000	1000	500
L. Ballast (tons)	39,000	–	5900	–
M. Pretension (tons)	–	–	2600	5000
* includes steel weight				
Sum K+L+M	40,000	1000	9500	5500
Sum A through M (tons) i.e. displacement	66,000	26,000	19,400	13,000

discussed as it is similar for all the three concepts and constructed separately and installed on top of the hull structure after it is secured. The deck structure for the first three units cannot be installed at the quayside due to their configuration (i.e. keel elevation). The fourth unit, MiniTLP, is typically unstable with the deck in free-floating mode, requiring the installation of tendons prior to the lift installation of the deck structure.

The typical Spar unit identified above is the more difficult to construct than others not only because of its stiffening arrangement but also because of its length. Since it could not be dry towed in one piece, it has to be constructed in two pieces, and then joined close to the final installation site. The trussed Spar is an improvement over a typical Spar because: (1) it can be constructed and transported in one piece, and (2) the trussed and plated lower segment is there to lower the centre of gravity of the unit and increase its added mass (i.e. to improve its heave natural period) and do not have to be constructed to tight tolerances.

The MiniTLP is easier to construct than the Spar or the Trussed Spar due to its orthotropically stiffened column and pontoon configurations. However, the complex load transfer system from the tendons to the pontoons and from the pontoons to the column requires substantial effort to meet the tolerances and minimise fabrication defects.

The BLS consists of ring- and stringer-stiffened buoyant column and a small diameter restraining leg. The fabrication of both components is simple and they can be joined at the quayside. Although it does not have the complexity of a column-to-pontoon joint details, the buoyant leg is connected to the restraining leg with a transition cone and the design details in this area need to be carefully developed to minimise stress concentration factors and achieve high fatigue lives.

Based on the past construction data and the characteristics of each unit, about 24 months will be required to design, construct and install a Spar or a Trussed Spar. The MiniTLP and the BLS can be designed, constructed and installed in less than 18 months.

2.7.3 Transportation and Installation

If a Spar unit is constructed and installed in the same region it can be wet towed, upended and secured with 12- or 14-point mooring system. The deck structure is most likely to be towed on a cargo barge and lift-installed. The deck structure for each concept will be precommissioned at the yard and the final commissioning performed at the installation site.

If the Spar unit is constructed in another region, such as a European or a Far Eastern yard, it will most likely be dry-towed due to both technical and commercial reasons.

A MiniTLP is most likely to be dry towed even if the construction and the installation sites are in the same region. The most likely foundation system for the MiniTLP will be drilled and grouted piles or the suction piles. The tendons (i.e. six for SeaStar and eight for Moses) are flooded, upended and secured. The unit is partially deballasted following connection of the tendons and further deballasted as the deck structure is lift-installed.

A BLS unit with the buoyant and restraining legs joined at the quayside can be wet towed to the installation site in the same region. If the construction site is at another region, the two components can be dry towed to a quayside in the same region to join the two units. Once at the installation site the lower three compartments will be ballasted in preparation for upending. Free-flooding of the restraining leg will upend the unit. Further ballasting will allow either float-over or lift installation of the deck and the unit will remain stable in the free-floating mode. Then the restraining leg will be connected to the pre-installed foundation system consisting of drilled and grouted piles, suction pile or a gravity base structure. An alternative for this installation method is the upending of the buoyant leg alone followed by the use of on-board drilling rig to join segments of small-diameter restraining leg and lower them to the seafloor.

Installation effort and cost will be somewhat similar for all the three options if the heavy lift vessel is used. BLS installation will be less than the others if the self-upended unit is joined with a float-over deck without the use of a heavy lift vessel.

2.7.4 In-Service Response and Utilisation

2.7.4.1 In-Service Response

The MiniTLP has the best in-service response characteristics as the tendons restrict heave, pitch and roll motions (i.e. only three-degrees of freedom: surge, sway and yaw). Platform natural periods are away from energy-intensive wave spectra and do not appreciably contribute to dynamic amplifications. However, these characteristics are achieved at a cost. About 40% of the displacement/buoyancy is allocated for tendon pretension and weight to preclude these tendons going into compression. The shallow draft unit is subjected to large lateral forces and the couple effect due to combined lateral and vertical forces develops large tendon tension variations. The tendon pretension has to be larger than this amplitude due to the extreme 100-year recurrence interval storm.

The BLS unit is heave-restricted (i.e. five-degrees of freedom) and the slender unit with the centre of buoyancy at about 61 m (200 ft) below the water surface [vs. about 24 m (79 ft) for the MiniTLP] is subjected to much smaller lateral forces and the couple effect. Thus, rotational excitation forces due to a 100-year recurrence interval storm are reasonably small and the rotational stiffness of the restraining leg limits the maximum 100-year storm condition roll and pitch angles of the unit to about 3.5°. The restraining leg bending stress due to this event is kept at a reasonable level by introducing the transition between the buoyant and restraining legs to control the curvature. These BLS characteristics indicate that pretension requirements are not controlled by pitch and roll motions. Indeed, pretension requirement is determined (i.e. about 15% of displacement) based on the magnitude of lateral forces due to wind, wave drift and current forces and the desirable offset limit for that water depth.

The Spar unit is not dynamically restrained and is identified to be a six-degrees of freedom system. Although the unit is designed to have deep draft to minimise heave forces, substantial heave motions occur and the surface-completed risers have to be decoupled from Spar motions. Typical relative motions (i.e. single amplitude) between the risers with their own buoyancy chambers in the centrewell and the Spar unit may be as much as 20 ft. Pitch and roll motions due to an extreme event are substantial and can reach 10°. Although the lateral forces are minimised due to deep draft characteristics of the unit and the dynamic forces do not result in appreciable dynamic surge (i.e. excursions) of the unit, quasi-static lateral forces due to wind, wave drift and current are substantial. A 12- or a 14-point mooring system is required for station-keeping.

2.7.4.2 Downtime and Utilisation

All of the above three structures have good motion characteristics throughout the year, except during periods when they are in the path of a large storm or a hurricane. Since all platforms in the Gulf of Mexico are shut down and the personnel evacuated when a large storm or a hurricane is imminent, the downtime is approximately similar for all of the platforms.

Spar and Trussed Spar exhibit heave motions and the decoupled risers and the export risers are susceptible to fatigue damage and will most likely be replaced more frequently on a Spar than heave-restrained platforms.

2.7.4.3 Consequences of Deck Payload and Water Depth Changes

All the structures are sensitive to both deck payload increases and increases in water depth. Typically, every thousand tons of additional deck payload will require 3500–4500 tons of additional buoyancy, substantially increasing the hull size. Increases in water depth will also require additional buoyancy to support the increased weight of mooring lines, restraining leg or the tendons. Additionally, highly pretensioned tethers of a MiniTLP would be more susceptible to high frequency dynamics, such as ringing and springing, in deeper water.

2.7.5 Post-service Utilisation

2.7.5.1 Utilisation Without Relocation

All the three platforms are ideally suited to support the surface-completed dry wells. When the production declines to unsustainable levels (i.e. cost of operation exceeding cashflow) the operator can either shut down the production facility or use the facility to produce from adjacent fields. The ability to drill horizontal wells 10 miles or more permit drilling of new wells into an adjacent field. Another possibility is to offer the facility to partially process the oil and/or gas from a nearby field with subsea completed wells and tied to the facility with flowlines.

2.7.5.2 Utilisation Following Relocation

The Spar and Trussed Spar will be the easiest platforms to relocate and put to service at another field. After a five- or a seven-year service life at a given marginal field, the Spar unit will most likely need a new set of mooring lines and, therefore, the new set of mooring lines cannot be attributed to the relocation cost.

Buoyant Leg Structure unit is stable following the ballasting (i.e. to reduce restraining leg pretension to zero) and disconnecting of the unit from its foundation. Thus, it can be relocated to another site within the region with relative ease. However, the restraining leg length has to be either shortened or extended to be compatible with the water depth at a new location. If the design of the unit is based on the restraining leg being installed by the onboard rig by joining the segments, the process of changing the restraining leg length is quite simple. If the restraining leg was joined to the buoyant leg at the quayside and the unit upended as a single unit, adjustment of the restraining leg length becomes more complex. The design has to incorporate connection elements allowing removal and addition of segments while on location.

An ultra-deep water BLS [Capanoglu, et al 1997; Shaver, et al 2001] is secured to a gravity base foundation with a 1525 m (5000 ft) wire rope and a 915 m (3000 ft) long chain. This unit can be deballasted to lift the gravity base and relocated to any site in the vicinity with a 1525–2440 m (5000–8000 ft) water depth range.

The MiniTLP is the least likely candidate for post-service utilisation at another location. Since the unit is unstable when the tethers are disconnected (i.e. free-floating mode), the installation process has to be reversed and the deck unit removed prior to relocation. Once at a new site, extended or shortened tethers are first installed and then the deck unit will be ready for installation.

2.7.6 Capital and Operating Expenditures

Capital expenditures (CAPEX) for the hull and the station-keeping system will differ as a function of both construction and installation costs. The unit with the simplest construction will require the least material, will suffer the least defects and repairs and will be delivered ahead of the other units. Since the water depth and the magnitude of applied excitation forces differ from one geographic site to another, their effect on the platforms discussed in this section will have a different relative impact on the constructed cost. Other reasons for variations in constructed cost of one platform over the others will be the market conditions, yard location and labour rates and the experience, or the lack of experience, of the yard with the platform configuration.

By relying on a recent bid data and the documented unit costs, capital expenditures were determined for each unit as a function of topside cost, which remains constant for each platform. These costs, expressed in percentages, are shown in table 2.6 to illustrate the approximate cost breakdown.

Operating expenditures (OPEX) are very much a function of the operational scenario and the management philosophy. The operator able to perform the tasks with the least number of personnel will have the least offshore OPEX. The add-on OPEX charges to an offshore facility to account for the cost of support base and the head office engineering and management support charges vary greatly from one operator to another. Thus, OPEX on the platforms discussed will vary less with the platform type and more with the operator and the operational philosophy.

Table 2.6 Estimated Capital Expenditure (CAPEX) Distribution

System Category	SPAR	BLS	MiniTLP
Facilities	17	28	21
Deck/Hull structure	41	27	24
Mooring system	22	–	–
Tethering system	–	19	23
Foundations	–	12	12
Towing/Installation	20	14	20
Total	100	100	100

2.7.7 Residual Value and Risk Factors

Although these platforms have substantial residual values, most often the residual value is not incorporated in determining the present value of investment. The use of conservative estimate of recoverable reserves and the neglect of residual value are intended to determine the minimum return on investment so that the operator with several fields to develop can rank them and ensure that company profits are maximised.

The risk factors exist in each critical phase of design, construction, transportation, installation and in-service operations. Assuming that risk factors for the design, construction, transportation and the installation are equally applicable to all of the platforms, it is necessary to review the in-service operations.

2.7.8 Summary Discussion

Each structure discussed above has unique advantages over the others. The selection of one concept over another primarily depends on actual field conditions (i.e. reservoir data, site water depth, foundation material and environmental characteristics, infrastructure) and the operator's marketing strategy and operational philosophy. Each concept can be optimised to meet specific reservoir and site characteristics. The differences in the in-service performance of each unit [Capanoglu, 1997] have an effect on not only the constructed cost but also the inspection, maintenance and repair costs. However, selection of one concept over another also depends on the operator's philosophy and the emphasis (i.e. weighing factor) placed on unique characteristics of each concept. Some of these weighing factors are:

- Spar and Trussed Spar can be re-deployed from one field to another with relative ease. They have the highest weighing factor to perform Early Production Testing (EPT) at a field or to produce from a truly marginal field. BLS has the second highest weighing factor while the MiniTLP has the lowest weighing factor.
- Spar and Trussed Spar exhibit substantial heave motions and both the production risers and the export risers require substantial inspection, maintenance and repair effort and cost. Thus, the Spar and the Trussed Spar have a lower weighing factor than a BLS or a MiniTLP.
- A MiniTLP has negligible pitch and roll motions and is desirable at a site susceptible to very harsh environment. Thus, a MiniTLP has the highest weighing factor here, BLS the second highest and the Spar and the Trussed Spar the lowest weighing factors.
- The BLS unit does not have complex stiffening details or column-to-pontoon nodes and can be constructed practically at any yard. It also has the least unit construction cost. Thus, BLS and the Cell Spar have the highest weighing factor for this item, MiniTLP the second highest and the Spar and the Trussed Spar the lowest weighing factor.
- The restraining leg of the BLS provides lateral support to production risers. If the number of risers is small (i.e. 4–6), they can be kept inside the restraining leg and protected from the environment. If the emphasis is on the well system (i.e. re-entry, fatigue lives, etc.), BLS will have the highest weighing factor. Since the other concepts do not provide support/protection to the risers, the weighing factors will be substantially smaller.

2.8 Future Field Development Options

2.8.1 Technological Innovations and their Impact

Technological innovations most frequently occur when the environment for innovation is suitable. A suitable environment is established when the number of engineers and scientists responsible for technological advancement and breakthrough is adequate. They are provided with adequate funds to validate their innovations including performance of model basin and field tests. Even after an innovation is validated, it is extremely difficult to get it accepted by the operators for various reasons, the most important being: "let somebody else be the first" syndrome. It took 15–20 yr before horizontal drilling and subsea trees were fully accepted. It took close to 20 yr before a viable subsea processing system was implemented as a pilot project on Troll field offshore Norway. Although Shell Oil's Brent Spar was installed in the North Sea in the 1970s, the first Spar-based production unit (i.e. Neptune field in the Gulf of Mexico) was not installed until the mid1990s. This time period between the development of a concept and its implementation is sometimes referred to as the "Valley of Death" [Schroeder and Pena, 2002].

Often it is also difficult to make a clear distinction between the technological advancement and innovation. In some instances an innovative concept or a component incorporate both technological advancement and innovation. Some of the more prominent advancements and innovations are briefly discussed here.

2.8.1.1 Hull Configuration and Vortex-Induced-Vibration (VIV) Phenomena

Large-diameter columns, whether defined as Deep Draft Caisson Vessels – DDCV (i.e. Spar), Tension Buoyant Tower (TBT) or Buoyant Leg Structure (BLS), are susceptible to Vortex-Induced-Vibration when subjected to high current, such as the loop current in the Gulf of Mexico. The DDCV units, provided with helical strakes to preclude or minimise VIV, are subjected to very high lateral forces due to current. The back calculation of applied forces indicates that the equivalent drag coefficient may be in the 1.3–1.8 range. Thus, the helical strakes do not appear to fully function as intended.

Technological advancement, which is also an innovation is the use of multiple cylinders joined together to form a single cylinder instead of the use of a single Caisson. One of the original intentions for the use of multiple cylinders was to replace the complex internal stiffening system with simple ring and stringer stiffening system and to accelerate fabrication schedule. The simple details not only enhance initial constructed quality by reducing defects and repairs, but also improve life cycle maintenance and repairs by lowering stress concentration factors (SCFs) and increasing fatigue lives of component members and joint details. However, there was a legitimate concern that theoretical analyses indicating a closed off centrewell water column would be at times subjected to large pressure differentials resulting in the rise of the water column.

This innovation occurred while preparing for the model basin tests to validate analytical study results. Theoretical analyses that indicated the rise of the isolated centrewell water column at certain periods also indicated that the openings between the cylinders would not only solve the water column problem but would tend to break up any vortices formed.

To study these concerns the small space between the multiple cylinders forming a single large-diameter column were provided with removable plate elements. Tests performed indicated that adequate open space between the cylinders would preclude centrewell water column rise [Shaver, et al 2001]. These tests and further analytical studies indicated that the multiple cylinders forming a single column with variable space in-between cylinders would reduce the VIV effect.

2.8.1.2 Tethering System Limitations and Potential Solutions

Positively buoyant structures, whether being identified as a three-degrees of freedom system (i.e. TLPs and TLWPs) or a five-degrees of freedom system (i.e. TBTs and BLSs), have pretensioned elements of various configurations. Typically, these elements provide adequate stiffness to keep the natural heave periods low (i.e. less than about 4.5 s) to preclude magnification of externally applied heave forces. As the water depth increases, it is necessary to increase the cross-sectional area so that the increased stiffness and the increased water depth offset each other to maintain a reasonable heave natural period. Increased cross-sectional area (i.e. increased weight) and the increase in water depth require substantial increase in pretension requirements, which in turn requires a larger hull displacement. Long and slender tethers with very high pretensions are susceptible to not only the high frequency dynamics (i.e. ringing, springing) but a higher probability of failure during its design life.

The potential solutions offered include the use of polyester and composite materials, which have nearly neutral buoyancy and do not require large increases in pretension due to increased weight with increasing water depth. The use of polyester, composite materials or the wire rope with the modulus of elasticity of 3–10 times lesser than that of a steel tubular combined with increased water depth push the natural heave periods upwards. To preclude dynamic amplifications of heave forces due to energy intensive wave spectra it is better to reduce the stiffness (i.e. lower Young's Modulus) and achieve a tethered heave period of 22–23 s for a unit with a free floating natural heave period of 25 s. It is an innovative approach to have a vertically tethered unit with a natural heave period close to its free floating natural heave period only because there was no precedent for this concept.

2.8.1.3 Steel Catenary Risers (SCRs) and their Response to High Current

Steel catenary risers (SCRs) are extremely susceptible to current-induced displacements and vortices. Three-and-a-half knot currents observed some 700 ft below the water surface in the Gulf of Mexico are different than the loop current observed closer to the surface. The depth and the speed of this current will dramatically shorten the fatigue life of SCRs, unless preventive actions are taken. The traditional method of placing fairings is effective but costly due to maintenance and replacement requirements.

An innovative restraining leg of a BLS may provide just the right solution to SCRs. These SCRs can be kept within the restraining leg to about 1000 ft below the water surface and exit through a transition cone. Since the function of the restraining leg is station-keeping, SCRs are protected from the environment at no additional cost.

2.8.2 Innovations Affecting Cost Efficiencies

2.8.2.1 Hull Configuration

The original intention to use multiple cylinders, instead of a single large diameter cylinder, was to achieve structural and constructional efficiency. Large-diameter hull requires complex internal stiffening system resulting in construction complications and substantial time to construct. The use of multiple small-diameter cylinders with ring- and stringer-stiffening results in a simple construction, as cylinders may be fabricated in parallel. A preliminary study produced more than 30% reduction in structural steel requirements, 25% reduction in constructed cost and a four-to-six-month reduction in construction period. Variable distance between cylinders tends to localise the VIV and the reduced lateral quasi-static forces also allow reduction in station-keeping system cost.

2.8.2.2 Tethering System Limitations and Potential Solutions

The use of materials with close to neutral buoyancy (i.e. polyester and the composite materials or wire rope with buoyancy elements) was originally contemplated to minimise tethering system weight increases with water depth (i.e. increased length) and increased area (i.e. stiffness for heave). An increase in tethering system weight increases hull displacement, which in turn attracts more lateral forces. Increased lateral forces require substantial increase in pretension to meet the target offset limits. All of these contribute to an increased CAPEX. Thus, avoiding tethering system weight increase results in cost savings.

2.8.3 Most Likely Field Development Innovations/Concepts

It is extremely difficult to predict future field development concepts and developments under conditions suitable for innovation. The circumstances affecting the oil industry at present are not suitable for innovation. Considering the current cuts in the research and development funds throughout the industry, the move by many talented engineers and scientists to better paying projects and the management and staff reductions within the oil companies do not offer much promise for an innovative concept to become commercial in the short term.

Further discussions covering technology gaps, innovation and potential application of such technologies are presented by Bhat, et al (2001), Botker, et al (2002), Kibbee, et al (2002), Paulo, et al (2001), Schroeder, et al (2002).

References

Bangs, A. S. and Miettinen, J. A. (2002). "Design of the truss spars for the nansen/boomvang field development", *Proceedings of 2002 International Offshore Technology Conference*, OTC Paper 14090, Houston, Texas, 6–9 May.

Beattie, S., Michael, P., and Steven, R. (2002). "Nansen/boomvang field development – construction and installation", *Proceedings of 2002 International Offshore Technology Conference*, OTC Paper 14090, Houston, Texas, 6–9 May.

Beims, T. (April, 1995). "Structures Help Marginal Field Profits", *American Oil & Gas Reporter*.

Bhat, S. U., Kelly, P. J., and McBee, H. (2001). "Architecture and technology gaps for TLPs and DDCVs in ultradeep water", *Proceedings of Offshore Technology Conference*, OTC Paper 13093, Houston, Texas, 30 April–3 May.

Botker, Stig and Johannessen, Thomas (2002) "Composite risers and tethers: the future of deep water TLPs", *Proceedings of Offshore Technology Conference*, OTC Paper 14176, Houston, Texas, 6–9 May.

Capanoglu, Cuneyt, C. (1997). "Comparative assessment of new generation deepwater marginal field structures", *JNOC Workshop II, Deepwater Offshore Field Development – Offshore Mechanics and Engineering*, OMAE '97, Yokohama, Japan, June.

Capanoglu, C. C. and Copple, R. W. (1997). "A new generation mobile offshore drilling unit (MODU) – buoyant leg structure with a retractable gravity base", *Black Sea 97, Deep Water Drilling and Production Technology Symposium*, Ankara, Turkiye, 2–3 October.

Capanoglu, C. C., Shaver, C. B., Hirayama, H., and Sao, K. (2002). "Comparison of model test results and analytical motion analyses for a buoyant leg structure", *Proceedings of International Society of Offshore and Polar Engineers*, ISOPE 2002, Kitakyushu, Japan, 27–30 May.

Clarkston, Bradley J., Dhuldhoya, Nimish, P., Mileo, Michael, A., and Moncrief, John, A. (2001). "Gulf of Mexico ultra-deepwater development study", *Proceedings of 2001 International Offshore Technology Conference*, OTC Paper 12964, Houston, Texas, 30 April–3 May.

Craig, Michael, J. K. (1995). "Minimum offshore structures cost less, pose higher risk", *OGJ Special, Oil & Gas Journal*, July 17.

de Naurois, Herve, J., Desalos, Arnaud, P., and Schroeder, Art, J. (2001). "Enabling/enhancing technologies: value-ranking process and results", *Proceedings of 2001 International Offshore Technology Conference*, OTC Paper 13089, Houston, Texas, 30 April–3 May.

Dorgant, P. L. and Balint, S. W. (2001). "System selection for deepwater production systems", *Proceedings of 2001 International Offshore Technology Conference*, OTC Paper 12966, Houston, Texas, 30 April–3 May.

Halkyard, J. E., Paulling, J. R., Davies, R. L., and Glanville, R. S (1991). "The tension buoyant tower: a design for deep water", *Proceedings of Offshore Technology Conference*, OTC Paper 6700, Houston, Texas, 6–9 May.

Halkyard, J. and Chao, P. (2002). "A deep draft semisubmersible with a retractable heave plate", *Proceedings of 2002 International Offshore Technology Conference*, OTC Paper 14304, Houston, Texas, 6–9 May.

Jeffris, R. G. and Waters, S. A. (2001). "Getting deepwater development right: strategies for the non-operator", *Proceedings of 2001 International Offshore Technology Conference*, OTC Paper 14201, Houston, Texas, 30 April–3 May.

Johannessen, T. B. and Wanvik, L. (2002). "Deep draft floater motions – verification study", *Proceedings of 2002 International Offshore Technology Conference*, OTC Paper 14303, Houston, Texas, 6–9 May.

Johnston, D. (1988). "Economics models: field development proposals require critical review of technical, economic models", Offshore Magazine, December.

Kibbee, Stephen, E. and Snell, David C. (2002) "New directions in TLP technology", *Proceedings of Offshore Technology Conference*, OTC Paper 14175, Houston, Texas, 6–9 May.

Koon, J. R. and Wybro, P. G. (2002). "Development of the prince field", *Proceedings of 2002 International Offshore Technology Conference*, OTC Paper 14173, Houston, Texas, 6–9 May.

Offshore Magazine, (September 2000). "2000 Worldwide Survey of Deep Water Solutions".

Offshore Magazine, (January 2001). "2001 Worldwide Survey of Minimal Offshore Fixed Platforms & Decks for Marginal Fields".

Paulo, C. A. and Euphemio, M. (2001). "Deepwater field development – combination of mature and nonconventional technologies", *Proceedings of 2001 International Offshore Technology Conference*, OTC Paper 12968, Houston, Texas, 30 April–3 May.

Shaver, C. B., Capanoglu, C. C., and Serrahn, C. S. (2001). "Buoyant leg structure preliminary design, constructed cost and model test results", *Proceedings of International Society of Offshore and Polar Engineers*, ISOPE 2001, Stavanger, Norway, 17–22 June.

Schroeder, Jr., Art, J., and Pena, E. (2002). "Accelerating the commercialization of deepwater technology – a global overview", *Proceedings of Offshore Technology Conference*, OTC Paper 14336, Houston, Texas, 6–9 May.

Young, W. S. (2002). "Typhoon development project overview", *Proceedings of 2002 International Offshore Technology Conference*, OTC Paper 14122, Houston, Texas, 6–9 May.

Young, W. S. and Matten, R. B. (2002). "Typhoon seastar TLP", *Proceedings of 2002 International Offshore Technology Conference*, OTC Paper 14123, Houston, Texas, 6–9 May.

Handbook of Offshore Engineering
S. Chakrabarti (Ed.)

Chapter 3

Ocean Environment

Subrata K. Chakrabarti
Offshore Structure Analysis, Inc., Plainfield, IL, USA

3.1 Introduction

The metocean conditions experienced by the offshore structures are the subject of this chapter. "Metocean" refers to the combined effect of the meteorology and oceanography. As such, the metocean condition refers to a number of meteorological and oceanographic conditions. These factors include:

- local surface wind,
- wind-generated local waves,
- swell (long-period waves) generated by distant storms,
- surface current also generated from the local storms,
- energetic deep water currents associated with low frequency, large basin circulation, and
- non-storm-related currents, which are site-specific, such as loop current in the Gulf of Mexico or coastal current in the Norwegian northern North Sea.

Offshore structures, which are placed in the ocean for the exploration and production of resources beneath the ocean floor, are at the mercy of the environment they are subject to by nature. These environments that the structures may face are the ocean waves, wind and current. Earthquakes and Tsunami waves may also occur in the water in certain parts of the world. For the survival of these structures, the effects of the environments on them must be known by the designer of these structures, and considered in their design. Additionally, the selection of rigs and handling equipment, and the design of risers and mooring systems are also critically dependent on the predicted site-specific metocean condition. This chapter describes the environment that may be expected in various parts of the world, and their properties. Formulas in describing their magnitudes are provided where appropriate so that the effect of these environments on the structure may be evaluated.

3.2 Ocean Water Properties

Important properties of ocean waters are the density and viscosity. The salinity, by itself, is of secondary importance in the offshore structure design. However, the mass density of water is a function of the salinity in the water. The temperature plays a major role in the values of these quantities and the values change with the change in the water temperature. The density values not only determine the forces on a structure placed in water, but the difference in the density in different layers of water with depth may contribute to the internal waves in the deeper region, which has a very important effect on a submerged structure.

3.2.1 Density, Viscosity, Salinity and Temperature

Generally, the largest thermocline occurs near the water surface irrespective of the geographic region of an ocean. The temperature of water is the highest at the surface and decays down to nearly constant value just above 0° at a depth below 1000 m. This decay is much faster in the colder polar region compared to the tropical region and varies between the winter and summer seasons.

The variation of salinity is less profound, except near the coastal region. The river run-off introduces enough fresh water in circulation near the coast producing a variable horizontal as well as vertical salinity. In the open sea, the salinity is less variable having an average value of about 35%.

The variation of density and kinematic viscosity of water with the salinity and density is shown in table 3.1. The dynamic viscosity may be obtained by multiplying the viscosity with mass density. For comparison purposes, the values of the quantities in fresh water are also shown. These fresh water values are applicable in model tests in a wave basin where fresh water is used to represent the ocean. In these cases, corrections may be deemed necessary in scaling up the measured responses on a model of an offshore structure.

3.3 Wave Theory

Ocean waves are, generally, random in nature. However, larger waves in a random wave series may be given the form of a regular wave that may be described by a deterministic theory. Even though these wave theories are idealistic, they are very useful in the design of an offshore structure and its structural members. The wave theories that are normally applied to offshore structures are described in this section.

There are several wave theories that are useful in the design of offshore structures [see, for example, Chakrabarti (1987)]. These theories, by necessity, are regular. Regular waves have the characteristics of having a period such that each cycle has exactly the same form. Thus the theory describes the properties of one cycle of the regular waves and these properties are invariant from cycle to cycle. There are three parameters that are needed in describing any wave theory. They are:

1. period (T), which is the time taken for two successive crests to pass a stationary point,

Table 3.1 Density and viscosity vs. temperature of fresh and sea-water

Water		Mass Density		Mass Density		Kinematic Viscosity		Kinematic Viscosity	
Temp.		Fresh water		Salt water @ 35%		Fresh water, × 10^{-5}		Salt water @ 35%, × 10^{-5}	
°F	°C	lbm/ft^3	kg/m^3	lbm/ft^3	kg/m^3	ft^2/s	m^2/s	ft^2/s	m^2/s
32	0	1.9399	31.074	1.9947	31.952	1.9291	0.1792	1.9681	0.1828
34	1.11	1.9400	31.076	1.9946	31.950	1.8565	0.1725	1.8974	0.1763
36	2.22	1.9401	31.077	1.9944	31.947	1.7883	0.1661	1.8309	0.1701
38	3.33	1.9401	31.077	1.9942	31.944	1.7242	0.1602	1.7683	0.1643
40	4.44	1.9401	31.077	1.9940	31.941	1.6638	0.1546	1.7091	0.1588
42	5.56	1.9401	31.077	1.9937	31.936	1.6068	0.1493	1.6568	0.1539
44	6.67	1.9400	31.076	1.9934	31.931	1.5530	0.1443	1.6035	0.1490
46	7.78	1.9399	31.074	1.9931	31.926	1.5021	0.1395	1.5531	0.1443
48	8.89	1.9398	31.073	1.9928	31.922	1.4538	0.1351	1.5053	0.1398
50	10.00	1.9396	31.069	1.9924	31.915	1.4080	0.1308	1.4599	0.1356
52	11.11	1.9394	31.066	1.9921	31.910	1.3646	0.1268	1.4168	0.1316
54	12.22	1.9392	31.063	1.9917	31.904	1.3233	0.1229	1.3758	0.1278
56	13.33	1.9389	31.058	1.9912	31.896	1.2840	0.1193	1.3368	0.1242
58	14.44	1.9386	31.053	1.9908	31.890	1.2466	0.1158	1.2996	0.1207
60	15.56	1.9383	31.049	1.9903	31.882	1.2109	0.1125	1.2641	0.1174
62	16.67	1.9379	31.042	1.9898	31.874	1.1769	0.1093	1.2303	0.1143
64	17.78	1.9375	31.036	1.9893	31.866	1.1444	0.1063	1.1979	0.1113

(Continued)

Table 3.1 Continued

Water		Mass Density		Mass Density		Kinematic Viscosity		Kinematic Viscosity	
Temp.		Fresh water		Salt water @ 35%		Fresh water, $\times 10^{-5}$		Salt water @ 35%, $\times 10^{-5}$	
°F	°C	lbm/ft^3	kg/m^3	lbm/ft^3	kg/m^3	ft^2/s	m^2/s	ft^2/s	m^2/s
66	18.89	1.9371	31.029	1.9888	31.858	1.1133	0.1034	1.1669	0.1084
68	20.00	1.9367	31.023	1.9882	31.848	1.0836	0.1007	1.1372	0.1056
70	21.11	1.9362	31.015	1.9876	31.838	1.0552	0.0980	1.1088	0.1030
72	22.22	1.9358	31.009	1.9870	31.829	1.0279	0.0955	1.0816	0.1005

2. height (H), which is the vertical distance between the crest and the following trough. For a linear wave, the crest amplitude is equal to the trough amplitude, while they are unequal for a non-linear wave, and
3. water depth (d), which represents the vertical distance from the mean water level to the mean ocean floor. For wave theories, the floor is assumed horizontal and flat.

Several other quantities that are important in the water wave theory may be computed from these parameters. They are:

- wavelength (L), which is the horizontal distance between successive crests,
- wave celerity or phase speed (c), which represents the propagation speed of the wave crest,
- frequency (f), which is the reciprocal of the period,
- wave elevation (η) which represents the instantaneous elevation of the wave from the still water level (SWL) or the mean water level (MWL),
- horizontal water particle velocity (u), which is the instantaneous velocity along x of a water particle,
- vertical water particle velocity (v), which is the instantaneous velocity along y of a water particle,
- horizontal water particle acceleration ($\dot{u}$), which is the instantaneous acceleration along x of a water particle, and
- vertical water particle acceleration ($\dot{v}$), which is the instantaneous acceleration along y of a water particle.

3.3.1 Linear Wave Theory

A wave creates a free surface motion at the mean water level acted upon by gravity. The elevation of the free surface varies with space x and time t. The simplest and most applied wave theory is the linear wave theory. It is also called small amplitude wave theory or Airy theory. For the linear wave theory, the wave has the form of a sine curve and the free surface profile is written in the following simple form:

$$\eta = a \sin(kx - \omega t) \tag{3.1}$$

in which the quantities a, ω and k are constants. The coefficient a is called the amplitude of the wave, which is a measure of the maximum departure of the actual free surface from the mean water level. The quantity ω is the frequency of oscillation of the wave and k is called the wave number. A two-dimensional coordinate system x, y is chosen to describe the wave propagation with x in the direction of wave and y vertical.

Therefore, this equation describes a simple harmonic motion of the free surface, which varies sinusoidally between the limits of $+a$ and $-a$ as shown in fig. 3.1. The point where the value of the profile is $+a$ is called a crest while the point with the value $-a$ is called the trough. This form is a description of a progressive wave, which states that the form of the wave profile η not only varies with time, but also varies with the spatial function x in a sine form. At time $t = 0$, the spatial form is a sine curve in the horizontal direction.

$$\eta = a \sin kx \tag{3.2}$$

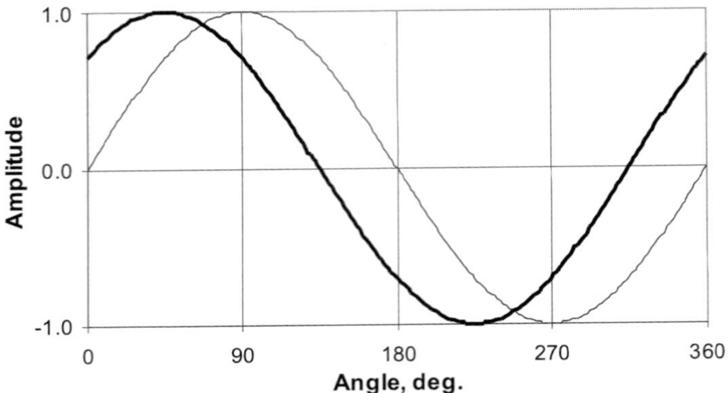

Figure 3.1 Free surface profile in wave motion at two times t_1 and t_2

This form is the same at any time as the wave is frozen at time t. The profile in fig. 3.1 is shown at two different times $t_1 = 0$ and t_2. Rewriting equation (3.1),

$$\eta = a \sin k(x - \frac{\omega}{k} t) \tag{3.3}$$

This form suggests that the frozen wave profile moves in the direction of the horizontal axis x with a velocity

$$c = \frac{\omega}{k} \tag{3.4}$$

The velocity c is called the celerity or speed of propagation of the progressive wave. We can write the frequency ω as

$$\omega = \frac{2\pi}{T} \tag{3.5}$$

where T is the period of the wave. It is clear that the waveform repeats at each cycle with the period T of the wave. Similarly we can write k as

$$k = \frac{2\pi}{L} \tag{3.6}$$

in which L is the length of the wave. Referring to fig. 3.2, L is then the distance between two crests or two troughs in the waveform. The wave propagation speed may also be written as

$$c = \frac{L}{T} \tag{3.7}$$

Of the three quantities, c, L and T, only one need be given. The other two can be expressed in terms of the given quantity. Flow field moves with wave speed c in the same direction as the wave profile.

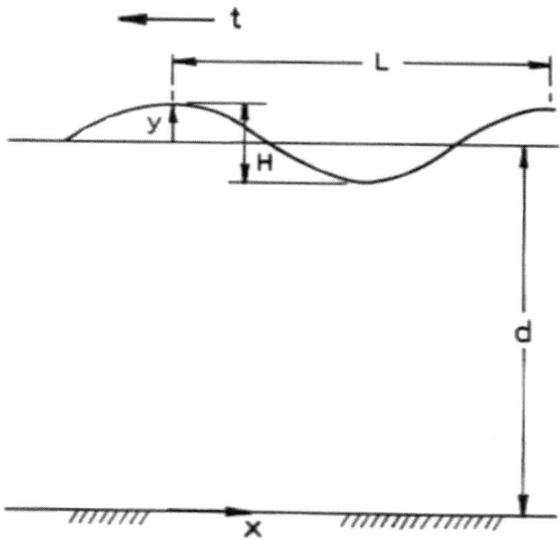

Figure 3.2 Two-dimensional wave motion over flat bottom

This description of the free surface flow is in two dimensions – horizontal (x) and vertical (y). It is bounded by the free surface on top and the sea floor at a water depth d. In order to describe the fluid particle motion within the wave field, the kinematic and dynamic properties of the particle should be established. For this purpose, it is assumed that the floor is flat so that the water depth is constant at all points along the x-direction. This is shown in fig. 3.2.

The formulas for the linear wave properties are listed in table 3.2. The two columns represent formulas for a finite water depth and for depths that may be considered infinite. These deepwater formulas are simpler to use and do not involve the water depth directly. The regions where they may be applied will be discussed here. The dispersion relationship, which is used to determine the wave length, relates the wave frequency to the wave number. There are several forms of this formula. For example, the wave length and wave period may be explicitly written in the formula by substituting T for ω and L for k including the factor 2π as appropriate. The kinematic properties of a wave are the particle velocities and accelerations, while the dynamic property is the pressure. For each of these quantities, the amplitude is obtained by setting the trigonometric sine or cosine function equal to one. The phase relationships for the amplitudes may be determined by examining the trigonometric functions. Note that the linear theory provides the kinematic and dynamic amplitudes as a linear function of the wave height (or wave amplitude). Therefore, if these amplitudes are normalised by the wave amplitude, then the normalised value is unique and invariant with the magnitude of the wave amplitude. The same applies to any linear responses of an offshore structure obtained by linear theory. The normalised responses as a function of wave period are called the transfer function or the response amplitude operator (RAO).

Table 3.2 Formulas for linear wave theory

Quantity	Finite depth	Deep water
Dispersion relationship	$\omega^2 = gk \tanh kd$	$\omega^2 = gk$
Wave profile	$\eta = \frac{H}{2}\cos[k(x - ct)]$	$\eta = \frac{H}{2}\cos[k(x - ct)]$
Horizontal velocity	$u = \frac{gkH}{2\omega}\frac{\cosh k(y + d)}{\cosh kd}\cos[k(x - ct)]$	$u = \frac{gkH}{2\omega}\exp(ky)\cos[k(x - ct)]$
Vertical velocity	$v = \frac{gkH}{2\omega}\frac{\sinh k(y + d)}{\cosh kd}\sin[k(x - ct)]$	$v = \frac{gkH}{2\omega}\exp(ky)\cos[k(x - ct)]$
Horizontal acceleration	$\dot{u} = \frac{gkH}{2}\frac{\cosh k(y + d)}{\cosh kd}\sin[k(x - ct)]$	$\dot{u} = \frac{gkH}{2}\exp(ky)\cos[k(x - ct)]$
Vertical acceleration	$\dot{v} = -\frac{gkH}{2}\frac{\sinh k(y + d)}{\cosh kd}\cos[k(x - ct)]$	$\dot{v} = \frac{gkH}{2}\exp(ky)\cos[k(x - ct)]$
Dynamic pressure	$p = \rho g\frac{H}{2}\frac{\cosh k(y + d)}{\cosh kd}\cos[k(x - ct)]$	$p = \rho g\frac{H}{2}\exp(ky)\cos[k(x - ct)]$

Table 3.3 Deepwater criterion and wavelength

Depth of Water	Criterion	tanh (kd)	Deep Water
Deep water	$d/L = > 1/2$	1	$L_0 = gT^2/2\pi$
Shallow water	$d/L < = 1/20$	kd	$L = T\sqrt{gd}$
Intermediate water depth	$1/20 < d/L < 1/2$		$L = L_0[\tanh(2\pi d/L_0)]^{1/2}$

Some simplifications may be made in the expression of the dispersion relation in table 3.2 if the water depth is shallow or deep. These simplifications are based on the fact that the hyberbolic function, tanh (kd), takes on simpler approximate forms in these two limits. This establishes the deep and shallow water criteria. The wavelength may be computed by simple formulas without resorting to the iteration in these two limiting cases. This is summarised in table 3.3. An approximate formula is also shown for the intermediate depth, which works well with small errors. The limiting criteria for the deep and shallow waters are stated in terms of the ratio d/L.

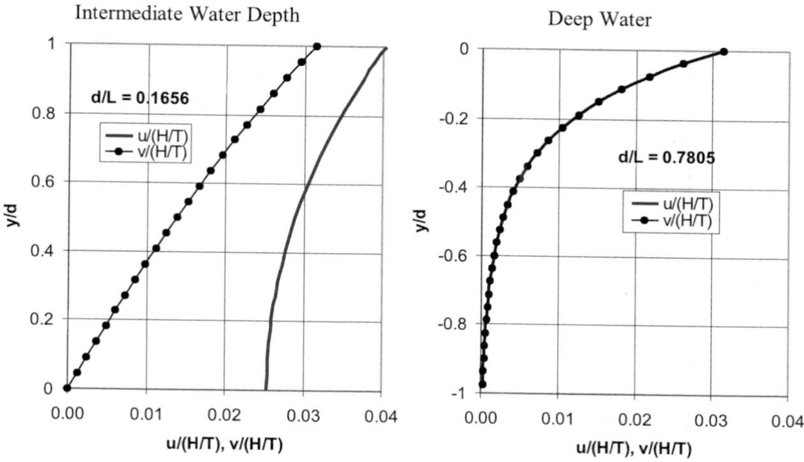

Figure 3.3 Particle velocity amplitudes with depth

An example of the linear velocity profiles for a finite and infinite water depth (deepwater) is shown in fig. 3.3. Linear theory defines the velocities up to the still water level and cannot determine them at the free surface. The vertical component of velocity is smaller in magnitude than the horizontal component. Their values approach each other as the water depth increases. Moreover, the vertical component of velocity is zero at the bottom for all depths in order to meet the bottom boundary condition. In deeper water, the horizontal velocity also becomes negligible with depth. If we examine the expressions of u and v in the table, we can see that the two velocities are orthogonal to each other so that when the horizontal velocity is maximum, the vertical velocity component is zero and vice versa. Since the path of the particles is defined by the particle displacements in these two directions, it is clear that the path will be defined by an ellipse with the major axis given by the horizontal particle motion double amplitude and the vertical axis will be the vertical motion double amplitude. For linear theory the particle path is closed indicating zero net transport of the fluid mass in the direction of the wave. In the ocean, the fluid motion in waves is associated with a mass transport (indicated by an open orbital path) giving rise to a wave-induced current. Some design guides suggest using a wave-induced current (e.g. 0.2 knots by API guidelines) for regular waves.

A sample calculation of linear wave properties from the equations in table 3.3 is provided below.

Consider,

water depth, $d = 100$ ft,
wave Height, $H = 20$ ft,
wave Period, $T = 10$ s,
elevation, $y = 0$ ft.

Then,

deep water wave length	$L_0 = gT^2/2\pi = 5.1248T^2 = 512.48\,\text{ft}$
the wave number	$k_0 = 0.01226$
	$d/L_0 = 0.1951$
which gives	$d/L = 0.2210$
so that	$k = 0.01389$
	$\sinh kd = 1.88$
	$\cosh kd = 2.13$

Finally the kinematic and dynamic amplitudes at the still water level are computed as follows:

horizontal velocity	$u_0 = \pi.20/10 \times 2.13/1.88 = 7.12\,\text{ft/s}$,
vertical velocity	$v_0 = \pi.20/10 = 6.28\,\text{ft/s}$,
horizontal acceleration	$\dot{u}_0 = 2\pi/10 \times 7.12 = 4.47\,\text{ft}^2/\text{s}$,
vertical acceleration	$\dot{v}_0 = 2\pi/10 \times 6.28 = 7.12\,\text{ft}^2/\text{s}$,
pressure	$p_0 = 1.94 \times 32.2 \times 20/2 = 4.34\,\text{lb/in.}^2$

The wave profile and the wave kinematics from the above calculations are plotted for one cycle in fig. 3.4. The wave profile is given a zero value at time zero. The x-axis is shown as θ, which is given in degrees representing $\theta = kx - \omega t$. The wave profile becomes maximum having a crest at $\theta = \pi/2$. The phase relationship among the parameters may be examined from this example.

The horizontal water particle velocity is in phase with the wave profile, whereas the vertical particle velocity is 90° out of phase with the horizontal. The particle accelerations are also out of phase with each other, the horizontal acceleration being in phase with the vertical velocity. Thus the forces proportional to the particle acceleration (e.g. inertia force) will be out of phase with forces proportional to the particle velocity (e.g. drag force). The dynamic pressure (excluding the hydrostatic component) on a fluid particle due to linear theory is in phase with the horizontal particle velocity. The pressure amplitude has a similar vertical profile as shown for the horizontal velocity in fig. 3.4.

In order to show how well the linear theory computes the wave particle properties in practice, an example is given in fig. 3.5. In this case the dynamic pressures were measured in regular waves at several submerged locations in a wave tank test. The pressures at different elevations were measured simultaneously in a water depth of 10 ft and a range of wave periods of 1.25–3.0 s. The amplitudes of these pressures with the water depth are compared with the pressures computed by the linear theory in fig. 3.6. The vertical axis is the non-dimensional depth (s/d where $s = y + d$). The pressure amplitude along the x-axis is normalised with the wave height. The symbols are the measurements and the solid line represents the linear theory results. Different symbols represent different wave heights at

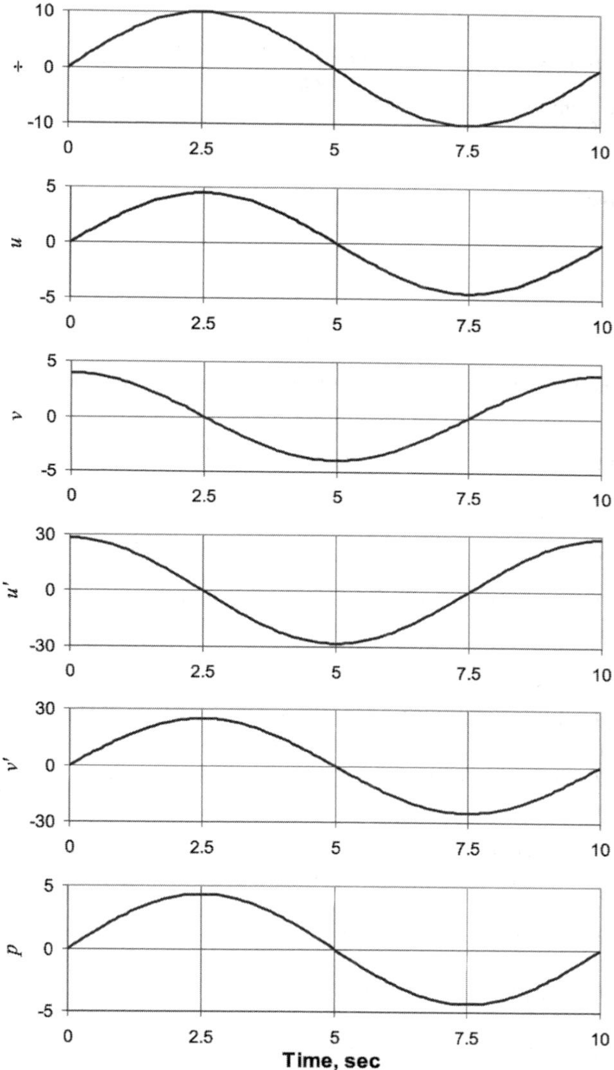

Figure 3.4 Time history of linear wave properties

the same wave period. The wave parameter is shown by the quantity d/T^2 for each wave. Note that the theory compares well in most cases with a systematic departure at the higher wave heights. Also, the measurements above the wave trough shows that the pressures approach a zero value at the free surface approximately in a linear fashion. This trend may be predicted by the second-order theory which is described next.

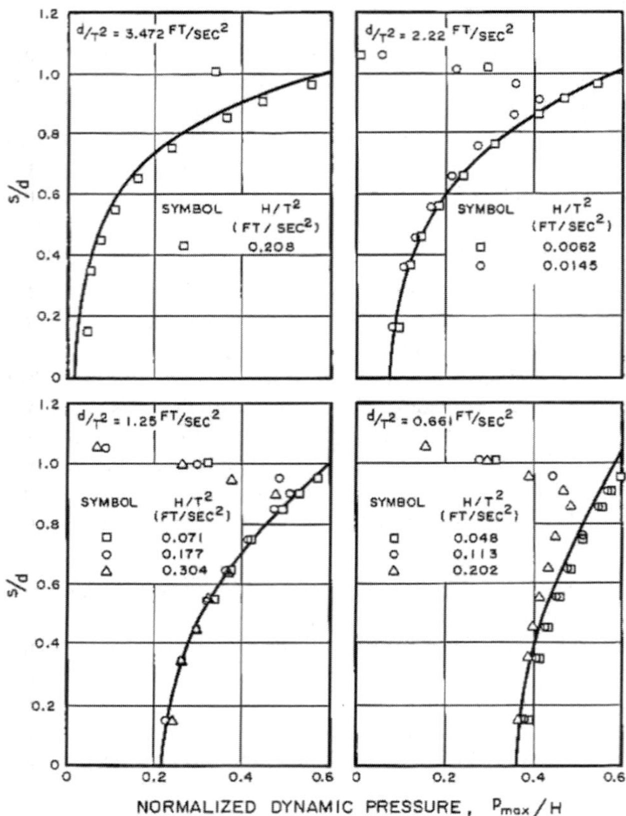

Figure 3.5 Comparison of linear wave pressures with tank measurements

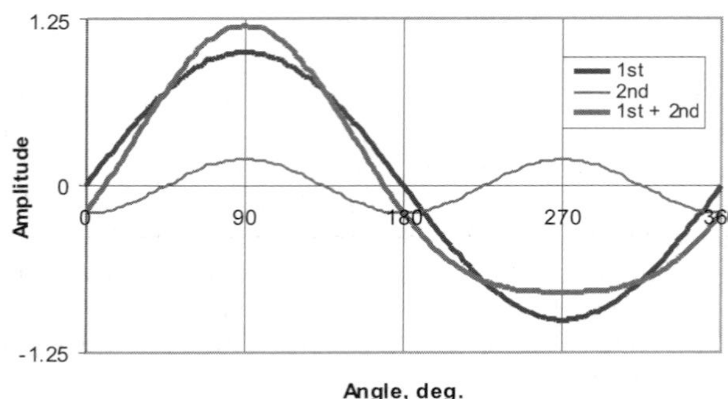

Figure 3.6 Time history of first- and second-order components

3.3.2 Second-Order Stokes Wave Theory

Stokes' second-order wave provides two components for the wave kinematics, the first one at the wave frequency and the second one at twice the wave frequency. The relevant formulas for the two components are shown in table 3.4. The second-order component is smaller than the first-order contribution. The wave profiles from the two components combine to provide a steeper crest and shallower trough. This is illustrated in fig. 3.7. Steeper waves in the ocean will have a similar form. Therefore, in selecting a regular wave theory in the calculation of response of an offshore structure, the higher wave heights resulting from larger storms require application of second-order theory or higher.

An example of the effect of the second-order theory is shown in fig. 3.6. The wave profile of a second-order wave is given along with the individual components. The second-order component is smaller in magnitude and has a frequency that is twice the specified wave frequency. When this component is added to the first-order contribution, the profile becomes peaked at the crest yielding a shallower trough. This form of wave profile is quite prevalent in the ocean where the height is finitely large. In this case the application of the second-order theory is justified. For the second-order theory the wave steepness (H/L) should be of the order of 0.1. The application of this theory vs. the first-order theory is based on the value of wave slope, which will be discussed later under applicability section.

In deep water the ratio d/L is larger than 0.5 and tanh kd is close to one. Generally, the second-order contribution is quite small in deep water. Considering the horizontal velocity at the still water level, the ratio of the second-order component to the first-order component is given by the quantity $1.5\ \pi H/(L\sinh^2 kd)$ where $L = gT^2/2\pi$. Numerical values of this ratio for a wave height of 75 ft and a range of periods in deep water are given in table 3.5 for three different water depths. It is clear that even for a high wave the second-order contribution is much less than 10%. In deeper waters, this value is even smaller.

A numerical example is given in a model scale. This example is chosen for a very steep wave in an intermediate water depth. The wave parameters chosen are:

wave height = 4 ft,
wave period = 2 s, and
water depth = 8 ft.

The amplitudes of the first and second-order horizontal velocity components are computed from the expressions in table 3.4 and are separately shown in fig. 3.7 as a function of water elevation. In order to show the second-order magnitudes clearly, a separate horizontal scale is chosen. Note that the second-order velocity amplitude is only about 3% of the first-order amplitude. Moreover, the second-order component decays with depth much faster so that in the deeper part of the water, the first-order component dominates. Table 3.5 and the example in fig. 3.7 show that the ocean waves in deep water may be represented better than shallow water waves by the linear theory and the higher order wave components have lesser effect on the offshore structure. The only time the higher order wave component may become important is when the structure responds to the higher frequencies present in the higher order waves.

Table 3.4 Formulas for Stokes' second-order wave theory

Quantity	First-Order Component	Second-Order Component
Dispersion relationship	$c^2 = \frac{g}{k}\tanh kd$	$c^2 = \frac{g}{k}\tanh kd$
Wave profile	$\eta = \frac{H}{2}\cos(kx - \omega t)$	$\eta = \frac{\pi H^2}{8L}\frac{\cosh kd}{\sinh^3 kd}[2 + \cosh 2kd]\cos 2(kx - \omega t)$
Horizontal velocity	$u = \frac{\pi H}{T}\frac{\cosh ks}{\sinh kd}\cos(kx - \omega t)$	$u = \frac{3}{4c}\left(\frac{\pi H}{T}\right)^2 \frac{\cosh 2ks}{\sinh^4 kd}\cos 2(kx - \omega t)$
Vertical velocity	$v = \frac{\pi H}{T}\frac{\sinh ks}{\sinh kd}\sin(kx - \omega t)$	$v = \frac{3}{4c}\left(\frac{\pi H}{T}\right)^2 \frac{\sinh 2ks}{\sinh^4 kd}\sin 2(kx - \omega t)$
Horizontal acceleration	$\dot{u} = \frac{2\pi^2 H}{T^2}\frac{\cosh ks}{\sinh kd}\sin(kx - \omega t)$	$\dot{u} = \frac{3\pi}{2L}\left(\frac{\pi H}{T}\right)^2 \frac{\cosh 2ks}{\sinh^4 kd}\sin 2(kx - \omega t)$
Vertical acceleration	$\dot{v} = -\frac{2\pi^2 H}{T^2}\frac{\sinh ks}{\sinh kd}\cos(kx - \omega t)$	$\dot{v} = -\frac{3\pi}{4L}\left(\frac{\pi H}{T}\right)^2 \frac{\sinh 2ks}{\sinh^4 kd}\cos 2(kx - \omega t)$
Dynamic pressure	$p = \rho g \frac{H}{2}\frac{\cosh ky}{\cosh kd}\cos[k(x - ct)]$	$p = \frac{3}{4}\rho g \frac{\pi H^2}{L}\frac{1}{\sinh 2kd}\left[\frac{\cosh 2ks}{\sinh^2 kd} - \frac{1}{3}\right]\cos 2(kx - \omega t)$ $- \frac{1}{4}\rho g \frac{\pi H^2}{L}\frac{1}{\sinh kd}[\cosh 2ks - 1]$

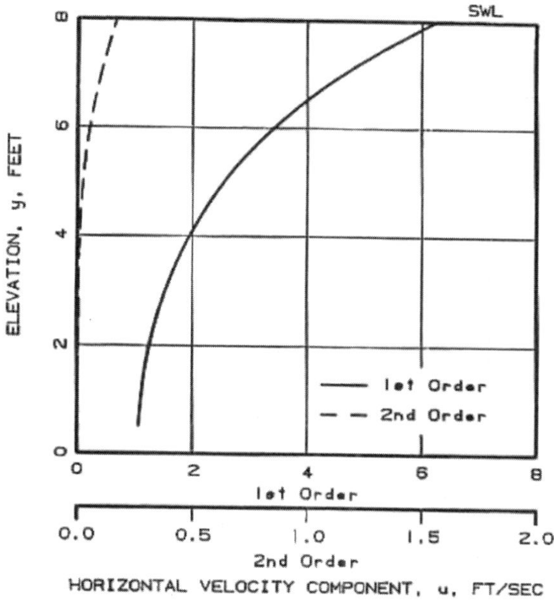

Figure 3.7 Example of first- and second-order horizontal particle velocity

Table 3.5 Ratio of second-order to first-order horizontal velocity

Period	Length	Ratio	Ratio	Ratio
T(s)	L(ft)	$d=500$ ft	$d=1000$ ft	$d=2000$ ft
6	184.5	1.54E-07	6.21E-15	1.00E-29
8	328.0	1.49E-04	1.03E-08	4.94E-17
10	512.5	3.00E-03	6.53E-06	3.09E-11
12	738.0	1.36E-02	1.92E-04	3.85E-08
14	1004.5	3.09E-02	1.35E-03	2.59E-06
16	1311.9	4.96E-02	4.48E-03	3.73E-05
18	1660.4	6.57E-02	9.68E-03	2.20E-04
20	2049.9	7.81E-02	1.61E-02	7.50E-04

3.3.3 Fifth-Order Stokes Wave Theory

This wave theory is applicable for deep-water high waves. As the name implies, the fifth-order theory comprises five components in a series form. Each component is generally an order of magnitude smaller than the previous one in succession. The horizontal velocity

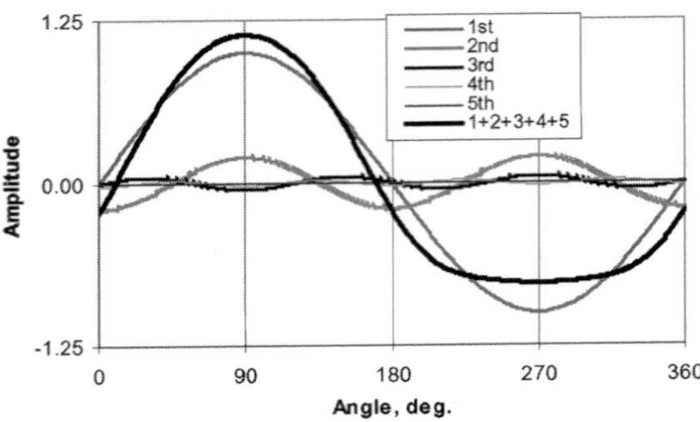

Figure 3.8 Example of five components of velocity for Stokes fifth-order theory

may be represented by a five-term series of the form:

$$u = \sum_{n=1}^{5} u_n \cosh nks \cos n(kx - \omega t) \tag{3.8}$$

Note that the frequencies of the higher components are multiples of the fundamental wave frequency. The higher components decay faster with depth and their effects in deeper depth are negligible. Therefore, even when a non-linear theory, such as Stokes' fifth-order theory, is applied in the design of an offshore structure, it has significant effect on the structure only near the free surface. Away from the MWL, the waves behave more like the linear wave. Because of this effect, many submerged structural components in deep-water steep waves can be designed simply with the linear theory without any measurable error. An example of the five components of the wave profile is shown in fig. 3.8. The total profile of the fifth-order wave shows clearly the vertical asymmetry with a steeper crest and shallower trough. This type of wave will have a significant effect on such structural responses as impact on the superstructure, airgap beneath the deck as well as the motion of a shallow draft floating structure.

3.3.4 Stream Function Theory

The stream function theory is a non-linear wave theory developed by Dean (1965). There are two types of stream function theory. One is called the symmetric or regular stream function theory, which like the previously described theories, is based on prescribed wave period, wave height and water depth. This theory describes a periodic wave of permanent form which are symmetric front-to-back but asymmetric crest-to-trough. Thus it can account for steep non-linear waves.

The other is the irregular stream function theory. This theory is applicable when the free surface wave profile is known. Thus, it is suitable in a design when the field or wave basin data is available. In the irregular stream function theory, no restrictions are placed on the wave form regarding its horizontal or vertical symmetry and one cycle of the wave profile from a random wave time history is analysed at one time in which the water surface elevation is known. Thus it is appropriate when a design based on a single extreme wave is desired. This wave profile may be chosen from a time realisation of a design wave spectrum. In this case all the system non-linearities may be maintained in a time domain design method.

3.3.4.1 Regular Stream Function Theory

The stream function theory is based on describing the stream function in a series form in a moving coordinate system. The inclusion of current whether uniform or shear is rather straightforward [Dalrymple, 1974]. For a current, the wave celerity c is replaced by $c - U$ where U is the uniform current velocity. For opposing current, U is negative in this formulation. According to the stream function theory, the wave is described by its stream function:

$$\psi(x, y) = (c - U)y + \sum_{n=1}^{N} X(n) \sinh nky \cos nkx \tag{3.9}$$

in which the summation is over N terms where N determines the order of the theory, and U = current velocity. The expression for ψ is written in a moving coordinate system (with speed c) so that time does not enter in the equation explicitly. The quantities c and $X(n)$ are unknown.

For regular stream function theory, the input is simply the wave height, period and water depth as with other wave theories. For irregular stream function theory, the free surface profile for one cycle of the wave becomes a constraint as well. Standard software is available to carry out the numerical computations.

In order to choose the appropriate order of the stream function theory in a particular application, one should consult the data on fig. 3.9. The plot is presented as H/T^2 vs. d/T^2

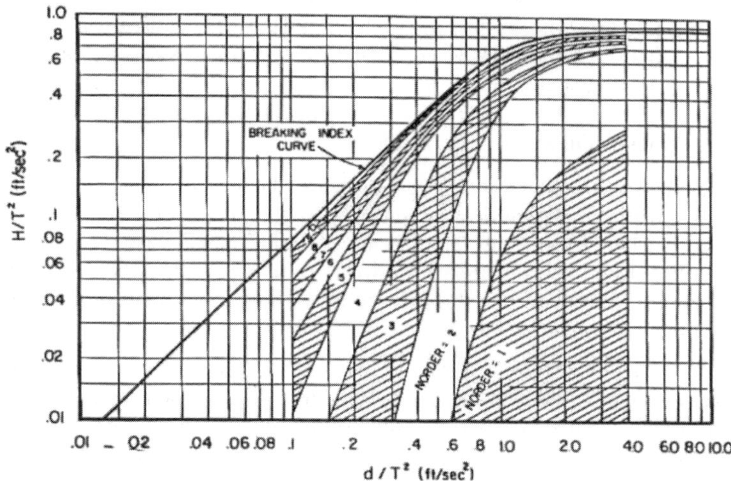

Figure 3.9 Order of stream function based on wave parameters [Dean, 1974]

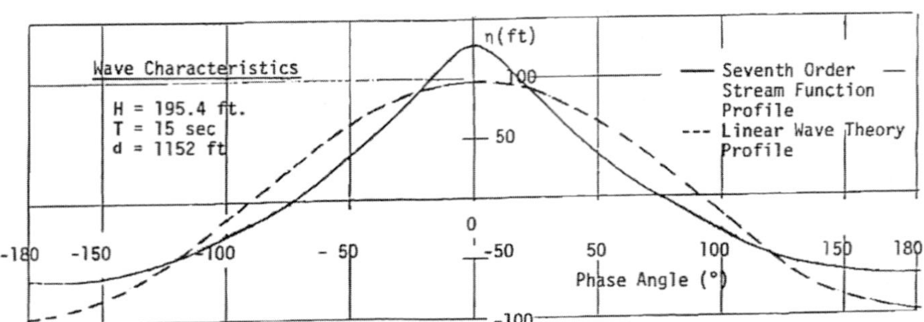

Figure 3.10 Example of non-linear regular stream function wave profile [Dean, 1974]

along the x-axis. Thus, depending on the values of these three input parameters, the order of the theory N may be determined from the figure.

Because the order of the stream function theory is open, it covers very small waves to very steep waves. Moreover, if the free surface profile is known (from the field or model tests) it can accommodate waves of large vertical asymmetry as well.

An example of the profile of a regular stream function wave is shown in fig. 3.10. The wave chosen is very steep with a height H of 195.4 ft and period T of 15 s in a water depth d of 1152 ft. A seventh-order theory was selected from the chart in fig. 3.9 for these parameters. For comparison purposes, the equivalent linear wave is superimposed on this profile. The horizontal asymmetry with a steep crest in the wave profile is clearly evident here.

3.3.4.2 Irregular Stream Function Theory

For the irregular stream function theory, the period of the wave is not needed, but is computed within the numerical algorithm as an unknown. If the wave is regular, then one cycle of data will represent the entire wave. For an irregular wave, generally, a steep wave cycle within the random wave representing an extreme wave is chosen for the subsequent design analysis of a structure. Upon convergence of the solution, the fit of the solution to the measured wave profile is generally excellent.

This is exemplified here. In a wave tank test the water particle kinematics in simulated near-breaking and breaking waves is measured, some in the presence of steady inline current. The measured particle velocities in the presence of steady inline current are compared with the tenth-order stream function theory and shown in fig. 3.11. The top plot shows the matching of the (a) wave profile as a constraint in the irregular stream function theory. The measured (b) horizontal and (c) vertical velocity profiles compare well with the theory in fig. 3.11, which was typical for these tests of near-breaking waves.

Even for the breaking waves, the horizontal velocity compared well. One area of exception in this comparison was the near-trough region of the vertical velocity, which did not predict well. Observation during the tests with breaking waves consistently showed that vorticity was generated in the flow as the flow came down near the trough from the breaking of the wave. Besides this exception, the irregular stream function wave theory has been found to

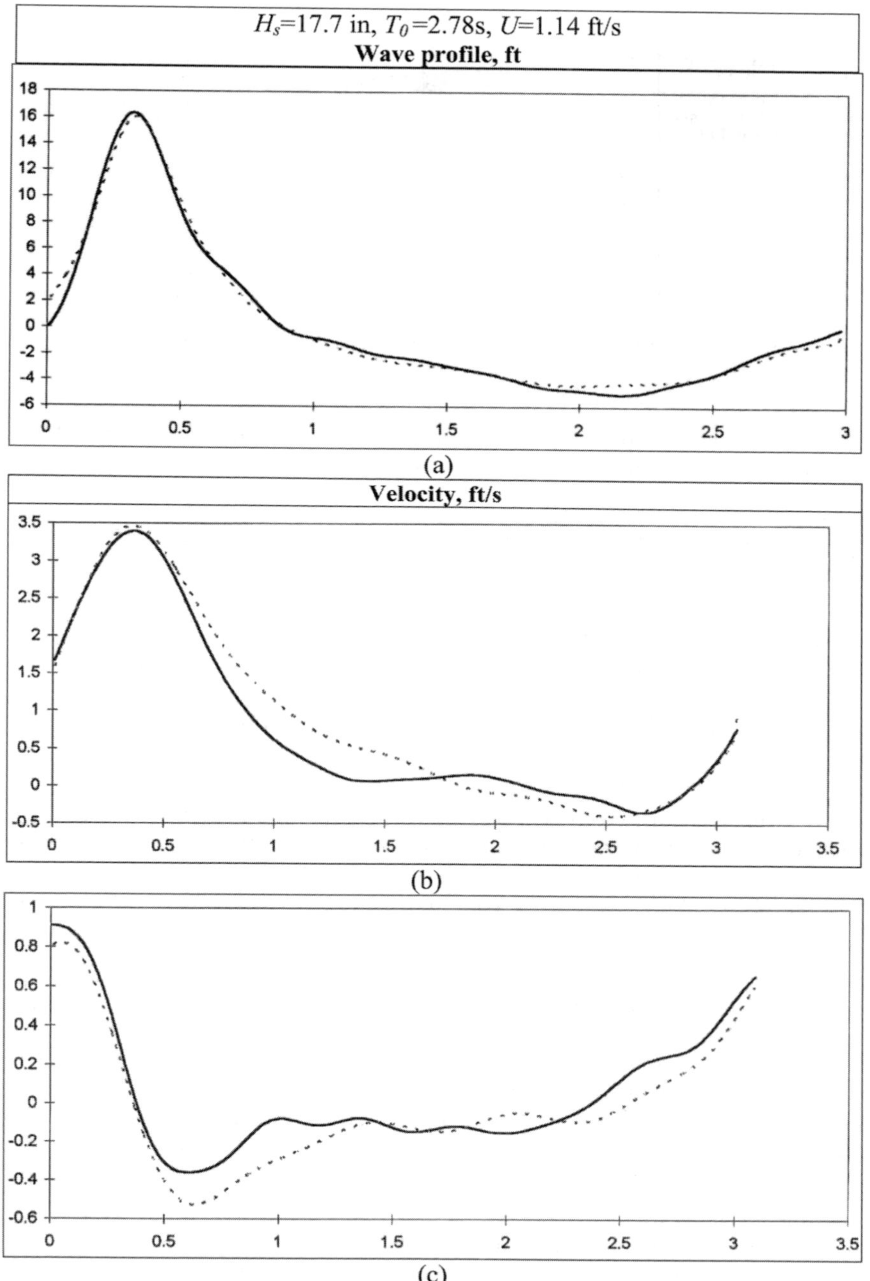

Figure 3.11 Comparison of irregular stream function velocities with measurements

match the wave kinematics and wave dynamics consistently well and its application is recommended in instances when the free surface time history is available.

3.3.5 Stretching Formulas for Waves at SWL

As stated earlier, the linear wave theory can only be applied up to the still water level. However, for structures at the free surface, the waves above the SWL may have a large influence on the response of the structure and may not be ignored. If linear theory is used, then the theory is modified to account for the free-surface effect. There are several possible modifications that have been proposed.

3.3.5.1 First-Order Stretching

Two stretching formulas use this technique. In these formulations, the effective water depth is modified using the instantaneous wave-free surface in computing the decay function. These are commonly known as the stretching formulas, since the water depth is stretched vertically at the crest and shortened at the trough. The methods provide the same horizontal particle velocity at the wave crest and the wave trough. These two formulas are included in table 3.6. They are identical at the free surface and their difference below the free surface is quite small. Wheeler's (1970) stretching is most commonly used in the offshore structure design.

3.3.5.2 Linear extrapolation

A couple of other stretching formulas are presented here. Their application for an offshore structure is not very common. But they have been used in designs. In the first case, the elevation y between the SWL and free surface is expanded in a Taylor series and the first two terms are retained [Rodenbush and Forristall, 1986]. Below the SWL, however, the standard forms for linear theory given in table 3.5 are used. This is termed linear extrapolation and is included in table 3.7.

The delta stretching method suggested by Rodenbush and Forristall (1986) is an empirical average between the Wheeler stretching and linear extrapolation (table 3.7). Here the horizontal velocity is extrapolated linearly to a height $\Delta\eta$, in which η is the free surface profile and $0 \leq \Delta \leq 1$. The velocity profile above a reference depth d' (where $-d' \leq \eta \leq \Delta\eta$) is mapped onto the water column d' $(-d' \leq \eta \leq \Delta\eta)$. For $y' > 0$, the wave velocity is computed by linear extrapolation. Wheeler stretching is the limiting case of this method when $\Delta = 0$ and $d' = d$. On the other end, when $\Delta = 1$ and $d' = d$, the linear extrapolation is reached. The values of Δ and d' are determined empirically. Rodenbush and Forristall suggested the values $\Delta = 0.3$ and $d' = 2\eta_{\text{rms}}$ (twice the standard deviation of wave elevation).

3.3.5.3 Second-Order Stretching

The stretching may be applied to non-linear waves as well. The extension is similar. A second-order stretching was suggested by Gudmestad and Connors (1986) using the first two components using linear superposition. The second-order corrections to the Chakrabarti stretching formulas are shown in table 3.8.

Table 3.6 Linear stretching formulas

Quantity	Wheeler Stretching	Chakrabarti Stretching
Horizontal velocity	$u = \frac{gkH}{2\omega} \frac{\cosh k\left(\frac{y}{1+\eta/d}\right)}{\cosh kd} \cos[k(x - ct)]$	$u = \frac{gkH}{2\omega} \frac{\cosh ky}{\cosh k(d+\eta)} \cos[k(x - ct)]$
Vertical velocity	$v = \frac{gkH}{2\omega} \frac{\sinh k\left(\frac{y}{1+\eta/d}\right)}{\cosh kd} \sin[k(x - ct)]$	$v = \frac{gkH}{2\omega} \frac{\sinh ky}{\cosh k(d+\eta)} \sin[k(x - ct)]$
Horizontal acceleration	$\dot{u} = \frac{gkH}{2} \frac{\cosh k\left(\frac{y}{1+\eta/d}\right)}{\cosh kd} \sin[k(x - ct)]$	$\dot{u} = \frac{gkH}{2} \frac{\cosh ky}{\cosh k(d+\eta)} \sin[k(x - ct)]$
Vertical acceleration	$\dot{v} = -\frac{gkH}{2} \frac{\sinh k\left(\frac{y}{1+\eta/d}\right)}{\cosh kd} \cos[k(x - ct)]$	$\dot{v} = -\frac{gkH}{2} \frac{\sinh ky}{\cosh k(d+\eta)} \cos[k(x - ct)]$
Dynamic pressure	$p = \rho g \frac{H}{2} \frac{\cosh k\left(\frac{y}{1+\eta/d}\right)}{\cosh kd} \cos[k(x - ct)]$	$p = \rho g \frac{H}{2} \frac{\cosh ky}{\cosh k(d+\eta)} \cos[k(x - ct)]$

Table 3.7 Linear extrapolation and Delta stretching formulas

Quantity	Linear Extrapolation	Delta Stretching
Horizontal velocity	$u = a\omega\left[\frac{\cosh ky}{\sinh kd} + ky\right] \times \cos[k(x - ct)]$	$u = a\omega\frac{\cosh ky'}{\sinh kd}\cos[k(x - ct)]$ where $y' = \begin{cases} (y + d')\frac{d'+\Delta\eta}{d'+\eta} - d' \\ \quad \text{for } y > -d' \text{ and } \eta > 0 \\ y \quad \text{otherwise} \end{cases}$
Vertical velocity	$v = a\omega\left[ky\frac{\sinh ky}{\sinh kd} + 1\right] \times \sin[k(x - ct)]$	$v = a\omega\frac{\sinh ky'}{\sinh kd}\sin[k(x - ct)]$

Table 3.8 Second-order stretching formulas

Quantity	Second-order Stretching
Wave profile	$\eta = a\cos[k(x - ct)] + \frac{1}{2}a^2k\frac{1}{2SC} + a^2kB'\cos[2k(x - ct)]$
Horizontal velocity	$u = a\omega\frac{\cosh ky}{S}\cos[k(x - ct)] + a^2k\omega\frac{C\cosh ky}{S^2}\sin^2 k(x - ct) + a^2k\omega[D'\cosh 2ky + D''\cosh 4ky]\cos[2k(x - ct)]$
Vertical velocity	$v = a\omega\frac{\sinh ky}{S}\sin[k(x - ct)] + a^2k\omega[2D'\sinh 2ky + 4D''\sinh 4ky]\sin[2k(x - ct)]$

The quantities in the table are defined below.

$$S = \sinh k[a\cos k(x - ct) + d] \tag{3.10}$$

$$C = \cosh k[a\cos k(x - ct) + d] \tag{3.11}$$

$$B' = \frac{1}{4CS^3(2S^2 + 1)}(4S^4 + 5S^2 + 1) \tag{3.12}$$

$$D' = \frac{(6S^4 + 7S^2 + 1)}{8S^4(2S^2 + 1)} \tag{3.13}$$

$$D'' = \frac{-1}{8S^4(2S^2 + 1)} \tag{3.14}$$

in which the wave amplitude $a = H/2$.

3.3.6 Applicability of Wave Theory

In the earlier sections, several wave theories applicable to offshore structures have been presented. The obvious question is, when is one of these theories suitable for application. A region of validity of the various theories that are applicable to offshore structures in relatively deeper waters is presented in fig. 3.12. The chart is taken from API RP2A Guidelines (2000), which was adopted from the report of Atkins Engineering Services (1990). Since the basic wave parameters are H, T and d, the regions are shown as functions

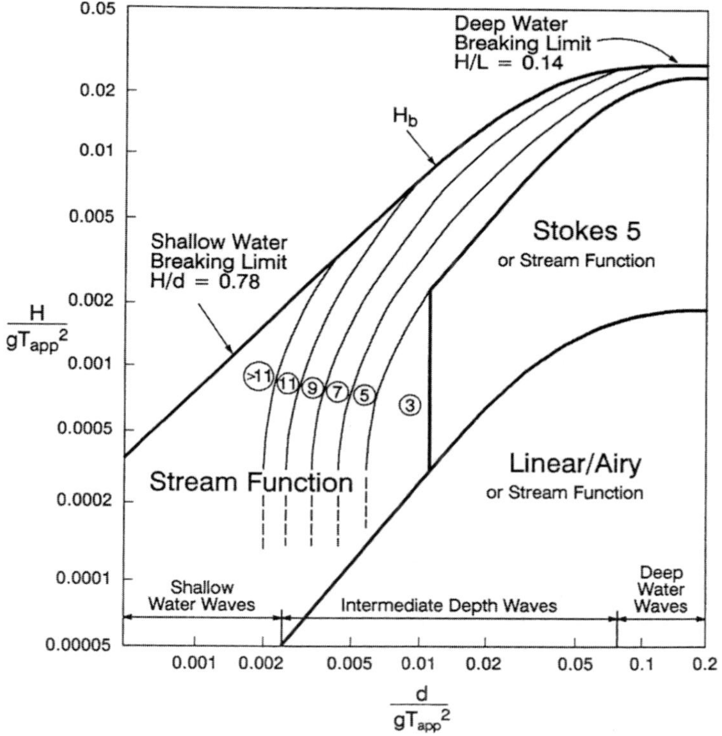

Figure 3.12 Region of application of wave theories [API, 2000]

of $H/(gT^2)$ and $d/(gT^2)$. The regular stream function theory is shown applicable over almost the entire region as long as the appropriate order N is determined from fig. 3.9. The solid boundaries among the various regions represent the fitting of theoretical equations, in particular, the free-surface boundary conditions. Another way to determine the limits of various theories is to examine the waveform or the steepness of the wave. For example, the limiting height of the Airy theory is given by

$$kH = 0.08 \frac{\sinh^3 kd}{\cosh kd\,(3 + 2\sinh^2 kd)} \tag{3.15}$$

and is a function of wave length (or period) and water depth. On the same basis, Stokes' second-order theory gives the limit as

$$kH < 0.924 \frac{\sinh^3 kd}{(1 + 8\cosh^2 kd)^{1/2}} \tag{3.16}$$

Similarly, the limiting zones between Stokes' third- and fifth-order theory are governed by the inequalities as follows:

$$\begin{aligned} &H/d < 0.725(kd)^2, && \text{in shallow water} \\ &H/d < 0.1, && \text{in deep water} \end{aligned} \tag{3.17}$$

Consider an example to find a suitable theory from the above chart on wave theory validity. Take a wave of height 30 m, period of 16 s in a water depth of 160 m. Then $H/(gT^2) = 0.012$ and $d/(gT^2) = 0.064$. From the chart (in the figure), the applicable wave theory is Stokes' third-order. A fifth-order Stokes wave may be applied here as well.

A rule of thumb for the practical application of wave theory is presented here with recommendation for the appropriate theory in specific cases. The theories for the cases shown are suggestions for normal applications. They should not be considered inviolate. There may be many instances where a different theory than listed here is more suitable in a particular situation.

Theory	Application
Linear theory	Low seastates (1 yr storm) Fatigue analysis Swell Large inertia dominated fixed and floating structures Linear radiation damping Long term statistics
Stokes' second-order theory	Slow drift oscillation of soft moored structures TLP tendon analysis
Stokes' fifth-order or Stream function theory	Storm waves Drag dominated structures Wave tank data (irregular stream function) Air gap Moorings and risers Non-linear damping near natural period

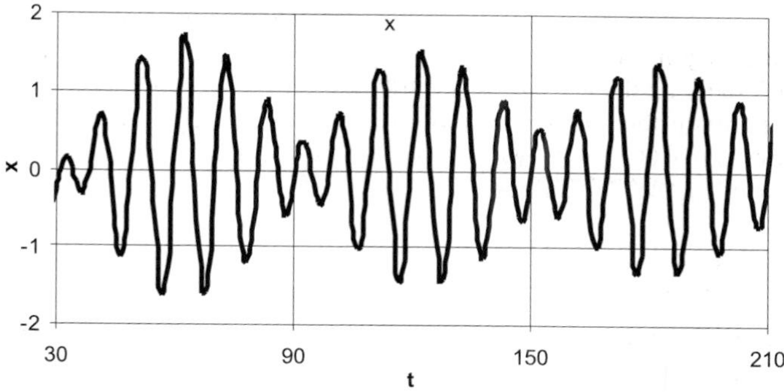

Figure 3.13 Example of regular wave group

3.3.7 Wave Group

Ocean waves are random whose description may only be possible in a statistical or spectral form. Although sea waves have a random appearance, closer inspection often reveals that the higher waves in a random ocean wave record appear in groups rather than individually. A sequence of high waves is called a wave group. One method of defining a wave group is to establish a threshold value and to consider a group to be a sequence of waves given by an envelope that exceeds this value. This threshold level may be the mean wave height, the significant wave height or a similar statistical wave height parameter.

The presence of such wave groups in an ocean wave has many engineering implications. From a practical standpoint, the number of waves in a group is important in the study of such topics as the long-period slow-drift oscillation of moored floating structures, capsizing of ships, resonance of structures, metal fatigue and stability of rubble-mound breakwaters. Ships, coastal and offshore structures have experienced exceptional damage resulting from groups of successive waves. In its simple form, a group of waves may be thought of waves having a narrow band of frequencies among them. A simple example in fig. 3.13 shows two superimposed waves of equal amplitudes having periods of 10 and 12. The beat period of 60 s is obtained as the difference between the individual periods in the wave group. Usually, the wave group in the ocean has a specific set of statistical properties.

In the analysis of slow drift oscillation the groupiness of waves is an important criterion. Since the group envelope period is generally large and the slow drift oscillation period can be quite long, the group period excites the slow drift causing large oscillation amplitude. This is why the model testing for slow drift is often done with random waves that duplicates the wave group spectra. This area will be discussed in Chapter 13.

3.3.8 Series Representation of Long-Crested Wave

In many applications a simpler representation of the random ocean waves may provide adequate information for its properties for a particular design case. Based on the linear wave theory, the random wave profile may be represented by a series of regular waves

Table 3.9 Series form for random ocean waves

Quantity	Formula
Wave profile	$\eta(x,t) = \sum_{n=1}^{N} a_n \cos[k_n(x - c_n t) + \varepsilon_n]$
Horizontal velocity	$u(x,y,t) = \sum_{n=1}^{N} \frac{a_n \omega_n}{\sinh k_n d} \cosh[k_n(y+d)] \cos[k_n(x - c_n t) + \varepsilon_n]$
Vertical velocity	$v(x,y,t) = \sum_{n=1}^{N} \frac{a_n \omega_n}{\sinh k_n d} \sinh[k_n(y+d)] \sin[k_n(x - c_n t) + \varepsilon_n]$
Horizontal acceleration	$\dot{u}(x,y,t) = \sum_{n=1}^{N} \frac{a_n \omega_n^2}{\sinh k_n d} \cosh[k_n(y+d)] \sin[k_n(x - c_n t) + \varepsilon_n]$
Vertical acceleration	$\dot{v}(x,y,t) = -\sum_{n=1}^{N} \frac{a_n \omega_n^2}{\sinh k_n d} \sinh[k_n(y+d)] \cos[k_n(x - c_n t) + \varepsilon_n]$
Dynamic pressure	$p(x,y,t) = \sum_{n=1}^{N} \rho g \frac{a_n}{\cosh k_n d} \cosh[k_n(y+d)] \cos[k_n(x - c_n t) + \varepsilon_n]$

summed together. In this case the linear superposition also applies to the kinematic and dynamic properties of the wave. The expressions for the wave profile, wave kinematics and dynamic pressure are included in table 3.9.

If such an approximation is applied in analysing an offshore structure, then the expressions in the table may be used in computing the responses of the structure.

3.4 Breaking Waves

Wind waves do not grow unbounded. For a given water depth and wavelength there is an upper limit of wave height or a given wave slope beyond which the wave becomes unstable and break. In deep water, this wave slope reaches a gradient of 1 in 7 according to Stokes (1880). The wave height limit in deep water is strictly a function of wavelength (or equivalently wave period)

$$H_b = 0.142 L_0 \tag{3.18}$$

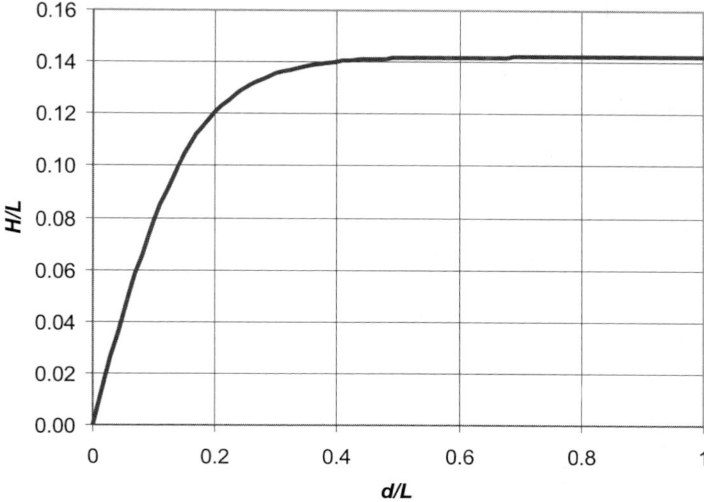

Figure 3.14 Breaking wave heights for different wave lengths and water depths

In a finite water depth, the wave height is limited by the formula involving wavelength and water depth.

$$H_b = 0.142L \tanh kd \tag{3.19}$$

This relationship is plotted in fig. 3.14 in a non-dimensional form in terms of the parameters H/L and d/L. Thus for a given water depth and wave period, the maximum wave height may be determined from this figure. In the limit of deep water, the ratio asymptotically approaches the limiting value of 0.142 shown in equation (3.18).

The theories described earlier for regular waves, including non-linear Stokes waves and stream function theory, do not predict the kinematics and dynamic properties of very steep waves well. These waves are not only vertically unsymmetric, but also have large horizontal unsymmetry. If the design is based on these single steep waves, then a numerical theory need to be utilised. There are current attempts in describing such waves by the numerical wave tank methods and the method of New Waves [see Tromans, et al (1991) and Kim, et al (1999) for details]. These methods have not reached the design stage yet and are not commonly used in the design of offshore structures.

3.5 Internal Waves

Internal waves are buoyancy waves caused by variations in density. They are large amplitude gravity waves, which propagate at the interface between a layer of warm water overlying a layer of cooler water. Although hidden from sight, the interior of the ocean is just as turbulent as its surface. Roughly 40 m below the surface, there is an abrupt change in both water density and temperature respectively called the pycnocline and the thermocline. Thus the pycnocline is a gradual interface between two fluids of different density. Disturbances travel along fluid interfaces, and disturbances, which travel along the pycnocline, are called internal waves. The evolution of internal waves and the interaction of internal waves with other waves and structures in the horizontal plane are not well understood and research in its propagation and evolution is ongoing.

These waves are difficult to measure, but have been found to occur even in very deep water. Internal waves in the ocean generally produce a small vertical displacement of the free surface and thus are undetectable from the floating structures. The height of these waves may be significant. Since internal waves are not directly visible, they can only be detected by their surface signature and by direct measurements of the pycnocline or thermocline. Where the density interface is shallow enough to permit the internal wave crests to interact with the sea surface, the waves can be detected by the resulting increased roughness of the surface ocean.

Internal waves are believed to be responsible for a great deal of damage. Large amplitude internal waves can create enormous local loads and bending moments in offshore structures. They have been reported to displace oil platforms as much as 200 m in the horizontal direction and 10 m in the vertical direction. There has been speculation that the loss of the submarine USS Thresher in 1969 came from an internal wave (called soliton) carrying the submarine rapidly into water deeper than its crush depth.

3.6 Sea Spectrum

The regular wave theories are applicable in a design where a single wave method is employed. This is often a common method in the design of an offshore structure. In this case an extreme wave is represented by a regular wave of the appropriate height and period. This method provides a simple analysis in determining the extreme response of an offshore structure. The random ocean wave, on the other hand, is described by an energy density spectrum. The wave energy spectrum describes the energy content of an ocean wave and its distribution over a frequency range of the random wave. Therefore, the random wave method of design may be important especially in the design of floating structures. The random wave is generally described by its statistical parameters. A few commonly used statistical parameters of a random wave are listed in table 3.10. The details of this method of design are covered in Chapter 5.

3.6.1 Spectrum Model

There are several spectrum formulas that are used in the design of offshore structures. These formulas are derived from the observed properties of ocean waves and are thus empirical in nature. The most commonly used spectrum formulas are Pierson–Moskowitz model, Bretschneider model, ISSC model, JONSWAP model and less used Ochi–Hubble spectrum model. These formulas are listed in table 3.11. In describing these spectra, one or more parameters are needed. These are representative statistical quantities for the storm represented by the spectrum and determine the total energy content of the storm as well as

Table 3.10 Common statistical parameters of ocean waves

Parameter	Symbol	Description
Short-term record length	T_s	Duration of the storm
Significant wave height	H_s	Average height of the highest one third waves in a short-term record
RMS wave height	H_{rms}	Root mean square value of the individual wave heights in a short-term record
Peak frequency	ω_0	Frequency at which the spectrum peaks,
Significant wave frequency	ω_s	Average frequency corresponding to the significant waves in the short-term record
Mean frequency	$\overline{\omega}$	Mean frequency of the individual waves in a short-term record
Wave standard deviation	σ	Standard deviation of the wave time history in the record

the frequency distribution. Both are equally important in determining the response of the structure. Different spectrum models for the same energy content distributes the energy differently across the frequency band. Thus the response of the structure for the same random wave energy (or equivalently, significant wave height) will be different if different spectrum models are used.

The P–M model shown in table 3.11 is a one-parameter spectrum model. It is written in terms of the peak frequency ω_0. The frequency corresponds to the frequency at which the energy density spectrum peaks. For this spectrum, the relationship between the peak frequency and the significant height of the wave may be obtained from

$$\omega_0 = 0.161g/H_s \tag{3.20}$$

The P–M spectrum model may also be described by the mean wind speed U_w, which is related to ω_0 by the relationship:

$$\omega_0 = \sqrt{\frac{2}{3}}\frac{g}{U_w} \tag{3.21}$$

In the formulas (table 3.11), α = Phillips constant, $\vec{\alpha}$ = modified Phillips constant, γ = peakedness parameter and σ = spectral width parameter, $\sigma = \sigma_A$ for $\omega \leq \omega_0$, $\sigma = \sigma_B$ for $\omega > \omega_0$.

The modified form of P–M, the Bretschneider and the ISSC spectra are called two-parameter spectra. In these cases, H_s and a statistical period are required as input. The periods respectively are ω_0, ω_s and $\bar{\omega}$, which are introduced in table 3.10.

Table 3.11 Formulas for common spectrum models

Model	No. of Parameter	Independent Parameters	Formula
Pierson–Moskowitz	one	U_w, or ω_0	$S(\omega) = \alpha g^2 \omega^{-5} \exp(-1.25[\omega/\omega_0]^{-4})$
Modified P–M	two	H_s, ω_0	$S(\omega) = \frac{5}{16} H_s \frac{\omega_0^4}{\omega^5} \exp(-1.25[\omega/\omega_0]^{-4})$
Bretschneider	two	H_s, ω_s	$S(\omega) = 0.1687 H_s \frac{\omega_s^4}{\omega^5} \times \exp(-0.675[\omega/\omega_s]^{-4})$
ISSC	two	H_s, $\overline{\omega}$	$S(\omega) = 0.1107 H_s \frac{\overline{\omega}^4}{\omega^5} \times \exp(-0.4427[\omega/\overline{\omega}]^{-4})$
JONSWAP	five	$H_s, \omega_0, \gamma, \tau_a, \tau_b$	$S(\omega) = \overline{\alpha} g^2 \omega^{-5} \exp(-1.25[\omega/\omega_p]^{-4}) \times \gamma^{[\exp\{-(\omega-\omega_p)^2/(2\sigma^2\omega_p^2)\}]}$
Ochi–Hubble	six	H_{s1}, ω_{01}, λ_1, H_{s2}, ω_{02}, λ_2	$S(\omega) = \frac{1}{4} \sum_{j=1}^{2} \left(\frac{4\lambda_j + 1}{4} \omega_{0j}^4\right)^{\lambda_j} \Big/ \Gamma(\lambda_j) \times \frac{H_{sj}^2}{\omega^{4\lambda_j+1}} \exp\left[-\left(\frac{4\lambda_j+1}{4}\right)[\omega/\omega_{0j}]^{-4}\right]$

While the JONSWAP spectrum has five parameters, only two are generally varied in its application – ω_0 and H_s. The suitable values of γ to use at various offshore locations will be described later. A suitable parameter for γ is in the range of 2–3 for the North Sea application.

If the peakedness parameter is not defined, the following can be applied:

$$\gamma = 5 \qquad \text{for } Tp/\sqrt{H_s} \leq 3.6; \quad \text{and}$$

$$\gamma = \exp\left(5.75 - 1.15 \frac{T_p}{\sqrt{H_s}}\right) \quad \text{for } T_p/\sqrt{H_s} > 3.6 \tag{3.22}$$

The value of $\vec{\alpha}$ for the North Sea application is commonly computed from

$$\vec{\alpha} = 5.058 \left[\frac{H_s}{(T_p)^2}\right]^2 (1 - 0.287 \ln \gamma) \tag{3.23}$$

The Phillips constant, α, is normally taken as 0.0081 and the width parameters σ_A and σ_B are 0.07 and 0.09 respectively, and the peaked parameter, $\gamma = 3.3$. In general, they are dependent on the significant height and peak periods. For a fully developed sea, the JONSWAP spectrum reduces to the Pierson–Moskowitz spectral formulation ($\gamma = 1.0$).

The Ochi–Hubble spectrum is a six-parameter spectrum describing combination of two superimposed seas – a locally generated sea and a swell. The swell is a narrow-band wave, which arrives at the site from a distant storm. The three parameters for each of these waves are given individually by significant wave height, peak frequency, and a parameter (H_{s1}, ω_{01} and λ_1, and H_{s2}, ω_{02}, and λ_2) respectively. The quantity Γ is the Gamma function. Generally, $\lambda_1 = 2.72$ and $\lambda_2 = 1.82 \exp(-0.02\, H_s)$. Note that $H_s = \sqrt{(H_{s1}^2 + H_{s2}^2)}$.

The significant wave height is related to the variance of the wave spectrum (also known as the zeroth moment) by

$$H_s = 4\sqrt{m_0} \tag{3.24}$$

Table 3.12 gives the values of H_s and corresponding ω_0 (and T_0) for the P–M spectrum. Examples of P–M spectrum for different values of H_s are shown in fig. 3.15. The table and plot should help the reader with a quick reference to the area of the maximum wave energy for a chosen significant wave height. Note that the peak frequency shifts with the value of the significant wave height.

The form of JONSWAP spectrum is such that the significant wave height does not directly enter in the equation. In order to describe a shape of a JONSWAP spectrum for a given H_s

Table 3.12 Significant height vs. peak period for P–M spectrum model

H_s	H_s	ω_0	T_0
ft	m	rad/s	s
5	1.52	1.018	6.171
10	3.05	0.720	8.726
15	4.57	0.588	10.688
20	6.10	0.509	12.341
25	7.62	0.455	13.798
30	9.15	0.416	15.115
35	10.67	0.385	16.326
40	12.20	0.360	17.453
45	13.72	0.339	18.512

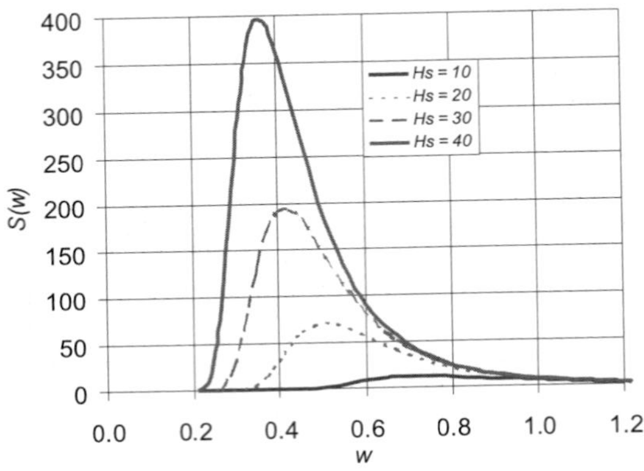

Figure 3.15 Fully developed P–M spectrum for different seas

an iteration is needed. In these cases, the following empirical relationship in equation (3.25) among H_s, γ, and ω_0 may be used to derive the value of the significant height for given values of peak period and peakedness parameter.

$$\omega_0^2 = (0.11661 + 0.0158\gamma - 0.00065\gamma^2)\frac{4\pi^2}{H_s} \tag{3.25}$$

A table is shown (table 3.13) based on this formula giving the values of peak frequency for different significant heights for five γ-values. This table may be used along with the equation in table 3.11 to generate JONSWAP spectrum for a given significant height. This is illustrated in fig. 3.16. The figure shows the JONSWAP spectrum for the same significant height of 40 ft (13.3 m) for different γ-values. Note that the energy contents are the same among the spectra, but the peak frequency shifts to higher values and becomes more peaked with higher γ-values.

3.6.2 Applicability of Spectrum Model

Two of the most common spectral models used in the offshore structure design are the P–M and JONSWAP spectrum. Note that the P–M spectrum is a special case of JONSWAP with the peakedness parameter value being one. If the site-specific spectral form is not known, table 3.14 summarises the most common form of spectral model applied for the different offshore locations of the world.

If the peakedness parameter is not defined, the values of γ in table 3.15 may be applied [partially adopted from DNV-OS-E01, 2001].

Table 3.13 Significant height vs. peak period for JONSWAP spectrum model

$\gamma=$		1		2		3.3		5		7	
T_0	ω_0	H_s	H_s	H_s	H_s	H_s	H_s	H_s	H_s	H_s	H_s
s		ft	m	ft	m	ft	m	ft	m	ft	m
5	1.26	3.29	1.00	3.64	1.11	4.04	1.23	4.48	1.37	4.88	1.49
6	1.05	4.74	1.45	5.24	1.60	5.82	1.77	6.46	1.97	7.03	2.14
7	0.90	6.46	1.97	7.13	2.18	7.92	2.42	8.79	2.68	9.57	2.92
8	0.79	8.43	2.57	9.32	2.84	10.35	3.15	11.48	3.50	12.50	3.81
9	0.70	10.67	3.25	11.79	3.60	13.10	3.99	14.53	4.43	15.82	4.82
10	0.63	13.18	4.02	14.56	4.44	16.17	4.93	17.94	5.47	19.54	5.96
11	0.57	15.94	4.86	17.62	5.37	19.56	5.96	21.70	6.62	23.64	7.21
12	0.52	18.97	5.78	20.97	6.39	23.28	7.10	25.83	7.87	28.13	8.58
13	0.48	22.27	6.79	24.61	7.50	27.32	8.33	30.31	9.24	33.02	10.07
14	0.45	25.82	7.87	28.54	8.70	31.69	9.66	35.15	10.72	38.29	11.67
15	0.42	29.65	9.04	32.76	9.99	36.38	11.09	40.36	12.30	43.96	13.40
16	0.39	33.73	10.28	37.28	11.36	41.39	12.62	45.92	14.00	50.01	15.25

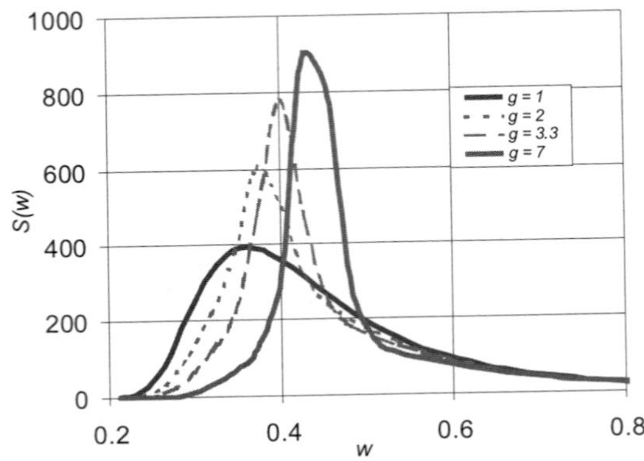

Figure 3.16 JONSWAP spectrum for different γ values for $H_s = 40$ ft

Table 3.14 Common form of spectral models applied to different regions

Location	Operational	Survival
Gulf of Mexico	P–M	P–M or JONSWAP
North Sea	JONSWAP	JONSWAP
Northern North Sea	JONSWAP	JONSWAP
Offshore Brazil	P–M	P–M or JONSWAP
Western Australia	P–M	P–M
Offshore Newfoundland	P–M	P–M or JONSWAP
West Africa	P–M	P–M

Table 3.15 Typical JONSWAP γ-values for various offshore locations around the world

Location	γ
North Sea or North Atlantic	3.3
Northern North Sea	Up to 7
Offshore West Africa	1.5 ± 0.5
Gulf of Mexico	1 for $H_s <= 6.5$ m
	2 for $H_s > 6.5$ m
Offshore Brazil	1–2

3.6.3 Simulation of Two-dimensional Sea

The frequency-domain design of an offshore system is based on wave spectrum. The frequency-domain analysis is applicable to linear systems. For a non-linear system, it is often desirable to design the offshore system with a time-domain design tool. In a time-domain analysis, the time history of the ocean wave is needed. The time history is computed from the spectrum model.

Based on a frequency band of width Δf as shown in fig. 3.17, the wave height (based on the blocked area in the figure) is derived from the formula

$$H(f_1) = 2\sqrt{2S(f_1)\Delta f} \tag{3.26}$$

where f_1 is the frequency within the Δf band and $S(f_1)$ is the mean amplitude of the spectral density within this band. The period for this band is simply

$$T = 1/f_1 \tag{3.27}$$

Thus the frequency band of the spectrum is represented with a height-period pair (H, T). A random phase is assigned to this pair by a random number generator to retain the randomness of the time history. If the entire spectrum is divided into N frequency bands of the width of Δf, then the time history of the wave profile is obtained from

$$\eta(x,t) = \sum_{n=1}^{N} \frac{H(n)}{2} \cos[k(n)x - 2\pi f(n)t + \varepsilon(n)] \tag{3.28}$$

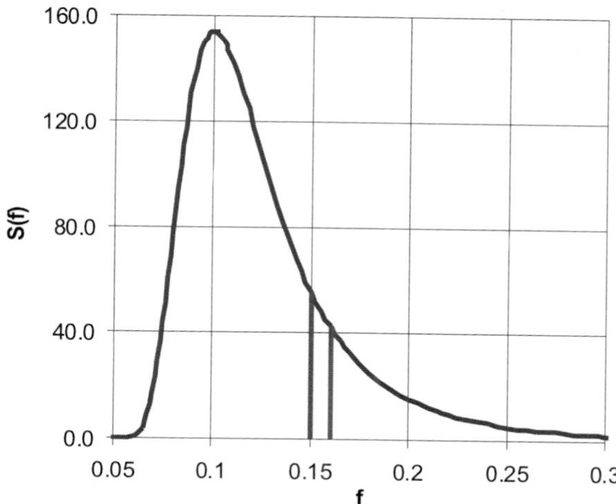

Figure 3.17 Time history simulation from given energy density spectral model

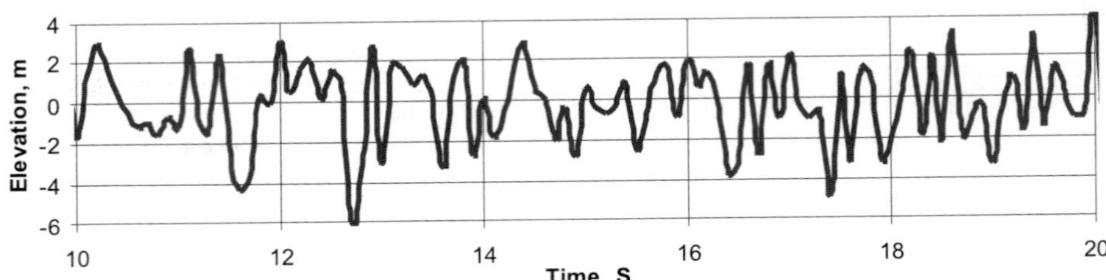

Figure 3.18 Time history simulation of two-dimensional random wave profile from an energy density spectral model

in which x is the location on the structure where the profile is computed and $\varepsilon(n)$ is the random phase angle of the component wave n. The quantity N should be large (200–1000) to ensure the randomness of the time history.

A sample two-dimensional time history generated from a spectrum is shown in fig. 3.18. The duration of the random wave time history record should be sufficiently large for data analysis, but should have a minimum of at least 200 cycles for statistical stability in data analysis.

Because of the method of generation of the time series, when inverted by the Fourier transform it yields a close match to the target spectrum. An example of the comparison for a JONSWAP spectral model is shown in fig. 3.19.

3.6.4 Directional Spectrum

The above formula assumes that the waves are unidirectional having incident from $-x$ to $+x$-direction. When waves have an angular spread, a spreading function should be introduced as well. Thus, the directional sea is a function of frequency ω as well as direction θ. The directional spectrum is obtained as

$$S(\omega,\theta) = S(\omega)D(\omega,\theta) \tag{3.29}$$

The common form of $D(\omega,\theta)$ over $-\pi$ and π is given as

$$D(\omega,\theta) = \frac{\sqrt{\pi}\Gamma(s+1)}{\Gamma(s+1/2)}\cos^{2s}\frac{(\theta-\theta_o)}{2\theta_{\max}} \tag{3.30}$$

Other forms are

$$D(\omega,\theta) = \frac{1}{\sigma\sqrt{2\pi}}\exp\left[-\frac{(\theta-\theta_o)^2}{2\sigma^2}\right] \tag{3.31}$$

$$D(\omega,\theta) = \frac{1}{\pi}2^{2s-1}\frac{\Gamma^2(s+1)}{\Gamma(2s+1)\sqrt{\pi}}\cos^{2s}\frac{(\theta-\theta_o)}{2} \tag{3.32}$$

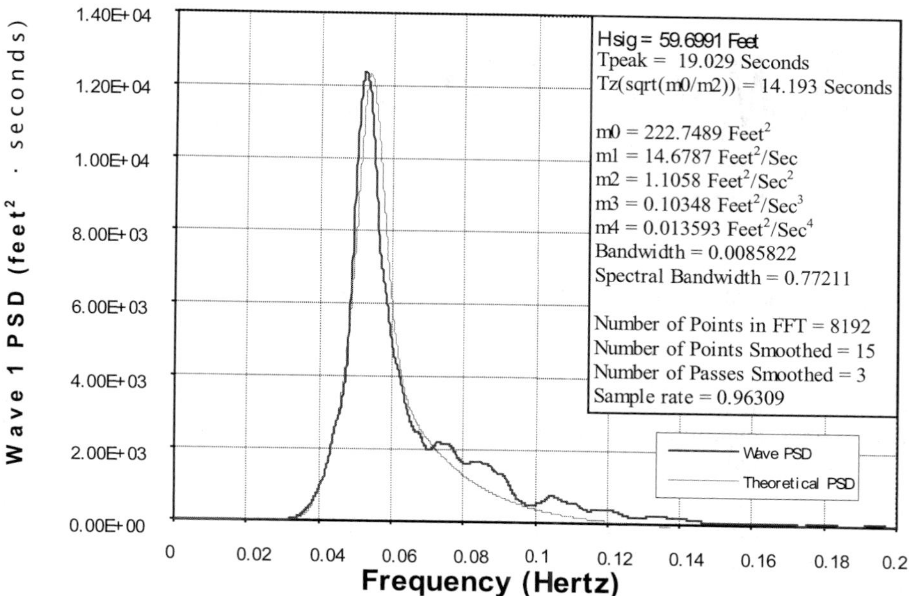

Figure 3.19 Comparison of generated JONSWAP energy density spectrum with target spectrum

where θ_o is the principal direction of wave, $\theta_{\max}$ is the maximum spreading angle and s is the spreading index. The coefficient $C(s)$ is defined as

$$C(s) = \frac{\Gamma(s+1)}{\Gamma(s+1/2)\sqrt{\pi}}, \quad s = 1, 2 \ldots \tag{3.33}$$

where Γ is the Gamma function. The values of the coefficient $C(s)$ are given in table 3.16 for the various spread functions of the directional seas.

Table 3.16 Spreading function coefficients for directional seas

Spread, s	Power, $2s$	Coefficient, $C(s)$
1	2	$2/\pi$
2	4	$8/3\pi$
3	6	$48/15\pi$
4	8	$384/105\pi$

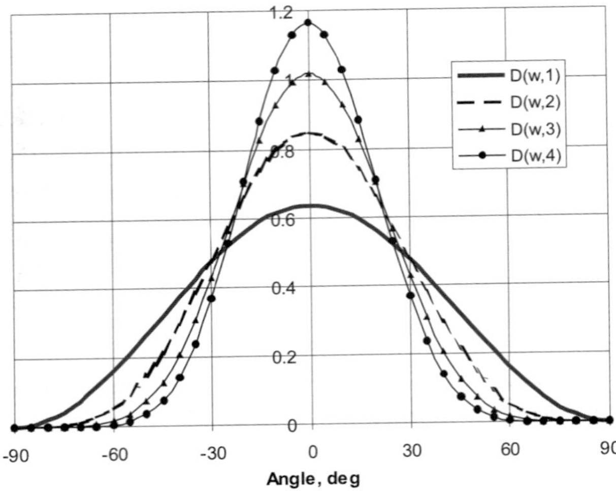

Figure 3.20 Spreading function for a directional sea

The most common value in use is $s=1$. A plot of $D(\omega, \theta)$ for different values of s is given in fig. 3.20. Note that directional spreading becomes narrower as the value of s increases. Thus, for a large value of s, the waves become essentially unidirectional.

3.6.5 Simulation of Directional Sea

The simulation of directional sea is similar to that of the unidirectional (i.e. monochromatic) waves (with additional dependence on the wave direction). The only difference is that the three-dimensional spectral density is divided into small sections not only along ω by $\Delta\omega$, but also along θ by $\Delta\theta$. Then based on the linear wave theory, the directional random wave profile may be represented by a series of regular waves summed together. Since the wave profile will be dependent on the horizontal coordinates x and y, the general form is

$$\eta(x, t) = \sum_{n=1}^{N} a_n \cos\left[k(x \cos\theta_n + y \sin\theta_n) - \omega_n t + \varepsilon_n\right] \tag{3.34}$$

where the amplitude a_n now includes the spreading function:

$$a_n = \sqrt{2S(\omega)D(\omega, \theta)\Delta\omega\Delta\theta} \tag{3.35}$$

in which $\Delta\theta$ is the increment in the wave spreading angle. The other quantities are computed the same way as before.

A comparison of the directional spectrum model and a simulated model in a wave basin is shown in fig. 3.21. Note that the measured spectrum is valid only in a small region where

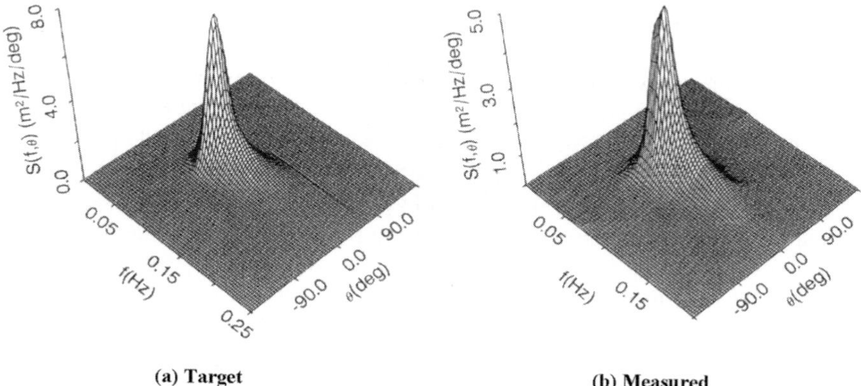

(a) Target **(b) Measured**

Figure 3.21 Directional sea spectrum generated in a wave basin [Cornett and Miles (1990)]

the target spectrum is desired. This is particularly true for a three-dimensional wave in a wave basin.

3.7 Sea States

A simplified description of the sea states from very small to extreme waves is presented in table 3.17. A simple description of the type of sea is given for sea states ranked 1–9. For each sea state the range of wind speed, significant wave height and wave period range is shown. The range of periods covers the range over which measurable energy of the random wave for the particular sea state exists.

Table 3.17 Definition of sea states

Sea states	Description of sea	Wind speed range	Significant wave height	Wave period range
		Knots	ft	s
1	Small wavelets	5–10	0.3–1.4	0.5–5
2	Large wavelets	10–14	1.4–3	1–7.5
3	Small waves	14–18	3–6	1.4–8.8
4	Small to moderate waves	18–19	6–7	2.5–10.6
5	Moderate waves	19–24	7–13	2.8–13.5
6	Large waves	24–30	13–22	3.8–15.5
7	Moderate gale	30–40	22–45	4.7–21
8	Strong gale	40–55	45–70	6.5–25
9	Hurricane type storm	55–70	70–115	10–30

3.8 Wave-driven Current

Current is a common occurrence in the open ocean. The current at the sea surface is mainly introduced by the wind effect on the water, variation of atmospheric pressure and tidal effects. But current is also present in the subsurface and seafloor region.

In the early years of offshore development, the common belief was that currents are confined to the upper waters of the ocean and practically no current exists below a water depth of about 1000 m. In recent days, however, it is recognised that a number of classes of currents exist in the deep waters, and some are known to extend to large depths. Improved definition and knowledge of these currents and their sources will lead to an improved design criteria for offshore structures extending in ultra-deep water. Examples of these classes of currents are tropical cyclones such as hurricanes, extratropical cyclones, and cold air outbreaks and currents arising from major surface circulation features.

The most common categories of current are:

- wind-generated currents,
- tidal currents (associated with astronomical tides),
- circulational currents (associated with oceanic circulation patterns),
- loop and eddy currents, and
- soliton currents.

The vector sum of these currents is the total current, and the speed and direction of the current at specified depths are represented by a current profile.

3.8.1 Steady Uniform Current

In most cases current is turbulent, but is generally approximated by the corresponding mean flow. In the design of offshore structures, it is customary to consider current as time invariant. For the design value, a 100 yr current is often chosen. The environmental conditions in design are obtained from the site-specific data. Therefore, it is difficult to be too specific in terms of the magnitude of current in offshore locations. However, some order of magnitudes of different types of current prevalent in the open oceans may be cited. They are discussed in further detail in Section 3.11.

A surface current speed with a 10-year return period should be used in the offshore structure design, based on the marginal distribution of current speeds at the location. In certain geographical areas, current loads can be the governing design loads. In areas where the current speed is high, and the sea states are represented with small wave heights, e.g. West Africa, an environmental condition represented by 100-year wind and current speeds combined with a sea state with a return period of 10-year should be considered.

If the current statistical data is not available, the wind generated current velocities at the still water level may be computed from the 1 h mean wind speed at a 10 m elevation as

$$U = 0.015U_W \tag{3.36}$$

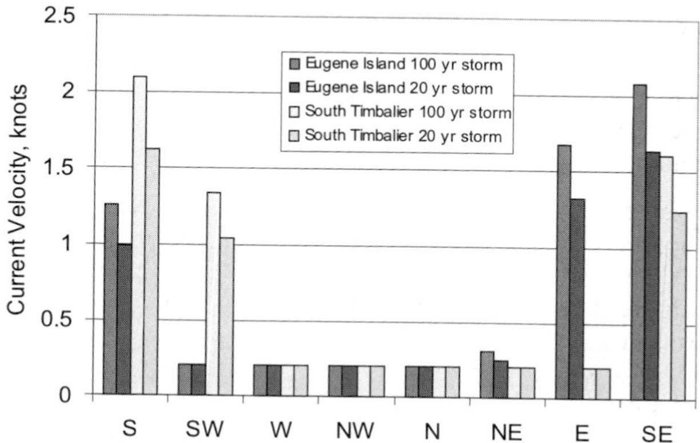

Figure 3.22 Current directional distribution at two locations in GOM

The surface current causes drift in ships and floating structures. The seafloor current may introduce eddies due to the irregularities in the seafloor, which may have a detrimental effect on the pipeline on the seafloor. Current at the seafloor is also responsible for scour under the pipeline and around the foundation of a gravity offshore structure causing danger to the survival of the structure.

An example of the directional current velocity distribution is shown in fig. 3.22 for Eugene Island and South Timbalier in the Gulf of Mexico (GOM). These values in knots are shown in 20th edition of API RP2A (2000) guidelines.

3.8.2 Steady Shear Current

While current is often uniform with depth, they may vary with water depth. The shear current is generally considered linear with depth or bilinear. In deep water, the current disappears near the bottom. Near the sea floor in shallow water the current profile is logarithmic due to bottom shear.

Besides introducing non-zero mean velocity in the water particle, current alters the shape and size of the waves. Moreover, the current is stretched under crest (or compressed under trough). An acceptable approximation for uniform current is to move the current profile up to the free surface straight up above the mean water level. For a linear shear current, a stretching formula similar to the Wheeler stretching for waves may be applied. For irregular waves, an extension for stretching current may be used based on the regular waves as stated above.

3.8.3 Combined Current and Waves

Modifications are needed in computing the wave kinematics and associated loading from waves propagating on a superimposed steady current. The wave period is modified to an apparent period by the free-stream current velocity. A current in the wave direction stretches the wavelength and opposing current shortens it. In a reference frame moving in the same direction as the steady current speed, U, waves encounter a structure at a (intrinsic) frequency lower than the wave frequency alone, i.e. $\omega_A = 2\pi/T_A$, where T_A is the apparent period seen by an observer moving with the current. The two frequencies are related by the Doppler shift as

$$\omega = \omega_A + kU \tag{3.37}$$

where k is the wave number. The last term in equation (3.37) is called the convective frequency. The Doppler effect is computed based on ambient current in the direction of waves. When current is not in the direction of waves, the component of current in the direction of wave should be considered in equation (3.37). In this computation the current profile with the water depth need not be necessarily uniform but the variation of current with the water depth should be known. The apparent wave period for a prescribed current profile is computed from the following three simultaneous equations [API RP2A, 2000]:

$$\frac{L}{T} = \frac{L}{T_A} + U_A \tag{3.38}$$

$$T_A^2 = \frac{2\pi L}{g \tanh kd} \tag{3.39}$$

and

$$U_A = \frac{2k}{\sinh 2kd} \int_{-d}^{0} U_c(y) \cosh[2k(y+d)] \tag{3.40}$$

Velocity U_A is a weighted mean velocity and is obtained by the integration of the local current over the entire water depth. It is called an effective in-line velocity. It is used to derive the apparent period, T_A. The quantity $U_c(y)$ is assumed to be the horizontal steady current at an elevation y. If the current is at an angle to the wave, the component in line with the wave should first be computed and this is what should subsequently be used. For a uniform current profile, the apparent period is given by the curves in fig. 3.23 [adopted from API RP2A (2000)]. For waves propagating in the same direction as the current, T_A is of the order of 10% greater than the wave period T. The apparent wave kinematics are computed using a two-dimensional wave theory since in a moving reference frame, the current does not enter in the calculation.

In deep water ($\tanh kd = 1$) in uniform current, the wave number is related to the wave frequency by the relation

$$k = \frac{4\omega^2/g}{[1 + (1 + 4U\omega/g)^{1/2}]^2} \tag{3.41}$$

Note that when U is positive (current in the same direction as the waves), the value of k is smaller so that the wavelength is larger (stretched wave). When U is negative (current opposing waves), k is larger and the wavelength is shorter. The total water particle kinematics is the sum of current and wave particle velocity, which should be computed using this interaction effect. For shear current, the variation of current with the depth should be included in this calculation.

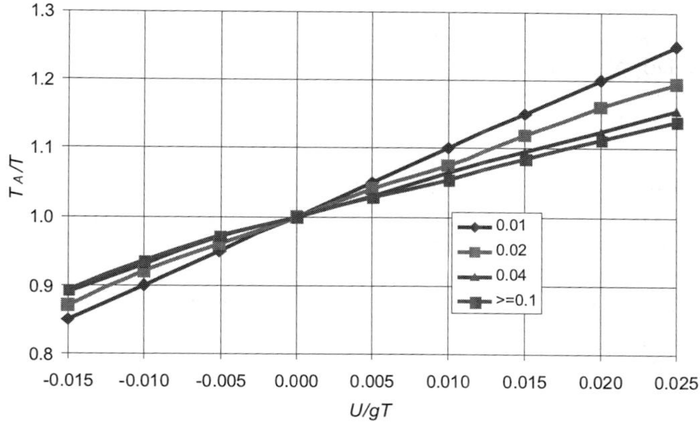

Figure 3.23 Apparent wave period due to Doppler shift in steady current

The difference of including the interaction effect and ignoring it in the superposition is illustrated by an example in which the horizontal water particle velocity was computed both ways and compared with the measurement. The comparison is shown in fig. 3.24. The wave parameters used in the example are amplitude, a of 0.25 in (6.3 mm), period $T = 1.12$ s in a depth of 0.37 ft (114 mm) as shown in the figure. The current profile with depth is shown on the left-hand side. It is uniform near the top with a shear near the bottom. The ambient current as well as current profile in waves is shown. It is found that the current increases in waves near the top with a corresponding decrease near the bottom. The horizontal water particle velocities are shown on the right-hand side. The velocities are computed using a shear current profile and an equivalent uniform current profile. The profile based on linear superposition of velocities without accounting for interaction is also shown. Note that the linear shear gives the best match with the measurement. Superposition without interaction provides the worst match.

3.9 Loop Current

Loop current is a part of the Gulf Stream system and loops through the Gulf of Mexico continually. It enters the gulf through the Yucatan Channel and exits through the Straits of Florida oscillating north and south.

The phenomena of most concern to deep water operators in the Gulf of Mexico are surface-intensified currents associated with the loop current, loop current eddies, and other eddies (both anticyclonic and cyclonic). While reliable information is available regarding the general speed distributions, translation speeds, sizes and shapes of these currents, the details of the velocity distributions and their variability are not well known. The information presented here on the loop current and its strengths is primarily obtained from Nowlin, et al (2001).

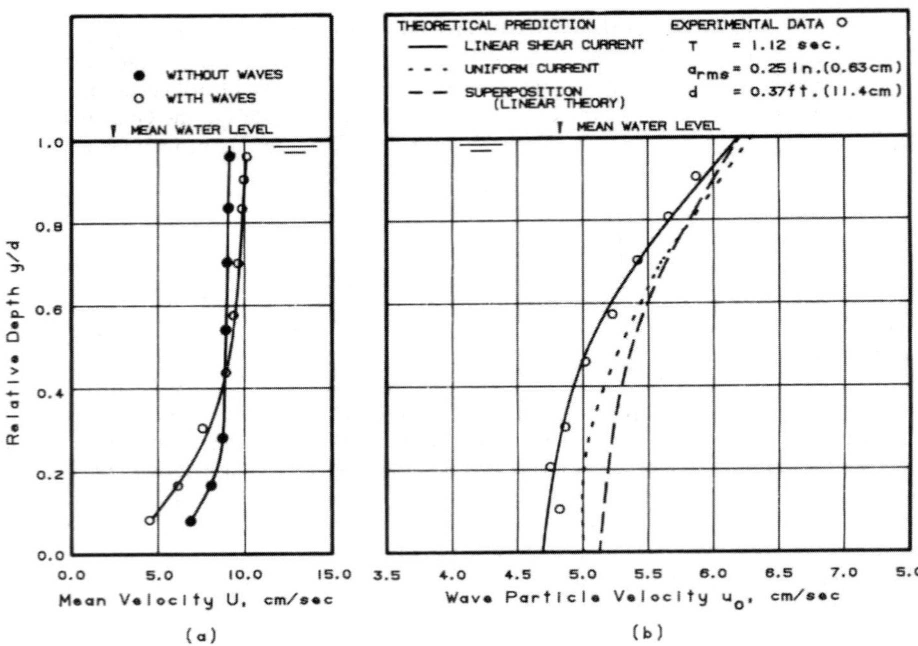

Figure 3.24 Comparison of horizontal water particle velocity in waves with uniform current [adopted from Ismail (1984)]

The Gulf's circulation is dominated by the loop current. The loop current is of warm subtropical water that enters the Gulf through the Yucatan Strait, extends northward, then loops around to the south and ultimately exits the Gulf through the Florida Strait. The strength of this loop current exhibits large variability and can be high. The loop current system passing from the Caribbean Sea to the Straits of Florida through the eastern Gulf has maximum speeds sometimes reaching the order of 3.0 m/s. The currents associated with the loop current may extend down into the water column to about 800 m. They may have surface speeds of 1.5–2.0 m/s or more; speeds of 10 cm/s are not uncommon at 500 m.

The loop current may extend far north, often reaching Mississippi Delta, where the circulation closes off and a large warm-core loop current eddy is shed. These eddies also possess strong currents, but unlike the loop current, they are not constrained to the eastern Gulf and typically drift westward. Often, the westward drift can interfere with offshore operations.

A few loop current cases reported by Nowlin, et al (2001) are listed here. Strong mid-water-column currents 1.25–2.0 m/s (2.5–4 knot) were observed in Ewing Banks along the 240 m (800-ft) contour. A similar event in deeper water had currents up to 1.5 m/s (3 knots) at 100–200 m beneath the surface. Several deep-water oil operators have reported cases of very high-speed subsurface currents at locations over the upper continental slope. Such currents may have vertical extents of less than 100 m, occurring in the depth range 150–400 m with maximum speeds above 1.5 m/s. Examining the data from Gulf of Mexico locations in water depths 1200–1500 m, Nowlin, et al (2001) have identified seven cases of subsurface current jets with speeds of 0.5–2.1 m/s. In the mid-1980s in the Gulf of Mexico depth-independent currents were observed to extend from depths near 1000 m to the bottom.

3.10 Wind and Wind Spectrum

The wind effect on an offshore structure becomes important when the superstructure (portion above the MWL) is significant. The wind generally has two effects – one from the mean speed and the other from the fluctuation about this mean value. The mean speed is generally treated as a steady load on the offshore structure. For a fixed structure, it is only the mean speed that is taken into account. The effect of the fluctuation of wind about the mean value has a little effect on the fixed structure. However, this is not the case for a floating structure. In this case, the dynamic wind effect may be significant and may not be ignored. It should be noted that even the mean wind flowing over a changing free surface produces a fluctuating load due to the variation of the exposed structure with the wave. This effect is sometimes considered. For a linear wave, this fluctuation may be determined in a simple straightforward manner if the exposed surface is assumed to vary sinusoidally.

In the following section the mean wind speed and the fluctuations about this mean value and its possible impact on an offshore structure are discussed.

3.10.1 Wind Speed

The accepted steady wind speeds in a design of an offshore structure are generally taken as the average speed occurring for a period of 1-h duration. The steady speeds are considered to be the mean speed measured at a reference height, typically 30 ft (10 m) above the mean still water level. A mean wind speed for a 100-year return period should be used in the design, based on the marginal distribution of wind speeds at the specific location. The directionality of the wind may be important in some applications.

Wind load on the structure should be treated as a steady component based on the above mean speed. Additionally, a load with a time-varying wind component known as the gust should be calculated, which generates low-frequency motion. The time varying wind is described by a wind gust spectrum.

3.10.2 Wind Spectrum

Just like the random waves, wind blowing over the deck structure is also random having a mean speed superimposed on it. This wind spectrum may be important to consider for certain types of offshore structures. There are several wind spectrum models available. Here we have adopted the expression of the wind frequency spectrum according to the guidelines of API-RP2A. However, the method described here may be applied to any practical frequency spectrum of wind [Ochi and Shin, 1988]. The variation of wind speed with elevation is estimated by the following expression:

$$U_w(1\ \mathrm{h}, z) = U_w(1\ \mathrm{h}, z_R) \cdot \left(\frac{z}{z_R}\right)^{0.125} \tag{3.42}$$

in which z = elevation of the wind centre of pressure above SWL, z_R = reference elevation taken as 10 m, $U_w(1\ \mathrm{h},\ z_R)$ = 1-h mean wind speed at the reference elevation. According to API-RP2A (the equations 3.3.2–5), the wind frequency spectrum about the 1-h mean value is described by

$$S(f) = \frac{(\sigma_w(z))^2}{f_p[1 + 1.5f/f_p]^{5/3}} \tag{3.43}$$

in which $S(f)$ = spectral energy density, f = frequency, f_p = peak frequency and $\sigma_w(z)$ = standard deviation of wind speed. Various values of peak frequency of the spectrum may be considered. The recommended range of f_p is indicated as:

$$0.01 \leq f_p\mathrm{coeff} = \frac{f_p \cdot z}{U_w(1\,\mathrm{h}, z)} \leq 0.10 \tag{3.44}$$

Generally, $f_p$coeff is taken as 0.025. The standard deviation of the wind speed is given by

$$\sigma_w(z) = \left| \begin{array}{ll} U_w(1\,\mathrm{h}, z) \times 0.15\left(\frac{z}{z_S}\right)^{-0.125} & \text{if } z \leq z_S \\ U_w(1\,\mathrm{h}, z) \times 0.15\left(\frac{z}{z_S}\right)^{-0.275} & \text{if } z > z_S \end{array} \right. \tag{3.45}$$

where z_s is the thickness of the "surface layer" and is taken as 20 m.

A wind spectrum model based on the above formula is shown in fig. 3.25. It is seen that unlike the wave spectrum, the wind spectrum is very wide-banded. The high

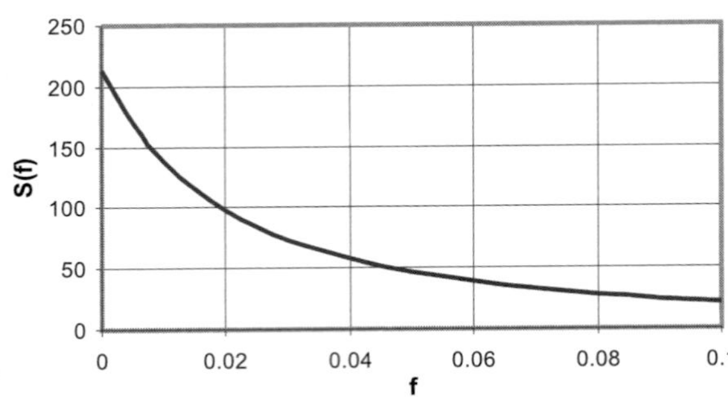

Figure 3.25 PSD of the wind speed

frequency part of the spectrum is generally not important for offshore structure application. However, floating structures are susceptible to low frequency area of the wind spectrum in which the structure experiences a slow drift oscillation. This area will be explored in the subsequent chapters on loading and floating structure design.

3.11 Offshore Environment by Location

The design criteria for a specific platform should reflect the risks involved. The early design guideline such as, API RP2A (1993) had recommended the Working Stress Design (WSD) based on a single maximum wave of given height and period, generally representing the 100-yr storm. API developed new recommendations (2000) for consequence-based design criteria for new platforms in the Gulf of Mexico.

Two classes of risks [API RP2A, 2000] were separately considered – those associated with Life Safety and those associated with consequences of failure. Risks involving environmental damage, economic loss and public concern were included as consequences of failure risks. In the Gulf of Mexico, the types of platform are often described in terms of three levels: L1, L2 and L3.

Life safety for consequence-based criteria considers three levels of exposure of personnel to the design event. Consequences of failure encompass damage to the environment, economic losses to the operator and the government, and public concerns. Consequences of failure are difficult to quantify. They are simply categorised as high, medium and low, and the platform characteristics whose damage or loss could result in high, medium or low consequences are described. These levels of exposure are described as the manning conditions as below:

- level 1: manned, non-evacuated,
- level 2: manned, evacuated,
- level 3: unmanned.

According to Level 1, new platforms with high life exposure and/or high consequences of failure must be designed to the 100-yr criteria as specified in the 20th edition of API RP2A. The type L1 is a manned platform that is not evacuated during a storm and thus has a high consequence of failure. For level 2 design, new platforms with minimal life exposure and moderate consequences of failure can be designed for a reduced 20th edition criteria that will result in a platform as reliable as those designed to 9th–19th editions of API RP2A. Type L2 is also manned, but is evacuated during a storm and, therefore, has a medium consequence of failure. This level is restricted to a maximum water depth of 400 ft. The level 2, i.e. manned, evacuated, represents the same design philosophy that has been practiced in the past in the Gulf of Mexico. Level 3 design requires that the new platforms with no life exposure and low consequences of failure can be designed to a reduced criteria that will result in a platform with an ultimate capacity equal to the 100-yr criteria as specified in the 20th edition of API RP2A. Type L3 has a low consequence of failure, as it is unmanned. This level is restricted to a maximum water depth of 100 ft. Consequently, the

Table 3.18 Environment for maximum wave height in GOM for various platform types [Ward, et al, 2000]

Level	Type	Wave period	Current velocity	Wind speed
		s	Knots	Knots
1	Manned, non-evacuated	13.0	2.1	80
2	Manned, evacuated	12.4	1.8	70
3	Unmanned	11.6	1.4	58

severity of environment for which these three types of platform are designed is based on these levels.

Oceanographic design criteria in terms of wave period, wind and current for L1, L2 and L3 platforms are shown in table 3.18. The wave heights are dependent on the water depth and are shown in fig. 3.26. Note that the design criteria are maximum for the level L1-type platform.

Typical seastates with a return period of 100 yr at various offshore locations around the world are shown in table 3.19, which are taken from DNV-OS-E01 (2001). Each short-term (representing a 3-h duration) seastate is characterised by a maximum significant wave

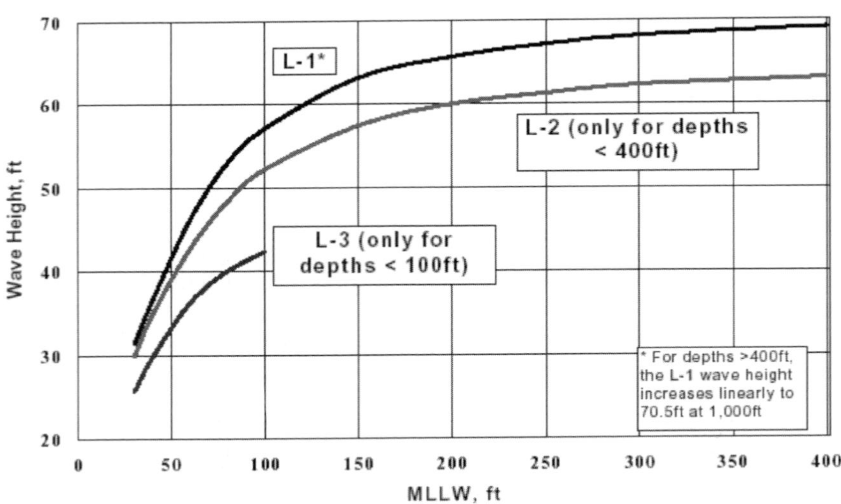

Figure 3.26 Water depth dependent wave height for different platform type for GOM [Ward, et al 2000]

Table 3.19 Typical sea states for various offshore locations around the world [adopted from DNV-OS-E01, 2001]

Description	Location	Type	H_s	T_p	U	U_w
			m	s	m/s	m/s
Norwegian Sea	Haltenbanken		16.5	17.0–19.0	0.90	37.0
Northern North Sea	Troll field		15.0	15.5–17.5	1.50	40.5
North Sea	Greater Ekofisk field		14.0	15.0–17.0	0.55	34.0
Mediterranean	Libya	shallow water	8.5	14.0	1.00	25.3
	Egypt		12.1	14.4	0.78	25.1
Gulf of Mexico		Hurricane	11.9	14.2	1.98	44.1
West Africa	Nigeria	swell	3.6	15.9	1.1	16.0
	Nigeria	squalls	2.7	7.6		16.6
	Gabon	wind generated	2.0	7.0	0.91	24.1
	Gabon	swell	3.7	15.5		16.0
	Ivory Coast	swell	6.0	13.0	0.90[(1)]	29.5
	Angola	swell, shallow water	4.1	16.0	1.85[(2)]	21.8
South America	Brazil (Campos Basin)		8.0	13.0	1.60	35.0
Timor Sea	Non-typhoon		4.8	11.5	1.10	16.6
Timor Sea	Typhoon		5.5	10.1	1.90	23.2
South China Sea	Non-typhoon		7.3	11.1	0.85	28.6
	Typhoon		13.6	15.1	2.05	56.3

(1) Ocean current going to east
(2) Ocean current going to 347.5° approximately parallel to the coast

height and a peak wave period, T_p or zero-crossing period, T_z. Some typical surface current speeds with a return period of 10 years are included for these locations. Also typical 1-h mean wind speeds with a return period of 100 years are shown in the last column of the table.

If the zero-crossing period is desired for the random wave, then the following relationship between the zero-crossing period, T_Z and the peak period, T_p can be applied:

$$T_z = T_p\left[\frac{5+\gamma}{11+\gamma}\right]^{1/2} \tag{3.46}$$

where γ is the JONSWAP peakedness parameter.

It should be stressed that the above is only a guideline and the environment for a specific site should be based on the local meteorological or hindcast data from the site-specific statistical data. Another example of the wave, wind and current data for a few offshore deepwater sites of the world are included in table 3.20 and fig. 3.27. The extreme design conditions are shown for the wave, wind and current. Again these are considered typical for the locations indicated in the figure. The water depths for the sites are indicated and the current profiles are shown.

Another example of a 100-yr environment for three different offshore locations is shown in table 3.21. These values of wave, wind and current are not site-specific. The statistical basis for the different environmental parameters is indicated in this table.

Table 3.20 Extreme design environment criteria for various locations

Location	Type	Water depth	H_s	Wind speed	Surface current	Seabed current	Current type
		m	m	m/s	m/s	m/s	
Gulf of Mexico	Hurricane	3000	12.9	42.0	1.1	0.1	bilinear
Gulf of Mexico	Loop	3000	4.9	32.9	2.57	0.51	bilinear
Brazil	Foz de Amazon	3000	6.0	20.0	2.5	0.3	bilinear
Northern Norway	Nyk High	1500	15.7	38.5	1.75	0.49	linear
West Africa	Girrasol	1350	4.0	19.0	1.5	0.5	bilinear
Atlantic Frontier	Faeroe–Shetland Channel	1000	18	40.0	1.96	0.63	linear

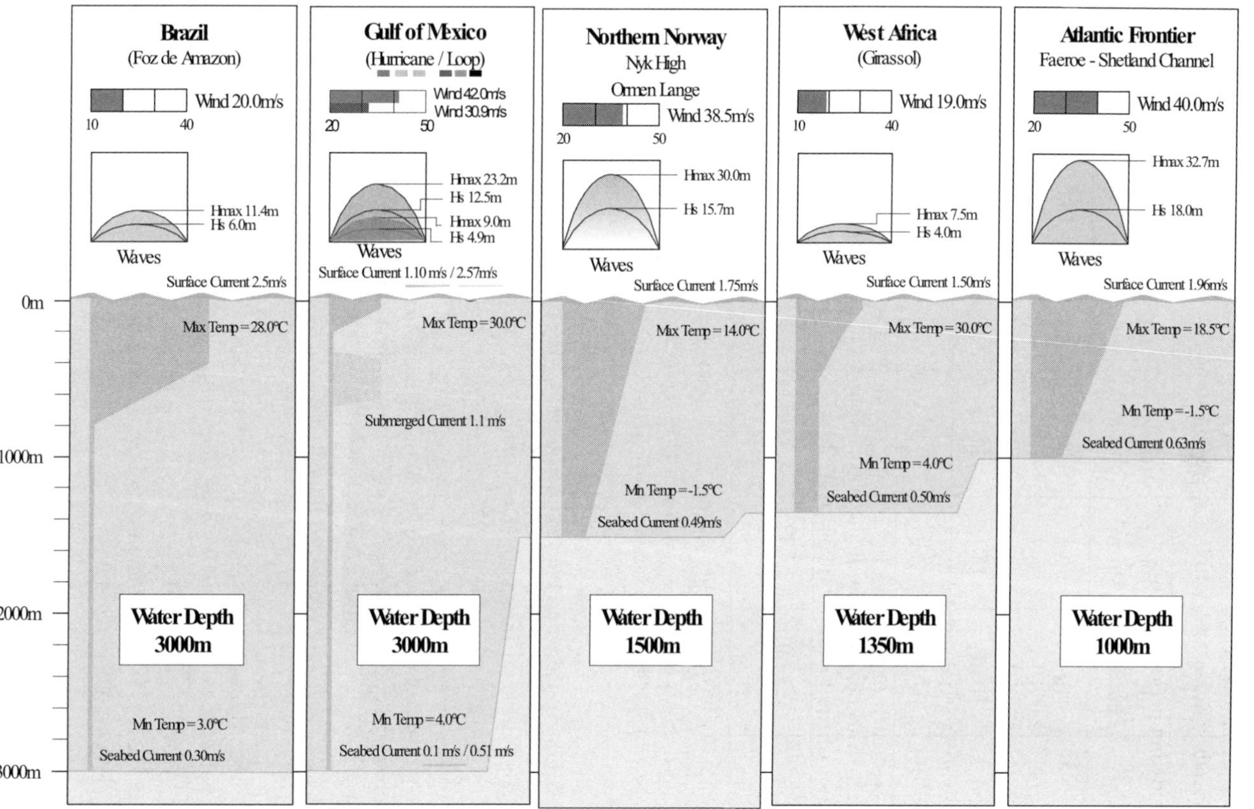

Figure 3.27 Environmental conditions at several deep water sites [Moros and Fairhurst, Offshore, April (1999), Courtesy BP]

Table 3.21 Typical 100-yr environment [Halkyard, et al 2000]

Region	West Africa		Brazil		Gulf of Mexico	
Dominant environment	100 yr current*	100 yr storm*	100 yr current*	100 yr storm*	Loop current	Hurricane
Sig wave ht, ft	10.5	12.1	20.7	24.9	15.0	44.3
Peak period, s	15.0	15.1	12.1	12.7	9.0	14.6
JONSWAP γ	3.0	3.0	3.0	3.0	2.4	2.4
Wind 1h @10 m kts	33.8	41.8	49.6	57.0	30.0	84.2
Surface Current kts	2.00	1.79	3.79	3.15	4.00	2.10

*Note: Current = 10 yr wind/wave and 100 yr current
Storm = 100 yr wind/wave & 10 yr current

References

Airy, Sir, G. B. (1845). Tides and Waves, Encyclopedia Mtrop., Article no. 192, pp. 241–396.

API (American Petroleum Institute). (Dec. 2000). "Recommended practice for planning, designing and constructing fixed offshore platforms", API-RP2A, 21st Edition, Washington, DC.

API (American Petroleum Institute). (1993). "Recommended practice for planning, designing and constructing fixed offshore platforms — working stress design", Washington DC, API Recommended Practice 2A-WSD (RP 2A WSD), 20th edition.

Atkins Engineering Services (1990). "Fluid loading on fixed offshore structures", OTH 90, 322.

Cornett, A. and Miles, M. D. (February, 1990). "Simulation of hurricane seas in a multidirectional wave basin", *International Conference on Offshore Mechanics and Arctic Engineering*, Houston, Texas, pp. 17–25.

Dalrymple, R. A. (1974). "A finite amplitude wave on a linear shear current", *Journal of Geophysical Research*, pp. 4498–4504.

Dean, R. G. (1974). "Evolution and development of water wave theories for engineering application", Vol. I and II, Special Report No. 1, Coastal Engineering Research Center U.S. Army, Fort Belvair, VA.

Dean, R.G. (1965). "Stream function representation of non-linear ocean waves," *Journal of Geophysical Research*, Vol. 70, pp. 4561–4572.

Det Norske Veritas. (June 2001). "Position mooring", Offshore Standard DNV-OS-E301, 65 pages.

Forristall, G. Z. (1986). "Kinematics in the crest of storm waves", *Proceedings of 20th Coastal Eng. Conference*, Taipei, Taiwan, ASCE, pp. 208–222.

Gudmestad, O. and Connors, J. J. (1986). "Engineering approximations to non-linear deepwater waves", *Applied Ocean Research*, Vol. 8, No. 2, pp. 76–88.

Halkyard, J., Choi, G., Prislin, I., and Steen, A., Hawley, P. (2000). "The box spar floating production, drilling, storage, and offloading vessel for the Gulf of Mexico and Brazil", OTC 11905, *Proceedings of Offshore Technology Conference*, Houston, Texas.

Ismail, N. M. (1984). "Wave-current models for design of marine structures", *Journal of Waterway, Port, Coastal and Ocean Engineering*, ASCE, Vol. 110, pp. 432–447.

Kim, C. H., Clements, A. H., Tanizawa, K. (1999). "Recent research and development of numerical wave tanks – a review", *J. of Offshore and Polar Engineering*, Vol. 9, pp. 241–256.

Moros, T.A. and Fairhurst, P. (April 1999). "Production riser design: integrated approach to flow, mechanical issues". *Offshore Magazine*, pp. 82–89.

Nowlin, W. D. Jr., Reid, R. O., DiMarco, S. F., Howard, M. K., and Jochens, A. E. (2001). "Overview of classes of currents in the deep water region of the Gulf of Mexico", OTC 12991, *Proceedings of Offshore Technology Conference*, Houston, Texas.

Ochi M. K. and Shin, Y. S. (1988). "Wind turbulent spectra for design consideration offshore structures", OTC 5736, *Proceedings of Offshore Technology Conference*, Houston, TX, pp. 461–467.

Rodenbush, G. and Forristall, G. Z. (1986). "An empirical method for calculation of random directional wave kinematics near the free surface", OTC 5097, *Proceedings of Offshore Technology Conference*, Houston, TX, pp. 137–146.

Stokes, G.G. (1880). "On the theory of oscillatory waves", *Mathematical Physics Papers*, Cambridge University Press, Vol. 1.

Tromans, P. S., Anaturk, A. and Hagemeijer, P. (1991). "A new method for the kinematics of large ocean waves – application as a design wave", *Proceedings of First Offshore and Polar Engineering Conference*, Vol. 3, pp. 64–71.

Wheeler, J. D. (March 1970). "Method for calculating forces produced by irregular waves", *Journal of Petroleum Technology*, pp. 359–367.

Ward, E. G., et al (2000). "Consequence-based criteria for the Gulf of Mexico: philosophy & results", OTC 11885, *Proceedings of Offshore Technology Conference*, Houston, TX.

Chapter 4

Loads and Responses

Subrata K. Chakrabarti
Offshore Structure Analysis, Inc., Plainfield, IL, USA

4.1 Introduction

This chapter will provide, in the first part, a generic description of how to compute loads on an offshore structure and how the structure responds to these loads in the second. Basic formulas are given, which may be referenced as needed in reviewing other chapters. A discussion is included regarding the design guidelines offered by certifying agencies.

Loads are differentiated between static and dynamic. The static loads on the structure come from gravity loads, deck loads, hydrostatic loads and current loads. The dynamic loads originate from the variable wind and waves. The environmental forces on an offshore structure can be steady or oscillating. The steady loads arise from steady winds and currents. The oscillating loads on the structure originate from the fluctuating structure motions and the waves acting on it. The steady current can also generate a fluctuating load in a (transverse) direction normal to the current direction. The corresponding dynamic current load in the inline direction is generally smaller. In certain offshore structure designs, this fluctuating load may have serious consequences.

The steady current and wind loads on the structure are computed by empirical formulas. The loads depend on the mean speed and the geometry of the member of the structure exposed to wind or current. The wave also produces a steady load known as wave drift load.

In computing wave forces on a structure, the structure is considered fixed in its equilibrium position. A distinction is made regarding small vs. large structures i.e. what is considered small and what is large. The Morison equation is used to compute forces on small structures. It is an empirical formula to include inertia and drag forces on a small structural member. The forces depend on the inertia and the drag coefficients. These coefficients are determined experimentally either in the laboratory or from the field measurements.

For large structures, the linear diffraction/radiation theory is used for the wave force computation. The diffraction part of the theory determines the exciting forces on the structure due to waves (first order) in its equilibrium position. The radiation part within the same general theory considers the moving structure in water and computes the added mass and damping coefficients. Once the exciting force and the hydrodynamic coefficients are known, the motions of the structure may be computed from the equations of motion for the six degrees of freedom of a floating structure. The drag effect on smaller members of the structure originating from the drag term of the Morison equation may be included in the motion analysis. Once the motions of the structure are known, the stresses on the mooring lines and risers attached to the structure may be obtained from their top deflection at the structure attachment point. The environmental loads on the mooring lines and risers by themselves may also be important in their design.

This chapter is mainly concerned with the methods of deterministic forces and motions of the structure. The subject of statistical and spectral representation of forces and motions due to random waves will be covered in Chapter 5.

In order to understand the effect of environment on the structure and the resulting forces experienced by the structure, certain non-dimensional parameters play an important role. The important parameters that determine the forces on a structure are included in table 4.1.

In the formulas in table 4.1, Re = Reynolds number, u_0 = water particle velocity amplitude, ν = kinematic viscosity of water, e = relative roughness parameter, K = surface roughness parameter, D = member diameter, KC = Keulegan–Carpenter number, L = wave length, U = current velocity, V_R = reduced velocity, and f_s = structure vibration frequency. The Reynolds number and the Keulegan–Carpenter number determine the importance of the drag force on the structure. The surface roughness influences the forces on a small structure. The rough surface of a structure is mainly contributed by the marine growth on the submerged part of a structure. In particular, the values of the hydrodynamic drag and inertia coefficients differ significantly with the roughness of the structure surface. The relative roughness is determined by the average size of the particles on the surface given by K normalised by the equivalent cross-sectional diameter of the structure member. A value of relative roughness of 0.02 is considered to be very rough. The diffraction parameter indicates if the scattering of waves from the structure surface is an important consideration. For large structures, the waves in the vicinity of the structure are diffracted causing a significant effect on the forces experienced by the structure from waves. In a steady flow, a flexible structural member is subject to cross-flow vibration. The reduced velocity determines if the structural vibration response in steady flows is important.

Table 4.1 Important non-dimensional quantites

Parameters	Formula
Reynolds number	$Re = u_0 D/\nu$
Keulegan–Carpenter number	$KC = u_0 T/D$
Relative surface roughness	$e = K/D$
Diffraction parameter	$\pi D/L$
Reduced velocity	$V_R = U/f_s D$

The relative importance of the above non-dimensional quantities may be examined by a simple analysis. A dimensional analysis may be performed to derive the dependence of force parameter on several non-dimensional quantities. The force on a structural member subjected to waves depends on a group of independent parameters as follows:

$$f = \psi(t, T, D, K, f_s, L, u_0, \rho, \nu) \tag{4.1}$$

in which t is time and ρ is the mass density of water. This dependence is considered inclusive in that most important parameters are included. However, not all parameters are important in all cases. The non-dimensional relationship may be obtained as:

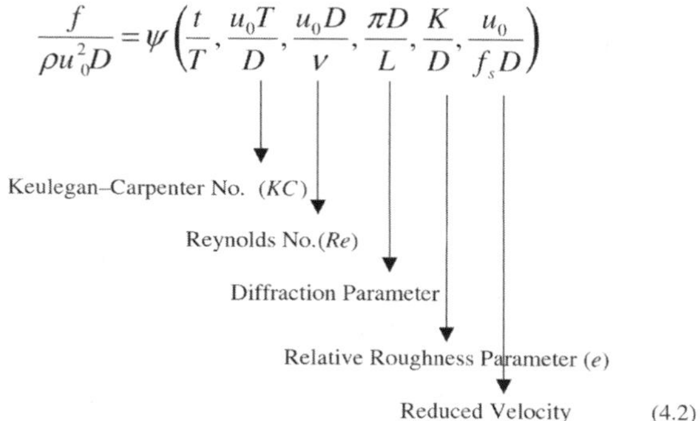

Note that:

$$KC \text{ large} \Leftrightarrow \frac{\pi D}{L} \quad \text{small} \tag{4.3}$$

The Keulegan–Carpenter number determines the relative contribution of the inertia and drag forces. The inertia plus the drag forces are determined by the Morison equation. The diffraction parameter establishes the size of the wave scattering from the surface of the structure. The general rule of thumb is that the Morison equation is applicable when KC is larger than 6 and the diffraction theory is applicable for large $\pi D/L$ (>0.5). For example, consider a deep-water wave of height, $H = 10$ m and period, $T = 10$ s. Then the wavelength, $L = 156$ m. In this case, by the linear theory, $u_0 = 3.14$ m/s at the still water level. For a structural member of diameter $D = 1$ m, we have $KC = 31.4$ and $\pi D/L = 0.056$ so that the Morison equation is applicable. If the member size is increased to $D = 10$ m, then $KC = 3.14$ and $\pi D/L = 0.56$. In this case, the diffraction theory should be used.

4.2 Gravity Loads

Gravity loads include dead loads, operating and equipment weights, live loads and buoyancy loads. The dead loads include the permanent loads of the structure and equipment and other fixtures that are not likely to vary during the service life of the structure. Live loads include the variable loads due to liquid and solid storage. A typical example representing the total payload breakdown for a fixed structure is given in table 4.2.

Table 4.2 Typical gravity loads of the deck

Item	% Weight
Deck structure	5.0
Helideck	1.5
Living quarter	3.5
Topside equipment and facilities	60.0
Drilling rig	12.0
Hook load	6.0
Jacket support structure	12.0
Total Payload	**100**

4.3 Hydrostatic Loads

A floating structure when at rest in still water will experience hydrostatic pressures on its submerged part, which act normal to the surface of the structure. The forces generated from these pressures have a vertical component, which is equal to the gravitational force acting on the mass of the structure. In other words, for a freely floating structure, this force is equal to the displacement weight of the structure. In other directions, the net force is zero.

4.4 Resistance Loads

When an offshore structure is towed in water, such as during the installation procedure or during the relocation of a drilling vessel, the forward motion of the vessel in the water generates a dynamic pressure distribution on the hull of the vessel. The result is a force on the vessel in the direction of motion, which opposes the forward motion. This is the normal force. There is another force experienced by the structure from this motion due to the viscosity of the water. This is generated due to friction of the fluid on the submerged surface of the structure and is a tangential force. Therefore, the total resistance (or, sometimes, called "drag") is the sum of the normal and the tangential forces.

The normal force, which is often called the "wave-making resistance", is a function of the Froude number. The frictional resistance is a function of the Reynolds number. Therefore, in a geometrically similar Froude model, the wave-making resistance will scale according to the scaling laws of Froude for forces (see Chapter 13). This force then is directly scalable from a model test. However, the frictional resistance cannot be scaled up due to its dependence on the Reynolds number, which is distorted in a Froude model. The frictional resistance is computed from the known values of the friction coefficients in the following manner. The frictional force is given by:

$$F_f = \frac{1}{2} C_f \rho A U^2 \tag{4.4}$$

in which A is the submerged surface area, U is the forward speed of the structure and the friction coefficient C_f is given by the Schoenherr formula based on Re number:

$$C_f = \frac{0.075}{(\log_{10} Re - 2)^2} \tag{4.5}$$

This value applies to a smooth surface. The wave-making resistance is not computed, but obtained from the model tests and scaled up by the Froude scale (see Chapter 13 for details of scaling method).

4.5 Current Loads on Structures

In the design of offshore structures, current is generally considered time-invariant represented by its mean value. The current strength, however, may have a variation with water depth. The existence of steady uniform and shear currents and their magnitudes has already been discussed in Chapter 3. The current introduces a varying pressure distribution around a member of the offshore structure generating a steady drag force on the structure in the direction of flow. Since the pressure distribution is not symmetric about the direction of flow, a transverse force is also generated on the structural member.

4.5.1 Current Drag and Lift Force

If a two-dimensional structure is placed in a uniform flow, then the force experienced by the structure will depend on the fluid density, the flow velocity and the frontal area of the structure encountering the flow. The force is found to vary with the square of the flow velocity:

$$f = \frac{1}{2}\rho C_D A U^2 \tag{4.6}$$

where ρ = fluid density, A = structure projected area normal to the flow, U = uniform flow velocity and C_D is a constant known as the drag coefficient. The drag coefficient C_D has been shown to be a function of the Reynolds number, Re based on mean current velocity and member diameter. For a circular cylinder across the flow, D is the diameter of the cylinder.

The drag coefficient for a smooth stationary circular cylinder in a steady flow has been obtained through laboratory testing (most of these with air as the flowing fluid) and is shown in fig. 4.1 [adopted from Schlichting (1968)]. For a Reynolds number less than 2×10^5, the flow is considered to be in the subcritical range. For an Re less than 50, the flow is strictly laminar and steady. The drag coefficient in this range decreases linearly as a function of the Reynolds number. The laminar flow is maintained up to an Re of about 200, beyond which the flow starts to become turbulent. The flow becomes fully turbulent for $Re > 5000$. The range of Re between 2×10^5 and 5×10^5 is termed critical where flow is in the transition mode. This area actually causes a low-pressure region behind the member (called the wake) to grow narrower with a corresponding decrease in the pressure gradient, causing a sharp drop in the actual value of the drag coefficient. This is known as drag

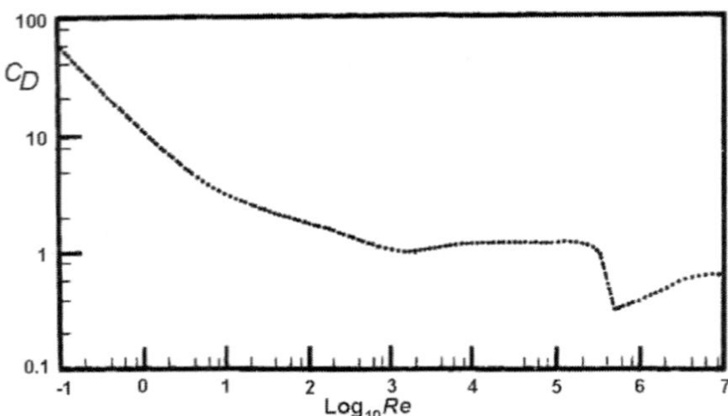

Figure 4.1 Drag coefficient for a smooth circular cylinder in steady flow

crisis. As the Reynolds number increases, the supercritical range is reached where the flow is strictly turbulent. In this range the drag coefficient slowly increases again. Beyond an Re of 3×10^6 the turbulent flow is called post-supercritical. Here the drag coefficient approaches a steady value because less dramatic changes occur in the boundary layer at still higher velocities.

This figure clearly demonstrates the difficulty associated with small-scale testing of structures in which the drag force is important. The Reynolds number in a small-scale model is much smaller than that in the full-scale structure (see Chapter 13 on scaling). Hence, the data from the model test is difficult to scale up to full scale due to the difference in the values of the drag coefficients in these two regions.

In practice, the surfaces of many structures in operational mode are not smooth. The roughness of the structure surface may be contributed from several sources. The appendages attached to a structural component may introduce irregularities causing flow tripping. An example of this type of structural component exists in ships and submarines. Another source of roughness present on a structural member is the marine growth of different kinds, which attaches to the surface of the structure near the ocean surface. Examples of such structures are the floating buoys and the submerged cylindrical structural members. It should be noted that this type of roughness on the surface increases the effective overall size of the structure. This means that there is an increase in drag force on the structure simply from the larger projected area to the flow. The API guideline recommends a 1.5 in. growth on members for depths from 0 to 150 ft below the surface. Moreover, if the surface of the structure is not smooth, the roughness moves the point of flow separation on the structural member and the corresponding wake behind the structure, resulting in a change in the drag coefficient as well. The roughness is quantified by the value of the roughness coefficient $e = K/D$ introduced earlier.

The main reason for the increase in the drag coefficient in the presence of roughness is that the roughness on the surface breaks up (i.e. trips) the flow earlier in the velocity field (i.e. lower velocity) creating a larger wake. Thus, the flow on a rough surface separates from the surface ahead of the point of separation for a smooth cylinder. The higher the roughness, the earlier is the separation of flow. This causes a larger pressure gradient between the

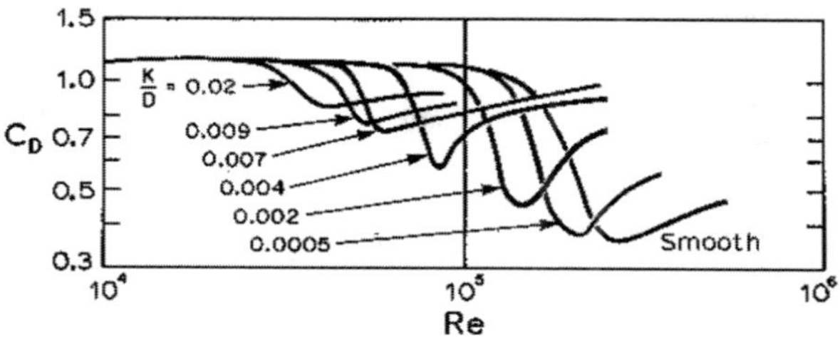

Figure 4.2 Drag coefficient for a rough circular cylinder in steady flow

upstream and downstream faces of the structure. The effect is a larger drag coefficient and a larger drag force.

The drag coefficients for a rough cylinder for a roughness coefficient value of up to 0.02 are shown in fig. 4.2 as a function of *Re* and the roughness coefficient, K/D. It has been shown that increased roughness moves the (drag crisis) transition area to lower values of *Re* and is much less pronounced. With sufficient roughness, the drag crisis can be eliminated. This is clear from the plots in fig. 4.2. Outside this region, the value of C_D is higher with the higher values of K/D. Note that at very low *Re* values where the flow is strictly laminar, the cylinder roughness does not appreciably alter the drag coefficient on the cylinder. This is simply because there is minimal flow separation in this range. Surface roughness plays an important role, not just in the drag coefficient, but in the VIV response as well.

In a recent experiment, the drag coefficients on a cylinder were obtained at high Reynolds number with various roughness, which is shown in fig. 4.3.

The API Semi-submersible Guidelines (1996) provide the following formulas for computing the steady current forces on several types of floating structures.

Ship bow or stern:

$$f = C_x S U^2 \tag{4.7}$$

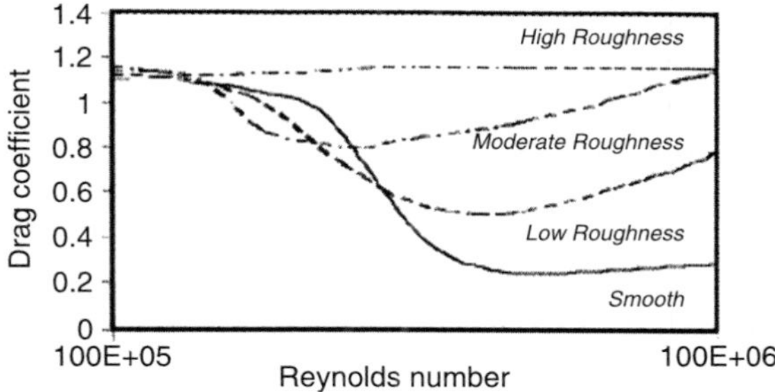

Figure 4.3 Effect of surface roughness on C_D at high Reynolds number [Allen, 2001]

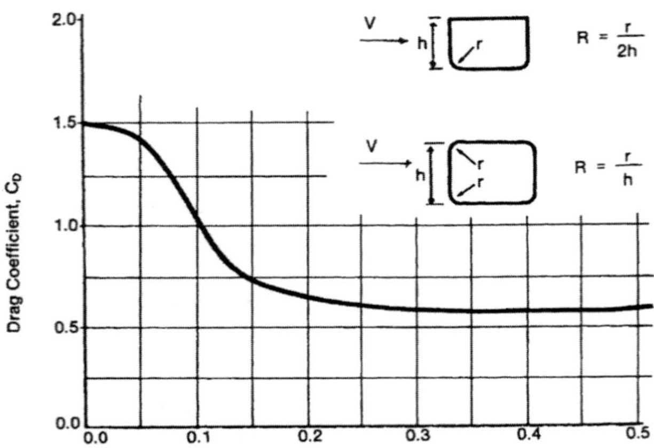

Figure 4.4 Drag coefficient for flat surface [API, 1996]

Ship beam:

$$f = C_y S U^2 \tag{4.8}$$

Semi-submersible:

$$f = C_s(C_D A_c + C_D A_f)U^2 \tag{4.9}$$

in which S = wetted hull surface area including appendages, A_c = projected area of submerged cylindrical members of semi, A_f = projected area of submerged flat surfaces of semi, $C_x = 2.89\ \text{Ns}^2/\text{m}^4$ ($0.016\ \text{lb}/(\text{ft}^2 \cdot \text{kt}^2)$), $C_y = 72.37\ \text{Ns}^2/\text{m}^4$ ($0.040\ \text{lb}/(\text{ft}^2 \cdot \text{kt}^2)$), $C_s = 515.62\ \text{Ns}^2/\text{m}^4$ ($2.85\ \text{lb}/(\text{ft}^2 \cdot \text{kt}^2)$), and C_D is 0.5 for circular cylinders. For flat surfaces, it is given by fig. 4.4.

It has already been noted that the wake formed behind the structural member and any associated vortex shedding from the boundary of its body surface is not symmetric with respect to the direction of flow. This is mainly because the formation (and shedding) of vortices on either side of the flow direction does not take place at the same time. In fact, it has been found experimentally that the shedding of the vortices alternates on the two sides. Thus the pressure distribution around the structure is not symmetric about the flow direction. Consequently, an additional force is generated transverse to the flow. This force is generated from the asymmetric pressure distribution due to the uneven formation of the vortices behind the member. Moreover, since this pressure field alternates in shape with the alternating vortex shedding, the transverse (also called lift) force is fluctuating. Moreover, the formation (and shedding) of vortices is somewhat irregular with respect to time. In other words, the change in the wake field even in a steady flow is not highly predictable. Therefore, the force generated in the transverse direction is irregular. This force is written in a

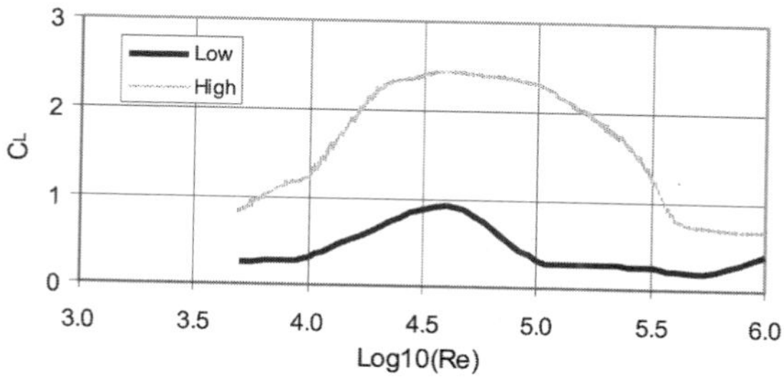

Figure 4.5 Lift coefficient for a smooth circular cylinder in steady flow

form similar to the inline drag force as:

$$f_L = \frac{1}{2}\rho C_L A U^2 \tag{4.10}$$

This form of the lift force requires that the lift coefficient, C_L in equation (4.10) is time-varying. Therefore, the lift coefficients are expressed either as rms values over one measured cycle or maximum values corresponding to the maximum lift force. Many experiments with circular cylinders transverse to steady flows have determined the values of lift coefficients. The lift coefficients obtained from experiments are shown in fig. 4.5 as functions of *Re* values. There is considerable scatter in the experimental data. Most of this scatter may be attributed to the free stream turbulence in the flow, flow over the ends of the cylinder, lack of rigidity in the mounting system and other physical sources in the experimental set-up. The two curves in fig. 4.5 provide the upper and lower ranges of experimental C_L and its value in a particular case can only be determined approximately.

4.5.2 Blockage Factor in Current

The values of C_D and C_L shown in the earlier figures in a steady flow are based on a single isolated member. However, a practical structure is often composed of relatively closely spaced small members connected in various orientations. These members may be circular or other geometrical shapes. Since the global load is computed from the loads on the individual members of a structure, it is important to account for the blockage in the flow field due to the presence of the structure itself. For example, if the structure is dense, then the steady flow actually slows down as it travels through the structure. In a practical design, this effect of blockage is accounted for with an overall factor, without describing the actual flow field, which is quite complicated.

The term blockage is used with reference to the entire structure. The current blockage factor introduced to account for the presence of the structure in the current flow field enables one to compute the true global load on the structure. The blockage factor is applied to the undisturbed current (i.e. steady flow) value in order to obtain an equivalent current velocity that accounts for the blockage by the structure. The blockage factor for steady current past a dense structure consisting of many circular members (e.g. an offshore jacket-type structure) may be estimated from an "actuator disk" model developed by

Taylor (1991):

$$f_{BF} = \left[1 + \frac{\sum_i (C_D A)_i}{4\overline{A}}\right]^{-1} \tag{4.11}$$

where the summation of the drag forces is computed from each member in the dense structure including the horizontal members in the flow, and the area $\overline{A}$ is the perimeter area of the projected structure normal to the flow. If the geometry of the structure changes significantly along the water depth, then the blockage factor may be computed at several levels of the structure for the total force computation.

In practice, one often encounters a vertical array of cylinders in a steady flow, which interfere with the flow field (such as, a group of drilling conductors or risers). The shielding from a vertical array of cylinders is given in terms of a blockage factor by the following simple empirical formula [API (2000)] based on the relative spacing of the cylinders with respect to the diameter:

$$f_{BF} = \begin{cases} 0.25S/D & \text{for } 0 < S/D < 4.0 \\ 1.0 & \text{for } S/D = 4.0 \end{cases} \tag{4.12}$$

The spacing S is the centre to centre distance of the conductors of diameter D, which includes any marine growth on the surface of the conductors.

According to the API guidelines, the current blockage factor for a structure having up to eight cylindrical members in close proximity in a row (with typical spacing of a jacket-type structure) is determined from table 4.3 based on the orientation of the structure relative to the current.

Table 4.3 Current blockage factor for jackets

No. of Cylinders	Current Heading	Blockage Factor
2	All	0.90
4	End-on	0.80
	Diagonal	0.85
	Broadside	0.80
6	End-on	0.75
	Diagonal	0.85
	Broadside	0.80
8	End-on	0.70
	Diagonal	0.85
	Broadside	0.80

4.6 Steady and Dynamic Wind Loads on Structures

Wind is often treated as a time-invariant environment, which has a mean value equal to its turbulent velocity. With this simplification, the effect of the wind on a superstructure (the portion of the offshore structure above the water as well as any equipment, deck houses and derricks) of an offshore structure is represented with a mean force. In this case the wind load is given by an expression similar to the current force in terms of a wind drag coefficient:

$$f = \frac{1}{2}\rho C_D A U_w^2 \tag{4.13}$$

where ρ = density of air, A = structure projected area normal to the wind flow, U_w = mean wind velocity, generally taken at an elevation of 10 m from the water surface and C_D is the wind drag coefficient. C_D is a function of the Reynolds number, *Re*. Average values of the wind drag (shape) coefficients for normal wind approach recommended by the API Guidelines (2000) are as follows:

Item	C_D
Beams	1.5
Sides of buildings	1.5
Cylindrical sections	0.5
Overall projected area of platform	1.0

The wind load may also be determined from the wind tunnel test on a specific platform and the deck structure.

Moreover, the fluctuation of the wind velocity over the open ocean is quite significant. The frequency spectral variation of the wind speed has been shown in Chapter 3. This is particularly true, because the wind spectrum has substantial energy at the low frequency, which is susceptible to the low-frequency oscillation of moored floating structures. This oscillation may be significant for the response of a floating structure, especially at its natural periods. This area of oscillating wind load is further discussed in Chapter 13.

Just like the ocean waves, the wind velocity fluctuation is represented by its power spectrum. Once the wind power spectral density (PSD) is known, the analysis tools developed for the waves may be equally applicable to the wind power spectrum in order to determine the dynamic response of a floating offshore structure due to wind. This method of computing response PSD is described in more detail in Chapter 5.

4.7 Wave Loads on Structures

As already mentioned, the forces on a structure due to waves are computed by two different methods depending on the size of the structure. In this regard, the structure is classified as small or large. The regions of small and large structures will be discussed later. The forces on small structures will be discussed first.

4.7.1 Morison Equation

Since the wave flow is not steady and, in particular, since the linear wave flow follows a simple harmonic motion, the flow around the cylinder will be more complex than the steady flow. In a simplified description we can say that the oscillatory flow over one cycle will change the low-pressure (wake) region immediately "behind" the cylinder every half cycle. As the flow changes direction, the low-pressure region will move from the downstream to the upstream side. Thus the force on the cylinder will change direction every half a wave cycle.

Combining the effects of water particle velocity and acceleration on the structure, the loading on the structure due to a regular wave is computed from the empirical formula commonly known as the Morison equation. Note that the force in the direction of waves is written for a unit length of a vertical pile shown in fig. 4.6:

$$f = \rho C_M \frac{\pi D^2}{4} \dot{u} + \frac{1}{2} \rho C_D |u| u \qquad (4.14)$$

in which D = pile diameter; f = horizontal force per unit length; and u, v = wave particle velocities in the horizontal and vertical directions respectively. The quantity w in fig. 4.6 is the resultant of u and v and dl = unit elemental length. The empirical constants C_M and C_D are hydrodynamic coefficients. Upper dot represents acceleration.

In the ocean environment of random waves, the force computed by the Morison equation shown earlier is only an engineering approximation. We can match the forces

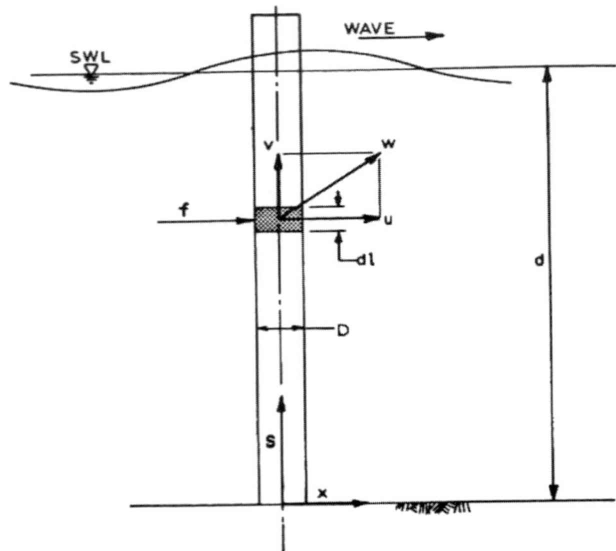

Figure 4.6 Morison force on a vertical pile

reasonably well in a particular half cycle of the wave profile using constant values of the hydrodynamic coefficients C_M and C_D. In the irregular waves, each half cycle is different in amplitude and shape. The half-wave cycle T_2 is defined as either under the crest or above the trough. For the linear wave these values are the same. But for a non-linear wave the two values are different. Usually, the crest period is smaller than the trough period. In such cases, the half-wave cycle approach is a way to make the wave cycle appear to be regular so that the regular wave theory may be justifiably applied. For this purpose, the period is assumed to be $2T_2$ where T_2 is the distance between the two zero-crossings. The half-wave cycle is chosen for the half period corresponding to the crest amplitude, which also corresponds to the value of u_0. It is not straightforward to compute this value for the half-wave cycle for a non-linear wave. It is suggested that the wave period, T should be used for the calculation of KC rather than $2 \times T_2$, as suggested in the API guidelines.

The hydrodynamic coefficients of a submerged member of a structure in an oscillatory wave flow are determined experimentally by testing an instrumented, scaled model of the member. The member, such as a cylinder, is held fixed in waves. Progressive waves are generated from a distance and allowed to flow past the cylinder. In this case, the flow is two-dimensional such that the free-stream velocity field varies both in the horizontal and the vertical directions. The waves, although generally regular, are not necessarily sinusoidal, and become particularly distorted by the refraction and shoaling effect from the bottom floor. The frequency of the regular waves is set as a constant while the amplitude and shape of the waves are obtained by measuring the free surface profile at the cylinder axis using a probe mounted near the cylinder. Water particle velocity is also measured directly at this point. The particle acceleration is seldom measured, but is computed by the numerical differentiation of the velocity profile.

Knowing the kinematics of the fluid flow at the instrumented section of the model and the corresponding inline loads, the mass and drag coefficients are computed from equation (4.14). In general, the coefficients have been established to be functions of the Keulegan–Carpenter number and the Reynolds number.

As in a steady flow, the waves produce a lift force in the normal direction to the flow. The Morison equation cannot describe this transverse force. As in a steady flow, in oscillatory flow the lift coefficient is a variant with time over a cycle. This is because the transverse force, unlike the inline force, is irregular and has multiple frequencies.

An illustration of the inline and transverse forces on a vertical cylinder is given in which the forces on the cylinder were measured in a progressive wave in a wave tank. The measured inline and transverse forces on the cylinder are shown in fig. 4.7 as a function of time. The wave was generated at a single frequency. The inline force, being inertia dominated in this case, follows the same frequency. However, the transverse force, although small, displays clear evidence of multiple frequencies.

4.7.2 Forces on Oscillating Structures

If a structure is free to move, it will oscillate due to the environmental loads. A modified form of the Morison equation is written to describe the force per unit length experienced

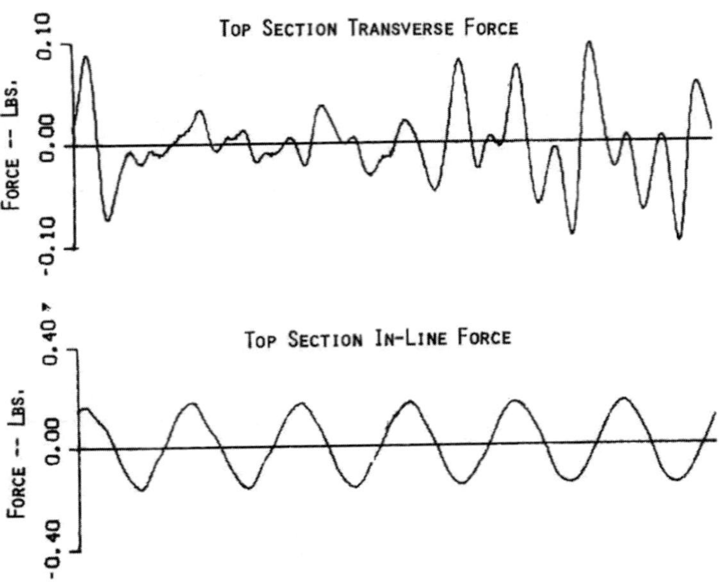

Figure 4.7 Measured inline and transverse force time history

by the structure due to its motion through the water by the following equation:

$$f = m\ddot{x} + C_A \rho \frac{\pi}{4} D^2 \ddot{x} + \frac{1}{2} C_D \rho D|\dot{x}|\dot{x} \tag{4.15}$$

The above expression is specific to a cylinder in which m is the mass of the cylinder per unit length, C_A is the added mass coefficient, x is the known harmonic displacement of the cylinder and dot represents derivatives of x. The first term on the right-hand side represents the cylinder inertia, while the last two terms are the hydrodynamic inertia and the drag forces due to the motion of the cylinder in water.

Since equation (4.15) is empirical, the values of the coefficients C_A and C_D are determined experimentally. The coefficient values are assumed invariant over a cycle for a given frequency of oscillation. There are two general methods of laboratory testing that determine the hydrodynamic coefficients for the cylinder. In the first method, the circular cylinder is oscillated harmonically in an otherwise still water. A known frequency and amplitude of oscillation are imposed on the cylinder through a variety of mechanical systems. Equivalently, the cylinder can be held fixed while the fluid is oscillated harmonically past the member. In both the cases the flow is one-dimensional, except in the vicinity of the cylinder. It can be demonstrated that the case of an oscillating structure in calm fluid is equivalent and kinematically identical to the oscillating fluid flowing past the stationary structure (except for the structural inertia). It is obvious that the kinematic

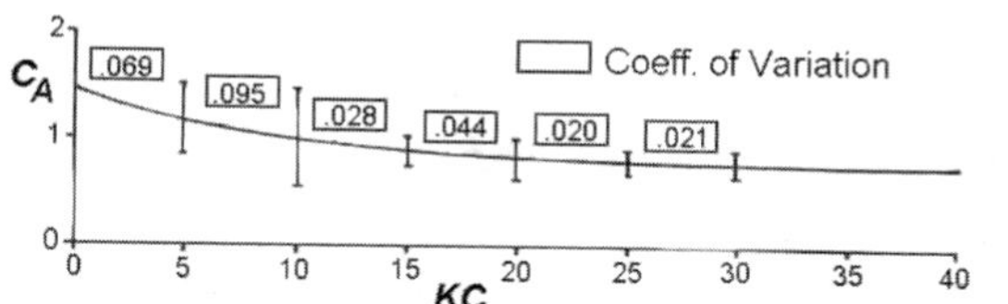

Figure 4.8 Inertia coefficients for an oscillating vertical cylinder

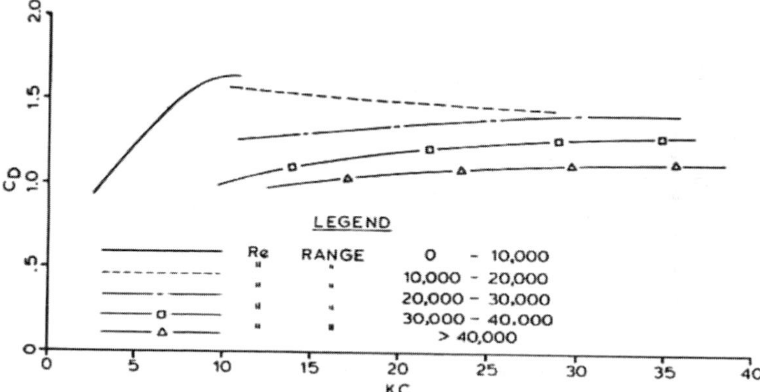

Figure 4.9 Drag coefficients for an oscillating vertical cylinder

fields in the mechanical oscillation of a structure are controlled and that the accuracy with which they can be described is excellent.

The results of one such experiment with a pivoted cylinder oscillated harmonically in water are reported here. The cylinder was instrumented over a small submerged section to measure the inline forces. The C_A- and C_D-values from the measurements are shown in figs. 4.8 and 4.9 as a function of the Keulegan–Carpenter number. In fig. 4.8, the C_A-values are less sensitive to Re and the scatter about the mean value is given in terms of coefficients of variation. Different symbols are used to distinguish the data points for different β-values ($\beta = Re/KC$) in fig. 4.9. The experimental β-values ranged from 700 to 2500.

Similar experiments were conducted by Sarpkaya (1976) in a large water-filled U-tube in which water was oscillated back and forth past a fixed horizontal cylinder placed in the horizontal arm of the U-tube. The forces on the cylinder were measured from which the inertia and drag coefficients were determined.

The results from one such experiment are shown in fig. 4.10 for the inertia and drag coefficients. The results clearly show the dependence of these coefficients on the quantities KC and β, or equivalently, KC and Re. The transverse force on the cylinder due to asymmetric shedding of vortices was also measured during these tests. The forces were irregular, having multiple frequencies. Therefore, unlike the drag and inertia coefficients, a lift coefficient over one cycle may not be determined. The lift coefficient is, generally,

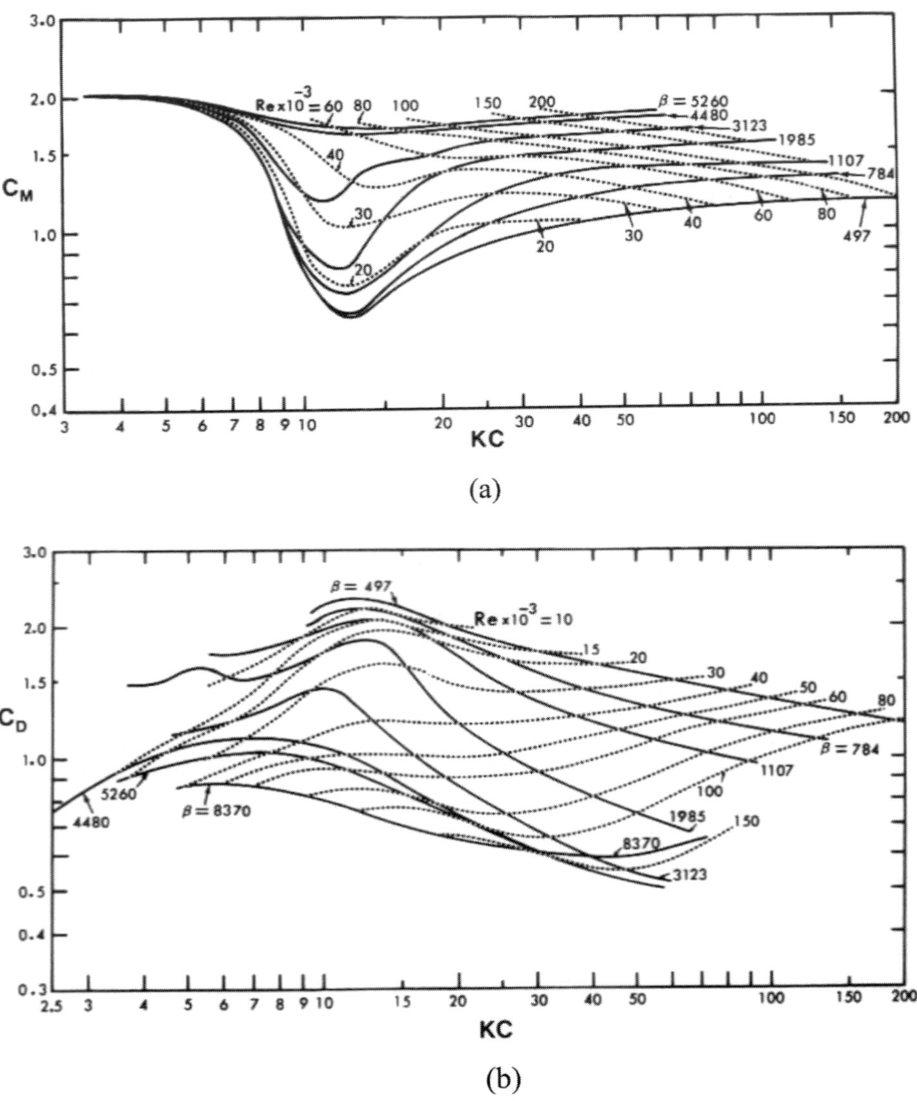

Figure 4.10 (a) Inertia and (b) drag coefficients from a fluid oscillation test [Sarpkaya (1976)]

presented as an rms or a maximum value. The rms lift coefficients from the U-tube tests are shown in fig. 4.11 as functions of *KC* and *Re*.

The lift force frequencies were multiples of one another and related to the number of vortices shed from the cylinder. The larger the number of vortices shed, the higher is the predominant frequency in the measured force. The ratio of the predominant lift force

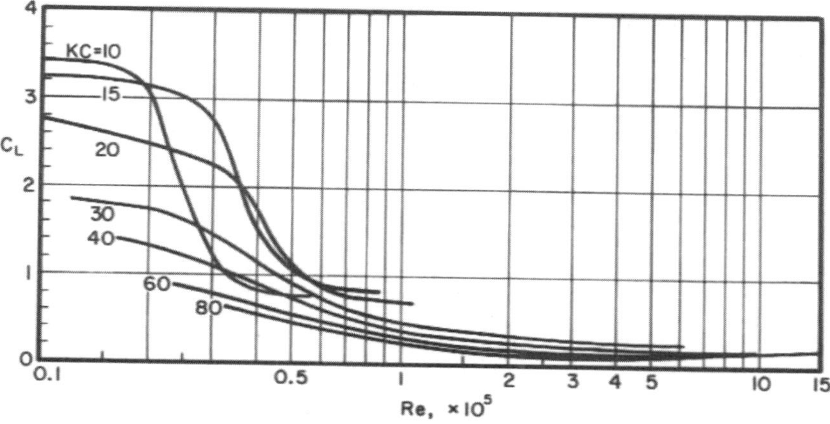

Figure 4.11 Lift coefficients from an oscillation test [Sarpkaya (1976)]

frequency to the imposed oscillatory frequency is plotted in fig. 4.12 as a function of the *KC* and *Re* numbers. This frequency ratio f_r is found to increase in value from 2 to 15 in the *KC* number range of 10 and 85. The dependence of the lift force frequency is strong on *KC* and weak on *Re* except in the high range of *Re*.

It is clear from this discussion that if the cylinder is free, flexible or mounted on springs and allowed to move, then the transverse force will cause the cylinder to oscillate in the flow. Moreover, the oscillations of the cylinder will, in general, be irregular having multiple frequencies. If one of these frequencies happen to fall close to the natural frequency of the cylinder (modal eigenfrequency), then this oscillation will experience a dynamic amplification (to be introduced later in this chapter). This is the source of the vortex-induced vibration and will give rise to a lock-in mechanism when the vibration frequency coincides with the natural frequency.

The above results on the C_M and C_D values are limited in scope for direct application to offshore structures. Then why are they included here? The main reason is to demonstrate and clearly show their dependence on *Re* and *KC*. Similarly, the dependence on the roughness parameter (K/D) has been established through similar experiments [refer to Chakrabarti (1987)].

It should be noted, however, that there are two hindering factors in their direct application to a design force calculation for offshore structure. The first one concerns the pure harmonic oscillation in the absence of a free surface such as the one found in water waves. It has been found that the C_M and the C_D values from a U-tube oscillation test are somewhat conservative compared to the values under waves. The second problem is more restrictive. It refers to the distortion in the Reynolds number, which is a couple of orders of magnitude higher in the prototype application. The dependence of these coefficients on such a large range of *Re* is quite strong. Therefore, field experiments are desirable to determine the appropriate values. This is not straightforward either due to many

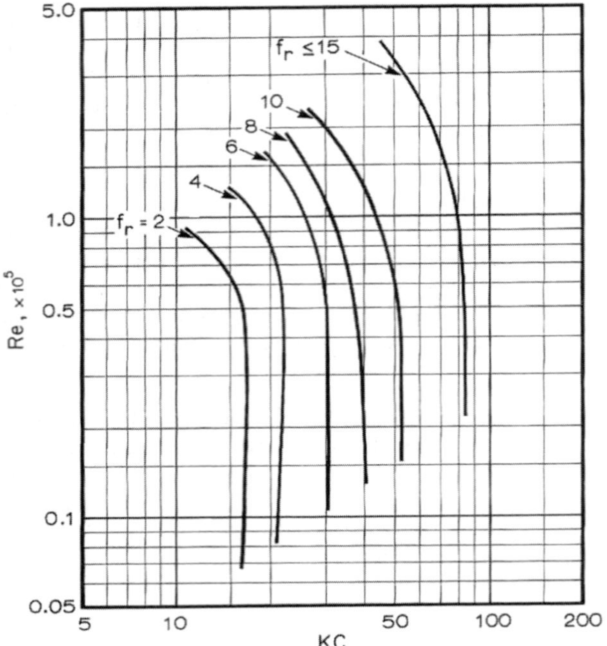

Figure 4.12 Lift force frequency as a function of *KC* and *Re* [Sarpkaya, 1976]

uncertainties in such experiments. This area and recommended values of the coefficients will be discussed later.

4.7.3 Wave Plus Current Loads

When a current is present, the total water particle velocity is modified by adding the wave particle velocity to the current velocity. If the current is inline, the magnitudes are added to give the total velocity. For a non-collinear current, the component of current inline with the wave is used. Additionally, the presence of the current alters the apparent wave period. This area has been discussed in Chapter 3 (see, e.g. fig. 3.23). The wave particle velocity is computed based on the apparent wave period. In this case the normal wave loading for a unit length of a cylindrical structure is based on the modified Morison equation:

$$f = \rho C_M \frac{\pi D^2}{4} \dot{u} + \frac{1}{2} \rho C_{DS} |u + U|(u + U) \tag{4.16}$$

where the drag coefficient corresponds to the combined wave-current flows.

Summarised below is essentially the recommended method (outlined in the Commentary on Wave Forces described in the Section 2.3.1 of the 20th edition) of the American Petroleum Institute design guide API-RP2A (2000) with some clarifications and explanation.

The API procedure for computing wave-current loading is based on the "design" approach in which a single wave height and period are selected to represent the extreme wave expected in a random sea. Wave kinematics are computed by the wave theory based on wave height, apparent wave period and water depth (as discussed in Chapter 3). Current does not enter into this calculation except for the altered wave period.

The wave–current interaction has been addressed in Section 3.8.3. The basic formulas of the combined wave and current are included there. For all circular cylindrical members, the drag coefficient C_{DS} is selected based on wave and current velocities and surface roughness. For members other than near-vertical (such as, horizontal and diagonal), the local wave particle velocity and the current velocity are used. In this case the inertia coefficient is assumed to be $C_M = 2.0$. The waves produce a net mass transport in the flow. The API suggests a minimum wave-induced current value of 0.2 knots irrespective of the direction of the free stream current.

As mentioned earlier, the hydrodynamic coefficients depend on several non-dimensional parameters (equation 4.2). In the presence of current, C_M and C_D additionally depend on the current-to-wave velocity ratio. In particular, the C_D value approaches a value closer to the steady current value, depending on the strength of the current. The current velocity strength is defined by the ratio r as the inline current velocity divided by the maximum amplitude of the wave–current velocity:

$$r = \frac{U}{u_0 + U} \tag{4.17}$$

Thus r is positive. If the current is in the opposite direction, then it is possible that the magnitude of this value will be obtained at the wave trough instead of at the wave crest.

For an inclined member in the structure, the velocity is normal to the cylinder axis, which includes both the horizontal and vertical components, or the total wave velocity. The wave kinematic factor provides the more appropriate value of wave particle velocity, considering the irregularity or directionality of waves.

In the presence of a current, the KC value is based on the maximum velocity including the current. In order to compensate for the current in the C_D value chosen, the KC value is modified by the following correction factor:

$$C_r = (1 + r)2\theta_* / \pi \tag{4.18}$$

where,

$$\theta_* = ATAN2\left(-r, \sqrt{(1 - r^2)}\right) \tag{4.19}$$

The correction factor is listed in table 4.4 for different values of r and θ_*. Note that the value of the correction factor increases with the value of r, which makes sense since the correction will be greater with increased current.

The corrected value of KC should be used to choose C_D for the force calculation. In the case of no current, $r = 0$. From equation (4.19), $\theta_* = \pi/2$ and the correction factor is equal to one. The Keulegan–Carpenter number as well as the value of C_D is based on the

Table 4.4 First-order correction factor

r	θ	Degrees	Correction Factor
0	1.5708	90.00	1.00
0.05	1.6208	92.87	1.08
0.1	1.6710	95.74	1.17
0.15	1.7214	98.63	1.26
0.2	1.7722	101.54	1.35
0.25	1.8235	104.48	1.45
0.3	1.8755	107.46	1.55
0.35	1.9284	110.49	1.66
0.4	1.9823	113.58	1.77

maximum wave velocity only. For the case of $r = 0.4$, the value of KC is modified by the correction factor, 1.77 as shown in table 4.4.

When $r > 0.4$, current is strong, so that the drag coefficient for all practical purposes is C_{Ds}, the steady-current value. There is no need to compute KC and hence the correction factor, C_r in these cases.

4.7.4 Design Values for Hydrodynamic Coefficients

From the experiments performed to date in the laboratory and in the ocean environment, it is not possible to prescribe the appropriate values of C_M and C_D for all practical cases. Engineering judgment is needed to choose these values in a particular case. This is particularly because the fluid flow problem past a submerged structure in waves is a very complex phenomenon and it is extremely difficult to duplicate the flow field in a confined environment. On the other hand, the field test is expensive and difficult to perform because of the measurement problem.

The certifying agencies have established their individual guidelines to follow based on experienced engineering judgment. Many offshore companies have also written standards from their own field experience. This section will describe results from some field experience and discuss the recommended values of the hydrodynamic coefficients from the design and certifying agencies.

4.7.4.1 Coefficients from Field Tests

Many ocean tests have been carried out in various parts of the world to measure wave forces on cylindrical members of an offshore platform. Some of the earlier tests made in the 1950s are relatively inaccurate, since no attempts were made to measure the particle kinematics. In a test with a vertical pile in the Gulf of Mexico, an average set of values for C_M and C_D were found to be 1.53 and 1.79 respectively. These values correspond to ocean waves in the presence of a current of magnitude of 0.55 ft/s.

Table 4.5 C_M and C_D values from field experiments

Field Project	Offshore Region	Re	KC	K/D	C_M	C_D
Wave projects I and II	Gulf of Mexico	1×10^5–3×10^5			1.3	1.2
		3×10^5–1×10^6			1.3	0.8
		1×10^6–3×10^6			1.3	0.7
		3×10^6–1×10^7			1.3	0.6
Ocean test structure	Gulf of Mexico		5–42	0	1.51	1.7–0.68
			3–42	0.02	1.25	2.0–1.0
Hurricane edith	Gulf of Mexico	3×10^5–3×10^6		0	1.5	0.7
				1/96	1.87	1.44
				1/32	1.99	1.41
				1/24	2.01	1.41
Christchurch bay	Offshore UK	1×10^5–1×10^6	2–30		1.22–1.85	0.73–1.0

Later, several tests were conducted in the Gulf of Mexico. Two series of tests named Wave Projects I and II were performed on vertical piles of diameters from 1 to 4 ft in a shallow water of 33 ft. During the Project II experiments, high waves were encountered from the Hurricane Carla (1961) on a pile of diameter 3.7 ft in 99 ft water depth. The average values of the hydrodynamic coefficients from a wave force test in Hurricane Edith (1979) were found to be $C_M = 1.06$ and $C_D = 0.70$ respectively. In 1979–80, there were extensive tests performed in the Christchurch Bay, UK on a large vertical pile and a small diameter wave staff equipped with force sleeve. In this case the particle kinematics was also measured. At about the same time another ocean test was conducted in the Gulf of Mexico on a platform called Ocean Test Structure. The forces and particle velocities were simultaneously measured in these tests. The instrumented section had the dimensions of 16 in. diameter and 32 in. long. The depth of submergence was 16 ft from the mean water level. The hydrodynamic coefficients from these ocean tests are summarised in table 4.5.

4.7.4.2 Design Guidelines of Certifying Agencies

From all the experiments in the laboratory and field measurements, the average values of C_M and C_D for a vertical cylinder have been found to vary from 1.5 to 2.0 and from 0.5 to 1.0 respectively. Based on these experiments and the field tests described earlier, various certifying agencies have made specific recommendations on the hydrodynamic coefficients

Table 4.6 Recommended C_M and C_D values by guideline agencies

Agency	Application	C_M	C_D
American Petroleum Institute		1.5–2.0	0.6–1.0
British Standard Institution	5th order @ splash zone	2.0 or Diffraction	0.8
British Standard Institution	1st order elsewhere	2.0 or Diffraction	0.6
British Standard Institution	Linear	2.0	1.0
British Standard Institution	Large cone/ pads	Diffraction theory	–
British Standard Institution	Low taper	2.0	–
Det Norske Veritas		2.0 or Diffraction	0.5 – 1.2
Norwegian Petroleum Directorate	Smooth cylinder	Variable	>=0.9

for use in offshore structure designs. Sometimes, the coefficient values shown are ranges, rather than single values and should be chosen based on the applications. For the C_M values, application of the linear diffraction theory is often recommended (table 4.6).

4.7.4.3 Example Design Procedures from API (2000)

For the sake of the following discussion, let us assume that the structural member is vertical. The horizontal steady current $U(y)$ is given at an elevation y. If the current is at an angle to the wave, the component in line with the wave should first be computed and this is what should subsequently be used. The expression for U in equation (4.16) gives the value of the effective in-line current at an elevation of y, $U(y)$, rather than a mean value over the entire depth. Thus r is a function of y and is always positive. For example, if the current is zero at a depth below the free surface, then the value of r is zero at that depth. When the member is at an angle, then a look-up table may be used. The table will provide the values of wave velocity amplitudes with depth.

The dependence of the steady-flow drag coefficient, C_{Ds}, on relative surface roughness is shown in fig. 4.13. The values apply to "hard" roughness elements at post-critical Reynolds numbers. The drag coefficient is referenced to the effective diameter D including the roughness elements. Natural marine growth on platforms will generally have $e > 10^{-3}$.

The values for C_D in waves plus current are given in terms of the steady drag coefficients. For extreme sea states, the drag coefficients may be obtained from fig. 4.14a. The drag coefficients for the lower sea states are shown in fig. 4.14b. The two curves apply to smooth and rough cylinders. The symbols refer to experiments performed. The reader should refer to the API guidelines for the appropriate references. The corresponding values of C_M are shown in fig. 4.15. The extreme sea states are on the left side, while the right side gives the values for the lower sea states.

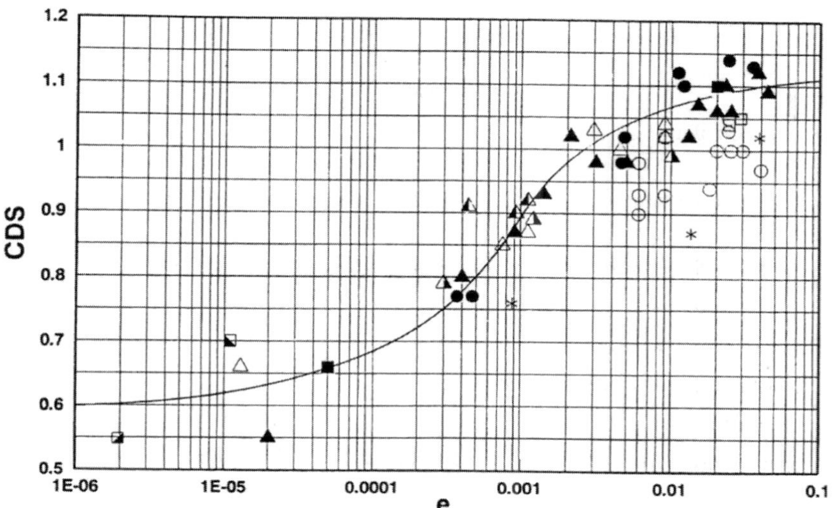

Figure 4.13 Steady drag coefficient vs. surface roughness [API, 2000]

The procedure for choosing the values of C_M and C_D for a cylindrical member in a structure (e.g. an offshore jacket structure or a jack-up) and computing the forces should be taken in the following simple steps:

- Compute effective member diameter:

$$D = D_c + 2K \tag{4.20}$$

where, D_c = cylinder diameter, and K= average 'hard' (marine growth) surface roughness height.

- Calculate the steady-flow drag coefficients, C_{DS} for circular members from fig. 4.13 for a relative surface roughness, $e = K/D$.
- Calculate the Keulegan–Carpenter number, KC.
- Calculate the drag coefficient, C_D for nearly vertical members from fig. 4.14a for the computed value of KC/C_{DS} from above. If KC is less than 12, then use fig. 4.14b for nearly vertical cylinders for prescribed values of KC and C_{DS}.
- If $KC < 12$ for nearly vertical members, the wake encounter effect on C_M for both smooth ($C_{DS} = 0.65$) and rough cylinders ($C_{DS} = 1.10$) may be represented by the sloped line in fig. 4.15b. Thus for $KC < 12$, use this figure for computing C_M.
- For $KC > 12$, compute C_M from fig. 4.15a for the values of KC/C_{DS}. If $KC/C_{DS} \geq 17$, then $C_M = 1.6$ for smooth and $C_M = 1.2$ for rough cylinders.
- For non-circular members, C_{DS} is independent of surface roughness.

For non-circular cylinders, the wake encounter is obtained from the same figures as the circular members as follows. Input the appropriate cylindrical member value for the inertia

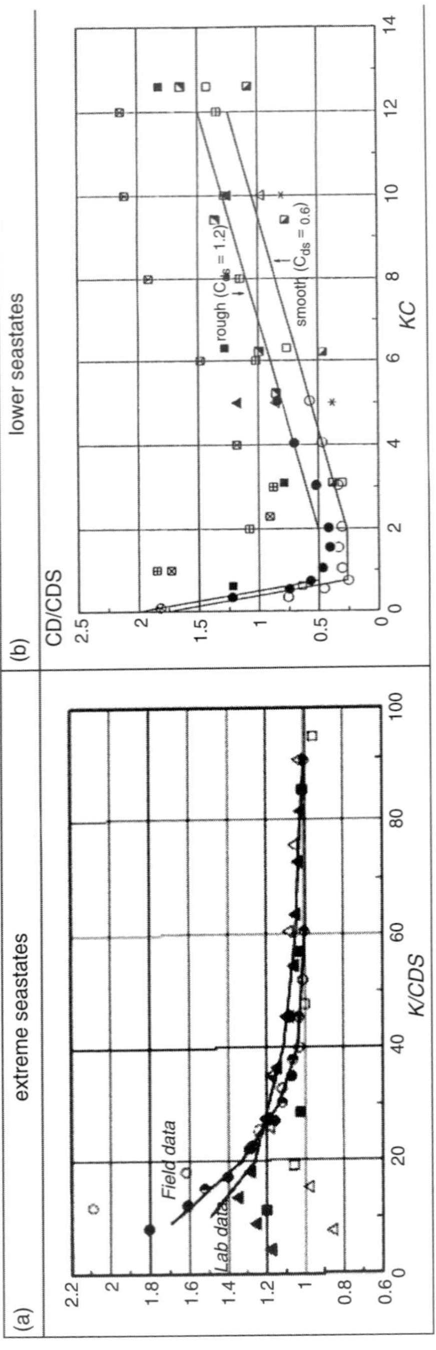

Figure 4.14 Drag coefficient vs. *KC* [API, 2000]

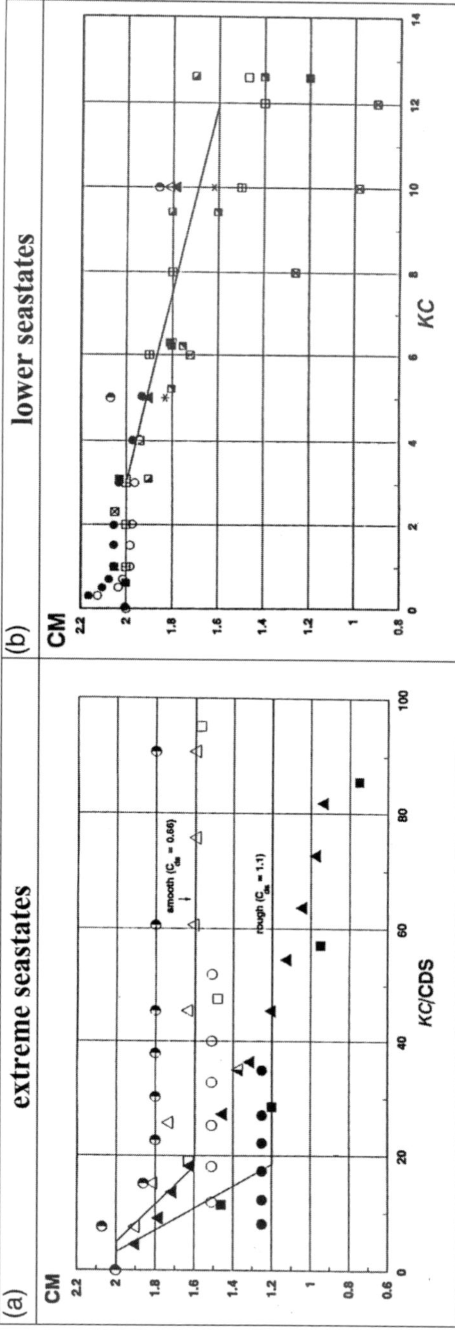

Figure 4.15 Inertia coefficient vs. *KC* [API, 2000]

coefficient C_{M0} for the member. Adjust this value using a factor f obtained from fig. 4.15 for the appropriate value of C_{DS}. Then, compute C_M for the non-circular cylinder,

$$C_M = fC_{M0}/2 \tag{4.21}$$

For members that are not vertical, no wake encounter is considered. Thus, for the horizontal and diagonal members, $C_D = C_{DS}$; $C_M = 2.0$ for circular members; and $C_M = C_{M0}$ for non-circular members.

As mentioned earlier, the preceding procedure is based on the recommendation of the API guideline [API (2000)]. If another certifying agency, such as Det Norske Veritas is consulted, then the recommended values for the hydrodynamic coefficients will differ slightly. It should be kept in mind that the Morison formula is empirical and the load calculation for a submerged structure in current and waves is only approximate for engineering design application.

4.7.5 Froude–Krylov Force on Structure

The forces on a submerged structure in waves appear from the pressure distribution on its surface. For a small structure these pressure distributions are difficult to compute because of the complex flow around the structure and formation of vortices in its vicinity. This is the reason for the use of the Morison empirical equation described above. However, if the flow remains essentially attached to the surface, then it is easier to compute this pressure field. When the structural dimension is large relative to the wavelength, then the flow around the surface of the structure remains attached. In this case, the incident wave experiences scattering from the surface of the structure, which should be taken into account in computing this pressure field. If the computation of the scattered wave potential is waived and its effect is incorporated by a force coefficient, then it is possible to compute a force simply by the incident wave potential. Thus, the calculation of the force is performed assuming the structure is not there to distort the wave field in its vicinity. The force is computed by a pressure–area method using the incident wave pressure acting on the submerged surface of the structure. This force is called the Froude–Krylov force.

While limited in its application to offshore structure design, the simplicity of the Froude–Krylov theory allows computation of forces in terms of a simple expression for a variety of shapes. The limiting criterion may be circumvented through the application of a force coefficient to account for the wave diffraction. For moderate-sized structures of simple contour shapes, this method provides a reasonable estimate of the forces. Its application is generally limited to estimating only approximate forces on structures. A few applications of this method may be cited as follows: concept design and rough estimate of structure sizes, sizing mooring lines, design of towing and transportation schemes, and model test planning such as scale selection and choosing load cells.

A few basic structural shapes for which a closed form expression may be obtained by the Froude–Krylov theory are a horizontal cylinder or half-cylinder, a vertical cylinder, a sphere or hemisphere and a rectangular barge. The expressions of forces and force coefficients are included in table 4.7. The forces are given in terms of the water particle acceleration at the centre of the structure wherever possible. The force coefficients shown are applicable over a small range of diffraction parameter ka ($= \pi D/L$), which is included

Table 4.7 Formulas for forces by Froude–Krylov theory

Basic Shape	X-Force	C_H	Y-Force	C_V	ka Range
Horizontal cylinder	$C_H \rho V \dot{u}_0$	2.0	$C_V \rho V \dot{v}_0$	2.0	0–1.0
Horizontal half-cylinder	$C_H \rho V [\dot{u}_0 + C_1 \omega v_0]$	2.0	$C_V \rho V [\dot{v}_0 + C_2 \omega u_0]$	1.1	0–1.0
Vertical cylinder	$C_H \rho V \frac{2J_1(ka)}{ka} \frac{\sinh(kl/2)}{(kl/2)} \dot{u}_0$	2.0	–	–	–
Rectangular block	$C_H \rho V \frac{\sinh(kl_3/2)}{(kl_3/2)} \frac{\sin(kl_1/2)}{(kl_1/2)} \dot{u}_0$	1.5	$C_V \rho V \frac{\sinh(kl_3/2)}{(kl_3/2)} \frac{\sin(kl_1/2)}{(kl_1/2)} \dot{v}_0$	6.0	0–5.0
Hemisphere	$C_H \rho V [\dot{u}_0 + C_3 \omega v_0]$	1.5	$C_V \rho V [\dot{v}_0 + C_4 \omega u_0]$	1.1	0–0.8
Sphere	$C_H \rho V \dot{u}_0$	1.5	$C_V \rho V \dot{v}_0$	1.1	0–1.75

Table 4.8 Numerical values of C_1–C_4

ka	C_1	C_2	C_3	C_4
0.1	0.037	15.019	0.042	12.754
0.2	0.075	7.537	0.085	6.409
0.3	0.112	5.056	0.127	4.308
0.4	0.140	3.825	0.169	3.268
0.5	0.186	3.093	0.210	2.652
0.6	0.223	2.612	0.252	2.249
0.7	0.259	2.273	0.292	1.966
0.8	0.295	2.024	0.332	1.760
0.9	0.330	1.834	0.372	1.603
1.0	0.365	1.685	0.411	1.482
1.5	0.529	1.273	0.591	1.156
2.0	0.673	1.105	0.745	1.034
2.5	0.792	1.031	0.867	0.989
3.0	0.886	0.999	0.957	0.977
3.5	0.955	0.989	1.015	0.978
4.0	1.000	0.087	1.945	0.985

in each case. If the values of ka are much different from this range, then caution should be used in using the values of the force coefficient included here.

The quantity V in the table is the submerged volume of the structure. C_H and C_V are the force coefficients in the horizontal and the vertical directions respectively. The subscript zero in the table refers to the amplitude of the particle velocity or acceleration at the centre of the structural shape. In other words, it is computed at the axis of the horizontal cylinder or half-cylinder, sphere or hemisphere. The quantities l_1 and l_3 are the length and the underwater depth respectively of the rectangular block. The numerical values of C_I–C_4 depend on the diffraction parameter ka and are given in table 4.8.

4.7.6 Wave Diffraction Force on Structure

When structures are large, the wave force computation is more streamlined. If the dimension of the structure is large compared to the wavelength, the structure alters the form of the incident waves over a large area in its vicinity. In this case the flow essentially remains attached to the body of the structure and the flow may be described well by the potential flow. Knowing the incident wave potential, there are several numerical procedures (e.g. fluid finite element method) that can be used to describe the potential function generated in the vicinity of the structure. One such method is known as the boundary element method (BEM). For structures in wave, the BEM is a well-established numerical technique for the analysis of many engineering problems, in particular, the linear and the second-order problems. Several commercial computer programs are applicable based on this method, which will be listed here for easy references.

The core technique in using these software is to describe the geometry of these structures with panels. There are two types of panels that may be used, based on the software. Most commercially available software uses the flat panels to describe the structure. These programs use what is known as the Lower Order Boundary Element Method (LOBEM). This method requires a large number of panels in describing the structural geometry for sufficient accuracy in the results. Each panel on the structure surface is, generally, described by the local coordinates of its corners. A preprocessor is required to generate the panel geometry. Since the finite number of flat panels are used to describe the complex surface of an offshore structure, the panel size must be small and a large number of them must be used for proper accuracy in the results. Typically 2000–3000 panels are adequate for the required engineering accuracy for structures e.g. semi-submersible, FPSO and TLP. If unsure about the accuracy, it is recommended that two different discretisations are used to compare the accuracy. The large number of panels makes the program slower and requires a large PC memory.

There is a more advanced paneling method developed in which curved panels are incorporated in describing the structure. This numerical method is known as the Higher-Order Boundary Element method (HOBEM). The advantage of this method is that it uses curved surfaces to follow the structural geometry better and requires fewer panels for the desired accuracy. The disadvantage is that it requires more sophisticated preprocessor and a higher user skill to panelise the structure. Either one of these software (i.e. the LOBEM or the HOBEM) provides similar accuracy with proper choice of the panels.

Note that, to first order, the forces on the submerged structure are computed in its equilibrium position. Therefore, the panels describing the structure should cover the submerged part of the structure up to the still (mean) water level. Generally, larger number of panels is required near the mean water level.

The details of the numerical method may be found, for example, in Chakrabarti (1987). Once the total potential comprising the incident wave potential and the scattered wave potential on the panel surface is known, the pressure at the centre of each panel on the structure surface is obtained from the linear term of the Bernoulli's equation (Chapter 3). Knowing the pressure distribution on the panels on the submerged surface of the structure, the forces and moments in six degrees of freedom, namely, surge, sway, heave, roll, pitch, and yaw, are computed by integration over the structure surface.

4.7.7 Added Mass and Damping Coefficients

Similar to the diffracted waves from the structure subject to incident wave, the motion of the structure introduces radiated waves. These waves generating from the structural motion produce radiated force on the structure. The surge motion is illustrated in fig. 4.16. The corresponding pressure field is computed using the appropriate body surface boundary condition describing the structural velocity.

This gives rise to six radiation potentials, hence six pressure fields on the surface for the six degrees of freedom. These pressures, when integrated over the body surface will give six

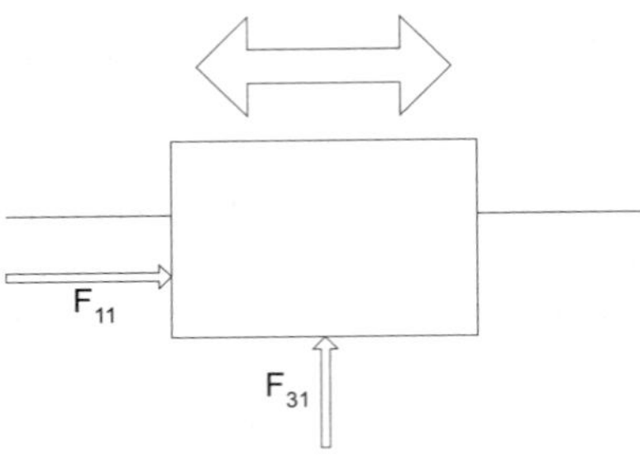

Figure 4.16 Forced Oscillation in Surge

forces (given by F_{11} in the surge in fig. 4.16). Moreover, the pressure in a particular oscillatory motion will generate a force in a different direction, such as the heave force F_{31} due to the surge motion in the illustration. In general, a 6×6 radiated force matrix results.

The component of the force that is in phase with the structural velocity acts as a damping term and is called the damping matrix. The out-of phase component is in phase with the structural acceleration and acts as an inertia term. This component is called the added mass term. Hence, the radiation problem provides a 6×6 matrix for the added mass and a 6×6 matrix for the damping term. The normalised values for these matrices are the non-dimensional added mass and damping coefficients.

4.7.8 Haskind Relationship for Accuracy Check

The method outlined above provides the pressure distribution over the submerged surface of the structure from the incident wave including the diffraction/radiation effect. This is known as the near field solution. However, the radiation problem may be solved simply by what is known as the far field solution. The Haskind relationship allows the computation of the exciting force from the radiation potential i.e. the added mass and damping coefficients. Thus a check may be made for the exciting force computed by the two different methods. The degree of agreement of the six exciting forces becomes an indication of the accuracy of the results. When some of these values are close to zero, it is expected that the percentage errors will be large and should not be considered critical. However, if the differences are genuinely large, then the number of panels used may not be adequate and finer resolution of the structure in terms of larger number of panels is recommended for acceptable accuracy. All commercial software performs this computation, which is, therefore, useful to examine before applying the results in a design.

4.7.9 Linear Diffraction/Radiation Theory Software

The first-order diffraction theory program typically computes the following responses of a large arbitrary-shaped floating structure:

- wave exciting forces at a given frequency in the surge, sway, heave, roll, pitch and yaw directions

- added mass coefficients as a 6×6 matrix including the cross-coupling terms
- damping coefficients as a 6×6 matrix including the cross-coupling terms
- first-order motions in six degrees of freedom
- exciting forces by Haskind's relation (generally for checking the program accuracy)
- free surface profile of wave at the structure surface (run-up or air gap)
- interaction of large neighbouring structures and resulting motions of floating structures
- steady drift force by the momentum principle
- steady drift force by the pressure–area method
- approximate slow drift force

These last two items regarding the drift force will be soon discussed. Additionally, the second-order bichromatic (two frequencies) diffraction theory generates the Quadratic Transfer Function (QTF) for the computation of the

- difference frequency slowly oscillating drift forces
- sum (high) frequency second-order forces

They will be discussed later in the chapter.

List of a few of the commercially available programs that perform the above computations is included in table 4.9. The capabilities of most of these programs are similar. The detailed capabilities of these programs and their sources and availability may be found in the web sites and contacts listed in the table. In some cases the program documentation is available for view/downloading.

Table 4.9 Commercially available diffraction theory programs

Software	Company	Web site	Contact
AQWA	WS Atkins of UK	www.ship-offshore.com/aqwa.htm	J.W. Manning
HOBEM	FCA International, Inc., TX, USA	fsfchou@yahoo.com	Frank Chou
HYDRAN	OffCoast, Inc., HI, USA	www.offcoast.com/offcoastinc/	H. Ronald Riggs
MORA	C. J. Garrison & Assoc, OR, USA	garrison@proaxis.com	C. J. Garrison
NBODY	OSA, Inc., IL, USA	http://members.aol.com/chakrab	S. Chakrabarti
NEPTUNE	ZENTECH, Inc., TX, USA	www.zentech.com	Rao Guntur
SESAM	DNV, Norway	www.dnv.com/software	DNV Software
WAMIT & HIPAN	Wamit, Inc., MA, USA	www.wamit.com	C-H Lee

The key capabilities of interest for a few of these programs are summarised in table 4.10. Note that the capabilities are quite similar.

The programs give very similar results as demonstrated in fig. 4.17 for the sway exciting force and the sway motion in a beam sea for a semisubmersible. The example represents a single unit for a Mobile Offshore Base concept and is taken from a BNI (2000) Report on MOB submitted to ONR.

4.8 Applicability of Morison Force vs. Diffraction Force

Earlier a dimensional analysis was shown to characterise the regimes of small structures vs. large structures. It is usually accepted that the diffraction effect becomes important when the diffraction parameter $\pi D/L$ or (ka) is greater than 0.5 where D or a is a charateristic dimension of a structural member in the plane of the wave, L is the wavelength and k is the wave number. In this case the Reynolds number or the wave steepness play a less important role. When the diffraction parameter is less than 0.5, then the potential flow assumption becomes invalid and the wave flow separates from the surface of the structure. Here, the Keulegan–Carpenter number becomes physically significant. In this case the Reynolds number is also an important parameter in determining the force.

The small vs. large structures may be determined from a chart similar to the one shown in fig. 4.18. In this chart the x-axis is the diffraction parameter while the vertical axis H/D is equivalent to the Keulegan–Carpenter number. For example, at the mean water level, $KC = u_0T/D = gT^2H/(2LD)$, which in deep water $L = gT^2/(2\pi)$ becomes approximately $KC = \pi\ H/D$. According to this chart then, when H/D is less than 2 (i.e. KC less than 6, regions I and II), the drag force is small. For region I, the inertia part of Morison equation suffices for the wave force calculation. If, in addition, $\pi\ D/L$ is greater than 0.5 (region II in fig. 4.18), then diffraction analysis should be made for the force calculation. On the other hand, for H/D greater than about two (regions II, V and VI), Morison equation should be used for the force computation of the submerged structural member. As KC number increases, the drag force contribution goes up (regions V and VI). For the region IV, both diffraction and drag effect may be important.

For fixed platforms, a design wave analysis based on the maximum design wave is often performed. The following steps for the static wave analysis closely follow the API-RP2A guidelines (2000):

- At the installation site for the platform, choose the design wave height, wave period, water depth and current from oceanographic data.
- Determine the apparent wave period based on the current strength.
- Compute wave kinematics using an appropriate wave theory. Refer to the applicability chart for wave theories given in Chapter 3.
- Reduce wave kinematics by the kinematic factor.
- Determine effective local current by the current blockage factor; vectorially add this current to the wave kinematics.
- Increase member dimensions due to possible marine growth.
- Choose the values of C_M and C_D for the members.

Table 4.10 Capabilities of a few diffraction theory programs

Capabilities	AQWA	WAMIT	MORA	NBODY	NEPTUNE	HYDRAN	HOBEM	HIPAN
Single rigid body	Y	Y	Y	Y	Y	Y	Y	Y
Multiple rigid bodies	Y	Y	N	Y	N	Y	N	Y
Flexible bodies, generalized mode approach	Y	Y	Y	Y	N	Y	Y	Y
Added mass, damping	Y	Y	Y	Y	Y	Y	Y	Y
Excitation loads and motions	Y	Y	Y	Y	Y	Y	Y	Y
Local pressure load	Y	Y	Y	Y	Y	Y	Y	Y
Free surface motion	Y	Y	Y	N	Y	N	Y	Y
Mean drift force	Y	Y	Y	Y	Y	N	Y	Y

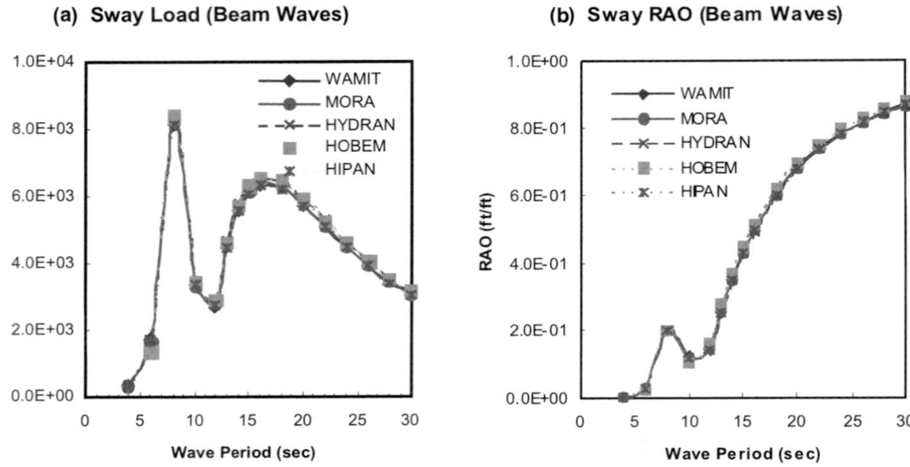

Figure 4.17 Comparison of sway first-order (a) loads and (b) motions

- For any conductor array present in the platform, reduce coefficient by the blockage factor.
- Use the Morison equation to compute local loads.
- Compute global loading on the platform as a vector sum.
- Use a structural analysis program to design the structural members of the platform.

For a large offshore structure, the design method uses the diffraction calculation to generate a transfer function. The design method in this case is described in Chapter 5.

4.9 Steady Wave Drift Force

While the diffraction theory gives the first-order force, inclusion of the second-order term from steeper waves may produce forces that are of the second order. The time independent portion of this force is called the steady wave drift force. Therefore, all quantities in the second order are included in the expression for the steady drift force. The steady diffraction forces arise from the potential flow. However, the viscous flow (with flow separation) also produces a steady second-order force. These quantities are described separately in the following sections. Generally, one or the other is significant for a member of an offshore structure. It is possible that different components of an offshore structure will contribute to the potential and viscous components respectively of the drift force.

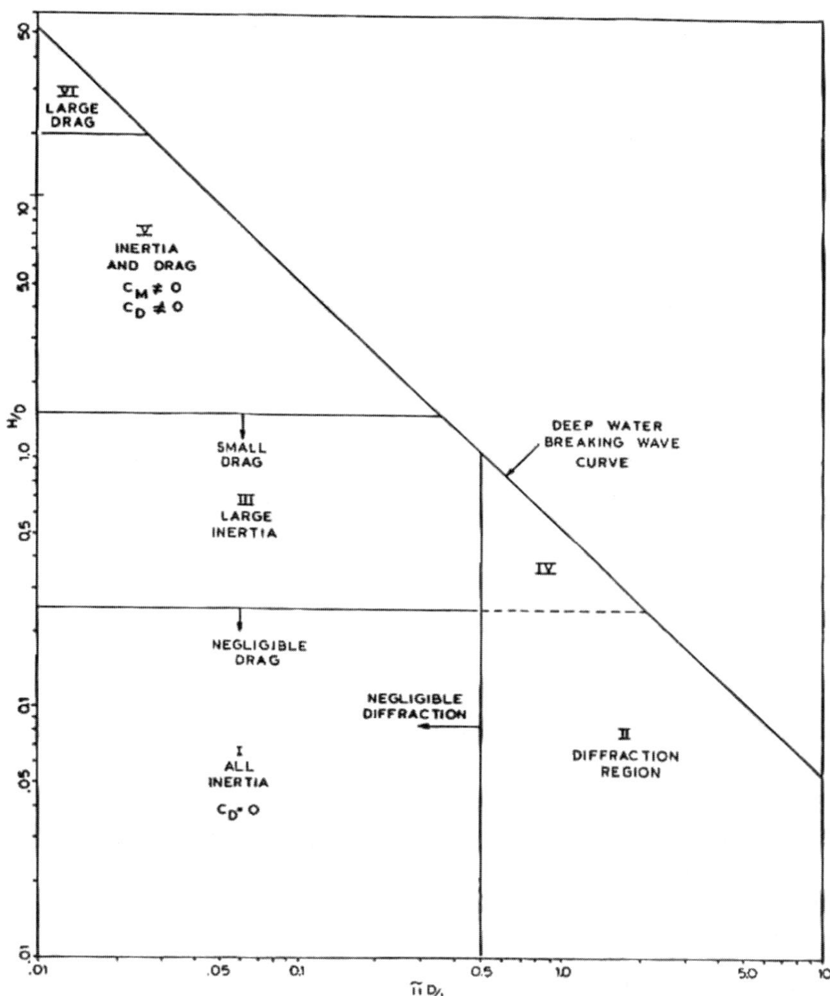

Figure 4.18 Limits of application for small vs. large structure

4.9.1 Steady Drift Potential Force

Recall that the linear diffraction theory computes the forces on the structure using the linear (Bernoulli's) pressure term in its equilibrium position up to the still water level. However, the structure motion, the wave free surface, and the non-linear (Bernoulli's) pressure terms will introduce non-linear forces on the structure. Considering these terms up to second order, both steady and oscillating second-order force components will result. The second-order terms correspond to the second power of wave amplitude. The steady drift

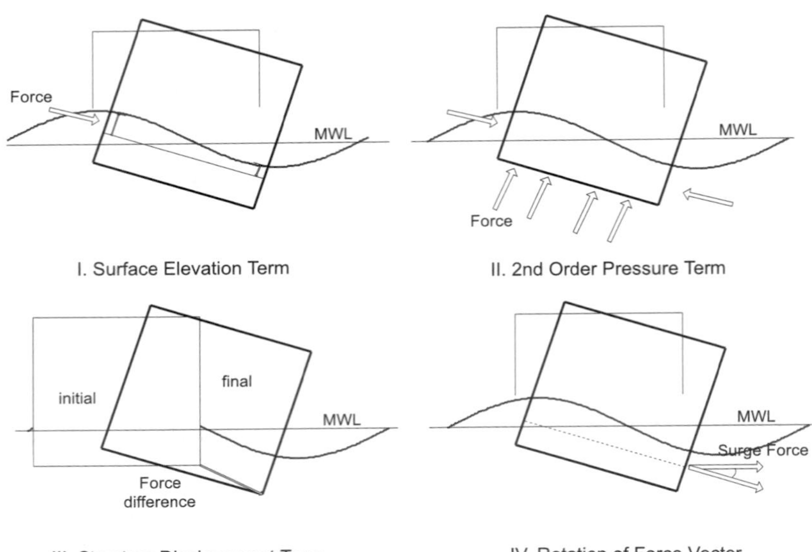

Figure 4.19 Definition of steady drift load contributions

force is the time averaged force over one cycle. The second-order terms introduce four individual contributions from the above effects as elaborated below:

1. **Free surface term:** This term arises because the wave-free surface at the structure continually changes the submerged part of the structure about the still water level. The angular motion of the structure has a similar effect. Therefore, the inclusion of these two terms as the relative elevation produces a higher-order force on the structure (see fig. 4.19 I).
2. **Velocity-squared term:** The first-order force considers only the linear pressure term in the Bernoulli's equation. If the velocity-squared term from the Bernoulli's equation is introduced, it produces forces on the structure that are of a higher order (see fig. 4.19 II).
3. **Body motion term:** The first-order force on the structure is computed in its equilibrium position. However, the motion of the structure in waves displaces the structure from its equilibrium position. If this displaced position of the structure is taken into account, it changes the pressure distribution on its surface. Therefore, the displaced position along with the first-order pressure distribution results in a second-order force (see fig. 4.19 III).
4. **Rotational term:** To first order the forces are computed along the axis of the structure at its equilibrium condition. However, the angular motion of the structure changes this direction of the forces along the x-, y- and z-directions. When the force is resolved from the rotated direction, second-order force results (see fig. 4.19 IV).

It is clear from the above discussion that there will be contributions of the second order and the higher. These contributions will be successively smaller in magnitude as the order of these terms increases. The steady force is limited up to second order only. In general there are six steady drift forces/moments in the six degrees of freedom.

The second-order steady drift force is the time-invariant component of the above four contributions. By its definition of second-order, the magnitude of steady drift force is expected to be small. Its magnitude is generally of the order of 5% of the first-order force, which is often the order of uncertainty in the force computation. Then, why is it important in a design of an offshore structure? It is generally not that important for an offshore structure, unless a moored floating structure possesses a soft mooring stiffness in a certain degree of freedom. Examples of such systems are the surge and the heave of a catenary moored FPSO or semi, surge of a TLP, etc. Since the initial stiffness characteristics of such systems are quite low, even a small steady force induces a large displacement. Thus, under this steady load, the structure takes on a displaced position. This large displacement may affect its riser to have a large initial angle, which may be operationally limiting. This is illustrated by the measured surge oscillation of a single point mooring tanker in regular waves as shown in fig. 4.20. Note that the high steady displacement (ΔL_S) of the tanker due to a steady drift load is higher than the first-order motion (ΔL_N). The long period slow drift motion is also evident here (which in regular wave arises as a transient response when the wave first hits the structure).

In random waves, the steady drift force may be determined from the steady drift force in regular waves. A transfer function is generated from the regular waves by normalising the steady force F_D at each frequency spanning the random wave frequencies by the square of the wave amplitude as:

$$\overline{F}_D(\omega) = \frac{F_D(\omega)}{a^2(\omega)} \tag{4.22}$$

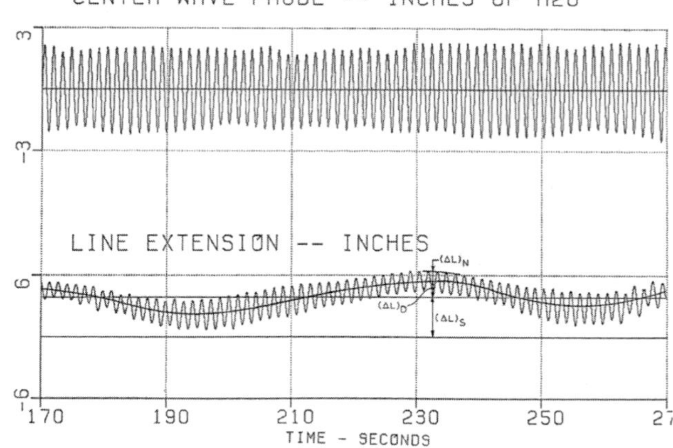

Figure 4.20 Measured surge motion of tanker model in regular wave

Then the steady drift force in random waves become

$$F_2 = 2\int_0^{\infty} \overline{F}_D(\omega)S(\omega)d\omega \tag{4.23}$$

4.9.2 Viscous Drift Force

When the structural member falls in the viscous regime, then a unidirectional current plus waves produce a steady drift force on the structural member as well. In the presence of current, the total velocity drag force (from the Morison equation) may be broken up into two simpler expressions depending on the strength of current relative to the wave velocity. Thus,

$$f_D = \pm\frac{1}{2}\rho C_D A\left[U^2 + \frac{1}{2}u_0^2(1+\cos 2\omega t) + 2Uu_0\cos\omega t\right] \quad \text{for } |U| \geq u_0 \tag{4.24}$$

in which the negative term corresponds to opposing current, and

$$f_D = \frac{1}{2}\rho C_D A\left[U^2 + \frac{1}{2}u_0^2(1+\cos 2\omega t) + 2Uu_0\cos\omega t\right]\text{sgn}(u+U) \quad \text{for } |U| \leq u_0 \tag{4.25}$$

in which sgn takes on the sign of $u+U$. The steady viscous drift force is obtained from these expressions (where the upper bar denotes mean value) as:

$$\bar{f}_D = \frac{1}{2}\rho C_D A u_0^2\left[\frac{1}{2} + \left(\frac{U}{u_0}\right)^2\right] \quad \text{for } \frac{|U|}{u_0} \geq 1 \tag{4.26}$$

and

$$\bar{f}_D = \frac{1}{2\pi}\rho C_D A u_0^2\left[\left(\frac{U}{u_0}\right)^2(2\psi - \pi) + 4\left(\frac{U}{u_0}\right)\sin\psi + \frac{1}{2}\sin 2\psi\right] \quad \text{for } \frac{|U|}{u_0} < 1 \tag{4.27}$$

in which the quantity ψ is given by

$$\psi = \cos^{-1}\left(-\frac{U}{u_0}\right) \quad \text{for } 0 \leq \psi \leq \pi \tag{4.28}$$

The normalised mean value for the viscous drift force on a cylinder in wave plus current is shown in fig. 4.21 vs. U/u_0. The drift force is asymmetric about the zero value and takes on a large magnitude at higher U/u_0 values.

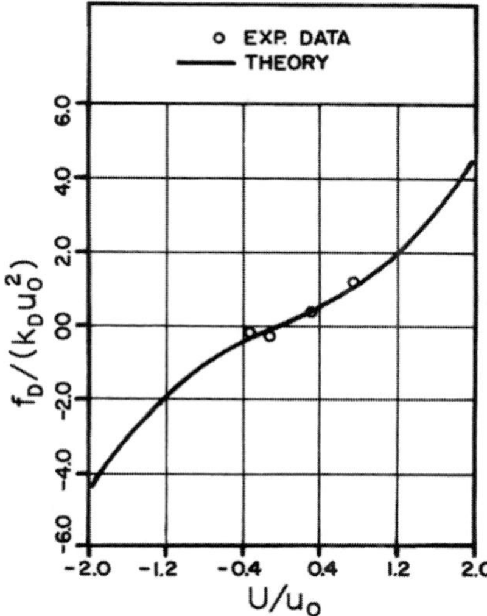

Figure 4.21 Mean viscous drift force on a fixed vertical cylinder

Table 4.11 Values of constant for free surface drift force

kH	0.1	0.2	0.3	0.4	0.5	0.6	0.7	0.8	0.9	1.0
C_1	1.0013	1.0053	1.0120	1.0215	1.0337	1.0487	1.0667	1.0877	1.1118	1.1392

If the drag force due to waves is considered applicable up to the free surface above the SWL, then a mean drift force results on a fixed vertical cylinder given by:

$$\overline{F} = \frac{1}{2}\rho C_D A \frac{g}{k^2}\frac{(kH)^3}{6\pi}\left[\frac{1}{\sinh 2\mathrm{kd}} + C_1(kH)\coth 2kd\right] \tag{4.29}$$

in which the values of the coefficients C_1 are given in table 4.11.

Since the use of the linear theory up to the free surface may be questioned, the modified forms had been proposed in Chapter 3. If Wheeler stretching is used for the modified expression for the free surface, a steady drift force results which is given below:

$$\overline{F} = \frac{1}{2}\rho C_D A \frac{g}{k^2}\frac{(kH)^3}{6\pi}\left[\frac{1}{\sinh 2kd} + \frac{1}{2kd}\right] \tag{4.30}$$

In this case, the free surface drift force on a vertical cylinder is a function of the third power of the wave height.

4.10 Slow-Drift Wave Forces

It should be clear from the discussion of the four contributions of steady drift force described in Section 4.9.1 that there is also an oscillating second-order contribution from these four components. Additionally, the second-order velocity potential will generate an oscillating component. These components will have a frequency that is twice the wave frequency. This is illustrated with the help of a simple example with the expression for the dynamic pressure due to waves. The pressure given by the Bernoulli's equation includes a velocity-squared term, which has the form $1/2\rho\ (u^2+v^2)$. For a linear wave, the horizontal velocity component is given in terms of its amplitude as,

$$u = u_0 \cos \omega t \tag{4.31}$$

so that,

$$u^2 = u_0^2 \cos^2 \omega t = \frac{1}{2} u_0^2 (1 + \cos 2\omega t) \tag{4.32}$$

Thus the first term contributes to the steady force, while the second term corresponds to the oscillating second-order force having twice the wave frequency. These contributions are generally not considered in the design of an offshore structure (unless waves are very steep or the structure natural frequency is very high, such as, the heave of a TLP).

Moreover, when waves have multiple frequencies (as in a random wave), the combination of frequencies will give rise to additional second-order oscillating terms. Their presence, again, may be illustrated by the velocity-squared term of the Bernoulli's equation. A simple derivation shows the source of these higher-order components. Choose a wave group having two frequencies, ω_1 and ω_2. The horizontal wave particle velocity component may be represented by the summation as follows:

$$u = u_1 \cos \omega_1 t + u_2 \cos \omega_2 t \tag{4.33}$$

Then, considering the second-order component as in equation (4.32) from this term:

$$p_2 = \frac{1}{2} \rho [(u_1 \cos \omega_1 t + u_2 \cos \omega_2 t)^2 + (v_1 \sin \omega_1 t + v_2 \sin \omega_2 t)^2] \tag{4.34}$$

Expanding the trigonometric functions to include the higher harmonics as in equation (4.32), the second-order pressure component becomes:

$$\begin{aligned} p_2 = \frac{1}{2} \rho \Big[& \frac{1}{2} (u_1^2 + u_2^2 + v_1^2 + v_2^2) \\ & + (u_1^2 - v_1^2) \cos 2\omega_1 t + (u_2^2 - v_2^2) \cos 2\omega_2 t) \\ & + (u_1 u_2 - v_1 v_2) \cos(\omega_1 + \omega_2) t + (u_1 u_2 + v_1 v_2) \cos(\omega_1 - \omega_2) t \Big] \end{aligned} \tag{4.35}$$

Therefore, when there are multiple wave components present, the structure experiences a steady and double frequency force (as described already) from multiple components in the random wave as well as a sum frequency force and a difference frequency force for each pair of frequencies in the random wave. This is only one source of low $(\omega_1 - \omega_2)$ and high frequency $(\omega_1 + \omega_2)$ force. There are other components similarly arising from the other three sources shown under steady drift forces. Additionally, the second-order velocity potential

Table 4.12 Quadratic transfer function for oscillating drift load (t/m²) due to paired wave frequencies [from Pinkster (1981)]

ω_1 (rad/s) → ω_2 (rad/s) ↓	0.354	0.444	0.523	0.600	0.713	0.803	0.887
0.354	1	21	17	32	16	39	30
0.444		1	6	30	10	12	5
0.523			6	19	19	18	10
0.600				1	18	13	6
0.713					5	13	8
0.803						2	9
0.887							2

based on two distinct frequencies will provide a slow drift and high frequency component. These contributions become important when the system natural frequency approaches one of the sum or difference frequencies.

For the oscillating potential drift force (for both the difference and the sum frequency), the linear diffraction theory (table 4.9) performs an approximate computation. This computation may not be adequate and a second-order bichromatic diffraction theory needs to be employed. Here the bichromatic means that the potential drift force is computed for a pair of frequencies chosen from the frequency distribution of a random wave. This requires solving the boundary value problem similar to the one described for the linear diffraction theory, but it should include terms up to the second order. Then, a quadratic transfer function (QTF) similar to the one shown in table 4.12 is generated. In this table, the frequencies appear along the horizontal and the vertical axes. Each value in the table represents a transfer function corresponding to the frequency pair. The diagonal terms are the steady forces at the corresponding frequency. It is clear that for a random wave, this table will be enormous if all combinations of frequency pairs are included in the computation. However, it should be noted that the importance of low and high frequency forces is in the area of the natural frequency of the structure. Therefore, it should suffice to choose the frequency pairs that give forces at the difference (or sum) frequencies close to the natural frequency of the particular structure. The order of frequency pairs may be limited to about 5 or 6. This computation should produce a matrix similar to the one illustrated in table 4.12 of order 5×5 or so and limit the number of QTF computations. Once this matrix is known, the time history of the second-order force may be generated.

For the slow drift force from a random wave, sum over all frequency pairs of a random wave yields the following expression for the force time history:

$$F(t) = \sum_{i=1}^{N}\sum_{j=1}^{N} P(\omega_i, \omega_j)\cos[(\omega_1 - \omega_2)t - \delta_i + \delta_j] + Q(\omega_i, \omega_j)\sin[(\omega_1 - \omega_2)t - \delta_i + \delta_j] \quad (4.36)$$

where $P(\omega_i, \omega_j)$ = in-phase difference force component, and $Q(\omega_i, \omega_j)$ = out-of-phase difference force component. Similarly, for the sum frequency force due to a random wave, the expression becomes

$$F(t) = \sum_{i=1}^{N}\sum_{j=1}^{N} P(\omega_i, \omega_j)\cos[(\omega_1 + \omega_2)t - \delta_i + \delta_j] + Q(\omega_i, \omega_j)\sin[(\omega_1 + \omega_2)t - \delta_i + \delta_j] \quad (4.37)$$

where P and Q are now the corresponding sum frequency force transfer functions.

4.11 Varying Wind Load

Wind is an important environmental parameter that influences the design of floating offshore structures, particularly in a harsh environment. Wind spectrum is broad-banded covering a large frequency band from low to high. Therefore, the wind-induced slow-drift oscillation is an important design criterion.

Based on the wind spectrum described in Chapter 3, a digital time series for the wind speed is obtained. The numerical simulation method is similar to the wave time history simulation in that a series of discrete frequencies are chosen at specified intervals over the frequency range and a sinusoidal representation is assumed at each of these frequencies. An inverse Fourier transformation is performed to change the wind spectrum from the frequency to the time domain. The wind velocity time series is expressed as a summation of these discrete frequencies:

$$U_w(t) = \sum_{i=1}^{M} U_i \cos(\omega_i t + \varepsilon_i) + U_w(1\text{h}) \quad (4.38)$$

in which U_w = wind speed, U_i = wind speed amplitude at each frequency, ω_i = circular frequency ($=2\pi f_i$), ε_i = corresponding random phase angle, t = time, M = number of discrete frequency components over the spectrum width, and $U_w(1\,\text{h})$ = one-hour mean wind velocity. The velocity amplitude is defined as:

$$U_i = \sqrt{2S(\omega_i)(\Delta\omega_i)}, \quad (i = 1 \ldots M) \quad (4.39)$$

in which $S(\omega_i)$ = wind spectral energy density (fig. 3.25), $\Delta\omega_i$ = frequency interval. The wind force is calculated from an expression similar to the drag force component of the Morison equation:

$$F_{\text{Wind}}(t) = \frac{1}{2}\rho_a C_s A \left|U_w(t)\right| U_w(t) \quad (4.40)$$

in which ρ_a = density of air, C_s = shape coefficient based on the geometry of the superstructure, and A = projected area of the superstructure. This force includes the steady component.

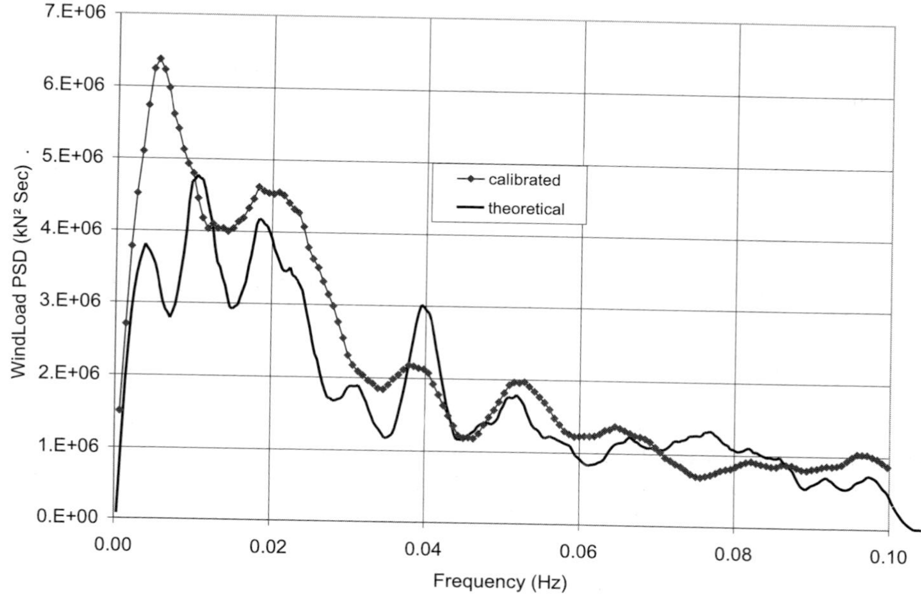

Figure 4.22 Power spectral density of the wind load

Figure 4.22 shows the computed wind load spectra using the computation in equation (4.40). Note that the wind spectrum is quite wide-banded having energy at low frequencies. It excites the slow drift of a floating structure.

The effect of the variable wind on the measured response of a floating structure is illustrated here taken from an actual model test. In this test, the response of the floating structure due to wind alone was measured. Figure 4.23 shows the pitch and surge motion of the structure. The pitch natural period is small and damping in pitch at this period is relatively large. The pitch response is small corresponding to the wind spectrum frequencies and has a small steady drift. On the contrary, the surge natural period is large having a low damping and hence the surge motion is large. The response in surge clearly shows a mean drift and a slow-drift motion of the structure induced by the wind spectrum similar in magnitude to the wave-drift motion. It is clear that the wind spectrum introduces a slow drift oscillation in the floating structure at a low frequency corresponding to the natural frequency of the floating system.

4.12 Impulse Loads

The forces on an offshore structure due to wave impact are of significant importance in the design of an offshore structure. This includes fixed structures, such as members of the drilling structures near the wave splash zone, as well as floating structures, such as wave

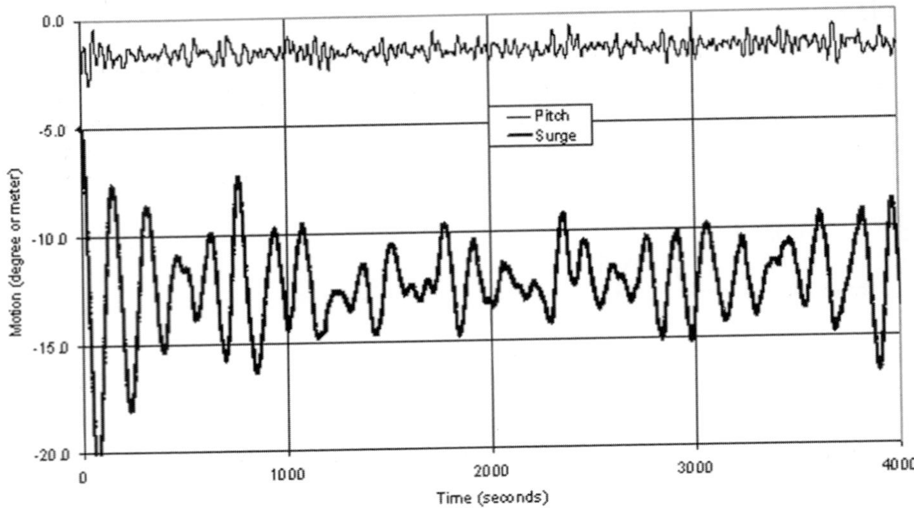

Figure 4.23 Slow drift oscillation due to wind

slamming on an FPSO. None of the methods described so far is suitable for the computation of the impulse load. Theoretical solution of the impact force is quite difficult due to the complicated flow field that generates around the structure from such a wave impact. Therefore, empirical equations are generally used for the computation of impact loads on structures near the wave-free surface.

Impact forces imposed on an offshore structure can be two to four times larger than the no-impact forces from the waves of similar amplitude. Also, the pressure due to impact may be ten times larger than the non-impact pressure and it rises at a fraction of time. The duration of impulse is quite small, typically about 2–4 ms.

4.12.1 Wave Slamming Load

The wave-slamming load on a horizontal cylinder may be computed by an empirical equation of the form:

$$F_0 = \frac{1}{2}\rho C_s A u_0^2 \tag{4.41}$$

in which the theoretical (potential flow) value of the slamming coefficient C_s is π. It has been experimentally found in a wave tank to be 3.6. However, the impact force is strongly related to the stiffness of the structural member. Sarpkaya (1978) found the coefficient to vary from 0.5π to 1.7π depending on the elasticity of the cylinder and the mounting system. The recommended practice is to use a C_s value of 3.5.

4.12.2 Breaking Wave Load

Ochi and Tsai (1984) experimentally showed that the magnitude of pressures on a structure from breaking waves is proportional to the square of the impact velocity. Two different breaking wave conditions were considered: the waves breaking in close proximity to the structure and the waves approaching the structure after they broke. These are called the breaking wave impact and the broken wave impact. The impact pressure is given by:

$$p = \rho \kappa_1 U^2 \tag{4.42}$$

in which the dimensionless constant was obtained as:

$$\kappa_1 = \begin{cases} 5.98 & \text{for breaking wave impact} \\ 2.74 & \text{for broken wave impact} \end{cases} \tag{4.43}$$

The impact velocity was shown to be proportional to the wave velocity:

$$U = \beta c \tag{4.44}$$

in which the constant β is,

$$\beta = \begin{cases} 0.48 & \text{for breaking wave impact} \\ 0.70 & \text{for broken wave impact} \end{cases} \tag{4.45}$$

The wave speed associated with wave breaking is given by,

$$c = 1.092 \frac{gT}{2\pi} \tag{4.46}$$

4.12.3 Wave Run-Up Load

For a vertical cylinder exposed to steep waves, the slamming coefficient is derived [from experiments of Campbell and Weynberg (1980)] as follows:

$$C_s = 5.15 \left(\frac{D}{D + 19s} + \frac{0.107s}{D} \right) \tag{4.47}$$

in which s is the penetration distance as defined in fig. 4.24 and D is the vertical cylinder diameter. The coefficient will change its value with the passing of the waves as the value of s changes.

The total force on the cylinder is then computed by integrating the force over the limits of the free surface intersection at the cylinder. The expression for the total force is given by:

$$F = \frac{1}{2} \rho D \int_{d+\eta_0}^{d+\eta_1} u^2 C_s(y) dy \tag{4.48}$$

in which η_0 and η_1 are defined in fig. 4.24.

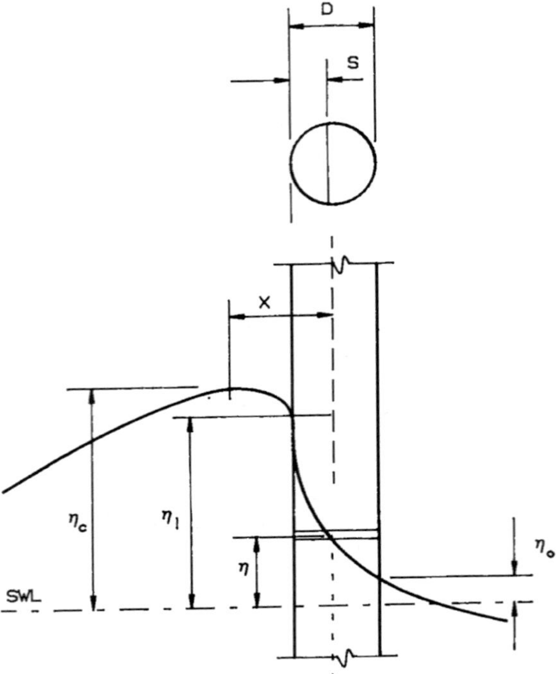

Figure 4.24 Wave impact on a vertical cylinder [Campbell and Weynberg (1980)]

4.13 Response of Structure

This section describes how the motion response of a floating structure or a floating multi-body system is computed. It should be clear that once the motions are known, the other responses, such as the mooring line forces, the component loads on a structure, the air gap at the deck of the structure, etc. may be determined. In order to explain how the motions of a complicated system may be determined, the formulas are given first for simple floating systems, such as a single-degree of freedom system. A floating structure that is constrained by the mooring or the anchoring system may be subjected to motions predominantly in one or two directions. A structure freely floating or held by soft mooring lines will be free to move in all the six degrees of freedom including three translations and three rotations. For two floating structures moored together, the total degrees of freedom increase to twelve. The methods of analysis in determining the motions in these cases are described in the following sections.

4.13.1 Structure Motion in One Degree

A floating structure held in place by its mooring system will experience motions when subjected to waves. If one is interested in motion in a particular direction and if the mooring system is treated as a linear spring in this direction, then the dynamic system is

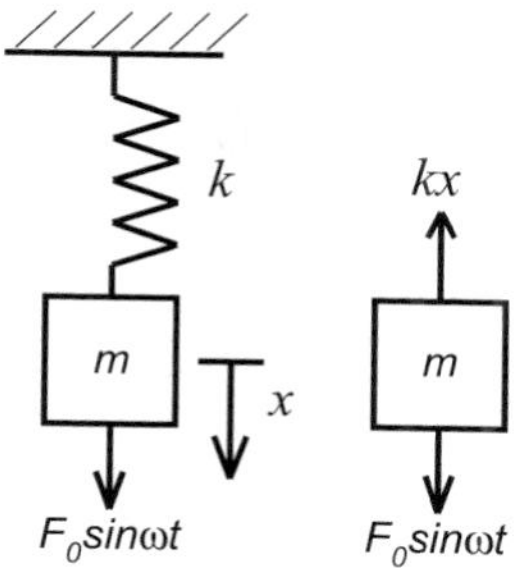

Figure 4.25 Forced spring–mass system

treated as a single-degree of freedom system in which a mass attached to a spring describes the system. The motions of many physical systems may be addressed by such a simplification. A few examples in offshore application are an oscillating articulated tower hinged at the bottom, the surge motion of a ship moored in head sea, and the heave motion of a semi-submersible. Here the basic properties of such floating systems in a single-degree of freedom will be given. These properties are useful in analysing the characteristics of an offshore structure.

Consider first a system having a mass m, and a spring having a spring constant k and no internal or external damping. The system is schematically shown in fig. 4.25. Such an undamped system will experience an inertia force from the motion of its mass and a restoring force from the spring, the sum of which will resist the excitation force. This is illustrated by the free body diagram in fig. 4.25.

The forcing function in this example is assumed to be a harmonic function of amplitude F_0 having a frequency ω. This is equivalent to the force obtained from the linear wave theory.

$$m\ddot{x} + kx = F_0 \cos \omega t \tag{4.49}$$

The natural period of this system is dependent on the mass and spring constant,

$$\omega_n = \sqrt{\frac{k}{m}} \tag{4.50}$$

This frequency ω_n is called the undamped natural frequency of the system. This implies that if the spring–mass system is displaced from its equilibrium position and released, it will undergo a harmonic oscillation whose period will coincide with the natural period of the system, and ideally in the absence of damping, the oscillation would take on a perpetual motion.

The displacement will be in phase with the force itself and the amplitude of displacement is obtained from:

$$X = \frac{F_0/k}{\omega_n^2 - \omega^2} \tag{4.51}$$

It is clear that the amplitude will "blow up", approaching infinity as the denominator becomes zero at the resonant frequency ω_n. It is customary to design a physical floating system in which the resonant frequency is removed as far as practical from the possible excitation frequency range that may be encountered from the environment.

A physical system, however, is never damping-free. The damping may be extremely small in which case the amplitude of oscillation at the resonant frequency will be quite large even for a small excitation. This is extremely important to note since many systems are excited from non-linear forcing effect at frequencies near the natural frequency of the system, which may be further from the excitation frequencies. Some of these offshore systems will be described later.

4.13.2 Transient Response of Structure

One of the common forms of damping for an offshore structure in waves is the viscous damping. In its simple form, the damping force is linearly proportional to the structure velocity $\dot{x}$:

$$F_d = c\dot{x} \tag{4.52}$$

where c is called the linear damping coefficient. This damping is very important for an oscillating structure since it helps limit the excursion of the structure in a resonance situation.

If a linear damper is present in the system, then a damping force is additionally introduced in the earlier system (fig. 4.25). This is illustrated with a free body diagram in fig. 4.26. Then the equation of motion for the displacement of the mass is described similar to equation (4.49) by:

$$m\ddot{x} + c\dot{x} + kx = F_0 \cos \omega t \tag{4.53}$$

Examine this equation first in the absence of the external force in which the mass m is displaced from its equilibrium position and released. It gives rise to a free oscillation of the structure about its equilibrium position and provides some very important properties of the system. The equation of free motion of the structure is given by:

$$m\ddot{x} + c\dot{x} + kx = 0 \tag{4.54}$$

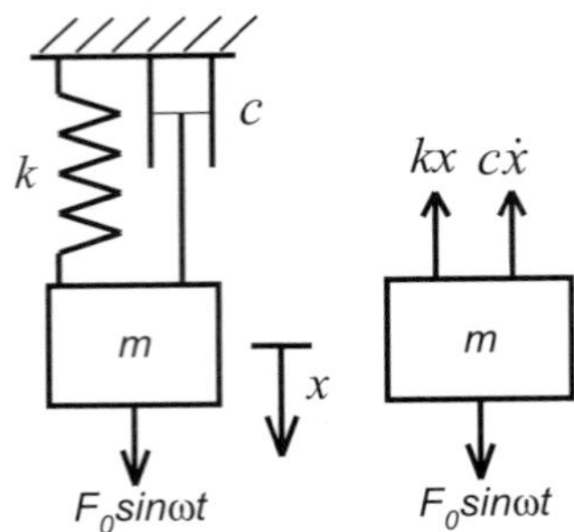

Figure 4.26 Damped single-degree of freedom system

In general, for a quadratic differential equation of this kind, there are two distinct solutions. However, when the two solutions coincide, the damping coefficient is given by:

$$c_c = 2\sqrt{km} \tag{4.55}$$

This damping has a special significance and is termed critical damping coefficient. The ratio of linear damping present in a system and this critical damping is generally known as the damping factor:

$$\zeta = \frac{c}{c_c} \tag{4.56}$$

The natural frequency of the linear system is given by the quantity ω_n shown in equation (4.50) so that the following relationship results from the above two equations:

$$\frac{c}{m} = 2\omega_n\zeta \tag{4.57}$$

It is a common practice to describe the amount of damping in a system by the value of the damping factor. The higher the value of the damping factor, the higher is the damping in the system. Systems with $\zeta > 1.0$ are called overdamped systems, while those with $\zeta < 1.0$ are termed underdamped. A critically damped or an overdamped system has an aperiodic or non-oscillatory motion.

For an underdamped system ($\zeta < 1.0$), the displacement of the system x may be written in terms of the above quantities by the formula

$$x = X\exp(-\zeta\omega_n t)\sin[\sqrt{1-\zeta^2}\omega_n t + \beta] \tag{4.58}$$

in which the unknowns are the displacement amplitude X and its phase angle β.

The solution has two parts: an exponential time decay function and an oscillating term, which has a frequency of $\sqrt{1-\zeta^2}\omega_n$. This frequency is different from the undamped frequency and is termed the damped frequency. Due to the exponential term, it is expected that the amplitude of oscillation will diminish in value as time passes, the rate of this decrease being a direct function of the damping factor. This relationship in equation (4.58) is plotted in fig. 4.27. When $\zeta = 1.0$, the oscillation disappears (see equation (4.55)) and the solution behaves as an exponential function (fig. 4.28). If the damping value is greater than 1.0, no oscillation of the solution is expected. This is why the unit value of damping ($\zeta = 1.0$) is referred to as critical damping.

Most of the practical systems that one encounters have a damping factor less than one. If a weight is hung in air from a coil spring and displaced, then the oscillation will continue for a long time before it comes to rest. This is an example of an underdamped system, which has a damping factor much less than 1. However, one does encounter systems whose damping factors are equal to or greater than 1.0. For example, the hinged storm doors in a house with attached dashpots should ideally be adjusted to a damping factor of close to one. This will allow the door to close slowly without oscillating when the door is opened and released. Another example of this type of motion is the heave motion of a floating barge. Since the heave of a flat bottom barge is highly damped, it has a large value of

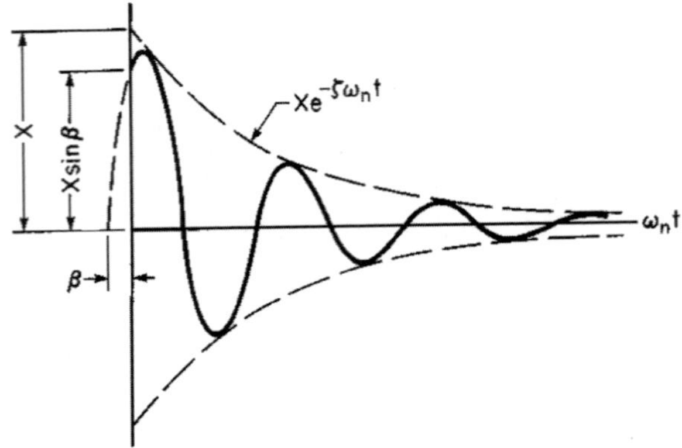

Figure 4.27 Decaying oscillation of a linearly damped SDOF system

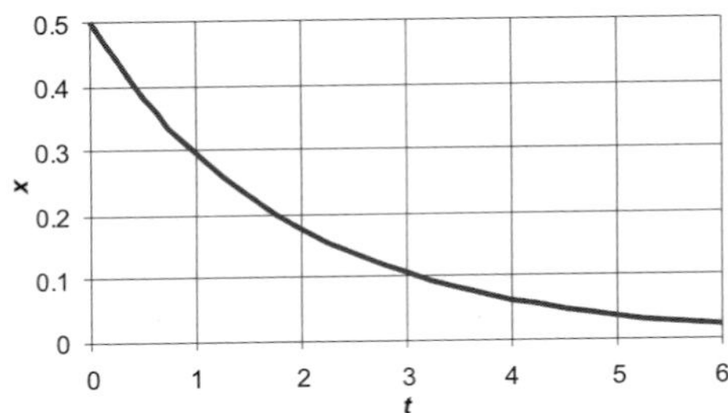

Figure 4.28 Free motion of an overdamped SDOF System

damping factor. In an experiment with a barge model, if the barge is displaced in the vertical direction and released, the measured vertical motion will have little oscillation before coming to rest in its equilibrium position.

The term $x_k \exp(-\zeta\omega_n t)$ represents the curve that can be drawn through the succeeding peaks of the damped oscillation. Strictly speaking, the curve does not pass exactly through the peaks, but the small deviation is usually neglected. If the natural logarithms of these peaks are determined, the quantity $\zeta\omega_n$ represents the slope, s of the line that can be drawn through the converted values. The frequency of the damped motion, ω_d, is also obtained from equation (4.58), and thus one obtains two equations in two unknowns:

$$s = -\zeta\omega_n \tag{4.59}$$

$$\omega_d = \omega_n\sqrt{1-\zeta^2} \tag{4.60}$$

The undamped natural frequency is close to the damped natural frequency so that the peak to peak distance is a good measure of the natural period of the system. The terms on the left-hand side of equations (4.59) and (4.60) are obtained from the free decay curve by a

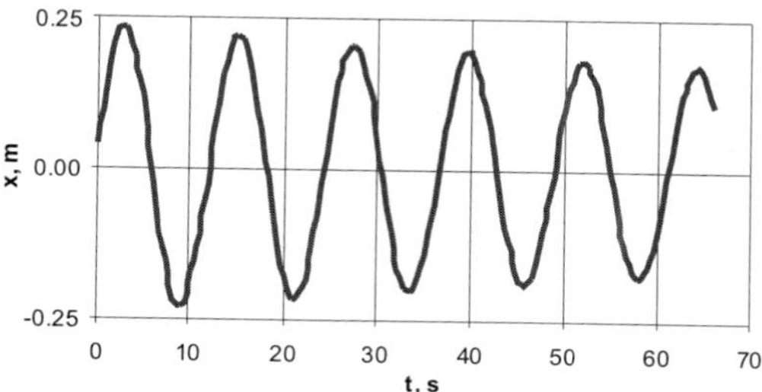

Figure 4.29 Extinction curve of the example problem

least square fit of the peak values. Once the values of ω_n and ζ are known from the above equations, the added mass m_a and damping coefficient c are computed as follows:

$$m_a = m - m_0 = \frac{k}{\omega_n^2} - m_0 \tag{4.61}$$

and

$$c = 2m\zeta\omega_n \tag{4.62}$$

in which m_0 is the mass of the structure in air.

Therefore, knowing the extinction curve for a moored floating structure, the damping of the system may be determined by a simple analysis. In a model test, the extinction curves in a particular direction of motion of the structure is obtained by displacing the structure and measuring its motion in this direction.

This is illustrated by an example based on fig. 4.29. The extinction curve represents the free oscillation of a moored tanker. The displacement of the tanker is $m_0 = 38.1$ kg/m (25.5 slugs) and the spring constant $k = 10.4$ kg/m (7.0 lb/ft). The least square analysis of the peaks of the free decay curve described above gives an added mass coefficient and a damping factor of 0.049 and 0.008 respectively. The natural period between the positive second and third peaks in fig. 4.29 is measured as 12.3 s. Then, $m_a = 1.75$ kg/m (1.17 slugs) and the added mass coefficient $c_a = m_a/m_0 = 0.046$. Also, the amplitudes of peaks 2 and 3 are 0.235 and 0.22 m (0.78 and 0.72 ft) respectively. Therefore, the logarithm decrement is given by $\delta = 0.5(\ln 0.235 - \ln 0.22) = 0.03$ which reduces to $\zeta = 0.01$ (or 1% damping). This illustrative example is based on only two peak values in the extinction curve to determine the two unknowns. In practice, for better estimates, a least square fit of the individual well-defined measured peak values (both the positive and the absolute negative peaks) are used to compute these two quantities. It is a common practice to determine these quantities in a test set-up of a floating system by displacing the float from its equilibrium position in a given degree of freedom and measuring the response in that direction.

4.13.3 Forced Linearly Damped System

For a linear wave, the exciting force (such as, by the linear diffraction theory) is given by a harmonic term. Then, the displacement of the structure in the direction of force is given by:

$$m\ddot{x} + c\dot{x} + kx = F_0 \cos \omega t \tag{4.63}$$

where x = structure oscillation, F_0 = exciting force amplitude, ω = forcing frequency, m = structure plus added mass, k = linear spring constant corresponding to the restoring force of the structure and c = linear damping coefficient. The solution in this case is harmonic in which the displacement leads or lags the excitation based on the damping and the frequency of the forcing function relative to the natural frequency of the system.

The displacement is commonly normalised by the static displacement $X_s(= F_0/k)$. The frequency is normalised by the natural frequency of the system. Then the displacement and phase are written as:

$$\frac{X}{X_s} = \frac{1}{\left[\left(1 - \omega^2/\omega_n^2\right)^2 + \left(2\zeta\omega/\omega_n\right)^2\right]^{1/2}} \tag{4.64}$$

and

$$\tan\beta = \frac{2\zeta\omega/\omega_n}{1 - \left(\omega/\omega_n\right)^2} \tag{4.65}$$

It is customary to plot these normalised forms of displacement and phase angle in terms of the normalised frequency. The displacement of the linearly damped system is shown in fig. 4.30. This is known as the frequency response curve. The normalised displacement is called the magnification factor, and reduces to its static value of X_s at zero frequency. The maximum displacement is found to occur at the damped natural frequency, the amplitude of which depends on the amount of damping in the system. This is also the point where the phase angle (fig. 4.31) goes through a rapid change from zero to 180°. For zero damping

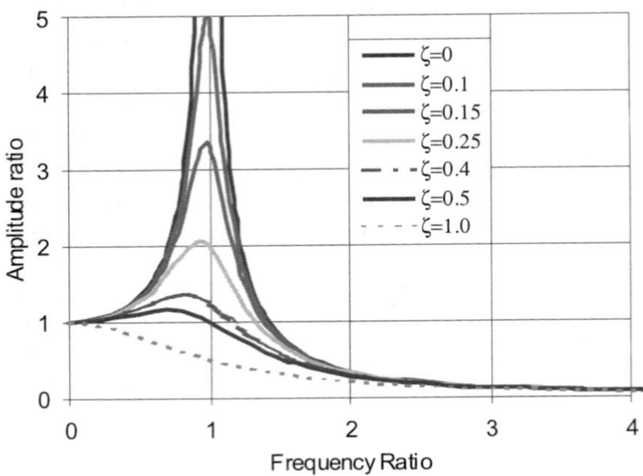

Figure 4.30 Normalized displacement amplitude of a SDOF system

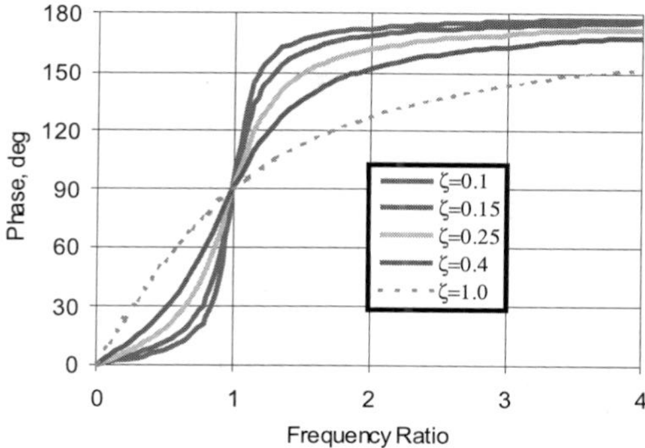

Figure 4.31 Displacement phase of a SDOF system

(i.e. undamped system, $\zeta=0.0$), the maximum amplitude occurs at the undamped natural frequency ($\omega=\omega_n$) and approaches infinity. This, of course, does not exist in nature, since some amount of damping is always present, so that the amplitude in practice is finite. Note that at resonance ($\omega=\omega_n$), the displacement amplitude may be simply computed from $X_0=X_s/(2\zeta)$ (see equation (4.64)). When the damping is large, the amplification disappears. For example, at $\zeta > 0.5$, there is practically no "hump" in the magnitude of displacement at the natural frequency.

The solution discussed above is the steady-state solution. However, when a spring–mass system is subjected to an external harmonic excitation at the start ($t=0$), its initial motion includes the transient oscillation (corresponding to free oscillation discussed in Section 4.3.1) due to the initial impact from the excitation. Therefore, the initial time history of motion will show the sum of the transient and forced oscillation. The transient dies out with time and the steady-state harmonic oscillation takes over. This decay time of free oscillation depends on the amplitude of system damping. An example of a floating moored ship-shaped structure subjected to a regular wave is shown in fig. 4.32. The surge motion of the structure shows the large transient oscillation superimposed on the sinusoidal oscillation at the imposed wave frequency. The excitation frequency of 10 s is far removed from the natural frequency of the system. The transient oscillation occurs at the damped natural period of the system. In this case, the natural period of the structure in surge is 100 s. This oscillation dies slowly since the damping in surge for the floating structure is small. Towards the end of the record, the oscillation is predominantly at the wave period.

However, the natural period in the heave for the same structure is much smaller. If the excitation period is close to the natural period, but not equal to it, then the initial response from the periodic excitation will take a form similar to fig. 4.33. Here, the natural period (e.g. in heave of the structure) is taken as 12 s and the wave period is 10 s. The response is shown for a small damping and it takes the form of a beating. The beating amplitude builds up slowly from a small amplitude to a maximum and then reduces to the small again. The beating period is based on the difference frequency between the excitation frequency and the natural frequency. In the example in fig. 4.33, the beating period is 60 s.

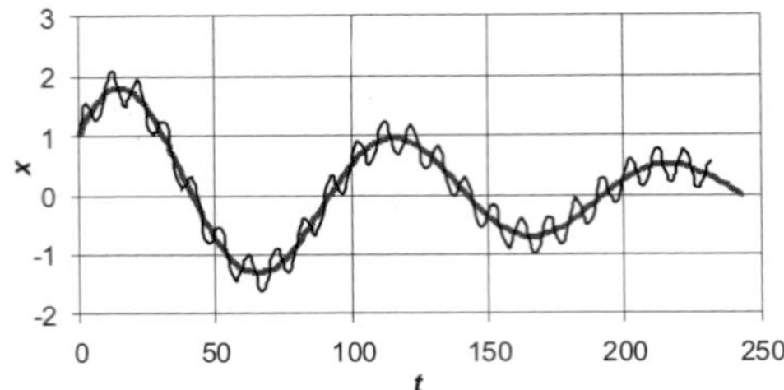

Figure 4.32 Initial surge motion of a floating structure in waves

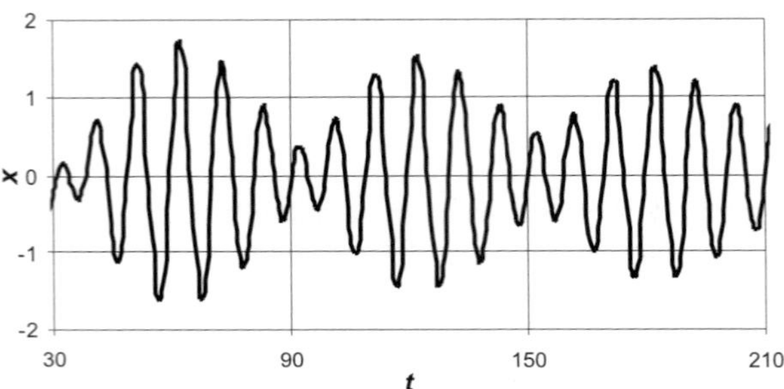

Figure 4.33 Beating effect from two close periods

4.13.4 Non-linearly Damped Structure Response

As already mentioned, the radiation damping is linear. In addition, the viscous damping has a linear contribution as well. However, the motion of the structure also introduces wake (low-pressure region) giving rise to (drag type) non-linear damping. In this case, the equation of motion (in equation 4.63) takes the form:

$$m\ddot{x}(t) + c_1\dot{x}(t) + c_2|\dot{x}(t)|\dot{x}(t) + kx(t) = F_0 \cos \omega t \tag{4.66}$$

in which c_1 is the linear damping and c_2 is the non-linear damping coefficient. The non-linearity in damping makes the solution difficult. This equation may be handled in a closed form as long as one linearises the non-linear damping term by the first term of the Fourier series expansion. Then, the linearised damping term is combined with the linear damping term giving c'. The equation of motion becomes:

$$m\ddot{x}(t) + c'\dot{x}(t) + kx(t) = F_0 \cos \omega t \tag{4.67}$$

The solution becomes harmonic (but non-linear) and obtained from the closed form expression:

$$x_0 = \frac{F_0}{\left[(k - m\omega^2)^2 + \left[\left(c_1 + \frac{8c_2}{3\pi}\omega_n x_0\right)\omega\right]^2\right]^{1/2}} \tag{4.68}$$

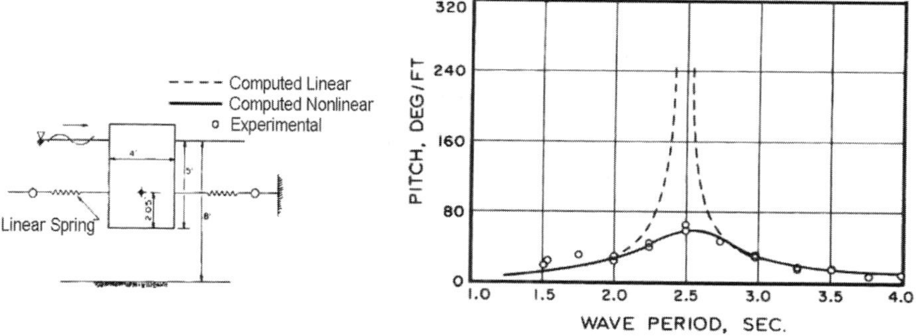

Figure 4.34 Pitch motion of a floating vertical caisson in waves

and the phase angle:

$$\tan\beta = \frac{\left(c_1 + \frac{8c_2}{3\pi}\omega_n x_0\right)\omega}{k - m\omega^2} \tag{4.69}$$

Since the amplitude of motion appears on both sides of the equation, the solution is obtained in an iterative way. The amplitude on the right-hand side of equation (4.68) is taken as zero in the initial estimate of the motion amplitude. This estimated amplitude is substituted on the right-hand side in the next iteration. This process is continued until a convergence is reached. The phase angle is then obtained from equation (4.69). There are many physical offshore systems where such non-linear damping is present. The above solution is still in closed form and provides a reasonable estimate of the response of the non-linearly damped system.

Experiments with a floating vertical cylinder in free surface waves have shown that this non-linear damping term is necessary for predicting the motions near the natural frequency. This is illustrated by an example in fig. 4.34, in which the pitch motion of a floating vertical cylinder held in place with a pair of horizontal soft linear spring lines was measured in a test with regular waves. The wave periods were chosen such that they excited the pitch natural period of the system. The motion was analysed with and without the presence of a non-linear damping term. The drag coefficient for the cylinder for the non-linear damping term was chosen from the available experimental data for the appropriate Reynolds number and the Keulegan–Carpenter number (see Section 4.7.1). It is clear that the non-linear damping term is needed in this case, at least near the natural period of the cylinder in pitch in order to match the measured data.

4.13.5 Motions of Floating Structure

Once the exciting forces and hydrodynamic added mass and damping coefficients are known, the motions of the floating structure are easy to compute. A floating offshore structure is anchored by some means to the ocean floor. An FPSO may be catenary moored. A semi-submersible in deep water is often anchored by taut mooring lines. A TLP is anchored vertically by tendons. If the stiffness of these mooring systems is considered linear, then the solution for the motions of the floating structure is straightforward. The equation of motion in this case takes on a familiar form for the six degrees of freedom. The motions are coupled because the added mass and damping matrices are 6×6. In many cases, however, this coupling is weak. The equations of motion have three terms related to the structure and include an inertia, a damping and a restoring force term. They are equated to the external force experienced by the structure from the waves.

The equations of motion for a moored floating structure is given by:

$$m_k \ddot{x}_k + \sum_{l=1}^{6} (M_{lk}\ddot{x}_l + N_{lk}\dot{x}_l + C_{lk}x_l) = F_k \exp(-i\omega t); \;\; k = 1{,}2\ldots 6 \qquad (4.70)$$

in which the quantity m_k is the structural mass matrix and M and N are the added mass and the damping matrix including cross terms. The matrix C includes hydrostatic as well as the external stiffness terms. The motion amplitudes for six DOF X_k are obtained from equation (4.70) by the inversion of the 6×6 matrix on the left-hand side.

Since the equation is linear, a frequency domain solution is possible. For a linear solution, the motions are harmonic, since the forces are. The solutions for the motion amplitudes X_k and phases ε_k

$$\sum_{l=1}^{6} [-\omega^2(m_{kl} + M_{kl}) - i\omega N_{kl} + C_{kl}]\, X_l \exp(i\varepsilon_l) = F_k; \quad k = 1{,}2\ldots 6 \qquad (4.71)$$

The calculation of these quantities is included in the aforementioned linear diffraction/radiation software.

Many offshore structures have large members as well as smaller components. These smaller components provide the non-linear drag forces. Normally, the wave exciting drag forces are small compared to the inertia forces and are not included in the analysis. The drag damping, on the other hand, may be a significant contribution to the overall damping of the system and should not be ignored. It is rather straightforward to include a velocity-squared drag term in the above equation (equation (4.70)) as shown in equation (4.66). In frequency domain analysis, this damping term is linearised by Fourier approximation so that a closed form solution may be determined. Most commercially available diffraction software mentioned earlier (table 4.10) include such terms through the input of what is sometimes called Morison elements.

When the wave drag forces are deemed important for structures that have mainly smaller members, then the relative velocity drag damping forces are included in the equation of motion. In these cases, the time domain solution is sought to preserve the non-linearity due to these terms. The non-linearity in the mooring system may also be maintained.

4.13.6 Interaction of Two Floating Structures

In practice, one often encounters two floating structures in the vicinity of each other. Examples of this type are a loading tanker in tandem to an FPSO, a supply vessel next to a TLP, a floating buoy-moored FPSO, a deck installation barge inside a platform, etc. For two large floating structures that move independently in the vicinity of one another in waves, the two-body motion problem is solved. A slight modification of the linear diffraction/radiation software for a single module handles such multiple modules as a multi-degree of freedom system. In this case, the interaction problem is a little more complex. The incoming wave, upon incident on one structure is scattered and becomes the incident wave, for the second structure. This generates a multiple scattering effect between the two structures. Similarly, for the radiated part, the motion of one structure creates radiated waves, which are incident on the second structure altering the radiated wave field just as the scattered waves. Therefore, the added mass as well as damping corresponding to each module in the two-module system generates a 12×12 matrix.

The equations of motion for the two floating structures include the mass and stiffness terms of each module and the combined added mass and damping. The right-hand side includes the exciting forces on the individual structure including interaction forces from all modules. Several software listed in table 4.10 can handle this problem.

4.13.7 Slowly-Varying Response

It is clear that for a linear system the forces at a given wave frequency yield responses of significant amplitudes at the same frequency. However, the motion of the structure at the free wave surface generates forces that may be different from the wave frequency. One such contribution arises from the combination of harmonics present in a random wave as already illustrated. Since the random waves are composed of a band of frequencies, these forces are expected to be irregular having frequencies different from the wave frequencies. As discussed before, an important contribution of such frequency combination is the slow drift forces. The motions from such forces will generally be determined from a time domain analysis.

4.13.8 Simplified Computation of Slow-Drift Oscillation

A simplified frequency-domain computation of slow drift motion and anchoring load may be made in the following way. Since the motion is critical near the natural frequency, the slow-drift force (equation (4.36)) is assumed only at the natural frequency of the floater as follows:

$$F(t) = \overline{F}_2 \cos \omega_n t \tag{4.72}$$

In the above, only the difference frequencies at resonance are considered i.e. $\omega_n = \omega_i - \omega_j$. Also, since this frequency is low for a floater, the two frequencies ω_i and ω_j are close to each other and the drift force amplitude $\overline{F}_2$ may be taken as the steady drift force at the mean frequency $(\omega_i + \omega_j)/2$. Then, the equation of motion in surge slow drift simply becomes:

$$m\ddot{x} + c\dot{x} + kx = \overline{F}_2 \cos \omega_n t \tag{4.73}$$

where x is represented by its amplitude and phase angle as

$$x = X\cos(\omega_n t - \beta) \tag{4.74}$$

At ω_n, the amplitude of motion becomes:

$$X = \frac{\overline{F}_2}{c\omega_n} \tag{4.75}$$

The mooring line load is then determined from the simple formula:

$$F_r = kX \tag{4.76}$$

This computation is illustrated by an example of a model test performed with a semi-submersible moored with linear springs. In this case the steady drift force on the semi-submersible was determined by a model test in the regular waves. The drift force in regular waves may be determined as the average mean value of the mooring line loads in the direction of waves as shown on the top curve in fig. 4.35. The normalised slow drift load on lines in several random waves is computed from equation (4.76) and compares well with the measurements vs. the spectral peak periods of the random waves.

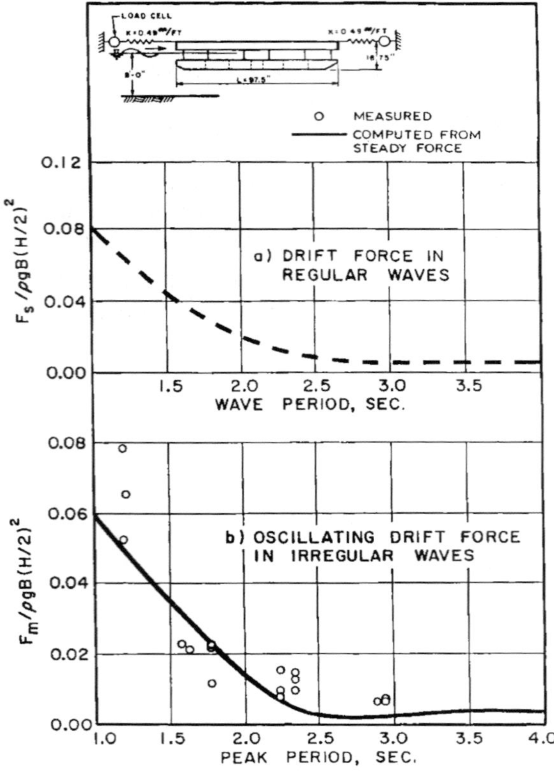

Figure 4.35 Computation of oscillating drift force on lines

4.13.9 High-Frequency Response

The sources and generation of the high-frequency forces have already been discussed along with the low-frequency forces. For stiff systems, which have low natural periods, the high frequency forces become important. The analysis for motions and response loads for such systems may be made similar to the low frequency effect described earlier. An example of such systems is the heave of a Tension Leg Platform. The heave motion of the TLP is small due to high tendon stiffness, but the tendon load becomes high.

This is illustrated with an example in which the tendon loads were measured in a TLP model test in regular and random waves. A random wave spectrum and corresponding measured tendon load spectrum are shown in fig. 4.36. This load is generally referred to as the springing load. In fig. 4.36 the wave spectrum is superimposed on the load spectrum in one plot for illustrative purpose. The y-axis is the energy spectral density with the respective units. Note that the tendon load at the wave frequency is quite small compared to the load at about twice the wave spectrum frequency. This is the sum frequency force on the tendons generated from the high-frequency force shown in equation (4.37). The high-frequency wave force will be small, but the tendon load response is quite high, because the TLP in heave has a small damping. This is illustrated in the next example.

The transfer function in regular waves for the tendon load was measured in the test. The tendon load transfer function was computed with a simple one-degree of freedom system in heave for various damping ratios. The computed results are given along with the measured load amplitudes in fig. 4.37. Note that the load amplitudes are normalised with the square of the wave amplitude. The computation was carried out for several damping factors. The damping ratio between 0.11 and 0.41% seems to correlate the measured tendon loads, which is extremely small. It is clear that the damping is the most important factor limiting the tendon load. Unfortunately, information on damping is limited.

The TLP also experiences a ringing type of response (fig. 4.38). The ringing loads in the tendon appear like an impact-type response. The source of ringing is the third-order effect

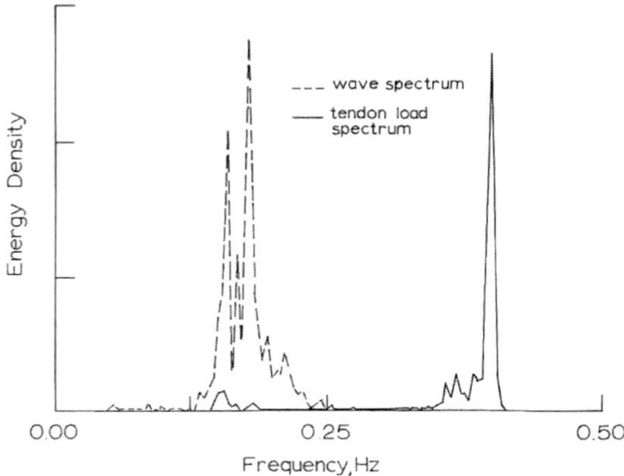

Figure 4.36 Measured TLP tendon spring load spectrum in random wave

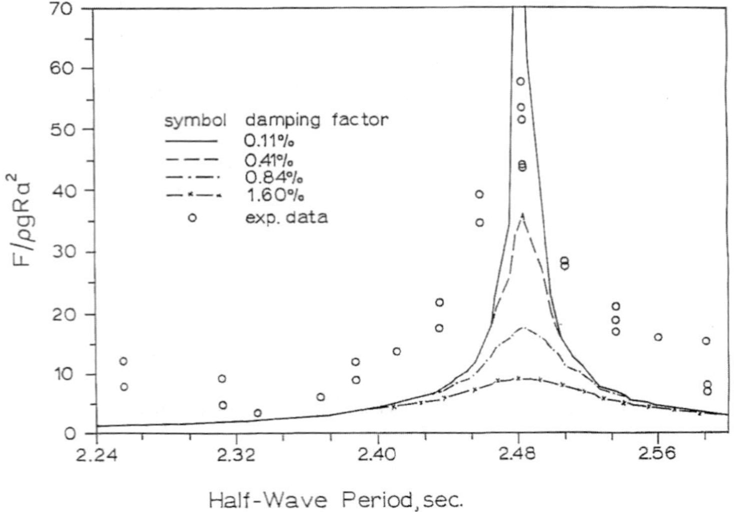

Figure 4.37 TLP tendon load response function

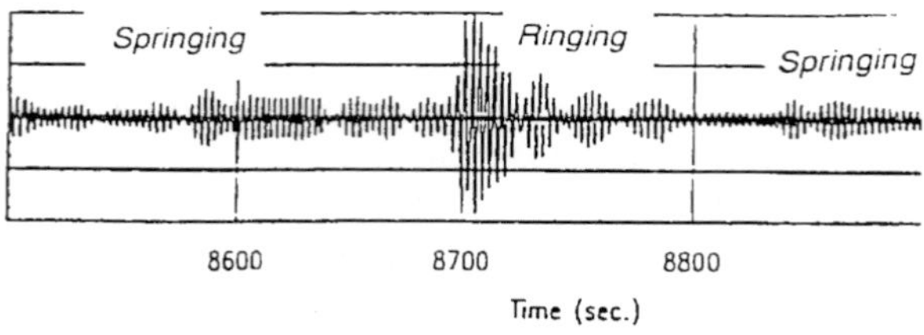

Figure 4.38 Field measured tendon load on a TLP

mainly from the run-up of waves on its columns. These loads are transient as found in fig. 4.38. For the design of the tendons, these loads are high and should be met by the tendon capacity. But they are not of much concern from the fatigue standpoint, since they are intermittent.

4.13.10 Hydrodynamic Damping for Floating Systems

There are several sources of damping besides material damping present in a floating system. These are of hydrodynamic type and are listed as follows:

- Wave Radiation damping
- Wave Viscous damping
- Wave drift damping
- Mooring line damping

The total damping of the cylindrical members in a TLP has been shown to be extremely small. The contribution from the radiation and material damping for a TLP is generally small. For example, the material damping ratio for a TLP generating from the tendon, bottom template, connectors etc. is of the order of 0.01%, whereas the wave radiation damping is about the same or slightly higher. The mooring line damping is absent for a TLP, but may be significant for other floaters (see Chapter 8).

For the members of a TLP the hydrodynamic values for the percent damping ratio are shown in table 4.13. The results were obtained experimentally on a small scale TLP model and reported by Huse (1990) and Troesch and Kim (1990).

Table 4.13 Percent damping factor for a TLP in model

Members ↓	Model Scale	Model Scale
Section →	Round	Square
Vertical column	0.049	0.049
Horizontal pontoon	0.176	0.278

Table 4.14 Key responses in the design of common floating systems

Structure	Key Components	Key Responses
Tension Leg Platforms	Deck Hull Tendons Risers Subsea templates Tendon anchors	Small heave and pitch Steady and slow drift in surge Springing of tendons Ringing of tendons Wind load on deck Riser VIV
SPAR	Deck Hard tank Midsection (shell or truss) Soft tank Mooring system Risers (w/buoyancy cans)	Large heave Steady and slow drift Large drag force on straked caisson Mooring line dynamics Caisson VIV Riser VIV
Semi-submersible	Deck Hull Mooring system Risers	Wave frequency motions Steady and slow drift Mooring line dynamics Riser VIV Wind load on deck

(Continued)

Table 4.14 Continued

Structure	Key Components	Key Responses
FPSO	Deck Ship-shaped Hull Turret mooring system Risers	Wave frequency motions Steady and slow drift in surge & pitch Mooring line dynamics Riser VIV Wind load on deck
Buoy Mooring (SBM)	Buoy Ship-shaped hull Catenary system Risers	Wave frequency motions Steady and slow drift Multi-structure interaction Mooring line dynamics Riser VIV

4.14 Applicability of Response Formula

This chapter discusses the various basic methods used to compute the loads on fixed and floating offshore structures and responses of floating structures. The design methods and formulas outlined in the preceding sections are the most common for both the fixed and the floating offshore structures. The application of these methods depends on the size of the structural components and distinction is made between a small component vs. a large component. Since an offshore structure is often composed of both these types, it is often desired to apply a hybrid method of computation based on the size of components comprising the structure.

4.14.1 Key Responses for Offshore Structures

An attempt is made here to select the most common structures and describe their key components. The key responses needed to analyse these structures are included in table 4.14. Some of the responses for the class of structures are highlighted. The responses that are present, but may not be important, are omitted.

It should be recognised that it is a very simplistic description of the uniqueness of these structures considering their dynamic responses. More details are given in Chapter 7.

References

Allen, D. W. and Henning, D. L. (2001). "Surface roughness effects on vortex-induced vibration of cylindrical structures at critical and supercritical Reynolds numbers", *Proceedings of Offshore Technology Conference*, Houston, TX, OTC 13302, May.

American Petroleum Institute (2000). "Recommended practice for planning, designing and constructing fixed offshore platforms", API-RP2A, 21st ed., Washington, DC, Dec.

American Petroleum Institute, (1996). "Recommended practice for design and analysis of station-keeping systems for floating structures", API-RP2SK, Washington, DC.

BNI, (2000). 'Mobile Offshore Base (MOB) Design Tools and Procedures – Benchmark Analysis Report', Bechtel National, Inc. San Francisco, CA.

Campbell, I. M. C. and Weynberg, P. A. (1980). "Measurement of parameters affecting slamming", Wolfson Unit Marine Technology Ind. Aerodynamics, University of Southampton, UK, Report 116.

Chakrabarti, S. K. (1987). "Hydrodynamics of offshore structures", Computational Mechanics Publication, Southampton, U.K.

Chakrabarti, S. K. (1990). "Nonlinear methods in offshore engineering", Elsevier, The Netherlands.

Chantrel, J. and Marol, P. (1987) "Subharmonic response of articulated loading platform", *Proceedings of the Sixth Conference on Offshore Mechanics and Arctic Engineering*, Houston, TX, pp. 35–43.

Hoerner, S. F. (1965). "Fluid-dynamic drag," published by the author.

Huse, E. and Utnes, T. (1994). "Springing damping of tension leg platforms", *Proceedings of 26th Offshore Technology Conference*, Paper 7446, pp. 259–267.

Huse, E. (1990). "Resonant heave damping of tension leg platforms", *Proceedings of 22nd Offshore Technology Conference*, Paper 6317, pp. 431–436.

Keulegan, G. H. and Carpenter, L. H. (1958). "Forces on cylinders and plates in an oscillating fluid", *Journal of Research of the National Bureau of Standards*, Vol. 60, No. 5, May, pp. 423–440.

Morison, J. R., O'Brien, M. P., Johnson, J. W., and Schaaf, A. S. (1950). "The force exerted by surface waves on piles", *Petroleum Transactions, American Institute of Mining and Metal Engineering*, Vol. 4, pp. 11–22

Ochi, M. K. and Shin, Y. S. (1988). "Wind turbulent spectra for design consideration of offshore structures", *Proceedings of the 20th Offshore Technology Conference*, OTC 5736, Houston, TX, pp. 461–467, May.

Ochi, M. K. and Tsai, S.-H. (1984). "Prediction of impact pressure induced by breaking waves on vertical cylinders in random waves", *Applied Ocean Research*, Vol. 6, No. 3, pp. 157–165.

Oil Companies International Marine Forum (OCIMF) (1977). "Prediction of wind and current loads on VLCCs", London.

Pinkster, J. A. (1981). "Low frequency second order wave exciting forces on floating structures", Publication No. 650, Netherlands Ship Model basin, Wageningen.

Sarpkaya, T. (1976). "In-line and transverse forces on cylinder in oscillating flow at high Reynolds number", *Proceedings of the Offshore Technology Conference*, Houston, Texas, OTC 2533, pp. 95–108.

Sarpkaya, T. (1978). "Wave impact loads on cylinders", *Offshore Technology Conference*, OTC Paper No. 3775, Houston, TX.

Sarpkaya, T. (1983). "A critical assessment of Morison's equation", *Proceedings of the 15th Offshore Technology Conference*, Houston, TX.

Schlichting, H. (1968). "Boundary layer theory", (6th ed.). McGraw-Hill Book Co., New York.

Taylor, P. H. (1991). "Current blockage: reduced forces on offshore space-frame structures", *Proceedings of the Offshore Technology Conference*, OTC 6519.

Thiagarajan, K. P. and Troesch, A. W. (1994). "Hydrodynamic heave damping estimation and scaling for tension leg platforms", *Journal of Offshore Mechanics and Arctic Engineering*, Vol. 116, No. 2, pp. 70–76.

Handbook of Offshore Engineering
S. Chakrabarti (Ed.)

Chapter 5

Probabilistic Design of Offshore Structures

Arvid Naess and Torgeir Moan
Centre for Ships and Ocean Structures,
Norwegian University of Science and Technology, Trondheim, Norway

5.1 Application of Statistics in Offshore Engineering

The design of structures to withstand the effect of environmental forces due to wind and ocean waves, typical of offshore structures, invariably has to take into account the irregular or random nature of these kinds of loads. To establish a rational design procedure, it is therefore necessary to use a statistical approach of some sort. The design procedures and codes adopted to accommodate the uncertainty in the wave loads are traditionally based on one of the three types of approaches, commonly referred to as the design wave, the design storm and the long-term design. These approaches will be discussed separately in section 5.4. Moreover, this design procedure should also account for the other uncertainties inherent in loads and resistances by using a structural reliability approach, as described in Sections 5.6–5.11. Before we embark on these topics, it is expedient to discuss in some detail the statistics of ocean waves and some of their effects.

5.2 Wave Statistics

The standard approach to the statistical modelling of ocean waves in engineering applications is to assume that the ocean surface constitutes a stochastic wave field that can be assumed stationary in time and homogeneous in space. In practice, stationarity is assumed only for limited periods of time, like three hours, which is then referred to as the short-term description of the wave field. This says that we are assuming steady state wave conditions for specified intervals of time with no transition periods between different sea states. This is clearly an approximation, but for most offshore design applications it seems to work well.

Usually, the wave field is assumed to be Gaussian, which has been verified to give a reasonably good approximation to reality. This assumption implies, for example that the wave elevation at a location on the sea surface can be represented by a Gaussian random variable. The Gaussian probability distribution therefore plays a central role in the assessment of loads and responses of offshore structures to ocean waves. It should be emphasised, however, that while the assumption about a Gaussian wave field is an acceptable one in many cases, especially for the calculation of the statistics of the dynamic response of compliant offshore structures, this is not always the case. It is a recognised fact that the crests of ocean waves are generally larger than the corresponding troughs. For problems relating to for example, a wave slamming into the deck structure of offshore platforms, this effect is of importance.

A Gaussian sea surface implies that the stationary wave field is fully described by a directional wave spectrum and a dispersion relation specifying the relationship between the (scalar) wave number and the wave frequency. For deep water waves, which is assumed here, this relationship is given by the formula

$$\omega^2 = g\,k \tag{5.1}$$

where ω denotes (circular) wave frequency, k denotes the wave number, and g equals the gravitational acceleration.

The wave field $X(t; x, y)$, relative to an inertial frame of reference (x, y, z), with $z = 0$ as the mean free surface, which is specified by a directional wave spectrum $S_X(\omega, \theta)$, can then be represented approximately as follows:

$$\begin{aligned} X(t;\, x,y) = \sum_{j=1}^{n}\sum_{k=1}^{m} & \sqrt{S_X(\omega_j, \theta_k)\,\Delta\omega_j\,\Delta\theta_k} \\ & \times \left[A_{jk} \cos\left(\omega_j t - \frac{\omega_j^2}{g}(x\cos\theta_k + y\sin\theta_k)\right)\right. \\ & \left. + B_{jk} \sin\left(\omega_j t - \frac{\omega_j^2}{g}(x\cos\theta_k + y\sin\theta_k)\right)\right] \end{aligned} \tag{5.2}$$

where $0 < \omega_1 < \cdots < \omega_n$ and $-\pi \le \theta_1 < \cdots < \theta_m \le \pi$ are discretisations of the frequency axis and the directional range respectively. $\{A_{jk}\}$ and $\{B_{jk}\}$ are independent sets of independent zero-mean Gaussian variables of unit variance.

As discussed in Chapter 2, the directional wave spectrum employed in offshore engineering is usually written as $S_X(\omega, \theta) = S_X(\omega)\, D(\omega, \theta)$, where $S_X(\omega)$ is chosen from a list of standard wave spectra, and $D(\omega, \theta)$ is a directional function . Typically for offshore applications, the spectral density function $S_X(\omega)$ is fully specified by the significant wave height H_s and for example, the peak spectral period T_p, cf. Chapter 2. As was also evident in Chapter 2, where various expressions for the directional function were given, in practice one usually adopts the approximation $D(\omega, \theta) = D(\theta)$, that is, the frequency dependence of the directional function is generally neglected.

In complex number notation, equation (5.2) can be rewritten in a more convenient and compact form as

$$X(t; x, y) = \Re\Big\{ \sum_{j=1}^{n} \sum_{k=1}^{m} \sqrt{S_X(\omega_j, \theta_k)\, \Delta\omega_j\, \Delta\theta_k} \times C_{jk} \exp\Big(i\omega_j t - i\frac{\omega_j^2}{g}(x \cos\theta_k + y \sin\theta_k)\Big)\Big\} \tag{5.3}$$

where $\Re\{z\}$ denotes the real part of the complex number z, $i = \sqrt{-1}$, and $C_{jk} = A_{jk} + iB_{jk}$.

The ocean surface wave elevation is usually specified relative to some chosen reference point, e.g. the origin of the inertial coordinate system. Let $X(t) = X(t; 0, 0)$ denote the wave elevation at the reference point. Hence

$$X(t) = \sum_{j=1}^{n} \sum_{k=1}^{m} \sqrt{S_X(\omega_j, \theta_k)\, \Delta\omega_j\, \Delta\theta_k} \left[A_{jk} \cos(\omega_j t) + B_{jk} \sin(\omega_j t)\right] = \Re\left\{ \sum_{j=1}^{n} \sum_{k=1}^{m} \sqrt{S_X(\omega_j, \theta_k)\, \Delta\omega_j\, \Delta\theta_k}\, C_{jk}\, e^{i\omega_j t} \right\} \tag{5.4}$$

Note that an equivalent version of equation (5.4) is the following expression, which is also widely quoted.

$$X(t) = \sum_{j=1}^{n} \sum_{k=1}^{m} \sqrt{S_X(\omega_j, \theta_k)\, \Delta\omega_j\, \Delta\theta_k}\, R_{jk} \cos(\omega_j t - \varepsilon_{jk}) \tag{5.5}$$

where $R_{jk}^2 = A_{jk}^2 + B_{jk}^2$, that is, $\{R_{jk}\}$ is a set of independent Rayleigh distributed variables, which is independent of the set of independent random variables $\{\varepsilon_{jk}\}$. Each ε_{jk} is uniformly distributed over the interval $(0, 2\pi)$, or equivalently, $(-\pi, \pi)$, cf. Chapters 2 and 12.

A simplified, approximate version of this formula is used extensively in offshore engineering in the following form:

$$\tilde{X}(t) = \sum_{j=1}^{n} \sum_{k=1}^{m} \sqrt{2\, S_X(\omega_j, \theta_k)\, \Delta\omega_j\, \Delta\theta_k} \cos(\omega_j t - \varepsilon_{jk}) \tag{5.6}$$

Note that $\tilde{X}(t)$ is never exactly a Gaussian random variable since the Rayleigh variables R_{jk} have been replaced by the constant $\sqrt{2}$, but due to the central limit theorem, it can be expected that $\tilde{X}(t)$ is approximately Gaussian when nm is large enough.

Each term in the sums of the expressions above is a periodic function. Hence there is a limit to how long the simulation time should be when such an expression is used. If we let $\Delta\omega_{\min}$ denote the smallest and $\Delta\omega_{\max}$ the largest of the $\Delta\omega_j$, $j = 1, \ldots, n$, then the time interval of validity of the expressions above will approximately be between $(0, 2\pi/\Delta\omega_{\max})$ and $(0, 2\pi/\Delta\omega_{\min})$, with $(0, 2\pi/\Delta\omega_{\max})$ as a first, simple estimate.

The expression for the wave elevation or any other quantity that is linearly related to it, can be obtained at any other relevant point from the expression given in equation (5.4) by using the linear transfer function that specifies the relationship. Let the linear transfer function relating the wave elevation at the reference point from a regular wave train propagating in the θ-direction to the desired quantity related to the point (x, y, z) $(z \leq 0)$, be denoted by $\hat{H}(\omega; \theta)$. Using this transfer function, the corresponding stochastic process, $Y(t; x, y, z)$ say, at the point (x, y, z) corresponding to the wave elevation process $X(t)$, is then calculated as follows:

$$Y(t; x, y, z) = \Re\left\{\sum_{j=1}^{n}\sum_{k=1}^{m} \hat{H}(\omega_j; \theta_k)\sqrt{S_X(\omega_j, \theta_k)\,\Delta\omega_j\,\Delta\theta_k}\; C_{jk}\, e^{i\omega_j t}\right\} \tag{5.7}$$

The spectral density $S_Y(\omega)$ of the stationary process $Y(t; x, y, z)$ is given approximately by the following equation

$$S_Y(\omega) = \sum_{k=1}^{m} |\hat{H}(\omega; \theta_k)|^2\, S_X(\omega, \theta_k)\,\Delta\theta_k \tag{5.8}$$

Let us look at two examples. For simplicity, we consider the case of no current. In the case of a regular wave train propagating in the direction θ, the linear transfer function relating the wave elevation at the reference point to the water particle velocity in the x-direction at the point (x, y, z) $(z \leq 0)$ is $\hat{H}_{XV_x}(\omega; \theta) = \omega \cos\theta\, e^{(\omega^2/g)z}\, e^{-i(\omega^2/g)(x\cos\theta + y\sin\theta)}$, when the x-direction corresponds to $\theta = 0$, cf. Chapter 2. According to equation (5.7), the water particle velocity $V_x(t; x, y, z)$ in the x-direction at the point (x, y, z) corresponding to the wave elevation process $X(t)$, is then calculated simply as follows

$$V_x(t; x, y, z) = \Re\left\{\sum_{j=1}^{n}\sum_{k=1}^{m} \omega_j \cos\theta_k \sqrt{S_X(\omega_j, \theta_k)\,\Delta\omega_j\,\Delta\theta_k} \times C_{jk} e^{(\omega_j^2/g)z} e^{i\left[\omega_j t - (\omega_j^2/g)(x\cos\theta_k + y\sin\theta_k)\right]}\right\} \tag{5.9}$$

Similarly, the water particle acceleration $A_x(t; x, y, z)$ in the x-direction at the point(x, y, z) is given by the following expression:

$$A_x(t; x, y, z) = \Re\left\{\sum_{j=1}^{n}\sum_{k=1}^{m} i\,\omega_j^2 \cos\theta_k \sqrt{S_X(\omega_j, \theta_k)\,\Delta\omega_j\,\Delta\theta_k} \times C_{jk}\, e^{(\omega_j^2/g)z}\, e^{i\left[\omega_j t - (\omega_j^2/g)(x\cos\theta_k + y\sin\theta_k)\right]}\right\} \tag{5.10}$$

The corresponding spectral densities are, according to equation (5.8), given as follows

$$S_{V_x}(\omega) = \sum_{k=1}^{m} \omega^2 \cos^2\theta_k\; e^{(2\omega^2/g)z}\; S_X(\omega, \theta_k)\,\Delta\theta_k \tag{5.11}$$

and

$$S_{A_x}(\omega) = \sum_{k=1}^{m} \omega^4 \cos^2\theta_k\; e^{(2\omega^2/g)z}\; S_X(\omega, \theta_k)\,\Delta\theta_k \tag{5.12}$$

5.2.1 The Gaussian Distribution

The $\{A_{jk}\}$ and $\{B_{jk}\}$ are independent sets of independent Gaussian variables, and the wave elevation $X(t)$ is a linear sum in these variables. The sum of independent Gaussian variables is again a Gaussian variable. Therefore, $X(t)$ is also a Gaussian variable for each point in time t. In fact, $X(t)$ is a stationary Gaussian process. From the fact that $E[A_{jk}] = E[B_{jk}] = 0$, it follows that $E[X(t)] = 0$. The variance is obtained from the spectral density by the formula

$$\sigma_X^2 = \mathrm{Var}[X(t)] = \int_{-\pi}^{\pi} \int_0^{\infty} S_X(\omega, \theta)\, \mathrm{d}\omega\, \mathrm{d}\theta \approx \sum_{j=1}^{n} \sum_{k=1}^{m} S_X(\omega_j, \theta_k)\, \Delta\omega_j\, \Delta\theta_k \tag{5.13}$$

The statistical distribution of the wave elevation $X = X(t)$ is therefore determined by the probability density function (PDF), denoted by $f_X(x)$, of a normal or Gaussian random variable of mean value zero and standard deviation σ_X, which is given by the equation [Papoulis, 1965]

$$f_X(x) = \frac{1}{\sqrt{2\pi}\sigma_X} \exp\left(-\frac{x^2}{2\sigma_X^2}\right) \tag{5.14}$$

The corresponding cumulative distribution function (CDF), denoted by $F_X(x)$, is given as follows

$$F_X(x) = \mathrm{Prob}\,[X(t) \le x] = \int_{-\infty}^{x} \frac{1}{\sqrt{2\pi}\sigma_X} \exp\left(-\frac{s^2}{2\sigma_X^2}\right) \mathrm{d}s \tag{5.15}$$

where $\mathrm{Prob}\,[E]$ denotes the probability of the event E.

A Gaussian random variable of mean value zero and standard deviation 1.0 is referred to as a standard Gaussian variable. Its PDF is often denoted by $\phi(x)$ and its CDF by $\Phi(x)$.

In fig. 5.1 is plotted, the PDF and the CDF of the standard Gaussian variable as given by equation (5.14) with $\sigma_X = 1.0$.

As with X, $V_x = V_x(t; x, y, z)$ and $A_x = A_x(t; x, y, z)$ are both zero-mean Gaussian random variables for every fixed time t. To fully specify their PDFs, we only need to calculate their variance, which is given by the expressions

$$\begin{aligned} \sigma_{V_x}^2 = \mathrm{Var}[V_x] &= \int_0^{\infty} S_{V_x}(\omega)\, \mathrm{d}\omega \\ &\approx \sum_{j=1}^{n} \sum_{k=1}^{m} \omega_j^2 \cos^2\theta_k\, e^{(2\omega_j^2/gz)}\, S_X(\omega_j, \theta_k)\, \Delta\omega_j\, \Delta\theta_k \end{aligned} \tag{5.16}$$

and

$$\begin{aligned} \sigma_{A_x}^2 = \mathrm{Var}[A_x] &= \int_0^{\infty} S_{A_x}(\omega)\, \mathrm{d}\omega \\ &\approx \sum_{j=1}^{n} \sum_{k=1}^{m} \omega_j^4 \cos^2\theta_k\, e^{(2\omega_j^2/gz)}\, S_X(\omega_j, \theta_k)\, \Delta\omega_j\, \Delta\theta_k \end{aligned} \tag{5.17}$$

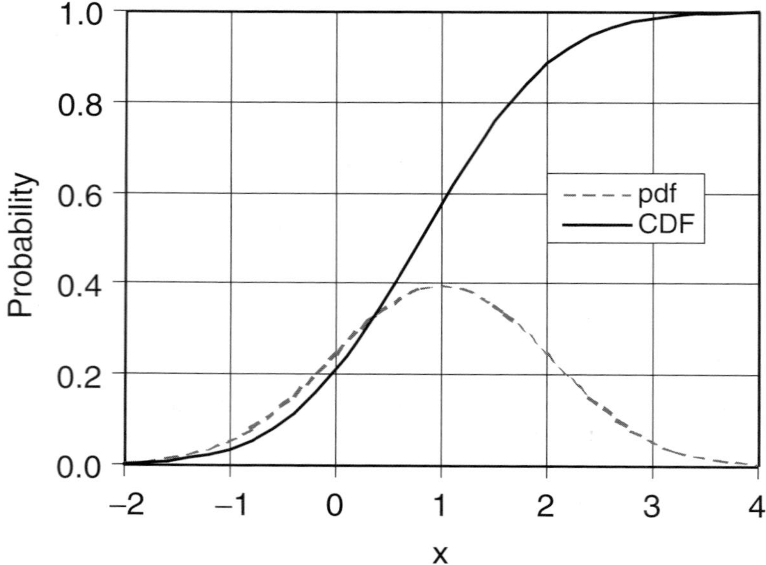

Figure 5.1 Normalized Gaussian Distribution

5.2.2 The Rayleigh Distribution

As we have just seen, the stochastic model adopted for the water waves leads to a normally distributed sea surface elevation. In many practical applications, it is not the instantaneous surface elevation that is of most interest, but rather the individual waves. So there are two immediate questions to answer: What is the statistical distribution of the wave heights? What is the distribution of the wave crests?

The answers we shall give will be based upon the notion of mean level upcrossing rate of a stationary stochastic process. The average number of times per unit time that the zero-mean, stationary Gaussian process $X(t)$ crosses the level a with a positive slope, which we shall denote by $\nu_X^+(a)$, can be shown to be given by the formula [Lin, 1967]:

$$\nu_X^+(a) = \frac{\sigma_{\dot{X}}}{2\pi\sigma_X}\exp\left(-\frac{1}{2}\frac{a^2}{\sigma_X^2}\right) \tag{5.18}$$

where $\sigma_X^2 = \mathrm{Var}[X]$ and $\sigma_{\dot{X}}^2 = \mathrm{Var}[\dot{X}]$. If $S_X(\omega)$ denotes the one-sided spectral density of $X(t)$, then

$$\sigma_X^2 = \int_0^\infty S_X(\omega)\,\mathrm{d}\omega \tag{5.19}$$

and

$$\sigma_{\dot{X}}^2 = \int_0^\infty \omega^2 S_X(\omega)\,\mathrm{d}\omega \tag{5.20}$$

The following expression is a widely adopted notation for the spectral moments

$$m_j = \int_0^\infty \omega^j S_X(\omega)\,\mathrm{d}\omega, \qquad j = 0, 1, 2, \ldots \tag{5.21}$$

It is quite common to express the spectral density in terms of frequency f in Hz. If $G_X(f)$ denotes the appropriate spectral density in Hz, then the following relation must be satisfied to preserve variance: $S_X(\omega)\,\mathrm{d}\omega = G_X(f)\,\mathrm{d}f$. Thus the spectral moments

$$\tilde{m}_j = \int_0^\infty f^j G_X(f)\,\mathrm{d}f, \qquad j = 0, 1, 2, \ldots \tag{5.22}$$

satisfies the relation $m_j = (2\pi)^j \tilde{m}_j$.

It is seen that $\nu_X^+(a)$ decreases quickly (with σ_X as measuring unit) on either side of the mean value zero, where it assumes its largest value

$$\nu_X^+(0) = \frac{\sigma_{\dot{X}}}{2\pi\sigma_X} = \frac{1}{2\pi}\sqrt{\frac{m_2}{m_0}} = \sqrt{\frac{\tilde{m}_2}{\tilde{m}_0}} \tag{5.23}$$

$\nu_X^+(0)$ is called the *mean zero-upcrossing rate*. A corresponding parameter one often comes across is its inverse, commonly referred to as the *mean zero-crossing period T_z*, which is then defined as

$$T_z = \frac{2\pi\sigma_X}{\sigma_{\dot{X}}} = 2\pi\sqrt{\frac{m_0}{m_2}} = \sqrt{\frac{\tilde{m}_0}{\tilde{m}_2}} \tag{5.24}$$

Hence it is important to know which spectral moments have been calculated.

Let X_p denote the size/height of an arbitrary peak of $X(t)$ relative to the zero level, that is, the wave crest height. X_p becomes a random variable. For a narrow band process, where there is typically only one peak between an upcrossing and the subsequent downcrossing of zero, a reasonable definition of the probability distribution of X_p is obtained by the relation

$$\mathrm{Prob}[X_p > a] = \frac{\nu_X^+(a)}{\nu_X^+(0)} \quad (a \geq 0) \tag{5.25}$$

The corresponding CDF then becomes

$$F_{X_p}(a) = 1 - \frac{\nu_X^+(a)}{\nu_X^+(0)} \quad (a \geq 0) \tag{5.26}$$

By substituting from equation (5.18), it is found that

$$F_{X_p}(a) = 1 - \exp\left(-\frac{1}{2}\frac{a^2}{\sigma_X^2}\right) \quad (a \geq 0) \tag{5.27}$$

The PDF is obtained as

$$f_{X_p}(a) = \frac{a}{\sigma_X^2}\exp\left(-\frac{1}{2}\frac{a^2}{\sigma_X^2}\right) \quad (a \geq 0) \tag{5.28}$$

Hence, the wave crest height X_p in a Gaussian seaway $X(t)$ is a *Rayleigh* distributed random variable [Papoulis, 1965] if $X(t)$ is a narrow band process. If the individual waves are defined as zero crossing waves, equations (5.27) and (5.28) can be adopted as valid expressions for the statistical distribution of wave crest heights also for non-narrow band wave processes.

The mean value and variance of X_p are $E[X_p] = \sqrt{\pi/2}\,\sigma_X$ and $\mathrm{Var}[X_p] = 2(1-\pi/4)\,\sigma_X^2$, respectively. In fig. 5.2, is plotted the PDF of the Rayleigh variable as given by equation (5.28) with $\sigma_X = 1.0$.

To avoid heavy wave loads on the deck structure of offshore platforms, the wave crest height becomes a parameter of particular importance. Already for some time, the knowledge that wave crests of real ocean waves tend to be higher above the mean water level than the subsequent trough which is below, has been taken into account in the practical design work. What this means in practical terms is that our Rayleigh approximation to the distribution of wave crest heights is not accurate enough for some applications. To account for the observation that the ocean surface process is typically positively skewed with higher crests and shallower troughs than expected under the Gaussian assumption, an empirical correction to the Rayleigh distribution of the wave

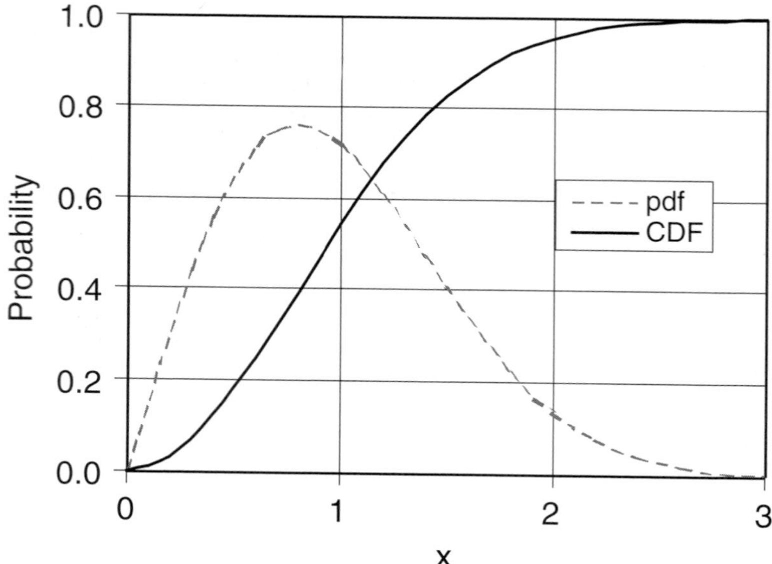

Figure 5.2 Normalized Rayleigh Distribution

crests was presented already some 30 years ago by Jahns and Wheeler (1972). Their proposed distribution can be written as

$$F_{X_p}(a) = 1 - \exp\left\{ -8\frac{a^2}{h_s^2}\left[1 - \beta_1 \frac{a}{d}\left(\beta_2 - \frac{a}{d}\right)\right]\right\} \quad (a \geq 0) \tag{5.29}$$

Here h_s denotes the significant wave height, d denotes the water depth, β_1 and β_2 are empirical coefficients. $\beta_1 = 4.37$ and $\beta_2 = 0.57$ are recommended by Haring and Heideman (1978). Note that in the narrow band limit, $h_s = 4\sigma_X$.

A more recent proposal is due to Forristall (2000), which is based on simulation data for a full second-order model of the wave elevation. From the analysis of these data, he cites a two-parameter Weibull distribution as the short-term model for the wave crest heights:

$$F_{X_p}(a) = 1 - \exp\left\{ -\left(\frac{a}{\alpha_F h_s}\right)^{\beta_F}\right\} \quad (a \geq 0) \tag{5.30}$$

where h_s is the significant wave height, and the two parameters α_F and β_F are given as follows:

$$\alpha_F = 0.3536 + 0.2892\, s_1 + 0.1060\, Ur \tag{5.31}$$

$$\beta_F = 2 - 2.1597\, s_1 + 0.0968\, Ur^2 \tag{5.32}$$

for the case of long-crested waves. Similar expressions are given for short-crested waves [Forristall, 2000]. s_1 is a measure of steepness and is given by

$$s_1 = \frac{2\pi h_s}{g t_1^2} \tag{5.33}$$

where t_1 is the mean wave period determined by the first two moments of the wave spectrum. The Ursell number Ur, which is a measure of the influence of the water depth on the non-linearity of waves, is given by the formula

$$Ur = \frac{h_s}{k_1^2 d^3} \tag{5.34}$$

where k_1 is the wave number corresponding to wave period t_1, and d is the water depth.

The statistical distribution of the wave height, denoted by H, is somewhat more involved than for the wave crest height since the usual definition of the wave height involves a wave crest and the subsequent wave trough, which occurs half a wave period later. This time delay introduces some difficulties, which are much simplified in the case of a narrow-banded wave process. This simplification is achieved by adopting the approximation that the wave crest and the subsequent wave trough have the same size. Hence, for a narrow-banded wave process, the approximation $H = 2\,X_p$ is often adopted. In fact, for many years the most commonly adopted CDF for the crest to trough wave heights in a random sea way has been the following Rayleigh distribution

$$F_H(h) = 1 - \exp\left(-\frac{h^2}{8\,\sigma_X^2}\right) \quad (h \geq 0) \tag{5.35}$$

Its simplicity and the fact that it agrees reasonably well with observations are probably the causes for its widespread use. However, high quality ocean wave data have shown significant differences between equation (5.35) and the empirical distributions [Haring, et al 1976; Forristall, 1978]. To improve the fit, Forristall proposed to use the following two-parameter Weibull distribution

$$F_H(h) = 1 - \exp\left\{-\frac{1}{\alpha_f}\left(\frac{h}{\sigma_X}\right)^{\beta_f}\right\} \quad (h \geq 0) \tag{5.36}$$

with parameters $\alpha_f = 8.42$ and $\beta = 2.126$.

The observation that the Rayleigh distribution above does not fit real wave data is not surprising. In fact, in his now classical paper of 1952, Longuet-Higgins shows that the CDF of the crest to trough wave heights in a Gaussian sea way of narrow but finite bandwidth, is given by:

$$F_H(h) = 1 - \exp\left(-\frac{h^2}{\overline{h}^2}\right) \quad (h \geq 0) \tag{5.37}$$

where $\overline{h}$ denotes the rms crest to trough wave height.

For an infinitely narrow band process, $\overline{h}^2 = 8\,\sigma_X^2$. In the case of finite bandwidth [Longuet-Higgins, 1980; Larsen, 1981; Boccotti, 1982] $\overline{h}^2 < 8\,\sigma_X^2$. Equation (5.35) therefore over-predicts the true values, which is exactly what has been observed. Longuet-Higgins (1980) showed that a good agreement can be achieved between equation (5.37) and the empirical distribution equation (5.36) if $\overline{h}$ is chosen properly. To account for the effect of finite bandwidth, Naess (1985b) proposed the following CDF for the crest to trough wave height

$$F_H(h) = 1 - \exp\left(-\frac{h^2}{4\sigma_X^2(1 - r')}\right) \quad (h \geq 0) \tag{5.38}$$

where $r' = R_X(T_d/2)/\sigma_X^2$, and the autocorrelation function R_X is given as

$$R_X(\tau) = \int_0^\infty S_X(\omega)\cos\omega\tau\, d\omega \tag{5.39}$$

T_d denotes the dominant wave period chosen so that $R'_X(T_d) = 0$, that is, T_d denotes the time when the autocorrelation function $R_X(\tau)$ assumes its first maximum after $\tau = 0$.

For most sea states of some severity and without significant swell, it has been observed that r' assumes values in the interval $-0.75 \leq r' \leq -0.65$ [Naess, 1985a]. Good agreement is obtained between the distribution given by equation (5.38) and the empirical distribution equation (5.36) based on the storm data from the Gulf of Mexico if the parameter $r' = -0.71$, which is a reasonable value to expect. Experience with the storm data from the North Sea indicates a typical peakedness parameter in the JONSWAP spectrum slightly less than 3, which corresponds to $r' \approx -0.71$ [Naess, 1985a]. A discussion of the accuracy of various proposed wave height distributions in mixed sea states, including those presented here, has recently been published by Rodriguez et al (2002).

5.3 Response Statistics

If the relation between the forces on a structure and its response, expressed by a specific quantity like a displacement response or a stress response, is described by a linear, time-invariant system model, then the relation is completely specified by a linear transfer function $H(\omega)$, which is a function only of the frequency. When such a model can be adopted, the response will be a stationary stochastic process if the force is modeled by such a process, and most importantly, the relationship between the force and the response is quite simple. To exemplify this, let $F(t)$ denote a stationary force process with a spectral density $S_F(\omega)$. Then the response, denoted by $Z(t)$, has a spectral density $S_Z(\omega)$ given by the simple relation

$$S_Z(\omega) = |H_{FZ}(\omega)|^2 \, S_F(\omega) \tag{5.40}$$

where $H_{FZ}(\omega)$ denotes the transfer function between $F(t)$ and $Z(t)$. The relationship between the mean values of load and response is given by the equation

$$E[Z(t)] = H_{FZ}(0) \, E[F(t)] \tag{5.41}$$

If the forces $F(t)$ on the structure can be approximated as first-order hydrodynamic forces, the relationship between the wave process and the forces is also fully described by a transfer function, $H_{XF}(\omega)$ say, where the index X may refer to the wave elevation $X(t)$ at some reference point as discussed previously in this chapter. In such a case the spectral density of the response can be expressed as

$$S_Z(\omega) = |H_{XF}(\omega)|^2 \, |H_{FZ}(\omega)|^2 S_X(\omega) \tag{5.42}$$

where $S_X(\omega)$ denotes the wave spectrum.

Also, if the wave elevation is assumed to be a stationary Gaussian process, the same is true of the response process. Hence, by using equations (5.41) and (5.42), it is straight-forward to calculate the statistics of the response. This also applies to the prediction of extreme response levels, as we shall see in the next section.

Example

The calculation procedure of transfer functions and response spectra will be illustrated by a somewhat simplified example of an offshore installation operation. In fig. 5.3, a crane vessel is shown, which is in the process of installing subsea equipment on a bottom-mounted frame. It is assumed that the vessel is able to maintain its horizontal position, for instance by the use of an automatic positioning system. It is therefore assumed that the vertical motion of the equipment module is due to the heave and roll motion of the vessel, if the crane is locked.

To be able to carry out the response calculations of this system, we have to know how to calculate the forces on the crane vessel. We shall limit ourselves to the wave forces, and assume that these hydrodynamic forces are linearly related to the waves, that is, as described by a linear transfer function. Assume for simplicity, the situation of long-crested, beam sea waves propagating in the x-direction. This means that a regular ocean wave represented as

$$\eta(t) = Re\{e^{i(\omega t - kx)}\} \quad (k = \omega^2/g) \tag{5.43}$$

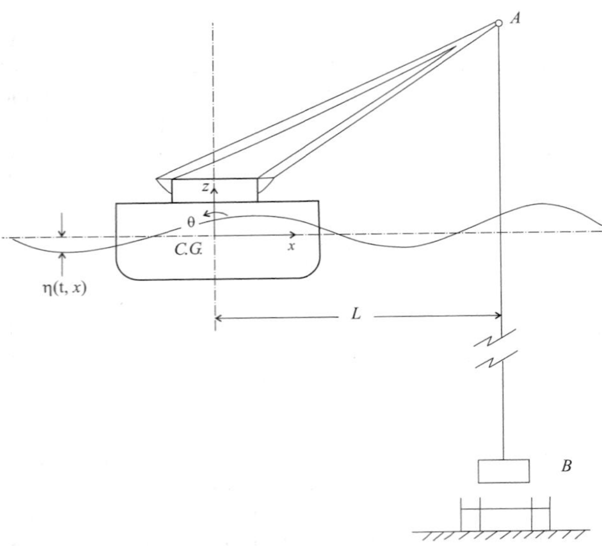

Figure 5.3 A crane vessel for installation of subsea equipment

gives a vertical heave force on the vessel, referred to the centre of gravity C.G. ($x=0$), which can be written as

$$f_z(t) = Re\{H_z(\omega)\, e^{i\omega t}\} \tag{5.44}$$

and a corresponding roll moment about C.G. given by

$$m_\theta(t) = Re\{H_\theta(\omega)\, e^{i\omega t}\} \tag{5.45}$$

It is thereby seen that $H_z(\omega)$ is the transfer function between the waves (referred to $x=0$) and the heave force, while $H_\theta(\omega)$ is the transfer function between the waves and the roll moment. It is assumed that the installation operation can only be carried out in small to moderate seas so that in particular the rolling motions of the vessel can be assumed to be small. Hence the following linear equations of motion for the heave and roll response $z(t)$ and $\theta(t)$ respectively, are adopted

$$M\,\ddot{z}(t) + c_z\,\dot{z}(t) + k_z\,z(t) + \int_0^\infty g_z(s)\,z(t-s)\,\mathrm{d}s = f_z(t) \tag{5.46}$$

and

$$I\ddot{\theta}(t) + c_\theta\,\dot{\theta}(t) + k_\theta\,\theta(t) + \int_0^\infty g_\theta(s)\,\theta(t-s)\,\mathrm{d}s = m_\theta(t) \tag{5.47}$$

$f_z(t)$ and $m_\theta(t)$ denote the general heave force and roll moment on the vessel, both with reference to C.G. M, I, c_j, k_j ($j = z, \theta$) are positive constants, and their interpretation needs no further elaboration. The reason for the Duhamel integral in each of the two equations of motion is due to the fluid structure interaction. This gives rise to an added mass term for the heave motion and an added mass moment of inertia for the roll motion. The motion of the vessel itself will also generate waves emanating from the vessel. This will induce added damping. Both these effects are taken care of by the two Duhamel integrals. This connection is brought to the fore by taking the Fourier transform of the two impulse response functions $g_z(t)$ and $g_\theta(t)$. It can be shown that

$$\int_0^\infty g_j(t)\, e^{-i\omega t} \mathrm{d}t = -\omega^2 A_j(\omega) + i\,\omega\, B_j(\omega) \quad (j = z, \theta) \tag{5.48}$$

where $A_j(\omega)$ and $B_j(\omega)$ are frequency-dependent added mass and damping terms respectively, due to the interaction effects between the fluid and the floating vessel. Both are real functions of the frequency ω, and computer programs to calculate these terms are standard commercial software.

On the basis of the equations of motion for the heave and roll motions, it is seen that there is a linear, time-invariant relationship between the wave elevation process at $x = 0$ and the corresponding heave and roll response of the crane vessel. Let us determine the associated linear transfer functions, denoted by $G_z(\omega)$ and $G_\theta(\omega)$. By exploiting the standard recipe for calculating transfer functions, we know that by putting $\eta(t) = e^{i\omega t}$, it is obtained that $z(t) = G_z(\omega)\, e^{i\omega t}$ and $f_z(t) = H_z(\omega)\, e^{i\omega t}$. By substituting this into equation (5.46) it is found that

$$\left(-\omega^2 M + i\,\omega\, c_z + k_z + \int_0^\infty g_z(s)\, e^{-i\omega s} \mathrm{d}s \right) G_z(\omega)\, e^{i\omega t} = H_z(\omega)\, e^{i\omega t} \tag{5.49}$$

Combining this equation with equation (5.48), it follows that

$$G_z(\omega) = \frac{H_z(\omega)}{-\omega^2 (M + A_z(\omega)) + i\,\omega\, (c_z + B_z(\omega)) + k_z} \tag{5.50}$$

Analogously, it is found that

$$G_\theta(\omega) = \frac{H_\theta(\omega)}{-\omega^2 (I + A_\theta(\omega)) + i\,\omega\, (c_\theta + B_\theta(\omega)) + k_\theta} \tag{5.51}$$

The vertical motion $y(t)$ of the crane top can be written as $y(t) = z(t) + L\,\theta(t)$, assuming small roll angles justifying the approximation $\sin\theta = \theta$. Assuming that the horizontal distance between C.G. and the crane top is kept constant, it is seen that the relationship between the wave elevation process at $x = 0$ and the crane top motions is linear and time-invariant, which is determined by the transfer function

$$H_{\eta y}(\omega) = G_z(\omega) + LG_\theta(\omega) \tag{5.52}$$

Let the wave elevation process at $x = 0$ be modelled as a zero-mean stationary Gaussian process $X(t)$ with a spectral density $S_X(\omega)$. Then the vertical motion of the crane top is also a stationary Gaussian process, which we denote by $Y(t)$. The spectral density $S_Y(\omega)$ of $Y(t)$ is given as follows:

$$S_Y(\omega) = |H_{\eta y}(\omega)|^2\, S_X(\omega) = |G_z(\omega) + LG_\theta(\omega)|^2\, S_X(\omega) \tag{5.53}$$

To finally arrive at the vertical motion response $\tilde{y}(t)$ of the equipment module at the end of a wire rope, an equation of motion of the module has to be established. Assuming that the motions are primarily in the axial direction of the wire rope, this can also be approximated by a linear, time-invariant model determined by a linear transfer function, denoted by $H_{y\tilde{y}}(\omega)$. Therefore, if $X(t)$ is a Gaussian process as above, the vertical response of the module $\tilde{Y}(t)$ also becomes a stationary Gaussian process with a spectral density $S_{\tilde{Y}}(\omega)$ given as

$$S_{\tilde{Y}}(\omega) = |H_{y\tilde{y}}(\omega)|^2 \, S_Y(\omega) = |H_{y\tilde{y}}(\omega)|^2 \, |G_z(\omega) + LG_\theta(\omega)|^2 \, S_X(\omega) \qquad (5.54)$$

Having access to the response spectral density, several statistical parameters that are important for the statistical analysis of the response can be calculated, for instance the standard deviation. In the next section, it will be demonstrated how the spectral density can be used to estimate extreme response quantities.

5.4 Design Approaches

The design provisions for offshore structures have as their common goal to secure a level of safety accepted by the professional community. The practical implementation of this typically takes the form of establishing load or load effect values that have a specified mean return period, e.g. 100 yr, or, equivalently, a specified annual probability of exceedance.

For the design of load carrying members of offshore structures which have to withstand the effects of the environmental forces generated by wind, waves and current, this would require a simultaneous probabilistic model of all three load-generating sources. This is not a practical approach at present, and therefore the procedure is usually simplified. For design problems where the wave loads are dominant, the Norwegian practice has been to assume that the simultaneous action of the 100-yr sea state, the 100-yr mean wind speed and the 10-yr current would represent a proper environmental condition from which characteristic values of extreme loads or load effects can be calculated, see e.g. NORSOK Standard (N-003) (1999).

In the subsequent sections, the focus is on the three dominant methods for calculating characteristic values of extreme loads or load effects due to wave forces.

5.4.1 Design Wave

For the design of offshore structures where the load effects to be considered are primarily of a quasi-static nature, i.e. for fixed platforms it has been common practice to carry out the design check for a so-called 100-yr wave. This has been understood to mean a wave with a wave height $H^{(100)}$ being exceeded on the average only once every 100 yr. Equivalently, $H^{(100)}$ can be interpreted as the wave height which is exceeded during a period of one year with a probability of 10^{-2}.

Design according to this format would in practice proceed as follows: (1) The design wave height is established on the basis of available data relevant for the offshore location in question. (2) A suitable range of corresponding wave periods, and, if relevant, wave directions is specified. (3) According to best practice, (1) and (2) are combined to provide a range of wave profiles for which the corresponding load effects on the structure are established, either by numerical calculations using state-of-the-art computer codes, or in some cases, model tests.

In general, design guidelines applicable in a specific case will specify how the design wave should be chosen. For example, to simplify the calculations, [NORSOK Standard (N-003), 1999] has suggested to take $H^{(100)} = 1.9 \cdot H_s$, if more accurate estimates are not available. The same document also recommends to vary the corresponding wave period T in the range $\sqrt{6.5 \cdot H^{(100)}} \le T \le \sqrt{11 \cdot H^{(100)}}$.

It should be noted that the approach will be different for floating platforms for which the structural response is very sensitive to the wave period [NORSOK Standard (N-003), 1999].

5.4.2 Short-Term Design

When the load effects to be taken into account in the design of load carrying members can be expected to have significant dynamic components, it has been good practice for many years to carry out the design check for load effects established during e.g. a 100-yr storm of specified duration, for example 3 or 6 h. Traditionally, the notion of a 100-yr storm was interpreted to mean a stationary storm condition with a significant wave height not exceeded on the average more than once every 100 yr. A corresponding spectral period was then chosen according to some specified rule or table.

In recent years, the environmental contour line approach has been advocated as a rational basis for choosing the appropriate short-term design storms leading to load and response extremes corresponding to a prescribed return period, or equivalently, a prescribed annual probability of exceedance, which otherwise has to be obtained from a long-term analysis.

Environmental contour line plots are convenient tools for very complicated structural systems where a full long-term response analysis is out of reach in practice. For such systems, extensive time domain simulations or model tests would be necessary for a wide range of sea states in order to determine the short-term distribution given the sea state. Environmental contour lines make it possible to obtain reasonable long-term extremes by concentrating the short-term considerations to a rather narrow area in the scatter diagram.

The contour line approach can be applied for an offshore site if the joint PDF for the significant wave height and the spectral peak period is available in the form of a joint model as described e.g. by equations (5.80)–(5.82). This joint model must be calibrated to fit the available data given for example in the form of a scatter diagram like the one in table 5.2. Contour lines corresponding to a constant annual exceedance probability can be obtained by transforming the joint model to a space consisting of independent, standard Gaussian variables and then utilise the Inverse First-Order Reliability Method (IFORM), see e.g. Winterstein et al (1993). In the standard Gaussian space, the contour line corresponding to an annual exceedance probability of q will be circles with radius $r = \Phi^{-1}(1 - q/2920)$, where Φ denotes the CDF of a standard Gaussian variable. 2920 is the number of 3-h sea states per year. Transforming these circles back to the physical parameter space provides the q-probability contour lines. Approximate contour lines can be obtained by determining

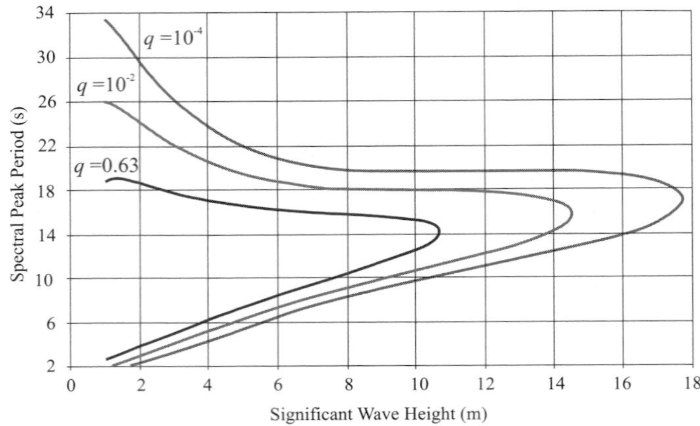

Figure 5.4 Contour line plot for the Statfjord area

the probability density for the point defined by the marginal q-probability significant wave height and the conditional median spectral peak period and then estimating the q-probability contour line from the line of constant probability density. Contour lines based on the joint model discussed in Section 5.4.3.3 are plotted in fig. 5.4, cf. Haver (2002). Even for the most complicated systems, simple methods may often be utilised in order to identify the most critical range of the q-probability contour line regarding a prediction of the q-probability response extreme. The bonus of this method is that thorough time domain analyses and/or model tests are required for only a limited number of sea states. As the most unfavourable sea state along the q-probability contour line is identified, a proper estimate for the q-probability response is taken as the p-fractile of the distribution of the 3-h extreme response value. It is important to note that the median 3-h extreme value for this sea state, i.e. $p = 0.50$, will not represent a proper estimate for the q-probability extreme value since this characteristic value will not account for the inherent randomness of the 3-h extreme value. The fractile level, p, will depend on the aimed exceedance probability target, q, and the degree of non-linearity of the system. For most practical systems, $p = 0.90$ seems reasonable for $q = 10^{-2}$, while $p = 0.95$ may be more adequate for $q = 10^{-4}$. More accurate estimates require full, long-term analyses.

5.4.2.1 Short-term extreme values

For a short-term design, we need to calculate the extreme loads and responses during a short-term storm condition. In this subsection the focus is therefore on the distribution of the extreme values of a stationary Gaussian process, which is specified by an appropriate wave spectrum that characterises a short-term wave condition. For North Sea locations, the JONSWAP wave spectrum would typically be used. This spectrum depends on the choice of four parameters, including the significant wave height and the peak spectral period, cf. Chapter 3.

Let $\hat{X} = \hat{X}(T)$ denote the largest value of the stationary process $X(t)$ during a time interval of length T, that is, $\hat{X} = \max\{X(t) : 0 \leq t \leq T\}$. $\hat{X}$ is clearly the same as the largest peak value during time T. A good approximation of the CDF of $\hat{X}$ is given by the following expression:

$$F_{\hat{X}}(\xi) = \exp[-\nu_X^+(\xi)T] \tag{5.55}$$

where $\nu_X^+(\xi)$ denotes the average rate of upcrossings of the level ξ by the process $X(t)$. In practice, the only approximation that has been made in the derivation of this equation is that the upcrossing events of high levels are statistically independent. For very narrow-band processes, this approximation tends to become less accurate, cf. Naess (1985a).

Invoking equations (5.18) and (5.23), it follows that for a Gaussian process,

$$F_{\hat{X}}(\xi) = \exp\left[-\nu_X^+(0)\, T \exp\left(-\frac{1}{2}\frac{\xi^2}{\sigma_X^2}\right)\right] \tag{5.56}$$

Note that the extreme value distribution of equation (5.56) is completely determined by the standard deviation σ_X and the mean number of zero/mean value upcrossings $\nu_X^+(0)\, T$.

Equation (5.56) can now be used to calculate various statistical descriptors of the extreme value $\hat{X}$. e.g. the mean largest value during time T will be given, to good approximation, by the formula

$$E[\hat{X}] = \sigma_X \left[\sqrt{2\ \ln\left(\nu_X^+(0)\, T\right)} + \frac{0.5772}{\sqrt{2\ \ln\left(\nu_X^+(0)\, T\right)}}\right] \tag{5.57}$$

while the variance is given as

$$\sigma_{\hat{X}}^2 = \mathrm{Var}[\hat{X}] = \frac{\pi^2 \sigma_X^2}{12\ \ln(\nu_X^+(0)\, T)} \tag{5.58}$$

Equations (5.57) and (5.58) clearly show how $E[\hat{X}]$ increases to infinity and $\sigma_{\hat{X}}$ decreases to zero as $\nu_X^+(0)\, T$ increases. This feature is illustrated in fig. 5.5, where plots of the underlying Gaussian distribution and the corresponding extreme value distributions for different values of the number of zero upcrossings.

Quantities of particular interest in design are the quantile values of the extreme value distribution . Let $\xi_p = \xi_p(T)$ denote the extreme value level not exceeded during time T with probability p $(0 < p < 1)$. That is, $F_{\hat{X}}(\xi_p) = p$. From equation (5.56) it is found that

$$\xi_p = \sigma_X \sqrt{2\ \ln\left(\frac{\nu_X^+(0)\, T}{\ln(1/p)}\right)} \tag{5.59}$$

The most probable extreme value, denoted by $\hat{\xi}$, is given to good approximation by the formula

$$\hat{\xi} = \sigma_X \sqrt{2\ \ln\left(\nu_X^+(0)\, T\right)} \tag{5.60}$$

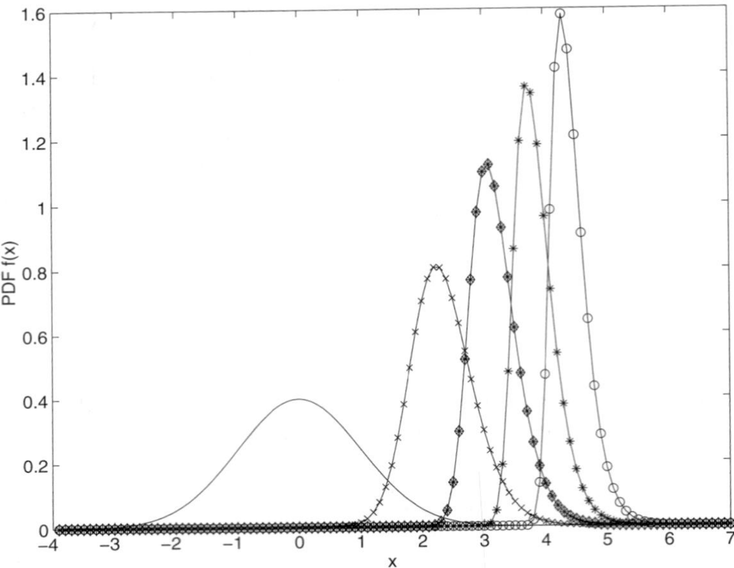

Figure 5.5 Plots of the extreme value PDF of equation (5.56) for $\nu_X^+ 0\ T\ =\ 10\ (\times)$, 100 (◆), 1000 ($*$), 10,000 ($\circ$). —: Underlying Gaussian PDF

Table 5.1 Table of ξ_p/σ_X-values

$\nu_X^+(0)\ T$	10	100	1000	10,000
$p = 0.37$	2.1460	3.0349	3.7169	4.2919
$p = 0.5$	2.3105	3.1533	3.8143	4.3765
$p = 0.9$	3.0176	3.7028	4.2797	4.7876
$p = 0.95$	3.2474	3.8924	4.4448	4.9357
$p = 0.99$	3.7156	4.2908	4.7975	5.2556

$\hat{\xi}$ is the value where the PDF of the extreme value $\hat{X}$ attains its maximum. It can be shown that $\hat{\xi} \approx \xi_{0.37}$.

Specific values of the quantiles ξ_p/σ_X for various numbers of zero upcrossings have been listed in table 5.1.

5.4.2.2 Extreme waves

Only the short-term distributions need to be discussed. The corresponding long-term distributions can be obtained by using the procedures described in the next subsection.

Due to experienced local damage on the deck structure of some fixed offshore platforms because of large wave crests hitting the deck, there has been an increasing attention in recent years towards the estimation of the extreme wave crest height during a storm, as has already been mentioned. If the wave surface elevation is modelled as a Gaussian process, the extreme value distribution for the wave crest height for a short-term condition is simply given by equation (5.56). Since it is known that the wave profile tends to be non-symmetrical wrt the still water level, using this equation would not represent best practice in estimating the short-term extreme wave crest height.

Approximate (empirical) distributions for the short-term extremes are obtained by the following expressions

$$F_{\hat{X}}(a) = \left(F_{X_p}(a)\right)^N \tag{5.61}$$

where $F_{X_p}(a)$ is given by equation (5.29) or (5.30), and $N = \nu_X^+(0)\,\tilde{T}$ = the expected total number of zero-crossing waves during the short-term duration $\tilde{T}$.

Similar to the wave crest height, approximate distributions for the extreme wave height $\hat{H}$ during a short-term condition is obtained by the equation

$$F_{\hat{H}}(h) = (F_H(h))^N \tag{5.62}$$

where $F_H(a)$ is given by equation (5.36) or (5.38), and $N = \nu_X^+(0)\,\tilde{T}$ denotes the number of zero-crossing waves during $\tilde{T}$ as before.

It has been observed that the mean value of the largest wave height $\hat{H}$ during time $\tilde{T}$ is often well fitted by the formula [Krogstad, 1978]

$$E[\hat{H}] = \overline{H}_{1/3}\sqrt{\frac{\ln N}{2}} \tag{5.63}$$

where $\overline{H}_{1/3}$ denotes the empirical significant wave height, which is the average of the one third highest waves.

Using equation (5.31), it can be shown that the corresponding theoretical value of the significant wave height, denoted by $H_{1/3}$, is given by [Naess, 1985a]

$$H_{1/3} = 2\sqrt{2(1-r')\,m_0} = \sqrt{\frac{1-r'}{2}}\,H_{m_0} \tag{5.64}$$

where $m_0 = \sigma_X^2$, and H_{m_0} is the significant wave height obtained from the wave height distribution given by equation (5.35), that is, $H_{m_0} = 4\sqrt{m_0}$.

The expression for the mean largest wave height based on equation (5.31) becomes

$$E[\hat{H}] = H_{1/3}\left(\sqrt{\frac{\ln N}{2}} + \frac{0.5772}{(1-r')\sqrt{2\ln N}}\right) \tag{5.65}$$

which has been shown to agree well with equation (5.63), cf. Naess (1985a).

5.4.3 Long-Term Design

In principle, the most satisfactory design approach is based on long-term statistics, but clearly it may not be the most economical method from a computational point of view as it may involve response calculations for sea states that contribute little or nothing to the design loads or load effects. However, for fatigue design, one or the other version of a full long-term analysis is usually required.

Clearly, the estimation of the extreme loads or load effects on an offshore structure subjected to the ocean environment over the design life of the structure has to take into account the changing weather conditions. This is done in a consistent manner by invoking the appropriate long-term statistical methods, which will be described in this section.

There are basically three different approaches to estimating characteristic long-term extreme values. These methods are based on either (1) all peak values, (2) all short-term extremes, or (3) the long-term extreme value. A more detailed description follows, where $X(t)$ denotes a zero-mean stochastic process, for example the wave elevation or a corresponding load effect, that reflects the changing environmental conditions. Therefore, $X(t)$ is a non-stationary process. Let T denote the long-term time duration, e.g. 1 yr or 30 yr, and let $\tilde{T}$ denote the duration of each short-term weather condition, assuming that $T = N\tilde{T}$, where N is a large integer. The long-term situation is considered to be a sequence of N short-term conditions, where each short-term condition is assumed to be stationary.

Let W denote the vector of parameters that describes the short-term environmental condition. W can be considered as a random variable. For simplicity, let us assume that $W = (H_s, T_s)$, where H_s is the significant wave height and T_s a suitable spectral period. In principle, the analysis is entirely similar if W contains more parameters, e.g. dominant wave direction, wind speed, etc.

5.4.3.1 All peak values

A peak value of $X(t)$, denoted generically by X_p, is defined here as the maximum value of $X(t)$ between two consecutive zero upcrossings. For each short-term condition, let $F_{X_p|H_sT_s}(\xi|h_s, t_s)$ denote the conditional CDF of the peak value. Battjes (1970) has shown that the long-term CDF $F_{X_p}(\xi)$ of the peak value X_p is given as follows:

$$F_{X_p}(\xi) = \frac{1}{\overline{\nu_X^+(0)}} \int_{h_s} \int_{t_s} \nu_X^+(0|h_s, t_s)\, F_{X_p|H_sT_s}(\xi|h_s, t_s) f_{H_sT_s}(h_s, t_s)\, \mathrm{d}h_s \mathrm{d}t_s \tag{5.66}$$

where $\overline{\nu_X^+(0)}$ denotes the long-term average zero-upcrossing rate given by

$$\overline{\nu_X^+(0)} = \int_{h_s} \int_{t_s} \nu_X^+(0|h_s, t_s) f_{H_sT_s}(h_s, t_s)\, \mathrm{d}h_s \mathrm{d}t_s \tag{5.67}$$

Here $\nu_X^+(0|h_s, t_s)$ denotes the average zero-upcrossing rate for the short-term stationary condition characterised by $H_s = h_s$ and $T_s = t_s$.

In practical applications, a commonly adopted statistical distribution for the peak values in a short-term condition is the Rayleigh distribution, that is

$$F_{X_p|H_sT_s}(\xi|h_s,t_s) = 1 - \exp\left(-\frac{\xi^2}{2\,\sigma_X(h_s,t_s)^2}\right) \tag{5.68}$$

Under the assumption that all peak values can be considered as statistically independent, which may not always be very accurate, the peak value ξ_q with a probability q of being exceeded per year is found by solving the following equation:

$$F_{X_p}(\xi_q) = 1 - \frac{q}{365\cdot 24\cdot 3600\cdot \overline{\nu_X^+(0)}} \tag{5.69}$$

It is seen that the short-term duration $\tilde{T}$ does not enter into this analysis. In the long run, the relative frequency of the various sea states is reflected in the PDF $f_{H_sT_s}(h_s,t_s)$, which can be approximated by using an appropriate scatter diagram if that is available. An example of such a scatter diagram is shown in table 5.2.

Let the scatter diagram be divided into m intervals for the h_s-values and n intervals for the t_s-values. It may often be an acceptable approximation to assume that $\nu_X^+(0|h_s,t_s) = T_z^{-1} \approx c\,T_s^{-1}$ for a fixed constant c. Equation (5.66) with the Rayleigh approximation for $F_{X_p|H_sT_s}(\xi|h_s,t_s)$ can then be expressed approximately in the following form

$$F_{X_p}(\xi) \approx \frac{1}{\overline{t_s^{-1}}}\sum_{i=1}^{m}\sum_{j=1}^{n}\left\{1 - \exp\left(-\frac{\xi^2}{2\,\sigma_X(h_i,t_j)^2}\right)\right\}\frac{N_{ij}}{t_j N} \tag{5.70}$$

where

$$\overline{t_s^{-1}} = \sum_{i=1}^{m}\sum_{j=1}^{n}\frac{N_{ij}}{t_j N} \tag{5.71}$$

Here N_{ij} equals the number of registrations in the h_s-interval $(h_i - \Delta h/2,\ h_i + \Delta h/2)$ and t_s-interval $(t_j - \Delta t/2,\ t_j + \Delta t/2)$; $i = 1,\ldots,m$, $j = 1,\ldots,n$. N equals the total number of registrations. Also note that the values for h_s and t_s that have been written on the scatter diagram in table 5.2 are upper class limits, that is, $h_i + \Delta h/2$ and $t_j + \Delta t/2$.

5.4.3.2 All short-term extremes

The conditional CDF $F_{\tilde{X}|H_sT_s}(\xi|h_s,t_s)$ of the largest peak value during a short-term condition, $\tilde{X}|_{H_sT_s}$, is clearly given by the expression

$$F_{\tilde{X}|H_sT_s}(\xi|h_s,t_s) = \left(F_{X_p|H_sT_s}(\xi|h_s,t_s)\right)^{k^{(st)}} \tag{5.72}$$

where $k^{(st)} = \nu_X^+(0|h_s,t_s)\,\tilde{T}$ is the number of peak values during the short-term condition specified by $H_s = h_s$ and $T_s = t_s$. The validity of equation (5.72) is again based on the assumption that all peak values are independent.

Table 5.2 Scatter diagram Northern North Sea, 1973–2001. Values given for h_s and t_p are upper class limits

h_s (m)	**t_p (s)**																	
	3	**4**	**5**	**6**	**7**	**8**	**9**	**10**	**11**	**12**	**13**	**14**	**15**	**16**	**17**	**18**	**19**	**20**	**> 20**
0.5	18	15	123	113	110	390	260	91	38	42	32	3	19	13	9	1	3	2	7
1.0	16	49	675	433	589	1442	1802	959	273	344	125	33	64	29	13	1	7	1	6
1.5	5	32	417	893	1107	1486	2757	1786	636	731	299	121	92	43	18	10	5	2	13
2.0	1	0	102	741	1290	1496	2575	1968	780	868	492	200	116	51	31	8	4	4	8
2.5	0	0	9	256	969	1303	2045	1892	803	941	484	181	157	58	23	19	5	1	8
3.0	0	0	1	45	438	1029	1702	1898	705	957	560	218	196	92	40	11	4	2	5
3.5	0	0	1	4	124	650	1169	1701	647	865	456	237	162	100	36	12	6	1	5
4.0	0	0	2	0	33	270	780	1369	573	868	427	193	157	91	51	13	3	0	1
4.5	0	0	0	0	3	90	459	1017	466	761	380	127	137	86	31	23	6	5	0
5.0	0	0	0	0	0	15	228	647	408	737	354	119	96	50	32	18	2	4	1
5.5	0	0	0	0	0	2	68	337	363	580	283	94	92	31	24	10	6	2	0
6.0	0	0	0	0	0	1	20	166	221	418	307	63	76	24	13	9	4	0	0
6.5	0	0	0	0	0	0	5	50	140	260	257	59	49	20	12	4	2	2	2

Table 5.2 Continued

h_s (m)	t_p (s)																	
	3	**4**	**5**	**6**	**7**	**8**	**9**	**10**	**11**	**12**	**13**	**14**	**15**	**16**	**17**	**18**	**19**	**20**	**> 20**
7.0	0	0	0	0	0	0	0	23	90	180	193	41	53	20	5	3	3	0	0
7.5	0	0	0	0	0	0	0	6	25	93	121	45	46	17	5	5	0	1	0
8.0	0	0	0	0	0	0	0	3	14	50	84	26	47	11	6	0	1	0	0
8.5	0	0	0	0	0	0	0	0	7	25	45	23	25	20	8	0	0	0	0
9.0	0	0	0	0	0	0	0	1	2	12	30	22	20	19	0	0	0	0	0
9.5	0	0	0	0	0	0	0	0	1	2	20	21	14	7	1	1	0	1	0
10.0	0	0	0	0	0	0	0	0	0	2	5	4	21	6	2	0	0	0	0
10.5	0	0	0	0	0	0	0	0	0	3	4	8	9	12	2	0	0	0	0
11.0	0	0	0	0	0	0	0	0	0	0	2	0	4	3	1	0	1	0	0
11.5	0	0	0	0	0	0	0	0	0	0	2	1	2	3	0	0	0	0	0
12.0	0	0	0	0	0	0	0	0	0	0	0	0	1	2	1	0	0	0	0
12.5	0	0	0	0	0	0	0	0	0	0	0	0	0	1	0	0	0	0	0
13.0	0	0	0	0	0	0	0	0	0	0	0	0	0	0	0	1	0	0	0

The long-term CDF of the short-term extreme peak values is now obtained as

$$F_{\tilde{X}}(\xi) = \int_{h_s}\int_{t_s} \left(F_{X_p|H_sT_s}(\xi|h_s, t_s)\right)^{k(st)} f_{H_sT_s}(h_s, t_s)\,\mathrm{d}h_s\,\mathrm{d}t_s \tag{5.73}$$

Assuming for illustration that $\tilde{T} = 3\text{h}$, an estimate of the value ξ_q which has a probability q of being exceeded per year is in this case determined by the equation

$$F_{\tilde{X}}(\xi_q) = 1 - \frac{q}{365 \cdot 8} \tag{5.74}$$

A relation analogous to equation (5.70) would be (with $\tilde{T} = 3\text{h}$ and $T_z^{-1} \approx c\,T_s^{-1}$)

$$F_{X_p}(\xi) \approx \sum_{i=1}^{m}\sum_{j=1}^{n}\left\{1 - \exp\left(-\frac{\xi^2}{2\,\sigma_X(h_i, t_j)^2}\right)\right\}^{\frac{60^2\times 3\times c}{t_j}} \frac{N_{ij}}{N} \tag{5.75}$$

5.4.3.3 The long-term extreme value

It can be shown that the CDF of the extreme value $\hat{X} = \hat{X}(T)$, that is, the global extreme value over a long-term period T, can be expressed as follows [Naess, 1984]

$$F_{\hat{X}}(\xi) = \exp\left\{-T\int_{h_s}\int_{t_s} \nu_X^+(\xi|h_s, t_s) f_{H_sT_s}(h_s, t_s)\,\mathrm{d}h_s\,\mathrm{d}t_s\right\} \tag{5.76}$$

where $\nu_X^+(\xi|h_s, t_s)$ denotes the average ξ-upcrossing rate for the short-term stationary situation characterised by $H_s = h_s$ and $T_s = t_s$.

From equation (5.18) it follows that for the case of a zero-mean Gaussian process, equation (5.76) would read

$$F_{\hat{X}}(\xi) = \exp\left\{-T\int_{h_s}\int_{t_s} \frac{\sigma_{\dot{X}}(h_s, t_s)}{2\pi\sigma_X(h_s, t_s)} \times \exp\left(-\frac{1}{2}\frac{\xi^2}{\sigma_X(h_s, t_s)^2}\right) f_{H_sT_s}(h_s, t_s)\,\mathrm{d}h_s\,\mathrm{d}t_s\right\} \tag{5.77}$$

where the standard deviations σ_X and $\sigma_{\dot{X}}$ in the long-term situation become functions of the environmental parameters h_s and t_s as indicated.

With $T = 1\text{ yr} = 365 \cdot 24 \cdot 3600\text{ s}$, the value ξ_q, which has a probability q of being exceeded per year, is now calculated from the equation

$$F_{\hat{X}}(\xi_q) = 1 - q \tag{5.78}$$

With reference to the scatter diagram above, equation (5.77) can then be expressed as a relation analogous to equation (5.70) in the following way (with $T = 1$ yr and $T_z^{-1} \approx c\,T_s^{-1}$)

$$F_{\hat{X}}(\xi) \approx \sum_{i=1}^{m}\sum_{j=1}^{n}\exp\left\{-\frac{60^2\times 24\times 365\times c}{t_j}\exp\left(-\frac{\xi^2}{2\,\sigma_X(h_i, t_j)^2}\right)\frac{N_{ij}}{N}\right\} \tag{5.79}$$

For the purpose of estimating extreme load effects, the use of scatter diagrams calls for a certain amount of caution. If the scatter diagram is too coarse, leading to poor resolution in the tail regions, the long-term extreme value estimates may become inaccurate. In such cases it is recommended to use a properly adapted smooth joint PDF of the parameters characterising the short-term sea states. For our purposes here, the joint PDF of $W = (H_s, T_s)$ is needed. For North Sea applications, the spectral period chosen is very often the spectral peak period T_p due to the fact that a commonly adopted spectral model is the JONSWAP spectrum, which is usually parametrised by the significant wave height and the spectral peak period. The following probabilistic model given by Haver and Nyhus (1986) has been frequently adopted:

$$f_{H_s}(h_s) = \frac{1}{\sqrt{2\pi}\alpha h_s} \exp\left\{ -\frac{(\ln h_s - \theta)^2}{2\alpha^2} \right\} \qquad h_s \leq \eta \tag{5.80}$$

and

$$f_{H_s}(h_s) = \frac{\beta}{\rho}\left(\frac{h_s}{\rho}\right)^{\beta-1} \exp\left\{ -\left(\frac{h_s}{\rho}\right)^{\beta} \right\} \qquad h_s > \eta \tag{5.81}$$

where the value of the transition parameter η separating the lognormal model for the smaller values of H_s from the Weibull model for the larger values, will depend on the geographical location.

This marginal PDF for the significant wave height is complemented by the conditional PDF of the spectral peak period T_p given by the value of H_s:

$$f_{T_p|H_s}(t_p|h_s) = \frac{1}{\sqrt{2\pi}\sigma t_p} \exp\left\{ -\frac{(\ln t_p - \mu)^2}{2\sigma^2} \right\} \tag{5.82}$$

where the parameters μ and σ are assumed to depend on the significant wave height h_s in the following manner

$$\mu = a_1 + a_2 h_s^{a_3} \tag{5.83}$$

$$\sigma^2 = b_1 + b_2 \exp(-b_3 h_s) \tag{5.84}$$

for suitably chosen constants a_i and b_i, $i = 1, 2, 3$.

The joint PDF for the environmental parameters is then obtained by multiplying the marginal PDF for the significant wave height with the conditional PDF for the spectral peak period, $f_W(w) = f_{H_s T_p}(h_s, t_p) = f_{H_s}(h_s) f_{T_p|H_s}(t_p|h_s)$. It is seen that the conditional distribution model for the spectral peak period is a lognormal distribution, which is also true for the distribution of significant wave heights, except in the tail where a Weibull distribution has been chosen. The following set of parameter values has been cited by Haver (2002) for locations in the Northern North Sea (Statfjord area): $\alpha = 0.6565$, $\theta = 0.77$, $\eta = 2.90$, $\beta = 2.691$, $\rho = 1.503$, $a_1 = 1.134$, $a_2 = 0.892$, $a_3 = 0.225$, $b_1 = 0.005$, $b_2 = 0.120$, $b_3 = 0.455$.

Finally it should be observed that the long-term analysis can be made more efficiently by calculating $F_X(x)$ or the probability of exceedance $Q_X(x) = 1 - F_X(x)$ by only accounting for sea states that contribute to $Q_X(x)$ for large values of x, e.g. Videiro and Moan (2000).

Table 5.3 Table of design alternatives

	Quasi-Static	Weakly Dynamic	Dynamic
Linear	Design wave long-term	Design wave + DAF$^{(*)}$ long-term	Long-term
Weakly non-linear	Design wave long-term	Design wave + DAF$^{(*)}$ long-term	Long-term
Non-linear	Design wave short-term$^{(**)}$	Design wave + DAF$^{(*)}$ short-term$^{(**)}$	Short-term$^{(**)}$

$^{(*)}$ A short-term analysis is required to determine an appropriate dynamic amplification factor (DAF)

$^{(**)}$ A long-term analysis is recommended, but may be too demanding computationally. An alternative is a long-term analysis based on selected storms.

Based on the earlier work by Jahns and Wheeler (1972) and Haring and Heideman (1978), Tromans and Vanderschuren (1995) proposed an alternative approach to the calculation of the long-term extreme load or response value. In their approach, the focus is on storm events, similar to what is done in a peaks-over-threshold analysis [Reiss and Thomas, 2001] Hence, the long-term situation is considered as a sequence of storm events. The method is based on the assumption that the distribution of the storm extreme response value can be approximated by a Gumbel extreme value distribution conditional on the most probable extreme response for that storm. The distribution of the most probable extreme value itself is assumed to follow a generalised Pareto distribution, which is determined by fitting to data. By invoking the rule of total probability, as exemplified by equation (5.73), the long-term extreme response value distribution can be calculated.

An indication of relevant design alternatives for various combinations of structural situations have been listed in table 5.3.

5.5 Combination of Multiple Stochastic Load Effects

In practical design it is very often necessary to account for the fact that the total load effect is the result of a combination of several load effect components. The challenge then is to provide a rational combination procedure for the estimation of the extreme combined load effect value which ideally has the prescribed return period of, say, 100 yr. In some cases it is possible to carry out a long-term extreme value analysis as described in the previous section, but often some degree of simplification has to be accepted in practical design work. To exemplify this situation, we shall investigate the case of the total surge response of a tension leg platform (TLP) in long-crested random seas.

5.5.1 Combination of First- and Second-order Responses

Let the total horizontal wave-induced response of the TLP structure subjected to a random, long-crested sea way be written as a sum of a linear and a quadratic response component. Specifically, let $Z(t)$ denote the total surge response process of the TLP. Then

$$Z(t) = \mu_2 + Z_1(t) + Z_2(t) \tag{5.85}$$

where the subscript 1 signifies the linear, first-order or wave frequency component while subscript 2 refers to the non-linear, second-order part of the TLP's surge response. Note that μ_2 = the mean drift offset, implying that $Z_2(t)$ denotes the zero-mean, slowly varying part of the second-order response. Strictly speaking, the assumed decomposition presupposes that the equation of motion for the surge response is linear, or linearised.

Assume that the quantity to be calculated is the expected largest surge response during a short-term storm condition, that is, $E[\hat{Z}]$, where $\hat{Z} = \hat{Z}(\tilde{T}) = \max\big(Z(t);\, 0 \leq t \leq \tilde{T}\big)$ denotes the extreme value of $Z(t)$ during a short-term time interval of length $\tilde{T}$.

There exist procedures for calculating the extreme values of the total response process $Z(t)$ as described in the previous section, and based on the level upcrossing rate of $Z(t)$. However, for the present case this is not a straight-forward matter since it involves extensive calculations of the second-order hydrodynamic coefficients of the TLP for the short-crested seas case, cf. Teigen and Naess (1999).

For routine design work, a realistic approach to the estimation of the extreme surge response is based on a simplified procedure. For example, API's Recommended Practice document [API RP 2FPI, 1993] for combination of wave frequency and low-frequency horizontal excursion of a moored floating structure is based on a specific, simplified combination formula, viz. Turkstra's rule [Turkstra, 1970; Naess and Royset, 2000]. In the present context, it assumes the following form

$$E[\hat{Z}] = \mu_2 + \max\left\{E[\hat{Z}_1] + \sigma_2,\ \sigma_1 + E[\hat{Z}_2]\right\} \tag{5.86}$$

where σ_i = standard deviation of $Z_i(t)$.

The main efforts in calculating the expected largest response according to this formula are in the calculation of $E[\hat{Z}_1]$ and $E[\hat{Z}_2]$. The state-of-the-art approach is to assume that the short-term random sea way can be described as a stationary Gaussian process. This implies that the linear response is also a Gaussian process, hence $E[\hat{Z}_1]$ can be calculated by equation (5.57). The slowly varying response $Z_2(t)$ is generally non-Gaussian, which invalidates the use of equation (5.57) without further justification. As a first estimate of $E[\hat{Z}_2]$, Naess (1989) has proposed the following formula based on a simplified, second-order response representation:

$$E[\hat{Z}_2] = \sigma_2 \ln\left(\frac{\tilde{T}}{T_{SD}}\right) \tag{5.87}$$

where T_{SD} denotes the natural period of the surge motion.

A more accurate estimate of $E[\hat{Z}_2]$ would lead to a result that falls somewhere between the Gaussian estimate provided by equation (5.57) and the simplified formula given by equation (5.87), cf. Naess (1989), Stansberg (1992), Stansberg (2000).

To finalise the calculation of $E[\hat{Z}]$ as given by equation (5.86), it is seen that μ_2 and σ_2 must be provided. The following formula results

$$\mu_2 = \frac{1}{M\,\omega_{SD}^2}\int_0^\infty C(\omega)\, S_X(\omega)\,\mathrm{d}\omega \tag{5.88}$$

where M = the total mass of the TLP, $\omega_{SD} = 2\pi/T_{SD}$, $C(\omega)$ = the wave drift coefficient [Faltinsen, 1990], and $S_X(\omega)$ = the wave spectrum.

Adopting Newman's approximation [Newman, 1974] for the calculation of the slowly varying response process $Z_2(t)$, it can be shown that σ_2 can be calculated as follows

$$\sigma_2^2 = 2\int_0^\infty \left(\int_0^\omega |\hat{L}(\omega-\omega')|^2\, S_X(\omega')\,\mathrm{d}\omega'\right) C(\omega)^2\, S_X(\omega)\,\mathrm{d}\omega \tag{5.89}$$

where $\hat{L}$ denotes the linear transfer function for the surge motion, that is,

$$\hat{L}(\omega) = \frac{1}{M\left(\omega_{SD}^2 - \omega^2 + 2\,i\,\kappa\,\omega_{SD}\,\omega\right)} \tag{5.90}$$

where κ denotes the damping ratio for the surge motion, and $i = \sqrt{-1}$.

An alternative combination format that is used extensively in practical design applications is the so-called SRSS-formula. SRSS is an abbreviation for Square Root of Sum of Squares. The SRSS equivalent to equation (5.86) is the following expression

$$E[\hat{Z}] = \mu_2 + \left\{E[\hat{Z}_1]^2 + E[\hat{Z}_2]^2 + 2\,\rho_e\,E[\hat{Z}_1]E[\hat{Z}_2]\right\}^{1/2} \tag{5.91}$$

where the parameter ρ_e denotes a correlation coefficient between first- and second-order wave-induced responses at large response levels. The standard form of the SRSS formula is obtained by neglecting this correlation coefficient, which, in general, cannot be justified for the combination of first- and second-order wave-induced responses. For applications considered here, the correlation coefficient ρ_e can be expected to assume values typically in the interval from 0.2 to 0.4.

5.6 Probabilistic Design of Offshore Structures

5.6.1 Introduction

In Sections 5.1–5.5 the statistical nature of waves and their effect on structural load effects have been dealt with. More precisely, methods to describe the fundamental variability of loads and the corresponding load effects have been outlined. Initial efforts in the 1970s on the rationalisation of safety measures were directed at establishing risk-based storm load criteria, e.g. [Marshall, 1969] based on the fundamental variability of loads. Later risk and reliability approaches with a wider scope have been adopted. Engineering decisions in general and about safety in particular, must be made in the presence of uncertainties arising from inherent randomness in many design parameters, imperfect modelling and lack of experience. Indeed, it is precisely on account of these uncertainties and the potential risks arising therefrom that safety margins provided by the specification of allowable stresses, resistance and load factors, and the like, are required in design. In the following sections the probabilistic treatment of uncertainty is therefore expanded to deal with all uncertainties in load effects and resistance that affect design of offshore structures. This is the topic of the probabilistic design.

Adequate performance of offshore structures is ensured by designing for serviceability and safety in a service life of 20–40 yr. Modern design of structures is based on dimensioning (design check) using calculated effects due to the different types of loads and the resistance, corresponding to the different failure modes. Separate sets of calculations are required to check that the structure will not attain each limit state for each structural component. Limit state criteria are classified into two groups, namely serviceability and safety requirements [ISO 2394, 1998]. Serviceability requirements refer to motions, deformations, vibrations etc. that can hamper the operation, but do not represent a threat to the safety. Safety means the absence of failures and damages and is ensured by fulfilling requirements to overall stability and ultimate strength and fatigue failure under repetitive loading to avoid ultimate consequences such as fatalities, environmental damage or property damage. The corresponding criteria are defined by limit states for ultimate failure and fatigue respectively.

Another issue in modern codes is the application of partial safety factors or load and resistance factor design (LRFD) instead of working stress design. While the main advantage of working stress design is its simplicity; however, it can lead to designs with less safety than normally considered adequate, particularly if loads counteract each other at a less consistent safety level.

The simplest example of ultimate limit state design check of a structural component subjected to two types of loads, may be written as:

$$R \geq S_1 + S_2, \tag{5.92}$$

where R is the resistance and S_1 and S_2 are extreme load effects in a reference period of 100 years, due to different types of loads, e.g. payload and wave load respectively. R and S refer to the same physical quantity, e.g. a stress or an axial force or a bending moment.

Since R, S_1 and S_2 are subjected to uncertainty and variability, the design format applied is:

$$\frac{R_c}{\gamma_R} \geq \gamma_{S1} S_{1c} + \gamma_{S2} S_{2c} \tag{5.93}$$

where subindex c refers to characteristic value and γ_R, γ_{S1} and γ_{S2} are resistance and load factors respectively. The characteristic resistance is obtained by, for example using the 95% fractile material strength while the characteristic load effect due to payload and waves correspond to a specified value and the load with an annual probability of exceedance of 10^{-2} respectively. Equation (5.93) implies that the failure probability

$$P_f = P[R \leq S] \tag{5.94}$$

for the present problem, becomes sufficiently small when load and resistance factors and characteristic values are properly chosen. The ultimate limit state criteria in modern design codes are based on the design format like equation (5.93). This approach is denoted semi-probabilistic while design based on direct calculation of the failure probability is a probabilistic approach. Then R and S are considered random variables which represent the uncertainties and variability in R and S. Some design codes permit direct probabilistic design for particular situations. Even more importantly, probabilistic approaches are used to calibrate the semi-probabilistic approaches.

An appreciation of the philosophy underlying such provisions is essential: in the presence of uncertainty, absolute reliability is an unattainable goal. However, the probability theory and the reliability-based design provide a formal framework for developing criteria for design, which insure that the probability of unfavourable performance is acceptably small.

More general expressions than equation (5.92)–(5.94) would be required to describe ultimate failure of beams, panels and shell structures under multiple loading. Also design formats relating to brittle fracture and fatigue need to be described.

Also, when S_1 and S_2 represent two time-varying loads, such as the still-water and wave loading on an FPSO, the fact that their maxima in a given period do not occur at the same time, necessitates particular analysis to determine the maximum of the combined load. By use of the stochastic process theory, the characteristic values of the combined loads may be expressed by: $(\psi_{S1}S_{1c}, S_{2c})$ or $(S_{1c}, \psi_{S2}S_{2c})$ where ψ_{Si} is a load reduction factor and the individual load effects are still defined separately with respect to the 10^{-2} exceedance probability.

So far the focus has been on achieving serviceability and safety by structural design. However, inspection, maintenance and repair (IMR) during fabrication and operation are crucial issues to maintain safety, especially in connection with deterioration phenomena such as fatigue and corrosion. But their effect on the reliability depends upon the quality of inspection, e.g. in terms of detectability vs. size of the damage. Hence, an inspection and repair measure can contribute to the safety only when there is a certain damage tolerance. This implies that there is an interrelation between design criteria (fatigue life, damage tolerance) and the inspection and repair criteria. For an adequate treatment of design and IMR issues relating to fatigue, a probabilistic approach is necessary [Moan, et al 1993]. To some extent these issues are reflected in the fatigue design criteria in offshore codes [NORSOK Standard (N-001), 2000; ISO 2394, 1998] where the fatigue design check depends upon the consequences of failure and access for inspection and vary between an allowable cumulative damage, Δ_d between 0.1 and 1.0 [NORSOK Standard (N-001), 2000]. The consequence measure is based on whether the structure fails in a condition with reasonable variable load, and 100-yr sea loads after fatigue failure of the relevant joint, and is thus linked to the Accidental Collapse Limit State, as subsequently discussed. The current treatment of both the consequence and inspection issue, however, could be improved, e.g. by taking Δ_d to be dependent on a more precise measure of reserve strength and an explicit measure of the effect of inspection.

In the following subsection, various limit states are briefly described, followed by an outline on how the reliability problem may be formulated in terms of random variables. In Section 5.7 uncertainties in loads and resistances are classified and characterised in probabilistic terms.

Section 5.8 deals with methods for reliability analysis of basic components, while system reliability is briefly outlined in Section 5.9. Section 5.10 outlines how the reliability for deterioration phenomena can be calculated and applied in connection with structural reassessment and inspection planning.

5.6.2 Limit States and Failure Criteria

Before pursuing modelling of uncertainties and methods for calculating P_f and how they can be applied, some introductory remarks on other limit states are made.

Overall stability check is formulated by a destabilising load effect, S_{destab}, and a stabilising load effect, S_{stab}, which is analogous to the resistance, R in equations (5.92)–(5.93).

The basic formulation for a structural component with one resistance R and a corresponding load effect S is by equations (5.92) and (5.93). This case could relate to the ductile collapse of a member.

Equations (5.92) and (5.93) may be reformulated by introducing a limit state function $g(\cdot)$ as defined by

$$g(R, S) = R - S \tag{5.95}$$

The design criterion, equation (5.93) is given by

$$g(R_d, S_{1d}, S_{2d}) \geq 0 \tag{5.96}$$

where $R_d = R_c/\gamma_R$, $S_{1d} = \gamma_{S1} S_{1c}$, $S_{2d} = \gamma_{S2} S_{2c}$ are the design values of resistance and load effects, respectively. Similarly, the expression for a failure event is given by

$$g(R, S_1, S_2) < 0 \tag{5.97}$$

As another example of an ultimate limit state, consider a steel beam-column subject to axial force and bending moment. The limit state function in this case may be formulated as

$$\begin{aligned} g(R_1, R_2, R_3, S_1, S_2) &= 1 - \left[\frac{S_1}{R_1} + \frac{S_2}{[1 - (S_1/R_2)]R_3}\right] \\ &= 1 - \left[\frac{X_1}{X_2} + \frac{X_3}{[1 - (X_1/X_4)]X_5}\right] = g(\mathbf{X}) \end{aligned} \tag{5.98}$$

where S_i and R_i are load effect and resistance parameters respectively. Clearly equation (5.98) is based on a Perry-Robertson approach, and represents one alternative. Many other problems can be formulated by a multiple set, $\mathbf{X}$ of random variables.

Brittle facture due to overload in welded metal structures has often been treated in a simple manner, by choosing material quality based upon environmental temperature, plate thickness and not a formal explicit design check analogous to equation (5.92). However, by use of fracture mechanics approaches, formal criteria can be established and treated in a semi-probabilistic or probabilistic manner, see e.g. Almar-Næss (1985).

Fatigue is an important consideration for structures in areas with more or less continuous storm loading and especially for dynamically sensitive structures. The fatigue design check is normally based on resistance defined by SN-data ($N = KS^{-m}$), that have been obtained by laboratory experiments, and the use of Miner Palmgren's hypothesis of linear cumulative damage. A simple expression for the cumulative damage can be obtained by assuming that the SN-curve is defines by $NS^m = K$ and the number, $n(s)$ of stress ranges is given by the two-parameter Weibull distribution. The distribution of stress ranges, s can then be described by

$$F_S(s) = 1 - \exp\left(-\left\{\frac{s}{A}\right\}^B\right) \tag{5.99}$$

where

$$A = s_0/(\ln N_0)^{1/B}, \quad P[S \geq s_0] = 1/N_0 \tag{5.100}$$

and B is the shape parameter and A is the scale parameter of the Weibull distribution.

The damage, D in a period, T with N_T cycles, is then

$$\begin{aligned} D &= \sum_i \frac{n_i}{N_i} = \sum_i \frac{n(s_i)}{N(s_i)} = \frac{N_T}{K} A^m \Gamma(m/B + 1) \\ &= \frac{N_T}{K}(s_0^m/(\ln N_0)^{m/B})\Gamma(m/B + 1) \end{aligned} \tag{5.101}$$

where A is given by equation (5.100). Guidance on the magnitude of the shape parameter, B is available. The scale parameter, A which is directly related to the extreme response value, s_0 required for ULS design checks can be estimated in connection with this kind of analysis. In this way, fatigue loading at least for initial design and screening to identify the importance of fatigue can be easily accomplished.

The failure function for reliability analysis based on the *SN* approach is given by

$$g(\cdot) = \Delta - D \tag{5.102}$$

where Δ describes the damage at fatigue failure.

The design check, $g(\Delta_d, D_c) \geq 0$, is normally based upon D_c determined by the best estimate of stress cycles, as specified by the Weibull parameters A and B; while K is taken to be K_c, the characteristic value of K, typically corresponding to 97.7% probability of exceedance (mean minus two standard deviation) and Δ_d is the acceptable fatigue damage, usually between 0.1 and 1.0 [NORSOK Standard (N-001), 2000]. When calculating the probability of failure A, K, Δ etc in $g(\cdot)$ are taken as random variables.

The failure function may also be expressed in terms of time by reformulating equation (5.102) as follows:

$$g(\cdot) = \frac{K \cdot \Delta}{\upsilon_0 \cdot A^m \cdot \Gamma(m/B + 1)} - \frac{N_T}{\upsilon_0} = T_f - \tau \tag{5.103}$$

where T_f and τ are the time to failure (resistance) and the time elapsed with fatigue loading (load effect). υ_0 is the mean frequency of the stress cycles (ranges).

The expression for the failure function may be generalised to cases where the *SN*-curve consists of two linear segments instead of a single segment in a logarithmic plot.

Instead of the *SN*-curve/Miner-Palmgren approach a fracture mechanics approach needs to be adopted to assess more accurately the different stages of crack growth including calculation of residual fatigue life beyond through-thickness crack, which is normally defined as fatigue failure. Such a detailed information about crack propagation is also required to plan inspections and repair. A model that includes the crack depth as a variable, also gives the opportunity to better compare predicted with observed fatigue behaviour. However, it is crucial that the fracture mechanics approach is calibrated to the *SN*-approach for the initial stage of the fatigue life – to ensure that the initial crack size and local geometry is properly represented.

The fracture mechanics approach is based on the Paris' crack propagation law:

$$\frac{\mathrm{d}a}{\mathrm{d}N} = \begin{cases} C(\Delta K)^m & \text{for } \Delta K > \Delta K_{\text{th}} \\ 0 & \text{for } \Delta K \leq \Delta K_{\text{th}} \end{cases} \tag{5.104}$$

where a is the crack depth, N is the number of cycles, C is the crack growth parameter, m is the inverse slope of the *SN*-curve, and ΔK_{th} is the threshold for $\Delta K =$ the stress intensity factor (SIF) range given by

$$\Delta K = S\left[Y_{m,\text{plate}}\frac{S_m}{S} + Y_{b,\text{plate}}\left(1 - \frac{S_m}{S}\right)\right]\sqrt{\pi a} \tag{5.105}$$

S is stress range and subscripts m and b refer to membrane and bending respectively. Compliance functions for semi-elliptical surface cracks in flat plates, $Y = Y_{x,\text{plate}}$ $(a/t, a/c, c/w, \phi)$, where index x is m or b [Newman and Raju, 1981] are used, where c is one half of the crack length, w is the plate width, and the angle ϕ measured from the surface defines a point on the elliptical crack boundary. For applications to tubular joints, $Y_{m,\text{plate}}$ and $Y_{b,\text{plate}}$ are corrected by a magnification factor M_k in order to account for effects from welds based on upper bound (conservative) values, see e.g. Almar-Næss (1985).

The failure function for fatigue, $g(\cdot)$ can be written as

$$g(\cdot) = a_f - a(t) \tag{5.106}$$

where a_f and $a(t)$ are the crack sizes at failure and the crack size at time t or $N = \upsilon_0 \cdot t$ cycles, where υ_0 is the average stress cycle frequency.

By assuming a constant geometry function $Y(a)$,

$$a(t) = \left[a_0^{1-m/2} + (1 - m/2)C\Delta\sigma^m Y^m \pi^{m/2} N\right]^{\frac{1}{1-m/2}} \tag{5.107}$$

where a_0 is the initial crack size.

As pointed out by Ayala Uraga and Moan (2002), $g(\cdot)$ in equation (5.106) needs to be reformulated when introducing $a(t)$ according to equation (5.107) when FORM/SORM is applied to determine P_f. In general, when $Y(a)$ is a function of a, $g(\mathbf{x})$ can be reformulated as a function of time as shown by Madsen et al (1986).

$$g(\mathbf{x}, t) = \int_{a_0}^{a_f} \frac{\mathrm{d}a}{G(a)\left[Y(a)\sqrt{\pi a}\right]^m} - C \cdot \upsilon_0 \cdot A^m \cdot \Gamma\left(1 + \frac{m}{B}\right) \cdot t \tag{5.108}$$

where a_0 and a_f are initial and final crack depths respectively. In case of through-thickness-crack (TTC), a_f = plate thickness. $G(a)$ is defined by

$$G(a) = \frac{\Gamma\left(1 + m/B; \left(\frac{\Delta K_{\text{th}}}{A \cdot Y\sqrt{\pi a}}\right)^B\right)}{\Gamma(1 + m/B)} \tag{5.109}$$

$\Gamma(\cdot)$ and $\Gamma(\cdot;\cdot)$ are the Gamma and the complementary incomplete Gamma functions respectively.

Fatigue and corrosion interaction can be accounted for based on the fact that the wall thickness reduction will increase the long-term stress ranges. Moreover, the crack propagation parameter, C is increased for steel structures in a corrosive environment. To account for the plate thinning effect, the scale parameter A in the Weibull distribution of long-term stress range, is taken to vary with time due to a corrosion rate r_c. A fixed annual thickness reduction rate due to corrosion is assumed throughout the service life. The wall thickness h at time t is given by $h(t) = h_0 - \alpha r_c(t - t_0)$, where h_0 is the initial value and t_0 is the coating protection time, varying randomly typically in the range of 5–15 years if no maintenance is carried out. The stress range due to axial forces is therefore $S(t) = S_0 \cdot h_0/h(t)$. The parameter A may be defined as the time-varying parameter since S increases as the thickness reduces with the annual rate mentioned above. α is a factor equal to 1 or 2 depending whether the corrosion is one-sided or two-sided respectively. Then the Weibull scale parameter A may be obtained as follows (Moan, 2004):

$$A = \begin{cases} A_0 & \text{for } 0 \le t \le t_0 \\ A_0\left[t_0/t + \frac{1}{(m-1)k \cdot t}\left\{(1 - k(t - t_0))^{-m+1} - 1\right\}\right]^{1/m} & \text{for } t > t_0 \end{cases} \tag{5.110}$$

where $k = \alpha \cdot r_c/h_0$.

Interaction between fatigue and corrosion is therefore, obtained by applying the time varying scale parameter $A(t)$ from equation (5.110) in the failure function defined in equation (5.108) or safety margins and inspection events defined below.

It is important to note that fatigue failure could be defined in various ways. SN-curves are obtained by referring to a visible crack or through-thickness-crack or loss of capacity for the (small-scale) specimen applied, and does not reflect the possible significant amount of material surrounding the crack especially in steel-plated structures.

If Δ_d is taken to be 1.0, it is indicated later that the implied $P_f(t)$ would be of the order of 10%. With a large number of possible crack sites, the likelihood of having one or more fatigue failure becomes significant when such a design criterion is used. Adequate safety can then be achieved by:

- providing a significant amount of surrounding material to ensure a sufficient residual fatigue life for the effective use of visual inspection to detect cracks, followed by repair, or,
- using non-destructive examination to detect even small cracks, which are then repaired.

Strength deterioration due to corrosion also needs to be considered, depending upon the possible protection, corrosion allowance and plating replacement policy.

Semi-probabilistic methods need to be calibrated by reliability analysis to account for the effect of inspection results, and possible repair (grinding, welding, replacement) on the safety. For instance by a given inspection plan (inspection method, frequency of inspection) the fatigue criterion may be relaxed. This effect would have to be estimated at the design stage. At this stage the outcomes of the inspections (no crack detection or crack detection) are not precisely known. However, the probability of various outcomes can be estimated. Another situation occurs when inspections are carried out during use of the structure. In such cases the outcome (e.g. no find of crack) of the inspection will be known.

By continuously updating the safety measure (P_f) depending on actions initiated during use, decisions regarding inspection scheduling, maintenance and repair can be properly made, to ensure an acceptable safety level.

While current design codes primarily refer to component requirements, the main safety issue is concerned with the global structure (system). Some fatigue requirements, e.g. NORSOK Standard (N-001) (2000), are made dependent upon consequence of the relevant fatigue failure, by making Δ_d dependent on whether the system is statically determinate or survives a storm with an annual probability of 10^{-2} after the relevant fatigue. A more refined criterion would be to let the fatigue requirement depend upon the failure probability given the fatigue failure of the relevant member.

There is currently no code which specifies ULS criteria for components, which depend upon the system failure probability, after component failure. If subsequent failures are highly correlated with the initial failure, this dependence is essentially deterministic.

The ultimate and fatigue design criteria in current codes are based on component failure modes (limit states) and a linear global model of the structure to determine the load effects in the components. However, an approach, which is based on the global (system) failure modes of the structure is desirable because significant consequences, e.g. fatalities will primarily be caused by global failure. A proper system approach is also necessary to obtain the optimal balance between the design and the inspection plan since it is normally based on a certain damage tolerance, especially when the inspections rely on detecting flooded or failed members. System failure is then expressed mathematically by load and resistance parameters relating to all failure modes for all components. The system failure probability is calculated by the probabilistic properties of these parameters. Broadly speaking, this may be achieved by a failure mode (or survival mode) analysis or by direct simulation methods.

Design checks of a structure with respect to a particular failure mode may be expressed by a model describing the limit state in terms of a function (called the limit state function) whose value depends on all design parameters. In general terms, attainment of the limit can be expressed as

$$g(\mathbf{X}) = 0 \tag{5.111}$$

where $\mathbf{X}$ represents the vector of (random) design parameters (also called the basic variable vector) that are relevant to the problem, and $g(\mathbf{X})$ is the limit state function. Conventionally, $g(\mathbf{X}) \leq 0$, represents failure (i.e. an adverse state).

For many structural engineering problems, the limit state function, $g(\mathbf{X})$, can be separated into one resistance function, $g_R(\cdot)$, and one loading (or effect) function, $g_S(\cdot)$, in which case equation (5.111) can be expressed as

$$g_R(\mathbf{R}) - g_S(\mathbf{S}) = 0 \tag{5.112}$$

where $\mathbf{S}$ and $\mathbf{R}$ represent subsets of the basic variable vector, usually called loading and resistance variables respectively.

5.7 Uncertainty Measures

5.7.1 General Description

For the purpose of quantifying uncertainties for subsequent reliability analysis, it is necessary to define a set of *basic variables*. The basic variables are described by probability density functions or distributions. Alternatively the statistical moments, e.g. mean value, variance (square of standard deviation) may be applied. Reference is made to textbooks, e.g. Thoft-Christensen and Baker (1982), Madsen et al (1986), Melchers (1999). In many cases, important variations exist over time (and sometimes space), which have to be taken into account in specifying basic variables. In probabilistic terms, this may lead to a random process rather than random variable models for some of the basic variables. However, simplifications might be acceptable, thus allowing the use of random variables whose parameters are derived for specified reference period (or spatial domain).

The values of the uncertainties are crucial to the estimates of the reliability. It is therefore important that authoritative uncertainty measures are applied. Data processed in connection with code calibration and assessed by an expert committee – with knowledge about the mechanics of the phenomena as well as the probabilistic characteristics, are most relevant.

Offshore-related code calibration studies are reported e.g. by Fjeld (1977), Lloyd and Karsan (1988) and Moan (1988).

The uncertainties of the main categories of basic variables may be described as follows:

5.7.1.1 Mechanical properties of structural materials

Relevant material properties include yield strength of steel, compression strength of concrete (cube or cylinder strength), Young's modulus and coefficient of friction.

Considerable quantities of data are available for the strength of materials such as structural steel, concrete and timber. Statistical analysis should be undertaken to obtain means and standard deviations, but care should be taken to ensure that the data are homogeneous. The strengths of most materials, including most concretes in compression, are well represented by log-normal distributions.

The actual characteristic values of material strength should normally be specified as a small percentage fractile (commonly taken as 5%) of the strength distribution. This may be evaluated from the data, taking into account the type of distribution, and the size of the sample.

5.7.1.2 Structural dimensions and geometry

The geometrical properties, for example include thickness of plates, lack of straightness in columns and geometrical imperfections in steel-plated structures.

Except for structures which are sensitive to geometrical imperfections (usually compression members), the variability of structural dimensions and geometry tends to be rather less important than variability in load and strength parameters. This is generally because the coefficients of variation of the former tend to be considerably less than 5%. Most dimensional variables (e.g. beam depths) can be assumed, and are indeed found, to be distributed with mean values corresponding to the specified nominal dimensions. However, there are some exceptions; see, e.g., Melchers (1999).

Most dimensional variables can be adequately modelled by normal or log-normal distributions. In the case of variables which are physically bounded a truncated distribution is appropriate.

5.7.1.3 Loads and imposed deformations

These quantities may be further classified according to whether they are:

Permanent – constant for the life of the structure (e.g. self-weight), or
Variable – in time throughout the life of the structure (e.g. wave loads)

A special case of a variable load or imposed deformation is an accidental load, which may or may not occur during the life of the structure.

Account may also need to be taken of whether loads and imposed deformations can vary in space as well as in time (e.g. superimposed floor loads) and whether they may be considered to be essentially static or dynamic in nature (the latter depends on the type of structure being considered).

Permanent loads. For permanent loads which are assumed to remain constant for the life of the structure (e.g. densities of the construction materials), the statistical distribution function should be assumed to be normal. The self-weight of the structure itself is generally not treated as a basic variable, as it is a function of two other types of more fundamental quantities (dimensions and densities).

Variable loads. Most loads come within this category (e.g. wave loads, variable deck load, ballast). For loads of this type, which vary with time throughout the life of the structure, it is necessary to know the distribution of the maximum load likely to occur in a specified time interval. When the design problem involves only one time-varying basic variable, it is necessary, if designing to a target lifetime reliability level, to know the distribution of the maximum value of the load during the proposed design life. On the other hand, if the specified target reliability is expressed on an annual basis, the relevant time interval is clearly 1 yr. When there is more than one time-varying load or basic variable, the situation becomes more complex. The necessary procedure for this situation is described in Section 5.8.5.

To estimate the extreme (maximum) values of the variable loads in a specified time interval, it is necessary to transform the distributions of the basic data collected. For wave loads, for example, it is convenient to use such statistics as the long-term variation of sea states in terms of scatter diagrams or some joint probability distribution. The load effects are obtained by a frequency domain analysis for a linear problem or a time domain analysis when important nonlinearities affect the load effects. This is a broad topic. The methods applied depend on which load effects are in focus. These issues are treated in Sections 5.1–5.5 of this chapter. Also, further remarks are given in Sections 5.7.2 and 5.7.3. In general, a Weibull distribution is found to accurately describe different types of long-term load effects in structures in areas with extratropical climates. Extreme values corresponding to one year or 100 yr periods can then be readily established based on the extreme value theory. However, for certain types of load effects such as forces obtained by the Morison's equation, different phenomena prevail at different load levels or wave heights. For instance, when forces are estimated by the Morison's equation, inertia forces dominate at moderate wave heights while drag forces may dominate at large wave heights, see Brouwers and Verbeek (1983). In such cases Weibull-tail or POT methods is recommended to use in connection with predicting extreme values.

Spatial variations. Spatial variations in the position of application of variable or permanent loads should be modelled separately from the variations in load intensity. In the case of superimposed deck loads, the load effects vary according to the spatial distribution of the loads. This should be taken into account by introducing an additional basic random variable X_s defined as:

$$X_s = \frac{\text{load effect due to spatially random load}}{\text{load effect due to uniformly distributed load}} \tag{5.113}$$

whenever possible, the statistical parameters of X_s should be evaluated from data, otherwise by subjective assessment. A normal distribution is usually appropriate.

5.7.1.4 Uncertainties in the theoretical model

Even in the best theoretical method to model a particular load effect (e.g. loads on a jacket calculated by the Morison's equation) or resistance (e.g. collapse of a steel beam by buckling), various effects are likely to be neglected either because they are not known or because they are too difficult to take into account. For this reason the ratio

$$X_m = \frac{\text{Actual behaviour}}{\text{Predicted behaviour}} = \frac{X_{\text{true}}}{X_{\text{pred}}} \tag{5.114}$$

tends to vary in a random way. If the theoretical method is good and takes into account all the major effects, but neglects a number of smaller effects, X_m can be considered to be an additional random variable independent of the other parameters. The magnitude of X_m describes the uncertainty in the mechanical model.

Such a model uncertainty is assessed by comparing prediction with the theoretical test results for full-scale structures or representative models. It is necessary to verify theoretical solutions by comparison with a sufficient number of tests to cover the ranges of the relevant parameters. The number should be an adequate sample for statistical analysis. The values of each of the basic variables should be measured (geometry, sizes, imperfections, material properties, etc.) and used to obtain the ratio X_m between observed and predicted behaviour. The mean value and coefficient of variation of this ratio for the sample may then be used in the statistical model, the distribution being assumed, for example to be normal or lognormal.

5.7.2 Representation of Uncertainty

The uncertainties used in reliability analysis should include all variables which may affect the failure probability. These would include "inherent" statistical variability in the basic strength or load parameter. Additional sources of uncertainty arise due to modelling and prediction errors and incomplete information; included in these "modelling uncertainties" would be errors in estimating the parameters of the distribution function, idealisations of the actual load process in space and time, uncertainties in calculation, and deviations in the application of standard and material specification from the idealised cases considered in their development. Frequently, the latter uncertainty measures must be estimated on the basis of professional judgement and experience. The key test in differentiating between the "inherent" and "modelling" uncertainties is in whether the acquisition of additional information would materially reduce their estimated magnitude. If the variability is intrinsic to the problem, additional sampling is not likely to reduce its magnitude. In contrast, uncertainties due to "modelling" should decrease as improved models and additional data become available.

Uncertainties can be represented by random variables. Each random variable, X, may be described by the probability density $f_X(x)$ or distribution function $F_X(x)$, or parameters such as mean values (μ_X), standard deviation (σ_X), etc.

Let X denote a basic resistance or load variable. Although the true mean and COV of X, μ_X and V_X, should be employed when evaluating reliability, these generally are not known precisely in structural engineering problems owing to insufficient data and information. What are available instead are the estimates $\hat{\mu}_X$ and $\hat{V}_X$, of the mean and COV of X, which are usually computed from idealised models and data gathered under carefully controlled conditions. Therefore, while $\hat{V}_X$ reflects basic statistical variability, it fails to encompass all sources of uncertainty that contribute to the total variability in X.

According to equation (5.114) a basic variable, X may be expressed by

$$X = X_m \cdot X_{\text{pred}}. \tag{5.115}$$

where X_m and X_{pred} are the model uncertainty and the predicted value respectively. X_{pred} may normally be a function of several variables. For instance, consider X to be the wave load effect for a submerged member in a jacket with static behaviour. Assume that the method for predicting X is the design wave method specified in the API RP2A/ISO code. This method is based on a regular wave with height H and an appropriately chosen wave period. Wave kinematics is described by Stokes' fifth-order theory and the wave forces on the wetted structure are obtained by the Morison's equation. X_{pred} may then be expressed for instance by, e.g. Moan (1995)

$$X_{\text{pred}} = k \cdot C(Cd, Cm; \text{wave kinematics}) \cdot H^{\alpha} \tag{5.116}$$

where

- k – influence coefficient – reflecting the transformation of load into its effect,
- C – variable that depends upon the coefficients appearing in Morison's equation, and wave kinematics
- α – exponent depending upon the dimensions of the platform members, and should be determined by the regression analysis for the range of H most relevant for extreme loading.

In this case, X_m needs to reflect the model uncertainty in all the factors, and it may be necessary to split the factor X_m into different contributions. Figure 5.6 shows how the model uncertainty for the predicted load on some fixed platforms can be estimated [Heideman and Weaver, 1992].

In reliability analysis, random variables are conveniently normalised with respect to their nominal values. In this way, the statistics are made applicable to a wide range of design situations. The statistics of the load or resistance variable can easily be computed for each design situation that is defined by nominal loads and resistances, since if

$$X = (X/X_n) \cdot X_n \tag{5.117}$$

then

$$\mu_X = \mu_{(X/X_n)} \cdot X_n \tag{5.118}$$

$$V_X = V_{(X/X_n)} \tag{5.119}$$

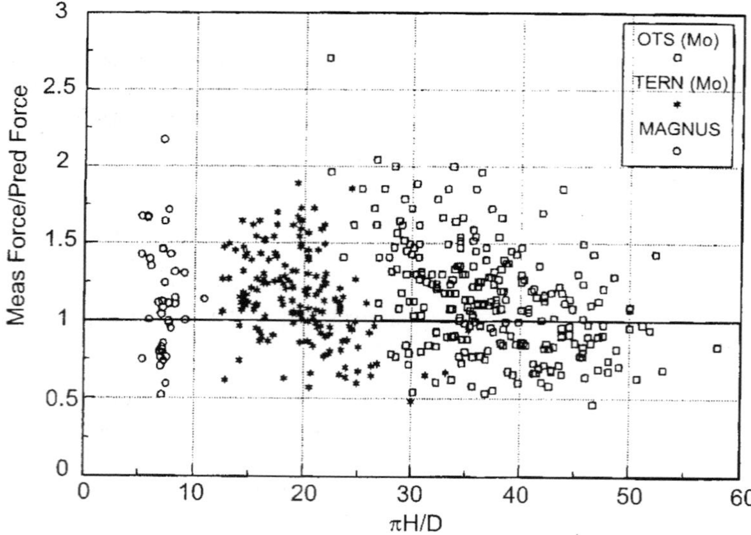

Figure 5.6 Model uncertainty for the API/ISO approach for predicting wave loading on jackets. Mean value and COV for the Tern jacket are 1.06 and 0.25 respectively [Heideman and Weaver, 1992]

Consider, for example, the ultimate capacity of a member with axial compression force. The predicted normalised ultimate capacity f_u/f_y (where f_y is the yield strength) is obtained as a function of member slenderness from curves determined experimentally and given in design codes. The true resistance, in terms of the axial force capacity, R is then expressed by

$$R = N_u = f_u \cdot A = R_{\text{pred}} \cdot X_R \cdot X_A \cdot X_{fy}$$

where

$$X_R = \left(\frac{f_u}{f_y}\right)_{\text{pred}}, \quad X_A = \left(\frac{A}{A_n}\right), \quad X_{fy} = \left(\frac{f_y}{f_{yn}}\right) \tag{5.120}$$

are the model uncertainty and parameter uncertainty in cross-sectional area (A) and yield strength (f_y) respectively [Moan, 1988]. The predicted value, R_{pred}, is based on nominal cross-section area A_n and yield strength f_{yn}. Typically X_R would have a mean value of 1.0–1.1 and a COV in the range of 0.05–0.10 depending upon the type of member and slenderness. X_A is close to 1.0 with a small uncertainty. The mean and COV of X_{fy} are 1.0–1.1 and 0.05–0.10 respectively.

More details about the assessment of uncertainties of loads and resistance are given in reports on code calibration such as Lloyd and Karsan (1988) and Moan (1988).

In most instances the basic resistance variable is taken as the strength of the structure member in question, and the basic load variable is the load effect (moment, shear, etc.) dimensionally consistent with the resistance. These variables can be directly used when the limit state is formulated as a linear combination of resistance and load variables.

5.7.3 Probabilistic Description of Response in Complex Structures

It is not yet feasible in a single load effect analysis to account for the stochastic features of sea waves and current, interaction between incident waves and structure, possible non-linear loads, dynamic behaviour, and the structural geometry to the detail e.g. required for fatigue analysis. Load effects are, therefore, determined in a hierarchy of analyses.

Extreme values for conventional ULS design check typically corresponding to an annual probability of exceeding 10^{-2}, are obtained by

- stochastic analysis considering all relevant sea stated in a long-term period of, say, 20 yr and typically using crude structural models to identify the load pattern (forces or nominal stress) that can be used to determine the extreme response
- deterministic analysis, using a hierarchy of refined models of the structure, using the simplified load pattern

Detailed ULS design checks of platforms is commonly based on refined FE models of the structure, considering functional loads and wave loads based on a regular (design) wave approach. The design wave for bottom-supported platforms, corresponding to a 100-yr return period is calibrated by using the 100-yr design wave height and varying the wave period within a reasonable range. For floating platforms the wave period is usually the most critical wave parameter. Then the most critical wave period (length) is first identified on the basis of response to regular waves and the wave height is calibrated by stochastic analysis. This calibration is carried out for example, for semi-submersibles by considering global response variables such as splitting force, transverse and longitudinal shear and torsion; and relevant short-term sea states – which themselves have to be determined as described below. The stochastic features of waves are especially of importance when dynamic effects due to rigid body motions or flexible structural modes, are significant.

The stochastic analysis is normally based on modelling the wave elevation as a Gaussian process, comprising regular waves with different frequency, phase lag and direction. For linear systems, the stochastic response is most efficiently and accurately obtained in the frequency domain. For a single response variable, Z the response is concisely described by the response spectrum, $S_Z(\omega)$, which can be obtained by equation (5.42) based on the amplitude transfer function from wave to response and wave spectrum respectively. The distribution of individual response peaks and their maximum is known to be a Rayleigh distribution and Gumbel distribution respectively for narrow-band Gaussian response, and the distribution parameters are readily obtained from $S_Z(\omega)$. Also for the wide-band response, there is a theoretical basis that makes the analysis efficient. For non-linear problems, a very limited theoretical basis is available. This applies only to very special cases, involving static response of simple offshore structures to waves. Time-domain simulations are in general required to calculate the non-linear response. It is generally important to ensure that the use of refined stochastic mechanics models are consistent with the current design practice. This means for instance that a stochastic analysis approach should be consistent with the design wave approach for structures with quasi-static behaviour. Moreover, dynamic effects should preferably be considered by their additional forces as compared to the quasi-static ones.

The stochastic long-term analysis becomes especially challenging when non-linearities are important, especially severe non-linearities associated with phenomena like wave slamming and green water effects. It is then necessary to apply time-domain simulation. See for instance Moan (2002). Since proper account of frequency-dependent mass and damping in general requires an integral–differential equation (convolution of e.g. mass with time and acceleration), such analysis becomes very time-consuming. Hence, it is computationally advantageous to use procedures described by Winterstein et al (1993) and Videiro and Moan (2000) to efficiently identify the relevant sea states which need to be considered in the estimation of expected maximum response. It has been shown that even a single sea state could represent the 100-yr value if the response is taken to be that corresponding to a 90–95% fractile.

5.7.3.1 Fatigue analysis

Fatigue crack growth is primarily a local phenomenon and requires that local stresses are calculated. Moreover, the loading over the platform service life is needed. Fatigue analyses are therefore carried out in a sequence of analyses:

(1) Long-term wave climate is the starting point fatigue analysis. This is the aggregate of all sea states occurring yearly (or for longer periods of time). Obtaining this data often requires a major effort, with significant lead times.

(2) Global scale space frame or coarse finite element models based on shell elements is performed to obtain structural response in terms of nominal cyclic stresses in the structure at large for each sea state of interest.

(3) Geometric stress concentrations at all potential hot spot locations within the relevant connections must be considered, since fatigue failure initiates as a local phenomenon.

(4) Accumulated stress cycles are then counted, and applied against suitable fatigue criteria (e.g. Miner's rule) to complete the analysis of fatigue damage.

In view of the scatter and uncertainty in fatigue analysis, the choice of target fatigue life requires careful evaluation of the consequences of fatique failure. Typically, the target fatique life is a multiple of the required service life.

Global analysis of nominal stress histories, considering all relevant sea states, is first carried out. Such analysis would require consideration of 70–400 directional sea states occurring over the life of the structure, with stresses and cumulative fatigue damage being examined at thousands of potentially critical locations throughout the structure. Commonly frequency-domain analysis methods, using coarse finite element models of the structures, are applied for each sea state, by assuming Rayleigh distribution and combining the stress ranges for each sea state by their probability of occurrence. Moderate non-linearities may be handled with "quasi-transfer" functions calibrated by time-domain analysis. In some situations, direct time-domain analysis combined with rainflow counting of cycles is necessary, for example in connection with the intermittent wetting in the splash zone. In the Gulf of Mexico such efforts are undertaken for fixed deep water structures, where structural dynamics amplify the effects of everyday small waves. For monumental North Sea structures, bigger consequences of failure and a more severe wave climate also created fatigue problems which must be analysed. The computer efforts involved make it necessary to improve the efficiency of the long-term analysis, for example, by using response surfaces established by a few representative sea states and appropriately used by interpolation for other sea states.

Equation (5.101) is convenient as a basis for an early screening of fatigue proneness, using a simple (conservative) estimate of the extreme response, s_0 and assuming the shape parameter, B of the Weibull distribution. The parameter B depends upon the environmental conditions, relative magnitude of drag and inertia forces and possible dynamic amplification. For a quasi-static response in an extratropical climate, like the North Sea B may be around 1.0 while "the effective" B may be as low as 0.4–0.6 for Gulf of Mexico platforms subjected to infrequent hurricanes, e.g. Marshall and Luyties (1982), Moan (2002). For structures with predominantly drag forces B will be smaller than for predominantly inertia forces. Structural dynamic effects may start to affect load effects relevant for fatigue when the natural period exceeds 2.0 s. By increasing the natural period from 2 s to 4 s, may, for example, increase B from 0.7 to 1.1 and from 0.9 to 1.3 for Gulf of Mexico and North Sea structures respectively. The implication is a factor of the order of 10 on fatigue damage.

The geometric stress concentration is determined by strain gauge measurements or by refined finite element analysis based on shell or solid elements.

Fatigue design criteria for marine structures are based on two alternative definitions of load effects and the corresponding strength (SN-curves). This is the nominal and the so-called hot spot stress method. Hot spot stress places many different connection geometries on a common basis by incorporating the microscopic notch effects, metallurgical degradation and incipient cracks at the toe of the weld into the *SN*-curve. While the nominal stress approach requires many *SN*-curves, a single *SN*-curve is applied for the hot spot approach. Stresses have to be calculated with due account of weld geometry.

Based on systematic studies of unstiffened tubular joints with different configurations and relative scantlings, parametric formulas for the stress concentration factor have been developed, see e.g. NORSOK Standard (N-004) (1998) and ISO 19900 (1994).

In this section the uncertainties in loads, load effects and resistances have been described. This information together with the limit states (or, failure functions) described in Section 5.6 form the basis for the structural reliability analysis, as outlined in the following sections.

5.8 Structural Reliability Analysis

5.8.1 Elementary Case

The term *structural reliability* should be considered as having two different meanings – a general one and a mathematical one.

- In the most general sense, the *reliability* of a structure is its ability to fulfill its design purpose for some specified time under specified conditions.
- In a narrow sense, it is the *probability* that a structure will not attain each specified limit state (ultimate or serviceability) during a specified *reference period.*

In this context the focus will be on structural reliability in the narrow sense, as the complementary quantity to *failure probability*, P_f defined by equation (5.94). Hence, the *reliability* may be determined as:

$$\Re = 1 - P_f = P[R > S] \tag{5.121}$$

which may be interpreted as a long-run survival frequency or long-run reliability and is the percentage of a notionally infinite set of nominally identical structures, which survive for the duration of the reference period T. $\Re$ may therefore be called a *frequentist reliability*. If, however, the attention is focused on one particular structure (and this is generally the case for one of its kind civil or marine structure), $\Re$ may also be interpreted as a measure of the reliability of that particular structure.

Depending upon how R and S are related to time, the failure probability will also be defined with respect to some reference time period, say year or service life time. This issue is explored later. First, the basic reliability problem – i.e. the calculation of P_f is examined.

When the load effect, S; and the resistance, R for a structural component are described by their probability density functions, $f_S(s)$ and $f_R(r)$, respectively, the failure probability, P_f can be calculated as

$$P_f = P[R \leq S] = \int_0^{\infty} F_R(s)\, f_S(s)\mathrm{d}s = F_M(0) \tag{5.122}$$

where M is given by $M = R - S$. If the probability density function $f_M(m)$ or distribution function $F_M(m)$ is known, P_f may be calculated.

The failure probability can generally be written as:

$$P_f = \Phi(-\beta); \tag{5.123}$$

with

$$\Phi(u) = \int_{-\infty}^{u} \phi(t)\mathrm{d}t = \int_{-\infty}^{u} \frac{1}{\sqrt{2\pi}} \exp\left[-\frac{1}{2}t^2\right]\mathrm{d}t \tag{5.124}$$

and $\phi(x)$ is the probability density of a normal variable with $\mu = 0$ and $\sigma = 1$. Tabulated values of $\Phi(u)$ and $\phi(u)$ may be found in handbooks on mathematics and textbooks on reliability analysis. Hence, the *reliability index* β is uniquely related to the *failure probability*, P_f. This relationship is given for some values of β in Table 5.4.

In some special cases, the failure probability can be expressed analytically. This applies to cases when both R and S have a normal distribution and both R and S have a lognormal distribution.

The reliability index, β for the two special cases of $P_f = P(R - S \leq 0)$ are:

(1) When both R and S are normal variables, β is:

$$\beta = \beta_N = \frac{\mu_R - \mu_S}{\sqrt{\sigma_R^2 + \sigma_S^2}} = \frac{\mu_R/\mu_S - 1}{\sqrt{(\mu_R/\mu_S)^2 V_R^2 + V_S^2}} \tag{5.125}$$

Table 5.4 Relation between β and P_f

β	P_f	β	P_f	β	P_f	β	P_f
1	0.16	2	0.023	3	$1.3 \cdot 10^{-3}$	4	$3.2 \cdot 10^{-5}$
1.1	0.14	2.1	0.018	3.1	$9.7 \cdot 10^{-4}$	4.1	$2.1 \cdot 10^{-5}$
1.2	0.12	2.2	0.014	3.2	$6.9 \cdot 10^{-4}$	4.2	$1.3 \cdot 10^{-5}$
1.3	0.097	2.3	0.011	3.3	$4.8 \cdot 10^{-4}$	4.3	$8.5 \cdot 10^{-6}$
1.4	0.081	2.4	$8.2 \cdot 10^{-3}$	3.4	$3.4 \cdot 10^{-4}$	4.4	$5.4 \cdot 10^{-6}$
1.5	0.067	2.5	$6.2 \cdot 10^{-3}$	3.5	$2.3 \cdot 10^{-4}$	4.5	$3.4 \cdot 10^{-6}$
1.6	0.055	2.6	$4.7 \cdot 10^{-3}$	3.6	$1.6 \cdot 10^{-4}$	4.6	$2.1 \cdot 10^{-6}$
1.7	0.045	2.7	$3.5 \cdot 10^{-3}$	3.7	$1.1 \cdot 10^{-4}$	4.7	$1.3 \cdot 10^{-6}$
1.8	0.036	2.8	$2.6 \cdot 10^{-3}$	3.8	$7.2 \cdot 10^{-5}$	4.8	$7.9 \cdot 10^{-7}$
1.9	0.029	2.9	$1.9 \cdot 10^{-3}$	3.9	$4.8 \cdot 10^{-5}$	4.9	$4.8 \cdot 10^{-7}$

(2) When both R and S are lognormal variables, β is:

$$\beta = \beta_{LN} = \frac{\ln\left[(\mu_R/\mu_S)\cdot\sqrt{(1+V_S^2)/(1+V_R^2)}\,\right]}{\sqrt{\ln(1+V_R^2)(1+V_S^2)}} \approx \frac{\ln(\mu_R/\mu_S)}{\sqrt{V_R^2+V_S^2}} \qquad (5.126)$$

If R and S have a lognormal and normal distribution respectively, it is not possible to determine P_f analytically. In this case, P_f may be obtained by numerical integration of the equation (5.122). However, more general and effective methods are available as subsequently indicated in Section 5.8.2.

Example: Implicit failure probability in design equations

Despite the simplicity of the expression for P_f by equation (5.123) with $\beta = \beta_{LN}$ (equation (5.126)), it has been extensively used in calibration of design codes, as described later. Here, it will be used to illustrate the failure probability implied by the design (equation (5.93)), with only one load effect, S. Let the random load effect, S and resistance, R be defined by:

$$\mu_S = B_S S_C, \text{ typically } \quad B_S \le 1; \quad V_S = 0.15 - 0.30$$

$$\mu_R = B_R R_C, \text{ typically } \quad B_R \ge 1; \quad V_R = 0.1$$

The B_S reflects the ratio of the mean load (which refers to an annual maximum if the annual failure probability is to be calculated, See Section 5.8.2) and the characteristic load effect (typically the 100-yr value) as well as possible bias in predicting wave load effects.

By inserting the design equation (5.93) for one load effect, S, into the approximate expression of equation (5.126)

$$\beta_{LN} \approx \frac{\ln(\mu_R/\mu_S)}{\sqrt{V_R^2+V_S^2}} = \frac{\ln(B_R\gamma_R\gamma_S/B_S)}{\sqrt{V_R^2+V_S^2}}$$

With $\gamma_R \cdot \gamma_S = 1.5$; a typical $B_S = 0.8$ for wave-induced load effects; $B_R = 1.1$ and $V_R = 0.1$, it is found that β_{LN} is about 2.7 and 3.2 for a V_S of 0.25 and 0.20 respectively. These reliability indices correspond to a P_f of $35{\cdot}10^{-4}$ and $7{\cdot}10^{-4}$ respectively. By inspection of the expression for β_{LN} it is seen that a percentage change of B_R and B_S has the same effect as a change of the safety factors, γ_R and γ_S. Hence, efforts should especially be devoted to estimate the possible bias in the model uncertainty for S and R. Moreover, it follows when V_S dominates over V_R, that β_{LN} is inversely proportional to V_S.

5.8.1.1 Tail sensitivity of P_f

It is interesting to compare the β (and, implicitly then P_f) obtained by assuming different distributions for R or S. This may be directly carried out by comparing reliability indices according to equations (5.125)–(5.126), valid when both (R, S) are normal and

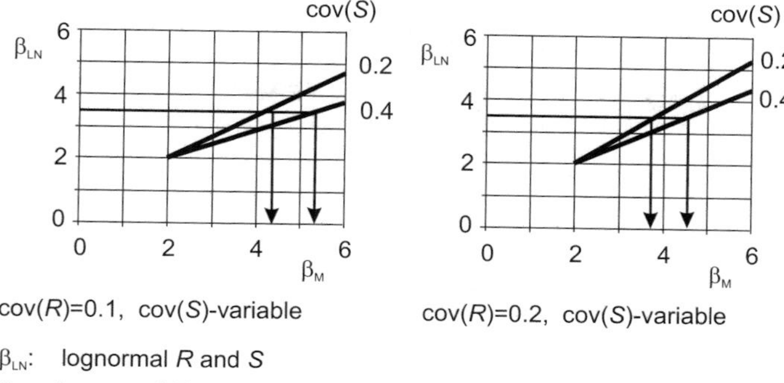

Figure 5.7 Comparison of reliability indices determined by different assumptions about the distributions of *R* and *S*.

lognormal, respectively. Figure 5.7 shows another comparison when βs are obtained under the following assumptions:

- *R* and *S* both with lognormal distribution
- *R* and *S* with lognormal and normal distribution respectively

With a typical $V_R = 0.15$, $V_S = 0.3$ and μ_R/μ_S given such that β_M is in the range 3–4, β_{LN} varies in the range of 2.6–3.3, implying that probabilities differ by a factor of 10.

These results clearly show the sensitivity of reliability to the shape of the distribution (e.g. the upper tail of *S*). In reliability analysis it is therefore crucial to use adequate distributions for the random variables. However, the effect of the "tail sensitivity" of reliability estimates on decisions about safety, will be reduced if the same reliability approach is applied to infer target values as used to estimate the values used to demonstrate compliance with the target values, see e.g. HSE (2002).

5.8.2 Generalisation of Reliability Analysis

The methods for calculating the failure probability outlined above, have resulted in a one-dimensional integral, equation (5.122). The failure probability may alternatively be expressed as the integral of the probability density function for *R* and *S* over the area which corresponds to failure, i.e. $S \geq R$. This area corresponds to all combinations of *R* and *S* where $S \geq R$. With reference to fig. 5.8 this integral may be evaluated as follows, assuming that *R* and *S* are independent variables:

$$P_f = \iint_{M \leq 0} f_{RS}(r, s)\,\mathrm{d}r\,\mathrm{d}s = \iint_{M \leq 0} f_R(r) f_S(s)\,\mathrm{d}r\,\mathrm{d}s \tag{5.127}$$

where $M = R - S$. Instead of doing the integration of $f_{RS}(r, s)$ in the (r, s) domain, the variables (R, S) may be transformed into a space (U_1, U_2) such that these variables are independent and have a standard normal distribution.

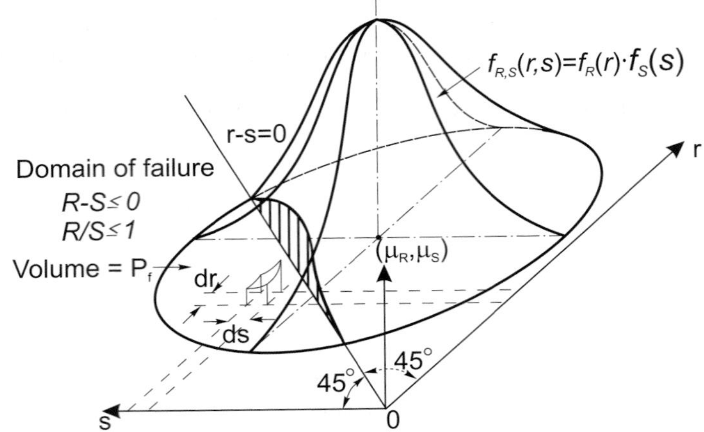

Figure 5.8 Illustration of P_f for a linear failure function and Gaussian R and S

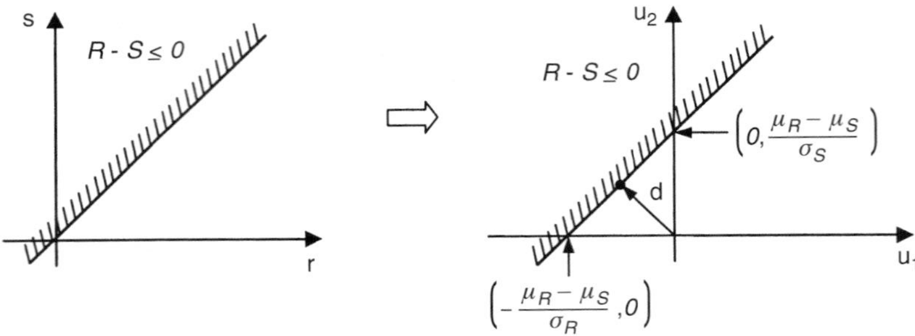

Figure 5.9 Illustration of P_f in the space of independent variables with standard normal distribution for a linear failure function and Gaussian R and S

The transformation of a variable, X with an arbitrary distribution $F_X(x)$ into U with a standard normal distribution $\Phi(u)$ is

$$\Phi(u) = F_X(x) \tag{5.128}$$

The integration in the (u_1, u_2) space (fig. 5.9) becomes very easy, since the variables U_1 and U_2 are independent and have a standard normal distribution. Actually, the result is equation (5.123) with β interpreted as the distance from the origin to the linear failure surface $M(u_1, u_2) = 0$.

The point $\mathbf{u}^*$ is denoted the design point and corresponds to the point with the maximum contribution to the probability of failure, i.e. it is the point in the failure domain $M(\mathbf{u}) \leq 0$ with the highest probability density. This point is located so that the distance to the origin is a minimum.

The notion of failure probability has so far been introduced with reference to a resistance R and a load effect S. Explicit expressions have been given for some special cases of R and S. In general, P_f is expressed by a two-dimensional integral, equation (5.127). This two-dimensional formulation may be generalised by considering multiple variables to describe the problem.

In general, the failure probability for a time-invariant reliability problem, may be formulated as

$$P_f = P[g(\mathbf{X}) \leq 0] \equiv \int_{g(\mathbf{x})\leq 0} \cdots \int f_{\mathbf{X}}(\mathbf{x})\mathrm{d}\mathbf{x} \tag{5.129}$$

where $\mathbf{X}$ is the set of n random variables used to formulate the problem. $f_{\mathbf{X}}(\mathbf{x})$ is the joint probability density function of $\mathbf{X}$. The n-dimensional integral is taken over the region where $g(\mathbf{x}) \leq 0$, corresponding to failure. Examples of failure functions, $g(\mathbf{x})$, are given in Section 5.6.2.

An important class of limit states are those for which all the variables are treated as time independent, either by neglecting time variations in cases where this is considered acceptable or by transforming time-dependent processes into time-invariant variables (e.g. by using extreme value distributions). The integral of equation (5.129) may be calculated by direct integration, simulation or FORM/SORM methods as described in, e.g. textbooks [e.g. Ang and Tang, 1984; Madsen, et al 1986; Melchers, 1999]. In this chapter, methods for structural reliability analysis are briefly outlined.

5.8.2.1 FORM/SORM

The basic steps in the reliability analysis by the First-Order Reliability Method (FORM) comprises the following:

- definition of failure function, $g(x_1, \ldots, x_n)$ and the relevant random variables
- establish uncertainty measures and distributions for the various variables
- transform the possibly dependent variables in the $\mathbf{X}$-space into a $\mathbf{U}$-space of independent standard normal variables according to a probability preserving transformation

$$\mathbf{X} = \mathbf{T}(\mathbf{U}) \tag{5.130}$$

  and correspondingly transform $g(\mathbf{X})$ into $g'(\mathbf{U})$. When the X_is are independent, equation (5.128) applies.

- calculate the failure probability, P_f

$$P_f = P[g'(\mathbf{U}) \leq 0]$$

Hence, according to Madsen et al (1986):

$$P_f = \int_{g'(\mathbf{u})\leq \mathbf{0}} \cdots \int f_{\mathrm{U}}(\mathbf{u})\,\mathrm{d}\mathbf{u} = \int_{g'(\mathbf{u})\leq 0} \cdots \int \phi(u_1)\ldots\phi(u_n)\,\mathrm{d}u_1\ldots\mathrm{d}u_n = \Phi(-\beta) \tag{5.131}$$

where β is the shortest distance from the origin to the surface $g'(\mathbf{u}) = 0$.

Again, the design point, $\mathbf{u}^*$ is the point on $g'(\mathbf{u}) = 0$ with the shortest distance to the origin. The design point is expressed as:

$$\mathbf{u}^* = \beta \cdot \boldsymbol{\alpha}^* \tag{5.132}$$

where the unit vector $\boldsymbol{\alpha}^*$ is positive outward from the origin.

The design values can be established as in the two variable case above. This FORM is an approximate method in which the failure function $g'(\mathbf{u}) = 0$ is approximated by its tangent in the design point. The good accuracy commonly achieved by FORM is due to the rotationally symmetric **U**-space and because the joint standard normal density function, whose bell-shaped peak lies directly above the origin, decreases exponentially as the distance from the origin increases. To determine the design point, a search procedure is generally required.

The parameters α_i ($i = 1, \ldots, n$) represent the direction cosines at the design point. These are also known as the sensitivity factors, as they provide an indication of the relative importance of the uncertainty in basic random variables on the computed reliability. Their absolute value ranges between zero and unity and the closer this is to the upper limit, the more significant the influence of the respective random variable is to reliability. In terms of sign, the convention adopted herein is that resistance variables are associated with negative sensitivity factors, whereas load variables have positive factors. The convention used in ISO 2394 (1998) is the opposite one.

The FORM method can be improved by a second-order reliability method (SORM) [e.g. Madsen, et al 1986].

The SORM is based on a quadratic approximation of the limit state surface at the design point. But experience has shown that the FORM result is sufficient for many structural engineering problems. When using the FORM, the computation of reliability (or equivalently of the probability of failure) is transformed into a geometric problem, that of finding the shortest distance from the origin to the limit state surface in standard normal space.

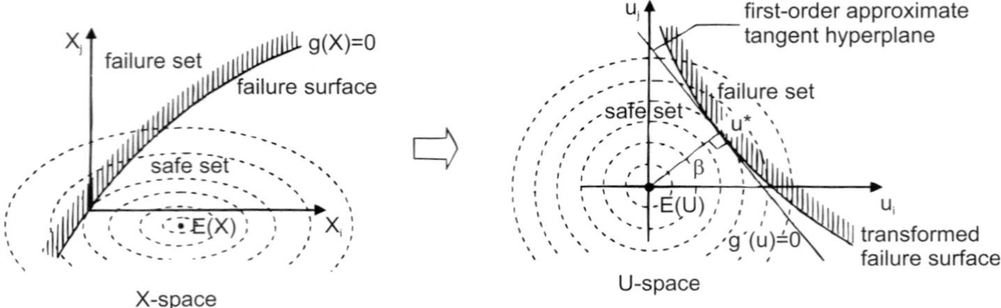

Figure 5.10 FORM analysis for general $g(\mathbf{x})$ function

5.8.2.2 Simulation methods

In this approach, random sampling is employed to simulate a large number of (usually numerical) experiments and to observe the results. In the context of structural reliability, this means, in the simplest approach, sampling the random vector $\mathbf{X}$ to obtain a set of sample values. The limit state function is then evaluated to ascertain whether, for this set, failure (i.e. $g(\mathbf{x}) \leq 0$) has occurred. The experiment is repeated many times and the probability of failure, P_f, is estimated from the number of trials leading to failure (n) divided by the total number of trials, (N), i.e.

$$P_f \approx n(g(\mathbf{X}) \leq 0)/N. \tag{5.133}$$

This so-called Direct or Crude Monte Carlo method is not likely to be of use in practical problems because of the large number of trials required in order to estimate with a certain degree of confidence, the failure probability. It is noted that the number of trials increases as the failure probability decreases. Simple rules may be found, of the form $N = C/P_f$, where N is the required sample size and C is a constant related to the confidence level and the type of function being evaluated.

Thus, the objective of more advanced simulation methods, currently used for reliability evaluation, is to reduce the variance of the estimate of P_f. Such methods can be divided into two categories, namely indicator function methods (such as Importance Sampling) and conditional expectation methods (such as Directional Simulation). Simulation methods are also described in a number of textbooks [e.g. Ang and Tang, 1984; Melchers, 1999].

Time-invariant reliability methods have been extensively used to calibrate ultimate strength code checks based on partial safety factors, to comply with a certain target reliability level [Melchers, 1999]. This application is centred around current design practice, in the sense that the $g(\cdot)$-function can be based on the relevant design equations in an existing code. The main result of the calibration is more consistent safety factors. Calibration of codes is further discussed in Section 5.8.4.

5.8.2.3 Time-variant *R* and *S*

So far, the failure probability has been expressed by time-independent random variables, R and S. In general, R and S are functions of time. For instance, the ultimate resistance may be a slowly decreasing function with time due to crack growth (fatigue) or corrosion. Load effects due to waves clearly vary with time, and a stochastic process model is required to describe load effects. However, good approximations of the reliability may be obtained also in the case of time-dependent resistances and load effects. This is especially the case if the time-dependence of the resistance can be neglected or is very slowly varying, and the load process is e.g. a stationary process. Both $f_R(r|t)$ and $f_S(s|t)$ may then be modelled as time-independent functions. In particular, the load process is replaced by the extreme value distribution, as shown in fig. 5.11.

The probability of failure in t_L may be determined by:

$$P_f(t_L) = P[R \leq S_{\max}(t_L)] \tag{5.134}$$

where $S_{\max}(t_L)$ is max S in the period t_L. This implies that the mean load effect is calculated by the expected maximum value in the reference period for the failure probability, e.g. the

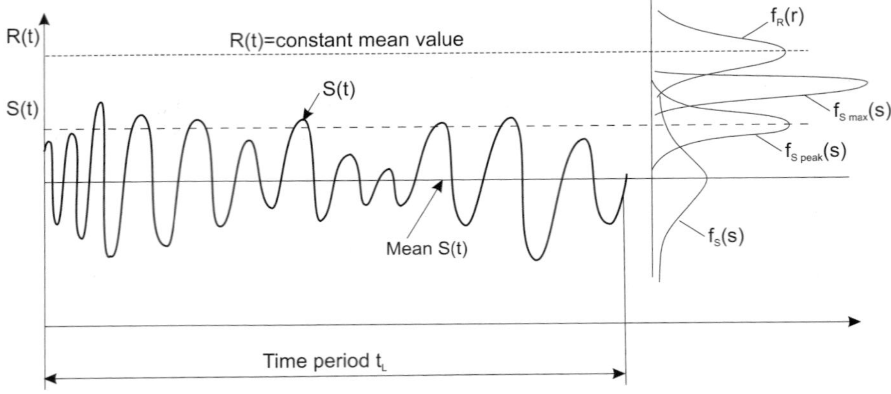

Figure 5.11 Time-dependent reliability problem

annual maximum if the annual P_f is to be calculated. The scatter of this expected maximum is combined with model uncertainties.

The relation between the failure probability per year and n years can be investigated by formulating the probability in n years as follows

$$P_f(T) = P[(M_1 \leq 0) \cup (M_2 \leq 0) \ldots \cup (M_n \leq 0)] \tag{5.135}$$

where M_i refers to the (linear) failure function for year i (annual failure). By assuming that the failure probability in individual years is equal and repeatedly applying the relationship

$$\begin{aligned} &P[(M_1 \leq 0) \cup (M_2 \leq 0)] \\ &= P(M_1 \leq 0) + P(M_2 \leq 0) - P[(M_1 \leq 0) \cap (M_2 \leq 0)] \end{aligned} \tag{5.136}$$

it is seen that the failure probability is bounded by $P(M_i \leq 0)$ and $n \cdot P(M_i \leq 0)$, depending upon the correlation between failure in individual years. Consider in this connection, for instance equations (5.114–5.116) that describe the load on fixed platforms. Here the model uncertainty, X_m will be identical from year to year while the annual maximum wave height relevant for the variable H (when annual failure probabilities are to be calculated), is independent from year to year. If only these two variables are considered as random with lognormal distribution, the correlation between the load (effects) in two years is approximately

$$\rho \approx \frac{V_{X_m}^2}{V_{X_m}^2 + \alpha^2 V_H^2} \tag{5.137}$$

For $V_{X_m} = 0.3$ and $\alpha = 1.6$ and $V_H = 0.15$, ρ becomes approximately 0.6.

While reliability problems involving a single, time-variant load effect, $X(t)$ and otherwise time-independent variables (**X**), can be readily formulated in terms of equation (5.107), this is not the case when the load effects are a vector process, $\mathbf{X}(t)$. Under such circumstances, the conditional failure probability, $P_f(\mathbf{x})$ for a given $\mathbf{X} = \mathbf{x}$ can be determined by estimating the rate of crossing from the safe to failure domain. Instead of formally considering the vector outcrossing, the scalar $g(\mathbf{x})$ can be sampled to check $P[g(\mathbf{x}) \leq 0]$ [Videiro and Moan, 2000]. P_f is then obtained by taking the expectation of $P_f(\mathbf{X})$ over $\mathbf{X}$.

5.8.2.4 Further generalisation of calculation of reliability

In the previous section it was shown that it would be necessary to calculate the probability of the unions and intersections in conjunction with expression (5.135). In general, this is necessary in connection with system reliability (Section 5.9) as well as Bayes' statistics (Section 5.10).

It is necessary to calculate probabilities like

$$\begin{aligned} P_s &= P[(M_1 \leq 0) \cup (M_2 \leq 0) \ldots \cup (M_n \leq 0)] \\ P_p &= P[(M_1 \leq 0) \cap (M_2 \leq 0) \ldots \cap (M_n \leq 0)] \\ P_{\text{general}} &= P[\cup_i \cap_j (M_{ij} \leq 0)] \end{aligned} \tag{5.138}$$

FORM/SORM and Monte Carlo simulation methods may be applied. Particular caution needs to be exercised when linearising the different failure functions in FORM analysis of the union of events.

A reliability method which gives acceptable estimates of the reliability for the structure or structural component shall be used. The choice of the method must be justified.

In general, it should be noted that FORM and SORM are approximate analytical methods. The advantage of the analytical methods is that they are fast. FORM and SORM have proved to be useful tools for evaluating reliability of marine structures.

But these methods need to be verified by simulation. Simulation can then be used to verify if this local estimate is sufficient as an estimate of the global reliability when there is more than one estimation point.

Reliability estimates by simulation methods are considered verified if a sufficient number of simulations are carried out. Simulations by basic Monte Carlo methods should be carried out with a number of simulation samples not less than $100/P_f$ where P_f denotes the failure probability. Simulations by other methods should be carried out such that the estimate of P_f is positive and the coefficient of variation in the simulations is less than 10%.

For FORM and SORM solutions, is recommended to check whether the estimation point has a sound physical interpretation.

Besides an approximation to the failure probability, the reliability analysis methods also provide importance and sensitivity measures.

Uncertainty Importance Measures. The uncertainty importance factor, α_i, indicates the importance of modelling the random variable X_i as a distributed variable rather than as a fixed valued variable, the median of the distribution being the fixed value. For the FORM analysis, it can be shown that the reliability index is increased by a factor $1/\sqrt{1-\alpha_i^{*2}}$

(called omission sensitivity factor) if the uncertainty in variable i is ignored and the variable is replaced by its median value (50% fractile). In a process of deciding on variables for which to collect further data to reduce the overall uncertainty, these parameters obviously give very useful guidance.

Sensitivity Measures. The parametric sensitivity factor gives the change in failure probability (through change in reliability index) to an increment of the parameter θ, whether a statistical distribution parameter or a deterministic parameter. Hence, the sensitivity of the approximate reliability index given by equation (5.126) is expressed by

$$\Delta\beta = \frac{\partial\beta}{\partial\theta}\cdot\Delta\theta$$

$$\text{with}\quad \frac{\partial\beta}{\partial\mu_R} = \frac{1/\mu_R}{\sqrt{V_R^2+V_S^2}} \tag{5.139}$$

$$\frac{\partial\beta}{\partial V_R} = \frac{\ln(\mu_R/\mu_S)}{\sqrt{V_R^2+V_S^2}}\cdot\frac{V_R}{V_R^2+V_S^2} = -\beta\frac{V_R}{V_R^2+V_S^2}$$

The FORM and SORM methods give parametric sensitivity factors and uncertainty importance factors for the reliability index. The directional simulation method provides parametric sensitivity factors of the reliability index.

5.8.2.5 Target reliability level

Reliability analysis is a decision tool. Hence, a calculated P_f is normally compared with an acceptable failure probability, P_{ft}, the target level for P_f. The target levels depend upon the consequences of failure, the ultimate consequences being loss of lives, pollution and property loss, see e.g. HSE (2002). Hence, if target levels for components are to be decided, they should depend upon the potential for progressive failure after a single component failure, i.e. the residual strength after component failure. However, this principle is difficult to implement, because failure of one component might imply failure of other components due to correlation between failure events, see e.g. HSE (2002).

The target levels may be set to be the average implied P_f in existing structures considered to have acceptable safety level. Alternatively, the target level can be reduced or increased compared to the implied level. It is particularly challenging to establish the target level for structures with a novel type of function or a novel layout. Consider for instance the introduction of FPSOs in the offshore industry. While ships have been used as trading vessels, the mode of operation in the offshore industry is different and warrant particular considerations.

Another issue is that the P_f is sensitive to the assumption of distributions made for loads and resistance variables, especially the character of the distribution tail. This fact shows that the target level needs to be determined based on the same reliability methodology as that which later will be applied to demonstrate compliance with the target level.

P_{ft}s are either referred to annual or intended service life (20 yr) values. It might be argued that the annual values are relevant if the focus is on fatality rate for individual personnel on board the platform, while service life values are relevant if the emphasis is placed on cost benefit considerations of the installation. By using the relationships expressed above, annual and service life values of P_f can easily be related for ultimate failure events relating to extreme loads. Fatigue failure probabilities naturally lend themselves for reference to service life. However, an alternative in connection with fatique failure, is to use hazard rate as discussed subsequently.

5.8.3 Fatigue Reliability

5.8.3.1 Elementary format

The elementary reliability format, equation (5.126), may also be used to obtain an estimate for the fatigue reliability based on *SN* formulation. It is then noted from equations (5.95) and (5.102) that equation (5.126) can be used with $R = \Delta$ and $S = D$. Moreover, if only the dominant variables s_0 and K are taken as random variables with lognormal distribution and other parameters are taken to be constants, D is characterised by

$$\mu_D = k \cdot \frac{\mu_{S0}^m}{\mu_K} \cdot \sqrt{\frac{1 + V_K^2}{(1 + V_{S_0}^2)^m}} \sqrt{1 + V_D^2}; \tag{5.140}$$

$$\ln(1 + V_D^2) = \ln(1 + V_K^2)(1 + V_{S_0}^2)^{m^2} \quad \text{or} \quad V_D^2 \approx V_K^2 + m^2 V_{S_0}^2 \tag{5.141}$$

where

$$k = N_T \cdot \Gamma(m/B + 1)/(\ln N_0)^{m/B}$$

The failure probability refers to the time period T, with N_T stress cycles.

Analogous with the exemplification of the implicit P_f in the design expression (5.93) for ULS, the implicit P_f in the fatigue design check $D \leq \Delta_d$, can be estimated using the β_{LN} in equation (5.126) as a measure of reliability. Based on the equality in the design equation and using $D_C = k \cdot \mu_{S_0}^m / K_C$, $\mu_D \approx k \cdot (1 + V_D^2)^{1/2} \cdot \mu_{S_0}^m / \mu_K$, equation (5.126) may be written as

$$\beta_{LN(\text{fatigue})} \approx \frac{\ln(\mu_\Delta/\mu_D)}{\sqrt{V_\Delta^2 + V_D^2}} = \frac{\ln\left[(\mu_\Delta/\Delta_d)(\mu_K/K_C)(1 + V_D^2)^{-1/2}\right]}{\sqrt{V_\Delta^2 + V_D^2}} \tag{5.142}$$

Typically $\mu_\Delta \approx 1.0$ and V_Δ =0.2–0.3 [Wirsching, 1983]. Here, it is assumed that $\mu_\Delta = 1.0$ and $V_\Delta = 0.25$. The characteristic value K_C is determined as the mean value minus two standard deviations. With the relevant uncertainty in *SN*-curves this implies a μ_K/K_C of the order 2.5–3.3. Here this ratio is assumed to be 3.0. The uncertainty in *SN*-curves typically corresponds to a V_K of 0.4–0.5. The uncertainty in V_{S0} consists of contributions from the load and global load effect calculation, and the total uncertainty would correspond to a V_{S0} in the range 0.15–0.30. Based on equation (5.141) V_D would be of the order of 0.75. Hence, $\beta_{LN\ \text{fatigue}} = 1.33 \cdot [0.88 - \ln(\Delta_d)]$. This means that $\beta_{LN} = 1.2$, 2.1 and 4.2, when $\Delta_d = 1.0$, 0.5 and 0.1 respectively; implying a P_f in the service life that is about 0.12, 0.017 and $0.13 \cdot 10^{-4}$ respectively. It is seen that P_f is very sensitive to the fatigue design criterion, i.e. Δ_d.

More refined fatigue reliability models can be established by using the fracture mechanics model or by modelling the random variables in a more refined manner. Such models are especially relevant in connection reliability analysis involving additional information due to inspection etc, as discussed in Section 5.10.

5.8.3.2 Hazard rate

Time-dependent reliability problems may be solved by means of the hazard function, which is based on the conditional probability theory. In principle, the hazard function or hazard rate, may be interpreted as the frequency of failure per unit of time. Let $F(t)$ be the distribution function of the random time-to-failure T, and let $f(t)$ be its probability density function. Then the hazard rate $h(t)$ is defined as, see for instance Melchers (1999),

$$h(t) = \frac{f(t)}{1 - F(t)} \tag{5.143}$$

This expression represents the probability that the structure of age t will fail in the interval $t + \Delta t$, given that it has survived up to time t, namely

$$\begin{aligned} h(t) &= P[T \leq (t + \Delta t)|T > t] = \frac{P[t \leq T \leq t + \Delta t]}{\Delta t \cdot (1 - P[T \leq t])} \\ &= \frac{\mathrm{d}P_f(t)/\mathrm{d}t}{1 - P_F(t)} = -\frac{\phi(\beta(t))}{1 - \Phi(-\beta(t))} \cdot \frac{\mathrm{d}\beta(t)}{\mathrm{d}t} \end{aligned} \tag{5.144}$$

5.8.3.3 Time variant interaction between strength degradation and overload failure

In the fatigue analysis, the cyclic loading is assumed to be steady, without account of sequence effect and the occurrence of extreme stresses that can cause fracture before the fatigue life is exhausted. For instance, *SN*-curves refer to constant amplitude loading and the approach used implies that variable amplitude loading is represented by an equivalent constant amplitude loading. However, the resistance to fracture or rupture is decreasing when the crack grows. The rupture criterion may be expressed in terms of the stress intensity factor K_I as follows:

$$K_I = Y(a(t)) \cdot S \cdot \sqrt{\pi \cdot a(t)} \geq K_{IC} \tag{5.145}$$

where $a(t)$ is the crack size at the time t, $Y(a)$ is a function of the geometry adjacent to the crack, and K_{IC} is a material parameter. S is the stress range that varies randomly with time.

Failure can occur when a high stress occurs at a crack size well below the critical one from a fatigue point of view. The problem of calculating the reliability under these circumstances is truly a *time-variant* reliability problem and is very time consuming, especially when a long-term variability of sea states is to be taken into account, Madsen et al (1986). However, it has been demonstrated by Marley and Moan (1992) that the reliability in many situations can be efficiently calculated by replacing the time-variant problem with a time invariant one. This is achieved by calculating the probability of fracture in the service life using the fatigue crack size at the end of the service period. This approach is conservative, but not much.

5.8.4 Design Values and Calibration of Partial Factors

5.8.4.1 Probability-based design values

In the previous section, methods to estimate the failure probability P_f have been briefly introduced. It is also shown that, P_f is uniquely defined by the reliability index β. In this section it will be shown that, for a specific case (design equation, uncertainty measures of R and S, ...), a unique relationship between β and partial factors, γ can been established, so that the design values for loads are given by $Q_d = \gamma Q_c$. Alternatively, the design values, Q_d can be expressed by certain fractiles, defined by sensitivity factors, α_i, and the reliability index, β. This relationship is also unique for each specific case.

These relationships can be demonstrated by using the FORM solution of the reliability problem. Here it is illustrated for the case which is defined by the random variables R and S and the failure function, $g(R, S) \geq R - S$.

The coordinates of the design point, $\mathbf{u}^*$ are

$$u_1^* = \alpha_1^* \cdot \beta \tag{5.146}$$

$$u_2^* = \alpha_2^* \cdot \beta \tag{5.147}$$

where α_i^* are the directional cosine with respect to the axis of u_i. (Note that with the opposite definition of sign for $\boldsymbol{\alpha}$, $\boldsymbol{u}^* = -\beta \cdot \boldsymbol{\alpha}^*$.) When R and S have normal distributions

$$\alpha_1^* = -\frac{\sigma_R}{\sqrt{\sigma_R^2 + \sigma_S^2}} \quad \text{(negative)} \tag{5.148}$$

$$\alpha_2^* = \frac{\sigma_S}{\sqrt{\sigma_R^2 + \sigma_S^2}} \quad \text{(positive)} \tag{5.149}$$

and

$$\alpha_1^{*2} + \alpha_2^{*2} = 1.0. \tag{5.150}$$

According to the transformation $\Phi(u) = F_X(x)$, the corresponding values, r^* and s^* of R and S are given by the following relations when R and S have *normal distributions*.

$$F_R(r^*) = \Phi\left(\frac{r^* - \mu_R}{\sigma_R}\right) = \Phi(u_1^*) \tag{5.151}$$

$$F_S(s^*) = \Phi\left(\frac{s^* - \mu_S}{\sigma_S}\right) = \Phi(u_2^*) \tag{5.152}$$

The design values can, hence, be written as:

$$R_d = r^* = u_1^*\sigma_R + \mu_R = \mu_R + \beta\alpha_1^*\sigma_R \quad (5.153)$$

$$S_d = s^* = u_2^*\sigma_S + \mu_S = \mu_S + \beta\alpha_2^*\sigma_S \quad (5.154)$$

Design values of resistance: $R_d = \mu_R - \beta\alpha_1'\sigma_R$ ($\alpha_1' = -\alpha_1^*$) may be specified in two alternative ways:

- $R_d = R_c/\gamma_R$ with the characteristic value R_c defined by e.g. $P[R \leq R_c] = 0.05$
- R_d directly defined by a fractile, by inserting equation (5.153) into equation (5.151), with $\alpha_1^* = -\alpha_1'$:

$$P[R \leq R_d] = \Phi(-\beta\alpha_1') \quad (5.155)$$

For example, with the following assumptions:

$$R_c = \mu_R(1 - k_R V_R), \quad (5.156)$$

the material factor, can be obtained from

$$\gamma_R = R_c/R_d = \mu_R(1 - k_R V_R)/(\mu_R + \beta\alpha_1^*\sigma_R) = (1 - k_R V_R)/(1 + \beta\alpha_1^* V_R) \quad (5.157)$$

where it is seen that γ_R depend upon the target safety level specified by β and α_1^*. Alternatively, the design value can be defined by

$$P[R \leq R_d] = \Phi(-\beta\alpha_1'). \quad (5.158)$$

Similarly, the design values for the load effects can be specified by $S_d = \mu_S + \beta\alpha_2'\sigma_S$ and can be written by:

$$S_d = \gamma_S S_C \quad (5.159)$$

where $P[S \geq S_c] = 10^{-2}$ and $\gamma_S = S_d/S_c = (\mu_S + \beta\alpha_2'\sigma_S)/S_c$, or

$$P[S \geq S_d] = 1 - P[S \leq S_d] = 1 - \Phi(\beta\alpha_2') = \Phi(-\beta\alpha_2') \quad (5.160)$$

A design format where the design values are specified by a characteristic value and a safety factor directly as a fractile, is denoted a *design value format*. If the reliability problem is solved to ensure that the implied $P_f = P_{ft}$(or, $\beta = \beta_t$) and the result is applied to specify design values according to the alternative definitions above, they yield the exact result. The design value format then does not offer any advantage as compared to a full reliability-based design. However, in practice the α_i-*factors are assumed* (based on experience) and the target safety level, in terms of β, is selected. Then, R_d and S_d can be easily obtained without reliability analysis.

Example. Considering a case with $k = 1.65$ for $P[R \leq R_C] = 0.05$; $\beta = 3.7$; $\alpha_1' = -0.8$; $V_R = 0.15$. These data imply the following values that specify the design value, R_d in an identical manner:

$$P[R \leq R_c] = 0.05, \quad \gamma = 1.35; \quad \text{(from equation (5.157))}$$

$$P[R \leq R_d] = 4.6 \cdot 10^{-3}$$

The discussion in this section shows that there are various ways to obtain the same safety level. For instance, one may make characteristic values of loads high and the partial factors small or vice versa.

Finally, the latter formulation was approximated by specifying β to be the target level and *assuming* the values of α_i i.e. the reliability problem was not explicitly solved.

As mentioned above, when a reliability analysis is carried out case by case, there is a unique relationship between P_f (or β) and the design value. However, this would imply that a safety factor will vary depending upon the case at hand. Such an approach, however, will not be convenient to use in practical design. The simplest way to achieve a practical design format is by applying the design value format with assumed, representative values of the α-factors. However, offshore codes have been established by determining a set of partial factors (γ_i) that yields a P_f or β as close as possible to the desired target level, P_{ft} or β_t.

The approach is pursued in the following sub-section, considering a simple example.

5.8.4.2 Multiple loads, single load effect – strength variable

Consider the format

$$\gamma_{f1} S_{1c} + \gamma_{f2} S_{2c} \le R_c / \gamma_m \tag{5.161}$$

where S_{1c}, S_{2c} are characteristic load effects due to load types 1 and 2, γ_{f1}, γ_{f2} are load factors, R_c is characteristic resistance and γ_m is a material factor.

Assume that the characteristic values are expressed by the (true) mean values of the random variables R and S_i as follows:

$$\mu_R = B_R R_c \quad \text{and} \quad \mu_{Si} = B_{Si} S_{ic}$$

where B_{Si} and B_R are "bias" factors.

Table 5.5 illustrates a typical set of uncertainty measures.

The aim in this example is then to determine one set of $\gamma_{f1}, \gamma_{f2}, \gamma_m$ which should be applied to all combinations of action effects S_{1c} and S_{2c}, $\gamma_m = 1.15$, and yield a P_f and β as close as possible to the target values. Here the reliability is assumed to be given by R and S_i with normal distribution, corresponding to β given by equation (5.125).

This task is commonly denoted "calibration of partial factors" or "code calibration". γ_{f1} and γ_{f2} are then determined so that $|\beta - \beta_t)|_{\min}$ for, say, for an assumed range of

Table 5.5 Probabilistic character of random variables

Variable, X_i	$B_{X_i} = \mu_{X_i} / x_{i(c)}$	Coefficient of variation, V_{Xi}
R	1.15	0.12
S_1 (permanent)	1.05	0.10
S_2 (variable)	0.80	0.30

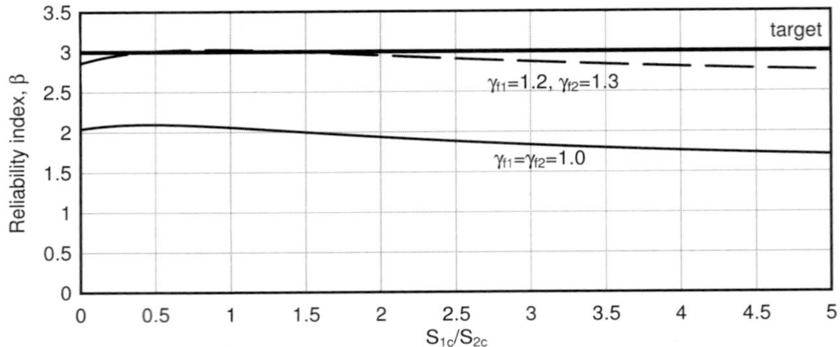

Figure 5.12 Safety index, β as a function of the load ratio, $k = S_{1c}/S_{2c}$ for an "optimal" set of load factors

$k = S_{1c}/S_{2c} = (B_{S2}/B_{S1})(\mu_{S1}/\mu_{S2})$: $0.2 < k < 5.0$. Here,

$$\Pi = \sum_{i=1}^{9}(\beta_i - \beta_t)^2; \quad \beta_t = 3.0; \quad \text{is minimised} \tag{5.162}$$

The solution is found to be:

$$\gamma_{f1} = 1.2; \quad \gamma_{f2} = 1.3. \tag{5.163}$$

The relative high γ_{f1} in this case is due to the large "bias" (B_{S1}) in S_1. A plot of β as a function of k is shown in fig. 5.12.

In the above calibration, one set of partial factors (γ_{f1}, γ_{f2}) was applied,with a given γ_m. It would normally be necessary to apply several sets to ensure a close fit to the target level, $\beta_t(P_{ft})$. Typically then one set of (γ_{f1}, γ_{f2}) could be (γ_{f1},1.0) and the other one (1.0, γ_{f2}) where γ_{f1} and γ_{f2} may be in the range 1.2–1.6.

5.8.4.3 General calibrations

Examples of more complete calibrations of codes may be found e.g. in CIRIA (1977), Fjeld (1977), Lloyd and Karsan (1988), Moan (1988) and Jordaan (1991). The calibration includes the following items:

- Establish code format, in terms of design parameters and safety factors ($\gamma_{f(ij)}$, $\gamma_{m(j)}$) e.g. for all types of components (j) (beams, girders, joints,...) and types of loads (i).
- Define characteristic values, $R_{c(j)}$, $S_{c(ij)}$
- Decide on possible limitation, e.g. load factors $\gamma_{f(ij)} = \gamma_{f(i)}$ (i.e. the same load factors for all types of components for a given type of load).
- Quantify the uncertainties (distribution, mean value, coefficient of variation) for all variables.
- Decide on a measure for failure probability, e.g. a simple lognormal format as used by API, or FORM etc.

- Decide on the target reliability level, commonly by calculating the implied P_f in cases which are considered to have an acceptable safety level.
- Determine (by iteration) the values $\gamma_{m(j)}$ and $\gamma_{f(ij)}$ that make the design format yield a failure probability for all load combinations and structural components as close as possible to the target P_f.

Structural reliability has been applied to calibrate ULS code requirements [Fjeld, 1977]. Later significant effort by API [Moses, 1987; Lloyd and Karsan, 1988] to develop a load and resistance factor design for jackets is noted. In certain situations when a new design falls outside the scope of existing codes, reliability analysis has been applied adhoc to establish design criteria. This was the case when the first offshore production ship was designed some years ago. It then became clear that application of the ship rules for trading vessels and existing offshore codes differed significantly, implying a difference in steel weight of the order of 20–30%. Moan (1988) conducted a study to establish ultimate strength criteria for this type of vessel, which complied with the inherent safety level in the existing NPD code for offshore structures. An evaluation of previous calibration efforts for offshore codes was provided by Moan (1995) in conjunction with the ISO effort to harmonise codes for offshore structures [ISO 2394, 1998; ISO 19900, 1994].

5.8.5 Probabilistic Calibration of Combination Values for Loads

Most structural loads vary with time. If a structural component is subjected to only one time-varying load in addition to its permanent load, the reliability may be determined simply by considering the combination of the dead load with the maximum time-varying load during some appropriate reference period. It is frequently the case, however, that more than one time-varying load will be acting on a structure at any given time. Conceptually, these load combinations should be dealt with by applying the theory of stochastic processes, which accounts for the stochastic nature and correlation of the loads in space and time.

Loads (or their effects) acting on structural elements typically are represented by various combinations of load process models. Permanent loads change very slowly and maintain a relatively constant (albeit random) magnitude. Wave and wind loads are variable loads, which are more or less continuous. Earthquake loads are also variable and occur in short periods.

The analysis of reliability associated with the ultimate limit states requires that the maximum total load during a reference period, typically taken as 100 yr, be characterized. When more than one time-varying load is present, it is extremely unlikely that each load will reach its peak lifetime value at the same moment. Consequently, a structural component could be designed for a total load which is less than the sum of the individual maximum loads.

To handle combination of loads – and their effect, various methods can be applied. Turkstra's rule [Turkstra, 1970] is the simplest, but may not be accurate enough. Efforts have been made to improve the method, cf. Naess and Royset (2000). To illustrate this principle, assume a linear relationship between the loads' effect S and the loads:

$$S = a_1 Q_1 + a_2 Q_2 = S(Q_1, Q_2) \tag{5.164}$$

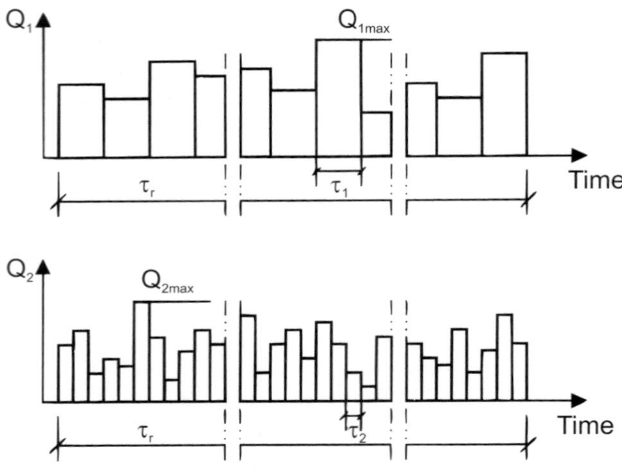

Figure 5.13 Load process models

The maximum load effect $S_{\max}$ from Q_1 and Q_2 during the reference period T can then be written as:

$$S_{\max} = \max S\{Q_1, Q_2\} = \max\{S(Q_{1\max}, E(Q_2));\ S(E(Q_1), Q_{2\max})\} \tag{5.165}$$

where $E(Q_1)$ denotes the mean value of Q_i. The maximum should be taken over all intervals T within the reference period.

A further refinement can be obtained by assuming that the loads can be described in a simplified manner as shown in fig. 5.13. This representation is due to Ferry Borges and Castanheta (1971). The following assumptions are made about the processes:

- $Q_1(t)$ and $Q_2(t)$ are stationary ergodic processes
- All intervals τ_1 are equal
- All intervals τ_2 are equal
- $\tau_1 \geq \tau_2$
- r_1 and r_2/r_1 are integers, where $r_1 = t_r/\tau_1$ and $r_2 = t_r/\tau_2$
- Q_1 and Q_2 are constant during each interval τ_1 and τ_2
- The values of Q_1 for the different intervals are mutually independent: the same holds for Q_2
- Q_1 and Q_2 are independent

It is noted that the methods presented subsequently will be approximately valid if the load process, Q_i is zero over a number of intervals, τ_1. r_i is then taken to be the effective number of intervals, where the load is non-zero.

Ocean wave conditions are frequency assumed to be stationary with a constant significant wave height in a 3 h period. However, in this case, the wave condition in the previous and subsequent 3 h period may not be completely independent. Assuming independence is, however, conservative.

In this case, $S_{\max}$ in the reference period may be expressed by

$$S_{\max} = \max_{\substack{r_1 \\ \text{events}}} \left[a_1 Q_1 + a_2 \max_{\substack{t, t+\tau_1 \\ r_2/r_1 \\ \text{events}}} [Q_2] \right] \tag{5.166}$$

In the following it is, for simplicity, assumed that $a_1 = a_2 = 1.0$.

In the interval τ_1 there are r_2/r_1 events of Q_2, and r_1 events $Q_1 + \max(Q_2)$ in t_r.

The distribution of $\max[a_1 + a_2 \max(Q_2)]$ is then (Ferry Borges and Castanheta 1971).

$$F_{S\max,t_r}(q) = \left\{ \int_{-\infty}^{\infty} F_{Q_2}(q - \nu)^{r_2/r_1} f_{Q_1}(\nu) \mathrm{d}\nu \right\}^{r_1} \tag{5.167}$$

This formula can be extended to the sum of more than two processes. However, it is complicated to calculate analytically. In general, numerical integration needs to be applied to determine the integrals, equation (5.167)

Still, more refined methods for treating the load combination problem are outlined by Wen (1990). However, it was found by Wang and Moan (1996) that the Borges–Castanheta model gave good results for combining still-water and wave load effects on an FPSO. By writing the design values

$$M_c = M_{wc} + \psi_{sw} M_{swc} \quad \text{or} \quad M_c = \psi_w M_{wc} + M_{swc}$$

it was found that the ψ_{sw} and ψ_w were in the range of 0.6–0.4 and 0.75–0.55 respectively. Factors were different for hog and sag, and was smaller for characteristic values which referred to long return periods.

5.9 System Reliability

5.9.1 General

The ultimate and fatigue design criteria in current codes are based on component failure modes (limit states) and commonly a linear global model of the structure is applied to determine the load effects in the components. However, an approach, which is based on global (system) failure modes of the structure is desirable because significant consequences, e.g. fatalities, will primarily be caused by global failure of offshore structures. A suitable system approach is also necessary to obtain the optimal balance between design and inspection plan since the effect of the inspection and repair strategy on the safety will normally depend on a certain damage tolerance, especially when the inspections rely on detecting flooded or failed members.

In the system approach the relevant structure is assumed to be composed of different physical components (members, joints, piles, ...) which may each have different failure modes, e.g. different collapse, fracture or fatigue failure modes.

System failure is then expressed mathematically by load- and resistance parameters relating to all failure modes for all components, and the system failure probability is calculated by the probabilistic properties of these parameters. Broadly speaking, this may be achieved by a failure mode (or survival mode) analysis, or direct simulation methods [Melchers, 1999; Karamchandani, 1990; Moses and Liu, 1992]. The failure mode analysis consists in

- identifying the sequences of component failures for each system failure mode, ES_i, considering members, joints and other components
- establishing a mathematical expression for the events of each sequence, ES_i, based on structural mechanics. The event sequence no. i, ES_i, may involve failure of n_i components such that

$$ES_i : E_{i_1} \cap E_{i_2} \cap \ldots \cap E_{i_{ni}} \tag{5.168}$$

where E_{i_j} is the event that component no. i_j fails given that the $i_1, \ldots, i_{j-1}$ components already have failed.

- establishing probabilistic measures for the random variables involved
- calculating the failure probability of the system.

In the following section, general system reliability analysis based on the failure mode approach is briefly reviewed.

Failure of redundant offshore structures may also be initiated by a fatigue failure or a fatigue-induced fracture. If repair is not accomplished, a second fatigue or overload failure may occur. Even if the damage is detected, it may not be repaired until some time later, depending upon load conditions. The increased stress in the remaining members will contribute to this second failure. While progressive overload failures are assumed to take place instantly (during the 18–20 s period of a storm wave), the fatigue failures occur at different points in time.

For each sequence (i) of fatigue failures, the failure functions for successive failures, in each mode may be established [Shetty, 1992; Dalane, 1993]. This task is even more complex than for overload failure, because the successive failures depend upon the (random) time between different failure events (memory effect).

Modelling the uncertainties in loads and resistances in the components of the system is a crucial task. The system approach in addition requires an estimate of the uncertainty of the system model as well as the correlation between variables in the different failure functions that represent the system.

Correlation in strength variables arises if joints belong to the same "batch", since the between-batch variability is predominant. Correlation in stress due to common hydrodynamic factors depends upon location in the same vertical truss plane, and closeness in space. Correlation in stress concentration factors depends upon geometric similarity.

While the failure probability of a series system with n components may vary by a factor of n depending upon the correlation, the failure probability of a parallel system may vary even more – depending upon correlation and component characteristics.

Typically load uncertainties predominate in the calculation of the probability of wave overload of jackets, while load and resistance uncertainties are of the same order of magnitude in fatigue problems. Also, the correlation between component failure modes is less in fatigue. The effect of the correlation depends upon the system, i.e. whether it is a series or parallel system.The small correlation between fatigue and overload failure events implies large systems effects for failure modes comprising fatigue events and a single overload failure.

Having established the limit state $g_{i_j}(\cdot)$ and uncertainty measures for all random variables the failure probability may be calculated, using:

$$P_{f\,sys} = P[FSYS] = P\left[\cup_{i=1}^{N} \cap_{j=1}^{n_i} \left(g_{i_j}(\cdot) \leq 0\right)\right] \tag{5.169}$$

by FORM/SORM, bounding techniques or simulation methods. Due to the effort involved, it is important to apply some kind of technique to limit the number (N) of failure modes considered in the analysis.

5.9.2 Analysis of Simple Systems

To illustrate the complexity of system analysis, a series and a most simple parallel system will be considered.

A *series system* fails when any component has failed and the failure probability can be expressed by equation (5.135). When the failure functions M_i are linear or have been linearized, $P_{f\,sys}$ can be expressed by FORM as follows [Madsen, et al 1986]

$$P_{f\,sys} = P[(M_1 \leq 0) \cup (M_2 \leq 0) \cup \ldots \cup (M_n \leq 0)] \leq 1 - \Phi_n(\boldsymbol{\beta}, \mathbf{R}) \tag{5.170}$$

where $\boldsymbol{\beta}$ is the vector of reliability indices for the failure functions and $\mathbf{R}$ is the correlation matrix for the failure functions, M_i. $\Phi_n(\cdot,\cdot)$ is the n-dimensional standard normal distribution function. However, commonly upper and lower bounds are also applied to calculate $P_{f\,sys}$ for series systems, see e.g. Madsen et al (1986).

Similarly, the failure probability of a (stand by) parallel system can be expressed by FORM as follows

$$P_{f\,sys} = P[(M_1 \leq 0) \cap (M_2 \leq 0) \cap \ldots \cap (M_n \leq 0)] = \Phi_n\,(-\boldsymbol{\beta}, \mathbf{R}) \tag{5.171}$$

The value of $\Phi_n(\cdot,\cdot)$ normally needs to be determined by some numerical method.

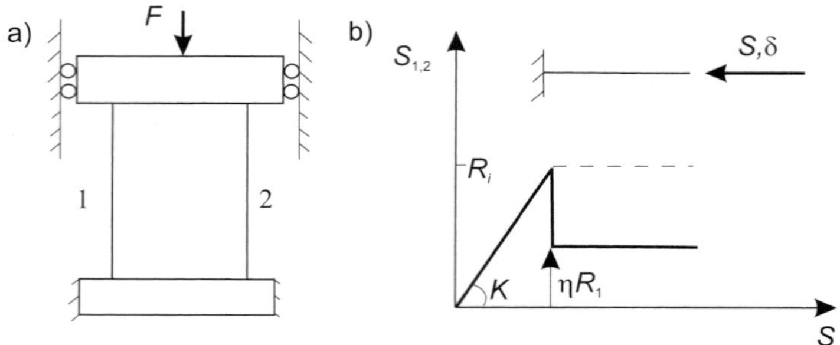

Figure 5.14 Simple parallel system: (a) physical system, (b) load-displacement relationship

This parallel system is an ideal one. Consider now the simplest relevant structural *parallel system*, shown in fig. 5.14, consisting of two parallel bars with component characteristics as shown in fig. 5.14b. If the failure mode approach is applied, the various failure modes have to be identified. The probability of a system failure, $P_{f\,sys}$ may schematically be formulated as:

$$P_{f\,sys} = [\text{failure of a sufficient number of components to cause a mechanism}]$$

The probability of failure may be expressed by

$$\begin{aligned} P_{f\,sys} = P[\{&(\text{failure of component (1)) AND (failure of component (1)} \\ &\text{before component (2)) AND (failure of component (2))}\} \\ &\text{OR } \{(\text{failure of component (2)) AND (failure of component (2)} \\ &\text{before component (1)) AND (failure of component (1))}\}] \end{aligned} \tag{5.172}$$

If the two components have identical stiffness, the axial force in each member before first failure is $S_1 = S_2 = F/2$. After failure, say, of member 1, the axial force in this member is ηR_1 and the load effect in member 2 is $F - \eta R_1$.

Mathematically, the probability of failure may be then formulated by

$$\begin{aligned} P_{f\,sys} = P\Bigg[&\left\{\left(R_1 - \frac{F}{2} \le 0\right) \cap (R_1 - R_2 < 0) \cap (R_2 - (F - \eta R_1))\right\} \\ &\cup \left\{\left(R_2 - \frac{F}{2} \le 0\right) \cap (R_1 - R_2 \ge 0) \cap (R_1 - (F - \eta R_2))\right\}\Bigg] \end{aligned} \tag{5.173}$$

Hence, computation of $P_{f\,sys}$ involves the calculation of the probability of intersections and unions.

It may be distinguished between two different situations

- The probability of failure is due to normal variability and uncertainties

- The probability of failure is due to damage initiated by gross errors, followed by a failure of the damaged structure due to normal uncertainties.

The first situation can be treated by structural reliability analysis. In the second situation, however, risk analysis needs to be used.

5.9.2.1 Simplified system analysis of framed offshore tower structures

Fortunately, accurate estimates of the system failure probability for jackets under extreme sea loading can be achieved with a very simple model, corresponding to a single system failure mode, i.e. by referring both the load and resistance to a given load pattern and using, the (overall) base shear as variable. This model has been validated for cases where the load uncertainties are dominant and the component forces are highly correlated [De, et al 1989; Wu and Moan, 1989]. However, it might be necessary to consider additional modes of failure relating to pile system failure and also possibly an upper storey "shear failure" in case of (abnormal) wave in deck loading.

This approach may be extended to include fatigue failure modes, using the basic overload case as the reference case. A first approximation to $P_{f\,sys}$, considering both overload and fatigue failure modes, may be achieved by,

$$P_{f\,sys} = P[\mathrm{FSYS}] \approx P[\mathrm{FSYS}(U)] + \sum_{j=l}^{n} P[F_j] \cdot P[\mathrm{FSYS}(U)|F_j] \qquad (5.174)$$

where FSYS(U) is the overload system failure; and F_j the fatigue failure of component j. It is noted that the first term then covers all pure overload failure sequences, i.e. all kinds of failure modes like: FSYS$|U_i, U_j, U_k, \ldots$. The main approximation in equation (5.174), is the negligence of sequences initiated by component overload followed by component fatigue failure or sequences initiated by fatigue failure and followed by more fatigue or overload failures. This approximation is non-conservative. On the other hand, disregarding the correlation between the failure modes that are included, is conservative.

The failure probability, $P_{f\,sys}$ (equation (5.174)) may most conveniently be referred to the service life or a period of a year. P(FSYS(U)) for the service life is estimated by the probability of a union of annual failure events. $P[F_j]$ is then conservatively, computed as the probability of fatigue failure in the service life. $P[\mathrm{FSYS}(U)|F_j]$ in principle should be calculated as the probability of overload failure in the remaining time of the service life after fatigue failure. However, if it is assumed that inspection is carried out and that fatigue failures (of the complete member) can be reliably detected, e.g. by an annual visual inspection, the latter failure probability may be calculated as an annual overload probability.

To estimate $P_{f\,sys}$ as an annual probability, the fatigue failure event F_j could be split into mutually exclusive events F_{jk}, which denote fatigue failure of component j in year k, given survival up to that time. The conditional probability of FSYS(U) given F_{jk} is then calculated as the probability of overload failure in the period from year k to the end of the service life. Alternatively, if inspections with reliable detection of member failure is assumed, the probability of overload failure could be referred to all annual value, as mentioned above.

The probability of ultimate failure of intact or damaged system may be achieved by a single mode approach, based on a single resistance (R_{sys}) and load effect (S_{sys}), as discussed e.g. by De et al (1989) and Wu and Moan (1989). While FORM/SORM and simulation methods as discussed above can be applied for this purpose, a simple approach-based lognormally distributed resistance and load effect is applied here. The *annual* reliability index β is

$$\beta_{sys} = \frac{\ln\left(\frac{\mu_{R_{sys}}}{\mu_{S_{sys}}}\right)}{\sqrt{V_{R_{sys}}^2 + V_{S_{shear}}^2}} = \frac{\ln\left(\frac{B_R}{B_S} \cdot REF\right)}{\sqrt{V_{R_{sys}}^2 + V_{S_{shear}}^2}} \tag{5.175}$$

where μ and V denote mean and COV, respectively. The mean values for strength and load are related to their respective characteristic (nominal) values by

$$\mu_{R_{sys}} = B_R R_{c,sys} \quad \text{and} \quad \mu_{S_{sys}} = B_S S_{c,sys} \tag{5.176}$$

where B_R and B_S are bias factors. Subscript c indicates characteristic (nominal) values. For S_{sys}, the characteristic value $S_{c,sys}$ is defined as the most probable 100 yr maximum load from the push-over analysis. For R_{sys}, the characteristic value $R_{c,sys}$ is defined according to the characteristic material properties and nominal values of geometry. B_R is of the order 1.05–1.10. $V_{R\,sys}$ is typically of the order of 0.1 and becomes negligible as compared to $V_{S\,shear}$. Relevant $R_{C,sys}$ for an intact as well as a damaged platform is considered. REF is defined as $R_{C,sys}/S_{C,sys}$.

The main contribution to the base shear load variable S_{sys} stems from wave loads. The load effect may be approximated by

$$S_{sys} = c\psi H^{\alpha} \tag{5.177}$$

where c is a constant, ψ is the model uncertainty, H is the wave height and α is a wave force exponent for the given structure. Expression (5.177) is based on the assumption of a fixed wave steepness (wave height to length). The uncertainty in S is then given by the model uncertainty and that in the wave height. Since the annual failure probability is addressed here, the wave height refers to the expected annual maximum value, $\mu_{H_{1\max}}$. If these variables are assumed to be lognormal,

$$\mu_{S,sys} \cong c \cdot \mu_{\psi} \cdot \mu_{H_{1\max}}^{\alpha} \tag{5.178}$$

$$V_{S,sys}^2 = V_{\psi}^2 + \alpha^2 V_{H_{1\max}}^2 \tag{5.179}$$

where $\mu_{S,sys}$ is the mean value of the annual maximum wave height. $V_{S\,shear}$ is typically of the order of 0.2–0.4. The characteristic value of characteristic value of S_{sys} with annual exceedance probability of 10^{-2} (the most probable maximum in 100 years) is given by

$$S_{c,sys} = c \cdot H_{100}^{\alpha} \tag{5.180}$$

H_1 denotes the most probable annual maximum wave height. This implies that: $B_S = \mu_\psi \cdot (\mu_{H_{1\,\max}}/H_{100})^\alpha$. It is noted that $\mu_{H1\,\max}/H_{100} \approx H_1/H_{100}$. In summary, the reliability index for the system may be obtained by:

$$\beta_{sys} = \frac{\ln[REF \cdot B_R(H_{100}/H_1)^\alpha/\mu_\psi]}{\sqrt{V_{R_{sys}}^2 + V_\psi^2 + \alpha^2 V_{H_{1\,\max}}^2}} \tag{5.181}$$

and the corresponding P_f can be calculated be equation (5.123) with $\beta = \beta_{sys}$. This simplified approach has especially been applied in connection with inspection planning, see e.g. Moan (1994) and Moan et al (1999).

5.10 Reliability Updating

5.10.1 General

Reliability measures may be updated based on additional information obtained about the as-built structure. The additional information can be incorporated by updating variables, updating events – such as probability of failure, and statistics updating at large. In this section updating distributions of variables and event updating will be illustrated with particular reference to additional information achieved by inspections to detect cracks.

The distribution of a variable can be updated by procedures developed by Shinozuka, Yang, Itagaki and others as reviewed by Yang (1994). Over the last decade FORM/SORM techniques have been used to develop general and efficient techniques for event updating, see e.g. Madsen et al (1986), Madsen and Sørensen (1990) and Jiao and Moan (1990). When the variables are updated, the failure probability can be easily calculated based on the new safety margin in which the updated variables are used to replace the original ones. However, if several variables are updated based on the same inspection event, the change of correlation between the updated variables due to the common information used, should be accounted for.

Updating of the probability of an event A given the information B, is based on Bayes theorem:

$$P(A|B) = P(A \cap B)/P(B) \tag{5.182}$$

where A and B are arbitrary events, "|" indicates a conditional event and the operator "∩" means intersection, i.e. that both of the events A and B shall occur.

Example

To illustrate updating of variables consider the case that the initial crack depth is assumed to follow an exponential distribution, with the mean crack size λ_0 as a parameter, and that the probability of crack detection, the POD curve is approximated by a cumulative exponential function, with λ and the minimum detectable crack size a_{th} as parameters.

By assuming a_{th} equal to zero, the density function for the initial crack depth, $f_{A0}(a)$, and POD curve, $P_D(a)$, are given by:

$$f_{A0}(a) = \frac{1}{\lambda_0} \exp\left(-\frac{a}{\lambda_0}\right) \quad (5.183)$$

$$P_D(a) = 1 - \exp\left(-\frac{a}{\lambda}\right) \quad (5.184)$$

The unknown parameters λ_0 and λ, can be determined by using the results of the first and second inspection, where the first inspection resulted in a no-finding. In principle, this method may be based on fabrication and fatigue cracks. However, if the fatigue cracks are used, they would have to be back-tracked to the time of the respective inspection due to the crack propagation. In the work by Moan, Wei, and Vårdal (2001), fabrication defects were used to estimate the two parameters in equations (5.183) and (5.184). By using the Bayes theorem and defining the events A and B to be that the crack size is in an interval $(a, a + \mathrm{d}a)$ and detection of a crack with size $(a, a + \mathrm{d}a)$, the updated probability density of crack depth may be obtained by:

$$f_{A,D}(a) = f_A(a)P_D(a) \Big/ \left(\int_0^\infty f_A(a)P_D(a)\mathrm{d}a \right) \quad (5.185)$$

$$f_{A,ND}(a) = f_A(a)(1 - P_D(a)) \Big/ \left(\int_0^\infty f_A(a)(1 - P_D(a))\mathrm{d}a \right) \quad (5.186)$$

where the subindices D and ND refer to the updated density functions of cracks detected and not-detected respectively. Similarly, the probability density $f_{A,ND1,D2}(a)$ for the crack size of cracks that were not detected in the first inspection, but detected in the second inspection, can be calculated. The expected values of cracks detected in the first and second inspection, denoted by $E[A|D1]$ and $E[A|ND1, D2]$ respectively, were used to determine the two unknown parameters, λ_0 and λ, based on about 3400 inspections in-service [Moan, et al 2001].

$$E[A, D1] = \lambda_0 \frac{2\lambda + \lambda_0}{\lambda + \lambda_0} \quad (5.187)$$

$$E[A, ND1, D2] = \lambda\lambda_0 \left[\frac{1}{\lambda + \lambda_0} + \frac{1}{\lambda + 2\lambda_0} \right] \quad (5.188)$$

The mean crack sizes in the first and second inspections were determined by correcting the rough data for measurement bias. The actual crack size used is $a = a_{rep} - 0.5 + \Delta a$; where a_{rep} is the reported crack size, 0.5 mm is the grinding increment carried out to size the crack, Δa is a parameter characterising the measurement uncertainty – and taken to be 0.3. Based on the reported mean values of 1.77 and 1.31 mm respectively in the two inspections, the unknown parameters: mean crack size, λ_0 and mean detectable crack size, λ, were determined to be $\lambda_0 = 0.94$ and $\lambda = 1.95$. General overviews of inspection reliability data are given e.g. by Silk et al (1987), Visser (2002), and Moan (2004).

5.10.1.1 Event updating

Reliability updating is relevant in connection with deterioration phenomena such as crack growth, corrosion etc. Here, crack growth serves as an illustration.

The crack depth may be represented by a random variable, $A(t)$ being a function of time t, see fig. 5.15. The distribution of the crack size at the initial time and a time t are given in the figure. The inspection may be used for updating of the distribution function for $A(t)$ into a distribution function representing the variable A_{upd} by accounting for the inspection. The updating is undertaken by defining failure and inspection outcomes by events.

The failure probability at the time, t (N-cycles) can be formulated as

$$P_f = P\left[a_f - A_N \leq 0\right] = 1 - F_{A_N}(a_f) \tag{5.189}$$

where a_f and A_N are the crack size at failure and after N cycles respectively, and $F_{A_N}(a)$ is the cumulative distribution function of the crack size, A_N. In general, $A(t)$ is given by Paris' law and is commonly not known explicitly and it would have been difficult to determine this distribution explicitly when taking into account all uncertainties that affect the distribution as well as the effect of inspections. The basic expression for the probability of fatigue failure, considering both A_f and A_N as random variables, is hence based upon

$$P_f = P\left[A_f - A_N \leq 0\right] = P[g(\mathbf{X}, t) \leq 0] \tag{5.190}$$

where $g(\mathbf{X}, t)$ can be reformulated as given by equation (5.108).

If inspection is made at time t_i, our belief in the probability of a large crack must be changed in accordance with the inspection result. The change in our belief of the probability of large cracks depends on the quality of the inspection method, the experience of the

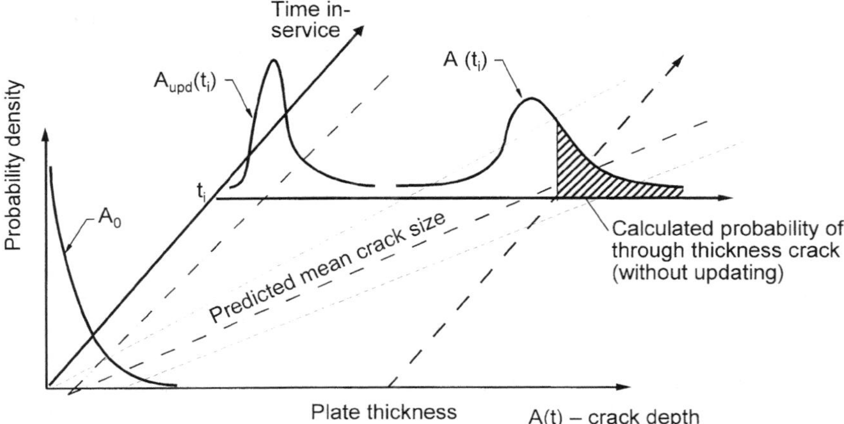

Figure 5.15 Initial distribution of crack depth A_0, simulated crack depth $A(t_i)$ at time t_i and updated distribution of crack depth $A_{\text{upd}}(t_i)$

operator, and on the conditions in which the inspections are carried out. The size of the detectable crack is also represented by a random variable A_d. The distribution of A_d depends on the *Probability of Detection* (POD) data valid for the inspection method in question.

The outcomes of inspections are assumed to be either no crack detection (ND) or crack detection (D) after N cycles, that can be described by events like:

$$I_{ND}: \quad a_N - a_d \le 0, \tag{5.191}$$

$$I_D: \quad a_N - a_d \ge 0. \tag{5.192}$$

These events can be reformulated analogous with the transition from equation (5.106) to equation (5.108).

The failure probability of a given joint (i), may be updated on the basis of a given inspection event

$$P_{f,up} = P\left[g(\mathbf{X}, t) \le 0 | IE_j\right] = P\left[(g(\mathbf{X}, t) \le 0) \cap IE_j\right] / P\left[IE_j\right] \tag{5.193}$$

where $g(\mathbf{X}, t)$ is the failure function for joint i in a time period, t and IE_j stands for a mathematical expression of inspection events such as, e.g. detection or no detection of a crack, e.g. equations (5.191)–(5.192). When $P_{f,up}$ according to equation (5.193) is computed by FORM, $g(\mathbf{X}, t)$ is linearized. The linearized failure function is denoted $M(t)$.

Bayesian statistics can be applied to update the reliability for instance in connection with proof testing of structures or by accounting for inspection results associated with deterioration due to corrosion or crack growth. In this context only the latter issue will be pursued.

To indicate the effect of updating more explicitly, it is convenient to express equation (5.193) in terms of reliability indices. Terada and Takashi (1988) showed that the following approximation is reasonable if the correlation, ρ between the events (the linearized failure function ($M \le 0$) and event ($IE \le 0$) is less than 1:

$$\beta_{up} = \frac{\beta_M - \rho A}{\sqrt{1 - \rho^2 B}} \quad \text{where} \quad \begin{aligned} \rho &= \alpha_M^T \alpha_{IE} \\ A &= \phi(-\beta_E)/\Phi(-\beta_{IE}) \\ B &= A(A - \beta_{IE}),\ 0 \le B \le 1 \end{aligned} \tag{5.194}$$

Figure 5.16 illustrates results for reliability updating based on consecutive inspections. Two situations are considered, namely based on an assessment at the design stage before actual inspections have been carried out and an assessment just after an in-service inspection has been carried out.

If the effect of inspections is estimated before they are carried out, two outcomes need to be considered, namely: (i) crack detection and repair or (ii) no crack detection. The exact outcome is not known, but the probability of the outcomes can be estimated based on the reliability method. The failure probability in a time period t, when a single inspection is made at time t and possible cracks detected are repaired, the failure

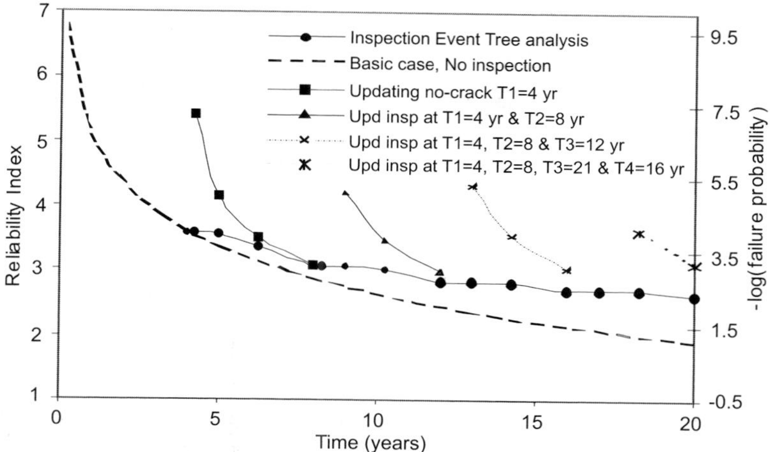

Figure 5.16 Reliability updating of a plated joint considering updating at the design stage (based on Event Tree and formulation of the type equation (5.195), and considering 1, 2, 3 and 4 consecutive inspections after 4, 8, 12 and 16 yr, respectively, with no crack detection, equation (5.196) [Ayala Uraga and Moan, 2002]

probability in the period $t \geq t_I$

$$P_F(t) = P[F(0, t_I)] + P[S(0, t_I) \text{ and } F(t_I, t)|I_D(t_I)] \cdot P[I_D(t_I)] + P[S(0, t_I) \text{ and } F(t_1, t)|I_{ND}(t_I)] \cdot P[I_{ND}(t_I)] \quad (5.195)$$

where $F(t_1, t_2)$ and $S(t_1, t_2)$ mean failure and survival in time period (t_1, t_2) and $I_D(t_I)$ and $I_{ND}(t_I)$ refer to crack detection and no detection respectively, at the time t_I. Equation (5.195) can be generalised to cover cases with several inspections, with two alternative outcomes. In this case an event tree of outcomes needs to be established. By comparing the results corresponding to no inspection and inspection at the design stage in fig. 5.16 the effect of inspection in a 20-yr service period is an increase of β from about 1.8 to 2.7.

If, on the other hand, no failure has occurred before time t_I and no crack is detected during an inspection at time t_I, the failure probability in the period $t \geq t_I$ is expressed by

$$P_F(t) = P[F(t_I, t)|I_{ND}(t_I)] \quad (5.196)$$

This expression may be generalised to account for the outcomes of possible previous inspections. Figure 5.16 also illustrates the effect of in-service inspections with known outcomes of no detection. It is seen that the more precise and favourable information obtained by no crack detection during an in-sevice inspection leads to a further increase of the reliability index.

It is observed that the two methods described above yield very different values just after inspection. The reliability index for the first method is continuous after updating but exhibits a change in slope after inspection and have almost a horizontal tangent. This means that the failure rate immediately after an inspection is small, for two reasons. The inspection has either not revealed a crack or repair of a detected crack has taken place.

The methodology described above can be extended to cover situations where a crack is detected and repaired, see e.g. Sigurdsson et al (2000) and Moan (2004).

The updating methodology is useful in connection with the extension of service life for structures with joints governed by the fatigue criterion. In such cases, the design fatigue life is in principle exhausted at the end of the planned service life. On the other hand, if no cracks have been detected during inspections, a remaining fatigue life can be demonstrated. However, it is not possible to bring the structure back to its initial condition by inspection only. This is because the mean detectable crack size typically is 1.0–2.0 mm, while the initial crack size is 0.1–0.4 mm. The fatigue damage, D actually caused in the service life can be estimated by updating the fatigue damage, based on the inspection results.

When a Bayesian updating of the remaining fatigue life is made, further improvement of the fatigue life can be achieved by grinding the crack toe to remove the possible crack.

By bringing the fatigue life back to the initial value, inspection can be kept at a minimum. Alternatively, if the members possess sufficient residual strength after development of through-thickness-cracks, less effort can be devoted to documenting the remaining fatigue life and an inspection strategy based on flooded member detection could be used to ensure adequate safety. If the global system is sufficiently damage-tolerant, another alternative is to utilise a systems (reliability) approach.

In the approach described, inspection results are used to update the initial estimates of fatigue loading, crack growth parameters and initial crack size based on an assumed probabilistic nature of these variables. Since only fundamental randomness and "normal" uncertainties are considered, the updated hazard rate will be very conservative if, e.g. the initial crack size had been accidentally large due to a gross fabrication defect. Additional information about loads, may be a second measurement of the crack size, etc. would then be necessary to achieve a better estimate.

5.10.1.2 Updating methods at large

The updating procedures mentioned above provide updating e.g. of mean value, standard deviation etc. of random variables associated with individual joints or groups of joints. In a given structure, there may be different groups of joints, requiring different methods to determine loads and resistance with different uncertainty measures. Likewise inspection results will apply to different groups. The challenge then is to correlate predicted and observed crack occurrences and no occurrences and then adjust model parameters and even the model itself to give a better prediction, see, e.g. Vårdal and Moan (1997).

5.10.2 Calibration of Fatigue Design Criteria

Since fatigue in offshore structures is mainly caused by wave loading, there is limited merit of calibrating fatigue criteria in the same sense as ULS criteria are calibrated to have the same reliability level independent of the relative magnitude of multiple loads. However, to achieve consistent design and inspection criteria, fatigue design criteria should be calibrated to reflect the consequences of failure and inspection plan. Moan, Hovde, and Blanker (1993) show that the allowable cumulative damage Δ_d in design can be relaxed when inspections are carried out. Figure 5.17 shows that a target level, corresponding to a design criterion $\Delta_d = 0.1$ and no inspection can be reached by a $\Delta_d = 0.25$ and $\Delta_d = 0.2$ for a

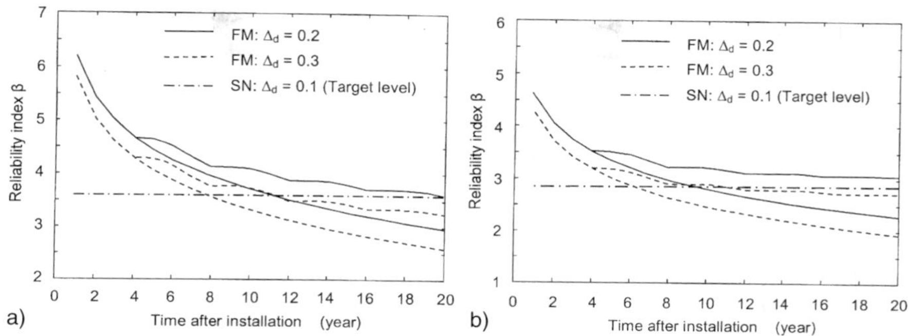

Figure 5.17 Reliability index for welded joints in offshore structures as a function of time – (a) in jackets and (b) in TLP tethers

tubular joint in a jacket and a tether butt weld respectively, when inspections are carried out every fourth year using NDE methods with a mean detectable crack depth of 1.5 mm.

While this calibration is done on a generic basis, it is important that information obtained for example by inspections during operation is used and the inspection plan is updated accordingly. For this reason structural reliability analysis of crack growth, needs to be done for each individual structure during the service life, while ultimate strength (design) criteria have been calibrated on a generic basis by code committees.

The fact that the inspection plan is adjusted up during operation also suggests that the fatigue design criteria could be based on a more optimistic view of the in-service performance, i.e. by assuming that no cracks will be detected during inspection. The effect of this assumption is indicated in fig. 5.16. If it turns out that cracks are detected, more frequent inspections would have to be implemented to maintain an acceptable safety level.

Fatigue design criteria for the North Sea depend upon the consequences of fatigue failure, i.e. whether ALS criteria are satisfied after the relevant fatigue failure or not. A more explicit account of the consequences of fatigue failure has been considered in connection with template jacket structures in the North Sea based on systems reliability considerations. Inspection planning for such structures is based on a simplified systems reliability approach, considering one failure mode at a time [Kirkemo, 1988; Vårdal and Moan, 1997; Moan et al., 1999]. The initial failure is considered to be fatigue, followed by a global overload failure, corresponding to each term in the sum of equation (5.174). The probability of system failure corresponding to mode (i) should then be less than the allocated target level, $P_{f\,sys(t)}$:

$$P_{f\,sys(i)} = P[FSYS(U)|F_i]P[F_i] \leq P_{f\,sys(t)} \tag{5.197}$$

With a given target level and an estimated $P[FSYS(U)|F_i]$ the allowable $P[F_i]$ for joint i can be obtained from equation (5.197). Based upon this information, inspections for this joint can be scheduled so as to ensure that the target reliability level is maintained during the service life. Further, description of reliability-based inspection planning may be found in Moan (2004).

5.11 Concluding Remarks on Probabilistic Design

It has been demonstrated that structural reliability analysis provides a measure of the safety which can be used to establish risk consistent criteria for design against ultimate and fatigue failure and inspection planning. It is shown that simple reliability measures based on e.g. lognormally distributed variables provide explicit and useful estimates. However, to properly account for all sources of uncertainty in connection with updated reliability based on inspection outcomes etc, methods based on FORM or Monte Carlo simulation are necessary to apply. The results of the reliability analysis are crucially dependent upon the uncertainties applied. It is strongly recommended to use uncertainty measures that have been authorised through code committees or other expert panels. Also, the emphasis should be placed on the most important variables. When some initial reliability analysis has been made and knowledge about the relative importance of different variables is available, efforts should be made to obtain more refined measures of the uncertainty associated with the dominant variables. Moreover, decisions regarding safety should be based upon comparison of the actual, implied reliability of a certain structure, as documented based on design and the plan for fabrication and operation, and the acceptable or target reliability level. The target value needs to be established by careful assessment of the implied safety level in structures that are considered to have acceptable safety, by using the same type of methodology that later will be used to demonstrate compliance with the target level.

Finally, it is noted that the structural reliability framework provides a basis for comparing the relative effect of uncertainties in loads, load effects and resistance on the reliability and, hence, as a basis for prioritising efforts to reduce the uncertainties.

References

Almar-Næss., A. (1985). "Fatigue handbook", Tapir Publishers, Trondheim, Norway.

Ang, A. and Tang, W. (1984). "Probability concepts in engineering planning and design, Volume II Decision, Risk and Reliability", John Wiley and Sons, New York.

API RP 2FP1 (1993). "Recommended practice for design, analysis and maintenance of moorings for floating production systems", (1st ed.). American Petroleum Institute, Washington, DC.

Ayala Uraga, E. and Moan, T. (2002). "System reliability issues of offsore structures considering fatique failure and updating based on inspection", *1st Int. ASRANet Colloquim*, Glasgow, Scotland, UK. Proc. Univ. of Glasgow, Das, P. K., July 8–10.

Battjes, J. A. (1970). "Long term wave height distributions of seven stations around the British Isles" (Report A.44). Godalming: National Institute of Oceanography.

Boccoti, P. (1982). "Relations between characteristic sea wave parameters". *Journal of Geophysical Research*, Vol. 87, p. 4267.

Brouwers, J. J. H. and Verbeek, P. H. J. (1983). "Expected fatigue damage and expected response for Morison-type wave loading". *Applied Ocean Research*, Vol. 15, pp. 129–133.

CIRIA (1977). "Rationalisation of safety and serviceability factors in structural codes". Technical Report Report 63, Construction Industry Research and Information Assoc., London, UK.

Dalane, J. (1993). "System reliability in design and maintenance of fixed off-shore structures". Dr. Ing. thesis, The Norwegian Institute of Technology, Trondheim.

De, S.R., Karamchandani, A. and Cornell, C. A. (1989). "Study of redundancy in near-ideal parallel structural systems", *Proc. 5th ICOSSAR*, Vol. 2, New York, pp. 975–982, ASCE.

Faltinsen, O. M. (1990). "Sea Loads on Ships and Offshore Structures", Cambridge University Press, Cambridge.

Ferry Borges, J. and Castanheta, M. (1971). "*Structural Safety*" (2nd ed). Laboratorio Nacional de Engenharia Civil, Lisbon.

Fjeld, S. (1977). "Reliability of offshore structures", *Proc. 9th OTC*, Vol. 4, Houston, pp. 459–472, OTC 3027.

Forristall, G. Z. (1978). "On the statistical distribution of wave heights in a storm". *Journal of Geophysical Research*, Vol. 83, p. 2352.

Forristall, G. Z. (2000). "Wave crest distributions: observations and second-order theory". *Journal of Physical Oceanography*, Vol. 30, pp. 1931–1943.

Haring, R. E. and Heideman, J. C. (1978). "Gulf of Mexico rare wave return periods", *Proceedings of the Offshore Technology Conference*, Number OTC 3230, Houston, Texas.

Haring, R. E., Osborne, A. R., and Spencer, L. P. (1976). "Extreme wave parameters based on continental shelf storm wave records", *Proceedings of the 15th conference on Coastal Engineering*, Honolulu, Hawaii: ASCE.

Haver, S. (2002). "On the prediction of extreme wave crest heights", *Proceedings of 7th International Workshop On Wave Hindcasting and forecasting*, Meteorological Service of Canada, Banff, Canada, Environment Canada.

Haver, S. and Nyhus, K. A. (1986). "A wave climate description for long term response calculations", *Proceedings of the 9th Int. Conference on Offshore Mechanics and Arctic Engineering*, Tokyo, Japan: ASME.

Heideman, J. and Weaver, T. (1992). "Static wave force procedure for platform design". *Proc. Civil Ingineering in the Oceans V*, College Station, Texas, pp. 496–517, ASCE.

HSE (2002). "Target levels for reliability-based assessment of offshore structures during design and operations", Report 1999/060, Health and Safety Executive, HMSO, Norwich, UK.

ISO 19900 (1994). "Standard for Offshore Structures for the Petroleum and Natural Gas industries".

ISO 2394 (1998). "General principles on reliability for structures".

Jahns, H. O. and Wheeler, J. D. (1972). "Long term wave probabilities based on hindcasting of severe storms", *Proceedings of the Offshore Technology Conference*, Number OTC 1590, Houston, Texas.

Jiao, G. and Moan, T. (1990). "Methods of reliability model updating through additional events". *Structual Safety*, Vol. 9, pp. 139–153.

Jordaan, I. J. and Maes, M. (1991). "Rationale for load specifications and load factors in the new CSA code for fixed offshore structures". *Canadian Inst. Civil Engng.*, Vol. 3, No. 18, pp. 454–464.

Karamchandani, A. (1990). *New Method in Systems Reliability*. Ph.d. thesis, Department of Civil Engineering, Stanford University, Stanford, CA.

Kirkemo, F. (1988). "Applications of probabilistic fracture mechanics to offshore structures". *Applied Mechanicas Review*, Vol. 41, pp. 61–84.

Krogstad, H. E. (1978). *Estimating the maximum wave height from the significant wave height* (Report Project No. P-201). Continental Shelf Institute, Trondheim, Norway.

Larsen, L. M. (1981). "The influence of bandwidth on the distribution of heights of sea waves". *Journal of Geophysical Research*, Vol. 86, p. 4299.

Lin, Y. K. (1967). "Probabilistic theory of structural dynamics", McGraw-Hill Inc., New York.

Lloyd, J. and Karsan, D. (198). "Development of a reliability-based alternative to API RP2A", *Proc. 20th OTC*, Vol. 4, pp. 593–600, Houston.

Longuet-Higgins, M. S. (1980). "On the distribution of the heights of sea waves: some effects of nonlinearity and finite bandwidth". *Annals of Statistics*, Vol. 85, pp. 1519.

Madsen, H., Krenk, S. and Lind, N. (1986). *"Methods of Structural Safety"*, Prentice-Hall Inc., New Jersey.

Madsen, H. and Sørensen, S. (1990). "Probability-based optimization of fatigue design inspection and maintenance", *Presented at Int. Symp. on Offshore Structures*, University of Glasgow.

Marley, M. and Moan, T. (1992). "Time variant formulation for fatigue reliability", Paper OMAE-92-1203, *Proc. OMAE*, Calgary, Canada.

Marshall, P. (1969). "Risk evaluation for offshore structures". *ASCE St. Div.*, Vol. 95, No. 12.

Marshall, P. and Luyties, W. (1982). "Allowable stresses for fatigue design", *Proc. BOSS'82*, McGraw-Hill, New York.

Melchers, R. E. (1999). "Structural reliability analysis and prediction", (2nd ed.). John Wiley and Sons Ltd, Chichester.

Moan, T. (1988). "The inherent safety of structures designed according to the NPD regulations". Report F8804, SINTEF, Trondheim, Norway.

Moan, T. (1994). "Reliability and risk analysis for design and operations planning of offshore structures", *Proc. 6th ICOSSAR, Structural Safety and Reliability*, Volume I, Rotterdam, pp. 21–43, Balkema.

Moan, T. (1995). "Safety level across different types of structural forms and materials – implicit in codes for offshore structures". Report STF 70 A95210, SINTEF, Trondheim.

Moan, T. (2002). "Wave loading", in Dynamic Load and Design of Structures, A. J. Kappos (ed.), Chapter 5, Spon Press, New York, pp. 176–230.

Moan, T. (2004). "Reliability-based management of inspection, maitenance and repair of offshore structures". *Structure and Infrastructure Engineering*, Vol. 1, No. 1, pp. 33–62.

Moan, T., Hovde, G. and Blanker, A. (1993). "Reliability-based fatigue design criteria for offshore structures considering the effect of inspection and repair", *Proc. 25th OTC, Offshore Technology Conference*, Volume 2, pp. 591–599, Houston, OTC 7189.

Moan, T., Vårdal, O. and Johannesen, J. (1999). "Probabilistic inspection planning of fixed offshore structures", *Proc. 8th ICASP, Applications of statistics and Probability*, Rotterdam, pp. 191–200, A.A. Balkema.

Moan, T., Wei, Z. and Vårdal O. (2001). "Initial crack depth and POD data based on underwater inspection of fixed platforms", *Proc. 8th ICOSSAR, Structural Safety and Reliability*, Newport Beach. Balkema.

Moses, F. (1987). "Load and resistance factor design-recalibration LRFD". Draft Report API PRAC 8722, API, Dallas.

Moses, F. and Liu, Y. (1992). "Methods of redundancy analysis for offshore platforms", *Proc. 11th OMAE*, Vol. II, pp. 411–416, ASME, New York.

Naess, A. (1984). "On the long-term statistics of extremes". *Applied Ocean Research*, Vol. 6, No. 4, pp. 227–228.

Naess, A. (1985a). "The joint crossing frequency of stochastic processes and its application to wave theory". *Applied Ocean research*, Vol. 7, No.1, pp. 35–50.

Naess, A. (1985b). "On the distribution of crest to trough wave heights", *Ocean Engineering*, Vol. 12, No. 3, pp. 221–234.

Naess, A. (1989). "Prediction of extremes of combined first-order and slow-drift motions of offshore structures". *Applied Ocean Research*, Vol. 11, No. 2, pp. 100–110.

Naess, A. and Royset, J. O. (2000). "Extensions of Turktra's rule and their application to combination of dependent load effects". *Structural Safety*, Vol. 22, pp. 129–143.

Newman, J. and Raju, I. (1981). "An empirical stress intensity factor equation for the surface crack". *Engineering fracture Mechanics*, Vol. 15, No. 1, pp. 185–192.

Newman, J. N. (1974). "Second order, slowly varying forces on vessels in irregular waves", *Proceedings of International Symposium on the Dynamics of Marine Vehicles and Structures in Waves*, The Institution of Mechanical Engineers, University College, London.

NORSOK Standard (N-001) (2000). "Structural Design (N-001)", Norwegian Technology Standards Institution, Oslo, Norway.

NORSOK Standard (N-003) (1999). "Actions and Action Effects (N-003, Rev. 1)", Norwegian Technology Standards Institution, Oslo, Norway.

NORSOK Standard (N-004) (1998). "Steel Structures (N-004)", Norwegian Technology Standards Institution, Oslo, Norway.

Papoulis, A. (1965). *Probability, Random Variables and Stochastic Processes*, McGraw-Hill Kogakusha, Ltd, Tokyo.

Reiss, R.-D. and Thomas, M. (2001). "Statistical Analysis of Extreme Values", Birkhauser Verlag, Basel.

Rodriguez, G., Soares, C. G., Pacheco, M. and Perez-Matrell, E. (2002). "Wave height distribution in mixed sea states". *Journal of Offshore Mechanics and Arctic Engineering*, Vol. 124, No.1, pp. 34–4.

Shetty, N. (1992). "System reliability of fixed offshore structures under fatigue deterioration". Ph.d. thesis, Imperial College, Univ. of London, London.

Sigurdsson, G., Lotsberg, I., Myhre, T. and Ørbeck-Nilssen, K. (2000). "Fatigue reliability of old semisubmersuibles", *Proc. Offshore Technology Conference*, OTC Paper no 11950, Houston.

Silk, M., Stoneham, A. and Temple, J. (1987). *The reliability of non-destructive inspection*, Adam Hilger, Bristol.

Stansberg, C. T. (1992). "Model scale experiments of extreme slow-drift motions in irregular waves", *Proceedings of the BOSS-92 Conference*, Vol. 2, pp. 1207–1222, London, UK: BPP Technical Services, Ltd.

Stansberg, C. T. (2000). "Prediction of extreme slow-drift amplitudes", *Proceedings of the 19th International Conference of Offshore Mechanics and Arctic Engineering*, Number OMAE00-6135. New Orleans, ASME, USA.

Teigen, P. and Naess, A. (1999). "Stochastic response analysis of deepwater structures in short-crested random waves". *Journal of Offshore Mechanics and Arctic Engineering, ASME*, Vol. 121, pp. 181–186.

Terada, S. and Takashi, T. (1988). "Failure-conditioned reliability index", *Structural Engineering Div*, ASCE, Vol. 114, pp. 943–952.

Thoft-Christensen, P. and Baker, M. (1982). *Structural Reliability and its Applications*, Springer-Verlag, Berlin.

Tromans, P. S. and Vanderschuren, L. (1995). "Response based design conditions in the North Sea: applications of a new method", *Proceedings of the Offshore Technology Conference*, Number OTC 7683, Houston, Texas.

Turkstra, C. J. (1970). *Theory of Structural Safety*, SM Study No. 2, Solid Mechanics Division, University of Waterloo, Ontario.

Vårdal, O. T. and Moan, T. (1997). "Predicted versus observed fatigue crack growth. validation of probabilistic fracture mechanics analysis of fatigue in North Sea jacket", *Proc. 16th OMAE*, Number Paper 1334, Yokohama, Japan.

Videiro, P. and Moan, T. (2000). "Reliability analysis of offshore structures under multiple long-term wave load effects", *Proc. ICASP 8, Application of Statistics and Probability*, Rotterdam, pp. 1165–1173, A.A. Balkema.

Videiro, P. and Moan, T. (2001). "Reliability based design of offshore structures", *Proc. Int. PEP-IMP Symposium on Risk and Reliability Assessment for Offshore Structures*, Invited, Mexico City, December.

Visser, W. (2002). "POD/POS curves for non-destructive examination", Offshore Technology Report OTO 2000/018, Health and Safety Executive, UK.

Wang, X. and Moan, T. (1996). "Stochastic and deterministic combinations of still water and ending movements in ships". *J. Marine Structures*, Vol. 9, pp. 787–810.

Wen, Y. (1990). "Structural load modelling and combination for performance and safety evaluation", Elsevier, Amsterdam.

Winterstein, S., Ude, T., Cornell, C. A., Bjerager, P., and Haver, S. (1993). "Environmental parameters for extreme response: Inverse FORM with omission factors", *Proceedings of ICOSSAR' 93*, Innsbruck, Balkema, Austria.

Wirsching, P. (1983). "Probability-based fatigue design criteria for offshore structures". Report API-PRAC Project No. 81–15, Dept. of Aerospace and Mechanical Engineering, University of Arizona, Tucson.

Wu, Y.-L. and Moan T. (1989). "A structural system reliability analysis of jacket using an improved truss model", *Proccedings of 5th ICOSSAR*, pp. 887–894, San-Franciso, ASCE, New York.

Yang, Y. (1994). "Application of reliability methods to fatique, quality assurance and maintenance", Schuëller, G., Shinozuka, M., and Yao, J. (eds.), *Proc. 6th ICOSSAR, Structural Safety and Reliability*, Rotterdam, Balkema.

Handbook of Offshore Engineering
S. Chakrabarti (Ed.)

Chapter 6

Fixed Offshore Platform Design

Demir I. Karsan Ph.D., P.E.
AMEC Paragon, Houston, TX, USA

6.1 Field Development and Concept Selection Activities

6.1.1 Introduction

Offshore platform design is preceded by a sequence of activities that result in the selection of a field development system that best fits the field characteristics and economics. Before feasible alternatives for producing oil and gas from an offshore field are identified and the most desirable production scheme is selected, exploratory work defining the reservoir characteristics have to be completed. First, a decision has to be made whether an offshore location has the potential for hydrocarbon reserves. Geologists and geophysicists do this assessment through a study of geological formations.

Next, a decision has to be made whether the field will be economically viable and if further exploratory activities are warranted. This decision involves preparing cost, schedule and financial return estimates for selected exploration and production schemes, comparing several of these alternatives and identifying the most beneficial. For this phase, due to the absence of detailed information with respect to the reservoir characteristics, future market conditions and field development alternatives, we rely on judgements made by experts based on their past experience and cost and schedule estimates based on data available to the owner.

If the preliminary economic studies are positive, the seismic data generation and evaluation done by geophysicists follow these. This results in reasonable information with respect to the reservoir characteristics such as its depth, spread, faults, domes, traps, permeability, etc. and an approximate estimate for recoverable reserves of hydrocarbons.

If the seismic indications are positive and the recoverable reserve estimates and global economic analyses are judged favourable, exploratory drilling activities may commence. Depending on the water depth, environment and the availability, a suitable exploration scheme is selected. Jack-up rig exploratory units are suitable for shallow water depths

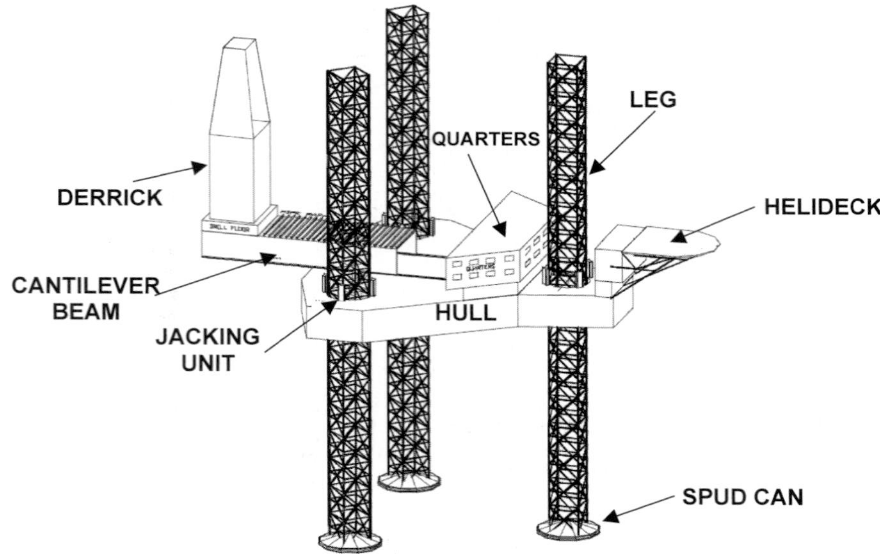

Figure 6.1 A jack-up drilling unit with spud cans

(fig. 6.1 and Section 6.3.3). In water depths exceeding 400 ft, ships or semi-submersible drilling units are generally utilised. Beyond 1000-ft depths, floating drilling units would require special mooring arrangements or a dynamic positioning system. Exploratory drilling may result in a discovery well.

Delineation drilling work follows a discovery well. This generally requires 3–6 wells drilled at selected points of a reservoir. These activities and production testing of the wells where oil and gas is encountered give a reasonably detailed information about the size, depth, extent and topography of a reservoir and its recoverable reserves, oil and gas content, pressure, viscosity (API Grade), liquid properties (oil water ratio, HC composition) and impurities such as sulphur, H_2S, etc.

Reservoir information enables us to estimate the location and number of wells that will be required to produce a field and volumes of oil, gas and water production. These information are needed to estimate the type of production equipment, facilities and the transport system that will be necessary to produce the field. Accuracy of reservoir data has a major impact on selecting a field development concept. In marginal or complex reservoirs, reliability of the reservoir data and the flexibility of a production system to accommodate changes in our reservoir appraisal becomes a very desirable feature.

6.1.2 Design Spiral and Field Development Timeline

The design spiral shown in fig. 6.2 illustrates the iterative process involved in selecting and designing an offshore production system. Each loop of the spiral indicates one design cycle.

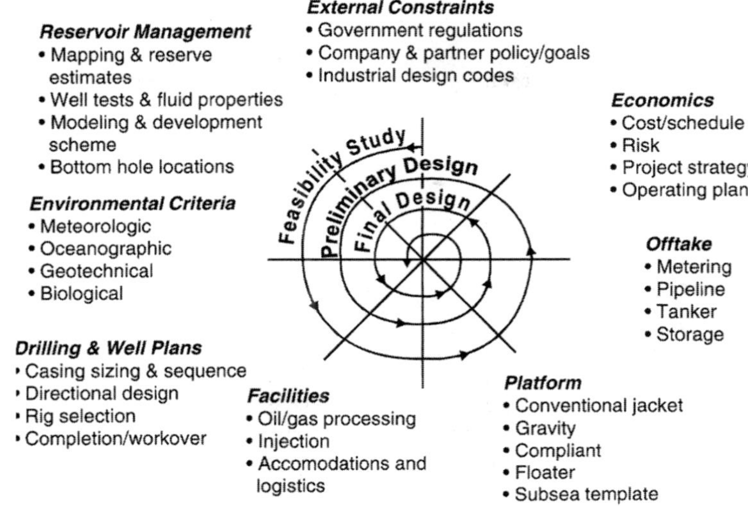

Figure 6.2 Offshore field development design spiral

The spokes of the spiral represents the activities of: gathering input parameters such as the reservoir and environmental data; selection and design of major system components such as the production drilling and the wells, platform type and its facilities, off-take system; and the decision criteria such as the economics and others including the external factors. The first loop around the spiral generally involves evaluating several field development options, which satisfy the input requirements and establishing their relative merits with respect to the decision criteria. In this loop, not only alternatives for field development systems, but also alternatives for each major system component are identified, developed and ranked. The next loop involves preparing a preliminary design for the selected system. In this phase, the selection activity is focused on the system components and detail elements. By the time we reach the last loop of the spiral, all the system components and the construction activities should be well defined. Beyond this point, a few changes to the system and its components could be made without suffering delays and cost overruns.

Figure 6.3 shows the design spiral shown in fig. 6.2 in a flow diagram format. The *Acquire* phase involves geological, seismic, concept and economic risk assessment activities that lead to the acquiral of a lease. The exploratory drilling, production test and data gathering and planning phases shown as the *Explore* and *Appraise* arrows follow this. The *Develop* phase includes engineering design, construction, production drilling, well completion, hook-up and commissioning. The *Operate* phase includes maintenance, production, repair and reassessment and transportation activities. The viable field development options are identified, developed and the most suitable option is selected parallel to the *Acquire*, *Explore*, *Appraise* cycle. All project activities that precede the start of the basic design phase are called the "*F*ront-*E*nd *L*oading – FEL" activities. FEL is the most important phase of a field development timeline. An ideal field development schedule should allow for a

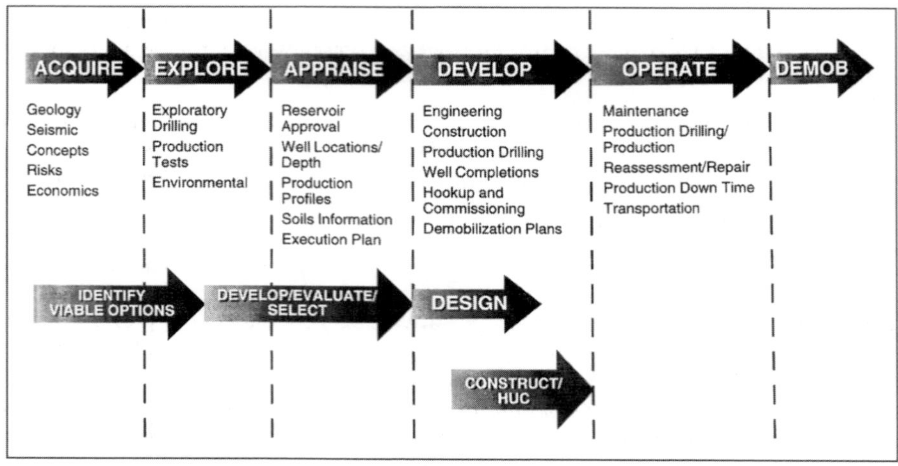

Figure 6.3 Field development timeline

sufficient lead-time to perform all FEL work with a high level of definition before basic design starts.

6.1.3 Factors That Drive Concept Selection

Figure 6.4 shows some of the major factors that drive the selection of a field development scenario. *Reservoir Management Plan* is affected by the reservoir and produced fluid

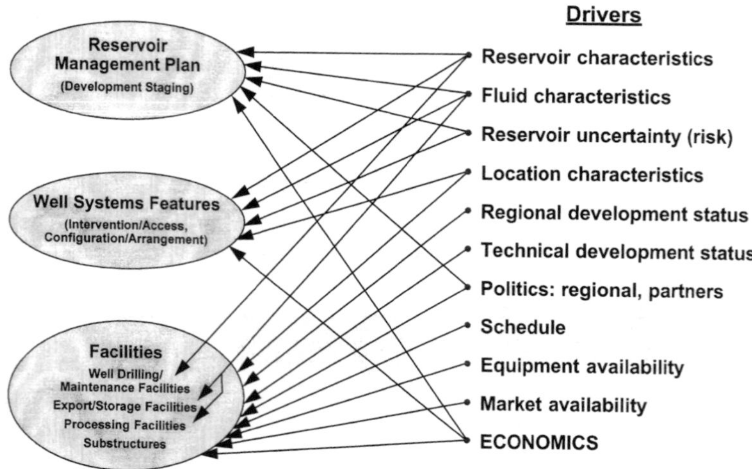

Figure 6.4 Factors that drive the field development concept selection [Morrison, 1997]

characteristics, reservoir uncertainty (size and topography), regional politics (national content, company and partner culture, etc.) and economics of the entire field development scheme.

Well system/completion design is affected by the same factors that affect the reservoir management plan except perhaps political factors. Platforms, facilities, (process and production), storage and export systems are affected by all factors listed in fig. 6.4.

6.1.3.1 Field Development Cost

Horizontal axis of fig. 6.5 references the field development timeline. Vertical axis shows the percent total installed cost of a project, 100% taken equal to cost at first oil/gas date. Experience shows that the FEL phase (identify viable options, develop/evaluate/select concept and conceptual design), which usually consumes only about 2–3% of the total installed cost (TIC) of the field development, has the highest impact on cost, schedule, quality and success. It is not uncommon to observe significant cost overruns when a full FEL is not performed. Reanalysis of a recently completed project, which did not perform a satisfactory FEL because of a tight schedule and political factors, indicated that a 50% TIC reduction could have been achieved if a satisfactory FEL was performed.

Our ability to influence cost and savings decreases as we march along the field development timeline. At the concept development stage, selecting and developing the right concept would have a major impact on the TIC. Savings in detailed design and construction phases would generally stem from good project controls and execution.

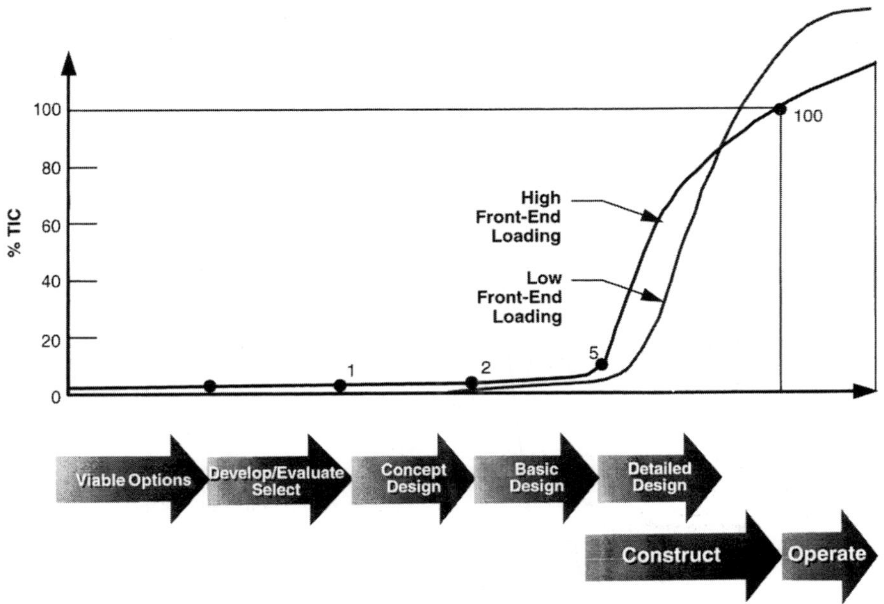

Figure 6.5 Field development cost build-up

The level of innovation (or the number of novel ideas and components) within a field development system has a significant impact on TIC and operability of the field. When first of a kind system or component is used, technical and construction issues may crop up during the implementation and operations phases that may affect the project capital expenses (CAPEX) and operational expenses (OPEX). Experience shows that a standard project with routine components and almost no novel ideas will experience very low operability problems. A recent survey of the offshore projects (Morrison, 1997) indicated that 90% of the projects with substantial innovations had major operability problems. Care must be taken not to introduce low value adding new ideas and components into a field development system.

Figure 6.6 shows the variation of the reliability of total installed cost (TIC) estimates in different phases of a project. Our understanding of economics and other features of a field development system improve as we move along the field development timeline. At the start of the FEL, a number of options would be available and identifying the right field development concept will have a profound impact on a project's success. During the *conceptual design phase* of FEL, general definition of each system component (well systems, platform(s), topsides facilities, transportation) and their subcomponents (hull, mooring system, tethers, living quarters, process, utility systems, pipelines, storage, risers, etc.) are made and a cost and schedule estimate is prepared. Selection and definition of the system components and subcomponents would also have a significant impact on our ability to reduce cost and/or schedule. At this phase, the accuracy of our TIC estimates would be around the ±25–40% range.

The *preliminary* or *basic design phase* includes a firm definition of the process through process flow diagrams (PFDs) and preparation of the field and equipment layouts. Piping and instrumentation diagrams (P&IDs), general platform drawings, materials and

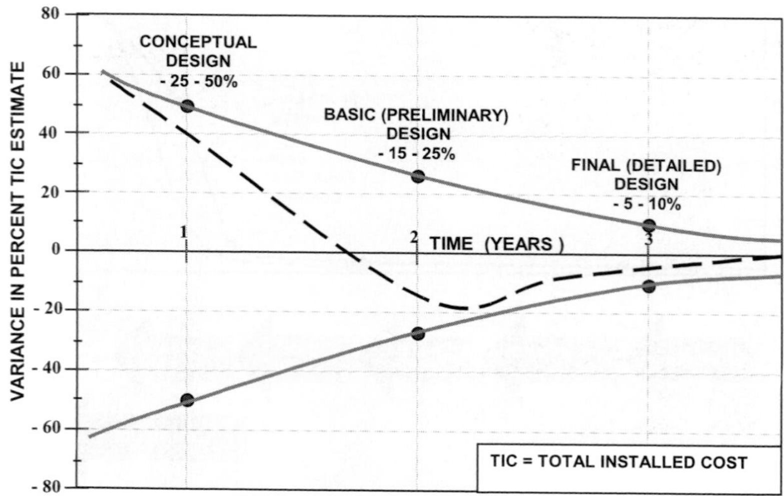

Figure 6.6 Reliability of total installed cost estimates at various project phases

equipment lists, data sheets, specifications and a final engineering, procurement and construction (EPC) cost and schedule estimate. Basic design phase allows some system optimisation but only at a subcomponent and specification level. At this phase, the accuracy of our TIC estimates would be around the ± 15–25% range.

In a well-organised project, the *final design* and *construction phases* should have the lowest impact on the TIC optimisation. This phase provides detailed engineering analysis and design, approved for construction (AFC) drawings and fabrication, transportation, installation, precommissioning and the hook-up and commissioning (HUC) of the field by selected constructors or an EPC (engineering, procurement, construction) contractor. Efficient project management [execution plan, cost, schedule and quality control (QC), verification, quality and safety assurance (V/QSA), purchasing and documentation] would have some impact on the TIC but not as profoundly as the FEL phase. Past experiences indicate that at the start of the construction phase, the accuracy of the TIC estimates would be in the ± 5–10% range.

Cash Flow Considerations

Figure 6.7 shows a *self-contained North Sea-type offshore platform.* The tower, deck structures and facilities are fabricated on land as separate units within about a 2- to 3-yr time

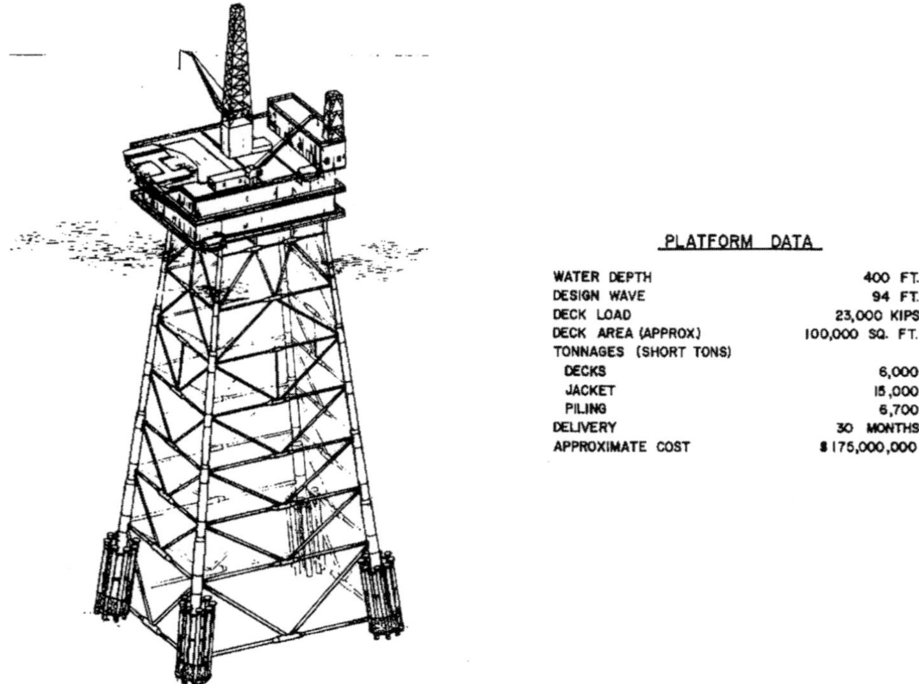

Figure 6.7 Self-contained North Sea production, drilling and living quarters (PDQ) platform

period. Tower is transported to the offshore location by a barge and launched in site or floated out. The deck and its facilities are transported to site either as a single integrated unit, or as modular units, lifted and set on the jacket using a heavy lift derrick barge. The jacket, deck and facilities units are hooked up.

A 2-to-3 yr drilling/field development programme generally follows self-contained platform installation. In some cases, some of the wells may be drilled before the tower gets in location or a successful exploratory well may be utilised, resulting in some early production. But, in general, about 3 to 4-yr from the platform installation date has to elapse before the peak field production rate can be reached.

Figure 6.8 shows a *North Sea multi-platform field development* through use of well protector/drilling platforms, a drilling/production platform and a living quarters platform.

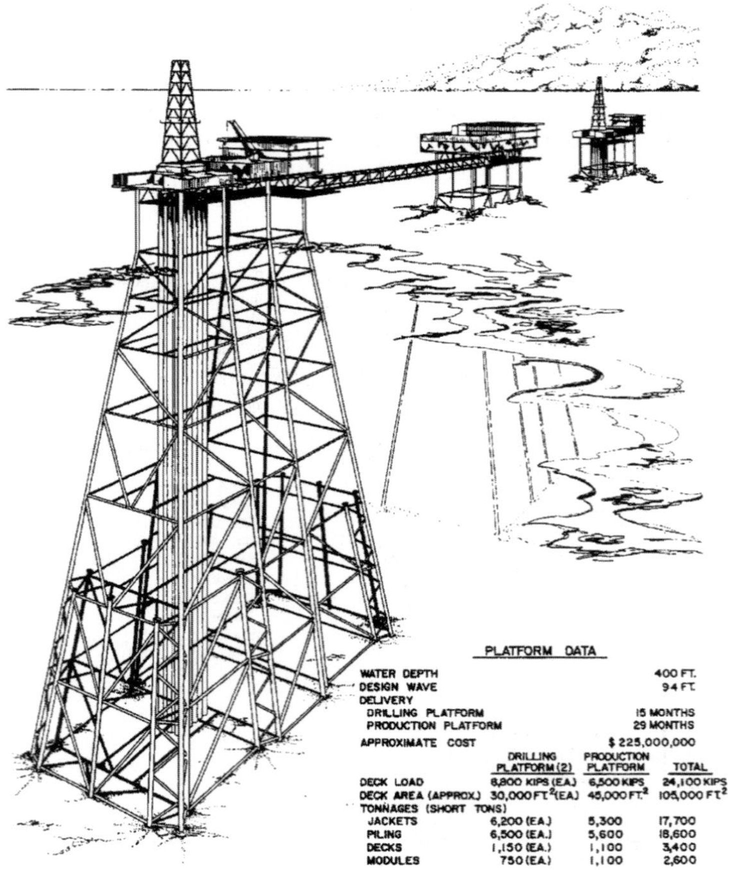

PLATFORM DATA

WATER DEPTH			400 FT.
DESIGN WAVE			94 FT.
DELIVERY			
DRILLING PLATFORM			15 MONTHS
PRODUCTION PLATFORM			29 MONTHS
APPROXIMATE COST			$ 225,000,000
	DRILLING PLATFORM (2)	PRODUCTION PLATFORM	TOTAL
DECK LOAD	8,800 KIPS (EA.)	6,500 KIPS	24,100 KIPS
DECK AREA (APPROX.)	30,000 FT^2 (EA.)	45,000 FT^2	105,000 FT^2
TONNAGES (SHORT TONS)			
JACKETS	6,200 (EA.)	5,300	17,700
PILING	6,500 (EA.)	5,600	18,600
DECKS	1,150 (EA.)	1,100	3,400
MODULES	750 (EA.)	1,100	2,600

Figure 6.8 North Sea multi-platform field development concept with separate production (P), drilling (D) and living quarters (Q) platforms

For the case of increased gas production in later years, addition of a gas treatment and compression platform may also be planned.

For this case, satellite-drilling platforms may be constructed and installed within a six-month time frame to allow early drilling to commence. By the time the production and additional drilling platform is installed say within a 3-yr time frame, a significant number of the satellite platform wells would have been drilled and completed, enabling a large volume of production to start.

This concept generally allows reaching the peak production rate about a year or so earlier than the self-contained platform concept.

Figure 6.9 shows the comparison of oil production rates for the two field development approaches shown in figs. 6.7 and 6.8. The vertical axis represents barrels per day oil production rate. The horizontal axis represents the years measured from the project start. A production rate of 6000 barrels of oil per day for each of the planned 36 wells is assumed for the production platform with satellite well protectors.

Due to earlier drilling and completion of 18 wells from a satellite well protector drilling (D) platform, a 100,000 barrels per day production rate would be reached immediately after the installation of production platform. Within one year from the production platform installation, all 36 wells would be completed and 200,000 barrels per day production rate will be reached.

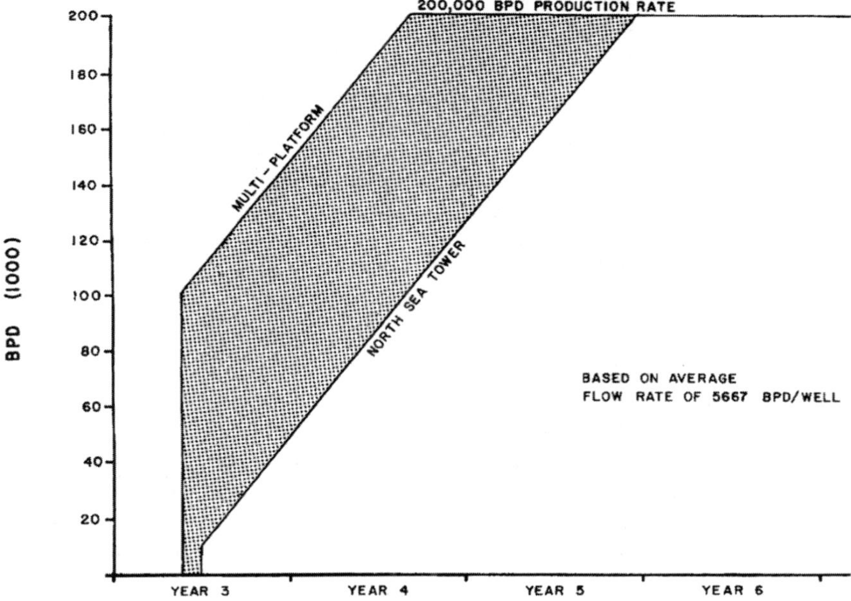

Figure 6.9 Oil production rate comparison, self-contained (PDQ) vs. multi platform field development approaches.

For the single self-contained PDQ platform case, a period of about three months from the tower installation has to elapse before 6000 barrels per day oil production rate may be achieved. For this case, a period of about two and a half years from the platform installation would be required to drill and complete all the 36 wells, reaching the peak production rate of 200,000 barrels per day.

The vertical axis of the chart in fig. 6.10 represents the cumulative cash flow from the project start to a given project year. The cumulative cash flow is normalised against the highest negative cash flow invested for the self-contained tower platform-based field development. The horizontal axis shows the years measured from the project start.

For the self-contained PDQ platform, the highest negative cash flow point is reached within the second quarter of the third year when some oil production will start. From this point on, positive cash flow from the produced oil will start offsetting the negative cash flow from early investment and operating costs. The zero cumulative cash flow position is reached within the fourth quarter of the fourth year, representing the date when all the field investments to date will be paid off.

The highest negative cash flow for the multi-platform concept is reached six months ahead of the self-contained platform, sometime within the first quarter of the third year, when the oil flow from the production platform will start. Due to the heavy up front investment on a satellite platform and the early drilling programme, the maximum cash invested will be about thirty percent more than that for the self-contained platform. However, rapid cash recovery from early drilled production wells would start offsetting the negative cash flow at a rapid rate. Within the first quarter of the fourth year, the zero cumulative cash flow (pay-off) point will be reached, about six months ahead of the self-contained platform concept. From this point on, the multi-platform concept would result in higher cumulative cash flow.

The up front available cash for investment may vary from one oil company to the other. If plenty of cash is available from other operations and some tax hedging is desired, an early high cash investment option may be preferred. However, if the company is cash starved or a higher corporate cash discount rate must be imposed due to many other competing investment options or interest rates, a lower up front cash investment option may be preferred. This is a somewhat simplistic presentation of the economic factors affecting the platform concept selection. Many other economic factors other than the cash flow, including the tax and discount rates, inflation and the time value of money also need to be considered, resulting in complex net present value calculations. Investment analysis specialists generally perform these calculations.

6.1.3.2 Multi-criteria Concept Selection

At the FEL stage, external factors such as national content, technology transfer, environmental pollution potential; the cultures, politics, economics and infrastructure of the host nation and the operating oil company and its partners may have major influences on the concept selection. Not so easily comparable criteria such as the economics, design completeness and maturity, and external factors have to be weighed against each other and used for concept ranking and selection. In a multi-criteria process, first the goal of the exercise is defined. Then the viable field development options are identified. This is followed by the identification of a multitude of selection criteria that are grouped and

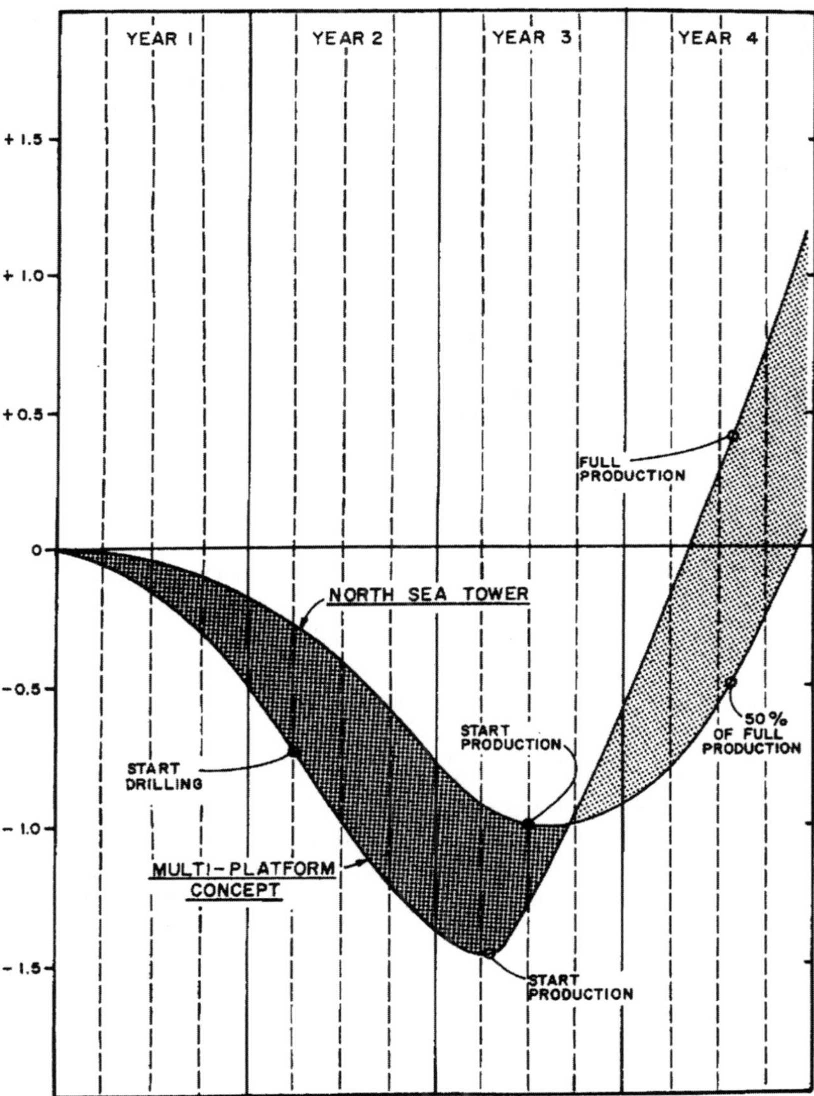

Figure 6.10 Cash flow comparison, self-contained (PDQ) vs. multi platform field development approaches

ordered in a hierarchical manner. This is followed by expert judgements where the importance of each criterion is compared to the others in a pair-wise manner. These comparisons are then passed through an analytical process to obtain weights for each ranking comparison and ranking of the alternatives. There are a number of such processes in current use (Saaty and Thomas describes the analytical hierarchy – AHP method, which is one of such tools in use for multi-criteria concept selection).

6.1.4 Field Development Design Phases

Front end loading (FEL) process ends with the completion of the conceptual design. At this point, the following information is available:

- A well-defined field development plan,
- Basis of conceptual design (field characteristics, operational and environmental parameters, foundation conditions, platform configurations, global materials selection, and other information and assumptions used for the concept development),
- Conceptual drawings showing major component configurations (platforms, topsides facilities layouts, well locations and well systems, reservoir maps and production profiles, storage (if needed) and offloading systems, pipelines to shore) and preliminary component sizes,
- In general, platform structure configuration is defined by a number of conceptual drawings that show side elevations and plans for legs and major bracing. Preliminary process flow diagrams (PFD) and major equipment lists may also be available,
- Concept cost estimate (±40%) and schedule for the entire development plan (including the capital and operational expenses, cash flow diagrams and net present value – NPV – of the total investment).

A conceptual design package that includes the above-listed information is prepared by the owner and given to the design contractor as input to the *Basic Design Phase*.

6.1.4.1 Basic Design Phase

Basic design phase (also called by some as the "define engineering phase" or the "preliminary design phase") defines the platform, production facility and structure configurations and dimensions in satisfactory detail to allow start of the detailed design. Basic design results enable a reliable cost-and-schedule estimate and ordering long lead major equipment and structural components (which enables a contractor to provide a reliable bid for detailed engineering, procurement and construction – EPC – contract for the platform). The basic design phase comprises the following tasks:

- Prepare detailed and final process flow diagrams. If detailed design is attempted without final PFDs, process changes during the final design phase may cause significant rework and schedule and cost overruns.
- Prepare piping and instrumentation drawings (P&IDs),
- Prepare final deck and facilities layouts, providing adequate space and clearance for all equipment and operations,

- Prepare equipment and material lists, data sheets and specifications,
- Prepare detailed engineering, transportation, pre-commissioning, Hookup and Commissioning – HUC – scopes of work,
- Prepare detailed design schedules and cost estimates,
- Perform global in-place analyses to confirm that major member and equipment sizes are adequate,
- Prepare basic design drawings for all major platform and deck structures and components (jacket, deck, piles and conductors). These should contain enough detail to enable a reliable – ±20% – field development cost and schedule estimate. Thus information is particularly important if owners wish to go out for bids and enter a lump sum EPC agreement with a contractor.
- Prepare basis of detailed design – BOD – document for the detailed design phase. BOD defines the detailed design requirements including the

  - Platform configuration,
  - Environmental parameters (metocean, seismic, ice, etc.),
  - Site-specific information (water depth, temperature, soil characteristics, mud slides, shallow gas pockets, etc.),
  - Non-generated loads (equipment and supplies – wet/dry –, empty area loads, live loads, dynamic vibrations from rotating machinery, mud pumps and operations, etc.),
  - Design life (operating and fatigue),
  - Accidental design loads (boat impact, dropped objects, fire and blast),
  - Load combination cases (extreme environmental, operational, serviceability, transportation, lift and launch),
  - Damaged stability and or redundancy requirements (missing member, flooded leg or compartment),
  - Preferred material classes,
  - Design regulations, codes and recommended practices,
  - Appurtenances and their locations (escape and evacuation equipment, escape ways, stairs, boat landings, barge bumpers, conductors and J-tubes, mud-mats, etc.),
  - Corrosion protection requirements (type-anode, impressed current- and life),
  - A narrative of construction methods and procedures that will have an impact on the platform configuration and sizes (skidding and load-out procedures, pulling points, lifting eyes, launch skids etc.),
  - Verification and certification requirements,
  - Any other owner requirements that will impact detailed design (jack-up drilling unit clearances, tender rig sizes and weights.

If a contractor is not already selected, a detailed engineering bid document is prepared. This document, which contains most information listed above, is issued to all qualified bidders. Most contractors will not be interested in bidding a lump sum EPC contract, if a basic design package is not available.

6.1.4.2 Detailed Design Phase

Detailed design (also called by some as the "execute engineering phase" or "final design phase") includes all engineering analyses; approved for construction (AFC) detailed design drawings, specifications, procedures and reports required to allow fabrication, transportation, installation, pre-commissioning, HUC, startup and operations of a platform.

- Engineering analyses include all design conditions:
  - In place (wave, current, wind, tide, settlement, earthquake, ice, accidents, redundancy, etc.)
  - Construction (fabrication, load-out, transportation, installation including, up righting, lifting, pile and conductor driving, temporary installation conditions, strength and stability)
  - Fatigue, corrosion
  - Accidental
  - Other relevant design cases (deck vibrations, drilling rig configurations, etc.).
- Design drawings include:
  - Deck plans and elevations
  - Main and secondary trusses
  - Connections (joints) and stiffeners
  - Welding details
  - Pile and conductor details
  - Pad-eyes and other lifting connections
  - Tie-down bracing
  - Launch trusses and braces
  - Appurtenances
  - Any other details required for platform fabrication
- Specifications cover:
  - Design and construction
  - Equipment and materials
  - Painting and corrosion protection
  - Hookup and commissioning
  - Others as needed
- Procedures define certain critical operations in steps to ensure that these are carried out in accordance with the design requirements. These may include:
  - Welding
  - Load-out, tie down and transportation
  - Lunch, up righting and placement to seabed
  - Pile, conductor installation
  - Grouting
  - Deck lift and installation
  - Pre-commissioning, HUC operations

- Escape and evacuation
- Others

- More detailed information on the basic and detailed design tasks of a fixed jacket-tower-type of a platform is provided in Section 6.2.

6.1.4.3 Construction, Hook-up, Commissioning (HUC) and Operations Phase

This phase includes all engineering services provided after the completion of the detailed engineering phase to the start-off of the oil and gas production operations on board the platform. Support is provided to fabricators, equipment and hardware manufacturers, fabricators, installers, HUC operations, and operations to assure that the platform and its facilities are constructed and operated as designed and specified.

6.2 Basic and Detailed Design of a Fixed Jacket-/Tower-type Offshore Platform

6.2.1 Introduction

In this section, we will discuss the basic and detailed design of a fixed steel jacket-/tower-type of an offshore platform (see fig. 6.11). Our initial efforts will concentrate on the problems surrounding the selection of the basic configuration and the major member sizes of a platform such that these will form a valid basis for the subsequent detailed engineering analyses and design activities. We will then spend some time on computerised structural analysis and code check. An analysis helps the designer answer questions about the adequacy or efficiency of his design. It does not, however, replace the experience required for the design. This is acquired through many years of hands-on experience gained from doing actual design and construction work. Remember that the lateral forces on a fixed offshore platform are generated because of the presence of the structure. Consequently, more steel may make the platform stronger but may also cause it to attract more lateral load.

In the detailed design section, we will step from one major component to the another. Some attempt will be made to look at a number of design details so that some of the important design issues can be identified, however within the limited space available for this topic, it is felt that an understanding of the forest is more essential than identification of the trees. A knowledge of the past, the gaining of experience, the understanding of cost, the correct application of analysis and common engineering logic and sense are the building blocks of a good design. For further details, API RP 2A API *Recommended Practice for Planning, Designing and Constructing Fixed Offshore Platforms* [API RP2A-WSD, 2000] its commentary and the extensive references it provides are recommended.

The operational function of the platform is also important to the designer, as an understanding of the function will help them more effectively proportion the structure. As an engineer matures in their field, they can begin to make suggestions as to alternate means to accomplish a given objective.

A good design is one which performs satisfactorily in service and which minimises the capital and maintenance investments.

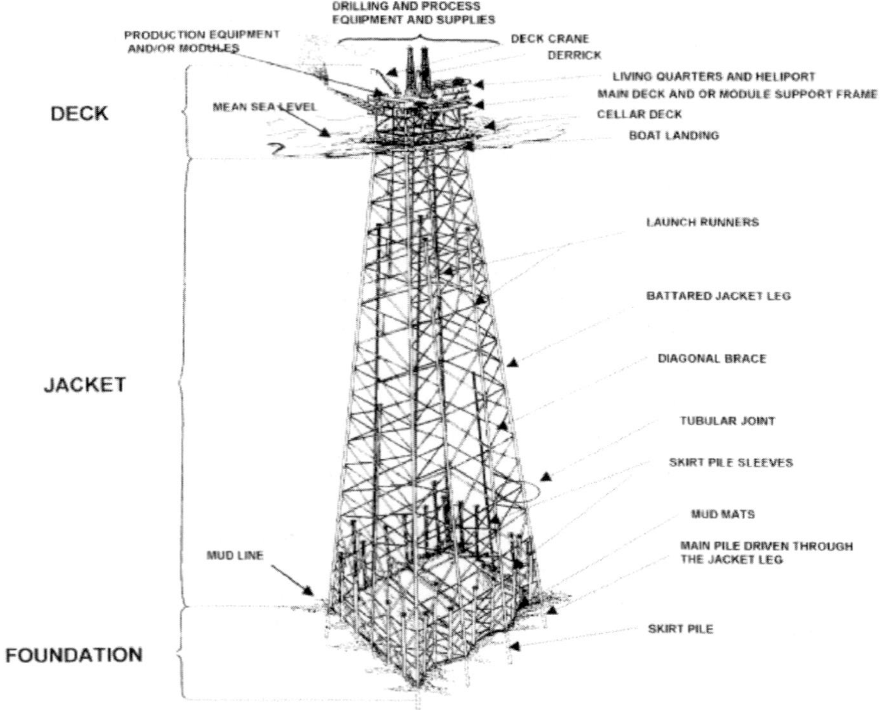

Figure 6.11 Elements of a steel jacket/tower structure

6.2.1.1 Major Structural Components

Figure 6.11 shows the principal structural elements of a steel Jacket type offshore platform. The three major substructures are:

Deck – supports the drilling and production equipment and life support systems of the platform.

A number of offshore platform deck types have evolved in response to operational requirements and the fabrication infrastructure and installation equipment availability. Operational requirements dictate the general deck size and configuration (number of deck levels and their layout, etc.). For example, need for a fully integrated drilling and production system would dictate vertical and horizontal layering of the deck structure in such a manner as to provide an efficient operation while also providing an acceptable level of human and environmental safety.

If fabrication facilities and skilled labour are not available in the area; the economics may dictate building deck in smaller pieces and modules and assembling these offshore, using low-capacity offshore lifting equipment available (fig. 6.17 a). This approach may result in increased steel weight; offshore construction time and cost, while the expense of investing in a major fabrication yard may be avoided. Alternatively, the owner may design the deck as an integrated single piece structure or as a module support frame (MSF) supporting few large modules, which can be built at a location where fabrication infrastructure and equipment are readily available (fig. 6.17 b). The "integrated deck" may then be installed in site using high capacity lifting cranes, or if not available, a float-over deck installation approach [Cotrell and Adrian, 2001]. In this approach, the fully integrated and pre-commissioned deck (or a large module) is loaded out onto a large transportation vessel(s) and transported to the installation site as a single piece. At the installation site, the deck is floated over and then lowered on the support structure by either ballasting the vessel or using quick drop mechanisms. Alternatively, for the case of a floating support structure, the support structure may be de-ballasted to pick up the deck. Other installation methods, such as the "Versatruss" are also in use for smaller integrated decks/modules [Cotrell and Adrian, 1997].

An integrated deck may be divided into a number of levels and areas depending on the functions they support. Typical levels are:

- m*ain* (*upper*) *deck*, which supports the drilling/production systems and several modules (drilling, process, utilities, living quarters, compression, etc.).
- *cellar deck*, which supports systems that need to be placed at a lower elevation and installed with the deck structures, such as pumps, some utilities, pig launchers/ receivers, Christmas trees, wellhead manifolds, piping, etc.
- *additional deck levels*, if needed. For example, if simultaneous drilling and production operations are planned, some process equipment may be located in a mezzanine deck.

A modular deck (fig. 6.17 a) may be divided into a number of pieces and modules depending on the functions they support and available installation equipment. The typical modular deck components are:

- module support frame (MSF), which provides a space frame for supporting the modules and transferring their load to the jacket/tower structure. MSF may also be designed to envelope a number of platform facilities, such as the storage tanks, pig launching and receiving systems, metering/proving devices and the associated piping systems,
- Modules. These provide a number of production and life support systems, such as the:

    Living quarters module (generally supporting a heliport, communication systems, hotel, messing, office and recreational facilities),

    Utilities module (generally supporting power generation and electrical and production control systems, including a control room),

    Wellhead module (generally supporting the wellheads, well test and control equipment),

    Drill rig module (containing the drill tower, draw-works, drillers and control rooms, drill pipe and casing storage racks and pipe handling systems). Drill rig module is located over and supported by the wellhead module.

Production module (containing the oil/gas/water separation and treatment systems and other piping and control systems and valves for safe production, metering and transfer of the produced liquids and gas to the offloading system). *A compression module may be added*, if gas compression for injection to the formation and/or high-pressure gas pumping to shore is needed. Since compression may be needed at later production stages, this module may be installed on the deck at a later date or on a nearby separate platform (generally bridge connected to the deck). Similarly, water injection and pumping modules may be added if these functions are needed at later field development stages.

In general, integrated decks result in more efficient and lighter structural systems, since additional module steel, which is only needed for installation reasons, is avoided. For demonstration purposes, the following paragraphs will elaborate only on the components and design of a mid-sized single-deck structure. The design of MSF and modules follow similar design principles and methods.

Major deck structure members are:

- Deck legs
- Main (longitudinal) and wind trusses (if the deck structure is selected to be an open girder system, this function is performed by a portal frame, see fig. 6.19 b),
- Deck beams,
- Deck plate and/or grating,
- Skid beams (if drilling using a tender rig is planned).

Jacket/Tower – in addition to providing support for the deck, jacket (may also be called steel template or the tower) provides support for conductors and other substructures such as boat landings, barge bumpers, risers, sumps, j-tubes, walkways, mud-mats, etc. The major jacket structure components are:

- Jacket legs
- Braces (vertical, horizontal and diagonal),
- Joints, which are the intersection points of legs and braces. Bracing stubs and cans may be provided to reduce stresses and improve the ductile behaviour of joints,
- Launch runners and trusses, if the jacket will be transported and launched to sea from a launch barge using skid and tilting beams,
- Skirt pile sleeves and braces (if skirt piles are needed),
- Appurtenances (boat landings, barge bumpers, conductor bracing and guides, risers, clamps, grout and flooding lines, j-tubes, walkways, mud-mats, etc.).

Foundation – Piles driven through the jacket legs or skirt pile sleeves fix the jacket to the seabed. Piles driven through the legs are generally welded to the jacket legs above the sea level. These may also be grouted to the jacket leg providing rigidity and composite action. Skirt piles are grouted to the sleeves. Mechanical pile to sleeve connectors may also be used.

Gravity foundations are seldom used, mostly for structures with high weight (such as the concrete gravity platforms). These are feasible at locations where the foundation soils are stable and non-consolidating (such as sand, gravel or highly compacted clays), large footprints are possible, construction infrastructure (such as graving docks and deep enough fjords) is available and geography permits (deep fjords, deep and short towing depths).

6.2.1.2 Major Classes of Loads

Depending on its function, location and construction method, an offshore platform is subjected to several types of loads. These are depicted in fig. 6.12.

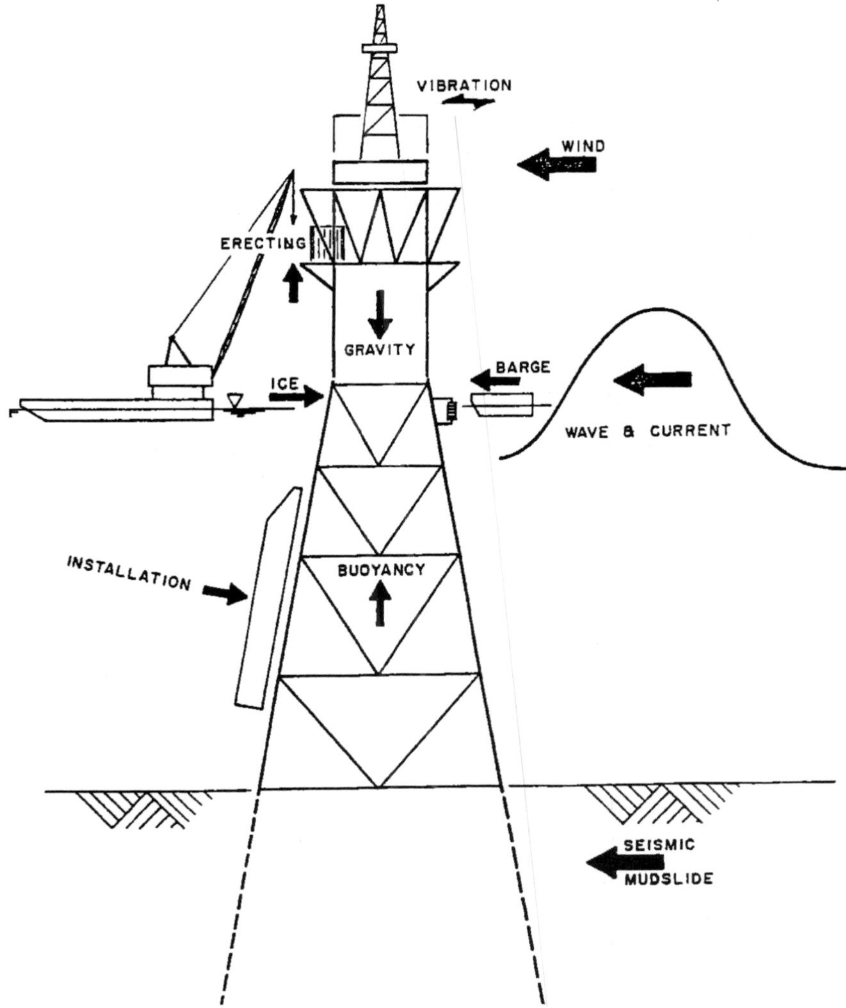

Figure 6.12 Types of loads acting on a fixed offshore platform [McClelland, 1986; Sect. 17]

Functional Loads – Several types of functional loads are imposed on the platform structure:

Deck and equipment loads – These include dry (self) and wet weight (self plus contents) of deck equipment and facilities, storage (mud, cement, pipe), operations (traffic in empty areas, drilling, pipe pull back and setback, unbalanced rotating equipment harmonics, etc.), buoyancy and self-weight of structural elements.

Environmental Loads – These loads depend on the location and environment of the platform site:

- *Meteorological and Oceanographic (Metocean) loads*. These include wind, wave, current and tide effects. Metocean effects are highly variable and their load effects generally control the jacket/tower structure design.
- *Ice loads*. If the platform will be placed in cold and/or arctic regions loads from sheet ice, ice floes and icing and snow on the platform must be accounted for. Cold temperatures also affect the type and quality of the structural materials.
- *Seismic loads*. If the platform will be located in a seismically active zone, loads and deformations caused by earthquakes must be taken into account.
- *Seabed settlements and movements*. Soft seabeds may be set to motion by waves, density differentials or gas pockets, causing mudslides or settlements. Depletion of reservoir or soil consolidation may also cause seabed settlements.

Construction Loads – The following types of loads may be experienced during construction:

- *Assembly and erection loads*. These include loads experienced during the roll-up of jacket bents, temperature and self-equilibrating loads from sequential welding of braces, loads experienced during the jacking/pulling of the jacket or deck over a transportation barge, lifting of the deck, modules and subcomponents, etc.
- *Transportation loads*. Decks and jackets are typically transported to the installation site using transportation barges or carriers. Barge motions and other environmental factors impose loads that are significantly different than a platform would experience in place. Transportation and tie down braces are installed to ensure that the structures stay on barge and resist all transportation loads.
- *Installation loads*. Decks may be lifted or floated over jackets. Jackets may be lifted or launched from barges using tilting beams, up righted and placed over the intended seabed location using derrick barges. Piles may be driven to the seabed using over or underwater hammers or drilled and grouted into the soils. Loads and stresses resulting on the platform and the piles from these operations must be accounted for.

Accidental Loads – These loads may occur due to human error, operational or equipment failures or uncertainties associated with the methods used to predict operational, environmental or construction loads:

- *Vessel impact loads* from construction equipment (barges, work boats, etc.), supply and crew boats, shuttle tankers, merchant vessels, fishing or pleasure boats cruising in the area.

- *Dropped objects*. These may be drilling supplies (drill pipe, casing, collars, BOP stack, etc.), supply packages, equipment on skids and modules that may be dropped by deck or construction vessel-mounted cranes. Drill pipe is lifted to the deck in large quantities and dropped drill pipes and collars are the major sources of injury and damage to the platform components and well systems.
- *Fires and explosions* caused by process equipment, vessel or pipe failures/leaks, blowouts and riser wall failures, etc.
- *Environmental events beyond those considered in the design*. Environmental parameters carry high level of uncertainty and there have been a number of instances where extreme environmental effects much higher than what is assumed for the design return period have been experienced in the past.

6.2.1.3 Platform – Detailed Structural Design Schedule

The tasks and schedule for the detailed structural design of an USA Gulf of Mexico platform in shallow water depths (generally less than 400 ft, where MMS certification is not required) is shown in fig. 6.13. Major tasks consist of *project administration*; *design engineering* and the *preparation of the design drawings*. Staffing and organising the design team; preparing design premises, specifications and schedule; project coordination; budget and schedule control; client and construction liaison; and reporting and documentation are the most important responsibilities of the project design manager and their lead engineers.

Design engineering (analysis and drawings) includes: Deck in-place and construction design; jacket in-place, transportation, launch and installation analyses and design; pile in-place and drivability analysis and design activities. Preparation of design drawings starts with the preparation of the design sketches followed by jacket and deck structural detail drawings, piping, electrical and instrumentation drawings, drawing checks revision and completes with the preparation of the as-built drawings. Not shown in the schedule are the construction, pre-commissioning and HUC support activities that extend into the construction schedule. Some of the design team follows into the construction phase and may even get involved in the maintenance and repair activities during the operations phase.

6.2.2 Selection of the Design Parameters

Environmental, construction and functional design requirements and parameters must be defined before the basic design phase starts. Some of these parameters may be further defined, and assumptions made during the concept development phase may be confirmed during the basic design phase. Definition of some of these parameters requires a site-specific metocean study, soil sampling, laboratory tests and a geotechnical study and, if relevant, seismic and ice environment conditions studies. Specialist companies or contractors perform these studies. Other parameters require detailed process calculations and layout studies that must be completed well before the detailed design starts.

Preparation of the Basis of Structural Design (BOD) Document

The review and finalisation of the BOD document provided by the owner (also known as the design premises document) is one of the most important responsibilities of the project

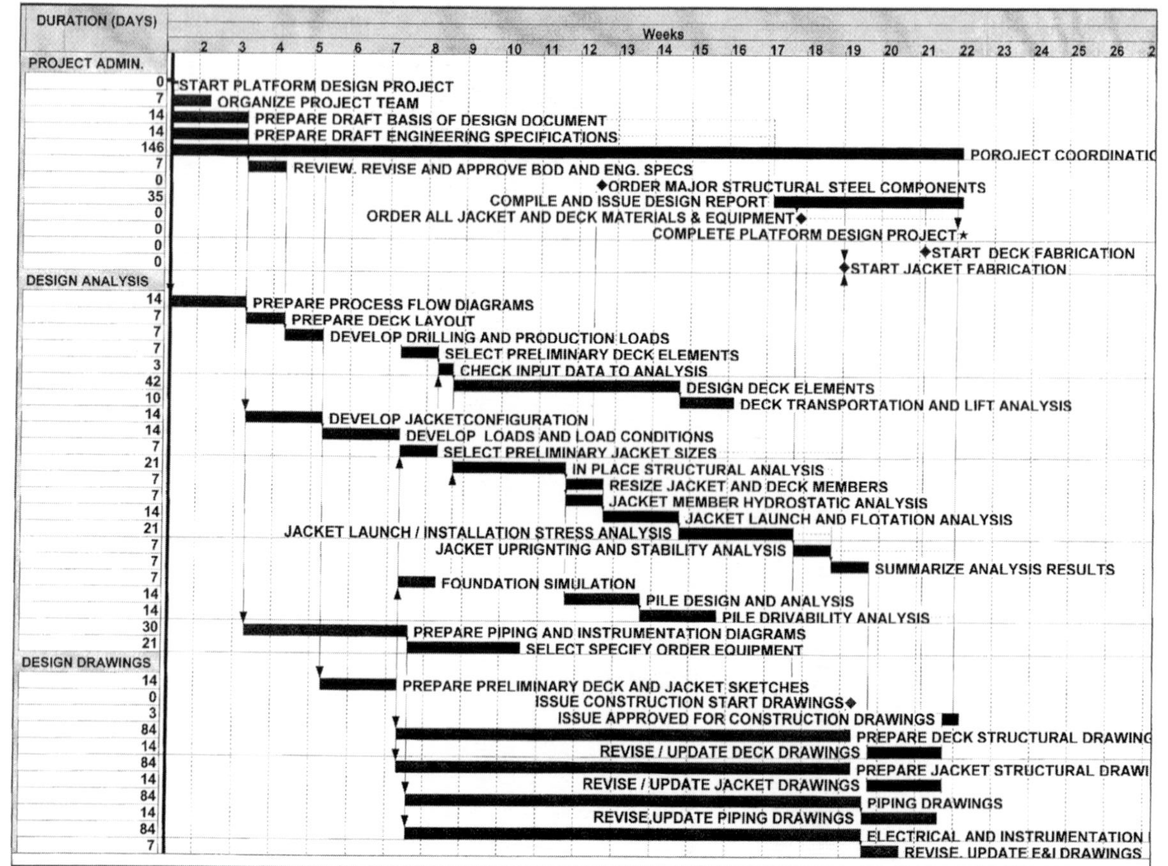

Figure 6.13 Fixed offshore platform design schedule

design manager and lead discipline engineers. This document, which is prepared before the design starts, should contain the following information:

Location, client name, document number issue and revision dates

Space should be provided to enter names and signatures of persons responsible for issuing, checking and approving the BOD document.

Description of the platform function and configuration

- Define the superstructure – deck – configuration including area requirements, number of levels, number of legs, number of bays, leg spacing, location of the wells, equipment, modules, deck plating/grating type, removable hatches, drainage ducts, hand-railing, walkways, stairways, lay-down areas.
- Define the substructure – jacket – configuration including number of legs, skirt piles, leg batter and primary bracing levels.
- Provide a table defining the number, size, location and spacing of well conductors, pump casings, sumps, risers and J-tubes.
- Define location, number, type and elevation of boat landings, barge bumpers, riser protectors and mud-mats.

Description of the site and environmental conditions

- Define water depth (mean water level–MWL).
- Define design winds, waves, tides and currents. Wave information should include significant wave height, associated period, direction vs. probability of occurrence (scatter) diagram; 100-yr extreme design (hurricane) wave height, period, direction and associated wind velocities and current profiles; operating wave height, period, direction and associated wind velocities and current profiles. If spectral static or dynamic analyses are to be provided, type of spectrum should be specified. Currents should be defined in terms of extreme and operating surface current velocity and its variation with depth from MWL. Tide and storm surge heights from the MWL should also be provided. Design winds associated with 100-yr extreme and operating wave conditions should be specified as the values of 1-min duration wind velocities 10 ft above the water level. Instantaneous wind gust velocities are also required for local deck design. In place of the above information, reference to a site-specific metocean report may be made.

Description of the foundation design parameters

- Provide site-specific information on foundation soil characteristics along the pile installation depth. This information should include type of soil (clay, silt, silty-clay, sand, gravel, etc.) the soil mechanics properties (angle of internal friction, cohesion strength, void ratio, water content, shear strength, etc.).
- Provide pile shaft friction and bearing strengths, for assumed approximate pile, as a function of depth from the seabed. Also provide P–Y (horizontal pile pressure vs. soil deformation) and T–Z (vertical pile shaft to soul friction force vs. soil deformation). In place of the above information, reference to a site-specific geotechnical consultant report may be given.

Description of design load conditions (combinations)

Define combination of loads to form critical design load combinations for member design. These include:

- Load combinations that create extreme member and foundation utilisation (extreme waves, associated current and wind attacking the platform from a number of critical directions, deck equipment and reduced live loads, structural weight, buoyancy, earthquake and ice loads, etc.),
- Operating loading conditions where higher safety factors and serviceability and operability considerations govern,
- Fatigue loading condition,
- Fabrication, transportation and installation loading combinations.

Also specify the percentage amount of each load to be used for each load combination. For example, in consideration of very low probability of all empty deck areas being full of live load during an extreme loading condition, for jacket and foundation design, a 70% carry down factor may be applied to most deck live loads. Safety or load and resistance factors applicable to each load type and combination should also be specified or a code reference should be given.

Provision of a table that lists all the load types and their percentage contribution to each design load condition is recommended (see Table 6.1).

Description of deck equipment and live loads

If drilling is to be conducted from the deck, define drilling loads (drilling rig weight, hook load, pipe, casing, collar, pipe setback, potable water weights), drilling package weights (chemical, engine packages, drilling quarters, material cranes; mud, drill water and fuel tanks), drilling supplies (mud, water, fuel).

Define dry (empty) and wet (full of supplies and produced liquids) production equipment weights.

- Define minimum distributed area loads. These are the minimum distributed loads the platform deck is required to carry in place of drilling and production equipment.
- Define empty area loads. These are the minimum live and storage loads that will be carried by the empty areas (walkways, lay-down areas, etc.) of the deck.

Description of the accidental design loads

These include loads from residual events that may be caused by accidents such as vessel impact, dropped object impact, fire and blast and unusual environmental conditions. Magnitudes of these loads are obtained from a global risk assessment study or based on past experience and code requirements. Platform is generally designed against these loads with safety factors set to unity.

Description of design regulations, codes and recommended practices

Provide a list of all regulations (NPD, MMS, DOE), Codes (AISC, ISO, ANSI), recommended practices (API, DnV, ABS) specified for the design.

Description of recommended material classes

Refer to material and welding codes and recommended practices (ASTM, API RP 2U and V, ASME, AWS) and preferred types of materials (2H for the joints, A 36, ST52, etc. for members, etc.).

Painting corrosion protection requirements

Specify the type of painting and corrosion protection system preferred and design life. Also specify any increase in member wall thickness or sacrificial wrap plate and painting requirements for the splash zone and above-water parts of the structure where the cathodic protection is not effective.

Provide other miscellaneous requirements and information

These may include types of deck cranes recommended, riser clamp types and locations and other preferences of the owner.

6.2.3 Selection of the Member Sizes

In most cases, a previous platform design for a similar environment and water depth would form the basis of a design for a new site. In this scenario, most initial member sizes and even a computer model may be already available or could be adapted to new requirements with some revisions. If a previous platform design to copy from is not available, most experienced designers will be able to guess the initial platform configuration and member sizes using rules of thumb based on their past experiences. These selected sizes are then verified and, if needed, re-sized through rigorous analyses in later stages of the design process. Member size selection is not a unique process and may vary from one designer to the other. This section outlines a design sequence and approach based on the author's past experiences.

Sequence of Member and Platform Bracing Configuration Selection

Preliminary member size selection generally starts with the selection of the main pile sizes. In most of the shallow water jackets, main piles are placed inside the jacket legs and the deck legs generally sit over the pile tops. Given the main pile diameter, the jacket leg inside diameter is selected to allow pile placement and driving. The deck leg diameter will generally be equal to the pile diameter. Once the main, pile, jackets and deck leg sizes are known, major jacket bracing, deck trusses beams and plating sizes can be selected. Detailed (computerised) structural analysis and design follow this.

For tower-type platforms with skirt piles only, starting with the determination of the skirt pile sizes is also recommended. Once the pile sizes and loads are estimated, the lower tower legs and braces can be proportioned to carry these loads and match the skirt pile geometry.

6.2.3.1 Pile Size Selection

Selection of the initial pile size requires the knowledge of the approximate axial and shear loads acting on the most critically loaded pile (fig. 6.14). In general, the highest axial pile load is expected to occur in a corner pile, or a corner pile cluster, under an extreme storm loading acting along the platform base diagonal direction. Experience suggests that this is also the most likely load condition that would result in the most critical pile axial

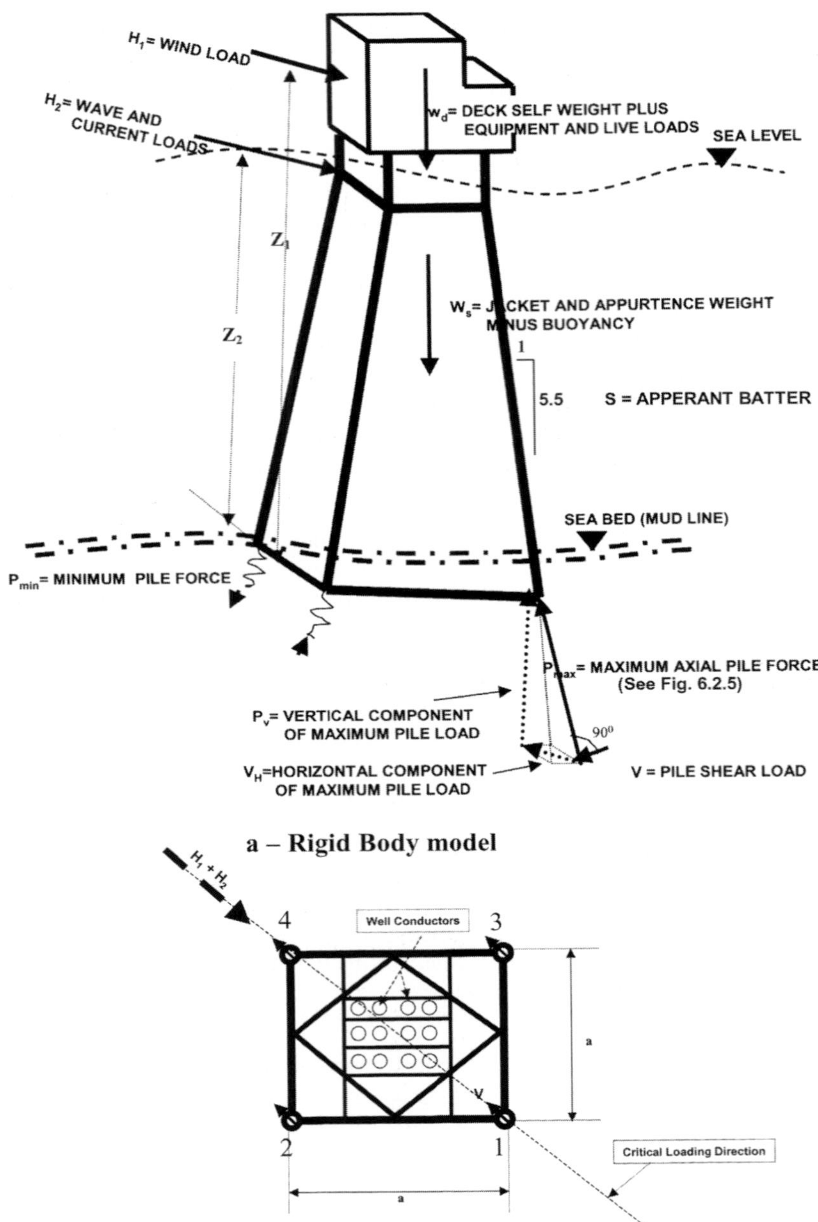

Figure 6.14 Rigid body model for a four legged Gulf of Mexico jacket platform

load and shear combination. The highest pile shear would most likely occur for the same loading but in a more centrally located pile. The estimation of the maximum pile load requires the knowledge of the approximate values for the following:

- Platform deck steel, equipment and supplies weight plus live load from the conceptual design phase and the BOD document.
- Wind load along the most critical loading direction plus its point of application (obtained using an approximate wind drag area for the above water portions of the structure and the maximum sustained wind velocity specified in the BOD).
- Jacket self-weight minus buoyancy (obtained from the conceptual design phase, a previous design or just an educated guess).
- Extreme wave plus current load plus its point of application (obtained from the conceptual design phase, a previous design, an approximate computer wave loading simulation, or an educated guess).
- If other types of loads (ice, earthquakes and mudslide) are to be considered, their approximate extreme load effects and points of application.
- Accurate values for these loads will become available, as detailed vendor information is received and computerised load simulations are performed in the later advanced stages of the design process.

Approximate extreme loads are applied on a rigid body platform model to estimate the maximum pile load. This estimation can be carried out either using manual calculations or a computer model. Figure 6.14 shows an approximate platform and load model for a jacket in 160-ft water depth in the Gulf of Mexico. Example 6.2.1 outlines a manual approach to estimating the pile axial load and shear given a platform and load model shown in fig. 6.14.

Initial pile size can be selected as follows:

a – Select pile outside diameter. Consideration should be given to the effect of the selected pile size on other structural components such as the deck and jacket legs. For a template-type structure with piles driven through its legs, once the pile diameter is known, deck and jacket legs will either be equal to or larger than this diameter. This will have a major influence on the wave loading and dead weight of the structure. Pile diameters should be kept to the minimum values dictated by soil strength, pile driving equipment capability and loads imposed on the pile.

b – Determine pile penetration. Pile penetration depth is calculated using the approximated extreme pile axial load times a safety factor. Axial load carrying capacity Q_d of a pile consists of two major components (see API RP2A-WSD (2000) and Section 14 of this handbook):

$$Q_d = Q_f + Q_p = \int_{Z=0}^{L} f(z)A_s(z)dz + q \cdot Ap \qquad (6.2.1)$$

where:

Q_f = Total pile shaft skin friction resistance
Q_p = Total pile end bearing resistance
$f(z)$ = Unit skin friction capacity (Force/unit area) at depth z

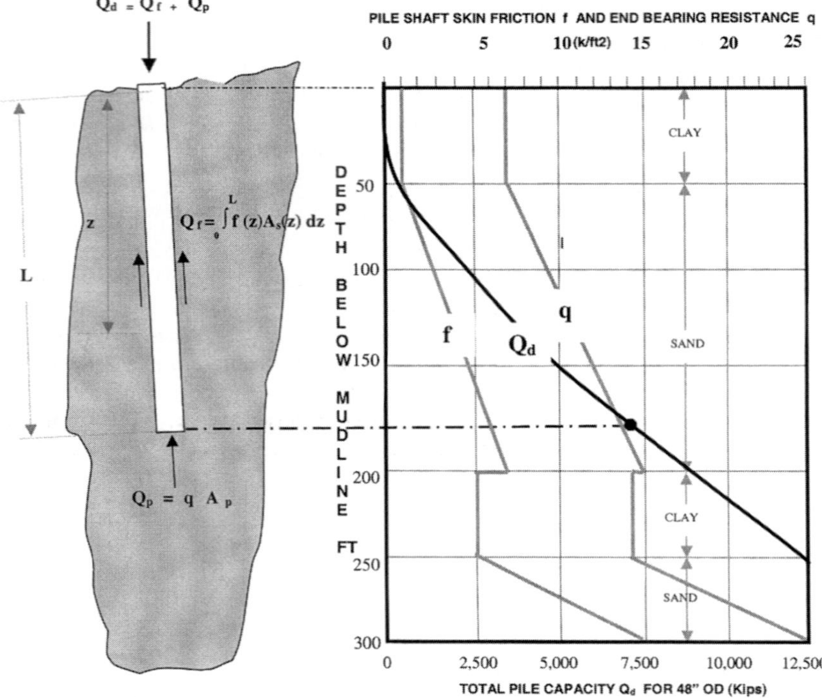

Figure 6.15 Pile capacity chart

$A_s(z)$ = Side surface area of the pile per unit length at depth z
q = Unit end bearing capacity (Force/unit area)
A_p = Gross end area of the pile
L = Pile length

Ultimate pile shaft friction and end bearing resistance values should be provided by a soil report from a geotechnical specialist (for further details, see Section 14). In accordance with the API Recommended practice 2A (2000), the pile safety factor can be taken as 2 for the operational loading and 1.5 for the extreme design environmental loading case. Figure 6.15 shows a typical pile capacity chart provided by a geotechnical consultant. Steps (a) and (b) are repeated until a reasonable pile penetration which can be achieved with the existing pile driving equipment is obtained. Pile outside diameters of 36–72 in. is in common use in the Gulf of Mexico.

c – *Select pile wall thickness*. The initial pile wall thickness is selected to resist maximum bending moment and axial load acting on the pile. If the pile is fixed to the jacket leg (grouted pile case), the most likely location of the maximum pile bending moment will be at the point where pile enters the jacket leg. If this connection is not rigid (un-grouted pile case) the maximum bending moment location will be below this point, at a distance determined by the flexibility of the pile to jacket connection and the pile sport points inside the jacket leg. The pile bending moment could be estimated using elastic laterally loaded pile analysis procedure [see Reese and Mattlock (1956); Bryant and Mattlock (1975)]. Alternatively, a laterally loaded pile analysis program may be used. P–Y data may be developed using API RP2A Section 6 foundation design requirements. Example 6.2.1 provides a manual method for determining the approximate pile penetration and wall thickness.

Most offshore piles are driven to ground using large steam, diesel or hydraulic hammers. In rocky or shale grounds where pile driving is not feasible, pile may be placed in pre-drilled shaft holes and grouted in. For hard to drive soil conditions, an intermittent drilling and driving sequence may be performed to reduce end bearing resistance allowing easier driving conditions. In deeper water depths and for skirt piles, underwater hammers are also in use. Dynamic pile driving loads and pile derivability may also influence the wall thickness selection. Generally, thicker wall pipe is used at pile tips to improve pile drivability. Two-inch or more wall thickness is commonly used at the pile tips. After the approximate pile wall thickness is determined, the dynamic stresses imposed on the pile by the hammer should be assessed and pile derivability should be confirmed using a dynamic pile driving analysis program (Smith, 1962).

Example 6.2.1

Select the initial pile dimensions for the platform model shown in fig. 6.14. Assume:

MWD = mean water depth = 160 ft
$H_1 = 100$ kip; $Z_1 = 230$ ft; $H_2 = 1500$ kip, $Z_2 = 120$ ft
$W_d = 8000$ kip; $W_s = 2000$ kip; Apparent batter $= s = 8$; $a = 85$ ft
M_B = base overturning moment $= H_1 \times Z_1 + H_2 \times Z_2 = 100 \times 230 + 1500 \times 120 = 203{,}000$ ft kip
V_B = base shear $= H_1 + H_2 = 100 + 1500 = 1600$ kip
P_B = axial force at base centre $= W_d + W_s = 8000 + 2000 = 10{,}000$ kip

Calculate the vertical component of the maximum pile load assuming that the foundation rotates as a plane around a line defined by the axial load P_B and the moment M_B:

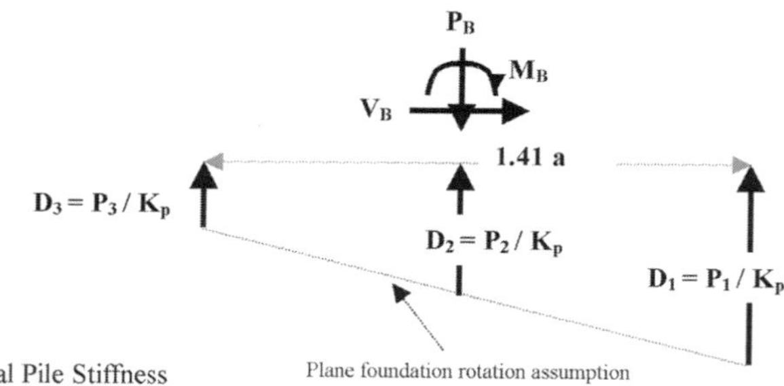

From the plane foundation rotation assumption:

$P_1 = M_B/(1.41)\ a + P_B/4 = 4194$ kip

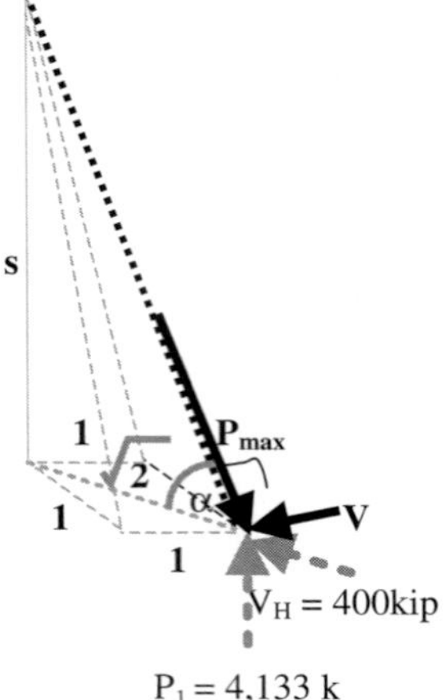

Figure 6.16 Resolution of the forces along the pile axis

Assuming equal shear load distribution to all four piles:

$V_H = V_B/4 = 400$ kip

Estimate the maximum axial force $P_{\max}$ and the associated shear force V acting on the corner pile:

$$\text{True batter} = s/\sqrt{2} = 5.67$$

$$\text{Cos}\ \alpha = \sqrt{2}/\sqrt{1+S^2} = 0.1754$$

$$\text{Sin}\ \alpha = 0.9845$$

$$P_{\max} = P_1\ \text{Cos}\ \alpha + V_H\ \text{Sin}\ \alpha = 4066 + 70 = 4133\ \text{kip}$$

$$V = V_H\ \text{Cos}\ \alpha - P_1\ \text{Sin}\ \alpha = 388 - 726 = -346\ \text{kip}$$

Note that, in this example, pile axial force acting along batter negated direct pile shear caused by V_H, showing the benefit of pile batter in resisting the horizontal forces. For this example, a lesser batter of $S = 15$ would have resulted in a pile shear of only −3 kip, while such a small batter would have resulted in a smaller foundation footprint (for a fixed deck size) and an increased P_1. For this case an apparent batter of 8 seems to result in an acceptable compromise.

Also note that, the base overturning moment M_B is linearly proportional to the water depth while the base shear V_B stays almost constant with water depth. As the water depth increases, there will be a need for increasing the number of piles through use of skirt piles (assuming that there is a limiting a single pile capacity dictated by the foundation and equipment capability). Increased number of piles will reduce V_H acting on an individual pile. Therefore, as the water depth increases, consideration could be given to reducing the leg batter.

As per API RP2A-WSD (2000) API RP2A Par 6.3.4 a factor of safety of 1.5 is required for design environmental conditions. Therefore the "design pile penetration" should be calculated using:

$P_{\text{Design}} = 1.5\ P_1 = 6200$ kip of pile axial load.

For corrosion allowance, protection against buckling from pile driving strains and drivability reasons, pile walls are generally selected not less than 1 in. thick. Forty-two to 54 in. OD piles, which are in common use in the Gulf of Mexico shallow water depths and represent the minimum diameters that can be cold-rolled from 1 to 2.0 in. wall thickness flat plate without significant rolling difficulty and high strain build up. Try an average pile diameter of 48 in. OD. Using the pile capacity Q_d curve for a 48 in. OD pile shown in fig. 3.2.1, a pile penetration of about 180 ft below the mud line is expected to generate a pile capacity of about 6200 kip. Use 185 ft length to account for pile's self-weight.

The next step is the selection of the pile wall thickness. Selection of the pile wall thickness requires the knowledge of the axial load and the bending moment acting on the critical pile cross section. The value of axial load in this section would be equal to P_1 at the pile top gradually decreasing to the end bearing resistance at the pile tip. The bending moment would be a function of the shear and bending moment at the pile top, the flexibility of the jacket to pile connection (generally assumed fixed if the pile is grouted inside the jacket leg) and the load deflection characteristics of the foundation soils (see fig. 6.31). Due to non-linear soil properties, solution of this problem would require the use of a non-linear and iterative solution using a special software. In such a solution, in addition to the pile and jacket properties, knowledge of the lateral load and corresponding deflection p–y (p = lateral resistance, psi and y = lateral deflection, in.) curves for the soils along the pile length is needed. For initial pile wall thickness selection, an elastic foundation subgrade modulus (κ)-based approach developed by Reese and Matlock (1956) could be used. For average soils a κ value of 4 lb/in.^{3} is commonly used. Selecting a wall thickness of 1.5 in. and assuming that the pile top is fixed to the jacket leg,

$I = 59{,}287$ in.^{4}; $S = 2470$ in.^{3}; $A = 219.1$ in.^{2}; $E = 29{,}000{,}000$ psi, from Reese and Matlock (1956), fig. 24:

Relative pile to foundation stiffness parameter (or length) $= T = (EI/\kappa)^{1/5} = 212.2$ in.

Moment in critical cross section $= M_F = F_M\,(V \times T)$. From Reese and Matlock (1956), for a fixed ended pile, M_F occurs at a cross section slightly below the pile to jacket leg connection point where $F_M = 0.93$. Therefore $M_F = 0.93\,(346 \times 212.2) = 68{,}282$ k in.

The next step is to calculate the maximum stress f in the critical section. Conservatively assuming that the axial force in the critical pile cross section is equal to $P_{\max}$:

$$f = f_a + f_B = M_F/S + P_{\max}/A = 27.64 + 18.86 = 46.5 \text{ ksi}$$

Use of $F_Y = 50$ ksi yield strength steel is common practice for most Gulf of Mexico piles. If this material is used, the combined critical stress f would be above the AISC allowable stress $F_a = 0.8$, $F_Y = 40$ ksi level. In this case either use of thicker wall pipe (use of 1.75 in. wall thickness pipe would result in an acceptable stress level of 39.9 ksi) or higher strength steel (use of X60 steel with 60 ksi yield strength would result in 48.0 ksi allowable stress) could be considered.

6.2.3.2 Deck Leg and Deck Structure Dimensions Selection

A number of platform deck structure systems are in use depending on the deck layout, operational requirements and fabrication and installation constraints (see Section 6.2.1). If the production equipment is to be packaged into modules, a two-way truss system called the module support frame (MSF) is utilised. The MSF consists of a system of horizontal girders and vertical and diagonal braces, which transfer the module loads to the top of jacket legs or piles (fig. 6.17a). Modular deck system is used when the deck weight or geometry exceeds the available crane lift capacity or modules containing some of the production equipment (such as a compression or pumping module) will be added at later dates. If the dry deck weight and geometry are within the available crane capacity, an integrated deck configuration may be considered. Integrated decks eliminate the need for module steel, which is only needed for lifting purposes and generally result in lighter structures. Hybrids of modular and integrated deck designs are also in use.

In hybrid deck configurations, some of the utility, wellhead and/or process equipment and storage may be placed inside the MSF, which would be installed in a manner similar to an integrated deck, followed by the installation of modules on its top. In all cases, the elements of the deck structure and the method of design are similar although the way they are loaded, fabricated, installed and commissioned differ.

Figure 6.18 shows the basic elements of a deck structure common to most offshore deck configurations.

One of the key decisions made prior to the deck member size selection is the determination of the deck elevation relative to the sea surface. The deck elevation must be selected to provide a satisfactory gap between the wave crest and the deck structure. Major Gulf of Mexico platform damage, including total platform and/or deck losses, has been experienced in the past due to platform decks being slammed by wave crests. To avoid the occurrence of this potentially catastrophic event, the standard industry practice is to

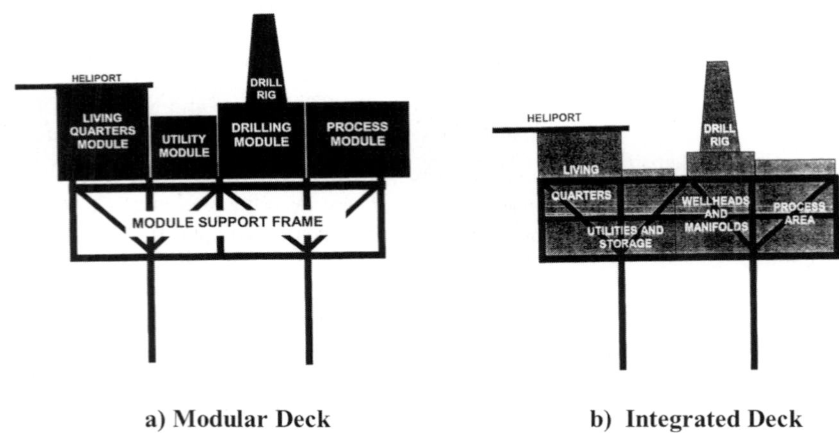

Figure 6.17 Modular and integrated deck concepts

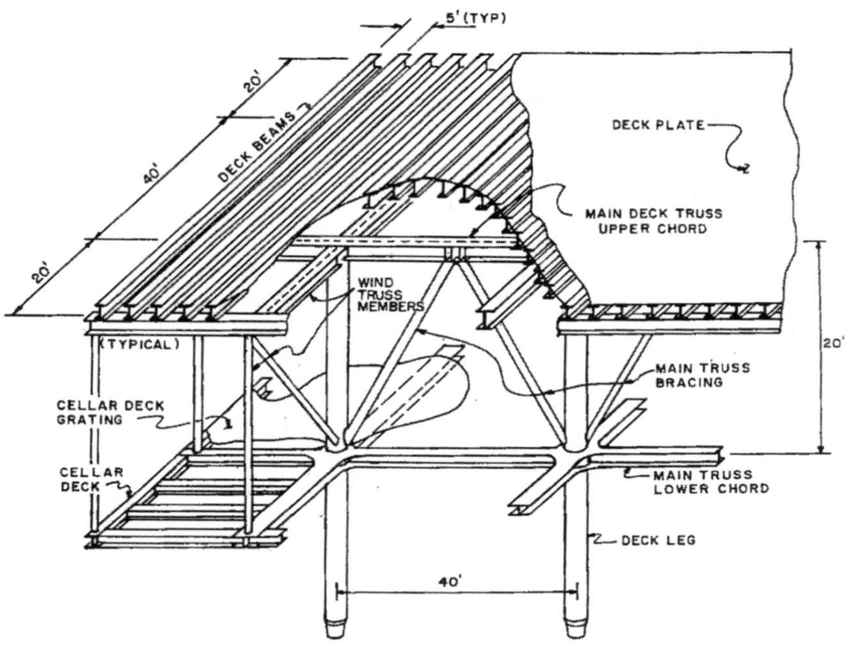

Figure 6.18 The basic elements of a deck structure [McClelland, 1986; Section 17]

provide a safety margin by providing an "air gap" between the elevation of the design wave crest (generally represented by the crest elevation of a nominal 100-yr return period wave) and the bottom of the deck steel or any piping or structural elements that may protrude below it. For the US GoM conditions, API RP2A (1993, 2000) recommends that this air gap should be a minimum of 5 ft, plus "any additional allowance provided for any known and predicted long-term seafloor subsidence". API RP2A goes further to recommend that "However, when it is unavoidable to position such items as minor subcellars, sumps, drains or production piping in the air gap, provisions should be made for the wave forces developed on these items". The local wave pressures generated by the wave crest hitting a horizontal or near horizontal deck member and surfaces could be far in excess of the drag and inertial pressures exerted on the platform members under normal wave progression. These "slam" pressures could involve high impulsive effects represented by wave slam coefficients of as high as three or more (as compared to the API recommended drag coefficients of 0.65–1.05 for the overall platform design). Therefore it would be a wise choice to do all possible to avoid wave crests hitting any deck area. API further goes to say "These provisions do not apply to vertical members such as deck legs, conductors and risers etc., which normally penetrate the air gap". Note that the 5-ft air gap requirement is for local GoM conditions and vary with the uncertainties inherent to the wave, tide and storm surge characteristics of an area. For regions with higher environmental uncertainties such as the Southern GoM or Far East, use of a larger air gap may be warranted. API RP2a Section 2.3.1 provides significant definition and details for the calculation of the minimum recommended wave forces acting on fixed platforms. It also provides useful charts and tables for guideline wave heights, storm tides and design deck heights for the Offshore US Gulf of Mexico.

In general, the skid beam spacing of a standard GoM platform drilling rig dictates the deck leg spacing for a drilling platform or module. Placing the deck legs directly underneath the skid beams facilitates direct transfer of the high rig and drilling loads onto the column tops, resulting in an efficient structural system. Most GoM platform rigs supplied by drilling contractors would have 40 ft skid beam spacing. Use of cantilevers to gain additional deck space is also commonplace. Most efficient cantilevers are achieved when these are about one half the lengths of the deck spans. Therefore, 80 ft by 80 ft four legged and 120 ft by 80 ft eight-legged GoM deck footprints are commonly encountered.

Deck structural component size selection starts with the estimation of the deck loads from the drilling rig, production equipment, live load, supplies, etc. The deck equipment and process designers calculate these loads in the conceptual design phase and they are included as part of the Basis of Design Document. At the early design stages, upper bound distributed loads representing the deck equipment and supplies may be used for the selection of the initial deck plate and beam sizes. As the design matures, more accurate equipment and operational loads are used for detailed design. For initial deck sizing purposes, 500 to 1000 Pounds per square foot (psf) loads for the main decks and 300 to 500 psf loads for the mezzanine and cellar deck design are commonly assumed.

Deck Leg Size Selection

Deck legs transfer the forces exerted on them by the main and wind truss members to the tops of the piles or the jacket legs. Deck legs are almost always made of tubular members. There are a few cases where deck legs may be made of stiffened steel boxes or rolled cross sections. Deck legs may be unbraced or may be supported by knee braces extending from the lower deck trusses/beams to a level dictated by the deck span and deck leg stability against buckling requirements.

Preliminary deck leg sizes may be selected as follows:

- For the case of a jacket with main piles driven through its legs, the deck leg outside diameter (D_o) is generally taken equal to the pile outside diameter. If the jacket does not have main piles, the deck leg outside diameter could be equal or less than the jacket leg diameter.
- For a tubular leg, radius of gyration (r) of its cross section can be calculated from 0.35D, which is true for thin-wall cylinders. For non-tubular cross-sections, this parameter could be calculated or picked up from a table. Depending on the end fixity conditions of the deck leg, a buckling length factor (k) is assumed. Depending on whether the deck leg is subject to side sway and/or rotations at its ends, the initial k-value assumption could be anywhere from 2 to 0.5. For initial leg size selection purposes, a k-value of 1.5 is commonly assumed [ASD, 1989]. Next, the slenderness ratio (kL/r) of the deck leg is calculated where L is the true deck leg length. With the slenderness ratio known, allowable axial stress value (F_a) can be obtained from the *AISC* (American Institute of Steel Construction) *Manual* [AISC ASD, 1988].

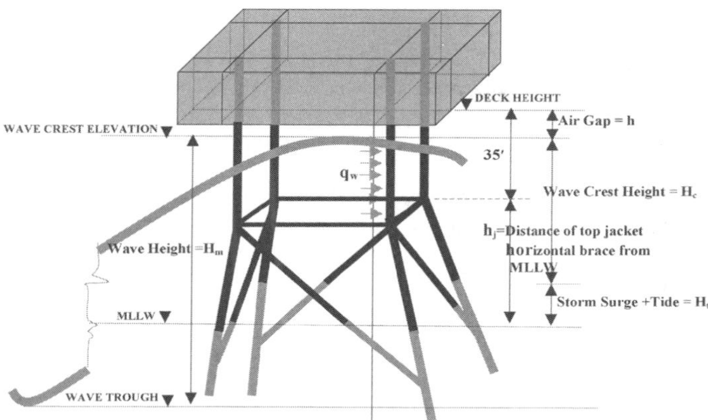

Figure 6.19 Deck model for preliminary air gap and deck leg sizing

- Using conservative values for deck equipment and structure weights and wind and wave forces, approximate values for maximum axial load and moment on the deck leg are calculated.
- Approximate deck leg wall thickness is selected, axial and bending stresses are calculated and *AISC* interaction ratios are computed. This procedure may be repeated for several wall thickness values until an interaction ratio of less than unity is obtained.

Example 6.2.2

For the deck configuration shown in fig. 6.19 and the platform loads used in Example 6.2.1, set up the deck bottom of steel elevation and estimate the deck leg size for a GoM platform. Assume the following:

- Mean lower low water (MLLW) depth = 160 ft (MLLW is the water depth corresponding to the height of the lowest tide level observed at a location averaged over a given time period, traditionally taken as a 19-yr period).
- Design wave height = $H_m = 62.5$ ft (wave height is measured from wave trough to the wave crest. Design wave height generally represents the maximum individual wave height that may be experienced within a 100-yr return period).
- Associated wave period = $T_m = 12$ s (Zero crossing period associated with H_m).
- Storm tide associated with the wave height = $H_t = 3.5$ ft (includes storm surge plus the astronomical tide).
- In-line current velocity at storm water level = $U_i = 2$ ft/s (associated with H_m).
- Assume a pressure of 350 psf at the wave crest, acting uniformly along the deck leg.

Calculate:

$$\text{Deck height above MLLW} = \text{Elevation of the bottom of the deck steel from MLLW}$$
$$= h + H_c + H_t = 5\,\text{ft} + 3.5\,\text{ft} + H_c = 8.5\,\text{ft} + H_c$$

Wave crest height H_c and the wave pressure distribution can be calculated by performing an analysis using a wave simulation theory in compliance with the API RP 2A Section 2.3.1, fig. 2.3.1–3 requirements. In the absence of such an analysis, for adequately deep water depths and only for preliminary deck height selection purposes (to be verified by detailed computer simulation later), H_c can be assumed to be about equal to 2/3 of H_m. In our example:

$H_c \cong 2/3\ H_m = 41.7$ ft and,

Deck height above MLLW = 50.2 Note that this height is reasonably close to the 51 ft height recommended by API RP2A fig. 2.3.4-8.

Distance of top of jacket horizontal brace from the MLLW $= h_j = hz + H_c + 35\text{ ft} = 15.2$ ft. Note that the elevation of the top of jacket horizontal braces should be selected to avoid frequent wave slamming and wet/dry (buoyant/non-buoyant) cyclic loads that may be experienced under commonly encountered waves. A 15.2 ft brace clearance is about equivalent to a wave height of 22.8 ft, which is outside the range of commonly encountered GoM waves.

Top of jacket horizontal bracing clearance of 15–20′ is in common use in offshore practice.

The most critical axial load on a deck leg will be created when the deck is fully loaded and the wind and waves are acting along the deck diagonal.

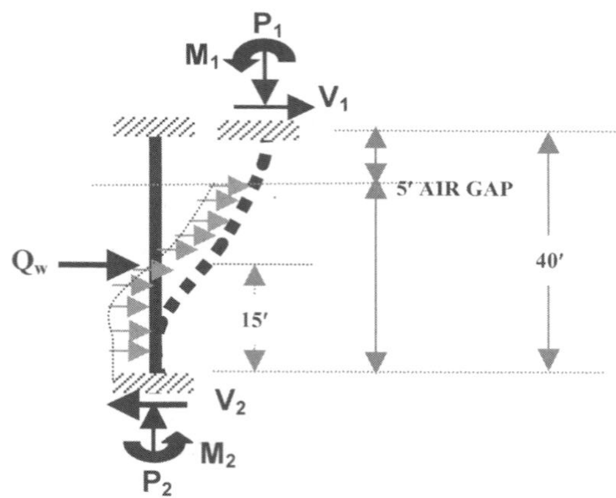

Figure 6.20 Deck leg loads

$V_1 = H_1/4 = 25$ kip

$P_2 \approx H_1(230\text{–}160\text{ ft}\text{–}15.2\text{ ft})/(\sqrt{2} \cdot 40\text{ ft}) + W_d/4 = 2097$ kip

$Q_w = 0.35\,\text{k/ft} \cdot (42\text{ in.}/12) \cdot 30\text{ ft} = 36.8$ kip

$V_2 = V_1 + Q_w = 61.8$ kip

Assuming fixed end restraint for the deck leg,

$$M_{\max} = M_2 = V_1 \cdot (35\text{ ft})/2 + Q_w(35/24)\{6(30/35)^3 - 16(30/35)^2 + 12(30/35)\}$$
$$= 561.45\,\text{kip ft} = 6738\,\text{kip in.}$$

(see Roark, Chapter 8, Item 34 for the above).

Select the deck leg outside diameter (OD) equal to the pile OD of 48 in. Try a wall thickness of 1.00 in. $A = 147.65$ in.^{2}; $S = 1699.57$ in.^{3} $r = 16.62$ in., use ASTM A572 Grade 50, $F_y = 50$ ksi steel.

Use a K buckling factor of 1.5; $KL/r = 1.5 \times 40 \times 12/16.6 = 43.37$, From AISC Manual for 50 ksi yield steel, for static loading conditions; $F_a = 25.26$ ksi, $F_b = 0.6 \times 50 = 30$ ksi.

First, check the static loading condition for operating conditions when no wind or wave load is present.

$$f_a = 2000/147.65 = 14.20$$

$$f_a/F_a = 0.55 < 1.00 \qquad \text{(AISC Eq. 1.6-1a)}$$

Next, check the dynamic loading condition with wind and wave when AISC 1/3 allowable stresses increase is applicable:

$f_a = 2097/147.65 = 13.78$ ksi; $f_b = 6738/1699 = 3.96$ ksi

$$f_a/1.33 \times 0.6F_y + f_b/1.33\,F_b = 0.35 + 0.10$$
$$= 0.45 < 1.00 \qquad \text{AISC ASD Eq.1.6-2 Interaction}$$

check:

$f_a/1.33F_a + C_m f_b/[1.33\,F_b(1 - F_a/F'_e)] < 1.00,$ AISC ASD Eq. 1.6-1a Interaction Equation, where $C_m = 0.85$ (column subject to side sway) and $F'_e = 79$ ksi

$$0.41 + 0.10 \times 0.85/[(1 - 18.78/79)] = 0.52 < 1.00$$

The above calculations indicate that a 0.75 in. pipe wall thickness would have resulted in an AISC IR of 0.80 and would be adequate. However, this wall thickness will yield a pipe D/t of 64, which may exhibit local pipe wall instability. Instead, selection of a 1.00 wall pipe also providing ¼ in. corrosion allowance in the splash zone would be a good starting size for the detailed three-dimensional structural analysis. (Note: use of 1/8 in. to ¼ in. corrosion allowance in the oxygen-rich splash zone, where cyclic wet and dry conditions prevail, is a common practice.)

Selection of the Deck Cover and Major Member Sizes

The main function of the deck structural system is to provide a suitable horizontal area above the water surface where the drilling and production operations are performed. In doing this, the deck framing transfers the loads from equipment supplies and operations exerted on the deck coverings to the deck legs, which in turn transfers loads to the jacket legs and the foundation. This load transfer can be accomplished by using a variety of deck structure systems. A popular system is the one where the loads are transferred through flooring and deck beams to longitudinal trusses made of tubulars and standard or built up beams (figs. 6.18 and 6.21a). Other deck framing systems, such as portal frames with stiffened plate girders directly framing into deck legs are also in use (fig. 6.21b). Selection of the type of deck framing very much depends on the deck span, operational requirements, the local construction practices and access clearance needed to install and maintain drilling and production equipment. Clearance for access is particularly important for integrated decks where the equipment is installed inside the deck structure. For such structures, K-braced trusses shown in fig. 6.21a generally allow reasonable access while also reducing the main deck span.

Hybrids of the above-described deck structural systems, where the main load bearing structure (truss lines 1 and 2 in fig. 6.21c) would be K-braced trusses and the transverse wind bracing structure (lines A and B) would be stiffened girders are also in use. Past studies suggest no major cost difference between the trussed or stiffened plate girder deck structures. However, the fabrication of the plate girders may pose dimensional control difficulties for construction yards with no plate girder fabrication experience.

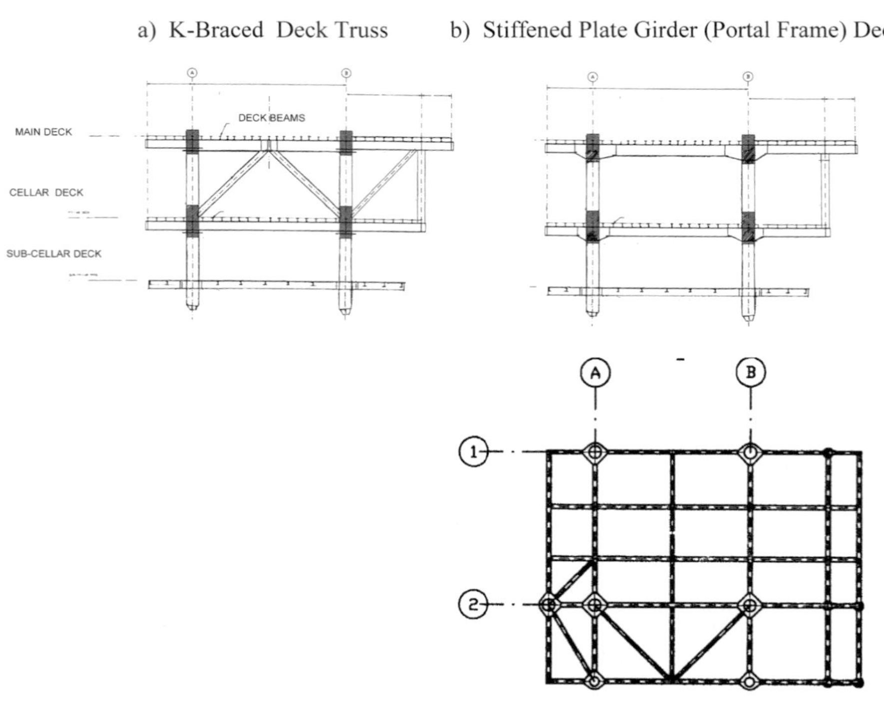

Figure 6.21 K braced truss and stiffened plate girder main deck structures

The flooring of the decks of a steel offshore platform may be non-existent, grated, checkered plate, timber or plain plate. For the cases where plate flooring is used, if properly welded to the deck beams, the deck plate also acts as a membrane that distributes the horizontal wind and wave loads to the main deck structure. This is particularly important for diagonal loadings, which would force the deck to behave as a horizontal portal frame, causing undesirable distortions. If deck plate is not present, using horizontal diagonal braces connecting the column tops at each deck level should be considered.

In designing the plate and beam flooring systems, the "lowest weight" solution for a specified deck load may not always result in the "lowest cost". The number of cuts, ease of fitting, length and difficulty of welds and ease of maintenance all influence the initial and maintenance costs of the system designed. Some deck systems may be designed using a two way framed (Orthotrophic) plate-beam approach. In this type of flooring design (fig. 6.22), the deck plate, deck beam and truss upper (lower) chord are replaced by a continuously welded plate/beam system. The plate carries the direct loads along its span while it functions as top flange of the deck beam and the main girder system. Two-way deck-covering designs generally result in lightweight but expensive structures because of the extensive cutting, fitting and welding requirements. Orthotrophic deck designs are popular in deepwater fixed or floating platforms [such as the Compliant Tower platforms (CPT), Tension Leg Platforms (TLP), SPARs and semi-submersible floating vessels (SSV)] where the deck weight may drive the platform cost.

The stacked deck plate – deck beam – main girder detail shown in fig. 6.18 results in heavier decks but requires much less labour to build it. If the deck width could be kept below 80 ft, the deck beams could be ordered from the manufacturers as single pieces, assembled stacked over the main deck girder, plated/grated over, lifted and installed over the pre-fabricated deck legs and braces, as a "Pancake". This results in much lower fabrication costs.

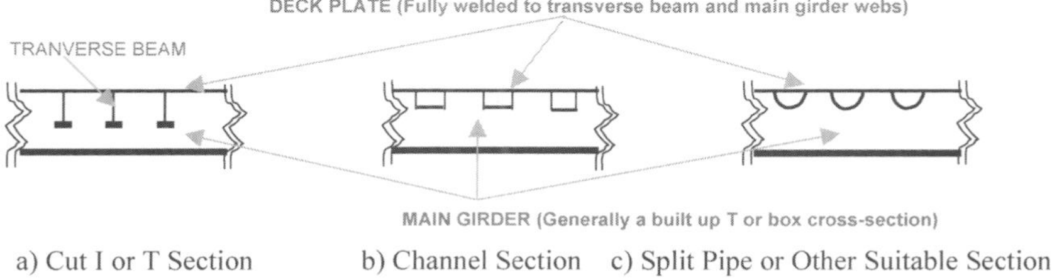

a) Cut I or T Section b) Channel Section c) Split Pipe or Other Suitable Section

Figure 6.22 Two-way (Orthotrophic) deck framing details

In the case of a plate floor system, selection of the deck plate size is accomplished by using an upper bound distributed deck load and selecting a deck plate thickness using the plate-bending formulas. These are readily available in many engineering handbooks and they generally degenerate to simple beam bending formulas for plates with high length to width ratios.

For the case of a stacked beam system, the deck plate loads are transferred to the deck beams. Deck beam spacing is generally dictated by the wellhead spacing. Beams are generally placed at equal spacing, spanning the space between the main deck trusses (or main deck girders). Selection of deck beam sizes can be accomplished by using simple or continuous beam formulas. An example to this process is shown in Example 6.2.3.

Example 6.2.3

Select the deck beam, main deck girder and main deck truss member sizes for the deck configuration shown in fig. 6.18.

Assume:

Main deck distributed equipment and live load = 750 psf

Main deck distributed self weight load = 50 psf

Cellar deck distributed equipment and live load = 350 psf

Cellar deck distributed self weight load = 35 psf

Use A36 (36 ksi yield) steel.

(a) Selecting the deck beam size (fig. 6.23)

$$Q_L = 0.75 \times 5 \text{ ft} = 3.75 \text{ kip/ft} \quad \text{(where the deck beam spacing is 5 ft)}$$
$$Q_W = 0.05 \times 5 \text{ ft} = 0.25 \text{ kip/ft}$$

Maximum deck beam moment M_1 will occur at point 1 when the cantilevers have no equipment and supplies load –

$$M_1 = (Q_L + Q_W)L_2^2/8 - Q_W\,L_1^2/2 = 750 \text{ kip ft}$$

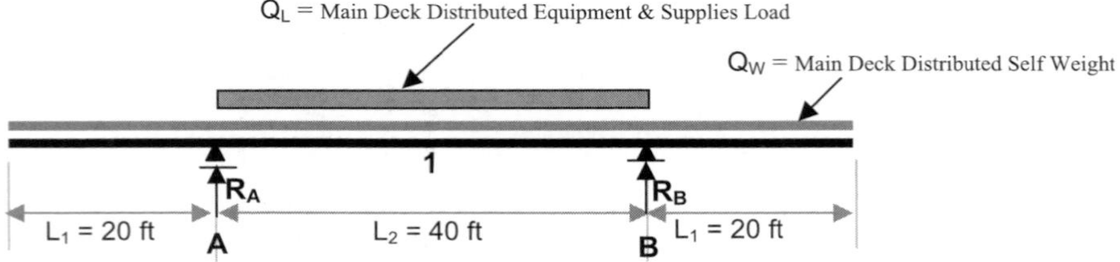

Figure 6.23 Deck beam model

Minimum deck beam moment M_A will occur at point A when the cantilevers are fully loaded with equipment and supplies load –

$$M_A = (Q_L + Q_W)L_1^2/2 = 800 \text{ kip ft} = 9600 \text{ kip in.}$$

Maximum deck beam shear V_A will also occur for this maximum loading condition

$$V_A = (Q_L + Q_W)L_1 = 80 \text{ kip}$$

Maximum deck beam support reactions would occur when the deck beam is fully loaded –

$$R_A = R_B = (Q_L + Q_W)(L_1 + L_2/2) = 160\,\text{kip}$$

For this case, the deck beam is designed by $M_A = 9600$ kip in. and $V_A = 80$ kip

Using AISC ASD (1989), Part 5 requirements:

Allowable beam bending stress $= F_B = 0.72F_Y = 25.92$ ksi

Required section modulus $= S_X = M_A/F_B = 371$ in.^{3}

Select WF 24×146 with $S_X = 371$ in.^{3}

Check shear stress f_v

$f_v \approx V_A/td$, where t and d are the thickness and the depth of the deck beam (0.625 and 24.25 in. respectively for the WF 24×146 section)

$f_v \approx 80/(0.625 \times 24.5) = 5.23$ ksi < Allowable Shear Stress $= 0.55\ F_Y = 19.8$ ksi

(b) Selecting main deck girder size

The maximum bending moment, shear and R_{112} support reaction for main girder on truss line *A* will occur when Bays 1, 2, and 4 are loaded with equipment and supplies.

For the 5 ft equal spaced equal loading case in a loaded bay (see fig. 6.24):

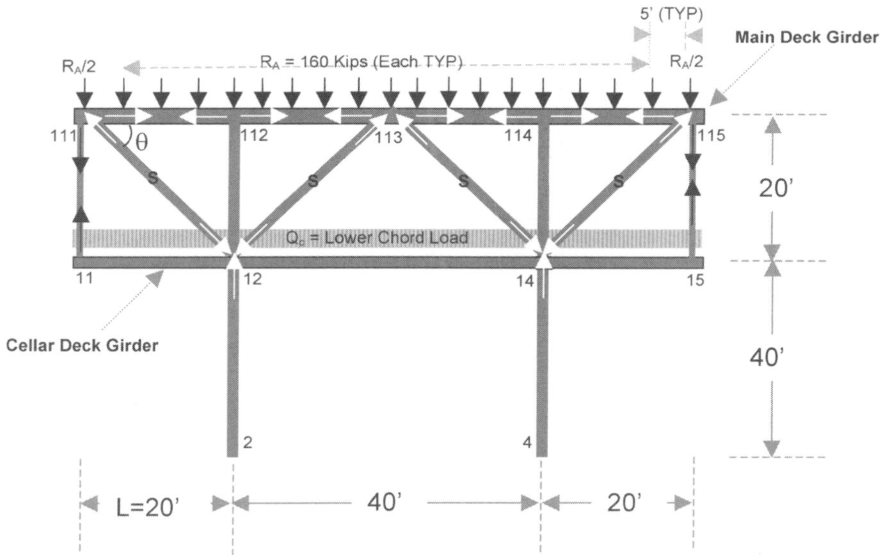

Figure 6.24 Main deck truss model

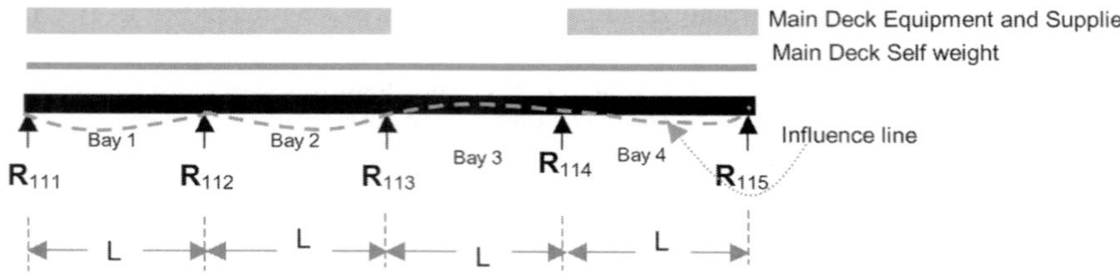

Calculate the moments using the Three-Moment Equation (a.k.a. Clapeyron's Equation, [Norris, 1960, Roark]) for equal length and equal stiffness four span continuous beam, where $M_{111} = M_{115} = 0$;

$$4M_{112} + M_{113} = -K_{1L} - K_{2R} \quad \text{(A)}$$

$$M_{112} + 4M_{113} + M_{114} = -K_{2L} - K_{3R} \quad \text{(B)}$$

$$M_{113} + 4M_{114} = -K_{3L} - K_{4R} \quad \text{(C)}$$

where K_{iL} and K_{iR} are the left and the right moment load terms for a loaded span i. For the symmetric loading case for all spans $K_{iL} = K_{iR} = K$.

For the 5 ft equal spaced loading case in a loaded bay (see fig. 6.24):

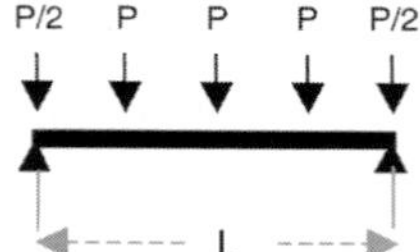

$$K = \frac{(n^2 - 1)}{4n} PL \qquad \text{where } n = \text{Number of equal spacings of load } P = 4$$

$$= 0.9375\ PL$$

(i) Calculate bending moments for the main deck self weight case,

$$P_w = R_{wa} = (0.25/4.0)R_A = 10 \text{ kip and } L = 20 \text{ ft} = 240 \text{ in.}$$

$$K_w = 0.9375 \times 10 \times 240 = 2250 \text{ k in.}$$

Using Equations A, B, and C for end moments at the supports 112, 113, 114:

$$4M_{112w} + M_{113w} = -2K_w$$

$$M_{112w} + 4M_{113w} + M_{114w} = -2K_w$$

$$M_{112w} = 4M_{114w}$$

Solving for Moments,

$M_{112w} = M_{114w} = -0.4286\ K_w = -964.35$ kip in. (Maximum bending moment location)

$M_{113w} = -0.2857\ K_w = -642.82$ kip in.

Support reactions:

$R_{111w} = R_{115w} = 1.5983\ P_w = 15.98$ kip

$R_{112w} = R_{114w} = 4.5358\ P_w = 45.36$ kip (Maximum Support reaction location)

$R_{113w} = 3.7372 P_w = 37.37$ kip

And shears:

$Q_{111Rw} = 1.0982\ P_w = 10.98$ kip (where Q_{111Rw} is the self weight caused shear force to the Right of Support 111)

$Q_{112Lw} = -1.9018\ P_w = -19.02$ kip (where Q_{112Lw} is the self weight caused shear force to the Left of Support 112) (Maximum shear force location)

$Q_{112Rw} = 1.6340\ P_w = 16.34$ kip

$Q_{113Lw} = -1.3660\ P_w = -13.66$ kip

(ii) Calculate bending moments and axial forces in the main deck for the Equipment and Supplies Loads case,

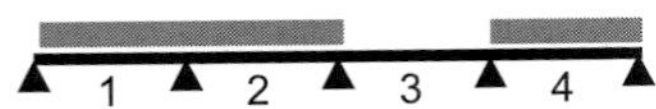

Load case q, Main Deck Equipment and Supplies Loads

$P_q = R_{qa} = (3.75/4.0)\ R_A = 150$ kip and $L = 20$ ft $= 240$ in.

$K_q = 0.9375 \times 150 \times 240 = 33{,}760$ k in.

Equations for end moments at the supports 112, 113, 114:

$$4M_{112q} + M_{113q} = -2K_q$$

$$M_{112q} + 4M_{113q} + M_{114q} = -K_q$$

$$M_{113q} + 4M_{114q} = -K_q$$

Solving for Moments,

$$M_{112q} = -0.4821K_q = -16{,}275.69 \text{ kip in. (Maximum bending moment location)}$$

$$M_{113q} = -0.0714K_q = -2{,}410.46 \text{ kip in.}$$

$$M_{114q} = -0.2321K_q = -7{,}835.70 \text{ kip in.}$$

Support reactions:

$$R_{111q} = 1.5480P_q = 232.20 \text{ kip}$$

$$R_{112q} = 4.8370P_q = 725.60 \text{ kip} \quad \text{(Maximum support reaction location)}$$

$$R_{113q} = 1.4644P_q = 219.66 \text{ kip}$$

$$R_{114q} = 2.3632P_q = 354.48 \text{ kip}$$

$$R_{115q} = 1.7824P_q = 267.36 \text{ kip}$$

And shears:

$$Q_{111Rq} = 1.0480P_q = 157.20 \text{ kip}$$

$$Q_{112Lq} = -1.9520P_q = -292.80 \text{ kip} \quad \text{(Maximum shear force location)}$$

$$Q_{112Rq} = 1.8850P_q = 282.75 \text{ kip}$$

$$Q_{113Lq} = -1.1150P_q = -167.25 \text{ kip}$$

$$Q_{113Rq} = -0.1506P_q = -22.59 \text{ kip}$$

$$Q_{114Lq} = -0.1506P_q = -22.59 \text{ kip}$$

$$Q_{114Rq} = 1.7176P_q = 257.64 \text{ kip}$$

$$Q_{115Lq} = -1.2824P_q = -192.36 \text{ kip}$$

The maximum support reaction R_{113} for main girder on truss line A will occur when Bays 2, and 3 are loaded with equipment and supplies –

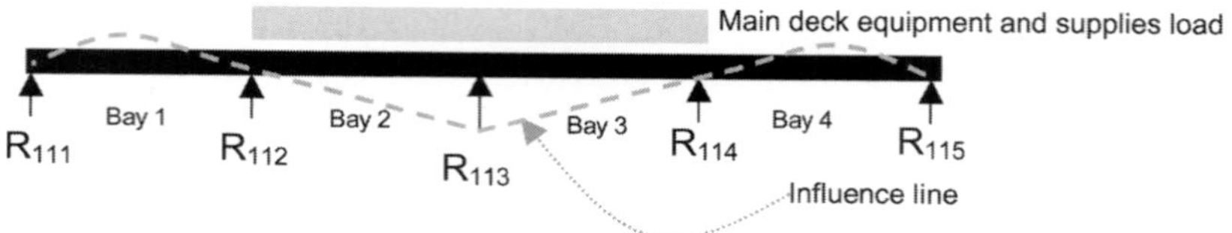

Using symmetry, the equations for end moments at the supports 112, 113, 114:

$$4M_{112q} + M_{113q} = -K_q$$

$$M_{112q} + 4M_{113q} + M_{114q} = -2K_q$$

Solving for Moments,

$$M_{112q} = M_{114q} = -0.1429\ K_q = -4{,}823 \text{ kip in.}$$

$$M_{113q} = -\ 0.4286\ K_q = -14{,}465 \text{ kip in.}$$

Support reactions:

$$R_{111q} = R_{115q} = -0.1340\ P_q = -20.10 \text{ kip}$$

$$R_{112q} = R_{114q} = 1.8622\ P_q = 279.33 \text{ kip}$$

$$R_{113q} = 4.5356\ P_q = 680.34 \text{ kip}$$

Calculate

Maximum Combined Bending Moment in the Main Deck Beam

$$\mathbf{M}_{112} = \mathbf{M}_{112w} + \mathbf{M}_{112q} = -964.35 - 16{,}275.69 = 17{,}240.04 \text{ kip in.}$$

Combined Beam Support Reactions

$$\mathbf{R}_{111} = \mathbf{R}_{111w} + \mathbf{R}_{111q} = 15.98 + 232.20 = 248.19 \text{ kip}$$

$$\mathbf{R}_{112} = \mathbf{R}_{112w} + \mathbf{R}_{112q} = 45.36 + 725.60 = 770.96 \text{ kip}$$

$$\mathbf{R}_{113} = \mathbf{R}_{113w} + \mathbf{R}_{113q} = 37.37 + 680.34 = 717.71 \text{ kip}$$

Maximum Axial Force in the Main Deck Beam (acting as the top chord of the deck truss):

Tension force due to Cellar Deck load in diagonal braces $S_{11,111}$ running from Joint 12 to 111:

$$\mathbf{S}_{11,111} \approx Q_c \times 10 \text{ ft} = (0.350 + 0.035) \times 40 \text{ ft} \times 10 \text{ ft} = 154 \text{ kip}$$

Compression force $S_{12,111}$ in diagonal brace running from Joint 12 to 111:

$$\mathbf{S}_{12,111} = \frac{\mathbf{S}_{111,11} + \mathbf{R}_{111}}{\sin 45^\circ} = -569 \text{ kip}$$

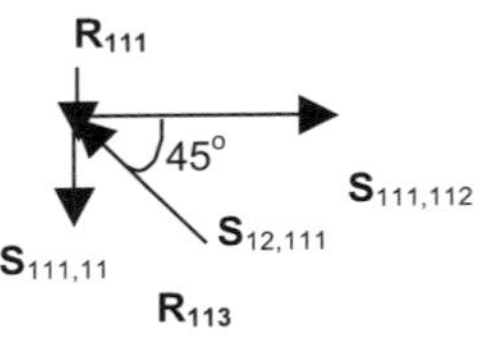

Tension force $S_{111,112}$ in the deck beam

$$\mathbf{S}_{111,112} = \mathbf{S}_{12,111} \times \sin 45^\circ = 402 \text{ kip}$$

Compression force $S_{12,113}$ in diagonal brace running from Joint 12 to 113:

$$\mathbf{S}_{12,113} = \mathbf{S}_{14,113} = \frac{\mathbf{R}_{113}}{2 \sin 45^\circ} = -508 \text{ kip}$$

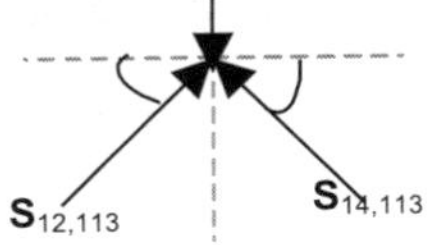

Maximum Combined Support Reaction in the Main Deck Beam

$$\mathbf{R}_{112} = \mathbf{R}_{112w} + \mathbf{R}_{112q} = 45.36 + 725.60 = 770.96 \text{ kip}$$

Maximum Combined Shear in the Main Deck Beam

$$\mathbf{Q}_{112L} = \mathbf{Q}_{112Lw} + \mathbf{Q}_{112Lq} = -19.02 - 292.80 = -311.82 \text{ kip}$$

Select and check the adequacy of the Main Deck Size

Using AISC ASD (1989) Manual Part 5 requirements for compact sections and A36 Steel:

Required section modulus $= S_x > \mathbf{M}_{112}/F_B = 666$ in.^{3} Also keeping in mind that the beam is under the combined action of moment and axial tension and, selecting a conservative beam size of WF 36 × 210;

$$S_X = 719 \text{ in.}^3; \quad A = 61.8 \text{ in.}^2$$

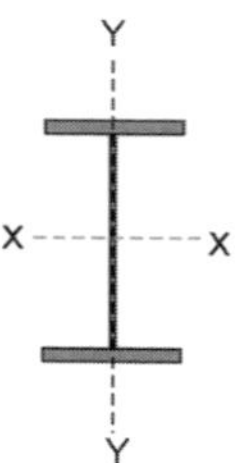

Calculate basic AISC bending and axial stresses:

$$f_b = \mathbf{M}_{112}/S_X = 23.98 \text{ ksi}; \quad f_a = \mathbf{S}_{111,112}/A = 6.50 \text{ ksi}$$

The beam is subject to combined bending and axial tension, therefore AISC Part 5 Section 1.6.1 Interaction Equation 1.6-1b applies –

$$f_a/(0.6F_y) + f_b/F_b = 0.30 + 0.88 = 1.18 > 1.00 \quad \text{(Not acceptable)}$$

At this point, we may consider three options,

(1) Use of a larger beam cross section. Try WF 36 × 245; $S_X = 895$ in.^{3}; $A = 72.1$ in.^{2}

$$f_b = \mathbf{M}_{112}/S_X = 19.26 \text{ ksi}; \quad f_a = \mathbf{S}_{111,112}/A = 5.58 \text{ ksi}$$

$$f_a/(0.6F_y) + f_b/F_b = 0.26 + 0.70 = 0.96 < 1.00 \quad \text{(Acceptable)}$$

However, WF 36 × 245 which is at the upper end of the beam sizes could be a long delivery item.

(2) Use of higher strength steel. Try ASTM A572 Grade 42 steel with $F_y = 42$ ksi, which is API RP2a Section I, table I.2 [RP2A-WSD, 2000] Group II, Class C material recommended for deck beams and legs for service temperatures above freezing. Using the initially assumed WF 36 × 210 –

$$f_a/(0.6F_y) + f_b/F_b = 0.26 + 0.75 = 1.01 \approx 1.00 \quad \text{(Marginally acceptable)}$$

(3) Use of a built up section. This would require additional iterations and is left to the reader for exercise purposes (Remember that the manufacturing of a built up section would require an experienced fabricator utilising special production techniqes, controlling welding/temperature generated distortions during the manufacturing operations and exercising strict quality control, thus avoiding out of tolerance dimensional variations).

Check shear stress f_v assuming use of WF 36 × 245 made of ASTM A572 Grade steel –

> $f_v \approx Q_{112L}/(td)$ where t and d are the thickness and the depth of the deck beam (0.830 and 36.69 in. respectively for the WF 36 × 210 Section)
>
> $f_v \approx 311.82/(0.830 \times 36.69) = 10.24$ ksi $<$ Allowable Shear Stress $= 0.40F_Y = 16.80$ ksi

The above calculations are satisfactory for the initial Main Deck Beam size selection. More detailed calculations should be made and code compliance should be further assessed during the three-dimensional structural computer simulation and stress analyses.

(c) Selecting main deck Truss brace dimensions

Use API 56 Grade X42 Steel with 42 ksi yield strength. (Note: API RP2A-WSD Section I, table I.3 recommends this material as Group II, Class C steel, suitable for jacket and deck braces and legs under service temperatures above freezing.)

Brace running from Joint 11 to Joint 111

Maximum tension force $\mathbf{S}_{11,111} \approx 154$ kip in brace running from Joint 11 to Joint 111

Area Required $\approx \mathbf{S}_{11,111}/(0.6\ F_Y) = 7.13$ in.^{2}

Select standard 10 in. (10.75 in. OD) 40 ST 40S (0.365 in. Wall Thickness) pipe

$A = 11.95$ in.^{2} $<$ 7.13 in.^{2} (Note that the hand analysis performed here did not consider any bending moment at the brace ends. Such moments will be present due to lateral wave and wind loads causing side sway distortions. Therefore, a conservative size selection is in order. Additionally, the tensile force present in this brace requires use of special welded connection details at both ends so that satisfactory load transfer between the brace and the deck beam webs is assured.)

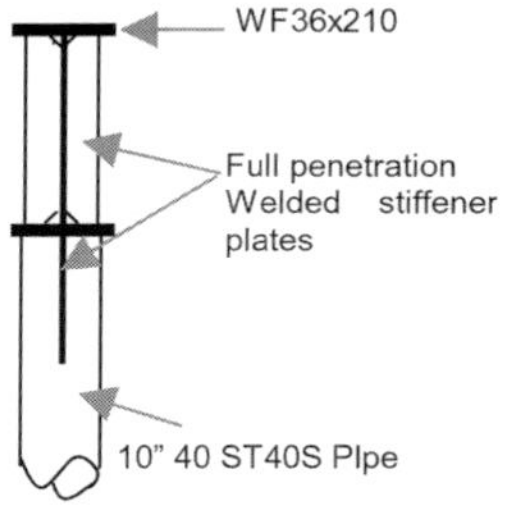

Brace running from Joint 12 to Joint 111

Maximum tension force $\mathbf{S}_{12,111} \approx -569$ kip in brace running from Joint 12 to Joint 111

Center to center length of the brace $= L' = L/\text{Cos}\,45° = 340$ inches. (Conservatively neglecting the length reduction due to deck beam and column width)

Since the beam-ends are restrained against side sway, a K buckling ratio of 0.8 can be conservatively assumed. Try an 18 in. (18 in. OD) 30 XS (0.50 in. Wall Thickness) pipe –

$$A = 27.49\ \text{in.}^2; \quad r_{xy} = 6.19\ \text{in.}; \quad f_a = -20.69\ \text{ksi}$$

$KL'/r_{xy} = 43.94$; From AISC Manual (1989) Part 5 Section 1.5.1.3, allowable axial compression stress F_a for $F_Y = 42$ ksi yield steel

$$F_a = \frac{[1 - (KL/r)^2/(2C_c^2)]F_Y}{5/3 + 3(KL/r)/(8C_c) - (KL/r)^3/(8C_c^3)}; \quad \text{if } C_c \le KL/r$$

(AISC Eq. 1.5-1)

where, $C_c = \sqrt{2\pi^2 E/F_Y} \le KL/r$;

$$\text{If,} \quad C_c > KL/r, \quad F_a = 12\pi^2 E/23\,(KL/r)^2$$

(AISC Eq. 1.5-2)

However,

If, $L/r > 120$,
$F_a =$ (Lowest F_a value from equations (1.5.1) or (1.5.2))/(1.6-L/200r)

(AISC Eq. 1.5-3)

In our example $E \approx 29{,}000$ ksi and $F_Y = 42$ ksi

$Cc = 126.94 < 43.94$ and $L/r = 54.9 < 120$, therefore eq. 1.5-1 governs and,

$F_a = 22.04$ ksi

$f_a/F_a = 0.94 < 1.00$ (Acceptable)

Braces running from Joint 12 to Joint 113 and Joint 14 to Joint 113

The compressive force of 508 kip acting on these braces is less than the 569 kip compression force acting on brace running from Joint 12 to111. For material uniformity purposes selecting these member sizes same as that for the member running from Joint 12 to 111 is suggested.

(*Note to Example 2.3*: The 16.51 in. flange width of WF 36 × 245 would present difficulties in framing diagonal brace diameters larger than 16 in., requiring side wing plates welded to the flange. If this additional work is not desired, an equal strength built up steel beam with wider bottom flange could be designed. The 18 in. OD brace size selected in our example would require such a solution sketched at the bottom. If wing plates are not desired, an equivalent area 16 in. OD pipe, such as the 16 in.Standard 60 XS 0.656 in. Wall Thickness pipe may be considered. An alternative solution would be to use of a built up deck beam with a 19 in. wide bottom flange.)

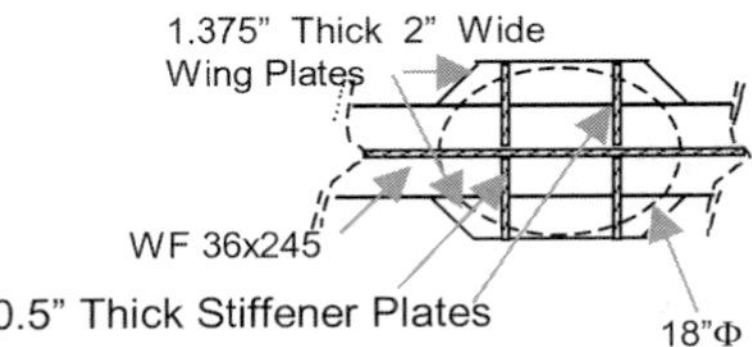

6.2.3.3 Jacket Bracing Configurations

Vertical, horizontal, and diagonal (mostly tubular) members (braces) connect jacket legs forming a stiff truss system. This system transfers the loads acting on the platform to its foundation (piles). There is a wide variation of platform bracing patterns, each with its advantages and some shortcomings. Several of these patterns are shown in fig. 6.25.

Type 1, K-brace Pattern results in fewer members intersecting at joints, reducing welding and assembly costs. It however lacks symmetry and redundancy. All K-braces in one level of a bay being either in full axial compression or tension do not provide tensile backup (robustness) and ductility. In case a brace would fail under a compression overload, the structure may unzip. K-brace pattern is used in locations where robustness is not important and seismic loads are not present (such as the shallow water Gulf of Mexico).

Type 2 and 5, V brace Patterns also result in fewer brace connections at a joint and also suffer from lack of redundancy and symmetry. Additionally, these patterns lack continuity of load flow from one bracing level to the other, resulting in larger horizontal brace dimensions. These patterns are seldom used and are not recommended.

Type 3, N-bracing Pattern has fever braces connecting to joints. It lacks symmetry and redundancy. All diagonal braces would be under compression or tension load depending on the horizontal load direction. Due to lack of tensile brace backup, buckling under compressive loading of one highly loaded diagonal brace can rapidly propagate to other braces causing platform collapse. Type 3 bracing pattern is seldom used and is not recommended.

Type 4, V plus X braced Pattern is in common use in most offshore locations. Braces run along the shortest diagonals of their bays with reduced buckling lengths. Adequate symmetry, redundancy and ductility are available. The only disadvantage of this pattern is higher number of brace connections at joints and the V braces at the transverse directions framing into horizontal braces. V braces in vertical plane carry high loads and would have larger diameters than the horizontal braces. Such a joint intersection would either require enlarged joint cans or larger than necessary horizontal brace dimensions. Replacing the V braces in the transverse direction by X braces (similar to the transverse direction of Type 6) results in higher ductility and better seismic resistance.

Type 6, Fully X-braced Pattern provides high horizontal stiffness, ductility, and redundancy. The joints are crowded and high volume of welding is present. This bracing pattern is popular in deepwater jackets where stiffness is needed to reduce sway periods and in seismically active regions where ductile behaviour is important.

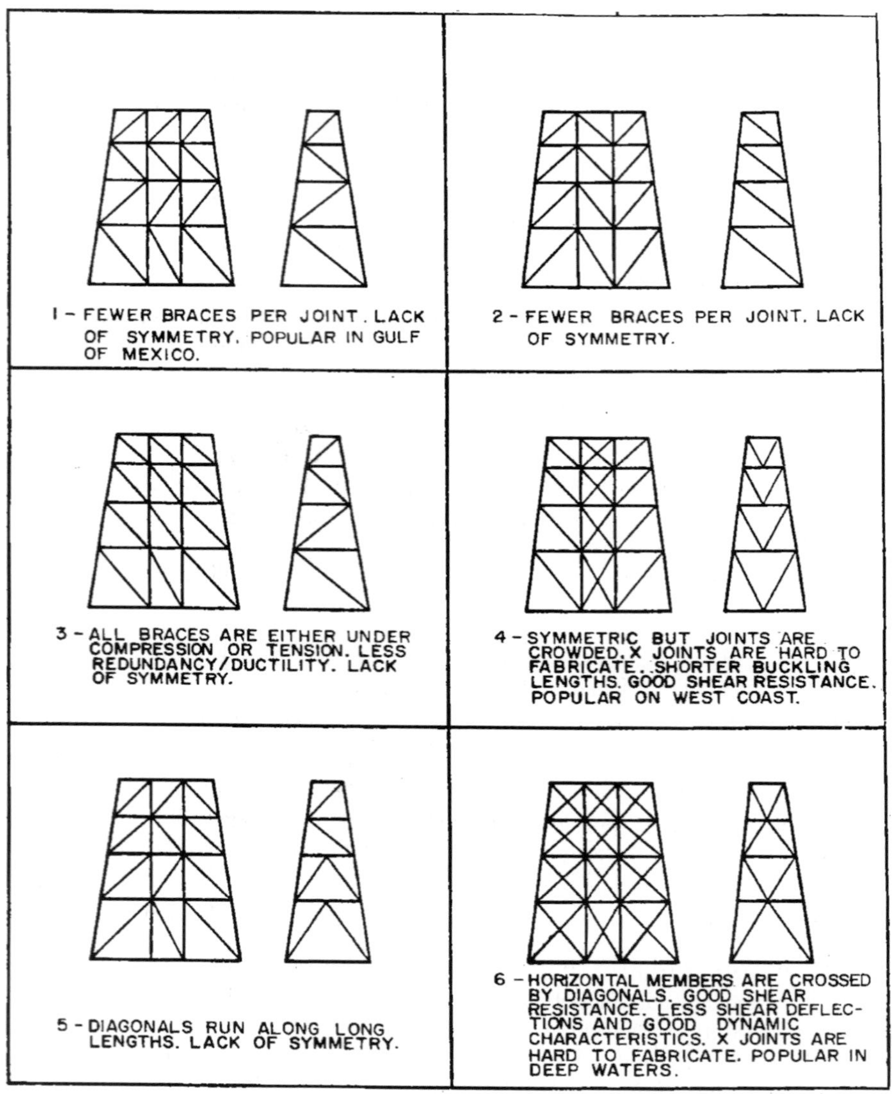

Figure 6.25 Jacket/tower bracing patterns

6.2.3.4 Jacket Leg Size Selection

For the commonly encountered case of main piles located inside the jacket legs, leg inside diameter (ID) is sized to accommodate pile driving and grouting operations. If all skirt piles are used (a common occurrence for deepwater jackets), jacket leg diameter may be controlled by buoyancy, stiffness and brace framing requirements. The selection of the leg ID should account for the pile guide/wear plates (welded inside the jacket legs to provide wear surfaces for the piles) and the dimensions and out of roundness of the pile. Large diameter pipes are commonly manufactured within 1% out of roundness (Out of roundness $= (OD_{max} - OD_{min})/OD_{average}$). Pile guide/wear plates located inside the jacket legs are usually 0.5 in. thick. For a commonly encountered Gulf of Mexico pile sizes of 42–72 in. OD, out of roundness will be about 0.5–75 in. Remembering that the out of roundness can be in different directions at different points of the pile lengths and allowing an extra inch for free play, the jacket leg ID should be a minimum of about 3–4 in. larger than the pile Outer Diameter (OD).

Jacket leg wall thickness is sized to resist the axial force and bending stresses and deformations exerted by intersecting braces. A thickness of 0.5 in. to 2.5 in. (leg thickness) is of common occurrence. Less than 0.5 in. wall thickness may result in corrosion problems. A 2.5 in. or thicker wall pipe is difficult to manufacture and may suffer through thickness cracks at brace weld points unless special steel is used. Increasing leg and brace thickness at the wave splash zone by about 1/8 to 1/4 inch (when cathodic corrosion protection is present) is commonly used as additional corrosion allowance. High stresses and weld requirements may dictate the use of specially manufactured (ductile, high through thickness strength, no laminations) steels at complex leg joints. In such cases, thick leg (cans) and brace ends (stubs) may be inserted at joints. Whether the pile will be grouted inside the leg or not also impacts the leg wall thickness.

6.2.3.5 Jacket Brace Size Selection

Guidelines for selecting jacket brace sizes can be outlined as follows:

- Choose a brace member diameter that has a slenderness ratio (kL/r) in the range of 70–90. Limiting the slenderness ratio to the 70–90 range is an industry-accepted practice, which is the end product of many factors. As the slenderness of a brace increases, its allowable axial stress (F_a) decreases. At a KL/r of 80, the allowable axial stress F_a for A-36 steel is 71% of that allowable for $KL/r = 0$. In 50-ksi-yield steel, at KL/r of 80, F_a is 63%. At high KL/r values, high yield pipe is less efficient than at lesser values.

  As brace diameter increases so do wave loads acting on it, which in turn results in a heavier structure. Lower slenderness ratios also encourage higher D/t ratios for tubulars that may compound local buckling problems. If the pipe area is satisfactory, other strategies, such as providing additional support points to cut down the buckling lengths (introducing small lateral braces or use of X-bracing) should be considered, before member diameters are increased.

- Check the capacity of the joint the member will be framing into [API RP2A-WSD, 2000; Section 4.3]. In some cases, the tubular joint capacity may control design and result in use of larger brace diameter or joint cans. One good approach to improve the joint capacity is to use brace to chord OD ratios (β) higher than 0.30.
- For sizes up to and including 18 in., use the wall thickness for seamless standard pipe as a starter. For sizes up to and including 29 in., which would most likely be rolled from

plate and seam and butt-welded, try $\frac{1}{2}$ in. For 30 through 36 inches start with 5/6 in. For larger than 36 in. OD, start with a wall thickness that satisfies $D/t > 31$ requirement. If the brace is at the splash zone, after selecting the brace size that satisfies all structural strength requirements, consider adding 1/8 in. to 1/4 in. to the wall thickness as corrosion allowance.

- Try to keep the pipe diameter to thickness (D/t) ratio of the members between 19 and 90. A D/t of 19 or less is a difficult pipe to buy or make. A pipe with a D/t of about 31 floats. For most steels, a D/t of 90 can present local buckling problems. In general, keeping D/t less than 60 is considered a good practice.
- For a water depth of "h_w" (in feet) begin to check for hydrostatic problems when D/t exceeds $250/(h_w)^{1/3}$.

Figure 6.26 represents a single braced (K or X) bay of a jacket truss. The horizontal stiffness of such a bay fixed at its base (see insert to fig. 6.27b below) is represented by the force required to cause an average unit deflection at its top two joints. This stiffness is determined by a number of geometric parameters. The most prominent parameters that control this stiffness are shown to be [Kumar, et al 1985] the jacket leg area (pile plus leg area, if grouted) relative cross-sectional areas of the diagonal braces compared to that of the jacket leg (A_3/A_1), the aspect ratio α $[(a+b)/2h]$, its batter S and height h. When there is no diagonal brace ($A_3 = 0$), the system behaves like a portal frame, with minimum horizontal restraint provided only by the bending stiffness of the jacket legs and the

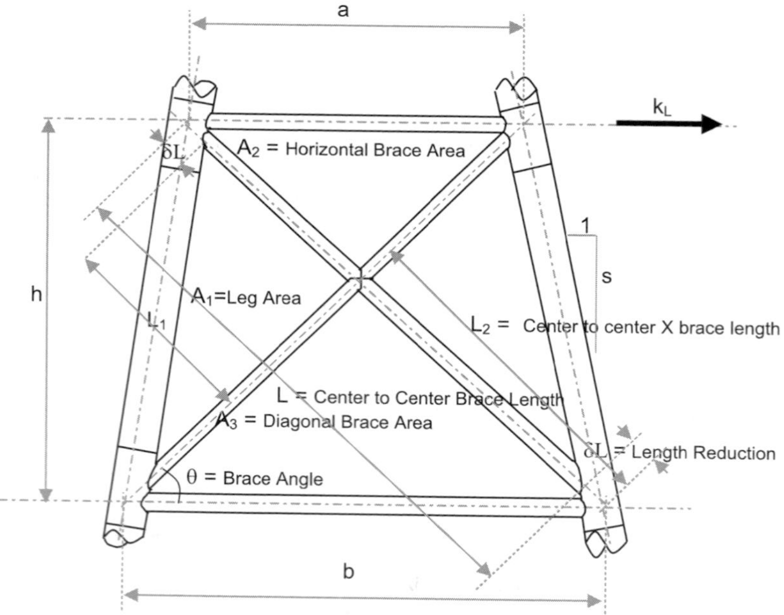

Figure 6.26 Geometry and stiffness parameters of a single X braced jacket bay

horizontal braces and the moment connections at their intersection points (joints). As the diagonal braces are introduced, with increasing brace cross-section area values ($A_3 > 0$) the system develops an increasingly stiff truss behaviour. At low A_3/A_1 values, the system stiffness would be low and high horizontal deflections would be experienced. As this ratio increases, the diagonal braces would become highly effective in transferring the shear forces from one jacket leg to the other. It can be shown that there exists a limiting $(A_3/A_1)_o$ ratio beyond which introduction of additional brace steel will not significantly contribute to the horizontal stiffness. Figure 6.27a shows the variation of the limiting $(A_3/A_1)_o$ value as a function of the aspect ratio for a single jacket bay module with $S = 12$. The behaviour stays similar when the S value is varied [Kumar, et al 1985]. Within the most commonly experienced jacket aspect ratio α ranges of 0.7 to 1.4 selecting an A_3/A_1 of 0.1 or more would generally assure rigid truss behaviour. Values above 0.2 would be acceptable from a structural strength point but should be considered inefficient steel use for commonly encountered jacket truss geometries. These ratios are for X braced trusses. For single diagonal (K) braced trusses higher A_3/A_1 ratios (0.2–0.4) should be considered.

Figure 6.27b from Karsan (1986) shows the results of a parametric study on the relation between the horizontal stiffness parameter k_L and the α aspect ratio. The study demonstrates that, for an X braced truss and for a constant A_3/A_1 ratio of 0.20 and a batter of 12, k_L reaches its maximum value at an aspect ratio of $\alpha = \sqrt{2}$ which corresponds to an approximate brace angle θ of 36°. $1 < \alpha < 2$ Range ($45° < \theta < 27°$) falls in a range of stationary horizontal stiffness values. Outside this range, rapid decreases in the horizontal stiffness would be observed. A similar behaviour is observed for other commonly used leg batter S values. Since a jacket is made of a stacked assembly of the truss modules shown in

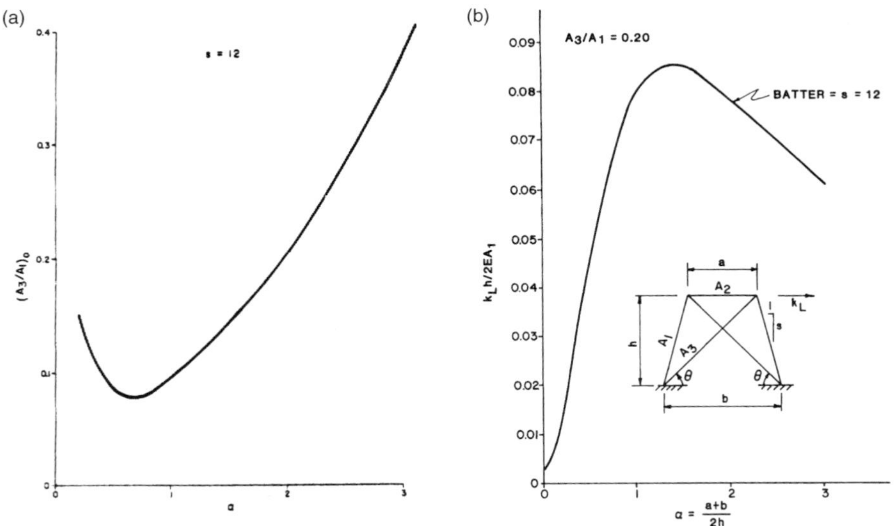

Figure 6.27 Variation of brace area ratio and horizontal stiffness with the aspect ratio. (a) Limiting brace area ratio $(A_3/A_1)_o$ as a function of the aspect ratio α [Kumar, et al (1985)]. (b) Variation of X braced single jacket bay horizontal stiffness k_L with the aspect ratio α [Karsan, 1986]

fig. 6.26, similar conclusions apply to the full jacket structure. Jacket horizontal stiffness becomes increasingly important for dynamic response reasons, as the water depth increases.

Example 6.2.3

Select the initial tubular brace dimensions for the X braced configuration shown in fig. 6.26. What should the initial brace dimensions be if a single diagonal K brace was used in place of the X braces?

Given:

$a = h = 45$ ft; $S = 12$; Jacket Leg 54 in. OD × 1.00 in. wall thickness; with 48 in. OD × 1.25 in. wall thickness pile sections ungrouted inside the jacket legs. Assume the brace bottom to be 80 ft below the mean water level.

Results: $b = a + 2h/S = 52.5$ ft

$$\alpha = (a + b)/2h = 1.08 \text{ (OK for aspect ratio)}$$

$$L = \sqrt{(a + a/s)^2 + h^2} = 66.34 \text{ ft}$$

$L_2 = Lb/(a + b) = 35.72$ ft and $\delta L = 2.92$ ft (left to reader to calculate using θ, S and jacket leg diameter)

For the X braced configuration:

For most critical horizontal loading conditions, one of the X braces would be in compression while the other would be in tension. This would result with an S shaped asymmetric fundamental buckling mode. Therefore, the buckling length could be taken equal to L_2–δL (where δL is a length reduction factor in recognition of the fact that the brace is fixed to the chord wall) and a buckling coefficient of $k = 0.8$ could be assumed.

$r \approx 0.35\ D_o$ (radius of gyration for thin walled tubes);

$k(L_2 - \delta L)/r = 0.8\ (L_2 - \delta L)/0.35\ D_o \leq 70$; Solving for D_o:

$D_o \geq 13$ in. The standard pipe diameter closest to this size is 14 in. OD. The β ratio for this diameter would be 14 in./54 in. = 0.25, which is on the low side. To satisfy the $\beta \geq 0.3$ requirement;

$D_o \geq 0.3\ D_{\text{chord}} = 0.3 \times 54 \approx 16$ in.

Since D_o is less than 18 in., standard seamless pipe will be used. Try a standard wall thickness of 0.5 in. For 16 in. OD, D/t ratio is 32 and satisfies the flotation and local buckling requirements.

For 80 ft water depth, if D/t is not extremely low, hydrostatic collapse will not be a real issue. For demonstration purposes, check for hydrostatic buckling potential:

$250/(h_w)^{1/3} = 58 > D/t = 32$ which shows no hydrostatic collapse potential. If potential is observed, more detailed calculations [as per API RP2A (LRFD 1993 and WSD 2000) requirements], and potential use of thicker wall pipe or ring stiffeners may be considered.

Now check the brace to jacket leg area balance:

Jacket leg area $= A_j = 166.5$ in.^{2}; Pile area $A_p = 183.59$ in.^{2} Since the pile will not be fully grouted inside the jacket leg, it will not add to the axial load capacity of the jacket leg:

$$A_1 = A_j = 166.5 \text{ in.}^2$$

$$\text{Brace area} = A_3 = 24.25 \text{ in.}^2$$

$A_3/A_1 = 0.15 > 0.1$ satisfying the efficient shear transfer requirement as per fig. 6.27.

If the piles were grouted inside the jacket leg A_1 will be equal to $A_j + A_p = 350$ in.^{2} and,

$A_3/A_1 = 0.07 < 0.1$ which suggests that use of a larger brace area (or a smaller pile plus jacket leg area) could result in a more efficient steel use.

For K braced configuration:

For this case, the buckling length will be equal to the brace length, extending from one jacket length to the other:

$L_2 - 2\delta L = 60.50$ ft. Using a k value of 0.8 and a limiting kL/r value to 70, $D_o \geq 24$ in. is calculated. The β ratio for this diameter would be 24 in./54 in. $= 0.44$, which is reasonable.

Since D_o is more than 18 in., rolled, seam welded pipe will be used. Try a wall thickness of 0.5 in. For 24 in. OD, D/t ratio is 48, which satisfies the flotation and local buckling requirements.

Check for hydrostatic buckling potential:

$$250/(h_w)^{1/3} = 58 > D/t = 48; \text{ which shows no hydrostatic collapse potential.}$$

Now check the brace to jacket leg area balance:

$$A_1 = A_j = 166.5 \text{ in.}^2$$

$$\text{Brace area} = A_3 = 74.51 \text{ in.}^2$$

$A_3/A_1 = 0.45 > 0.44$ which is slightly higher than 0.2 to 0.4 ratio required for an efficient K braced truss steel utilisation.

The above calculations are recommended for initial member size estimating purposes only. Full three dimensional simulation and analysis of the jacket and deck structures under actual design loading conditions will result in a better determination of final member sizes (see next Section 6.2.4).

6.2.4 Computer Simulation and Detailed Analysis

Once estimates for the platform geometry and member sizes are available, these are input to a structural analysis computer program to perform a three-dimensional structural analysis. There are a number software programs available for offshore platform design and analysis [SACS, Sesam, StruCad]. A typical platform analysis program would compute the structural deflections, member loads, stresses, utilisation ratios and support reactions, given initial member sizes and platform loads. Generally, repeated structural computer analyses are performed to revise over or under utilised member sizes and the platform loads and load combinations at each step (see the design spiral in Section 6.1.2). These analyses are repeated until acceptable member sizes are obtained. In some cases, platform geometry is also revised. The recent advances in the computer hardware and ever increasing personal computer (PC) capabilities now allow executing these software interactively using personal computers. The following paragraphs summarise the approach that is common to most fixed offshore platform design software.

6.2.4.1 Deck and Jacket Geometry Simulation

Three types of Cartesian co-ordinate axis systems are used for coding various aspects of the platform geometry and loads.

Global Axis System: A set of orthogonal axes, (**X**, **Y**, **Z**) which locate the direction of the gravity and the environmental effects (e.g. deck equipment weight, self-weight, buoyancy, wave and wind loads, earthquake and ice loads). Normally, the **X** and **Y** axes form the water plane (tangential to the Earth) and the **Z**-axis runs through the Earth's centre (fig. 6.28a).

Structural Axis System: A set of orthogonal axes (X, Y, Z) in which the joints and members of a structure are defined. Usually, one of the X or Y axes is chosen to lie in the Global **X–Y** Plane. A co-ordinate transformation matrix relates this axis system to the Global Axis System (fig. 6.28b).

Member Axis System: At each end of the member, a set of right-handed Cartesian axes are oriented such that plus "x" coincides with the Member Axis and points out from the lower to the highest numbered joint of the member. *y* and *z*-axes lie in a plane perpendicular to the Member Axis and are oriented to be parallel to the principle axes of the member. Usually, one of the *x* or *y* axes is taken to lie in the structural X–Y Plane. Co-ordinate transformation matrices relate each structural member's axis system to the Structural Axis System (fig. 6.28c).

Joints

Platform joints are generally coded using the structural axis system. The structural axis system should be selected at a point of symmetry of the structure. Symmetry can be used to advantage minimising coding and sub-structuring (a technique where a repeating joint

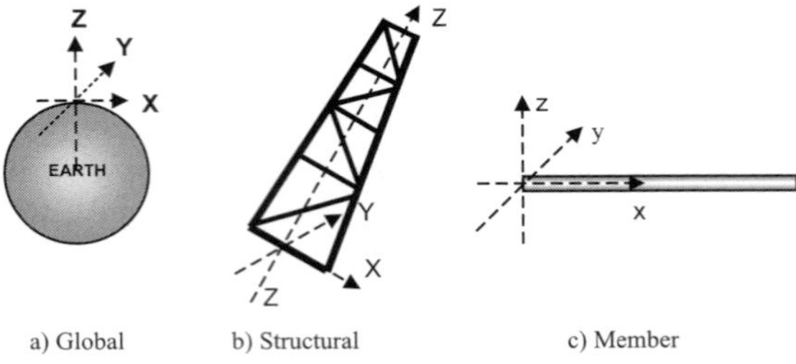

Figure 6.28 Axis systems

pattern or structural component is coded once and copied and attached to other parts of the structure, such as a skirt pile bracing pattern that is symmetric to four corners of a jacket).

Figure 6.29a shows the three-dimensional computer model for an eight-legged platform in 160 ft water depth. The model shows the deck and jacket structures coded as single units. Separate jacket and deck modelling may be useful if detailed analyses and member sizing will be done by separate groups or design teams or if the computer program has size limitations. With today's ever-increasing computer hardware capabilities, such a separation may not be necessary. It is recommended that the joints are numbered by Level (such as assigning 100 series to the joints at seabed elevation, 200 series at second level, etc.) and by row number (such as 10, 20, 30 series assigned to first, second, third joint rows along the Y axis direction and 1, 2, 3 assigned to first, second third joint rows along the X axis direction). In this way, the analyst can keep a simple picture of the structure in mind when reviewing the analysis results (one can quickly remember that Joint 231 is at second platform level along Z axis third row of joints along the Y-axis and first row along the X-axis). Most structural analysis software provide means of automatic joint coordinate generation if the joints can be specified as a sequence (such as equally spaced joints along a line, platform levels that are repetitions of a starting level along a constant batter, joints that are at the intersection of two lines etc. (fig. 6.29b). Foundation piles are coded as special substructures with their cross section properties and P–Y and T–Z curves attached to a joint at the structure base. If an elastic pile simulation is desired, pile stiffness characteristics can be input as a pile length and a set of six springs attached to a dummy member or a 6 by 6 pile stiffness matrix added to the pile connection joint stiffness (see McClelland (1986), Section 17 and Section 14.3 of this Handbook).

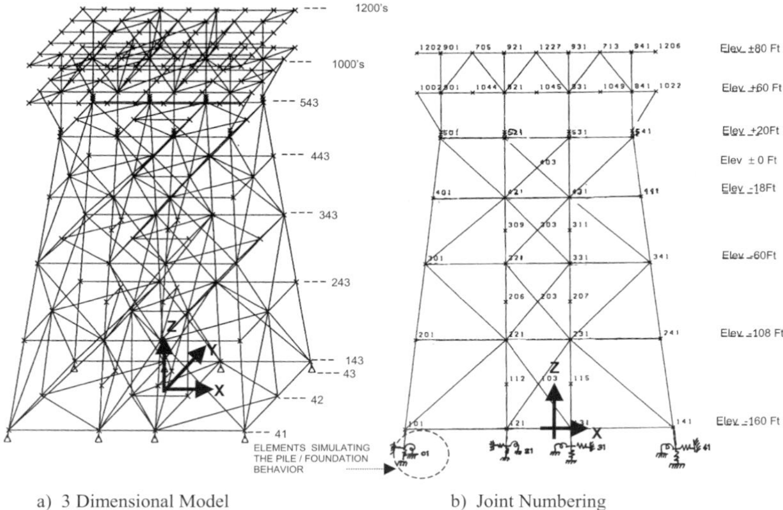

Figure 6.29 Computer model of a platform structure

Members

Members connect two structural joints forming a structural element. Members are coded by specifying the joints they connect. Cross-sectional properties along the member axes, end connectiveness; buckling lengths, buckling ratios and other member dependent characteristics such as the marine growth thickness (to be added to the member diameter to simulate marine growth) are also input. End length reduction (overlap length with the connecting braces that should be subtracted from the joint to joint length to find its true length) can also be input, if needed. Members with common properties and function are coded in design groups which eases coding, data interpretation and sizing as a group.

Figure 6.30 shows the members and member-group assignments for the structural model shown in fig. 6.29. There are a number of computer programs available in the market that have been developed for the specific purpose of analysing and designing fixed offshore platforms [SACS-User's Manual; Sesam Users's Manual; StruCad User's Manual]. These programs contain lists of weights, geometric and structural design parameters of structural steel shapes (WF, I, T, Channels, angles, etc) which are commonly available from international steel manufacturers and/or suppliers. These programs are also equipped with formulae and subroutines that enable the calculation of the weights and structural design

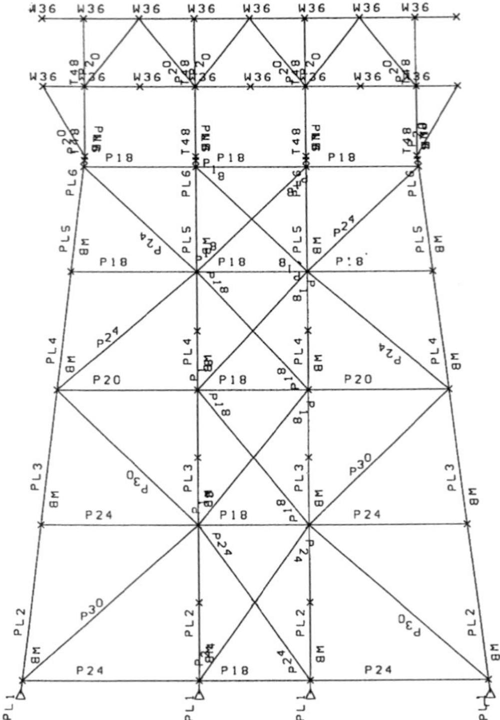

Figure 6.30 Members and member groups

parameters of structural cross-sections definable by a number of parameters (such as circular and rectangular tubulars, built up beams, orthotrophic beam plate assemblies, etc.). For further details, readers are referenced to the user and application manuals of these software programs.

6.2.4.2 Foundation Simulation

The soil stiffness depends on the applied pile loading levels. Therefore, the pile-soil system behaves non-linearly (fig. 6.31). Since the platform structures are generally analysed using linear matrix analysis methods, the pile-foundation system is either linearised around the expected pile load levels or an iterative procedure is used to assure load and deformation compatibility at the platform to pile connection points. Linear foundation modelling is generally useful for dynamic analysis purposes (earthquake, impact, blast, or harmonic loadings) or for large platform models with excessive loading conditions to be analysed.

The iterative (or pile-platform coupled analysis) approach involves direct coding of the pile and foundation characteristics ("P–Y" lateral load per unit pile length versus lateral foundation deformation and "T–Z" axial load per unit pile length versus axial foundation deformation) along the pile axis. The linear elastic structural jacket analysis and the non-linear pile-soil analysis are executed successively, feeding back the unbalanced axial force, shear and bending moments generated from establishing deflection compatibility at the

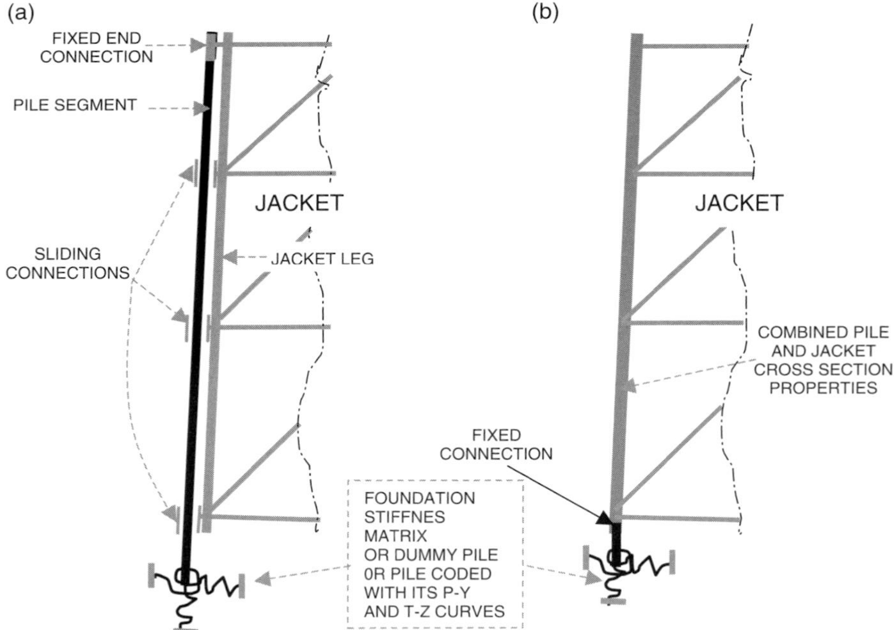

Figure 6.31 Foundation simulation. (a) Pile not grouted inside the jacket leg. (b) Pile grouted inside the jacket leg

pile-jacket interface at each iteration, until full force and deflection compatibility is achieved. Such modelling will differ for the grouted and un-grouted pile models (see figs. 6.31a and b). Certain structural strategies, such as condensing the jacket stiffness and loads at the pile to jacket connection joints are utilised to speed up the iteration process.

If a computer program with coupled interaction analysis capability is not available or if one desires to perform elastic analyses, a pile linearisation approach may be used. The principle behind the elastic pile simulation is to define a flexibility matrix such that it will yield the same displacements as the pile head displacements under a given set of platform loading. A special purpose "beam-column" analysis computer program is used [Bryant and Mattlock, 1975; Reese and Mattlock, 1956] to provide information needed to determine the soil-pile simulation model. The force-deformation relationship of the pile head in two dimensions can be described in a matrix form:

$$\begin{vmatrix} X_T \\ Y_T \\ \phi_T \end{vmatrix} = \begin{vmatrix} d_{11} & d_{12} & d_{13} \\ d_{21} & d_{22} & d_{23} \\ d_{31} & d_{32} & d_{33} \end{vmatrix} \times \begin{vmatrix} P_T \\ V_T \\ M_T \end{vmatrix} \tag{6.2.1}$$

where P_T, V_T, M_T are the axial and shear forces and the bending moment and X_T, Y_T, ϕ_T are the corresponding deflections along the pile x and y axes and rotation at the pile top respectively. The d_{ij} flexibility coefficients should be selected such that the deflections at the support locations of the structure will be equal those which will be obtained at the pile head from a pile "beam column" analysis using support reactions of the structure. One method in common use for calculating these coefficients [McClelland (1986), Section 17 on this subject by the author] is copied below for the reader's convenience:

- Since the axial load effects on the lateral deformation and the rotation will be secondary, the coefficients d_{12} and d_{21}, d_{13}, d_{31} would be very small and are assumed to be zero. The non-zero flexibility coefficients remaining are d_{11}, d_{22}, d_{23}, and d_{33}.
- Approximate pile reactions are calculated using a rigid body model for the structure (see fig. 6.14 and Example 6.2.1).
- An axial versus the pile settlement analysis is performed by using the axial pile reaction P_T and d_{11} is computed from:
  $d_{11} = d_p/P_T$ where d_p is the axial settlement under P_T.

- Three beam-column analyses of the soil pile system is performed.

  Analysis One: With the full axial load, shear and moment, to obtain Y_{T1} and ϕ_{T1} deflection and rotation. From equation (6.3.1);

$$Y_{T1} = d^*_{22}V_T + d^*_{23}M_T \tag{6.2.2}$$
$$\phi_{T1} = d^*_{32}V_T + d^*_{33}M_T \tag{6.2.3}$$

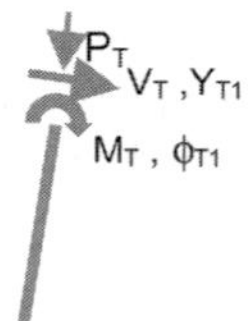

Asterisks are used to denote intermediate quantities of d_{ij} which are used for calculating their final adjusted values through equations (6.2.4)–(6.2.8).

Analysis Two: With full axial load and shear but moment equal to zero, obtain the Y_{T2} deflection. From equation (6.2.1);

$$d_{22}^{*} = Y_{T2}/V_T \qquad (6.2.4)$$

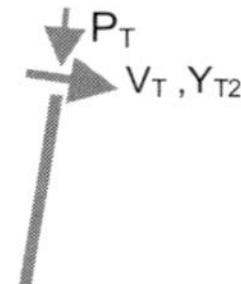

Analysis Three: With full axial load and moment but shear equal to zero obtain ϕ_{T3} rotation. From 6.2.1

$$d_{33}^{*} = \phi_{T3}/M_T \qquad (6.2.5)$$

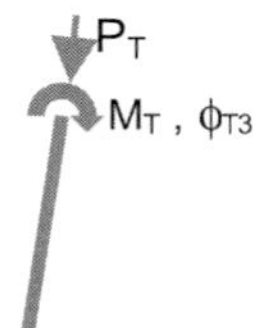

Substituting d_{22}^{*} and d_{33}^{*} into equations (6.2.2) and (6.2.3) and solving for d_{23}^{*} and d_{32}^{*};

$$d_{23}^{*} = (Y_{T1} - d_{22}^{*} V_T)/M_T \qquad (6.2.6)$$

$$d_{32}^{*} = (\phi_{T1} - d_{23}^{*} M_T)/V_T \qquad (6.2.7)$$

In an elastic system, off diagonal flexibility coefficients are equal. Taking the average value:

$$d_{23} = d_{32} = (d_{23}^{*} + d_{32}^{*})/2 \qquad (6.2.8)$$

Due to this averaging, values of d_{22} and d_{33} have to be recalculated if equation (6.2.1) has to be satisfied. Substituting d_{23} and d_{32} into equation (6.2.1) and solving for d_{22} and d_{33} can accomplish this:

$$d_{22} = (Y_{T1} - d_{23} M_T)/V_T \qquad (6.2.9)$$

$$d_{22} = (\phi_{T1} - d_{32} V_T)/M_T \qquad (6.2.10)$$

The d_{ij} flexibility coefficients calculated using the above procedure would result in an accurate representation of the in-plane soil-pile behaviour in the vicinity of a given V_T, M_T pair. If needed, the support stiffness matrix can be calculated by inverting the above flexibility matrix. The above approach is provided for demonstration purposes. For the three-dimensional case, similar calculations could be carried out to generate linearised d_{ijk} foundation flexibility coefficients. If the maximum shear value along the pile x axis is close to the same along the y axis (would generally be the case for diagonal or multiple loading conditions), the out-of-plane stiffness deflection and rotational flexibility coefficients may be assumed equal to the values calculated for the in-plane loading case. The fact that the above simulation is load dependent, it should be carried out for different loading combinations. The foundation-pile system would behave near linear and demonstrate higher stiffness for lower load levels and small amplitude dynamic loadings. Therefore development of different foundation flexibility matrices for extreme, operational, and dynamic loading conditions should be given consideration.

The aforementioned procedure is used for grouted piles. Formulation for the ungrouted piles would be identical except that the pile head constraints provided by the jacket would be reduced due to the existence of the annular gap between the pile and the jacket leg. In the normal working loading level, the location of contact points between pile and jacket leg are at the bottom two levels of horizontal bracing and/or where pile spacers have been installed. The jacket, through the horizontal couples at these two contact points, experiences the bending resistance of the pile. The axial supports of the jacket will be at the jacket top instead of at the mud-line. All these differences should be considered in the foundation simulation of ungrouted piles.

Conductor Simulation

Conductors are pipes (generally 20 in. to 30 in. OD) that are driven to ground for the support and protection of the well drilling and production operations. During the well drilling operations, the conductor guides and protects the drill string against external effects. During the well production phase, the conductor houses and protects the well casing and the production tubing(s). The total weight of the conductor and other pipe and systems supported by it (such as the production casing strings grouted to its inside, BOP, X-mas trees, etc.) is transferred to the seabed through shaft friction between the conductor and the soil, in a manner similar to a pile. Near the wave action zone, the conductor picks up significant wave loads and transfers these to the upper horizontal jacket bracing levels through conductor guides. Near the seabed level, the conductor provides restraint against the jacket horizontal motions, similar to a pile, if proper bracing strength is provided.

Figure 6.32 summarises the forces exerted by a conductor on the jacket caused by the wave and current forces and the platform horizontal motions acting on it. Before taking advantage of the beneficial restraint provided by conductor against the platform motions, the designer should assure that the platform motions are large enough to close the gap provided between the conductor and the conductor guide. This gap can be as much as 1 in. and taking advantage of restraint provided by a conductor may not be warranted for operating and dynamic loading conditions. If full advantage of this restraint is needed, lower level conductor guides may be provided with wiper type guides closing this gap. The designer should also remember that some conductors may be installed at later stages of the field development and not all the conductors may be present at all stages of the platform life.

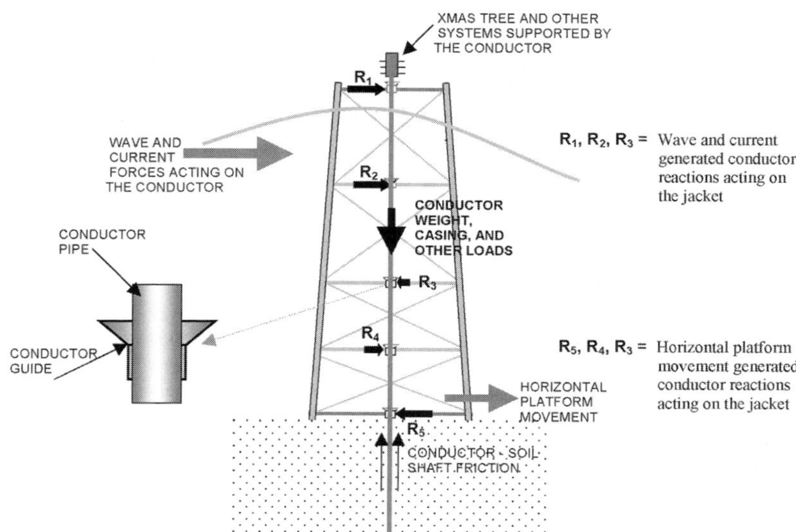

Figure 6.32 Wave and horizontal platform movement generated conductor forces acting on the jacket structure

Appurtenances

A number of non-structural elements (appurtenances) attached to the jacket structure exert self-weight and gather wave and wind loads while adding very little to none to the structural strength. The wave and current exerted pressures depend on the relative positions of the wave crest and an appurtenance and it would be very unpractical to manually input these loads. Therefore, these appurtenances are simulated as non-structural subsystems exerting buoyant weight and picking wave loads only. Some of these appurtenances are listed below:

Boat Landings: Generally, two boat landings each located in opposite faces of the platform are installed to provide supply boat and small watercraft access for all current/wind directions. The boat landings are located near the mean water surface with suitable depth and elevation to provide boat access at low and high tide levels. Their positions and sizes subject the boat landings to high current and near wave crest pressures. Wave and current loads transferred on a jacket structure by boat landings can be substantial. In some parts of the world (such as the North Sea) or in deepwater where the wave load effects are substantial and the main mode of personnel transportation is by Helicopters, not providing boat landings could be given consideration.

Barge Bumpers: These are generally made of steel pipe lengths placed at a suitable distance from and welded or clamped onto the jacket legs. Barge bumpers are generally fitted with truck tires or rubber fenders providing soft impact surfaces against collisions from larger supply vessels, workboats or material barges mooring at or near the platform. These appurtenances also transfer weight and substantial wave and current loads on the jacket.

Risers and Riser Clamps: Risers are pipes clamped onto the jacket legs or braces, transferring the oil and gas from (or to) the seabed pipeline to (or from) the deck piping. Riser and riser clamps transfer substantial wave and current loads on the jacket structure. Additionally, the riser clamps transfer the seabed pipeline and riser expansions caused by the oil temperature variations to the jacket depending on the relative stiffness and support conditions of the riser-jacket system.

Riser Protection Cages or Fenders These elements are located near the mean sea surface level with functions similar to the barge bumpers, protecting the risers against vessel impact.

Sump, pump and Other Surface Piercing Pipes: These are pipes that are used to dump treated water, treated bio-gredable refuse to sea or for seawater intake for fire fighting or cooling purposes.

Bridges connecting platforms: For the multiple platform case, there may be bridges connecting the living quarters, drilling, production, compression, pumping platforms to each other, transferring their substantial self weight, piping and wind loads to their deck support points. These could also transfer substantial horizontal loads or deflections between the platforms depending on their support conditions.

Anodes (Cathodic Protection): The submerged portions of the steel jackets are usually left uncoated and cathodic protection is provided to protect these areas against corrosion. Cathodic protection is a process where an electric field is created at the jacket surface so that current flows from an external source into the metal preventing the formation and loss of iron ions. Cathodic protection can be provided either by supplying a current flow into the jacket steel from an electric generator (impressed current) or by attaching sacrificial anodes to selected points on the jacket braces.

In the sacrificial anode system, a metal or alloy reacting more vigorously than the jacket steel is attached to structure providing the needed current flow, similar to a dry cell battery. These anodes are made of materials such as magnesium, aluminum or zinc, which are anodic with respect to the protected steel structure.

Sacrificial anodes, which are connected directly to the structure, are generally cast over tubular steel cores, which are welded to the structure. The sizes of these anodes are substantial (such as 4-in. square 3–6 ft long, sometimes weighing as much as 4% of the total jacket weight). These can pickup substantial wave loads and impose significant dead weight on the jacket structure. Additionally, there had been observations of the anodes fatiguing and shaking off their connection points to the jacket under hard pile driving conditions (see Salama, 1988). It is a common practice to provide wrap plates at their welding points to the jacket braces (see fig. 6.33), especially if the brace thickness does not provide adequate strength against fatigue and shaking off.

6.2.4.3 Checking the Platform Geometry Input

Geometry input to a platform analysis computer program should be carefully and thoroughly verified before a structural analysis is attempted. There had been a number of instances where errors in the input geometry resulted in wasted analysis, drafting and construction resources, major project delays and major loss of property. Modern

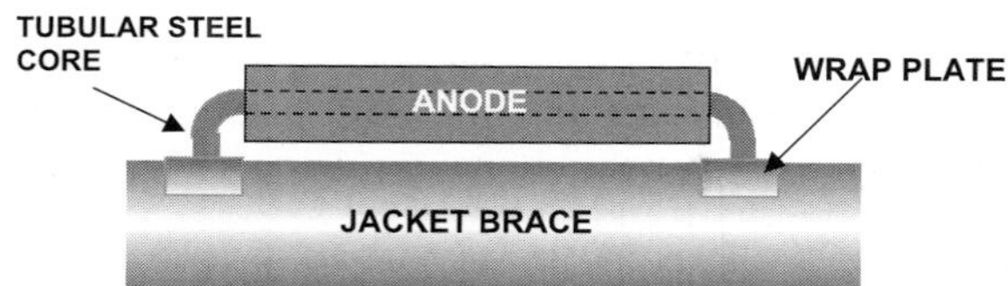

Figure 6.33 Sacrificial anode attachment to a jacket brace

interactive software provides extensive graphics (see figs. 6.28a, b and 6.29) and three dimensional verification capabilities, which should be used to thoroughly check and remedy input errors such as the:

- Missing joint and/or incorrect joint coordinate,
- Joints intended to be on a straight line but not,
- Joints intended to be at the intersection of two or more members but not,
- Missing members,
- Member dimensions and/or properties missing or incorrect,
- Pile simulation and/or end connection (specifically for ungrouted piles) incorrect,
- Support flexibility matrix and directions incorrect,
- Others (remember the old man Murphy!)

6.2.4.4 Load Simulation

Major classes of loads acting on an offshore platform are shown in fig. 6.12 and were summarised by type in Section 6.2.1. These loads may also be classified based on the manner they are simulated for computer analysis:

a. *Non-generated (Manually Coded) Loads*

These type of loads are calculated or specified externally and are manually input to the computer program. Some examples to manually input loads are:

> Dry and wet weights of deck equipment and supplies (these are either specified by the vendors or manually calculated by the engineers/designers),
> Live load on deck,
> Most accidental loads (such as dropped object and vessel impacts, explosion pressures etc.)
> Seabed settlements,
> Some wind loads that cannot be internally generated,
> Drilling loads (drill rig positions, pipe on rack, pipe on setback, pipe pull loads).

Prior to the manual load input, load analyses are performed transferring the dry and wet (contents) equipment and supplies, piping and storage tank, drilling and supplies loads provided by the equipment vendors and designers to the forces and reactions at their deck and jacket member or joint support points. Manual load input is usually done interactively using a menu driven PC screen. The loads are assigned to load groups so that they can be combined into sets making up the load combinations. Coding a load manually requires the input of its location at a joint or along the member it acts on as well as its direction (along

X, **Y**, **Z** axes or skew angles from these); it's magnitude (starting point and length, if a distributed load) and comments identifying its type and purpose.

b. *Generated Loads*

The computer program internally generates these types of loads; either by using the geometric properties of the structure or through input of a few parameters that can define an entire system of loads acting on a platform. Some examples to the generated loads are:

Dead weight of the structure
Buoyancy of structural members (whether a member is buoyant or flooded is specified)
Wave and current loads
Wind loads (some wind loads have to be input manually)
Earthquake loads
Certain types of dynamic/inertial loads (transportation and launch accelerations etc.)
Fire and heat loads
Some fire loads and explosion pressures

Methods in use for input of several of these load types are described in the following pages.

Wave Input Parameters and Wave Load Generation.

Wind driven ocean wave theory and calculation of the wave load effects on an offshore platform is presented in Chapters 2 and 3 of this Handbook. This section provides a short description of how these wave parameters are input to a computer program to generate wave loads acting on fixed platform members. Wave loads acting on a shallow water depth fixed offshore platform are calculated assuming that the platform does not move and there is no dynamic interaction between the structure and the waves. As the water depth increases, the structures become flexible and dynamic interaction between the jacket and the waves will result in wave load amplification. If dynamic wave structure interaction is expected, a wave energy spectrum with full range of wave frequencies and corresponding wave energy levels should be used. Where dynamic effects are not significant, a simpler wave formulation defined by the Linear/Airy, Stream Function, Stokes Fifth Order wave theories can be input. A wave theory selection depends on the water depth and wave steepness. API RP2A Section 2.3 and fig. 2.3.1–3 and Chapter 2 of this Handbook provide a guideline for selecting the applicable wave theory.

Figure 6.34 shows the parameters needed for computer analysis input. The required input parameters are:

- Wave height – H (measured from wave through to crest)
- Wave period or wave length (wave length can be calculated given the wave height and period)
- Distance from the global axis system origin to the wave crest
- Still water depth (for the extreme wave loading case this is equal to mean water depth, plus maximum astronomical tide, plus the storm tide heights)

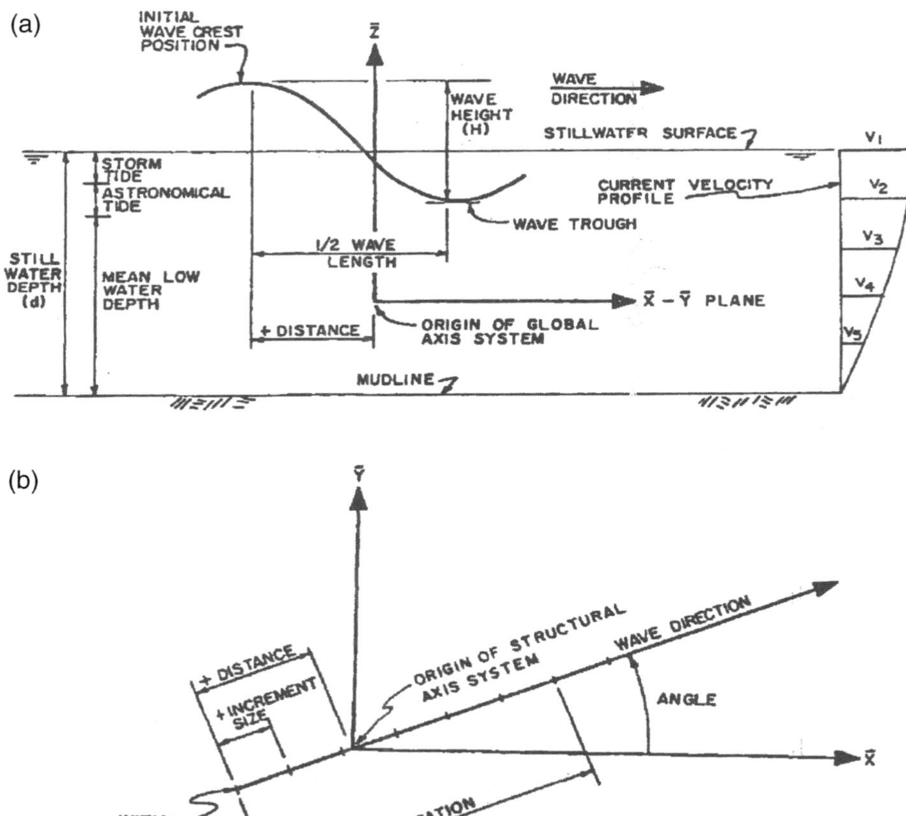

Figure 6.34 Wave and current input parameters (McClelland, 1986; Section 17). (a) Wave and current parameters; (b) wave attack direction, crest position and incrementation

- Current profile as a function of water depth (current velocity values to be added to the wave particle velocities for drag force calculation)
- Wave direction (wave attack angle measured counterclockwise from the **X** axis in the **XY** plane)
- Wave distance increments by which the wave crest distance will be moved along the wave attack direction to search and find the most critical member loads. See fig. 6.34b.

It is customary to calculate the global wave pressures exerted on the cylindrical members of a jacket through use of the Morrison's Equation. Below the Morrison's Equation and the description of its terms and recommendations for the calculation of its parameters are summarised directly from the API RP 2A Section 3 [API RP2A-WSD (2000); LRFD (1993)], for reader's convenience:

$$F = F_D + F_I = C_d(w/2g)AU|U| + C_m(w/g)V(\mathrm{d}U/\mathrm{d}t) \quad \text{[API (2000), Eq. 3–4]}$$

where:

F = Hydrodynamic load vector per unit length acting normal to the axis of the member
F_D = Drag force vector per unit length acting normal to the axis of the member in the plane of the member axis and U
F_I = Inertia force per unit length acting normal to the axis of the member in the plane of the member axis and $\mathrm{d}U/\mathrm{d}t$
C_d = Drag coefficient
W = Weight density of the water
g = Gravitational acceleration
A = Projected area normal to the cylinder axis, per unit length (=D for circular cylinders)
U = Component of the water velocity vector caused by wave plus current, normal to the axis of the member
$|U|$ = Absolute value of U
C_m = Inertia coefficient
V = Displaced volume of the cylinder (= $\pi D^2/4$ for tubular members)
$\mathrm{d}U/\mathrm{d}t$ = Component of the local acceleration vector of the water normal to the axis of the member.

In addition to the above defined drag and inertia generated pressures, other hydrodynamic effects, such as the lift and slam pressures, buoyant and non-buoyant forces (on members near the sea surface) and the axial Froude-Krylow force acting along the member axis should be taken into consideration for individual member design (for further details on these topics, refer to Chapters 2 and 3 of this Handbook).

It is common practice to compute the local water velocity vector U and the acceleration vector $\mathrm{d}U/\mathrm{d}t$ using the two-dimensional wave kinematics defined by an appropriate wave theory. Because of the out-of-plane wave energy dissipation and the approximations inherent to various wave theories, the horizontal components of U and $\mathrm{d}U/\mathrm{d}t$ are generally overestimated. This approach is conservative and does not have a significant impact on a shallow water depth jacket design. However, its impact on deepwater platform design may be significant. For deep water, the velocity and acceleration vectors may be reduced by a wave kinematics factor ranging from 0.85 to 1.0 (for details see API RP 2A Section 3 and its Commentary, (API RP2A-LRFD, 1993; WSD, 2000).

The current speeds are added to the wave velocities at each point to obtain the design U value in the API equation (3)–(4). The fact that the drag force is a squared function of U, small associated current velocities may result in significant increases in the drag force. The current direction is diverted and its speed is reduced in the vicinity of a platform because of the blockage offered by the platform. The API allows reduction of the current speed by multiplying it with a "current blocking factor" before it is added to the wave velocity. This factor varies from 0.70 to 0.90 depending on a platform's number of legs and the wave attack direction (see Chapters 2 and 3 of this Handbook). The current profile is also stretched or compressed to match the wave surface. For most design cases, a linear stretching of the current profile to the wave surface is satisfactory (for details, please see API RP2A-WSD, 2000; and LRFD, 1993, Sections 3 and their commentaries)

The values of the C_d and C_m coefficients depend on the tubular member dimensions as well as its surface roughness, the wave particle velocity, wave period and a number of hydrodynamic parameters (such as the Reynolds, Kuelagan-Capenter, Froude numbers. For details see Chapters 3 of this Handbook). The API had conducted a series of Gulf of Mexico offshore wave load measurement and data gathering studies where the measured total platform base (foundation) shears were compared against the same calculated using API recommended procedures. These studies resulted in the observation that, the following C_d and C_m coefficients for all unshielded jacket tubulars can be used, in conjunction with the guideline wave heights and wave force calculation procedures outlined in API RP 2A Section C3 [See Chapter 3 and API RP2A-LRFD, 1993; WSD, 2000]:

	C_d	C_m
Smooth unshielded cylinders:	0.65	1.6
Rough unshielded cylinders:	1.05	1.2

In most jackets with member length to diameter ratios more than 6, the size of the tubulars will not influence the wave pattern and the drag loads will dominate the wave loading. As this ratio decreases below six, the inertial loads will start dominating the wave pressure and a diffraction wave analysis that considers the presence of the structure must be used. For this case the maximum wave pressure will shift to a location to the right of the wave crest. Figure 6.35 shows the average wave pressure distribution on a jacket in 160 ft water depth from a 48 ft high 12 s period wave, plus a 3 ft/s surface current decreasing as a 5th order parabola to zero at mud-line. API recommended $C_D = 0.65$ and $C_m = 1.6$ values are used. In this example, wave pressure decreases from 309 psf at the wave crest to less than 100 psf at about 100 ft from the seabed. This demonstrates why the selection of the member sizes at the wave action zone is so important. For very short stubby members [such as a semi-submersible vessel (SSV) or a Tension Leg Platform (TLP) column] the maximum wave pressure will be at a point closer to the wave through. Given the wave input parameters, the computer program will calculate the wave pressure loads along the length of each jacket member and perform a structural analysis under these loads to calculate the internal member forces maximum member utilisation ratios and the support reactions. The program should also calculate the sum of all wave loads acting on the platform. These sums can be used to check and verify that the wave loads are calculated correctly.

Coding of Geometry and the Loads from Load-out, Transportation, Launch and Lift Operations

A number of key platform components are subjected to critical loadings during the construction operations. Some jacket members and joints may be subjected to high bending and punching shear loads while braces and bents are assembled into a jacket in the fabrication yard. Analysis of such assembly loading conditions would require detailed

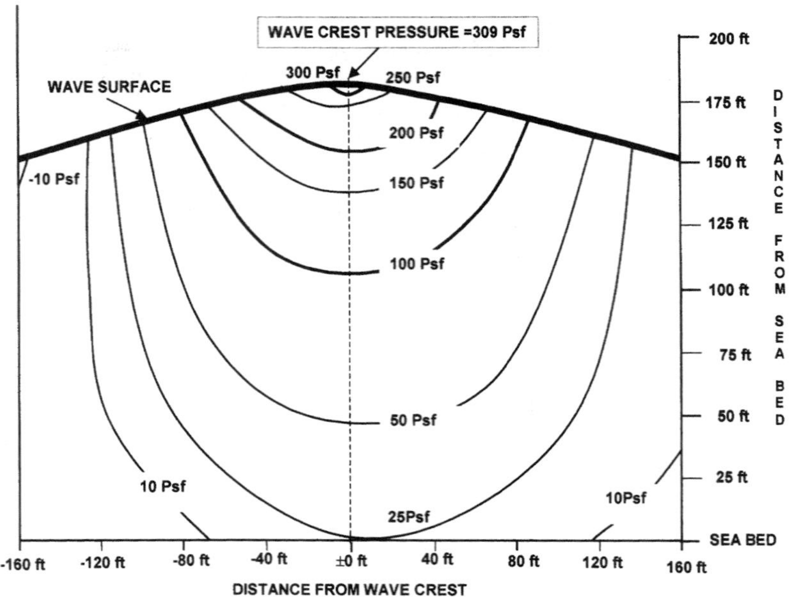

Figure 6.35 Wave pressure distribution in shallow water depth

knowledge of the jacket assembly plan and procedures and sequential simulation of the jacket geometry and loads representing the assembly plan.

Once the jacket fabrication is completed, it is loaded out and tied down to a vessel ready for transportation to its offshore installation site. Lighter, three to six legged, jackets may be assembled horizontally or in a vertical configuration with all legs standing up and horizontal and diagonal braces are fitted in and welded to these legs. The jacket is than lifted and loaded onto the transportation vessel (generally a barge) using multiple cranes or a large single crane, which may be barge mounted. The heavier and/or the eight-legged jackets are generally assembled using a bent roll-up process. Following the welding together of each jacket leg, these are placed on support cradles or skid beams and assembled into jacket bents by fitting and welding in their braces. This approach allows fast low cost welding of the bent braces while lying low on the ground. Assembled bends are then rolled up and tied together by welding in the remaining jacket braces higher off the ground, completing the jacket assembly. If crane capacity is available, jackets fabricated using the bent rollup method may be lifted, loaded up and tied down to the transportation barge.

If lifting is not possible, the jacket would be pushed along its skid-beams and loaded onto the transportation barge. The skidded load-out may be made onto a grounded or a floating barge. For the grounded barge load-out case (see fig. 6.36), the sea floor is dredged and filled with fine sand to ensure that there are no hard spots. The barge is grounded over the sand in-filled seafloor and then fully ballasted (generally to a ballast weight equal or more than the jacket transportation weight) pre-loading the sea surface. In general, skidding

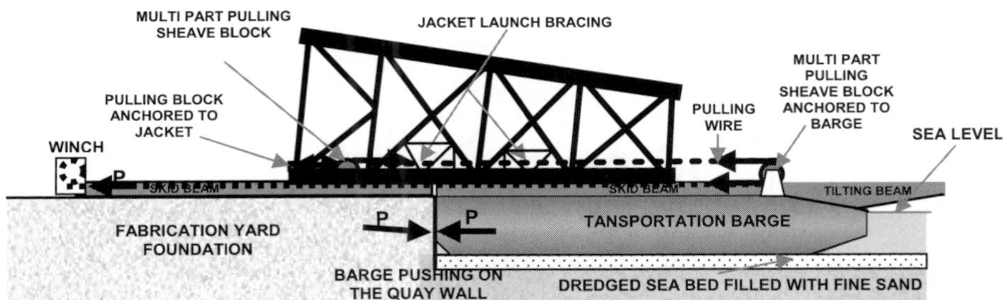

Figure 6.36 Jacket load-out to barge using the grounded barge method

onto a grounded barge is simple and poses no unusual stress conditions on the jacket structure, provided that the seabed surface is properly prepared and preloaded. During the jacket load out process, it is important to pull or push the jacket onto the barge while generating a compressive reaction force between the barge and the quay wall, eliminating the potential pull out damage to the quay wall. Figure 6.36 shows winches installed on shore pulling the jacket onto the barge using steel wires that go through multi-part wire sheave blocks one anchored to the barge stern and other attached to the dead end points provided at the land end of the jacket. Hydraulic long stroke jacks mounted on the barge pushing the jacket onto the barge and the barge stern on the key wall or similar arrangements are also in use. Rotating the jacket in place structural analysis model into a horizontal configuration generates the jacket load out structural model. Jacket's progression from the fabrication yard skids onto the barge mounted skid beams is simulated by a set of vertical springs and horizontal forces that represent the various support conditions and the friction forces generated while being skidded onto the barge. The static friction forces between the barge and the fabrication yard skids could be substantial, especially if rust and debris accumulates in between during the fabrication and or storage period.

Load out to a floating barge method may be used if the seabed soils are not strong enough to support the jacket weight or excessive seabed preparation is needed. This approach is done using small skidding increments (such as 10 ft per skid increment) and ballasting the barge in a manner that provides adequate jacket support for each skidding increment. Each skidding and ballasting increment could take anywhere from 30 to 60 min and offloading a large deepwater platform onto a floating barge may consume several days. The nearby sea traffic, which may create waves and ripples causing barge movements, must be restricted or stopped during the load out period. Tidal variations and wind and wave conditions in the area should also be taken into consideration. In this method, it is customary to check a number of accidental loading conditions where the jacket may be left cantilevering off the fabrication yard skids because of unplanned barge motions.

During its transportation to the field on a material barge, the jacket and the transportation tie down braces, their connections and the transportation barge are subjected to significant dynamic accelerations and inclined self-weight loads (fig. 6.37). These motions and resulting dynamic loads must be simulated in incremental loading sequences to determine and dimension the highest stressed components. The jacket horizontal bracing, which is of

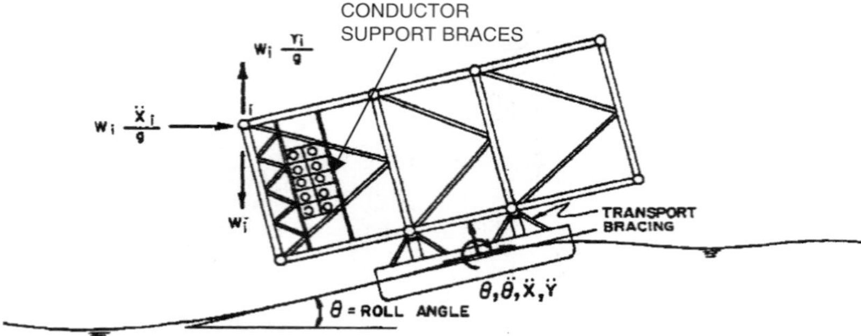

Figure 6.37 Forces experienced during the transportation of a jacket to its installation site [McClelland, 1986; Section 17]

secondary importance for the jacket in-place strength, plays a major role during the transportation stage. Proper design of this bracing, including the conductor support trusses could save jackets from sustaining damage or disintegration while under tow to site. Potential dynamic excitation of some jacket braces from drag and lift forces caused by high wind velocities that may be experienced during long trans-oceanic tows must also be verified. Some bracing may be needed only for jacket transportation phase; some of these braces may have to be removed, before jacket is installed on site, to reduce in place wave loads.

During its launch to sea, the jacket will be subjected to significant inertial and drag loadings (fig. 6.38). In general, the most critical loading would occur as the jacket starts tilting around the launch beam hinge and rapidly descends to sea. At this position, the tilting beams would exert high concentrated loads on the stiff bracing levels. These require a launch bracing system specially designed to distribute and reduce the launch forces. As the jacked hits the water plane and rapidly descends to sea, the leading jacket braces may experience high slam, drag and inertial forces.

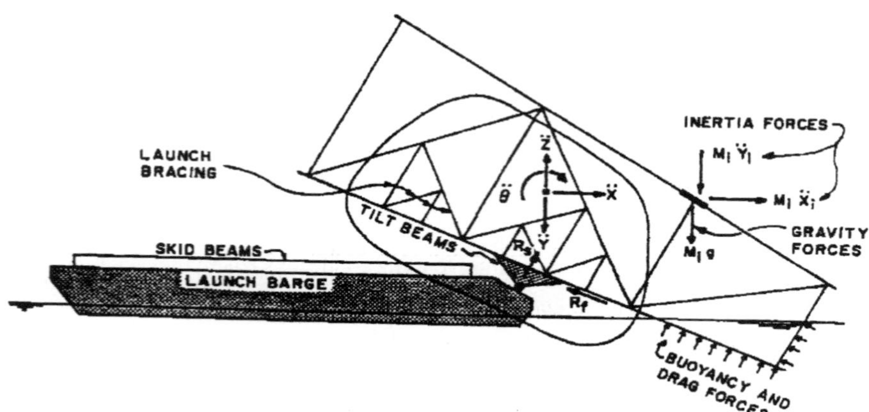

Figure 6.38 Forces experienced during the launch of a Jacket (Ref. 5.7 Section 17)

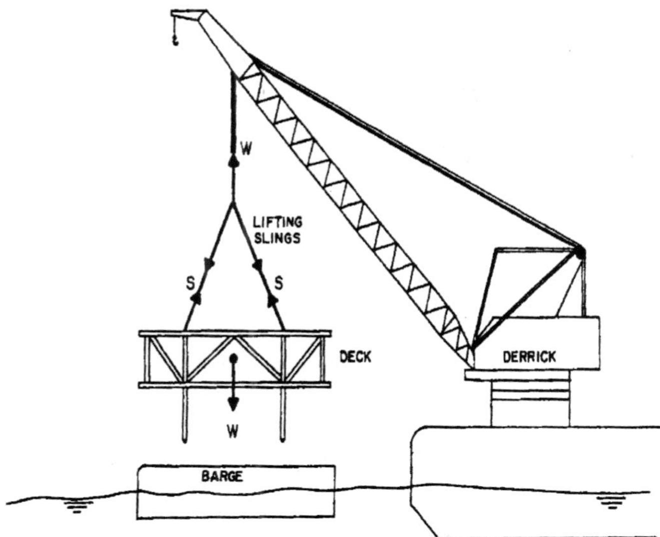

Figure 6.39 Forces experienced during the lift of a deck [McClelland, 1986; Section 17]

Crane lift of a deck or a jacket from a transportation barge is another critical loading condition that requires simulation, analysis and design (fig. 6.39). In such lifting operations, deck and jacket members and connections are loaded in directions different than their in place loading directions. Additionally, redundant or shorter/longer lifting sling lengths than planned may result in substantially different loads than those calculated for idealised conditions (for the case of a four sling lift, if one sling is shorter than planned, three instead of four slings may carry the entire deck load). Such an unplanned load distribution may also be caused by a center of gravity that may be at a location somewhat different than calculated. Lifting pad-eyes and lugs are components with high consequence of failure. A single pad-eye failure may result in the loss of the entire deck, jacket and the crane. Such critical components and their connections to the structures lifted must be designed for higher safety factors. Safety factors of four or more against ultimate capacity are commonly used for pad-eyes, their connections to the structure and the associated lifting gear.

6.2.4.5 Load Combination

Loads that belong to the same class (by type or by direction) are input in groups to facilitate their combination into a load condition. Loads are multiplied by load or carry down factors and combined to form critical load conditions that design the platform components. Typical load combinations are:

Operating Conditions: Different design codes (or recommended design practices) define operating load combinations in different ways.

API RP2a WSD version [API RP2A-WSD, 2000] defines an "Operating Load Condition" where the "Operating Environmental Conditions" are combined with the "Dead loads and maximum (or minimum) live loads appropriate to normal operations of the platform". The Operational environmental condition, while not clearly defined, is generally taken to be represented by the critical combinations of wind wave and current effects that may occur with a recurrence period of one to five years. Most designers commonly use one-year return period. Dead loads and maximum live loads include the platform weight and buoyancy, deck and equipment dry and wet weights, live loads in empty areas and any other operational static or dynamic effects (including drill-pipe set back and pull) acting along a number of directions creating critical load cases. A maximum loading condition that will yield the maximum compressive pile loads and a minimum loading condition that will yield the minimum axial pile load (which could be tension, if not enough deck dead and live load is present) are recommended. Due to the large number of possible operating deck load variations, many operating load combinations may have to be simulated.

Operating load condition strength checks for the steel plate and prismatic sections (Angle, Channel, Wide Flange built up beam and similar non-tubular component) are generally done using the basic AISC allowable stresses [AISC ASD, 1989]. API recommends a special set of design formulae and safety factors for the tubular member and connection (joint) strength and fatigue design. In some cases these formulae, differ significantly from the AISC recommended basic allowable stresses. For tubular member and connection design the reader is referred API RP2A-WSD (2000), Sections 3, 4 and 5.

API RP2a LRFD version [API RP2A-LRFD, 1993] does not define a clear operating condition load case. However, a high gravity and operating load factored "Factored Gravity Loads" case Q is defined –

$$Q = 1.3D_1 + 1.3D_2 + 1.5L_1 + 1.5L_2 \qquad \text{(API-LRFD Eq. C.2-1)}$$

where:

Q = The combined factored gravity load effect.

D_1 = *Dead Load 1*, which includes the weight of the structure, including the piles, grout and solid ballast, equipment and other objects permanently attached to the platform, hydrostatic forces including the buoyancy, weight of water included in the structure (such as the ballast water)

D_2 = *Dead Load 2*, dead load of movable equipment (such as the drill rig, forklifts and production equipment and modules that move or may be added or removed depending on the mode of operation)

L_1 = *Live Load 1*, weight of consumable supplies and fluids in pipes and tanks

L_2 = *Live Load 2*, short duration force exerted on the structure from operations such the lifting of drill string, crane lifts, machine operations, vessel mooring and helicopter loadings.

Additionally, API RP2a-LRFD suggests (optional) that an owner defined Operating Wave and current load can be added to the "Factored Gravity Loads" as follows:

$$Q = 1.3D_1 + 1.3D_2 + 1.5L_1 + 1.5L_2 + 1.35(W_o + 1.25\, D_{no}) \qquad \text{(API Eq. C.3-3)}$$

Here:

W_o = Static component of the owner defined operating wind, wave, current load (Return period not defined by API but generally taken as one year)

D_{no} = Dynamic (inertial) component of W_o (for further details see API RP2A-LRFD (1993), Section C.3.3).

API RP2a LRFD also recommends use of deck load carry down factors (generally applicable to L_1 and L_2) for design of the lower parts of the platform such as the jacket braces and piles. This is in consideration of the very low probability of all live load being present on all live load areas all the time. For other details of the application of the API LRFD equation C.2.1 please see API RP2A-LRFD (1993). A deck load carry-down factor of 0.70 is in common use.

In conjunction with the above load combinations, the API RP2a LRFD recommends a set of ϕ resistance factors, which depend on the type of loading and the platform component. Please see API RP2A-LRFD (1993), Sections D and E for these ϕ factors.

NPD (Norwegian Petroleum Directorate) Rules for "Load Bearing Structures" [NPD, 1992] specifies "The Ultimate Limit State – ULS" "Case-a" load combination:

$$Q_a = 1.3\,P + 1.3\,L + 0.7\,E + 1.0\,D \qquad \text{[NPD, 1992; Section 27]}$$

where:

P = Permanent Loads (Generally equivalent to D_1 plus D_2 loads of the API LRFD equation C.2.1)

L = Variable Functional Loads (Generally equivalent to L_1 plus L_2 loads of the API LRFD equation C.2.1)

E = Environmental Loads. For this load case the NPD and the API RP2a WSD have a major difference. While the API WSD suggests use of a one to five year return period for the operational environmental loads, the NPD specifies a return period of 100 yr, while reducing these by multiplying them with a factor of 0.70.

D = Deformation loads (deformation loads generally refer to loads that are applied on the structure as fixed displacements or rotations, such as temperature deformations or foundation settlements, which relax and reduce as the deformation is allowed and do not follow through to structural collapse, when satisfactory ductility and/or redundancy are available)

NPD recommends that a γ_m Material factor of 1.15 is used for the design of steel and concrete materials (note: Resistance Factor $\phi_m = 1/\gamma_m$), 1.25 for geotechnical properties and 1.30 for pile design.

Extreme Environmental Conditions: These include combinations of extreme environmental loads (storm wind, wave and currents; earthquakes; ice sheets, floes and bits, mudslides etc.). These are added to the platform weight and buoyancy, deck and equipment wet weight, live loads and any other static or dynamic effects that may be expected to be present under extreme environmental conditions. Pile compression and pull back conditions are simulated. Extreme environmental conditions are generally defined by all offshore codes of practice as those that have a 100 yr return period (0.01/year probability of occurrence). A significant number of extreme environmental load combinations may have to be simulated depending on the type and direction of environmental loads and the platform geometry.

API RP2a WSD version (2000) defines an extreme environmental load condition where the "Design Environmental Conditions" are combined with the "Dead loads and maximum (or minimum) live loads appropriate for combining with these extreme conditions". The "Extreme Environmental Effects" are clearly defined, as represented by the critical combinations of wind wave and current effects that may occur with a recurrence period of one hundred years. Dead load and maximum live loads are defined same as the operating load condition.

Strength checks for these load combinations are done using operating conditions allowable stresses increased by a factor of 1/3 for dynamic loadings. For further details and special component design, please see API RP2A-WSD (2000), Section 3, 4 and 5.

API RP2a LRFD version (1993) defines a "Wind Wave and Current" factored load combination which is representative of the extreme environmental conditions –

$$[Q = 1.1\ D_1 + 1.1\ D_2 + 1.1\ L_1 + 1.35\ (W_e + 1.25\ D_n) \quad \text{(API-LRFD Eq. C.3-1)}$$

where D_1, D_2, and L_1 are same as the API equation C.2-1 and –

W_e = "The force applied to the structure due to the combined action of the extreme wave (typically 100-yr return period) and associated current and wind,accounting for the probability of occurrence of winds and waves (both magnitude and direction)".

D_n = "The inertial load acting on the platform at the time when the total dynamic response (static and inertial) is a maximum".

When inertial forces due to gravity loads act in a direction opposing the inertial forces due to wind wave and current loads API LRFD recommends use of the following additional load combination:

$$Q = 0.9\ D_1 + 0.9D_2 + 0.8\ L_1 + 1.35\ (W_e + 1.25\ D_n) \quad \text{(API-LRFD Eq. C.3-2)}$$

For other details of the application of the API-LRFD equation C.3.2 please see API RP2A-WSD (1993). The ϕ resistance factors used for the API LRFD Extreme Environmental Cases are the same as the operating cases. Background of the LRFD design method is provided in Chapter 4, Section 4.6 of this Handbook.

NPD (Norwegian Petroleum Directorate) Rules for "Load Bearing Structures" [NPD, 1992] covers the extreme environmental loading conditions as the "Ultimate Limit State–ULS, Case-b" by specifying the following load combination:

$$Q_a = 1.0\ P + 1.0\ L + 1.3\ E + 1.0\ D \qquad \text{[NPD, 1992; Section 27]}$$

where P, L, E and D has the same definition given for the NPD ULS, Case-a

The NPD resistance/material factors used for ULS Case-b is same as the ULS Case-a.

Accidental (Progressive Limit State) Loading Combinations: These conditions include the combination of accidental loads and platform weight and buoyancy, deck and equipment weight, live loads and any other static or dynamic effects that may be present during an accidental loading condition. The recurrence period of accidental loads (dropped objects, vessel impact, fires and explosions, loss of buoyancy, blowouts etc.) may be calculated using a total risk assessment approach, a HAZOP study or from a prescriptive design rule [such as the NPD PLS design case and Karsan, et al (1996); Moan, et al (1993)] or from the cost of consequence versus the cost of prevention considerations. Code checks for these conditions are generally done using mean material strengths or resistance factors set to unity. For such cases the value of allowable safety/resistance factors will depend on the return period selected for the accidental load effect.

API RP2a WSD and LRFD versions [API RP2A-LRFD (1993) and WSD (2000)] list a number of accidental loading conditions "such as: collision from boats and barges; impact from dropped objects; explosion and fire". These recommend that "Consideration should be given in the design of the structure and in the layout and arrangement of facilities and equipment to minimise the effects of these loads". No clear procedure is recommended for the calculation of the magnitude or the frequency of occurrence of design accidental effects. Determination of the frequency of occurrence and magnitude of an accidental design event requires that a quantitative field development risk assessment criteria (RAC) are established and a set of design events are selected to keep these risks below the risk acceptance criteria. RAC is based on acceptable levels of loss of life, environment damage and production loss and their probabilities of occurrence. For further details please see Karsan, et al (1996); Moan, et al (1993).

NPD (Norwegian Petroleum Directorate) Rules for the "Load Bearing Structures" [NPD, 1992] covers the accidental loading conditions under their "Progressive collapse Limit State–PLS". NPD requires that all load coefficients are set equal to unity for the PLS load condition:

$$Q_a = 1.0\,P + 1.0\,L + 1.0\,A \qquad \text{[NPD, 1992; Section 27]}$$

where A is the load effect from the accidental event under consideration. Section 28 of NPD (1992) requires that the value should be selected such that its annual probability of exceedence would be 10^{-4}/year (or, it is a 10,000-yr event). NPD would accept an accidental event recurrence period different than 10^{-4}/year, provided that this value is the result of a risk assessment process. It is worthwhile to note here that it is not normal practice to combine two extreme events in one load condition unless the two are correlated (such as the winds and waves). This is because of the low probability of both extreme events occurring at the same time (such as the extreme waves and a major hydrocarbon fire).

Other Loading Combinations: These include:

(a) Conditions where *serviceability*, deformations, clearance, or *operability* may be of concern. *NPD Rules* define an additional Serviceability Limit State (SLS) load combination, which is achieved by setting all coefficients in their ULS-a equation to unity.

(b) *Temporary loading cases* – These include the load conditions, which the platform goes through during fabrication, load-out, transportation to site, installation hook-up and commissioning (HUC) phases. These are generally short duration loadings and load combinations and safety and resistance factors that are dependent on the types of loads expected and consequences of failure are used. For further details on these loading conditions, please see API RP2A-LRFD (1993), WSD (2000) and NPD (1992).

Table 6.1 provides a list of loadings and load combinations summarised above. It is highly recommended that such a table is prepared and incorporated in the Basis of Design Document before a computer run is attempted. Such a table will ensure that all the relevant loads, load combinations, load factors and corresponding allowable stress ratios or resistance factors are taken into consideration and discussed and agreed among the design team, client and, if applicable, the certifying authority or CVA. For the extreme environmental loading, it is customary to reduce the API recommended deck live loads by an appropriate carry down factor to account for the low probability of experiencing the full deck live load when the extreme loading is experienced. A factor of 0.70 on full deck live load is commonly used.

Checking the Platform Load Input

Load input to a platform analysis computer program should be carefully and thoroughly verified before a structural analysis is attempted. There had been a number of instances where errors in load and load combinations resulted in waste of analysis, drafting, construction resources, major project delays and loss of property and life. Modern interactive software provides convenient load summary tables that allow checking the load input. Some examples to these errors are:

- Missing load and/or incorrect load application, point, direction and magnitude
- Incorrect loading direction, combination and factors,
- Incorrect safety/material factor applied to the load condition,
- Others (remember the old man Murphy!).

A computer output table that provides a list of loads, their percentage participation in a load combination and contribution to load vector in global axes directions is very useful. Sums of all loadings along the global axes should also be provided and verified. Such tables enable the software program user to check the magnitudes and directions of loads to ensure that there are no gross coding errors.

6.2.5 Solution of the Load Deflection Equation $P_i = k_{ij} d_j$

The geometry and load conditions input to the structural analysis program are processed and analysed to compute the platform deflections, internal member forces and support reactions. While the engineer is not expected to know all the theory and computational details undertaken by the program, it is important that he/she has a global understanding of the methods used and a feeling for what to expect as output. The method used and a feel for what to expect as output and a good understanding of the relations between the support conditions, stiffness and how the structure should deform under a given load condition is

Table 6.1 API recommended and NPD required load cases and their safety/resistance factors

AUTHORITY	LOAD MULTIPLIER / FACTOR										SAFETY OR RESISTANCE FACTOR	COMMENT
	D_1	D_2	L_1	L_2	W_o	D_{no}	W_e	D_n	D	A		
API RP2a-WSD	1.00	1.00	1.00	1.00	1.00	1.00	-	-	-	-	Generally Basic AIC Allowables (Ref. 3) used . For additional tubular member requirements, see Ref. 1-Sect 3.2. Foundation SF =2.0	API Recommended. US Government MMS Required. A factor of 1.1 is generally applied to D1 and D2 to account for stiffeners etc.
API RP2a-LRFD, Factored Gravity Loads Case	1.30	1.30	1.50	1.50	0.00	0.00	-	-	-	-	Resistance factor depends on load and member type. See Ref 2 Sect. D & E Resistance Factors	API Recommended. US Government MMS Required
API RP2a-LRFD, Normal Gravity Loads Case	1.30	1.30	1.50	1.50	1.35	1.69	-	-	-	-	Resistance factor depends on load and member type. See Ref 2 Sect. D & E Resistance Factors	Optional
NPD-ULS Case-a	1.30	1.30	1.30	1.30	-	-	0.70	0.70	1.00	-	Steel & Prestressed Concrete =1.15; Geotechnical = 1.25; Piles = 1.30	Norwegian Govermnet regulatory requirement
NPD Servicability	1.00	1.00	1.00	1.00	1.00	1.00	-	-	1.00	-	1.00	Norwegian Govermnet regulatory requirement
API RP2a-WSD	1.10	1.10	0.70	0.70	-	-	1.00	1.00	-	-	Generally Basic AIC allowables (Ref 2) increased by 1/3. For additional tubular member requirements, see Ref.1-Section 3.2. Foundation SF=1.50	API Recommended, US Govt. MMS Reqd. A carry-down factor of 0.70 is traditionally applied to L1 and L2
API RP2a-LRFD, Maximum Gravity Loading Case	1.10	1.10	1.10	-	-	-	1.35	1.69	-	-	Resistance factor depends on load and member type. See Ref 2 Sections D & E Resistance Factors	API Recommended. US Government MMS Required
API RP2a-LRFD, When Gravity Opposes Inertial Loads	0.90	0.90	0.80	-	-	-	1.35	1.69	-	-	Resistance factor depends on load and member type. See Ref 2. Sections D & E Resistance Factors	API Recommended. US Government MMS Required
NPD-ULS Case-b	1.00	1.00	1.00	1.00	-	-	1.30	1.30	1.00	-	Steel & Prestressed Concrete =1.15; Geotechnical = 1.25; Piles = 1.30	Norwegian Govermnet regulatory requirement
API RP2a-WSD	1.00	1.00	0.70	0.70	-	-	-	-	-	1.00	Generally mean load and material strength used in conjunction with the extreme accidetal load effect.	A carry-down factor of 0.70 is traditionally applied to L1 and L2
API RP2a-LRFD	1.00	1.00	0.70	0.70	-	-	-	-	-	1.00	Resistance factor depends on load and member type. See Ref. 2 Sect. D & E Resistance Factors	A carry-down factor of 0.70 is traditionally applied to L1 and L2
NPD-PLS	1.00	1.00	1.00	1.00	-	-	-	-	-	1.00	1.00	Norwegian Govermnet regulatory requirement

Notes: Reference 1: API RP2A-WSD (2000)
Reference 2: API RP2A-LRFD (1993)
Reference 3: AISC (ASD; 1989)

important. So is the ability to investigate the causes of over utilisation of a member or connection and to arrive at an effective solution to redesign the member and verify this through calculations.

Having defined the structural geometry and loads, we can now proceed with the computation of the joint deflections and member end forces of the skeleton structure. The load-deflection relation of a linear elastic structure in three-dimensional space can be expressed as:

$$pi = k_{ij}dj \quad i = 1, 6N - m \quad j = 1, 6N - m \tag{6.2.4.1}$$

k_{ij} = Element of the structural stiffness matrix K setting up a linear relationship between the force at the ith independent degree of freedom and the deflection of the jth independent degree of freedom, in the structural axis system. Physically, k_{ij} can be visualised as the reaction at ith degree of freedom when the jth degree of freedom is deflected by a unit amount while all the other degrees of freedom are kept at zero.

N = Number of joints of the structure (six degrees of freedom each, unless a support or settlement restraint is prescribed).

m = Number of support restraints of the structure (generally more than six for a properly supported structure).

Furthermore, the support reactions can be calculated from:

$$R_i = c_{ij}dj + pi \quad i = 1, m \quad j = 1, 6N - m \tag{6.2.4.2}$$

where,

R_i = Reaction at ith support of the structure.

c_{ij} = Support stiffness coefficients establishing a relation between the ith reaction and the deflection at the jth independent degree of freedom.

p_i = Sum of the fixed end forces from members connecting to, plus direct forces acting on, support i.

From the Maxwell-Betti's reciprocity theorem, it can be shown that k_{ij} is symmetric. It can also be shown that, for a properly supported stable structure, the K matrix is positive definite.

The values of k_{ij} and c_{ij} are directly calculable from the structural geometry. p_i loads are calculated from the loads acting on the structure. The main structural analysis work consists of calculating the d_i deflections and R_i support reactions when k_{ij}, c_{ij} and p_i are known. Once the d_i deflections are known, internal forces and stresses for each member can be calculated.

Structural Member Stiffness

From the definition of k_{ij} in equation (6.2.4.1), it can be seen that when a unit deflection is imposed on the jth structural degree of freedom, every member end that is connected to it will undergo a unit deflection while deflections of all other degrees of freedom stays equal to zero. Using this observation, the structural stiffness matrix element k_{ij} can be obtained from the sums of the stiffness terms of all members that follow the deflection of that degree of freedom:

$$k_{ij} = \sum_{b=1}^{B} k_{ij}^{b} \tag{6.2.4.3}$$

where B is the number of members taking part in the deflection of the jth degree of freedom. k_{ij}^{b} is the stiffness element of the bth member connecting to the jth and ith structural degrees of freedom.

It is convenient to compute the k_{ij}^{b} terms in the member axis system (see 6.2.4.1) then convert these to the structural axis system by use of the directional cosine matrix between the two systems. Once k_{ij}^{b} are calculated k_{ij}^{b} in the structural axis system is obtainable from:

$$k_{ij}^{b} = l_{ic} \cdot k_{ca}^{b} \cdot l_{aj} \quad i = 1, 12; \quad j = 1, 12 \quad a = 1, 12; \quad c = 1, 12 \tag{6.2.4.4}$$

where l_{aj} is the directional cosine matrix element for transferring from the member axis system's "a" direction to the structural axis system's "j" direction.

Load Vector

Elements of the load vector **P** are made of external loads acting on the structure. At each structural joint, one load and one moment can be applied along each axis direction resulting in six load vector elements per joint (three forces and three moments). These loads can be applied to the joints or to the members of the structure. Thus, each load vector element p_i is made of two components:

$$p_i = p_i^o + p_i^s \tag{6.2.4.5}$$

where

p_i^o = Loads applied directly on the degree of freedom i

p_i^s = Load transferred to the degree of freedom i by members connecting to it. These are caused by the loads within the member spans and are calculated as fixed end (frozen body) reaction forces and moments of each member.

In order to calculate p_i, one must visualise structural deflections to take place in two steps. In the first step, all the structural degrees of freedom can be considered as restrained against deflection (e.g. $d_i = 0$, $i = 1, 6N$). Using this configuration p_i^s are calculated as the fixed end reactions of the members loaded in their spans. In the second step, the structure is loaded with the $p_i = p_i^o + p_i^s$ loads and allowed to deflect to reach its final deflected configuration.

Similar to the calculation of the member stiffness matrix, it is easier to calculate the member fixed end forces in the member axis system and then transfer these to the structural axis system using a coordinate transformation matrix where,

$$p_i = l_{ij} \cdot p_j^- \quad i = 1, 12; \quad j = 1, 12 \tag{6.2.4.6}$$

p_j^- are the fixed end forces in the member axis system.

Methods of Solution

$$p_i = k_{ij}dj \quad i = 1, 6N - m \qquad (6.2.4.1)$$
$$j = 1, 6N - m$$

Relation is a set of linear equations with $6N - m$ unknowns. Many methods applicable to the solution of linear equations can be used to solve this problem. Many researchers have studied the solution and stability of linear sets of equations. With the advance of computing technology, many methods that take advantage of speed and volume of computer operations, evolved. A full treatment of this subject will not be attempted in this section. We will try rather to give an overall view of the methods that are in common use. For further details on structural and support stiffness matrix and load vector generation and methods of solution, readers are referenced to Faddeva (1959), Livesly (1975), Martin (1966), Nair (1978) and similar publications on these topics.

The methods of solution can be divided into three major categories:

a. Methods Involving Direct Elimination

For a single load condition (i.e. one load vector **P**) it has been shown that no method exists which requires fewer arithmetic operations than the Gaussian Elimination. The well-known Gaussian elimination method essentially consists of a sequence of operations performed in order to transform the original matrix to an upper triangular one. This is achieved by a series of divisions and subtractions in such a way as to eliminate progressively the unknowns one by one. Once the elimination is completed. The last row contains only the coefficient relative to the last unknown, all other coefficients being equal to zero. Once the last unknown is computed from the last row, all the other components can be found by backward substitution. Many variants of the Gaussian Elimination that take advantage of certain structural properties are available. Each variant may result in a more expedient solution for a certain type of a structure or problem. For the unbraced chain structures such as the continuos beams, building frames and bridge structures, the degrees of freedom can be numbered such that the non-zero elements of the stiffness matrix will be condensed in a narrow band along the diagonal axis of the matrix. For this type of stiffness matrices, banded Gaussian elimination methods will result in a faster solution. The algorithm used for a banded solution is still the same, but the advantage is taken of the fact that all the stiffness elements outside the band are zero and division, multiplication and subtraction operations for these elements need not be considered. In computerised solutions, this results in substantial savings in the ranges of the program's "DO loops".

Other elimination methods take advantage of the topological characteristics of the structure under investigation. One such method consists of separating the structure into "substructures" each of which is considered a separate system. Each substructure may be thought of as a super-element with many degrees of freedom, few of which are coupled to those of the rest of the structure. In this way an entire super-element (i.e. a sub-matrix made from a set of degrees of freedom belonging to the super-element) is eliminated in a manner similar to the Gaussian elimination. This method requires inversion of the sub-matrices or condensation of all stiffness into the interface nodes before the elimination procedure. For structures subjected to a large number of loading conditions, a judicious selection of the size of the super-elements depending on the geometry, structure topology and the number of load conditions, may result in more economical solutions than the direct Gaussian Elimination Method. The recent unlimited PC memory and processing speed availability is progressively making the direct Gaussian Elimination the most convenient solution method, even for very large number of load conditions.

b. Methods Involving Iteration

Iterative methods are based on an estimate of the solution (deflection vector d_i) which is improved through an iteration procedure making use of the stiffness matrix and the load vector. One example of these methods uses a set of equations where:

$$d_i = p_i - 1/k_{ii}[k_{ij}d_j] \qquad (6.2.4.7)$$

$$i = 1, \quad 6N - m$$

$$j = 1, \quad i-1, i+1, 6N - m$$

Starting with the estimate of the d_i vector, an improved value can be obtained by substitution into equation (6.2.4.7) to obtain an improved estimate for d_i. The process is repeated until the answer converges towards prescribed accuracy limits. Like many other iterative procedures, these methods may become unstable due to certain numerical properties of the stiffness matrix and should be used with caution so that run-away computer costs can be avoided.

Matrix Inversion

A matrix f_{ij} can be defined such that:

$$\delta_{in} = f_{ij} \cdot k_{jn} \qquad (6.2.4.8)$$

$$i = 1{,}6N - m$$

$$j = 1{,}6N - m$$

$$n = 1{,}6N - m$$

where δ_{in} is the Kronecker delta (e.g. δ_{in} is a matrix with all elements equal to zero except ones along its diagonal axis). f_{ij} is termed the inverse of k_{jn} and it is equal to the flexibility matrix of the structure where:

$$d_i = f_{ij} \cdot p_i \qquad (6.2.4.9)$$

Inversion of the stiffness matrix needs more operations than the elimination and sub structuring methods and should be avoided for large structures with many degrees of freedom and few load conditions.

Verification of the Structural Analysis Results and Member Sizes

As a result of the computer analysis, the following output is generally made available by most standard structural analysis software:

1. As a first step, joint deflections and rotations, member end forces and moments, support reaction forces and moments are output.
2. In second step, these member forces and reactions are used to calculate the basic member stresses and the code compliance checks.

Most computer programs have the options for calculating the interaction ratios as defined in the American Institute of Steel Construction (AISC) Manual, American Petroleum Institute Recommended Practice 2A (API-RP-2A) and stress and other limitations defined by the Norwegian Petroleum Directorate (NPD) and Det norske Veritas (DnV) regulations/codes. Currently ISO standards for the same are under development. Most programs have options for sorting these stresses and interaction ratios for all load conditions and printing out summary tables showing the maximum stresses, maximum deflections, interaction ratios and their locations. Other specialised computer programs may interface with this output to perform joint analysis using API or NPD/DnV joint design rules, or perform deterministic or probabilistic fatigue analysis and code checks.

Deflections and Rotations

Figure 6.40 shows a typical joint deflection and rotation (d_j) output from a structural analysis program. The engineer should check this output carefully in order to ensure that the deflected shape is acceptable and is in agreement with the restraints imposed on the structure. Very large or very small deflections than what is anticipated or an unexplainable deflected shape may point out to input errors such as erroneously coded loads, geometry or supports. In general, structural deflections that are in excess of few inches at the platform top should be taken causes for investigating whether the structural and/or foundation stiffness is adequate.

Support Reactions

Another direct output from the solution of the force-deflection equations is the table of support reactions (R_i). Using these reaction quantities, piles and conductors of the structure can be checked and, if required, resized

Member End Forces and Stresses

Once the d_i deflections of the structure are known, the member end forces can be calculated from:

$$p_i^a = k_{ija} \cdot d_j^a \qquad i = 1,12 \qquad j = 1,12 \qquad (6.2.4.10)$$

where p_i^a, k_{ija}, d_j^a are the end forces vector, stiffness matrix and the end deflections of structural component (member) a. Before equation (6.2.4.10) can be used d_j^a member end deflections in structural axis system must be transferred to the member axis system using the directional cosines vector for these two axis systems.

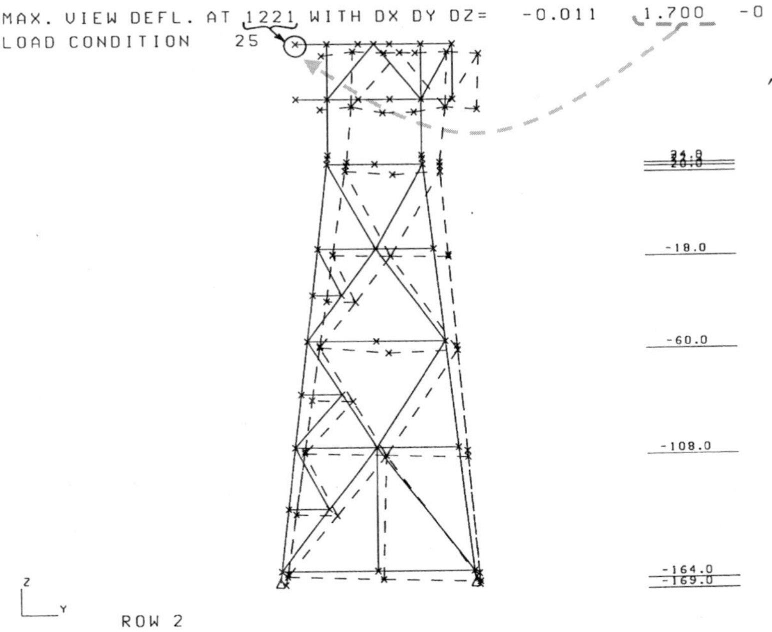

Figure 6.40 Undeflected and deflected shape plots output from a fixed offshore jacket platform structural analysis software program

Using p_i^a and the member in-span loadings, forces at any point along the member axis can be calculated. These forces and the member cross-section properties are then used to calculate the basic member cross-section stresses.

Code Check

Basic member forces and stresses calculated by the computer program should be checked against their allowable values set forth by an industry-accepted code of practice for the design of fixed offshore structures. Selection of a code may depend on many factors among which are the government regulations pertaining to the location where the structure will be installed, the type of structure, materials and types of loadings, etc.

Effective use of an industry code of design requires a good understanding of the basic statistical parameters effecting the loads and safety factors. Some of these statistical parameters and their use are summarised in Section 4.0 of this Handbook.

The current industry codes of practice use two different methods for design of steel offshore structures. Most U.S. practices use the allowable stress (safety factor) design method (also known as the *W*orking *S*tress *D*esign, WSD or *A*llowable *S*tress *D*esign, ASD). In this method, actual expected maximum values of loads are used for calculating the existing internal forces and stresses. These stresses are then checked against allowable safe loads (or stresses) i.e.

$$S \geq \frac{R}{\text{Safety Factor (SF)}} \qquad \text{where SF} = \text{Safety Factor} \qquad (6.2.4.11)$$

relation is used.

The current industry accepted code of practice in the U.S.A. is the American Petroleum Institute Recommended Practice 2A (API-RP2a). API RP 2a provides both WSD and LRFD methods for structural design RP2A-LRFD (1993) and WSD (2000). This RP bases most of its stress limitations on the American Institute of Steel Construction (AISC) Manual for the Design Fabrication and Erection of Structural Steel for Buildings [AISC ASD (1989) and LRFD (1988)].

The fundamental AISC stress and stability check for uniform cross section members subjected to combined axial and bending effects is performed by use of allowable stress defined under AISC code section 1.5 and interaction equations 1.6-1 a, b and 1.6.2 under section 1.6 where:

$$f_a/F_a + \frac{C_{mx} f_{bx}}{(1 - f_a/F'_{ex})F_{bx}} + \frac{C_{mx} f_{by}}{(1 - f_a/F'_{ey})F_{by}} \leq 1.0 \qquad \text{(AISC Eq. 1.6-1a)}$$

$$f_a/0.6F_y + \frac{f_{bx}}{f_a/F_{bx}} + \frac{f_{by}}{f_a/F_{by}} \leq 1.0 \qquad \text{(AISC Eq. 1.6-1b)}$$

When $f_a/F_a \leq 0.15$ the AISC Equation 1.6-2 below may be used in lieu of the above formulae:

$$f_a/F_a + \frac{f_{bx}}{f_a/F_{bx}} + \frac{f_{by}}{f_a/F_{by}} \leq 1.0 \qquad \text{(AISC Eq. 1.6-2)}$$

where:

f_a is the axial stress and f_{bx} and f_{by} are the bending stress acting along member principal axes x and y,

F_y is the yield strength of steel,

F_a is the allowable axial stress and F_{bx}, F_{by} are the allowable bending stress along member principal axes x and y,

F'_{ex}, F'_{ey} are the true buckling (Euler) stress of the beam along the member principal axes x and y divided by safety factor (generally equal to 23/12) and C_{mx}, C_{mx} are the moment amplification factors along the member principal axes x and y.

Further definition and additional formulae for calculating the parameters of the above equations and other stress and design limitations are described in detail in AISC ASD (1989).

The above AISC equations 1.6-1a, 1b and 2 converge to the following API RP2a equations for the cylindrical members:

$$f_a/F_a + \frac{C_m\sqrt{f_{bx}^2 + f_{by}^2}}{(1 - f_a/F'_e)F_b} \leq 1.0 \qquad \text{(API Eq. 3.3.1-1)}$$

$$f_a/0.6F_y + \frac{\sqrt{f_{bx}^2 + f_{by}^2}}{f_a/F_b} \leq 1.0 \qquad \text{(API Eq. 3.3.1-2)}$$

When $f_a/F_a \leq 0.15$ the following formula may be used in lieu of the foregoing formulae:

$$f_a/F_a + \frac{\sqrt{f_{bx}^2 + f_{by}^2}}{f_a/F_b} \leq 1.0 \qquad \text{(API Eq. 3.3.1-3)}$$

In addition to generally using the formulae for calculating the above parameters equations given by the AISC Code [AISC ASD, 1989], API further defines some of these parameters to parallel the past offshore industry experiences and research. These and other stress limitations and interaction equations are further defined in API RP2A-WSD (2000), Section 3.

Other codes of design (mostly European codes and the API RP2a – LRFD Document) have an approach more closely following a constant safety index β approach (see Chapter 4 of this handbook). In this approach, factors are applied to loads, depending on the type and reliability of knowledge of a certain type of load and sum of these factored loads is checked against the characteristic strength of the material divided by a material factor:

$$S_d = \sum_{1}^{N} (\gamma_{fi} \cdot S_{ki}) \geq Rd = \frac{Rk}{\gamma_m} \qquad (2.4.12)$$

(DnV Rules (Latest edition) and NPD Regulations (1992))

where S_d is the combined load effect, S_{ki} represents the characteristic values of individual loadings such as dead load, live load, wave loads, wind loads, etc. acting on a component.

This characteristic value is generally taken as the mean or biased two standard deviations from the mean value. γ_{fi} is a load factor representing the level of uncertainty of the information for a given load S_{ki}. Rd represents the design resistance of a given component, while Rk represents the characteristic resistance. γ_m is a material coefficient depending on the uncertainty level of our knowledge bias and scatter (Coefficient of Variation = COV) of our knowledge with respect to a given resistance (strength) case. Further details of this method are described in NPD (1992), DnV Rules (Latest edition) and Chapter 4 of this Handbook.

The API recommended Practice for Planning, designing and constructing Fixed Offshore platforms – Load Resistance Factored Design API RP2A-LRFD (1993) utilises a similar format where:

$$\phi \cdot R_n \geq \gamma_d \cdot D + \gamma_L \cdot L + \gamma_w (W_e + \gamma_{\text{dyn}} \cdot D_n) = Q$$
$$\text{and} \tag{6.2.4.13}$$
$$\phi \cdot R_n \geq \gamma'_d \cdot D' + \gamma'_L \cdot L' + \gamma'_w (W_e + \gamma_{\text{dyn}} \cdot D_n) = Q$$

Are checked in these code check equations. R_n represents the nominal component design resistance, ϕ the resistance factor < 1.0. γ_d, γ_L, γ_w, γ_{dyn} represent load factors applied to the global loads where $\gamma > 1.0$; W_e the static component of environmental load effect, D_n the inertial component of load induced by environmental effects; D and L the expected dead and live load effects. Primes indicate the value to be used in checking the conditions when gravity loads are beneficial (e.g. $\gamma' < 1.0$), such as a pile pullout case.

Data Management

Structural analysis of an average size offshore structure with many hundreds of members and as many as thirty to fifty load combinations result in the output of tens of thousands of quantities of deflections, stresses, interaction ratios and reaction forces. Review and sorting of these quantities by visual inspection may get quite involved, time consuming and error prone. To overcome this difficulty, most computer programs contain special post-processing sub-routines where the information is summarised and output in conveniently formatted tables. Some common output summary tables are:

Member Lowest and Highest Stress and Interaction Ratio Tables:

Members with stresses or interaction ratios above a specified maximum or below a specified minimum are flagged out. Using this table, over or undersigned members can be identified and resized.

Maximum Support Reaction Tables:

Load conditions that cause maximum axial load, bending moment or shear for a given support is tabulated. These tables can be used for the design of the piled supports.

Member Maximum Stress or Interaction Tables:

These types of tables identify the load condition for which the stress, interaction ratio or any other stress related quantity of a given parameter is a maximum. Figure 6.41 shows a graphic output of maximum interaction ratios from an offshore platform analysis software program.

Member Group Maximum Stress or Interaction Ratio Tables:

These tables identify the load conditions for which the stress or interaction ratio of a given member group is maximum. Such information enables sizing of a group as a whole and is useful in the design of symmetric or functionally similar members without the need for running many load directions.

Re-sizing and Re-analysis

At the end of a structural analysis, certain structural elements may show stresses or stability characteristics beyond what is permitted by a selected design code. Certain other elements

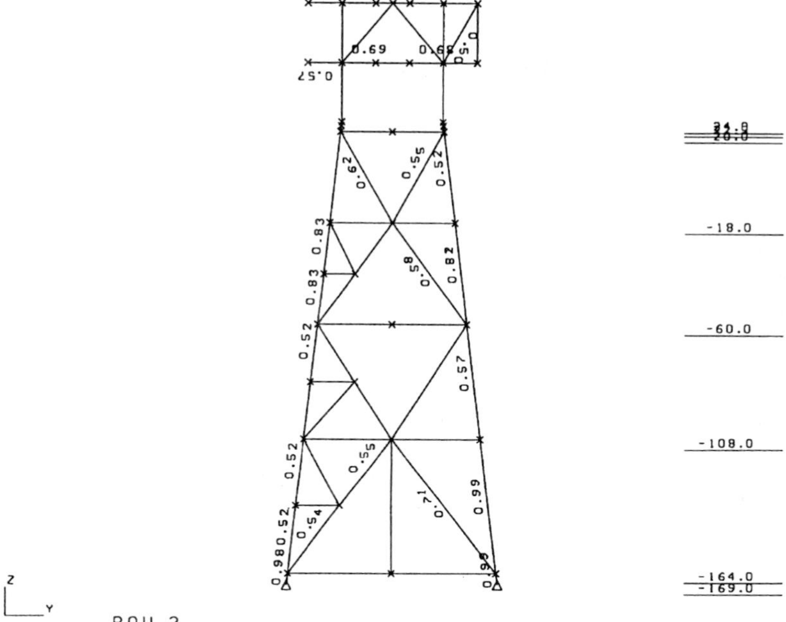

Figure 6.41 Maximum interaction/utilization ratios of a jacket structure from a typical offshore platform analysis software program

that are not critical under static conditions may found to be critical under other design conditions such as fatigue, earthquake, installation, impact, accidental, etc. In cases where this over utilisation is a local condition such that any change in member size will not have a major effect on the overall stiffness and stress distribution of the structure, critical members can be re-sized without any need for the re-analysis of the structure. In most cases, changing the size of a major structural member will have a significant effect on the overall stress distribution and the entire structure may have to be re-analysed to assess the true effect of member resizing. A structure may have to be re-analysed several times before arriving at its final geometry and member sizes.

Table 6.2 shows the weight breakdown for a typical shallow water depth (depth less than 300 ft) Gulf of Mexico jacket structure after all analysis and design work is completed. As expected, a majority of the weight is almost equally divided between the jacket legs and the braces (36% each). Piles that are about 20% of the jacket weight are the easiest to fabricate and the lowest cost per ton elements of a jacket. Jacket legs are easier to fabricate than braces and are expected to cost about twice as much as the piles per ton. Braces are more expensive to fabricate than the jacket legs and require additional end countering, fitting and welding. Other framing and appurtenances are about 8% of the weight. These are, however, the most costly components and require extensive cutting, fitting and welding. These components, while not major weight items, may become major cost escalating elements and their design requires attention.

Table 6.2 Relative weights and costs of jacket components

Jacket Component	Percent of Total Steel Weight	Cost Factor	Comment
Jacket legs	36%	2.0	Long, thick wall pipe with thick cans seam and girth welded
Jacket braces	36%	3.0	Long thinner wall girth and seam welded pipe segments with significant end profiling and welding
Vertical diagonals	18%		
Horizontals	12%		
Horizontal diagonals	6%		
Appurtenances	8%	4.0	Short seamless standard pipe with significant cutting, fitting and assembly work.
Boat landings and barge bumpers	2.5%		
Walkways	1.0%		
Anodes	1.5%		
Other misc. steel	3%		
Piles	20%	1.0	Long thick wall girth and seam welded pipe

6.3 Special Topics

6.3.1 Tubular Connections

6.3.1.1 Tubular Joints

Tubular braces connect to jacket legs and to each other at the "joints" forming a three-dimensional jacket truss structure. The external forces acting on the member spans are transferred to their end support points (joints) from where they are distributed to the other connecting members and/or to the global jacket structure. Deck and jacket geometry and load modelling and simulation, the methods for developing and solving the resulting load deflection matrix equations and calculation of the member end forces acting on each joint have been summarized in Section 6.2.4. Joints must be designed to safely transfer the member end forces, while also providing adequate strength and ductility to resist fracture and fatigue when subjected to extreme and or repeated loads caused by the ocean environment during the structure's design life.

Through a mix of model test programs, analytical simulations and field-gained experience, offshore industry developed a set of guidelines for the geometric requirements, strength and fatigue design of the hollow tubular joints. The earlier tubular joint design methods were based on rational engineering mechanics formulations derived from strength of materials and shell design principles [Marshall, et al. 1974]. As the model test results and field experience became available, these methods were upgraded to fill in gaps and match the observed test results [Boone, et al 1983; Yura, et al 1989]. Due to size, geometry and test hardware limitations, not all-joint parameters were fully represented in the tubular joint model tests. As a result, simpler, easier-to-test joint geometries, such as the non-overlapping T and K joints were first to be tested and formulated. As sophisticated non-linear structural analysis computer software programs became available, these software were calibrated against the available model test results [Pecknold 2000, 2001, 2003, 2005]. This enabled accurate formulation of the chord stress effects and more complex joint geometries, such as the overlapping and stiffened tubular joints and stress concentration factors for fatigue design [Eftymiu, 1988]. Currently, there are several

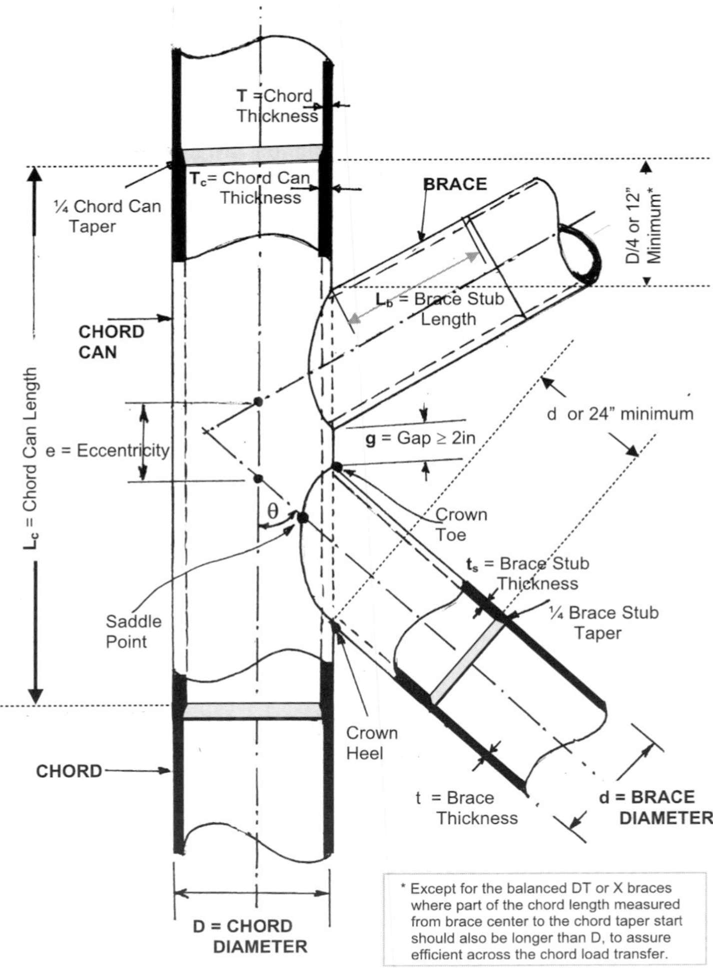

Figure 6.42 Geometric characteristics of a tubular joint

tubular joint design methods recommended by a number of design codes in use. The most prominent of these are the current API RP 2a method (1993, 2000), the method under development by the International Standards Organisation (ISO/CD 19902) and the results from the current analytical research collaborated by the API, EWI and the University of Illinois in Urbana [Pecknold, et al 2000, 2001]. The University of Illinois method is in the process of being adopted by the API as their RP2A 22nd Edition, under development [Karsan, et al 2005].

Geometric Requirements

Figure 6.42 shows a typical tubular joint geometry. Through years of industry experience and analytical and experimental work, a number of tubular joint geometry requirements have evolved. These are listed below.

(a) If the thickness of a chord (fig. 6.42) is found not adequate for carrying the brace loads imposed on it, a joint "Can" with an adequate thickness of T_c may be inserted. If an increased wall thickness can is utilised, its length should extend past the outside edge of the bracing a minimum distance of one quarter the chord diameter or 12 in. (excluding the chord taper) whichever is greater. A joint can taper of 1 to 4 is in common use and recommended. If a pile will be installed inside the jacket leg with joint cans, increasing the can diameter to accommodate increased wall thickness while keeping its internal diameter same as the jacket leg should be given consideration.

(b) Where an increased wall thickness or specially manufactured steel "Stub" is used for brace end, it should extent past the brace crown heel a minimum of one brace diameter or 24 in. (including the Stub taper). A brace Stub taper of 1 to 4 is in common use and recommended.

(c) Clear distance (gap) between the crown toes of two braces forming a "K-braced" joint configuration should not be less than 2 in. for non-overlapping braces. The method for determining the gap (g) depends on the joint loading, as well as the geometry. The method for determining gap is described in following paragraphs (see fig. 6.43).

(d) The joint offset (eccentricity = e) may be as much as one-fourth the chord diameter (D/4). Moments caused by joint offsets that exceed the D/4 limit should

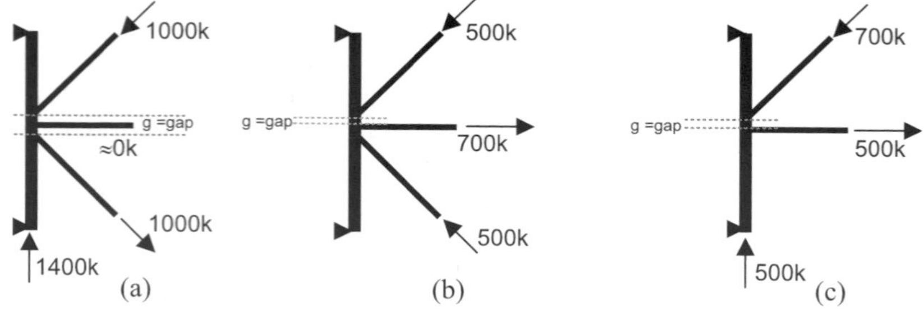

Figure 6.43 Examples to balanced K joint loading (all angles ≈ 45°)

be assessed through structural analysis by coding the brace-chord intersection points as joints in the platform geometry simulation. This offset may be an important design parameter for the K and X joints where all members are of similar outside diameter and should be assessed even if the D/4 requirement is satisfied.

(e) If there should be a circumferential (girth) weld on a joint can, this should be located at a location where it will not be crossed by a brace to chord weld. If this weld crossing cannot be avoided, the girth weld should be located at the crown heel of the lightest (or the least loaded) brace location (such as the crown heel of a small horizontal brace) (see fig. 6.44).

(f) The longitudinal seam weld on a chord should be offset a minimum of 12 in. from the point of intersection of any brace, measured along the brace surface.

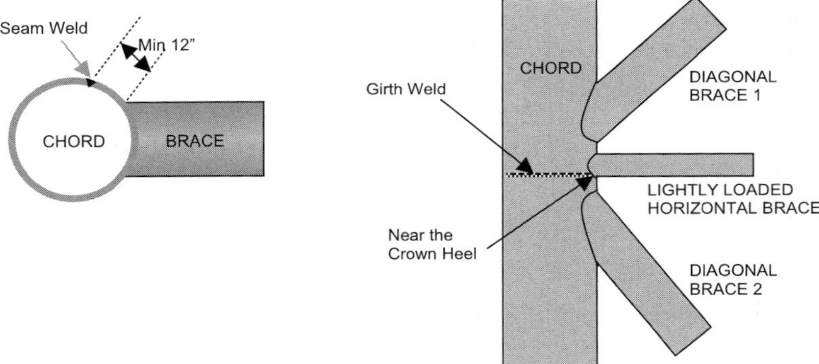

Figure 6.44 Chord seam and girth weld locations

(g) A tangential intersection of brace footprint and can seam should always be avoided, as this sets up fatigue cracks to grow with a substantial part of their length residing in a local brittle zone of the seam weld.

(h) Adequate through thickness and ductility characteristics are required of all joint chord (or can) material. The through thickness and ductile characteristics, free of lamination and inclusions are important for protection against lamellar peeling of the chord steel due to high strains imposed while cooling after welding or under extreme environmental loading conditions. Ductile chord material characteristics are also essential for the absorption of high joint deformations if a structure is subjected to overload caused by an accident or earthquake. Minimum material characteristics for steel tubular pipes recommended for use in offshore platform joints are described in API RP 2A, Section 8.4 [API RP2A, 1993, 2000].

Tubular Joint Classification

Tubular joints are classified into Balanced K, T/Y and Cross Double T/X based on their axial load transfer mode within a plane formed by the brace and the chord tubulars.

Balanced K Joints

In this class of joint, all brace axial loads are balanced such that no residual shear force is transferred to the chord member end points (see fig. 6.43). Load is transferred from one brace to the other(s) along the gap and through the chord wall.

The gap (g) of a balanced K joint is measured as the clear distance between the toes of the braces that balance the brace loads. For example, the gap brace configuration shown in fig. 6.43(a) is measured from the toe of the top diagonal brace to the toe of the bottom diagonal brace, disregarding the horizontal brace, which does not participate in load balancing. In fig. 2.3.3 (b) and (c) the horizontal brace contributes to load balancing and the gap is measured from the toe of the top diagonal brace to the toe of the horizontal brace.

Unbalanced T or Y Joints

In T or Y class joints, brace axial load is transferred directly into the chord member as shear and axial loads. The chord transfers these loads to its end points (See fig. 6.45). The T/Y joint chord behaves like a beam-column subjected to axial, shear and bending loads.

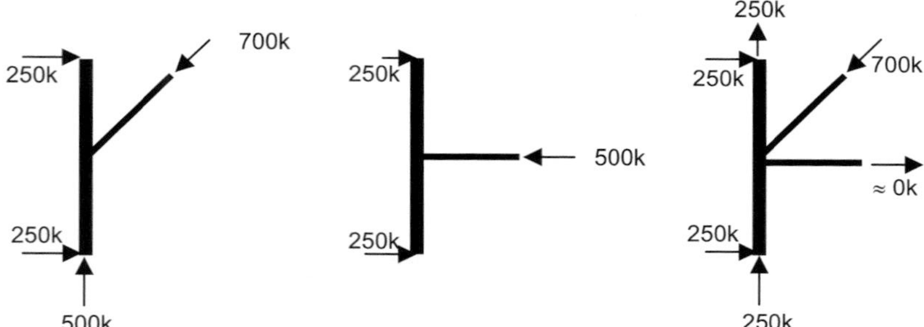

Figure 6.45 Examples to unbalanced T or Y joint loading

Balanced Double T (DT) or Cross X Joints

In this class of Joint, the brace axial loads are balanced by brace loads of equal and opposite magnitude located in the opposite side of the chord. No residual shear or axial force is transferred into the chord member end points (see fig. 6.46). Shear is transferred from one brace to the other(s) across the chord circumference.

Static Strength of Simple (Non-Overlapping) Tubular Joints

Strength Parameters

The following non-dimensional geometric parameters are used in the design of tubular joints subjected to overload and/or fatigue loadings:

$\tau = t_s/T_c$ Brace stub thickness to chord can thickness ratio

$\beta = d/D$ Brace diameter to chord diameter ratio

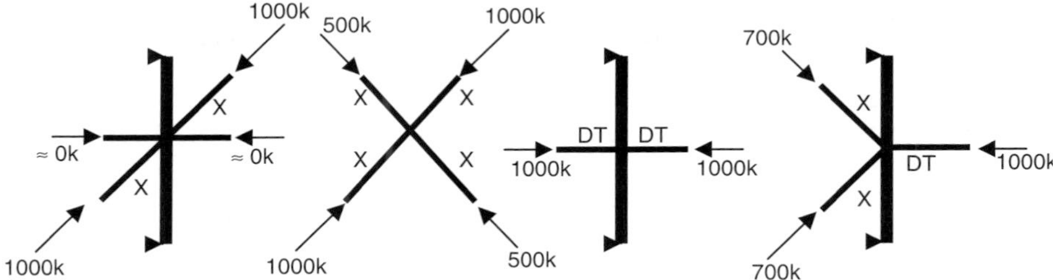

Figure 6.46 Examples to balanced double T or cross-X joint loading (all angles ≈ 45°)

$\gamma = D/2T_c$ Chord diameter to twice the chord can thickness ratio

$\xi = e/D$ Brace eccentricity ratio

Static Strength Formulation

The earlier tubular joint strength formulation was based on limiting the punching shear stress created on the chord wall by the brace load (API RP2A-WSD (2000), Editions 17 and earlier). As the joint behaviour was better understood, this formulation gave way to axial load and moment capacity equations, which are more representative of the tubular shell behaviour and correlate well with the test and non-linear finite element analysis results. The parametric formulation shown in fig. 6.47 is currently in use by all codes of design.

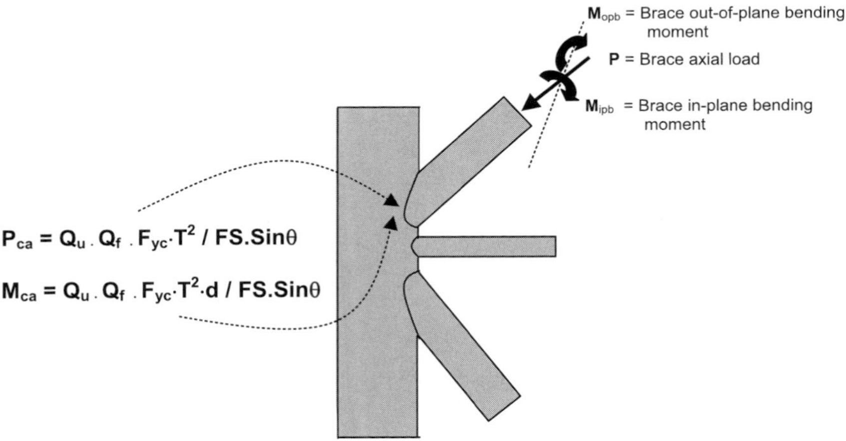

Figure 6.47 Tabular joint capacity equations

In fig 6.47,

F_{yc} = Yield strength of the chord member at the joint. The API RP-2A limits this strength to 80% of the yield strength for steels with yield strength less than 72 ksi.

P_{ca} = Allowable capacity of the chord subjected to the brace axial load P.

M_{ca} = Allowable moment capacity of the chord subjected to the brace bending moment M [Note that the brace bending moment could be in the brace-chord plane (in-plane-bending moment M_{ipb}) or outside the brace-chord plane (out-of-plane-bending moment M_{opb})]. Generally, moments less than 10° out off the brace-chord plane could be assumed to be in the same plane, as M_{ipb}.

Q_u = The ultimate strength factor which varies with the joint and the load type. Development of formulations for the Q_u factor has been the subject of extensive research [Boone, et al 1983; ISO/CD 19902; Marshall and Toprac, 1974; Pecknold, et al 2000, 2001; Yura, et al 1989]. The most recent advanced formulation developed by Prof. Pecknold et al of the University of Illinois in Urbana, Champlaign (UIUC) is described below.

Q_f = The chord load factor which depends on the load present in the chord and the joint geometry. Different Q_f formulations have been recommended by different design codes. The Pecknold et al developed formulation is described below.

θ = The angle between the brace and the chord centre lines (see fig. 6.42)

FS = Factor of Safety (1.6 for normal operating load cases 1.2 for extreme loading cases)

Ultimate Strength Factor Q_u

The recent advanced tubular joint design formulation by Prof. D. Pecknold et al (2000, 2001) is in the process of being adopted as an API Recommended Practice and ISO standard. The Q_u factor formulation by Pecknold et al is provided in table 6.3.

Chord Load Factor Q_f

Q_f is a factor that accounts for the effects from the presence of the axial load and the moments in the chord.

$$Q_f = [1 + C_1(FS \cdot P_c/P_{yc}) + C_2(FS \cdot M_{ipb}/M_p) - C_3 \cdot A^2]$$

where:

$$A = [(FS \cdot P_c/P_{yc})^2 + (FS \cdot M_c/M_p)^2]^{0.5}$$

and,

$FS = 1.6$ for normal operating loadings and $FS = 1.20$ for extreme loadings
P_c and M_c are the axial load and bending resultant in the chord (i.e. $M_c^2 = M_{ipb}^2 + M_{opb}^2$)
P_{yc} is the axial yield capacity of the chord ($P_{yc} = \text{Area}_c \cdot F_{yc}$),

M_p is the plastic moment capacity of the chord. Use the average value of the chord loads and bending moments on either side of the brace intersection. Chord axial load is positive in tension, chord in-plane bending moment is positive when it produces compression on the joint footprint. The chord thickness at the joint should be used in the above calculations.

C_1, C_2 and C_3 co-efficients depend on the joint and load type as given in table 6.4.

For further details see Karsan (2005) and Pecknold (2000, 2001, 2003, 2005).

Table 6.3 Ultimate strength factor Q_u values by Pecknold et al (2000, 2001)

<table>
<tr><th>Type of joint geometry</th><th>Axial tension*</th><th>Axial compresion</th><th>In-plane bending</th><th>Out of plane bending</th></tr>
<tr><td>K</td><td colspan="2">$(16+1.2\ \gamma)\ \beta^{1.2}\ Q_g$
but $\leq 40\ \beta^{1.2}\ Q_g$</td><td rowspan="3">$(5+0.7\gamma)\beta^{1.2}$</td><td rowspan="3">$2.5+(4.5+0.2\gamma)\beta^{2.6}$</td></tr>
<tr><td>T/Y</td><td>30β</td><td>$2.8+(20+0.8\gamma)\beta^{1.6}$
but $\leq 2.8+36\ \beta^{1.6}$</td></tr>
<tr><td>X (Cross)</td><td>23β for $\beta \leq 0.9$
$20.7+(\beta - 0.9)$
$(17\gamma - 220)$
for $\beta > 0.9$</td><td>$[2.8+(12+0.1\gamma)\ \beta]Q_\beta$</td></tr>
</table>

* For X joints under tension, the expression given for $\beta > 0.9$ applies for the co-axial braces. If braces are not co-axial ($\xi > 0.2$, where e to be taken as the distance between opposite brace centre lines), use 23 β for the full range of β.
where,

Q_g is a geometric factor defining the effect of the K joint gap g:

$Q_g = 1.0+0.2(1-2.8g/D)^2 \geq 1.00$ for $g/D \geq 0.05$

$Q_g = 0.13+0.65\phi\gamma^{0.5}$ for $g/D \leq -0.05$

where $\phi = t \cdot F_{yb}/(TF_{yc})$

Q_β is a geometric factor defining the effect of the brace to chord diameter ratio β:

$Q_\beta = 0.3/[\beta(1-0.833\beta)]$ for $\beta > 0.6$

$Q_\beta = 1.0$ for $\beta \leq 0.6$

F_{yb} and F_{yc} are the yield strengths of the brace and chord steels, respectively
The overlap Q_g should preferably not be less than 0.25/30. The Q_g values for the $0.05 < g/D < -0.05$ range may be found by linear interpolation between the limiting values of the above two expressions.

Table 6.4 C_1, C_2 and C_3 chord load effect coefficients

Joint type	C_1	C_2	C_3
K joints under brace axial loading	0.2	0.2	0.3
T/Y joints under brace axial loading	0.3	0	0.8
X joints under brace axial loading*			
$\beta \leq 0.9$	0.2	0	0.5
$\beta = 1.0$	−0.2	0	0.2
All joints under brace moment loading	0.2	0	0.4

*Linearly interpolate values between β=0.9 and β=1.0 for X joints under brace axial loading.

Load Transfer Across Chords

If a thicker chord "can" is specified for a joint with balanced DT or X braces, the chord "can" has to have adequate length to be considered as a uniform thickness cross-section transferring the balanced axial brace load from one face of the chord to the opposite face.

Analytical work shows that the chord length should be a minimum of 2.5 times the chord diameter (2.5D) with a minimum distance of 1.25 times the chord diameter (1.25D) measured from the brace centre at each side, to allow effective load transfer across the chord. If the joint can length does not meet this requirement, the chord's brace load transfer capacity has to be de-rated according to the chord length and could be as low as that is determined by the un-thickened chord length as follows:

$$P_{ca} = P_{cao} + (L_c/2.5D) \cdot (P_{cac} - P_{cao}) \leq P_{cac} \qquad \text{for } \beta \leq 0.9$$

$$P_{ca} = P_{cao} + [(4\beta - 3)L_c/1.5D) \cdot (P_{cac} - P_{cao}) \leq P_{cac} \quad \text{for } 0.9 < \beta \leq 1.0$$

where,

P_{cao} is the axial load capacity calculated for the nominal chord thickness
P_{cac} is the axial load capacity calculated for the chord with can thickness
L_c is the thickened chord length calculated as follows (refer to fig. 6.47 below):

For Brace 1 $L_c = 2b + d_1$ or $= b + d + f$, whichever is smaller
For Brace 2 $L_c = 2f + d_2/\sin\theta$ or $= b + d + f$, whichever is smaller
For Brace 3 $L_c = 2a + d_3$ or $= 2c + d_3$, whichever is smaller

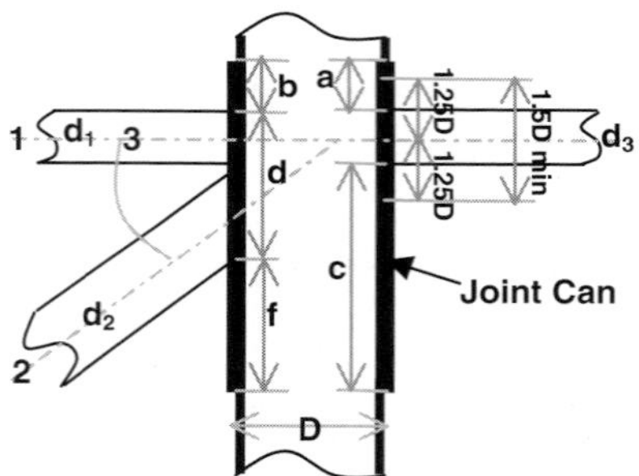

Figure 6.47 Chord length determination

Mixed Strength Cases

In real life joint loading cases, portions of a brace's axial load may fall under the balanced K, DT/X and the unbalanced T/Y brace classes. In such a case, the total joint capacity should be calculated as a combination of each loading class, proportionate to the load that falls in each loading class.

Based on the example given in fig. 6.48;

Brace 1 falls 100% (700 k/700 k) into DT brace class.
Brace 2 falls 33% (500 k/1500 k) into K, 33% (500 k/1500 k) into X and 33% (500 k/1500 k) Y brace classes.
Brace 3 falls 100% [(360 k + 140 k)/1500 k) into K brace class
Brace 4 falls 29% (200 k/700 k) into K and 71% (500 k/700 k) into X brace classes.

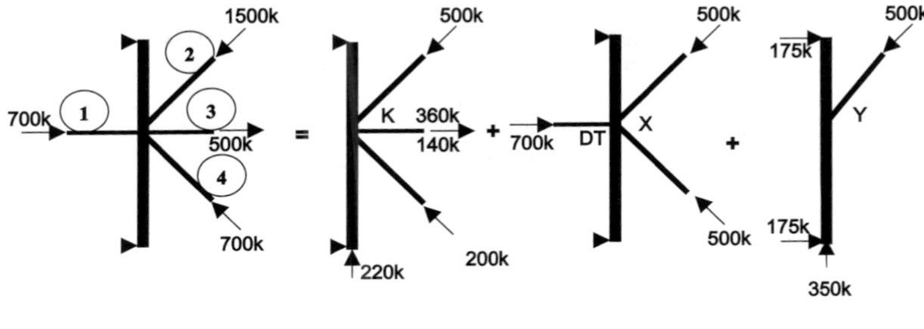

Figure 6.48 Example for decomposing brace load into its K, DT/X and T/Y loading components (all angles ≈ 45°)

Therefore, as an example, the axial and moment loading capacity of the Brace 1 should be calculated by adding 33% of each of its capacities calculated using the Q_u and Q_f coefficients for the K, X and Y joint classes.

Tubular Joint Strength Check

The strength check for a brace chord subjected to a combination of brace axial force and bending moments and the chord stresses should be checked using the following Interaction Equation [API RP2A-LRFD, (1993), WSD (2000)]:

$$\text{Brace Interaction Ratio} = \text{IR}_a = |P_a/(P_{ac})| + [M_{aipb}/(M_{caipb})]^2 + |M_{aopb}/(\cdot M_{caopb})| \leq 1.0$$

where P_a, M_{aipb}, M_{aopb} are the axial force and the in-plane and out-of-plane bending moments acting on brace a connecting to chord. P_{ac}, M_{caipb} and M_{caopb} are the axial load and the in-plane and out-of-plane bending moment capacities of the chord, calculated based on the joint classification class of brace a.

If the Load Resistance Factored Design (LRFD) Method is used, replace FS in all tubular joint formulations given in this section with unity.

Static Strength of the Overlapping Tubular Joints

The most common cause for joint overlap is excessive brace eccentricity ($\xi = e/D$) caused by large joint β ratios. In such cases, providing 2 in. minimum gap between the braces may result in a large brace eccentricity that may violate the geometric requirements for a simple joint formulation. In such cases the braces may be overlapped resulting in efficient load transfer between the overlapping braces (See fig. 6.49). In an overlapped joint, part of the overlapping brace's load is transferred to the through brace before it is transferred to the chord wall. The brace with the largest diameter should be made the through brace. If the diameters are same, the brace with the thickest wall should be made the through brace. While overlapped braces are efficient for static load transfer, because of their high stress concentration factors and need to re-weld over an already welded connection, they may result in fatigue problems.

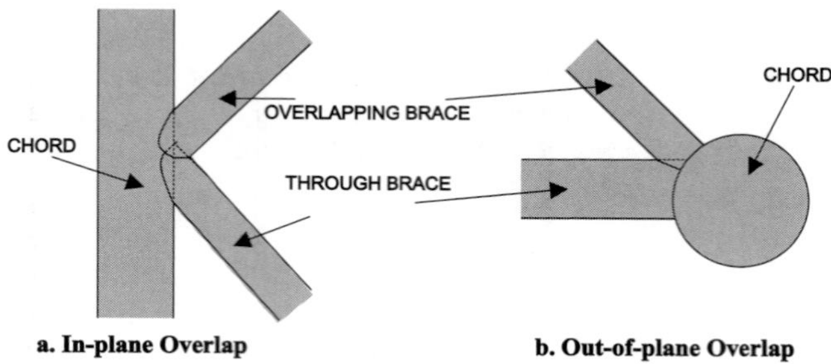

Figure 6.49 Overlapping tubular joint

Overlapped joints may be designed using the simple joint equations with the following exceptions and additions:

> Because of the reduced brace to chord contact area, shear transfer parallel to the chord surface may become critical and should be checked.
>
> If the axial force in the through and the overlapping braces act in the same direction (both towards or away from the chord), the chord to through brace connection strength should be checked for the through brace load plus a portion of the overlapping brace load. The portion of the overlapping brace load should be taken as the ratio of the overlapping brace area that bears on the through brace to its full area. If axial forces act in opposite directions, the chord to brace connection of each brace should be checked separately without any overlapping brace load deduction consideration.
>
> The combined sum of the in-plane or the out-of-plane moments acting on the overlapping and through braces should be used to check the through brace intersection capacity.

The overlapping brace to through brace connection strength should also be checked assuming the through brace as an imaginary continuous chord supporting the overlapping brace.

Other Joint Types

Please see ANSI/AWS D1.1-98, API RP2A-WSD (2000) and ISO/CD 19902 for methods and formulae for the design of other tubular joint configurations, such as grouted tubular joints, connections made of rectangular or square tubular cross sections, or tubulars to wide flange beam or closed square cross section connections.

Fatigue Strength of Simple Tubular Joints

Tubular joint fatigue failure may occur under cyclic environmental or other dynamic loading conditions. Fatigue is an important design consideration at locations where the dynamic loading environment is harsh and persistent (such as the North Sea) or where a dominant high load design event, such as a hurricane or typhoon that controls the joint design, does not exist (such as the Arabian Gulf or West of Africa). In the earlier offshore platform designs, steel strengths were low and the connections were generally riveted or bolted. These resulted in larger member cross-sections, highly redundant connections and lower cyclic stress. Since fatigue strength of steel is not strongly correlated with its yield strength, fatigue was a lesser problem for the members and connections in earlier structures. Fatigue became an important design consideration with the increasing use of higher strength steels and welded connections. Jack-up platform connections that use steel yield strengths as high as 100 ksi are especially vulnerable to fatigue.

Fatigue design of the welded tubular connections relies on input, analysis and material parameters with high levels of uncertainty. Because of these uncertainties, lower bound fatigue design curves, extensive quality control and testing during steel and pipe manufacture and welding, high safety factors and extensive in-service non-destructive inspection are used to assure their safety. Steel tubular joint fatigue design involves the following steps:

Prediction of the Fatigue Design Wave Parameters and their Likelihood of Occurrence

Fatigue design wave parameters are derived from a wave climate hindcast performed by a Meteorological and Oceanographic (MetOcean) specialist or from available records. This information is presented in the form of a wave scatter diagram that defines the wave energy spectrum physical parameters (significant wave height, mean zero crossing period, direction and spreading) and their annual likelihood of occurrence. Wave scatter diagrams are presented in a number of formats. Table 6.5 shows a two-dimensional wave scatter diagram that tabulates the significant wave heights, zero crossing periods and their annual probability of occurrence. For example, for the offshore location characterised in table 6.5, waves with heights 9–12 ft and zero crossing periods 6–8 s are expected to occur 1.45% of the time. A "two-dimensional scatter diagram" is useful for offshore locations where wave characteristics do not vary with direction. For locations where the wave scatter varies with direction, probability of wave height and period occurrence should be predicted for the entire 360° wave attack range and presented in a "directional scatter diagram". Use of a directional wave energy scatter diagram where the spreading and dissipation of wave energy around a wave attack direction could also be considered. Types of wave-scatter diagrams and their applicability are described in detail in the API RP 2A Commentary Section C5.2 [API RP2A-WSD, 2000]. Also refer to Chapters 2 and 3 of this Handbook for the physical wave parameters and spectra definitions and formulation.

Table 6.5 Two-dimensional scatter diagram for significant wave heights and zero crossing periods $[P\ (H_m,\ T_n)]$

H_m Significant Wave Height (ft)	T_n Zero Crossing Period (Seconds)								Total Probability
	0-2	2-4	4-6	6-8	8-10	10-12	12-14	14+	
0-3	0.114	0.225	32.551	16.810	0.105	0.001	0.000	0.000	49.806
3-6	0.000	0.002	7.951	30.110	2.575	0.003	0.000	0.000	40.641
6-9	0.000	0.001	4.255	3.250	0.006	0.006	0.000	0.000	7.518
9-12	0.000	0.000	0.080	1.450	0.150	0.003	0.001	0.000	1.684
12-15	0.000	0.000	0.000	0.006	0.125	0.075	0.015	0.000	0.221
15-18	0.000	0.000	0.000	0.000	0.050	0.010	0.010	0.000	0.070
18-21	0.000	0.000	0.000	0.000	0.010	0.045	0.000	0.000	0.055
21-24	0.000	0.000	0.000	0.000	0.000	0.005	0.000	0.000	0.005
24-27	0.000	0.000	0.000	0.000	0.000	0.000	0.000	0.000	0.000
27-30	0.000	0.000	0.000	0.000	0.000	0.000	0.000	0.000	0.000
30-40	0.000	0.000	0.000	0.000	0.000	0.000	0.000	0.000	0.000
40+	0.000	0.000	0.000	0.000	0.000	0.000	0.000	0.000	0.000
Total Probability	0.114	0.228	44.837	51.626	3.021	0.148	0.026	0.000	100.000

Calculation of the Nominal Cyclic Stress Ranges and Number of Cycles at the Brace Ends

Structural response should be assessed for the wave characteristics included in each probability of occurrence box of the scatter diagram. Generally, upper bound conservative wave parameter values are used. (In table 6.5, waves with heights 9 to 12 ft and zero crossing periods 6 to 8 s could be conservatively represented by a single wave spectrum with 12 ft significant height and 6s zero crossing period.) Background on types and development of wave spectra and their use in platform response analysis are described in detail in Chapter 2.

For fatigue design purposes, the wave height to nominal member end stresses relation is calculated through use of linearised wave height to brace end nominal stress range transfer functions. Methods for developing structural transfer functions (Response Amplitude Operators = RAO) and performing spectral analysis to result in response spectra are described in detail in Chapter 3. The RAO assumes that there exists a linear relationship between the wave height and the resulting wave forces and the nominal brace end stress ranges for each given wave period. Such a linear relation assumption generally holds true for structures made of tubulars with high diameter to wavelength ratios (Diffraction Ratio = $\pi D/L > 0.5$) where inertial loads dominate (In this regime, wave forces are generally proportional to the product of the structural mass and the wave particle acceleration). Wave forces on structures such as jackets, where the member diameter to wave height ratio is relatively small ($\pi D/L < 0.5$) drag forces that are proportional to the square of the wave particle velocities dominate. For such structures, the wave height to wave force relation is therefore not linear.

For drag force dominated structures, RAOs must be linearised in the region of each finite wave height corresponding to each spectral wave period. This linearisation is generally achieved by assigning a finite wave height to each spectral wave period defined by a constant wave steepness that is appropriate for the region. For the Gulf of Mexico, a wave steepness ratio $[H/(gT^2)$ [where H is the significant wave height and T is the spectral period] of 1/20 to 1/25 is used. H should not be allowed to be less than one (1) foot nor should it be allowed to exceed the 100-yr design wave or breaking wave height. The structure is then analysed subjected to a deterministic wave defined by the combination of this period and the finite wave height. When this analysis is performed, the mean sea level (no tide or storm surge) should be used. There is generally no need to consider adding the current velocity to the wave particle velocities with the exception of the higher wave heights where these may enhance the cyclic wave loads. In this analysis, the deterministic wave crest should be raised to a height described by the significant wave height of the spectrum under consideration to capture the finite wave height effects action on members in the wave action zone. The drag and inertia coefficients recommended in Section 6.2.4.4 are average values for global design wave force generation. The actual wave force experienced by an individual member depends on its dimensions, location and wave parameters (type, height, period). The drag and inertia coefficients used in fatigue wave load generation should be calculated based on the individual member dimensions and parameters of each fatigue wave under consideration (Reynolds, KC, Froude numbers). For details, see Chapter 3 and API RP2A-WSD (2000) Section 5.2.2. For deeper water jackets and Compliant Tower Platforms (CPTs), where dynamic amplification could be significant and wave particle and structural velocities and accelerations may add to or subtract from each other, a time domain analysis should be performed. In such dynamic analyses, the choice of structural damping coefficient is an important design parameter. API RP2A-WSD (2000) recommends a damping coefficient value of 2% of critical damping or less for spectral analysis purposes. Lower damping values should be used for lower wave heights and higher structural modes of vibration where the internal structural energy dissipation would be low. The calculated nominal brace end stress range is divided into the finite wave height to obtain the brace end stress range transfer function. For additional details of this analysis see API RP2A-WSD (2000) Commentary C5.2.2. The nominal brace end stress range response spectra is calculated from (for details see Chapter 3):

$$S_{xx}(\omega) = |H_{xw}(\omega)|^2 S_{ww}(\omega)$$

where,

ω = Spectral wave frequency = $2\pi/\mathrm{T}$
$S_{ww}\ (\omega)$ = Wave power spectra density function value at wave frequency ω
$H_{xw}(\omega)$ = Wave height to nominal member end stress range transfer function (RAO) at wave frequency ω
$S_{xx}(\omega)$ = Nominal member end stress range power spectra density function value at wave frequency ω

Once the $S_{xx}(\omega)$ power spectrum function is calculated, the standard deviation σ_{xx} of the brace end stress range is calculable from:

$$\sigma_{xx}^2 = \int_0^{\infty} S_{xx}(\omega)\,\mathrm{d}\omega$$

In this case, since the stress range distribution is a zero mean process, its standard deviation σ_{xx} is equal to the square root of the root mean squared value stress range caused by the particular wave spectrum. The probabilistic distribution of the stress ranges within an $S_{xx}(\omega)$ power spectrum of brace end nominal stress range is generally assumed to be a Raleigh Distribution. Modelling the statistics of stress range with a Raleigh Distribution is very convenient, because the entire distribution can be described once σ_{xx} is known:

$$P(X) = [X/\sigma_{xx}^2]e^{-(x^2/2)\sigma_{xx}^2}$$

where

X = The nominal stress range at the brace end
$P(X)$ = Raleigh probability distribution of the stress range X
e = Base of Natural Logarithm = 2.7183

Once $P(X)$ is generated, the cumulative probability of occurrence of various nominal brace end stress ranges could be obtained through:

$$P(X_i, X_j) = \int_{X_i}^{X_j} P(X)\mathrm{d}X$$

where $P\ (X_i,\ X_j)$ represents the probability of the nominal brace end stress range being between the stress range values of X_i and X_j. For further details, please refer to Chapter 4 Statistical Design Section of this Handbook. The lifetime number of stress cycles can be calculated from:

$$n[(X_i + X_j)/2] = DL \sum_{m=1}^{M} \cdot \sum_{n=1}^{N} P(X_i, X_j) \cdot P(H_m, T_n)(31.56 \cdot 10^7/T_n)$$

where:

$n\,[(X_i + X_j)/2]$ = Number of Stress Ranges between X_i, X_j expected during design life DL of the structure

DL = Design Life of the structure (yr)

$P(X_i, X_j)$ = Probability of occurrence of stress ranges between the values of X_i and X_j occurring given a wave spectrum with significant height H_m and zero crossing period T_n (see above Raleigh's distribution equation).

$P(H_m, T_n)$ = Annual probability of occurrence of a wave spectrum with significant height H_m and zero crossing period T_n from a scatter diagram (see table 6.5 as an example). For directional scatter, summation over the wave attack directions is also required.

$31.56{\cdot}10^7$ = Number of clock seconds in one year.

The statistical analysis of stress distributions are generally presented in a semi logarithmic diagram where the stress ranges are plotted against the logarithm (to the base 10) of the number of cycles to failure. Research observations suggest that these generally fit the format of a Weibull distribution (For further details, see API RP2A-WSD (2000) Commentary fig. C5.1–3).

Calculation of the Stress Concentration Factors (SCF) and the Hot-Spot Stresses Range (HSSR)

Welded tubular joints have high Stress Concentration Factors (SCF) and are the most fatigue sensitive components of an offshore tubular jacket platform. Nominal stress ranges (X) calculated for the brace ends must be increased by a SCF to account for the stress amplification caused by the tubular joint and weld geometry. The HSSR is calculated as a product of the SCF and the Nominal Brace Stress Range X.

$$\mathrm{HSSR} = \mathrm{SCF} \cdot X$$

The SCF may be calculated from finite element analyses, model tests or empirical equations derived from these. Various empirical SCF formulations have been developed by researchers with increased levels of sophistication and accuracy through the historical development of the fatigue analysis methods for tubular joints [Eftymiu, 1988; Gulati, et al 1982; Hart and Sablok, 1992; Kuang, et al 1977]. The HSSR versus number of cycles that lead to fatigue failure at this stress range level curves (S–N curves) are experimentally derived. The cyclic HSSR are measured at a point near the toe of the weld. Therefore, it is important that the stress ranges calculated using the analytically derived SCFs are also calculated relative to such points. Although there has been some variance in the strain gauge locations used in tests, API RP2A-WSD (2000) Commentary C5.3 documents that, for the database used in the API RP2A fatigue curves, "typical hot spot strain gauges were centered within 0.25 in. to $0.1(Rt)^{0.5}$ of the weld toes, with a gauge length of 0.125 in. and oriented perpendicular to the weld. Here R and t refer to outside radius and thickness of the member instrumented, either a chord or brace."

Due to a significant number of close form SCF formulations developed by researchers and many being still under development and some are being calculated using FEA, formulations of these are not given in this Handbook. Please see API RP2A-WSD (2000), Eftymiu (1988), Gulati, et al. (1982), and Hart and Sablok (1992) for more details and other references. For more detailed and accurate SCF assessment, formulations given in Eftymiu (1988) or detailed finite element analysis (FEA) are recommended. If FEA is used, plane curved shell element models, with welds simulated as solid bodies, should be used. Accuracy of the FEA results should be verified through use of successively smaller finite element dimensions until a solution convergence is assured.

Selection of the Applicable S–N Fatigue Design Curve

The number of HSSR cycles that lead to fatigue failure are experimentally derived and presented as the Stress range versus Number of cycles to failure curves (S–N Curves). Fatigue failure is generally characterised as a crack that goes through the total thickness of the connection component under consideration. Some may also define failure as the separation of a brace from its chord. The weld geometry (profile) and its application, including the material and the welding process used (weld chemistry, heat input, stress relieving in the weld and the heat affected zone – HAZ – through use of appropriate heating and cooling temperatures and rates); weld deposit surface geometry and surface treatment (buttering, hammer peaning and grinding) affect the fatigue life. Environment parameters (such as weld being above or under water, ambient sea-water temperature, chemical content such as H_2S or other corrosive agents in water, cathodic protection, surface, coating etc.) also have varying effects on the fatigue life. The weld geometry, weld technique and environmental effects are generally included when design S–N curves are specified. Basic S–N curves are presented as a log/log relation:

$$\mathrm{Log}_{10}(N_i) = \mathrm{Log}_{10}(k) + m\,\mathrm{Log}_{10}(X_i)$$

where N_i is the number of cycles to failure under stress range X_i (in units of MPa), k is a constant and m is the inverse slope of the S–N curve. The selection of the S–N curve parameters for welded connection design is a hotly debated research topic that attracts the intense attention of the fracture mechanics specialists as well as the welding and structural design experts. The API (2000), ISO (19902, Section 14), AWS (D1.1-98) and other research organisations such as the TWI (The Welding Institute) and EWI (Edison Welding Institute) have proposed numerous formulations. The shape of the S–N curve at high cycle-low stress range (Generally $N_i > 10^7$ cycles) is also a hot topic of discussion. Some codes agree in the existence of a fatigue endurance limit (stress range below which fatigue damage does not occur) provided that adequate protection against corrosion and mechanical wear (erosion) exists. Most European codes are more conservative and do not specify a fatigue endurance limit but agree that the fatigue damage accumulation rate will be significantly less in higher cycles by specifying a higher negative S–N curve slope for the low stress-high cycle region. Conducting fatigue tests at low stress-high cycle region is difficult due to time and test hardware limitations. Many test results depend on the way the load is applied (constant amplitude vs. variable amplitude), how the cycles are counted (Max to max, raindrop method, etc.) and the sequence of loading if variable amplitude tests are conducted. Table 6.6 shows the fatigue design curves recommended by the API

Table 6.6 API-OTJTG Recommended Basic S–N Curves for API RP 2A 22nd Edition

S–N Curve*	Log_{10} (k)	m
Welded Joints (WJ)		
$N < 10^7$ Cycles	12.48	3
$N \geq 10^7$ Cycles	16.13	5
Cast Joints (CJ)		
$N < 10^7$ Cycles	15.17	4
$N \geq 10^7$ Cycles	17.21	5

*For steels with yield strengths less than 500 mPa (72 ksi), S in MPa

Offshore Tubular Joints Task Group (API-OTJTG), which are in the process of being adopted as the recommended API practice for fatigue design of tubular offshore joints.

Notes for table 6.6:

i. Weld profile geometry are graphically defined in the following API figure (API RP2A-WSD, 2000; fig. C5.4-1). The Welded Joints (WJ) Curve applies to welds with no profile control (fig. C5.4-1 Case "b" below). Where profile control is practiced (Case "a" below) an enhanced curve should generally apply (see table 6.7).

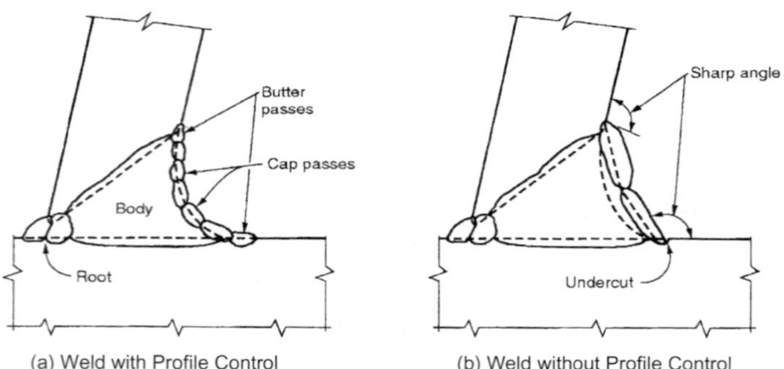

API fig. C5.4-1 Weld Profile Conditions [API RP2A-WSD, 2000]. (Reproduced courtesy of the American Petroleum Institute)

ii. The S–N curves given in table 6.6 are applicable to joints in air and submerged coated joints. For Welded Joints exposed to seawater with adequate cathodic protection, the $m = 3$ portion of the S–N curve should be reduced by a factor of 2.0 on life, with $m = 5$ portion remaining unchanged.

iii. The WJ curve given in table 6.6 is based on 16 mm or less chord wall thickness. A thickness correction should be made for higher than 16 mm chord thickness using the $X_i = X_o \cdot (16/t)^{-0.25}$ relation, where X_i is the reduced allowable chord stress range, X_o is the allowable chord stress range from the S–N curve and t is the chord thickness in mm for which the fatigue life is predicted. If the weld has profile control (fig. C5.4-1

Case a above), the exponent of the X_i equation may be taken as 0.2. If the weld toe is ground or peened, an exponent value of 0.15 should be assumed. No thickness effect correction is needed for the Brace S–N curve.

iv. The CJ curve given in table 6.6 is based on 38 mm or less chord thickness for the Cast Joints. For CJ with chords thicker than 38 mm, use the $X_i = X_o \cdot (38/t)^{-0.15}$ correction.

v. For welds with profile control as shown in fig. C5.4-1 Case "a", further fatigue life improvement could be achieved by a number of methods, including controlled burr grinding of the weld toe, or hammer peening. The recommended improvement factors are given in table 6.7.

Table 6.7 Factors on fatigue life for weld improvement techniques

Weld improvement technique	Improvement factor on X_i	Improvement factor on N_i
Profile per fig. C5.4-1a	$\tau^{-0.1}$	varies
Weld toe burr grind	1.25	2
Hammer peening	1.56	4

It is important to note here that, for large critical connections or one of a kind designs, or steels with higher than 500 mPa yield, full or adequate large size fatigue tests on representative samples are recommended. These tests should not be carried out for statistical sampling and new curve development purposes (which would require significant numbers of tests and statistical analysis) but for calibration and if necessary, adjustment of a selected S–N fatigue design curve.

Calculation of the Fatigue Damage Ratio (Miners Fatigue Damage Ratio)

Under normal offshore operating conditions, a tubular joint is subjected to variable stress ranges. The fatigue damage caused by different stress levels that may occur at different times of a structure's design life is customarily calculated using the Miners cumulative fatigue damage ratio.

$$D = \sum (n_i/N_i)$$

where,

D = Fatigue damage ratio representing the fatigue damage experienced by the structure during its design life DL

n_i = Number of cycles applied at stress range X_i

N_i = Total number of cycles that will cause fatigue failure at stress range X_i (from the S–N fatigue design curve)

Demonstration that the Required Safety Factors Against Fatigue is Available

If all data and analytical procedures used for fatigue analysis were deterministically accurate, a D fatigue damage ratio of unity would represent failure. In real life, because of the high level of data uncertainty and conservative data and procedures used in its calculation, it is observed that actual joint failures occur for D values in the range of 0.5–20 with a 95% level of confidence. If a tubular connection is located at above water, easily inspectable location of the structure and periodic code prescribed inspections are carried out and if the consequence of a failure is tolerable, a D ratio of unity should be considered acceptable. However, joints in below water, hard to inspect or non-inspectable locations with high consequence should be designed for lower $D_{\text{allowable}}$ values:

$$D_{\text{allowable}} = D/FS$$

where FS is a code prescribed factor of safety and $D_{\text{allowable}}$ is the allowable fatigue damage ratio. The API Offshore Tubular Joints Task Group (API-OTJTG) recommended fatigue design safety factors for the API RP 2A 22nd Edition are given in table 6.8.

Generally, a platform is considered critical when manned and not evacuated prior to Hurricane exposure (API Category L-1 structure). Most North Sea platforms are not evacuated and fall in this class. Manned evacuated or unmanned structures (API Category L-2 and L-3) are generally considered not critical. Most Gulf of Mexico platforms fall into this class. A safety factor of unity may be used for unmanned minimum consequence conventional jacket structures (API Category L-3) with redundant framing. Frequent inspection and a reliability based inspection plan may be used to justify a safety factor lower than those listed in table 6.8.

Norwegian Petroleum Directorate (NPD), Regulations Relating to Load bearing Structures [NPD, 1992] requires that the fatigue safety factors listed in table 6.9 are adhered to.

Table 6.8 Fatigue design safety factors recommended by the API OTJTG for API RP2A 22nd edition

Failure consequence	Inspectable	Not inspectable
Not critical (low)	2	5
Critical (high)	5	10

Note: In general, Gulf of Mexico platforms are unmanned or evacuated prior to Hurricane exposure.

Table 6.9 Fatigue design safety factors recommended by NPD (1992)

Damage consequence	Access for inspection and repair		
	No access or in the splash zone	Accessible	
		Below the splash zone	Above the splash zone or internal
With substantial consequences	10	3	2
Without substantial consequences	3	2	1

Note: Most North Sea platforms are manned and are not evacuated when exposed to extreme environmental conditions.

Simplified Fatigue Analysis

The Gulf of Mexico (GOM) normal offshore operating conditions are characterised by low wave heights with seldom-experienced extreme wave heights due to hurricanes. Fatigue is not a major design consideration in shallow water GOM (Shallow water depth is defined as less than 400 ft, where platform fundamental sway period is less than 3 s). In general, overloads caused by Hurricanes dimension the GOM platform joints.

API RP 2A 21st Edition (API RP2A-WSD, 2000; Section 5.1) recommends an alternative simplified fatigue analysis method for shallow water GOM platform design. In the API Simplified Fatigue Analysis Method, joint hot spot stresses calculated for a fatigue design wave (generally equal to the extreme wave height at site) are limited to Allowable Peak Hot Spot Stresses, which depend on the water depth, location (in waterline or not) and the intended fatigue life of the platform.

The API Allowable Peak Hot Spot Stress values have been derived for the joints characterised by the API recommended fatigue S–N curves. The method assumes the relation between the wave heights vs. number of waves exceeding this wave height in GOM as the superposition of two Weibull Distributions, one characterising the normal operating environment and the other representing the hurricane regime. An exponential relation between the wave height and the stress range is also assumed. The derivation of the API Allowable Peak Hot Spot Stresses is outlined in API RP 2A 21st Edition Commentary Section C.5.1 and described in detail in Geyer and Stahl (1986). Also refer to API RP 2A 21st Edition figs. C5.1-1 and C5.1-2 where the API Allowable Peak Hot Spot Stress values for GOM are specified. These values should not be used for offshore locations other than GOM but could be re-calibrated and extended to other locations in the world, if the wave height distribution for the site is known.

6.3.1.2 Pile to Jacket Connections

Welded Pile to Jacket Leg Connections

Safe transfer of the jacket loads to the foundation piles is an important design consideration. If piles are to be driven through the jacket legs, the axial load transfer could be achieved through use of a welded connection between the jacket leg top and the pile (fig. 6.50). The fillet welded shim plate connection detail (fig. 6.50a) is easier to fabricate but results in higher SCF and may not be desirable for dynamic loading conditions, where fatigue may be a problem. The crown block plate connection detail shown in fig. 6.50b is difficult to fabricate due to need for thicker plate and precision fitting and welding requirements but results in smoother load transfer and better fatigue behaviour. The thickness of the crown block plate should be selected to make up for the gap between the pile outside and the jacket leg inside diameters plus more to provide adequate load bearing area on the leg top. This typically results in 2 in. plus crown plate thickness. In both details, bending and axial load compatibility between the jacket and the piles could be established by grouting the pile inside the jacket leg. If grouted, pile area and bending stiffness will add to the jacket leg strength, resulting in thinner jacket legs.

Grouted Pile to Sleeve Connections

If the piles are driven through the skirt pile sleeves, the skirt pile to jacket connection could be achieved through grouting the piles inside the pile sleeves. In such connections, the jacket

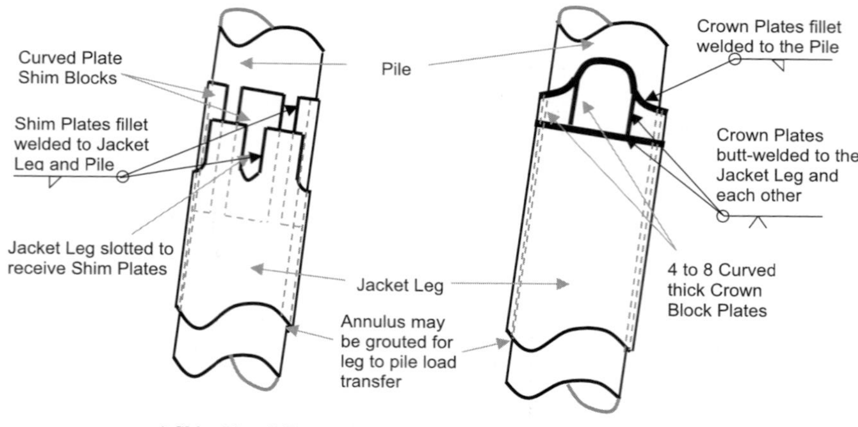

Figure 6.50 Pile to jacket leg connection details

load is transferred to the pile by the sleeve across the grout. Tests on the strength of plain pile to sleeve grouted connections with no shear keys demonstrate high scatter and uncertainty because of the inadequate confinement due to flexibility of the large pile and sleeve diameters and difficulty of fully displacing the water in the annulus with surface pumped grout. The low strength and higher safety factor requirements for plain grouted connections generally result in longer pile sleeves. This difficulty may be overcome by deploying shear keys welded on the inside wall of the sleeve and outside surface of the pile (fig. 6.51).

Tests on shear keyed connections show that the two separate mechanisms occur during the failure of a grouted tubular connection. First a slippage between the steel and grout, and second, a crushing of the grout against the shear keys [Karsan, et al 1984]. Such a connection normally fails in a ductile manner, so that the ultimate strength is taken as the sum of the two separate sources of strength. The statistical analysis of the test results yielded following mean strength equation for this combined strength:

$$f_{bu} = 167 + 1.72 f_{cu} \cdot (h/s)$$

where:

$f_{bu} = P_u/(\pi/D_p L) =$ Ultimate axial load transfer value of P at failure (psi)
$P =$ Total axial force transferred through the connection (psi)
$P_u =$ Ultimate value of P at failure (lbs.)
$D_p =$ Outside diameter of the pile (in)
$f_{cu} =$ Unconfined grout compressive strength (psi)
$h =$ Height of shear connector (in. See fig. 6.51b)
$s =$ Spacing of shear connectors (in. See fig. 6.51a)
$L =$ Length of grouted connection

The first 167 psi term in the above equation represents the strength of the plain connections with no shear keys. Statistical analysis of the plain strength suggests that a high safety

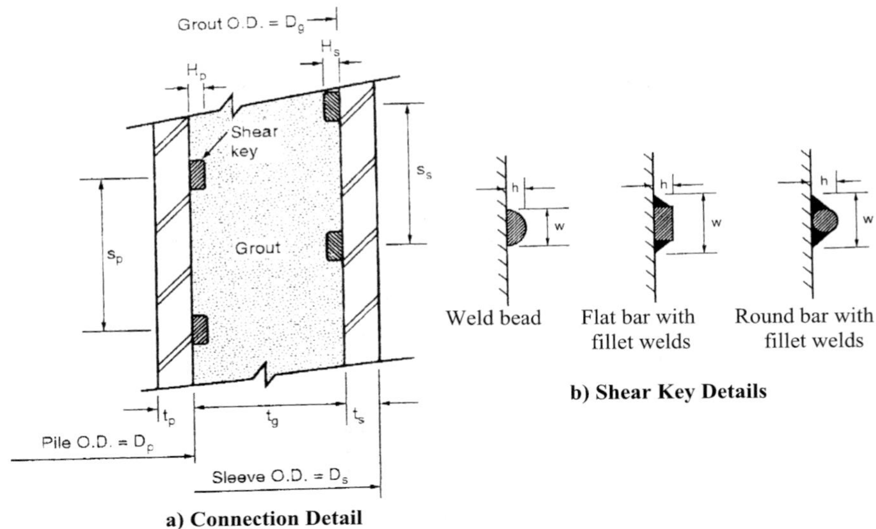

Figure 6.51 Shear keyed grouted pile to sleeve connection (Reproduced courtesy of the American Petroleum Institute)

factor near 8 is needed when h is equal to zero. Thus a 20-psi value defaulting in the old API recommended plain grout strength is recommended for this strength case. The 1.72 $f_{cu} \cdot (h/s)$ term in the equation represents the strength contribution from shear keys. This strength is found to be more reliable and a safety factor of 3.45 was found appropriate for this term, resulting in the API equation (7.4.4.1) [API RP2A-WSD, 2000]:

$$f_{ba} = 20 + 0.5 f_{cu} \cdot (h/s)$$

where f_{ba} (psi) is the API recommended axial load transfer capacity of the grouted shear keyed connection (psi)

If the API LFD Format [API RP2A-LRFD, 1993] is used, the above equation becomes:

$$f_{ba} = \Phi_{ba}[32 + 0.90 f_{cu} \cdot (h/s)]$$

where Φ_{ba} is the resistance factor recommended to be taken equal to 0.90.

The above equations are based on statistical curve fit to experimental data. The following limitations are recommended:

$2{,}500 \text{ psi} \leq f_{cu} \leq 16{,}000 \text{ psi}$
$D_s/t_s \leq 80 \leq D_s$ and t_s are the sleeve outside diameter and wall thickness)
$D_p/t_p \leq 40 \leq D_p$ and t_p are the pile outside diameter and wall thickness)
$7 \leq D_g/t_g \leq 45$ (D_g and t_g are the grouted annulus outside diameter and thickness)

Shear key spacing ratio, $2.5 \leq D_p/s \leq 8$ (the lower limit is only applicable to helical shear keys.)
Shear key ratio, $h/s \leq 0.10$
Shear key shape factor, $1.5 \leq w/h \leq 3$ (see fig. 6.51b)
$f_{cu} \cdot (h/s) \leq 800$ psi

Experimental validation of the connections outside the range of the above limitations is recommended. Test data indicates that the presence of moment and the transverse shear loads in the composite pile/sleeve member does not result in reduced axial load transfer. There are some indications that the presence of moment may even result in increased axial load transfer capacity. To date, no test results are reported in the literature indicating that the cyclic fatigue loads on shear keyed grouted pile to sleeve connections as a significant design issue.

6.3.2 Deep Water Jackets and Compliant Tower Platforms

In its installed configuration, a jacket platform can be visualised as a cantilevered beam fixed to the seabed through use of a multitude of large diameter piles (see fig. 6.11). The jacket responds to the dynamic ocean environment as a function of its geometry and mass that is distributed along its length with the deck payload lumped at its top. Time dependent forces are exerted on the system by water particle motions due to wave and current. These forces are proportional to the dimension of the platform elements located in the "wave action zone", which generally represents a depth of approximately 200-ft below the sea level. As the size and number of its wave load catching elements such as jacket legs, conductors, boat landings, barge bumpers, skirt pile guides, anodes, etc. increase, so does the wave and current generated drag and inertial forces.

In shallow water depths, the jacket structure exhibits a low first natural period of vibration (typically less than 4 s in less than 800-ft water depths). This short period is far removed from the mean period of an extreme design wave energy spectrum. The wave mean period generally lies in the 12–14 s range and no appreciable dynamic amplification of the wave loads would occur (fig. 6.52). Deeper water depths, larger number of conductors and higher platform mass result in increases in the platform period, which in turn result in increased dynamic wave loads.

Design of jackets for deepwater environment generally requires that the wave loads acting on the jacket are minimised and kept below reasonable levels and the technology and facilities are available for its construction and installation at the intended offshore site.

6.3.2.1 Deepwater Jackets

In-Place Design Considerations for Deepwater Jackets

Total design base shear V_B exerted by waves and current on a jacket structure could be approximated as a simple product of two effects:

$$V_B = V_S \cdot \text{DAF}$$

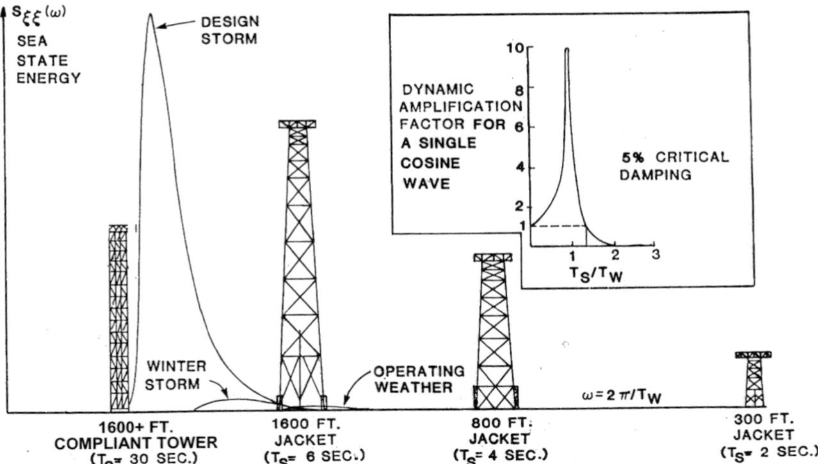

Figure 6.52 Wave energy spectrum vs. platform periods in the US Gulf of Mexico

Where V_S is the static base shear with no dynamic amplification and DAF is the Dynamic Amplification Factor. Minimisation of V_B not only requires a reduction in DAF but could also be achieved through a reduction in V_S by selecting appropriate structural sizes and configurations in the wave action zone.

Reducing the Static Base Design Shear V_S

The drag forces (see Chapter 2, Section 2.4.2) dominate maximum design wave pressures acting on slender jacket members. For combined storm wave and current loadings, the squared relation between the wave particle velocity and the wave drag pressure results in increased hydrodynamic forces with increased current velocity. For most cases, combination of the maximum expected current velocity with the maximum expected design wave particle velocity is not warranted. Accurate estimation of the current velocity and profile and the design wave height, period and direction associated with the maximum environmental design event return period could result in significant static base shear reduction.

Noteworthy reductions in the static design shear could also be achieved through the selection of smaller diameter leg and bracing members at the wave action zone. Reducing the sizes and number of the wave load catching components such as the boat landings and barge bumpers, skirt pile guides and similar appurtenances and, in certain cases, doing away with them altogether should be given serious consideration.

The number and size of conductors, risers and other utility piping in the wave action zone have a major effect on the total static base shear. For example, presence of forty 26-in OD conductors would more than double the shear load calculated for the case of a standard eight-legged Gulf of Mexico platform with 80 in. OD legs (fig. 6.53). While the number of conductors is a function of the production rates and the designers generally have little

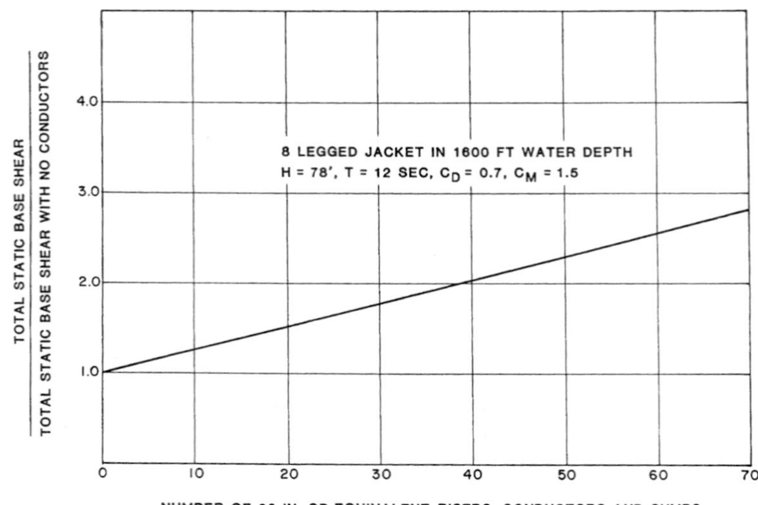

Figure 6.53 Increase in the total base shear with number of 26 in. OD risers, conductors and sumps [Karsan, 1986]

choice in the selection of their number, their impact on the static base shear should be appreciated by the production operators who set up the design requirements.

Static wave shear may also be reduced through use of the so-called "Wave Cancellation Effect" [Nath and Harleman, 1967]. When two vertical cylinders of equal diameter spaced at a horizontal distance D from each other are subjected to a simple Airy wave of length $L = 2D$ (See Chapter II Section 2.3.1, Linear Wave Theory), the wave particle velocity field acting on each cylinder will be in opposing directions (see fig. 6.54a). While each cylinder will be subjected to a pressure field equal to the maximum wave crest and through pressures, the sum of the forces acting on the two cylinders would be much less than twice the maximum wave crest shear force for a single vertical pile. The response amplitude operator (RAO) for such a configuration would have the cusped shape shown in fig. 6.54b.

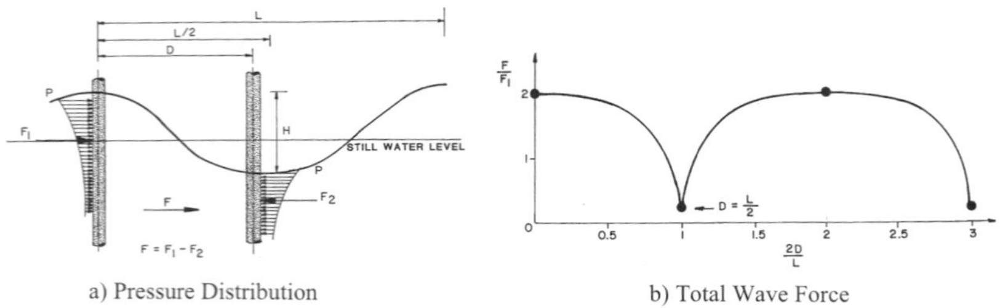

a) Pressure Distribution b) Total Wave Force

Figure 6.54 Wave force cancellation (Karsan, 1986; Nath and Harleman, 1967). (a) Pressure distribution; (b) Total wave force

Advantage of the wave cancellation effect could be realised in the deepwater jacket design if the drag generating elements (legs, conductors, boat landings, etc.) are positioned so as to filter out the forces from wave components with periods equal or near the first natural period of the platform. In this way while the DAF would be higher for these resonance periods, the V_S would be lower and their end product would result in manageable values of base shear V_B.

Wave force cancellation effect should be used in consideration of a number of uncertainties inherent to the calculation of the wave forces and the platform natural period. Field measurements performed on several installed offshore platforms indicate that the field measured fundamental structural period could be as much as one half second different than the calculated periods. The theoretical approximations made in the calculation of wave forces and spectral analysis also contribute to uncertainty. It would therefore be prudent to take advantage of the wave cancellation effect in a conservative manner. Model tests and use of conservative strategies such as shifting of the calculated period down, by as much as one-half second from the expected wave cancellation period and accounting for wave velocity shielding should be considered.

Reducing the Dynamic Amplification Factor (DAF)

Dynamic amplification of the wave forces occurs when the structural sway periods fall into spectral wave periods with high energy content. DAF is significant for structural periods within 6–20s range in the Gulf of Mexico. Efforts should be made to keep the structural sway period outside this range while also keeping the platform dimensions within reasonable and economic ranges.

Figure 6.55a shows a simplified deepwater jacket platform structural model that can be used to assess its dynamic response. The jacket structure can be visualised as a cantilevered beam with a cross sectional unit bending stiffness of $EI(z)$ and mass per unit length of $m(z)$ at a distance z from the seabed. Its large diameter, heavy walled legs act as the flanges and the diagonal bracing system behaves as the shear web of the beam (If piles are driven through and grouted inside the jacket legs, their area will add to the jacket leg area). The foundation can be simulated as a set of springs that simulate the rotational and horizontal stiffness of its piles driven into the soil. The rotational stiffness C_f of the foundation is of primary interest since this quantity multiplied by the distance L from the deck center of gravity (CG) to seabed is a significant contributor to the deck motions and the platform's fundamental mode of vibration. The pile axial stiffness determines the foundation rotational stiffness (see Section 6.2.4.2 Foundation Simulation). The dynamic effect of deck is simulated as a mass M_T concentrated at its centre of gravity (CG).

For a cantilevered beam of length L, with known first mode shape of $Y(z)$, the period of the fundamental mode of vibration can be estimated using the Raleigh's expression [Rogers, 1959]:

$$T = 2\pi\sqrt{\frac{\int_o^L Y^2(z)\cdot m(z)\cdot \mathrm{d}z + M_T\cdot Y_T^2}{\int_o^L Y''^2(z)\cdot EI(z)\cdot \mathrm{d}z + C_T\cdot Y_T'^2}}$$

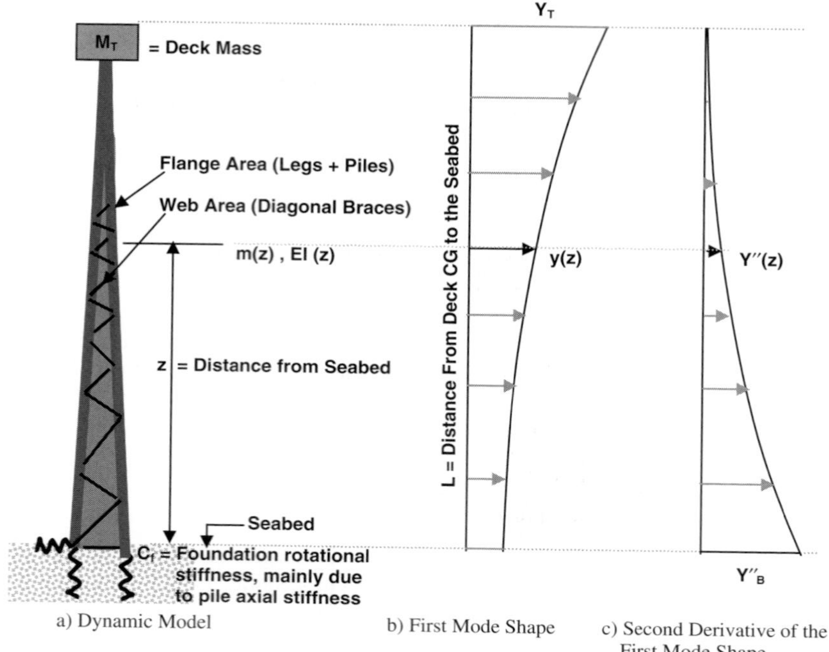

Figure 6.55 Dynamic model and first sway mode of an offshore platform

In the above expression $Y(z)$ represents the fundamental sway mode of vibration of the platform. $Y(z)$ can be derived from a direct dynamic analysis of the platform or could be approximated by its deflected shape when subjected to transverse loads proportional to its mass distribution. The above expression represents the ratio of platform's total mass distribution weighed by the square of its fundamental mode shape, divided by the jacket's total stiffness distribution weighed by the square of the second derivative of its fundamental mode shape. This observation suggests the following approach of reducing the platform's fundamental period of vibration:

Increase the "Stiffness Weighed by the Square of the Second Derivative of the First Mode of Vibration Term"

The shape of the second derivative of the fundamental mode of vibration shown in fig. 6.56c indicates that, increasing the bending stiffness $EI(z)$ at lower portions of the jacket, coupled with strong emphasis on increasing the foundation's rotational stiffness, would yield the highest payoff in decreasing the platform's fundamental period of vibration.

As a starting point, the brace areas should be selected such that the brace to jacket leg area ratios are equal or close to the $(A_3/A_1)_0$ limiting brace area ratio defined in Section 6.2.3.5 and fig. 6.27a.This will eliminate most shear deformations while resulting in optimum brace steel weight. Next, attention should be given to selecting the brace aspect ratios near the ideal value of $\sqrt{2}$ indicated in fig. 6.27b. This is specifically important for lower parts of the jacket structure.

Leg areas in the lower portions of the jacket could be increased either through increasing the leg diameter and/or wall thickness or through use of thick walled grouted in piles or pipe inserts. The welding difficulties and lamellar tearing problems inherent to thick leg cans may favour use of grouted insert piles or pipe inserts.

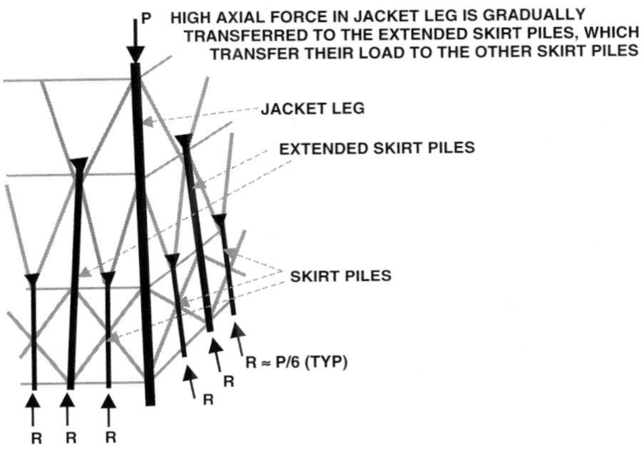

Figure 6.56 Extended skirt pile concept

As the water depth increases, the increased overturning moment would necessitate use of a large number of skirt piles. Consideration should be given to extending some of these skirt piles further up from the platform base. This approach would accommodate smooth axial load transfer from the jacket legs to the skirt piles while also increasing the bending stiffness of the lower parts of the jacket structure (see fig. 6.55). The base shear stays almost a constant while the water depth increases. This results in decreased shear per pile, which reduces the need for batter. As water depth increases, reducing the jacket leg batter should be given consideration.

Decrease the "Mass Weighed by the Square of the First Mode of Vibration Term"

The shape of the fundamental mode of vibration shown in fig. 6.55b indicates that, decreasing the unit mass $m(z)$ at the top portions of the jacket, coupled with emphasis on decreasing the deck mass M_T, would yield the highest payoff in decreasing the platform's fundamental period of vibration:

Reduce jacket top and deck leg diameters to a minimum compatible with the deck load and wave plus wind overturning moment imposed on them.

Consider reducing the number of deck and jacket legs, conductors, risers and other water surface piercing utility piping.

Use buoyant bracing and if possible, buoyant jacket legs, with emphasis on the near top locations (thereby reducing the added water mass).

Reduce deck weight without impairing the operational requirements. Use of two-way orthotrophic deck structures (see fig. 6.22) could be given consideration. Due to operating requirements, reducing the deck payload weight may not always be possible. However, realistic rather than conservative deck payload values should be considered when natural period calculations are made.

Fatigue Design Considerations

Fatigue is a major design concern for deepwater platforms that are subject to dynamically amplified stress ranges. The fatigue analysis procedures for deepwater platforms are same as those described in Section 6.3.1.1 – Fatigue Strength of Simple Tubular Joints. Fatigue behaviour of deepwater platforms can be improved by:

Development of accurate site specific database, wave parameters, spectral characteristics and associated wave scatter diagrams,
Generating accurate stress range RAOs by performing time domain analyses and using realistic structural damping values,
Using accurate SCF formulations or performing special finite element analyses for calculating the SCFs of critical joints,
Paying attention to the quality of the tubular joint welding. Using appropriate fatigue life improvement techniques, including weld profiling, buttering, grinding and peening justifying use of higher fatigue S–N curves,
Considering use of specially contoured cast steel joint nodes with low SCF values for fatigue sensitive platform joints.

Construction Considerations

Due to the fact that deepwater jackets would most likely be fabricated, transported to site, launched and installed as a single unit, construction operations would have a significant impact on their design feasibility.

Traditionally, Gulf of Mexico jackets have been fabricated using the well-known "bent-roll-up" method (fig. 6.57-1). In this method, each bent made of a vertical jacket truss plane containing two or more legs and interconnecting framing is fabricated laying flat on the fabrication yard surface. These bents are then rotated around one jacket leg and rolled up to a vertical configuration through use of numerous crawler cranes with large height and lift capacities. The rolled up bents are then connected with "in-fill" bracing using crawler cranes and scaffolding supported from the rolled up bents. In comparison to the nodal construction in air method popular in North Sea yards, the roll-up method eliminates the need for lifting and welding prefabricated joint nodes high up in the air and results in reduced welding and fitting man-hour costs.

Large base widths and bent weights for deepwater jackets would create serious brace handling problems in the rollup operations. Stresses generated during bent roll-up and assembly and load out to a launch barge could change some of the major member sizes and may require introduction of additional bracing. In these cases, use of temporary lifting

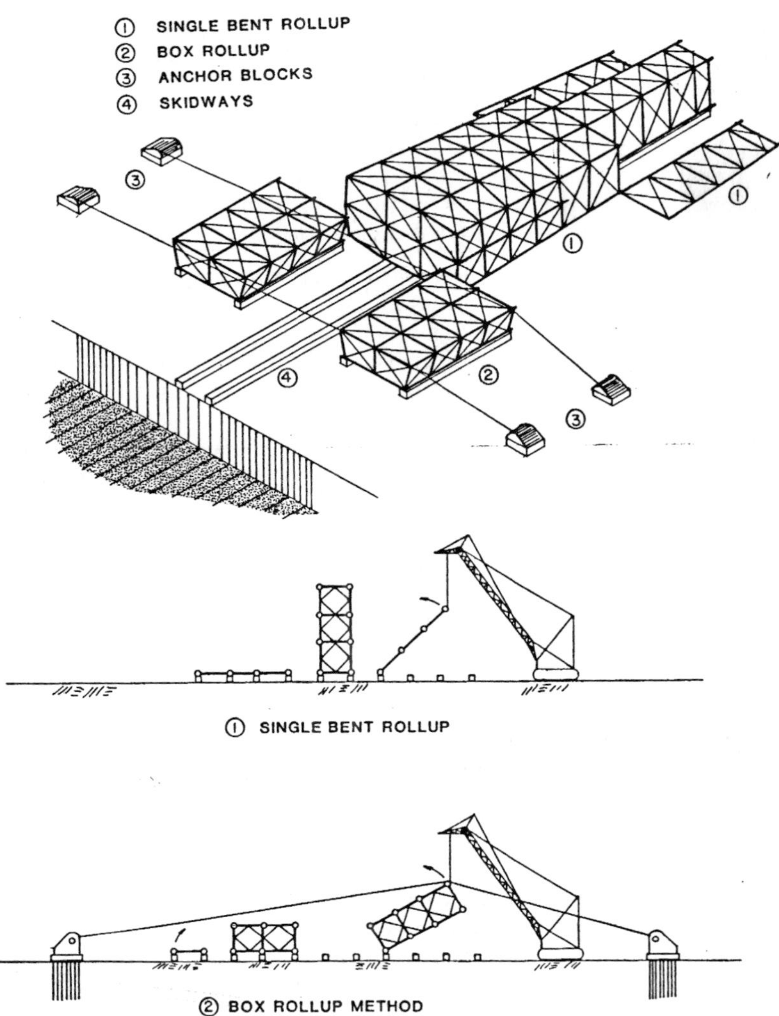

Figure 6.57 Bent and box roll-up construction techniques for deepwater jackets

"strong back" trusses or "box roll-up" techniques could be used (fig. 6.57-2). In the "box roll-up" method, first two adjacent bents are fabricated into a "box" through rolling up of its shorter and lighter bents. Following the installation of its in-fill bracing, the entire box is rolled up similar to a bent, using crawler cranes, assisted by linear jacks or heavy duty winches. (If parts of the structure will be used as support points for bent roll-up, bracing patterns may be effected.)

The size of a deepwater platform and the long construction period, which may take several years, presents challenging dimensional control problems. The differences between day and night temperatures, seasonal temperature variations, weld heat effects, construction sequence and lift deformations require active dimensional control, analysis and adjustments so that the platform could be successfully assembled within the dimensional tolerance requirements. Currently, there are several yards available in the Gulf of Mexico, with yard length, offloading key depth and access to open water that can undertake the construction of jackets for water depths as deep as 1,600 ft

Dynamic and gravity loads acting on the structure during transportation and launch operations would dictate the sizes of some jacket members in the cantilevering end support areas. Skid beam spacing would also be affected by the transportation phase. Transportation and installation of the jacket as a single piece offshore requires the availability of a large transportation and launch barge. Currently, there is only one barge available that is capable of transporting and launching deepwater jackets up to 1600 ft length and near 50,000 ton weight. This barge was used for the installation of Shell's Bullwinkle platform in 1353 ft water depth in the Gulf of Mexico in 1988 [Digre, et al 1989]. Shell's Bullwinkle platform is the deepest water depth jacket installed till date.

Providing skirt pile configurations to facilitate pile installation and designing of local areas to resist the forces generated during the pile stabbing and installation operations are also major design factors, which could impact the jacket design.

6.3.2.2 Compliant Tower Platforms

Preceding paragraphs provided strategies for extending the feasibility of the fixed offshore jacket/tower platforms to deeper water depths. As the water depths increase, fixed platforms get more flexible, resulting in longer fundamental natural vibration periods. Wave force components with periods near the platform natural periods are amplified, resulting in undesirable dynamic response. This problem is prevalent when the natural periods of the structure fall within the 6–20 s ranges, where the ocean waves contain significant energy (fig. 6.52). Compliant platforms with fundamental natural periods outside this period range are used in deeper water depths to resolve this problem. Compliant Tower Platforms (CPT) "move" with the waves, and therefore, the inertial forces generally resist the applied wave forces. This results in less force transmitted to the platform and its foundation.

Several means of providing platform compliance have been proposed in the past. Earlier versions, such as the Roseau (Reed) Tower [Tour Roseau, 1984] proposed to provide the platform compliance through use of a very slender "reed" like tower section, which flexes with the waves. In the Roseau Tower the platform is rigidly fixed to the seabed. While the earlier model tests and analytical work suggested the feasibility of the Roseau tower, it was not put to use due to potential fatigue of the tower joints under high cyclic loads.

In 1981, Abbot et al [Déserts, et al 1986] proposed a means of providing platform compliance by providing long ungrouted piles that run through the centrally located pile guides and tied to the platform at a location near its top. In this approach, the tower is relatively rigid and lowly stressed. The axial shortening and elongation of its piles (like large pogo sticks) provide a rotational hinge like behaviour near its base, while the tower stays almost straight. The pile founded flexible tower concept was first applied to the Exxon's Lena Guyed Tower Platform (GTP) installed in the Gulf of Mexico in 1983.

Figure 6.58 shows the forces acting on a pile founded Compliant Tower Platform (CPT). The tower stiffness and the dynamic inertial forces resist the external wave, wind and current forces. The foundation piles provide most tower stiffness restraint. Piles are welded to the tower near the top and slide through the tower legs or sleeves provided along the tower length. This provides a hinge (gimbals) action near the seabed, providing compliance under external loads. The inertial resistance is provided by the deck load and the tower mass (mass of steel plus water enclosed plus external water mass that moves with the tower members = hydrodynamic mass). In some designs, an Added Mass Stabiliser (AMS) may also be provided below the wave action zone (about 200 ft below the water level) to enhance the inertial resistance. The AMS may consist of a shroud that would force a large

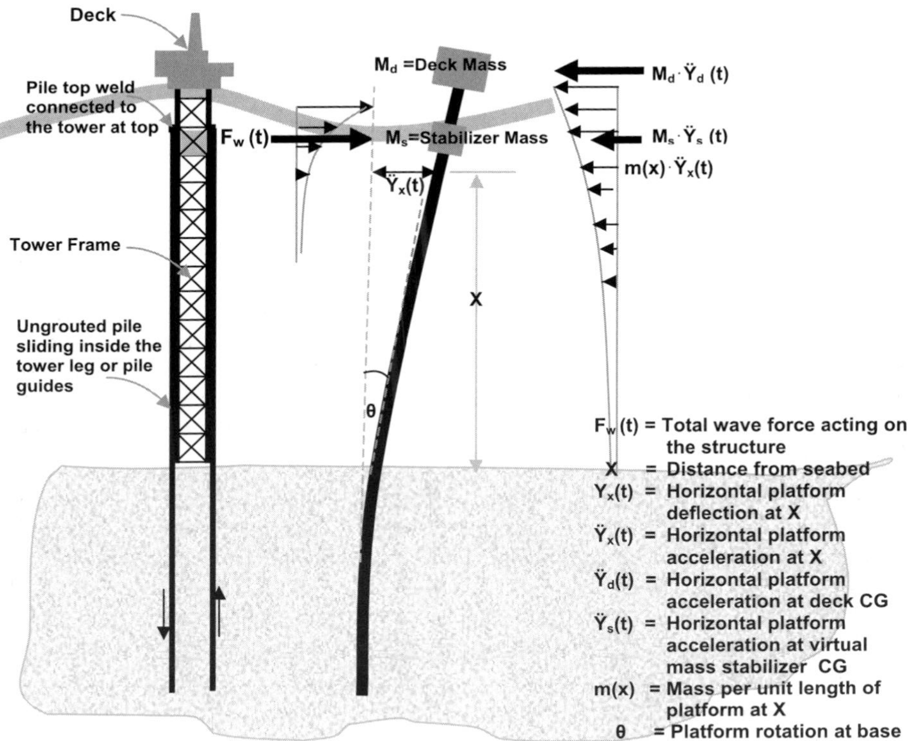

Figure 6.58 Dynamic response model for the pile founded compliant tower platform

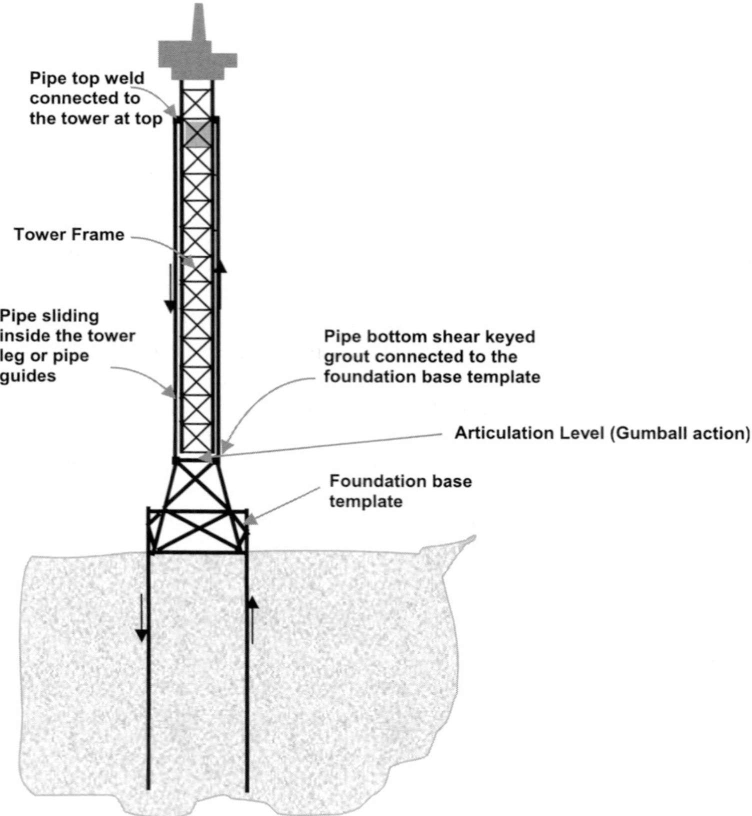

Figure 6.59 Articulated compliant tower platform (CTP)

mass of seawater to move with it (inside and around) generating additional inertial forces resisting the external forces.

In deeper water depths, long support pipes that are connected to the tower top and a base template that is piled to the seabed may replace the piles. The long support pipes behave in a manner similar to the piles, providing gimbals action at a conveniently selected height above the seabed (fig. 6.59).

As the tower depth increases, higher modes of vibration with shorter natural periods are excited, limiting the water depth applicability of CTPs. Provision of an articulation point higher up from the seabed would increase the water depth range of CTPs. Current water depth limitation for CTPs with near seabed articulation points is about 2500 ft. Taller foundation base templates or use of multiple articulation points may further increase the water depth ranges of CTPs [Karsan, et al 1988].

6.3.3 Structural Design of Jack Up Rig Platforms (by Vissa Rammohan, Vice President, Stress Offshore, Inc.)

6.3.3.1 Introduction

The current day design engineer of jack-up rigs encounters two distinct objectives. These may generally be classed as the upgrade of existing units and the design of new units. The bulk of the engineering effort on jack-up rigs today is expended on modifications and upgrade of the existing units. However, the development of new designs is also an important component of jack-up rig engineering. Hence, both of these aspects of jack-up design are addressed here.

Unlike jackets, the normal design (both upgrade and new design) of a jack-up rig is generally not tailored for a specific offshore location. By its very definition, a jack-up rig changes the geometry of its structure to suit a given location. Also, the environmental forces on the unit are a function of its location and the time of year. Hence, the design engineer is required to cope with a structure that is not fully defined subjected to forces whose maximum values and whose combinations are functions of its temporary location. The factors to be considered in the design are determined by the target capabilities of the unit. The parameters of these capacities would include the desired deck loads during operation, the total load during tow, the maximum water depth for operation, the corresponding combination of maximum wave height, current velocity and wind speed and design soil conditions. A given design would incorporate a range of these parameters. For example, a reduction in water depth would allow the unit to withstand greater wave heights or stronger current forces or higher wind speeds. The detailed specification of a jack-up would thus include a series of tables listing several useful permutations of the different parameters. Table 6.10 is a sample of one set of permutations.

There are also a number of cases where jack-up rigs are used as permanent production platforms at a specific site for long periods of time. In these cases, the geographic and environmental design parameters are selected similar to a fixed offshore platform. (See the preceding sections and Society of Naval Architects and Marine Engineers.)

6.3.3.2 Structural Elements of Jack-Up Platforms.

A simple structural approximation of a jack-up rig is one of a "3-D" portal frame. The legs constitute the "columns" of the frame and the hull forms the horizontal element. For the stability and structural integrity of such a structure, the connection between the hull and each leg should necessarily be capable of resisting large moments generated by the horizontal environmental load effects. This "joint fixity" or moment connection is achieved in different ways on different types of jack-up platforms.

The Hull is generally made of a stiffened plate "box" structure, often consisting of an upper deck, an intermediate (or equipment) deck and a bottom deck. Figure 6.60 illustrates a typical structural configuration for an upper hull. A flat plate (1) is "stiffened" with a series of closely spaced (usually 24–36 in.) bulb flats or angle sections (2). These sections span across "frames" (3) (beams) spaced between 6–9 ft Normally, these frames would be continuous across a section through the upper hull and form a closed structure around the box section. The frames span between bulkheads (4) that are strategically positioned at

Table 6.10 Typical table of environmental criteria for a jack-up rig

Storm conditions leg length 410 ft; elevated hull weight 16,200 kips				
Item	1	2	3	4
Water depth (ft)	100	150	200	250
Significant wave height (ft)	57	55	55	53
Wave period associated with the Significant wave height (s)	15	15	15	15
Wind speed (knots)	100	100	100	10
Current (uniform)(knots)	1.5	1.5	1.5	1.0
Penetration (ft)	25	25	25	25
Air gap (ft)	5	5	5	5
Pin point below mudline (ft)	10	10	10	10
Maximum unity check of members	1.00	1.00	1.00	1.00
Average pinion shear (kips)	857	860	847	875
Maximum leg reaction (kips)	11,517	11,569	11,461	11,637

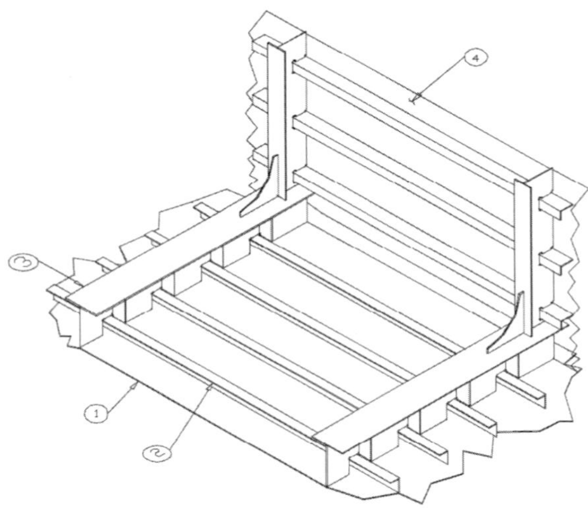

Figure 6.60 Typical structural configuration for a jack-up platform upper hull

locations of high loads and also at locations to optimise the span of frame girders. Figure 6.60 illustrates an example of bulkhead locations for a typical jack-up rig in plan.

The bulkheads provide a load path to the "legwells" in the hull. The legwells are framed by a series of bulkheads that provide a means to transfer forces to the leg structure. Various mechanisms such as "gear units", "guides", "rack chocks" and pins are utilised to transfer these forces. The forces thus transferred to the legs are then carried down to the "foundations" of the legs. These foundations can either be a rigid plated structure at the mudline commonly referred to as a mat or a spud can at the lower end of each leg, which distributes loads into the soil or seabed.

A significant structural component of a jack-up drilling rig that is not a part of its global structural system is the drill floor support structure. This structure is designed to enable the drill floor to be skidded longitudinally and transversely over a wellhead pattern on the seabed. Most current day jack-up rigs are equipped with deep cantilever beams that skid

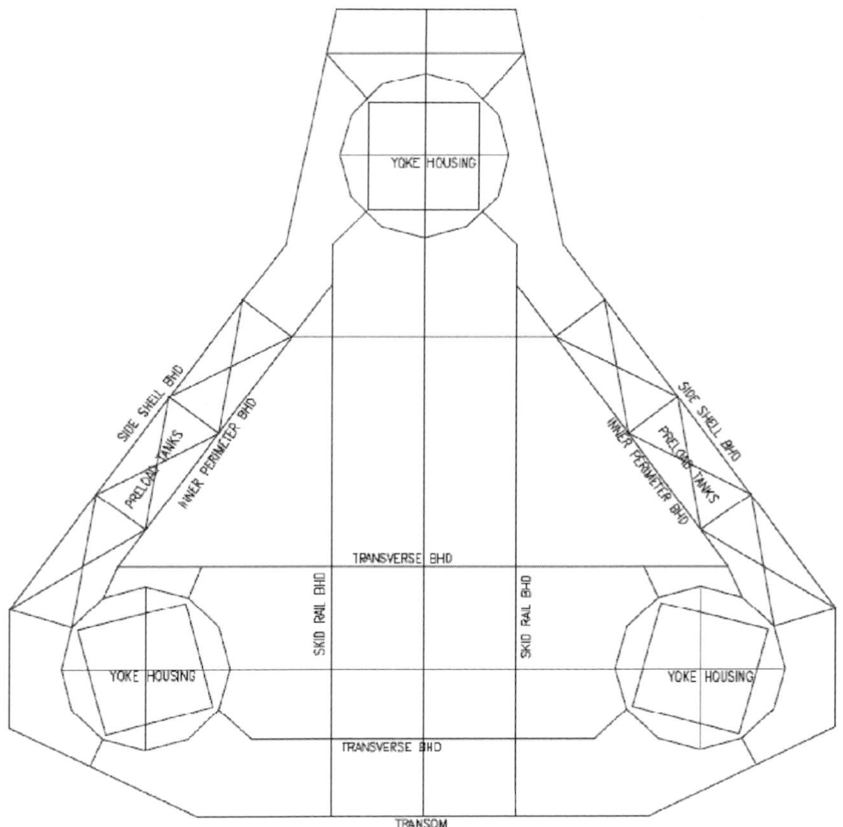

Figure 6.61 In-plan bulkhead locations for a typical jack-up rig

fore and aft over the transom of the vessel. The drill floor support structure is positioned atop these cantilever beams.

The first step in the jack-up rig design is the definition of its configuration. This is based on operational and economic requirements and past design experience. Decisions made at this stage have a significant impact on the behaviour of the structure. The geometry of the configuration developed should have the necessary capacity to accommodate needed equipment, preload tanks and quarters. Preliminary estimates of weights should be made and a naval architect should assess the configuration for the "afloat" mode of the jack-up rig. A configuration for the legs should be developed. The system for connecting the legs to the hull so as to achieve efficient moment transfer should be chosen. A classification society should also be chosen [American Bureau of Shipping, 2001; Det Norske Veritas-Rules for Classification of Mobile Offshore Units]. A preliminary assessment should then be made to ensure that the chosen configuration complies with the requirements of the chosen classification society. After this, the basic design can be developed. The efforts of the structural engineer are important from this stage on. Hull scantlings are the individual elements that makeup the structure. The individual items noted on fig. 6.1 are the scantling items such as bottom plating (1), stiffeners (2), side shell (3) and transverse girder (4). These items are usually designed based on predefined loads and the minimum requirements of the classification society. Jacking units should be chosen and their connection to the hull should be developed. Preliminary sizing of the earlier configured legs should be estimated. Other structural elements such as the spud cans, mat and gear unit brace beams are designed based on the loads that result after reaching the limits of the primary structures (Legs, gear units).

6.3.3.3 Global Structural Analysis of a Jack-Up Platform

Due to the numerous complexities associated with jack-ups, it should be remembered that a structural analysis would be based on a number of simplifying assumptions and approximations. Though software is available to execute a non-linear dynamic analysis, the designer may opt for a simple static analysis using wave forces generated from a hydrodynamic analysis applying a linear wave theory (Such as the Stoke's Fifth Order Potential Wave Theory) to a hydrodynamic model generated for this purpose. The following steps should serve as a general guideline for the analysis of a jack-up platform:

- Define the environment including water depth, wind speed, wave (type, height, period) and current velocity and its variation with depth. This can be a location specific environment (North Sea, Persian Gulf) or a world wide criteria (see table 6.10). The worldwide criterion is a reference benchmark that does not necessarily reflect any particular location. Some of the storm parameters (100 knot wind) are defined per code. The results of these environments are then used as reference for the actual unit location.
- With the exception of very heavy loads (such as cantilever, transom and hold-down reactions, heliport support members, etc.), this may be accomplished by summing all the equipment weight on a deck, a proportion of the variable load on that deck and dead load and distributing this load uniformly over the entire deck. This may be done for all decks. Loads from the drill floor may be applied as concentrated forces at

appropriate locations. Usually, the weight is assumed to be balanced equally among the three legs. This is normally achieved by moving the liquids among the various tanks to reach a balanced condition.

- Generate a hydrodynamic model of the jack-up platform. This may be a simple model consisting of three "stick" elements that have the same hydrodynamic properties as the trussed leg. The ideal source of the drag values of the unit would generally be determined via wind tunnel models. This takes into account the actual geometry of the unit and the effects of shielding. Usually the product of these studies is a single drag value for the legs and hull. The main problem with this source of parameters is cost and time. These tests should include the effect of biofouling on the drag coefficient. Alternately, a "building block" method using the different members of the leg with a suitable allowance for biofouling may be applied, this being a less accurate method of simulating the drag coefficient as compared to a wind tunnel test. Once the value of C_d on the legs is established, Morison's equation may be applied to determine hydrodynamic forces on them.
- Generate a Global Structural Model: Figure 6.62 illustrates a typical finite element analysis model of a jack-up platform structure. Figure 6.63 illustrates the length of leg that should be used in this model for a given water depth. For a jack-up platform whose legs have independent spud can foundations, the legs are usually assumed to be pinned at a depth of about 10 ft below the mudline. For a mat supported jack-up, the structure of the mat may be modelled using plate elements and the legs could be fixed to this structure. Per the ABS Rules [American Bureau of Shipping, 2001], the minimum crest clearance to be provided is 4 ft (1.2 m) above the crest of maximum wave or 10% of the combined height of the storm tide plus the astronomical tide and height of the maximum wave crest above the mean low water level, whichever is less between the underside of the unit in the elevated position and the crest of the design wave. The crest elevation is required to always be measured above the level of the combined astronomical and storm tides. It is most important that the wave NEVER be allowed to impact on the hull. The error in loading on the legs would be usually less than 10%. If the wave were to hit the hull, the design loads could increase by more than 500%, generally resulting in loss of the unit.

6.3.3.4 Simulation of the Major Structural Components

Legs

Current practice is to simulate each structural element of the leg in the global analysis model as a stick element. The section properties of each element should be accurately determined. The section of the rack should be included in determining the section properties of the chords. If the hydrodynamic loads were determined on the entire leg through wind tunnel tests, a "dummy" element with very low stiffness may be provided inside the leg and connected to all the chords with a load spider in each bay. Loads from the hydrodynamic analysis may be applied to this member and, through it, distributed into the leg structure.

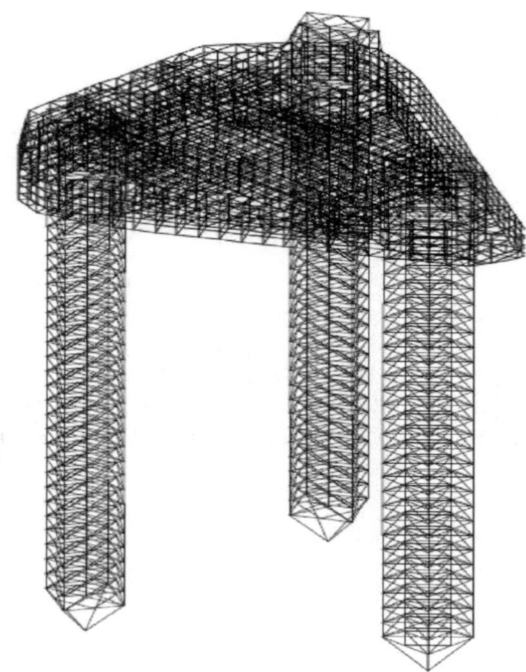

Figure 6.62 Typical jack-up platform structure finite element analysis model

An important aspect of the global analysis is the simulation of the connection between the leg and hull/jacking units. This is a function of the system chosen. Examples are: the single rack system, the opposed rack system, "rack-on-rack" system and the hydraulic jacking system. Figure 6.64 illustrates the simulation of a single rack system. The angle of the elements is equal to the pitch angle of the teeth on the pinion of the elevating gear unit. The stiffness of this element is best determined through tests on the gear unit or as provided by the manufacturer of the elevating unit. The same figure also illustrates the simulation of the top and bottom guides of the unit. The top guide is normally provided atop the "gear box" of the jacking unit and the bottom guide is at the level of the bottom of the hull. These guides (on each leg chord) together provide the necessary moment connection between the leg and the hull. The leg should be modelled in detail to at least one bay above the top guide. The portion above this may be simulated with a "dummy" member so that wind loads on the protruding portion of the leg may be applied.

Hull Structure

Plate elements are often used to simulate the hull. The important bulkheads that should be included are: side shell, inner perimeter, yoke housing, skid rail bulkheads, all structural bulkheads connecting to the yoke housing beneath the gear units and structural transverse bulkheads. It is not necessary to simulate small openings, such as doorways and hatches;

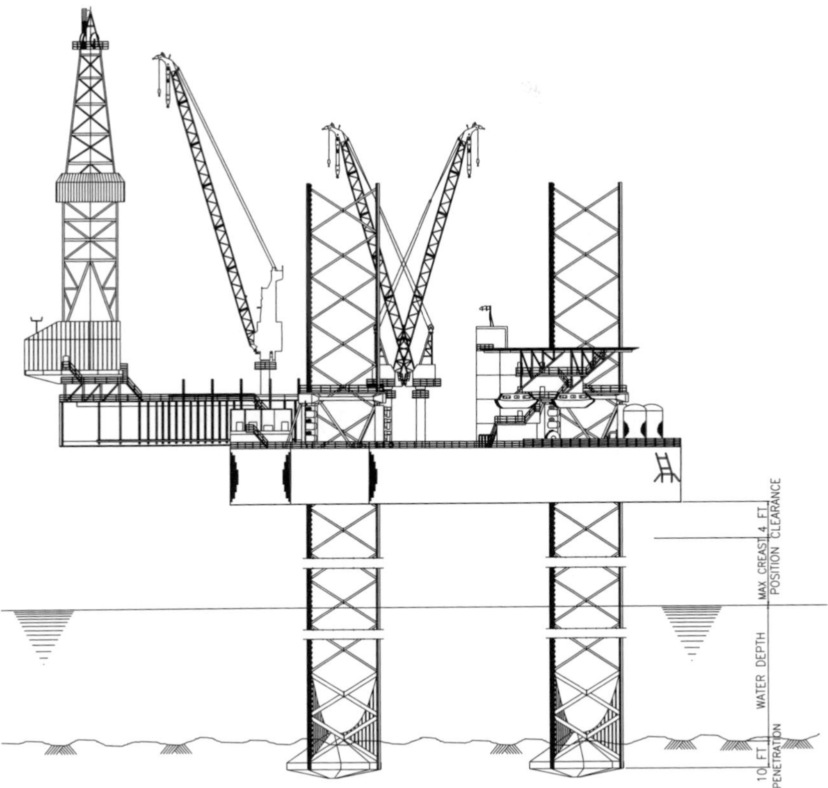

Figure 6.63 The length of leg used for a given water depth

large openings should however be included. The size of elements should be optimised keeping in mind computer run time and the relevant information to be gained by an overly detailed model and need for an accurate distribution of loads and calculation of global stresses. Typically, 5 ft × 5 ft elements are adequate in the areas that are not highly stressed. Smaller (triangular and 1 ft × 1 ft quadrilateral) elements should be modelled at areas of high stresses and close to intersections with yoke housings, other bulkheads and deck plating. The engineer may use his discretion in including girders and frame elements. Inclusion of stiffeners usually does not improve the results of the analysis significantly. All decks should be included in the model. This normally means that three decks will have to be simulated. The fineness of the mesh may be the same as that for bulkheads. The engineer often faces a need for a finer mesh near yoke housings and in areas of heavy loads. Inclusion of deck girders in the model facilitates a more realistic application of gravity loads on the upper hull. Structural components such as the helideck, cantilever beams, spud cans and the drill package are not normally included in the global analysis model of the hull and legs. Detailed models of each of these components are built separately and analysed.

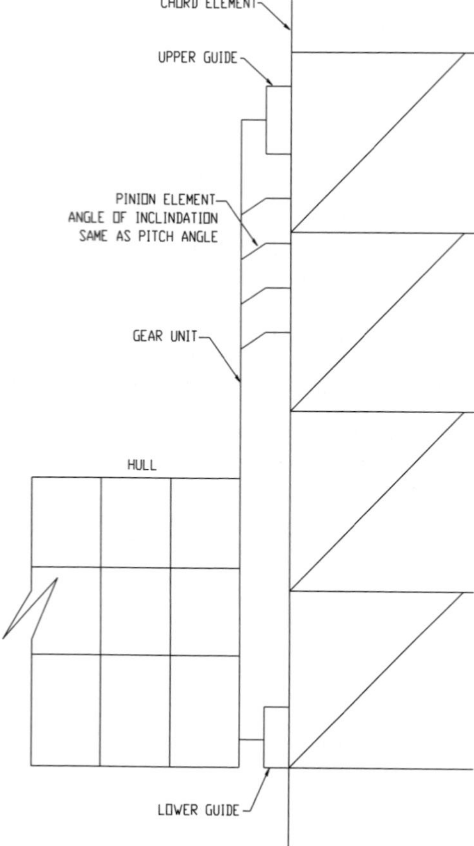

Figure 6.64 Simulation of the hull of leg connection

However, the reactions from these components should be included in the global structural analysis if significant.

Foundations

Foundations for jack-ups can be broadly classified into two types: independent leg spud can foundations and mats. Independent leg spud cans are simulated by providing a "pinned" support for the leg 10 ft beneath the mudline; such an approximation has proved over time to be sufficiently conservative in predicting moments in the leg at the level of the hull. This provides a representation of the average restraint of the leg for various penetrations and spud can fixities.

For a mat foundation, the mat structure may be modelled with plate elements and the legs could be fixed to this structure (For further details, see Whitley)

6.3.3.5 Description of the Jack-up Platforms Components

Foundations

Although spud cans are the commonly used jack-up platform foundation types, a design engineer may be called upon to assess the capacities of a jack-up with a mat foundation. Hence a description of mat foundations is provided below:

Mats

Figure 6.65 shows a plan view of a mat foundation. The basic purpose of the mat is to more uniformly distribute the concentrated loads from the legs into the soil of the seabed and to reduce the pressure loads on the soil. The mat is also required to resist lateral loads and it achieves this by a combination of cohesion or friction between the bottom of the mat and the soil and lateral passive soil pressure acting on the vertical surface of the mat that has penetrated into the soil. A scour skirt may be added to the mat to increase its lateral soil resistance.

The foundation mat provides a moment connection to the base of each leg that serves to reduce the bending moment in the leg at the hull level. This helps reduce the overall weight of each leg. Since the legs are fixed to the mat, the legs stay in the same position relative to the hull. This allows use of a jacking system that is simpler than one on an independent leg unit.

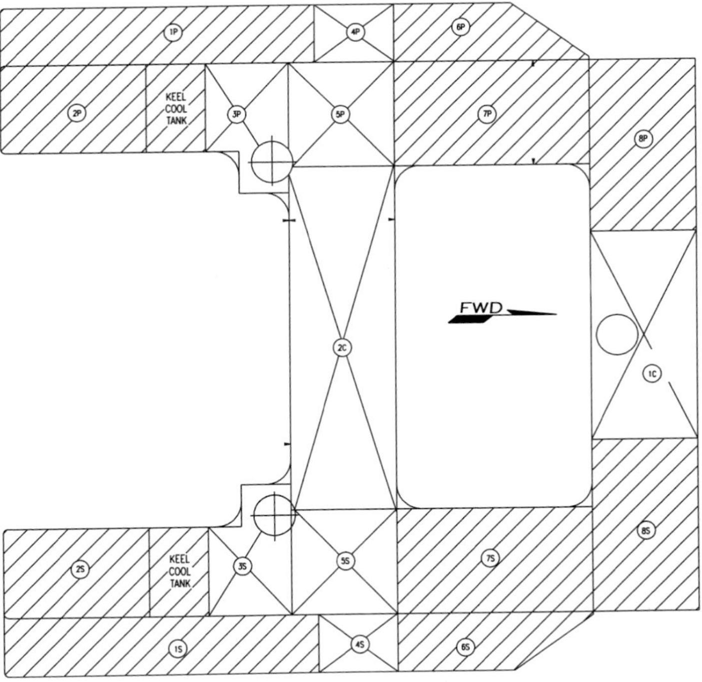

Figure 6.65 Jack-up platform foundation mat

Another purpose of the foundation mat is to provide buoyancy during the afloat condition. Thus, the depth of the mat is determined by two basic considerations (1) structural integrity and (2) adequate buoyancy to float the entire lightweight of the unit. Whitley carries details on the factors to be considered in the design of a mat foundation.

Spud Cans

This is the most common type of jack-up platform foundation in use. Spud cans typically consist of a conically shaped bottom face. The purpose of a spud can is to transfer the jack-up leg loads into the seabed below. The structure of the spud-can should thus have the capacity to resist the resulting shear and bending stresses exerted on it by the leg and the foundation soils. To determine the maximum force on a spud can during the design phase, the total weight of the upper hull during the worst design storm condition and its center of gravity is first established. This weight is then distributed over all the legs of the jack-up platform. From the applied environmental forces, the overturning moment is determined next. The direction of this overturning moment should be so as to cause the maximum compressive force on one leg. An appropriate load factor should then be applied to this force. The area of contact between the spud can and the soil should be sufficient for the weakest chosen soil condition to support this force. Due to uneven distribution of critical contact stresses on the spud can, the can structure and the legs should be checked for an eccentricity moment M. This could be conservatively approximated as being equal to:

$$M = (1/2) \times F_v \times R$$

where:

F_v = Maximum design axial load on the leg
R = Equivalent radius of the spud can

Other criteria that are applied to design the structural strength of the spud can are:

(1) Assume that the entire reaction acts as a concentrated load on the tip of the spud can.
(2) Assume that the entire reaction acts on a circle centred on the tip of the spud can, whose radius is (i) ¼, (ii) ½, (iii) ¾ and (iv) 1 times the equivalent radius of the can. The lower plating should be designed for the resulting distributed loads.
(3) Spud cans are usually designed to be flooded during operation. To facilitate access to the inside of the can, during the floating condition of the jack-up platform, vents may be provided to a certain height above the top of the can. The upper plating should be designed for a hydrostatic head corresponding to the height of this vent in case the can is not flooded.

Legs

Trussed legs are the most common type on modern jack-up rigs, the other type being cylindrical legs. Legs are subjected to the following forces:

(1) Elevated condition:
    (a) Compression forces due to gravity loads on the hull.
    (b) Compression forces due to the reactive couple caused by overturning moments on the jack-up.

(c) Bending moments at the hull due to the horizontal displacement of the hull and the moment connection between the leg and the hull.
(d) Horizontal forces on the leg due to wave, current and wind action. These forces are usually considered as collinear.
(e) Bending moments due to $P-\triangle$ effect on the leg.
(f) High local stresses due to force transfer and from the pinions, "rack chocks, hull upper and lower guides".
(g) Fatigue causing cyclic stresses due to constant wave action on the legs. Static and Fatigue analysis of welded tubular jack-up leg connections can be performed in the same manner defined for the fixed offshore platforms described in Section 6.

(2) Afloat condition:
(a) Gravity loads on the leg.
(b) Wind force.
(c) Inertia forces due to vessel motions.
(d) Restraining reactions from guide units or other locking devices in the hull that create high moments in the leg.
(e) Fatigue causing cyclic stresses in the lower bays of the legs due to the constant pitch and roll motions of the floating vessel. Static and Fatigue analysis of welded tubular jack-up leg connections can be performed in the same manner defined for the fixed offshore platforms described in Section 6.
(f) Effect of different leg positions on the legs and hull

Trussed jack-up legs would typically have either three or four chords. These chords are connected together by a system of horizontal and diagonal braces, normally made of circular cross-sections. These connections can be designed against overload and fatigue by using procedures outlined in Section 6.3.1. Gear racks are an integral part of the chords. Figure 6.66 illustrates a typical jack-up platform leg. Figure 6.67 shows two common cross sections of chords – "tear drop" and "opposed rack". Racks are normally cut from very high strength steel ($Fy > 90$ ksi). Braces are made of high strength steels ($Fy \cong 80$ ksi). The factors to consider in designing a brace would be: (a) high buckling stresses, (b) low hydrodynamic drag, (c) high tensile strength, (d) optimal section modulus, (e) feasibility for rolling and (f) weldability.

Due to high stresses induced into the lower braces during tow and possibility of eccentric loading on the spud can, these braces will be heavier and sometimes would be of built-up I or H sections.

The leg to hull interface is also affected by the number, location, stiffness and type of elevating units. If the stiffness of the elevating system is "soft" the pinions tend to share the load more evenly. As the stiffness increases, the load would become more concentrated in the lower elevating pinions.

Hull Component Design

The jack-up platform hull is generally built of stiffened plate (see fig. 6.1). The structure is configured so as to efficiently transfer loads acting on the various hull locations into the

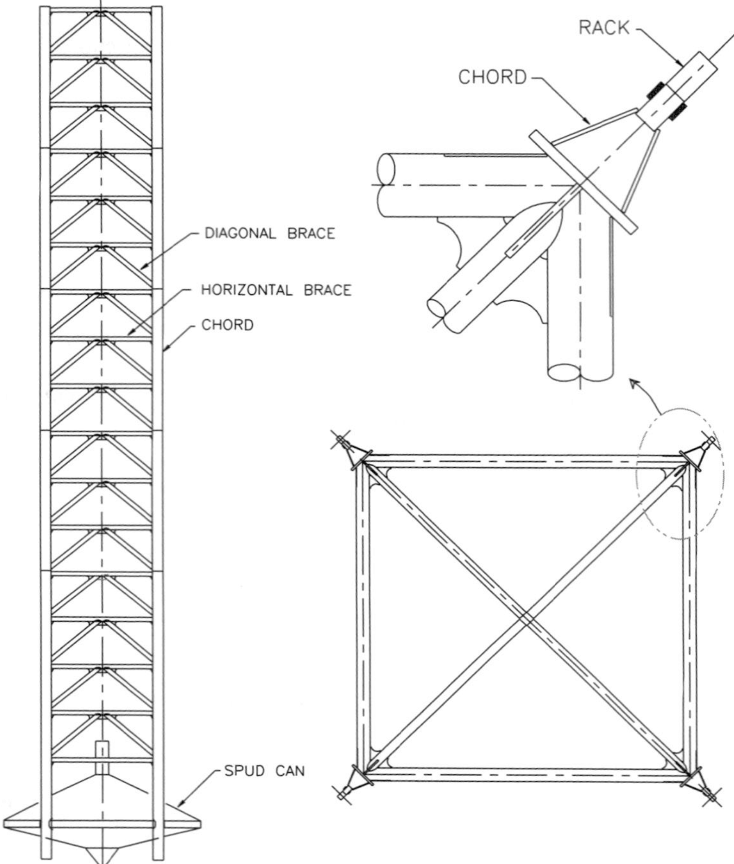

Figure 6.66 Components of a jack-up platform leg

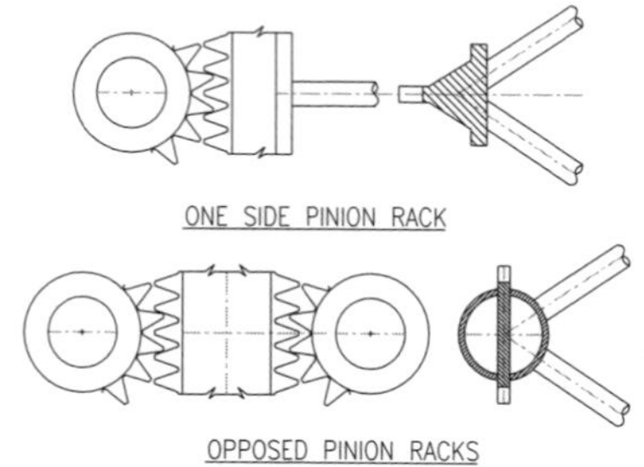

Figure 6.67 Two common types of gear racks

legs. Axial and horizontal loads are transferred into the legs through the hull leg interface connections and chords. Hence a bulkhead terminates at each leg chord location. The structural components of the hull may be listed as follows:

(1) Support points for jacking and holding units.
(2) Devices to provide a moment connection between the legs and the hull. These may be in the form of upper and lower guides or a "rack on rack" locking device.
(3) Longitudinal and transverse bulkheads, the side shell and the inner perimeter bulkhead to transfer loads from various parts of the hull to the legs.
(4) Deck plating often in the form of a main deck, a machinery deck and inner bottom plating. This plating is often stiffened with angles at between two and three foot spacing. The angles themselves span between hull transverse frames spaced at regular intervals of between six and ten ft.
(5) Cantilever beams – these are two deep girders that normally skid longitudinally over two bulkheads called the "Skid Rail Bulkheads". The drilling structure is located atop these beams. The purpose of the cantilever beams is to provide the capability for the drill floor to be moved over the well bay of a platform.

Several additional loading situations need to be considered. While these conditions are unusual, they can have a major effect on the unit. It is possible for one of the spud cans to suddenly penetrate further than the other spud cans due to a local soil failure. On a three leg unit, this could cause the total unit to tilt from its usual vertical orientation. If the tilt is large ($>5°$) the loads inducted on the legs can exceed their design capacities. This usually occurs in the foundation preloading operation. This is the reason why this operation should take place with a small (<5 ft) air-gap under calm weather conditions. The small air-gap allows for the hull to be supported by the water after a relatively small tilt.

Another unusual event is the collision of an object or boat on the leg(s). While to date no unit has failed due to this type of event, numerous collisions have occurred. The Rules [API RP2A-LRFD (1993), and WSD (2000)] provide some guidance for the magnitude of these events. However, if a ship is large enough, it could bring the unit down.

References

Abbot, P. A., Dailey, J. A., Karsan, D. I., and Mangiavacchi, A. I. U.S. Patent 4,417,831, Mooring and Supporting Apparatus and Methods for a Guyed Marine Structure.

American Bureau of Shipping ABS (2001): Rules for Building and Classing Mobile Offshore Drilling Units, 2001, Part 3-Hull Construction and Equipment.

American Welding Society, Structural Welding Code – Steel, ANSI/AWS D1.1-98, ISBN 0-87171-252-2.

API recommended practice for planning and constructing fixed offshore platforms, load resistance factored design, API RP2A-LRFD, (1st ed.), July 1, 1993.

API recommended practice for planning, designing and constructing structures and pipelines for arctic conditions, API RP2N, (2nd ed.), December 1, 1995.

API recommended practice for planning and constructing fixed offshore platforms – working stress design, API RP2A-WSD, (21st ed.), December, 2000.

Boone, T. J., Yura J. A., and Hoadley, P. W. (1983). Chord stress effects on the ultimate strength of tubular joints. Phase I Report to API, February.

Bryant, L. M. and Mattlock H. (1975). "Three dimensional analysis of framed structures and nonlinear pile foundations", *Proceedings of Offshore Technology Conference*, OTC 2955.

Clough, R. W. and Penzien J. (1975). Dynamics of structures, McGraw Hill.

Cotrell, Adrian (1997). "Giant scissors jack detaches Gulf deck", Offshore Engineering, October.

Cotrell, Adrian (2001). Float-over puts Malampaya on final straight, *Asian oil and Gas*, August.

Déserts, L. D., Michel, D., and Sedillot, F. G. (1986). "Oscillating offshore platform on flexible piles", *U. K. Patent Application GB 2162 883A*, C. G. Doris, February 12.

Det Norske Veritas – Rules for Classification of Mobile Offshore Units, Part 3, Ch. 1 – Structural Design General.

Digre, K. A., Brasted, L. K., and Marshall, P. W. (1989). "The design of the Bullwinkle Platform". OTC 6050, *Offshore Technology Conference Proceedings*, Vol. 3.

DnV Rules for Classification of Fixed Offshore Installations, Structures, Part 3, Chapter 1, Latest Edition, Det norske Veritas, Veritasveien 1, 1322 Hφvik, Norway.

Eftymiu, M. (October 1988). "Development of SCF formulae and generalised influence functions for use in fatigue analysis". *Recent developments in Tubular Joint Technology*, OTJ'88, London.

Faddeva, V. N. (1959). Computational methods for linear algebra. Dover Publications, Inc.

Geyer, J. F. and Stahl, B. (May 1986). "Simplified fatigue design procedure for offshore structures", OTC 5331, *Offshore Technology Conference Proceedings*.

Gulati, K. C., Wang, W. J., and Kan, D. K. Y. (1982). "An analytical study of stress concentration factor effects in multibrace joints under combined loadings", OTC 4407, *Offshore Technology Conference Proceedings*, May, 1982.

Hart, W. H. and Sablok, A. (1992). "Weld profile and plate thickness effects as applicable to offshore structures". Final report, API Project 87-24.

International Standards Organisation ISO/CD 19902, Section 14, Design of Tubular Joints.

Karsan, D. I. (1986). "Design of jackets in deepwater Gulf of Mexico waters", *ASCE Journal of Waterway, Port, Coastal and Ocean Engineering*, Vol. 112, No. 3, pp. 427–446, May.

Karsan, D. I., Ross, C. G., and Olsen, Conrad-Eck (1996). Risk assessment and safety assurance Program for The Heidrun Field Development, OTC 8103.

Karsan, D. I., Valdivieso, J. B., Suhendra, R. (1980). "An economic study on parameters influencing the cost of deepwater fixed platforms", *Proceedings of the 1986 Offshore Technology Conference*, OTC 5301, p. 79.

Karsan, D. Krahl, and I N. W. (May 1984). "New API equation for grouted pile-to structure connections". OTC 4715, *Offshore Technology Conference Proceedings.*

Karsan, D. I., Hanna, S. Y., and Yeung, J. Y. (1988). "Tension restrained articulated platform – TRAP", USA Patent 4,781,497, Dated Nov. 1.

Karsan, D. I., Marshall, P. W., Pecknold, D. A., Mohr, W. C. and Bucknell, J. (2005). "The new API RP2A 22nd edition tubular joint design practice", OTC 05, Paper no: 17236, Houston, Texas, May 2005.

Kuang, J. G., Potvin, A. B., Leick, R. D., and Kahlich, J. L. (1977) "Stress concentration in tubular joints", *Journal of Society of Petroleum Engineering*, August

Kumar, A., Nair, V. V. D., and Karsan, D. I. (1985). Stiffness properties of fixed and guyed platforms, *ASCE Journal of Structural Engineering*, Vol. 111, No. 2, p. 239, February.

Livesly, R. K. (1975). Matrix methods of structural analysis, (2nd ed.). Pergamon Press.

Lloyd, J. R. and Karsan D. I. (1988). "Development of a reliability-based alternative to API RP2A", *Proceedings 1988 Offshore Technology Conference*, OTC 5882, p. 593.

Manual of Steel Construction, Load Resistance Factored Design, LRFD method (1988). American Institute of Steel Construction, AISC, (1st ed.), July.

Manual of Steel Construction, Allowable Stress Design Method (ASD) (1989). American Institute of Steel Construction, AISC (9th ed.), July.

Marshall, P. W. and Luyties, W. H. (1982). "Allowable stresses for fatigue design", *Proceedings of the BOSS'82 Conference held at MIT*, Cambridge, Mass, August 2–5.

Marshall, P. W. and Toprac, A. A. (1974). "Basis for tubular joint design", *Welding Journal*, Vol. 53, No. 5, May.

Marshall, P. W., Bucknell, J. and Mohr, W. C. (2005). "Background to new RP2A fatigue provisions", OTC 05, Paper no: 17295, Houston, Texas, May 2005.

Martin, H. C. (1966). Introduction to matrix analysis methods of structural analysis, McGraw Hill.

Mc Clelland, Bramlette (Editor), (1986). Planning and Design of Fixed Offshore Platforms (1986). M. D. Reifel, Van Nostrand Reinhold. Library of Congress Cat. No. 84-27078.

Moan, T. G., Karsan, D. I., and Wilson, T. J. (1993). Analytical Risk Assessment of Floating Platforms Subjected to Ship Collision and Dropped Objects, OTC 71223.

Morrison, D. G. (1997). "Low cost designs for facilities in shelf and deepwater development", *Proceedings of the OMAE' 97 Conference*, Yokohama-Japan.

Nair, V. V. Damodoran (1978). "Aseismic design of offshore platforms", *Proceedings of ASCE Geotechnical Engineering Division Specialty Conference*, June 19–21.

Nath, J. H. and Harleman, D. R. F. (1967). "The dynamics of fixed towers in deepwater random waves", *Proceedings of the Conference on Civil Engineering in the Oceans, ASCE*, pp. 99–122.

Norris, C. H. and Wilbur, J. B. (1960). Elementary structural analysis, McGraw Hill & Co. International Edition, N.Y. Library of Congress cat No: 60-10615, 1960.

Norwegian Petroleum Directorate (NPD) (1992). Regulations relating to load bearing structures, 7 February.

Pecknold, D. A., Ha, C. C., and Mohr, W. C. (2000). "Ultimate strength of DT tubular joints with chord preloads", *Proceedings of the Offshore Mechanics and Arctic Engineering Conference*, New Orleans.

Pecknold, D. A., Park, J. B., and Koeppenhoefer, K. C. (2001). "Ultimate strength of gap K tubular joints with chord preloads", *Proceedings of the Offshore Mechanics and Arctic Engineering Conference*, OMAE01/OFT-1214, Brazil.

Pecknold, D. A., Chang, T- Y. and Mohr, W. C. (2003). "Static strength of T tubular joints with chord preloads under brace axial and moment loads" Report to the American Petroleum Institute, EWI Project no: 42705-CAP, Edison Welding Institute, 2003.

Pecknold, D. A., Marshall, P. W. and Bucknell, J. (2005). "New API RP2A tubular joint strength design provisions", OTC Paper no: 17295, Houston, Texas, May 2005.

Provisional Rules for Fixed Offshore Installations (1987). Lloyds Register of Shipping, June.

Reese Lymon C., Mattlock Hudson (1956). "Non dimensional solutions for laterally loaded piles with soil modulus assumed proportional to depth", The University of Texas, September.

Roark, R. J. Formulas for stress and strain, McGraw Hill & Co. International Edition, N.Y. Library of Congress cat No: 64-66027.

Rogers, G. L. (1959). "Dynamics of framed structures", Chapter 5, pp. 186–189. John Wiley & Sons Inc, Library of Congress Catalog No. 59-11808.

Saaty, Thomas L. Decision making for leaders, the Analytical Hierarchy Process (AHP) for decision in a complex world, RWS Publications, 4922 Ellsworth Avenue, Pittsburh, PA 15213.

SACS-User's Manual, Structural Analysis Computer System, Engineering Dynamics Inc. 2113, 38th Street, Kenner, LA 70065, USA.

Salama, M. M., Ellis, N., Hanna, S. Y., and Karsan, D. I. Pile driving dynamic loads on offshore structures, *Proceedings of the 1988 Offshore Technology Conference*, OTC 5704, p. 207.

Sarpkaya, T. and Isaacson, M. Mechanics of wave forces on ocean structures, Van Nostrand Reinhold, 115 Fifth Ave., N.Y., NY 10003.

Sesam User's Manual, Structural Analysis software, Det norske Veritas, Vertasveien 1, 1322 Hϕvik, Norway

Smith, A. E. L. (1962). "Pile driving analysis by wave equation". *Transaction ASCE*, Vol. 127, Part 1, Paper no. 3306, pp. 1145–1193.

Society of Naval Architects and Marine Engineers-Technical and Research Bulletin 5-5, Guidelines For Site Specific Assessments For Mobile Jack-Up Units.

StruCad User's Manual, Structural Analysis of Linear and Non-linear Systems, Zentec Engineering Inc., 8582 Katy Freeway, Suite 205, Houston, Texas 77024, USA.

Tour Roseau – The Reed Tower (1984). *Ocean Industry*, April.

Whitley J. O. Jr. "Some aspects of the structural design of a three column mat supported, self elevating, mobile drilling platform" – Bethlem Steel Corporation.

Yura, J. A., Zettlemoyer, N., and Edwards, I. F. (1989). Ultimate capacity equations for tubular joints, OTC 3690, May.

Handbook of Offshore Engineering
S. Chakrabarti (Ed.)

Chapter 7

Floating Offshore Platform Design

John Halkyard
Technip Offshore, Inc., Houston, TX, USA
(with contributions from John Filson, and Krish Thiagarajan)

7.1 Introduction

This chapter will discuss the design of floating structures, in particular those used in offshore oil drilling and production. Floating structures have been used since the 1950s for drilling, and they have become increasingly popular for production, particularly in deep water. Floaters pose new design challenges, for example:

- Weight control and stability become key design drivers,
- Dynamic responses govern the loads on moorings and equipment,
- Fatigue is an important consideration,
- In some areas, the new environmental challenges make design difficult, e.g.
    - Large currents in the deepwater of the Gulf of Mexico
    - High seas and strong currents in the North Atlantic
    - Long period swells in West Africa
- Installation of the platforms, mooring and decks in deep water present new challenges,
- New materials for risers and moorings are required in ultra-deep water

Floating platforms can be characterised as one of two types:

1. "Permanent" facilities
2. "Mobile" facilities

"Permanent" facilities are those, which are designed to be moored in place for typically 20–30 years. Inspections are performed in-place. They must be capable of surviving extreme environmental conditions including 100-year events. These are used primarily for the production and processing of oil and gas.

"Mobile" facilities include those used for drilling or marine construction and installation. The mooring and stationkeeping requirements for these are more constrained by operational considerations. The survival criteria are less than those for the permanent facilities. Inspections and maintenance is performed during scheduled drydocking.

Permanent facilities used for oil production will be the primary subject of this chapter. Many of the design criteria used for the mobile facilities, especially the MODU Rules, are used to design these permanent facilities.

Design of floaters and other offshore structures is often based on published industry standards and classification rules. There are a number of international conventions and regulations governing floating systems design which must be reviewed prior to operating a vessel (Rogers and Bloomfield, 1995). Most useful to the designers are the industry standards and classification rules; examples of these standards are used liberally in this chapter, especially in Section 7.8.

Industry standards and rules are written to reflect past design practices that have proven successful. They do not necessarily cover new or "novel" structures. It is a standard practice when designing a new structure to fall back on standards used for more established structures, but this is not foolproof. As the industry moves into ever deeper water and newer environments, it becomes even more important to question the standards developed for shallow water or mobile facilities. The best practice is to use standards as a guide, but to perform a rigorous amount of front end engineering based on "first principles" before embarking on the detailed design of a new concept. There is an ever increasing amount of tools available for response and stress analysis. World class model testing facilities exist to check the responses of new concepts. Analysis and testing should be performed early in the design evolution to avoid surprises.

The other important point to make here is that as the industry moves to deeper water the floater cannot be considered as simply a piece of real estate to hold a payload and to support risers. The dynamics of the floater can be affected by the risers and mooring systems. An inadequate hull mooring system design can invalidate the use of certain types of risers or riser components. The layout of equipment may result in an eccentric weight, which must be compensated by a large amount of ballast, increasing the total amount of displacement needed. This means the floater designer must understand all of the systems supported by the hull, and be prepared to include their effects in his modeling and design. It may be said that "the best hull is the hull which best supports the risers". A common mistake is to select and design the hull before the well layout and the riser makeup has been finalised, let alone analysed.

Finally, this technology is continuously changing and it is incumbent upon the user of this handbook to research current technology. The following web sites are particularly useful for this purpose:

www.otcnet.com – The official web site of the annual Offshore Technology Conference. There is an online search capability for all past OTC Papers, which reflect the current state of technology.

www.offshore-technology.com – This web site chronicles offshore projects.

www.spe.org – Society of Petroleum Engineers web site (includes eLibrary link).

www.omae.org – American Society of Mechanical Engineers, Offshore, Oceans and Arctic Engineering (Annual Conference).

www.sname.org – Society of Naval Architects and Marine Engineers.
www.isope.org – International Society of Offshore and Polar Engineers (Annual Conference).
www.asce.org – American Society of Civil Engineers.
www.eagle.org – American Bureau of Shipping (Classification Society).
www.dnv.org – Det Norske Veritas (Classification Society).
www.api.org – American Petroleum Institute (publications).
www.coe.berkeley.edu/issc/ – International Ship and Offshore Structures Conference (Summarises R&D in the field).
www.shipstructure.org – Interagency Ship Structures Committee (particularly structural issues).
http://ittc.sname.org/ – International Towing Tank Conference (particularly hydro-dynamic issues).

7.2 Floating Platform Types

7.2.1 General

The history and description of the various platform types has been discussed in Chapter 1. This chapter will deal with the design of floating systems, which are particularly suited for deepwater: the FPSO, Semi-submersible, TLP and Spar.

The TLP and the spar are the only floaters used today with dry trees (Ronalds and Lim, 2001). There is little standardisation of floater units. Shell Offshore and their partners achieved significant cost savings when they designed the multiple TLPs following similar design practices (i.e. Ram-Powell, Mars, Brutus). Kerr McGee achieved some saving by designing the Nansen and Boomvang Spars identically (Bangs et al, 2002). However, for the most part, each deepwater field has been developed with a "fit for purpose" design.

Floater types might be distinguished by several characteristics such as functions, stability, motions, load or volume capacities, transportability, reusability. Figures 7.1 and 7.2 illustrate the relative profile and plan view of the floaters considered here. Each has a significant difference in terms of design drivers, performance, construction and installation. FPSOs have a relatively shallow draft, but a large waterplane area. They provide a large area for process facilities, and large storage volumes. Semi-submersibles have a small waterplane area and a moderate draft. Spars have a very deep draft and a moderate to small waterplane area. Since each platform requires significantly different design considerations, the design of each is covered in a separate subsection herein. Hull stability and structural design are relatively generic and are described under specific subsections.

7.2.2 Functions

Table 7.1 shows the functions currently performed by the floater types. While the drilling and workover with dry trees has been limited to TLPs and Spars, semi-submersibles are used to drill and workover wet tree wells positioned under the hull. FPSOs have been designed with drilling and workover capability for benign environments, but they have not

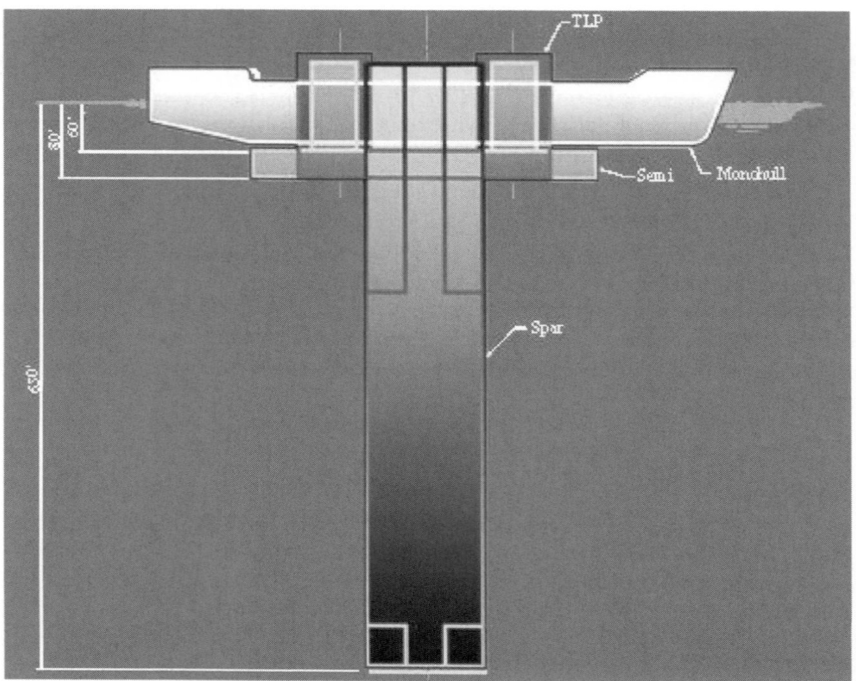

Figure 7.1 Relative cross-sectional shapes of floaters (Courtesy Technip Offshore, Inc.)

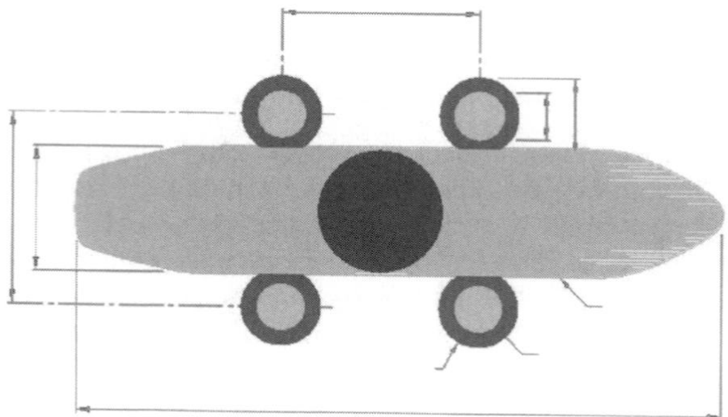

Figure 7.2 Relative plan area of floaters (Courtesy Technip Offshore, Inc.)

been implemented. Chapter 2 deals with new (as of the time of this writing) platforms which might perform all of these functions. It should be mentioned that, while this chapter is devoted almost exclusively to oil field and offshore construction applications, the U.S. Navy has spent considerable effort developing designs for large Mobile Offshore Bases (MOBs) using semi-submersible principles (see Taylor and Palo, 2000).

Table 7.1 Existing functions for floaters

	FPSO	Semi-submersible	Spar	TLP
Production	Yes	Yes	Yes	Yes
Storage	Yes	No	Yes	No
Drilling	No	Possible	Yes	Yes
Workover	No	Possible	Yes	Yes
Water Depth Limitation	No	No	No	Yes

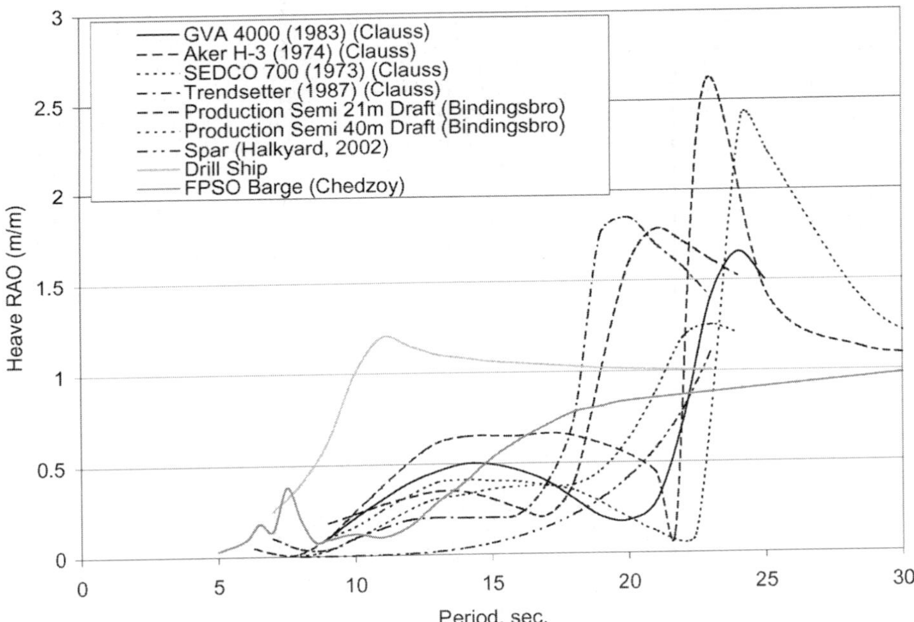

Figure 7.3 Example heave RAOs of various floaters

7.2.3 Motions

Figure 7.3 illustrates the differences in heave RAOs for various floaters. TLPs are not included here because their heave and pitch responses are not significant, except in the range of resonance where the effect is primarily on tendon tension. Heave response is the most critical response for the support of the risers, and for the operability of a drilling platform.

Several semi-submersible designs are included to show the influence of optimisation on the heave response. The most critical wave periods are: Brazil, 12–14 seconds; Gulf of Mexico, 13–16 seconds; North Atlantic and North Sea, 15–18 seconds; West Africa: 16–22 seconds. Semi-submersibles achieve different heave responses by varying the ratio of the pontoon to the column volume. The wave forces acting on the bottom of the columns due to wave pressure are partly cancelled by the inertial forces acting on the pontoons by the accelerating fluid. These forces completely cancel at a period unique to a volume ratio and spacing of the columns. A lower response at, like 14 seconds, is usually associated with a reduction in the natural period which would result in higher responses at longer periods. The Trendsetter Semi-submersible responses are particularly noteworthy. These were accomplished by adding a central column, which provides further wave-force cancellation in the hurricane wave period range. Clauss (1998) showed that the shape of a semi-submersible could be further optimised if the cross-sections of the columns and the pontoons could be adjusted.

Deep draft semi-submersibles and those with heave mass traps have been shown to have suitable motions for dry trees, but as of this writing, they have not been employed (Le Guennec et al., 2002; Halkyard et al, 2002). The response for two ship shaped hulls of differing length and displacement are shown in fig. 7.3. The "Drill Ship" response is typical of early generation drillships such as the SEDCO 445 and SEDCO 470 series. Larger drillships, and particularly VLCC class FPSOs, which are on the order of 300 m or more in length respond as the FPSO Barge indicated. These responses are for head seas. The ship-shaped hulls in the beam sea respond more like the "Drill Ship".

Pitch responses of the various floaters are shown in fig. 7.4. The RAO is presented in terms of pitch angle over the maximum undisturbed wave slope. The FPSO barge follows the wave slope for waves with lengths on the order of the ship length. The pitch responses of the semi and the spar are also to draft. A conventional drilling semi-submersible shows the effects of cancellation of pitch moments from the horizontal loads on columns and the vertical inertial forces on pontoons.

The surge response at vessel center of gravity for various floaters is shown in fig. 7.5. Again cancellation effects are evident for the semi-submersibles. The irregular variations in the response of the FPSO barge are due to the diffraction/radiation effects.

Acceleration at the deck level is an important consideration for equipment design and operations. Figure 7.6 illustrates the acceleration RAO.

7.2.4 Concept Selection

The process of floating facility selection and design can be a long and complicated process. The beginning is a perceived need for oil or gas recovery from a reservoir. The first indications are only that there might be oil or gas based on a geologic feature, but until it is drilled, there is no way to be sure of it. Even after drilling one well, there are often many uncertainties about the accessibility of a body of hydrocarbons and their quality. There are often several "appraisal" wells drilled following the "discovery" well to ascertain this. It is during the appraisal drilling that most operators begin to worry about how a particular oil field will be produced. The options for

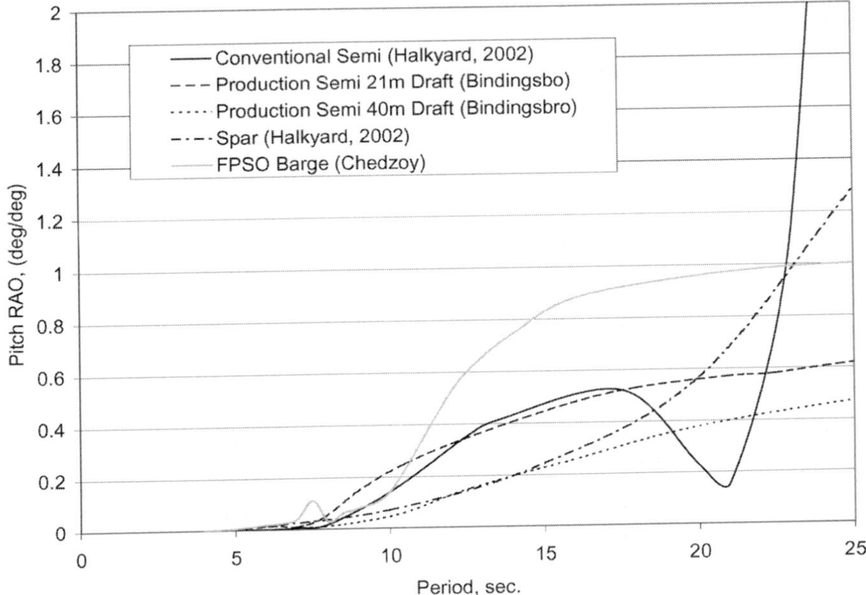

Figure 7.4 Example pitch RAOs of various floaters

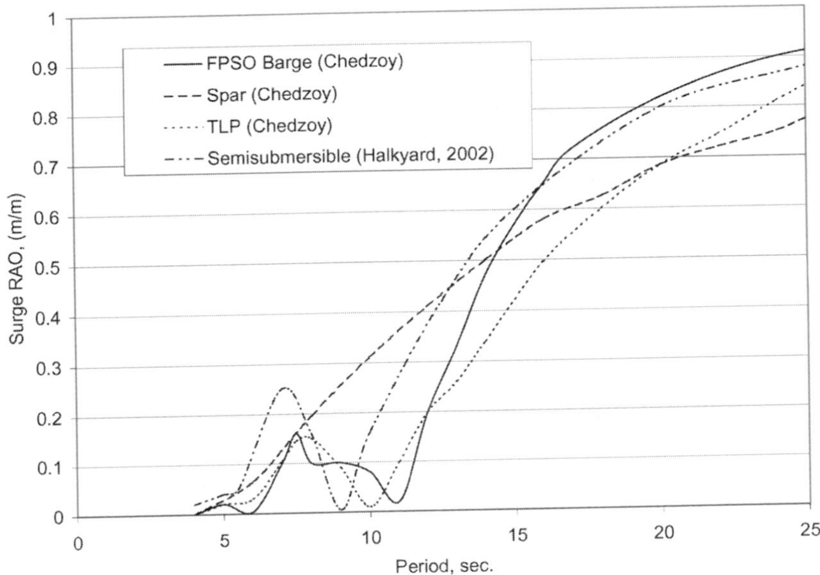

Figure 7.5 Example surge RAOs of various floaters

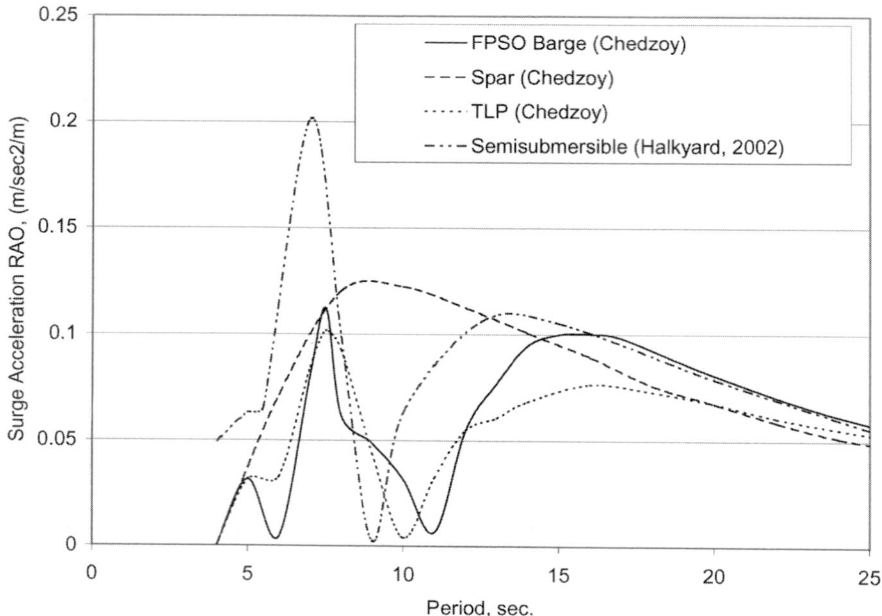

Figure 7.6 Example surge acceleration RAOs of various floaters

developing an oil field are innumerable, and it is not uncommon for the process of deciding which option to select to take years. The most important fundamental decisions are:

- How are the wells located and structured?
- How will the drilling and completion of the wells be performed?
- How will the well flow be delivered to the platform, processed and exported to market?

Floater concept selection should not be approached without an appreciation of all of the components required for a particular field development: drilling, substructures, processing and export. Selection of a particular substructure can have profound impacts on the other components. Equally important in the selection of a concept is an appreciation for the method of construction and installation. The local conditions, quay availability, weather, etc., can limit the options for transportation and installation. The availability of suitable fabrication yards and transport and construction vessels may limit the options.

Drilling costs typically make up around 50% of the total value of a deepwater project. The floater itself, installed and moored, usually represents less than 15%. The discovery and the appraisal wells are drilled from MODUs, but this can be an expensive approach to drilling the "development" wells (the ones that will be the basis for production). A fundamental decision is whether the wells should be subsea ("wet tree") or surface ("dry tree") wells. If a subsea development is utilised, a floating platform may not be needed at all. The lowest cost developments utilise a subsea well tied back to an existing platform. A subsea development was recently commissioned in the Gulf of Mexico which actually linked the subsea wells from several different oil fields owned by different operators to one pipeline connecting to a shallow water fixed platform (Rijkens, 2003). Subsea production has several limitations, e.g.:

- Each well requires an MODU for installation and maintainence.
- An existing structure is required to receive the oil.
- Flow assurance and well conditions may limit the range for tying back to an existing structure.

Reservoir recovery percentages are historically lower from the subsea wells because it becomes uneconomic to continue to operate the well when flow rates fail to meet the threshold values, and it is too expensive to mobilise a MODU to redrill or service the well.

Most floating production systems in use today actually support wet tree developments. Floating Production Storage and Offloading systems (FPSO), the most prolific type of floating platforms today, are primarily serving oil fields in remote parts of the world where there is no infrastructure to transport or use the oil. They receive hydrocarbons from one or more subsea wells, process the oil and offload the oil to tankers bound for the oil consuming part of the world. Many of these FPSOs are converted tankers. They do not use dry trees because their motions do not allow it, and because the cost of converting a used tanker to accommodate dry trees would be prohibitive. Many of these FPSOs are leased because the oil fields only last 5–7 years.

There are also numerous semi-submersible FPSs, which produce from subsea wells and deliver product through pipelines. These have proven to be more cost effective than fixed platforms in moderately deep hostile environments like the North Sea. In Brazil, there is a large infrastructure for servicing wet trees and there are many FPSs supporting wet tree developments.

As mentioned earlier, dry trees may allow relatively inexpensive intervention and maintenance of the wells, leading to higher productivity. Also, the drilling and/or the completing wells from a floating platform may be at a significantly lower cost than using a MODU for this purpose. Much of the drilling may be deferred until additional reservoir data is available from early production. In addition to deferring the cost, the information gained can greatly improve the productivity of future wells.

7.3 Design of Floaters

7.3.1 Functional Requirements

Floating offshore structures must satisfy numerous functional requirements which for the floater designer can be simply reduced to:

- Buoyancy must equal weight plus vertical loads from the moorings and risers,

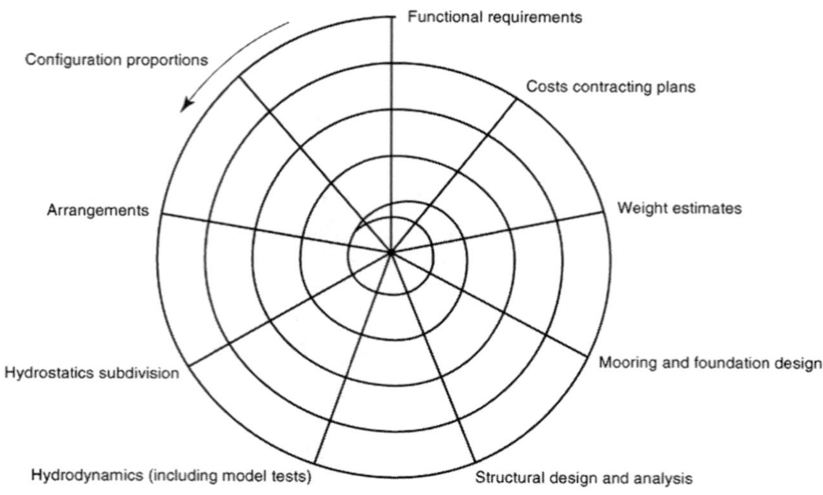

Figure 7.7 Design spiral (API RP2T, Reproduced Courtesy of the American Petroleum Institute)

- Space available must equal or exceed the space required for the functions to be performed,
- Motions, stationkeeping and stability must meet minimum criteria.

A unique feature of the floating offshore structures is the interrelationship between the floater itself, risers and mooring. The design process consists of a "spiral" in which weights, buoyancy, mooring loads, stability, etc. become better defined with each pass around the spiral (Fig. 7.7).

The process begins by listing the functional requirements, which should include, as a minimum, the weight and size of the payload to be supported. The weight and center of gravity of the payload is a primary input to the design process. Even if all the information is missing, all of the equipment should be listed and allowances made for everything. Since the floaters under consideration here are intended primarily for drilling or oil and gas production, their primary function will be to support the topsides (equipment plus structure) and risers. No configuration work should be started before clearly assessing the present and future types and sizes of risers to support, and how their behavior might be effected by the configuration. For example, steel catenary risers in deep water may be sensitive to heave motions, which will depend on the floater draft among other things. The "Hydrodynamics" step should include riser responses. Riser assessment should include both strength and fatigue.

Topsides Weight and Dimensions

"Topsides" is typically the term applied to the weight of the facilities structure, which must be carried by the hull. Different terminology is applied to various subsets of the "topsides" weights:

"Maximum Topsides Weight", or simply "Topsides Weight" may refer to the sum of the fixed and variable payload under maximum operations, and it might include "Topsides Primary Structure". This is important because it sets the full amount of buoyancy required (which must also include the "Other Loads").

"Topsides survival weight" may refer to the weight of the topsides in a shut-in (non-operating) condition typical, for example, of the condition assumed in a Gulf of Mexico Hurricane. Many of the variable loads are excluded. There may even be some so-called "Fixed" payload, which is removed for this condition. For example, many floaters in the Gulf of Mexico have removable workover rigs, which are removed during hurricane season.

"Lightship weight" usually refers to the weight of the outfitted hull without the topsides, but not including the mooring equipment and components which are not installed at the shipyard.

"Transport weight" includes the weight of the hull, deck or both which must be transported from the yard to an installation site.

"Lift weight" usually refers to the weight a derrick barge or other installation devices must lift in order to set the deck on the platform. This could typically include the Fixed Payload plus the "Topsides Primary Structure".

There are numerous load cases to consider depending upon the presence of the fixed and the variable loads. When sizing or analyzing a structure, it is useful to define a matrix of cases, which typically includes the various weight conditions and the associated expected environments.

Tables 7.2 and 7.3 show examples of some load cases and performance criteria, which may represent the basis for selecting the size of a floater.

In order to estimate facilities weight, the properties of the produced fluids needs to be defined. As a minimum, the flow rates for oil (BPD), gas rate (MMSCD) and produced water (BWPD) are required to perform an initial sizing.

Table 7.2 Example of load case definition

Operation	Design Storm		
	1 yr	10 yr	100 yr
Well Drilling	Yes	No	No
Well	Yes	No	
Intervention			Yes
(work over) Well	Yes	Yes (Winter Storm)	No (GOM Hurricane)
Production	Yes	Yes (Winter Storm)	Shut-in (GOM Hurricane)

Table 7.3 Example of performance criteria

Criteria	Well Drilling	Well Intervention (Work Over)	Well Production	Process Facilities	Survival
Mean Heel Angle (deg)	2	2	2	2	6
Max. Pitch Angle (deg)	6	6	6	6	10
Lateral Acceleration (g) incl gravity					
RMS	0.06	0.06	0.06	0.06	.13
Extreme	0.20	0.20	0.20	0.20	.50
Riser Stroke (Draw Down) (ft)	+15	+15	+15	+15	+30
	−10	−10	−10	−10	−15
Air Gap (No green water) (ft)	5	5	5	5	0 or show no damage

Figure 7.8 shows the typical range of process facility weights for the Gulf of Mexico floaters as a function of oil production rate. These weights represent the maximum fixed plus the variable payload weights. These weights exclude drilling, if any.

Drilling or workover weights can vary from a few thousand tons for a minimal workover rig to 10,000 tons for a deepwater full drilling rig. The principal variables are the water depth, reservoir depth and pressure capabilities.

Primary deck structure is typically 30–40% of the weight of facilities and drilling equipment (including variable weights).

Data on the past offshore projects with production, topsides weights and substructure information may be found on the web site: www.offshore-technology.com.

In addition to the weight, the deck area required for the topsides needs to be determined.

Before sizing a floater, a high level general arrangement of the topsides should be prepared to insure adequate area. Consideration needs to be given to the number of decks and the location of the center of gravity (vertically and horizontally). These considerations will be platform specific. For example, multi-column platforms (semi-submersibles and TLPs) may be more amenable to a single level deck, figure 7.9. Single column platforms (spars, mini-TLPs) may require multi-level decks to avoid excessive cantilevering of the deck structure, figure 7.10. If there is a well bay, i.e. for the vertical risers and/or drilling, the space for the risers and trees must be accommodated at every deck level. The drilling and the workover rigs are typically installed on the upper deck over the well bay.

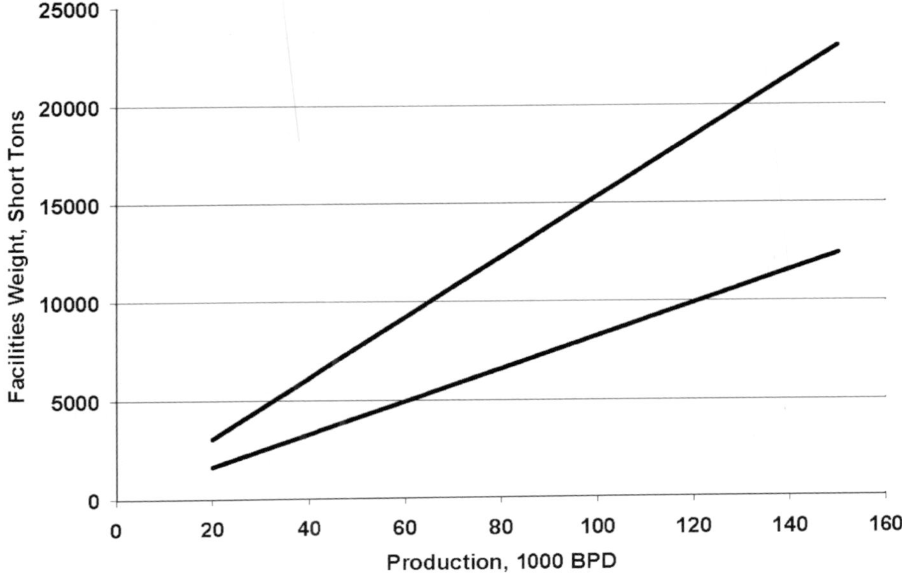

Figure 7.8 Range of topside facility weights for typical floaters, Gulf of Mexico

Figure 7.9 Asgaard "B" semi-submersible with single level integrated deck (Øyvind Hagen, STATOIL)

Environmental Conditions

Wave, wind and current criteria are site-specific and should be derived from measured data, hindcast or other theoretically sound oceanographic theories. It is a common practice to specify a design criteria based on a "100-year" return period, which statistically is considered equivalent to an event with a probability of occurring equal to 0.01 in any given

Figure 7.10 Three level topsides deck (Diana DDCV, Courtesy of EXXONMOBIL)

year at a particular site. This means it has a probability of about 0.18 of occurring over a 20-year lifetime of a structure. Often, the estimate of the "100-year" event has to be extrapolated from only a few years of data.

Typically the 100-year wave does not occur with the 100 year current, especially in regions with strong inertial or geostrophic currents such as the North Atlantic/Norwegian Sea or the Gulf of Mexico. The design criteria in these regions typically call for combining a 100-year Wave and Wind with a 10-year Current. Such simplified environmental specifications may be supplemented by more sophisticated "response-based" analysis wherein an extensive database of hindcast weather conditions is used to determine statistical distributions for a particular response, e.g. mooring tension (see, e.g. Standing, et al, 2002; Leverette, et al, 2004).

Table 7.4 shows typical environmental conditions provided by the DNV OS E-301 (Offshore Standard – Position Mooring). Here the wave and the wind are 100-year conditions, the current is 10 year. The spectral shape parameter, γ, refers to the three parameter Jonswap spectrum discussed in Chapter 3. Recent data on swell in West Africa suggests that a more appropriate spectral shape is an Ochi-Hubble shape with six parameters representing a combination of sea and swell (Chapter 3). The current speed is at the surface. Wind driven current is typically considered constant over a "mixing length" proportional to the wind speed. The mixing length for a 40 m/sec, 1-hr. wind speed is about 60 m.

Water Depth and Geotechnical Properties

Water depth is specified at the platform location. The topography of the seafloor should be specified for the design of moorings and risers, particularly catenary or flexible

Table 7.4 Guidance on environmental criteria (Derived from DNV OS E-301)

Location	Hs	Tp	γ	Uw (1-hr. avg.)	Uc
	m	sec		m/sec	m/sec
Norwegian Sea	16.5	17.0–19.0	1.9	37	0.9
Northern North Sea	15	15.5–17.5	2.3	40.5	1.5
North Sea	14	15.0–17.0	2.3	34	0.55
Gulf of Mexico	11.9	14.2	2.8	41.4	1.98
West Africa (swell)	3.6–4.1	15.5–16.0	1.0	16	0.9–1.85
West Africa (Squalls)	2.0–2.7	7.0–7.6	1.3	22–30	1.6
Brazil	8	13	1.6	35	1.6
South China Sea (non Typhoon)	7.3	11.1	2.8	28.6	0.85
South China Sea (Typhoon)	13.6	15.1	2.8	56.3	2.05

risers. Sometimes the seafloor topography, well layout or soil properties dictate that anchors be placed in specific areas or corridors. These kinds of constraints should be established early.

Seafloor geotechnical properties are required to design anchors and to analyse the interaction of the catenary risers with the seafloor. The following should be specified as a minimum: sediment type, shear strength profile, frictional coefficient.

Risers

Risers consist of production, drilling, import and export lines. Satellite wells also require umbilicals. The number and type of risers has a profound effect on the selection of a floater concept. One of the most fundamental issues in planning a deepwater field is the decision on whether to use wet or dry trees.

From the standpoint of the floater functional requirements, the main riser information required is:

- Wet Trees (Flexible Risers or Steel Catenary Risers, SCRs)
    - Number, size and azimuthal angle for each riser
    - Top tension and vertical angle for each riser
    - Interface with hull: j-tube, flex joint hangoff, or other
    - Allowances for future risers (hull must provide sufficient buoyancy for this)
- Dry trees (Top Tensioned – "rigid" risers)
    - Number and size of each riser
    - Top tension (calm water)

- Method of support
- Minimum spacing between risers

An important consideration for a floater is the method of providing riser tension, in particular whether the tension is to be carried by the vessel or by another means of buoyancy such as the spar buoyancy cans discussed in Chapter 9. Figures 7.11 and 7.12 show two

Figure 7.11 Hydraulic tensioner system for dry tree semi (from Wanvik and Koos, 2000)

Figure 7.12 Passive spring riser tensioner for Prince TLP (from Koon, et al 2002)

methods used to transmit riser tension to a floater. Figure 7.11 shows an arrangement of hydraulic tensioners on a separate substructure for a dry tree semi-submersible concept. The total stroke is in excess of 20 ft. Figure 7.12 shows a novel tensioner method used on a mini-TLP. This is a simple coiled spring. The total stroke in this case is only a few feet.

Whatever tensioning system is used, the design needs to consider the range of tensions and strokes which can be encountered in all weather conditions, and in the case of one or more failures. Failures might include:

- Broken mooring line
- Flooded compartment in hull
- Flooded compartment in buoyancy can (if applicable)
- Failure of one or more hydraulic components (if applicable)

It is important to select, at the beginning of the design phase, the appropriate combination of events and the associated safety factors to use in the design of the risers. A matrix of cases should be developed. Table 7.5 shows some example cases.

Analysis of the risers requires global responses of the floater hull. The coupling between the vessel and riser responses is often ignored, but it becomes increasingly more important as the water depth increases.

Table 7.5 Example of some riser load cases

Case ID	Riser	Fluid	Pressure at Surface	Environment	Damage	Allowable Design Factor
1	Production	Tubing: 0.95 SG Annulus: SW 1.02 SG	5000 psi	10 year Winter Storm	None	Basic
2	Production	Tubing: 0.95 SG Annulus: SW 1.02 SG	5000 psi	100 year Hurricane	None	1.33 × Basic
3	Production	Tubing: 0.95 SG Annulus: SW 1.02 SG	5000 psi	1 year Winter Storm	Flooded Compartment in Hull	1.33 × Basic
4	Production	Tubing: 0.95 SG Annulus: SW 1.02 SG	5000 psi	100 year Hurricane	Flooded Compartment in Hull	1.0 × Serviceability Limit State

7.3.2 Configuration Proportions

The following is a general discussion of selecting proportions. Detailed sizing methods are described for the specific concept type.

Design Criteria

The main dimensions of the floater are determined from considerations of the space required, total displacement (buoyancy), stability and seakeeping (motions). Various "criteria" for proportions may be used depending on the type of floater, as summarised in table 7.6.

The initial configuration requires "guesses" for many of the weights, which make up the total displacement, i.e. the hull weight and the mooring and the riser vertical force.

The configuration may be driven by the use of design rules, i.e. the Classification Society Rules, which will dictate a minimum freeboard, stability margins, etc. Consideration should be given at this stage to the fabrication and installation steps in order to insure that the vessel can be economically installed.

Arrangements

Space requirements for deck and hull equipment need to be accommodated. A preliminary layout of the decks on the topsides and the hull should be prepared to indicate adequate space. Also, the arrangement of compartments for the ballast and storage of liquids needs to be determined. An elevation plan should be developed to estimate wind loads.

Hydrostatic Subdivision (Stability)

Floaters should be designed so as to float in a safe and stable condition even if a buoyant space is accidentally flooded, either by human error or collision. Once an overall configuration is determined the internal subdivision of tanks needs to be determined based on this criteria. Stability rules are discussed herein. Generally, an initial sizing may consider external dimensions only. The internal compartmentation may be addressed after the initial sizing is performed.

Table 7.6 Example "Rules of thumb" for configuration sizing

Floater Type	Criteria
TLP	Tethers minimum top tension in most extreme environment.
	Heave and Pitch Periods < 4 seconds.
Semi-submersible	Metacentric height (GM) greater than 15 ft under normal operating conditions.
	Positive GM while at transit draft (MODU)
	Heave period greater than 20 seconds.
Spar	Maximum heel angle 5 degrees in 100 year storm.
	Heave period ~2 times peak storm wave period.
Ship Shaped FPSO	Provisions for process, quarters, turret and oil storage govern.

Hydrodynamics

The motions of the floater are affected by its configuration. Once the buoyancy and the stability are acceptable, motion predictions may be necessary to determine if the motions are acceptable for the desired operation. This may be an analytical procedure and/or it may include scale model testing. The activity of determining motions is traditionally referred to as "seakeeping analysis", and more frequently in offshore circles as "Global Response Analysis". This analysis is often coupled with the generation of loads for structural analysis, mooring analysis and riser analysis. For floaters with moorings and multiple risers, this analysis may be very complex and take several months. At the initial stages, simplified methods for determining motions are desirable. This may involve, for example, solving for natural periods and perhaps responses in heave alone. Some simplified methods are described in the subsections on specific hull forms.

Structural Design and Analysis

The initial configuration must provide buoyancy for the payload and the hull weight. The first configuration must necessarily be determined before the hull structure is designed using a "guess" for the hull weight. In most cases, the hull steel weight may be estimated to be within 10% by an analysis based on hydrostatic pressures alone (local strength). The dynamic loads are important for local strengthening and fatigue checks which are normally performed in the later stages of a design, but these refinements normally add a small amount to the weight, usually less than 15%. The initial hull weight estimates for configuration should be performed by a spreadsheet or simple programming using a parametric method, whereby approximate weights for varying proportions may be quickly determined, along with the hydrostatic balance of weight and buoyancy and the stability.

Mooring and Foundation Design

The mooring system holds the floater in position and in the case of the tension leg platform is also an integral part of the platform stability. An analysis needs to be performed to determine whether the platform can withstand the survival environment(s) while keeping its offset (distance away from a calm water neutral position) and the loads on the mooring system within an acceptable margin. The offset margin is usually determined by the properties of risers supported by the vessel and the water depth. A separate *riser analysis* may be needed to definitively determine this, however in most cases a maximum offset is prescribed based on past experience. Mooring safety factors and the methods of analyses are discussed below. The mooring analysis is often combined with the Global Response (motions) analysis discussed above. Mooring vertical forces can be a significant factor in the total displacement and need to be included as a "guess" in the initial configuration.

7.3.3 Weight Control

Since weight is the primary design criteria for floaters, it is important to maintain a comprehensive and consistent breakdown of the weights of a floater including the "payload" and the other weights and loads. There should be a continuous tabulation of weights and their locations (x, y and z) for purposes of evaluating the floater's performance.

There is no industry standard for weight breakdown, nor even a consistent definition of "payload". This is partly because of the different types of floaters, and also because of the differing conventions adopted over the years by various fabrication and shipyards where weight control is a significant management function, often involving whole departments.

Weights are usually divided into two main categories: "fixed" and "variable". The fixed weights are normally weights, which are physically attached to a platform when it is installed. A "payload" refers to the functional weights which the substructure (the structure providing the buoyancy) must support. Weights associated with the substructure, e.g. the vertical mooring loads and ballast, are typically not considered "payload".

The following, table 7.7, can be considered as an example weight breakdown.

Not all of the items listed apply in all cases, and there could be additional items. The term "Module" refers to a collection of equipment which is set on a frame. The frame might

Table 7.7 Example weight breakdown

1. Payload – Fixed
    - 1.1. Process Equipment or Modules
    - 1.2. Drilling Equipment or Modules
    - 1.3. Utilities Equipment or Modules
    - 1.4. Accommodation Modules
    - 1.5. Outfitting (piping, electrical, non structural steel, etc.)
    - 1.6. Secondary steel
    - 1.7. Tankage
    - 1.8. Flare tower
    - 1.9. Helideck
    - 1.10. Cranes
    - 1.11. Riser tensioning equipment on deck
    - 1.12. Mooring equipment on deck
2. Payload – Variable
    - 2.1. Personnel
    - 2.2. Operating Supplies
    - 2.3. Mud (active, bulk)
    - 2.4. Tubulars (drill pipe, casing)
    - 2.5. Moveable equipment (risers, BOP)
    - 2.6. Working loads (hook load, drilling riser tension)
    - 2.7. Liquids (fuel, drill water, potable water)
    - 2.8. Drilling loads (hook, riser guideline loads)
    - 2.9. Production Liquids and Consumables
    - 2.10. Subsea Equipment to install
    - 2.11. ROV and support equipment
    - 2.12. Production Riser tensions
    - 2.13. Trees
    - 2.14. Spares
3. Topsides Primary Structure (or "Deck Steel")

(Continued)

Table 7.7 Continued

4. Hull Primary Steel
5. Hull Marine Outfitting
    5.1. Structural Components (where applicable)
        5.1.1. Riser/Tendon Porches
        5.1.2. Strakes (on hull for vortex suppression)
        5.1.3. Riser Supports (Guides)
        5.1.4. Fairlead Supports
        5.1.5. Fairings (around external piping)
        5.1.6. Chain Jack Supports/Foundation
        5.1.7. Chain Locker and Hawse Pipes
        5.1.8. Winch Foundation
        5.1.9. Topsides Interface
        5.1.10. Truss/Hull Interface (Spar Hard Tank)
        5.1.11. Truss/Hull Interface (Spar Soft Tank)
        5.1.12. Tendon Porches
        5.1.13. Riser Porches
    5.2. Riser Support
        5.2.1. Tensioner support
        5.2.2. J-Tubes
        5.2.3. Internal Riser Sleeves (Guide Tubes)
    5.3. Miscellaneous Mechanical Equipment (pumps, etc.)
    5.4. Electrical
    5.5. Steel Outfitting
        5.5.1. Hatches and Ladders
        5.5.2. Boat Landing
    5.6. Mooring
        5.6.1. Fairlead and Bending Shoe
        5.6.2. Chain Inboard of Fairleads
        5.6.3. Tail Chain
        5.6.4. Chain Jacks and Sheaves (on hull)
        5.6.5. HPU
        5.6.6. Winch
    5.7. Piping
        5.7.1. Piping
        5.7.2. Bilge/Ballast Pumps
        5.7.3. Caissons
        5.7.4. Liquid in Piping Systems
    5.8. Corrosion Protection
        5.8.1. Anodes
        5.8.2. Painting
    5.9. Marine Growth
    5.10. Cargo Handling Systems (e.g. for FPSO)
6. Other loads
    6.1. Mooring vertical force
    6.2. SCR Vertical force
    6.3. TTR Vertical force

include piping and electrical components and the whole assembly might have been tested and commissioned prior to integration with a deck. Sometimes there is a separate weight category for "module steel" to identify the steel in the module, which is not part of the primary deck structure.

7.3.4 Stability (Krish Thiagarajan, University of Western Australia, Perth, WA, Australia)

Buoyancy

Stability and weight estimation are closely related. In the initial design stages, the stability of a floater must be evaluated simultaneously with its geometry refinement and weight estimation.

Stability is the ability of a system to return to its undisturbed position after an external force has been removed. When a floating vessel is in static equilibrium, it is under the influence of two forces: weight and buoyancy. While weight is the product of mass and gravitational acceleration, buoyancy is given by the weight of the displaced volume of water (∇) due to the presence of the body. Consider the examples shown in figs. 7.13 and 7.14.

In the preliminary design stages, the intact stability at small angles of heel should be sufficient. For more detailed design, it is important to consider the stability at large angles.

There are several types of stabilities that need to be assessed for a floating vessel. Since FPSOs typically are long compared to their width, the stability in the transverse plane can be different from that in the longitudinal plane. Further, a floating vessel should be stable in the static condition (e.g. due to a steady wind force), and in the dynamic condition (e.g. when a sudden gust blows along with a steady wind). The vessel should also have a reserve stability such that it can sustain a moderate environment in a damaged condition, e.g. when one of its compartments is flooded.

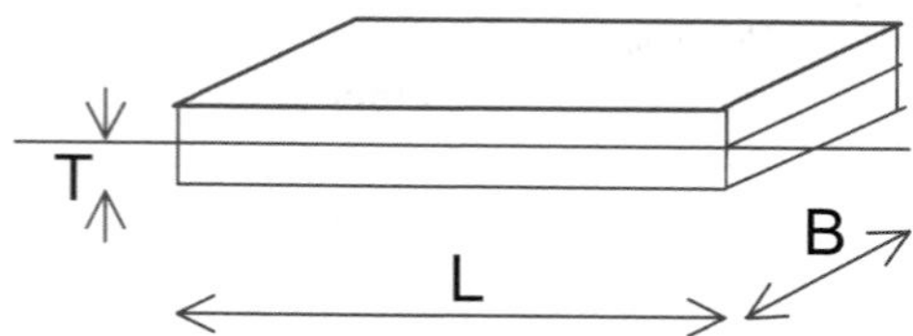

Figure 7.13 Displacement of a prismatic structure, $\Delta = LBT$

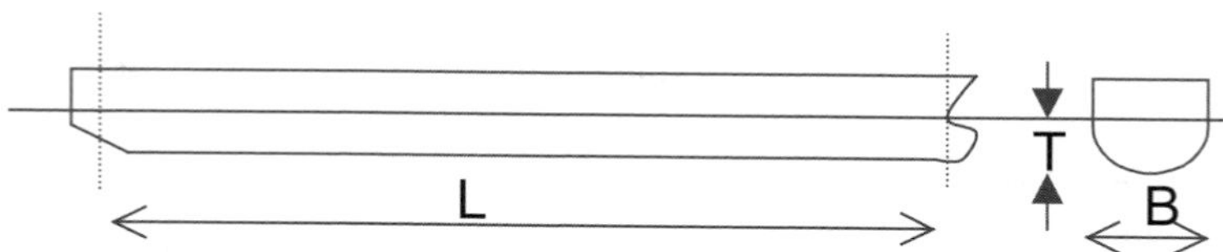

Figure 7.14 Displacement of a ship, $\Delta = C_B LBT$ (C_B is the block coefficient)

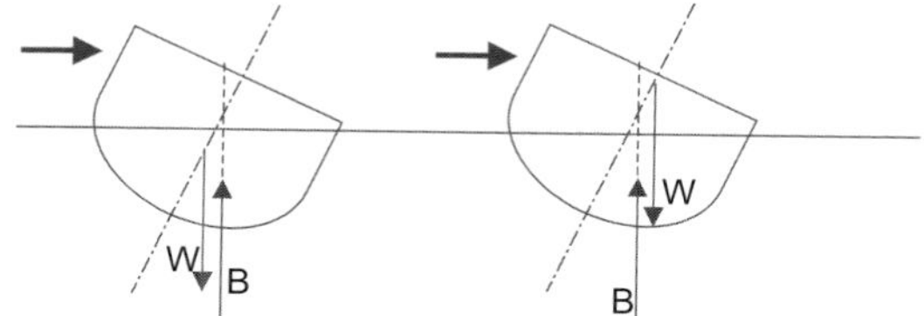

Figure 7.15 Positive and negative stability

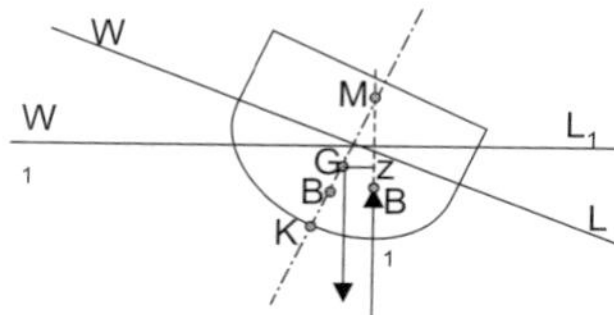

Figure 7.16 Metacentre defined

Basics of transverse stability

The points of action of weight (the center of gravity) and buoyancy (center of buoyancy) determine if a vessel is stable or otherwise. Consider the two cases of fig. 7.15. Case 2 is negatively stable because the net moment due to the two forces tends to destabilise the system further.

To better appreciate the difference between the two cases, it is necessary to look at the point of intersection between the line of action of buoyancy and the centreline, the *metacentre M*, fig. 7.16.

Here, K = keel point, G = point of action of weight, i.e. *centre of gravity*, B = point of action of buoyancy, i.e. *centre of buoyancy*; The position of B shifts with the amount of heel (B to B_1), M = Point of intersection of line of buoyancy and centreline, i.e. *Metacentre*, and GM = Distance between G and M, i.e. *metacentric height*.

For a floating system to be positively stable, GM > 0. The metacentre can be likened to the centre of oscillation of a suspended pendulum. Then GM becomes the length of the string, and for the pendulum to swing in a stable oscillation and return to its original position, the centre must be above the pendulum. For a submerged object to be stable, the centre of gravity must be below the centre of buoyancy. But since the point of action of buoyancy is fixed along the line of gravity and does not change, the metacentre is B itself. The criterion GM > 0 thus still holds well.

Evaluating the metacentric height follows from the above figure. It can be seen that:

$$GM = KB + BM - KG \tag{7.1}$$

where *KB* and *KG* are the distances from the centres of buoyancy and gravity respectively to the keel of the vessel. *BM* is the distance between the centre of buoyancy and the metacentre, and is given by (see e.g. Comstock, 1979):

$$BM = \frac{I_{xx}}{\nabla} \tag{7.2}$$

I_{xx} is the second moment of waterplane area about the x-axis. Typical GM values for a semi-submersible is 6 m, and a FPSO in ballast around 3 m.

Longitudinal stability

A floating ship or an FPSO is very stable longitudinally compared to the transverse plane. This is because GM_l is very large compared to GM (transverse). Figure 7.17 illustrates the point.

Evaluating the longitudinal metacentric height is similar to the transverse case, i.e.

$$GM_l = KB + BM_l - KG \tag{7.3}$$

$$BM_l = \frac{I_{yy}}{\nabla} \tag{7.4}$$

where I_{yy} is the second moment of waterplane area about the y-axis. For a typical vessel, since BM_l is an order of magnitude larger than $(KB - KG)$, we can assume

$GM_l \approx BM_l$.

Dynamic stability

The dynamic stability criteria for a ship or an FPSO are set based on the stability requirement to withstand a sudden environmental change, e.g. a gust of wind. Vessels that are intact are required under the ABS certification to be able to withstand a 100-knot (51 m/s) wind in a storm-intact condition. This applies equally to the column-stabilised vessels (e.g. semi-submersibles). In a damaged condition, the vessel should have sufficient stability to withstand a 50-knot (25.7 m/s) wind. Yet another requirement under the ABS regulations, is the requirement of at least 40% residual positive stability. Let us demonstrate this by considering the righting moment (product of the buoyancy and the righting arm GM, see fig. 7.18) distribution of a 30,000 t displacement cargo vessel. The heeling moment caused by a steady 100-knot wind is also shown. The heeling moment is computed from the couple produced by the wind load (which varies with heel angle) and the center of pressure of drag loads, assuming the vessel is free floating. If permanently moored, the heeling moment is

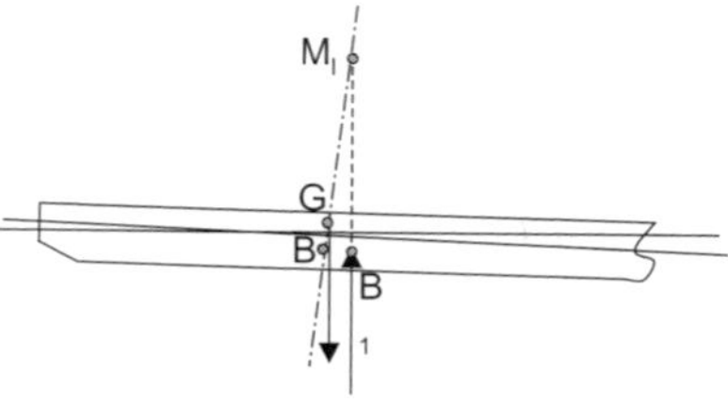

Figure 7.17 Longitudinal stability

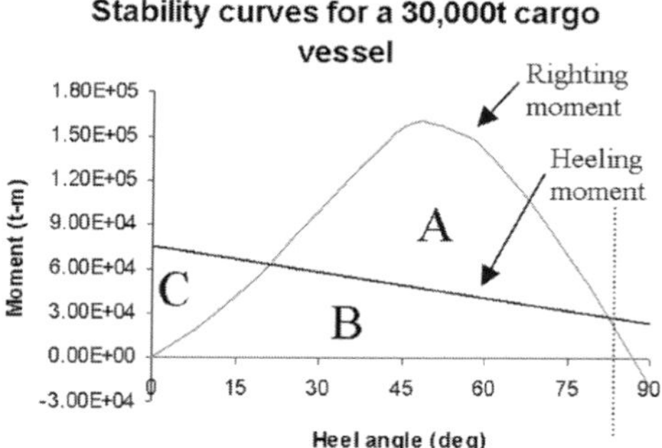

Figure 7.18 Dynamic stability curves for a 30,000 t cargo vessel

based on the worst location of the reaction loads: fairload locations or center of drag pressure.

If A, B and C are the designated areas, such that (A+B) is the area under the righting moment curve up to the dashed line (called the second intercept) and (B+C) is similarly the area under the heeling moment curve up to the second intercept. Then the ABS rule requirement implies:

$$(A + B) > 1.4(B + C) \tag{7.5}$$

The 40% excess is a safety limit, and denotes the work required to be done by an external force (in addition to the heeling moment) to capsise the vessel. For semi-submersibles, the excess requirement is 30%.

Stability in damaged condition

A vessel should be compartmentalised sufficiently to withstand flooding from the sea of any one main compartment. Further, in the damaged condition, the vessel should have sufficient stability to withstand a 50-knot wind (ABS MODU rules). The final waterline in the damaged condition is to be below the lower edge of any opening through which downflooding may occur.

Flooding of a compartment results in sinkage as well as trim. There are two methods of assessing stability in this condition:

Lost buoyancy method:

- Flooded volume treated as lost underwater volume
- Loss of water plane area calculated
- Sinkage and trim estimated
- Iterations carried out to get final position of vessel

Added weight method:

- Flooded water treated as added weight
- New displacement and KG evaluated

- Corrections for water plane area lost and displacement adjusted up to sinkage condition
- Repeat calculation to get convergent results

Both the methods give equivalent results.

Other considerations

Partially filled tanks affect stability. Half-filled tanks shift liquids when the vessel heels, thus moving G. This creates an adverse effect of decreasing stability. If the liquid cargo has density ρ_c, then the metacentric height is corrected to include effects of all partially filled tanks:

$$GM(\text{new}) = GM - \sum \frac{\rho_c}{\rho} \frac{i_{xx}}{\nabla} \tag{7.6}$$

where i_{xx} is the second moment of the partially flooded compartments waterplane.

For crane vessels operating offshore, when a load w is lifted from the deck, the metacentric height changes by:

$$GM(\text{new}) = GM - \frac{wh}{\rho\nabla} \tag{7.7}$$

where h is the distance of the crane tip from the deck level.

Effect of Flooded Column

The stability rules are intended to prevent catastrophic loss of a vessel, even if a compartment floods. In spite of this, there have been some dramatic losses, fig. 7.19. The semi-submersible P-36 was lost after an explosion in one of the columns led to flooding of multiple compartments (Filho, 2002). This type of accident is not envisioned by the classification rules, which are mainly concerned about flooding from collision damage. Each maritime catastrophe leads to an investigation and a rules review, which often results in new standards. As a result of the P-36 accident, some rule-making bodies have proposed

Figure 7.19 P-36 converted semi-submersible after flooding in one column (Barusco, 2002)

a requirement to have a reserve buoyancy on the deck of semi-submersibles, and to prevent storage of hydrocarbons in columns meant for stability.

Also, the above discussion is based largely on the ABS Modu Rules. The reader is cautioned that other, more restrictive, rules may apply for vessels carrying passengers, or in other parts of the world (see Wijngaarden and Heemskerk, 2000) which should be reviewed for each project.

7.3.5 Coordinate Systems and Transformations

Coordinate systems and transformations are central to much of the analysis surrounding floaters. For example, vessel motion computations made using one computer program are often used to provide input to another computer program, which computes riser dynamic responses. Often these programs use different coordinate systems. Also, the motions at one point often needs to be transferred to another point, e.g. from the center of gravity (computed by a vessel motion program) to a riser hangoff locations (required as input to the riser analysis). This section presents some common coordinate systems and their associated transformation matrices.

The coordinate system commonly used in Naval Architecture is shown in fig. 7.20.

Translational and rotational motions are referred to in terms of surge (x), sway (y), heave (z), roll (x-axis), pitch (y-axis) and yaw (z-axis). Directions on the floaters, even though they are not underway, are often referenced in naval architectural parlance:

The *bow* is pointing forward,
The *stern* is facing aft (backwards)
The *starboard* side is on your right as you face forward,
The *port* side is on your left as you face forward

On the ship-shaped vessels, these directions are self-evident. Such is not the case on spars and other symmetrical types of floaters. Typically, when performing motion

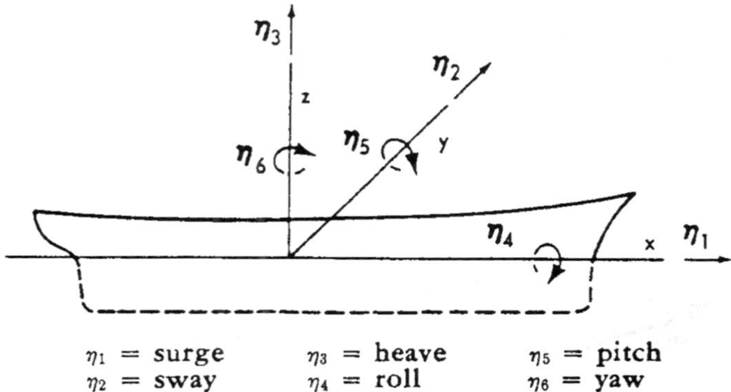

Figure 7.20 Naval architecture convention and nomenclature for translatory and angular displacements (Society of Naval Architects and Marine Engineers, 1988)

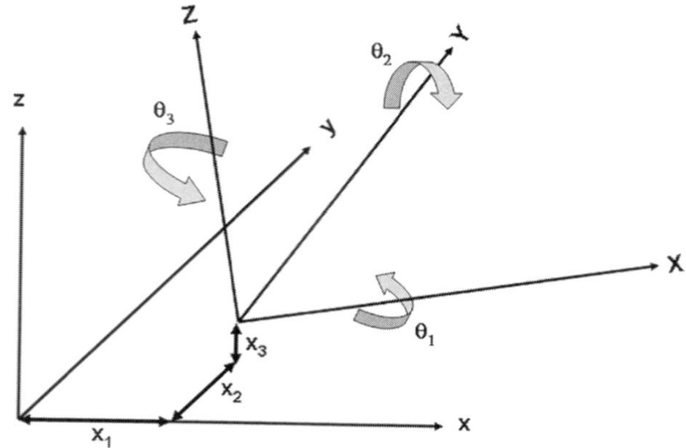

Figure 7.21 Coordinate system #1

calculations, it is typical to refer to the bow as the direction facing the oncoming waves or current, otherwise it is always necessary to clearly define the coordinate system and directions.

A common problem in floating system design is the translation of motions from one point on the body to another when the body is translating and rotating relative to a fixed coordinate system. For large rotations the coordinate transformation depends upon the order in which the rotations are taken. It is common to use the so-called Euler angles to define these rotations.

There are several conventions of Euler angles depending on the axes about which the angles are taken about (see, e.g. Landau and Lifchitz, 1976). A common convention is to assume the order of rotation to be first about the yaw (z) axis, then about the roll (x) axis and then about the pitch (y) axis. These rotations are relative to the coordinate system shown in fig. 7.21. Figure 7.21 shows the two coordinate systems: (x, y, z) represents a global coordinate system fixed in space, and (X, Y, Z) represents a coordinate system fixed to a rigid body. The coordinates of a point, P, may be referred to in either coordinate system using Euler angles as follows:

Body coordinates in terms of global coordinates:

$$\vec{X} = [A] \cdot (\vec{x} - \vec{x}_i) \tag{7.8}$$

Global coordinates in terms of body coordinates:

$$\vec{x} = [A]^T \cdot \vec{X} + \vec{x}_i \tag{7.9}$$

Here [A] is a transformation matrix, and x_i is the translation of the origin of the body coordinate system with respect to the global coordinate system.

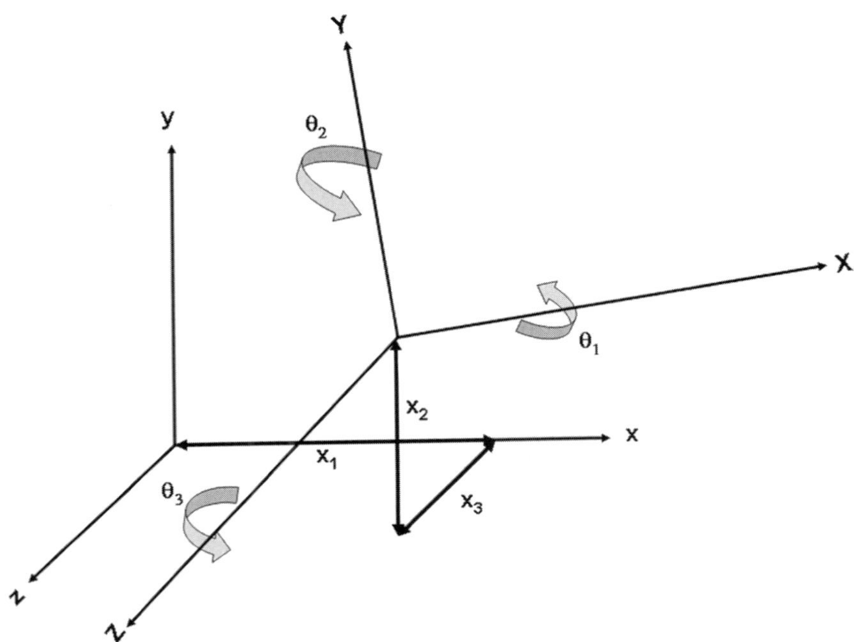

Figure 7.22 Coordinate system #2

Sometimes the coordinate system is defined with the vertical axis as "y" as illustrated in fig. 7.22 (Paulling and Shih, 1985).

The coordinate transformation matrices for both these coordinate systems are presented in table 7.8.

Table 7.8 Coordinate transformation matrices

	Coordinate System #1 (fig. 7.21)	Coordinate System #2 (fig. 7.22)
A_{11}	C_2C_3	C_2C_3
A_{12}	C_2S_3	S_3
A_{13}	$-S_2$	$-S_2C_3$
A_{21}	$S_1S_2C_3 - C_1S_3$	$S_1S_2 - S_1C_2S_3$
A_{22}	$S_1S_2S_3 + C_1C_3$	C_1C_3
A_{23}	C_2S_1	$C_1S_2S_3 + S_1C_2$
A_{31}	$C_1S_2C_3 + S_1S_3$	$S_1C_2S_3 + C_1S_2$
A_{32}	$C_1S_2S_3 - S_1C_3$	$-S_1C_3$
A_{33}	C_2C_1	$C_1C_2 - S_1S_2S_3$

The terms in this table are the cosine and sine functions:

$C_1 = \cos(\theta_1)$
$C_2 = \cos(\theta_2)$
$C_3 = \cos(\theta_3)$
$S_1 = \sin(\theta_1)$
$S_2 = \sin(\theta_2)$
$S_3 = \sin(\theta_3)$

These transformations are used frequently for the analysis of moorings and risers. In particular, computation of the floating platform motions usually result in rigid body motions at a particular point on the body, typically the centre of gravity or the centroid of the mean waterline. In order to perform riser or mooring analysis, it is necessary to translate these motions to the global coordinates of the point of attachment: the fairlead or riser hangoff point. Global coordinates in terms of the body coordinates may be used for this purpose:

$$\vec{x} = [A]^T \cdot \vec{X} + \vec{x}_i \tag{7.10}$$

7.4 Floating Production Storage and Offloading Systems

FPSOs are the most prolific floating production platforms. They were developed in the 1970s to produce from smaller, remote fields where pipelines or fixed structures would be uneconomic. The early FPSOs were restricted to mild environments, which allowed the use of tanker conversions. The turret mooring was introduced in 1986, which, in turn, opened the door for the FPSOs into more severe environments. The first new build and North Sea FPSOs were also introduced in that year (see Ronalds and Lim, 1999, for a review of FPSO history). The North Sea and Brazil are today the primary FPSO markets. In spite of their presence in the world's offshore oil producing regions there was not a single FPSO in the Gulf of Mexico as of 2004. The US government has only recently approved their use in the Gulf of Mexico (Minerals Management Service, 2002), however as of 2004 no oil company has made a proposal to install one.

The FPSOs are generally ship shaped floaters with provisions for storing and offloading of oil simultaneously. They may be designed to weathervane so that they always face the weather, minimising roll and heave motions. In benign environments such as West Africa and South East Asia, the FPSO may be spread moored to face one direction at all times. Some FPSOs for Brazil have been designed to semi-weathervane by using a spread mooring with slack aft moorings, giving the vessel the options of some limited weathervaning (Kaster et al, 1997; Portello et al, 2003).

Oil production is through either flexible risers or riser towers with flexible jumpers. The motions of the FPSOs generally prohibit the attachment of rigid vertical risers or steel catenary risers.

The FPSOs have a large area for setting a deck on the top of the hull. However, many FPSO hulls are conversions and the deck structure may not be designed to carry a processing facility. This needs to be checked carefully before committing to a surplus tanker for conversion.

The converted FPSOs often offer the shortest and the cheapest path to initiating production. Their main limitations include a lack of ability to operate dry trees, and technical feasibility of mooring in very deep water in harsh environments. The turret assembly can become very complicated and difficult to integrate with the hull.

7.4.1 FPSO Hull Design

There are four principal requirements that drive the size of a typical FPSO (fig. 7.23):

1. Provision of oil storage capacity compatible with the production rate and offloading arrangements, i.e. shuttle tanker turnaround time
2. Provision of topsides space for a safe layout of the process plant, accommodation and utilities.
3. Provision of displacement and ballast capacity to reduce the effects of motions on process plant and riser systems.
4. Provision of space for the production turret (bow, stern or internal), and the amount of hull storage capacity lost as a consequence (new-build or conversion).

Figures 7.24 and 7.25 show FPSOs with a bow turret and internal turret, respectively. Kizomba A, Figure 7.26, is an example if an FPSO with a spread mooring and no turret.

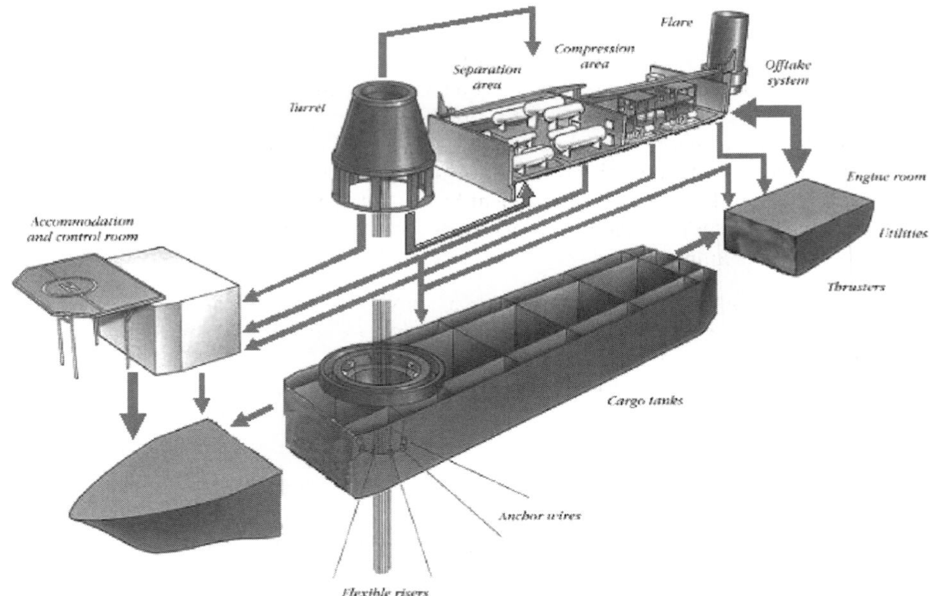

Figure 7.23 Typical new-build FPSO configuration (courtesy J. Ray McDermott, Inc.)

Figure 7.24 Xikomba with Bow Turret (SBM)

Figure 7.25 Balder FPSO with Turret Midships (ExxonMobil)

As water depths increase, the sensitivity of the FPSO mooring and riser systems to wave frequency motions increases rapidly. The FPSO hull form can be optimised to counter this by varying primarily the length, width, depth, draft and mass distribution.

The hull length can be adjusted with respect to the expected wave lengths to ensure that the extreme environmental conditions do not generate wave lengths located in the range of the peak heave and/or pitching responses, thus reducing the turret vertical motions

Figure 7.26 Kizomba A, Spread Morred FPSO in West Africa (ExxonMobil)

and associated dynamic loads. The hull form can also be adjusted in order to control rolling and pitching behaviour thus reducing riser bend-stiffener design requirements, particularly for transverse conditions. This optimisation also reduces the need for special topsides process equipment design requirements such as baffles for separators.

The hull form can also be optimised to reduce the wave and current actions by varying the length, width, draft and shape of bow and stern. This reduces the vessel loading with respect to incoming waves under transverse conditions, reducing mooring loads and vessel roll. The aforementioned hull form optimisation can have a beneficial contribution to the long term operational performance of the FPSO, but this should be carefully assessed against the fabrication cost and Life of Field (LOF) costs.

In the more recent FPSO developments, there have been cases where the cargo carrying capacity of the FPSO is not sized against the initial production flow rate, but one further into the production life where a reduced and more economical storage can be achieved. The initial storage requirements are met by supplementing the FPSO with a Floating Storage Unit (FSU) based on a cheap second hand tanker.

The above highlights that a full LOF approach should be taken in deciding both the vessel design and the field configuration. The following sections discuss an example FPSO design for 800,000 BBL storage in deep water.

7.4.2 Hull Structure

There is a marked difference in the hull arrangement between the new purpose-built FPSO hulls and the converted tanker hulls. The overall configuration for tankers is driven by the need to transport large cargo volumes at a low cost. Tankers have consequently evolved to a length to breadth ratio of about 6:1, which gives a good compromise between the enclosed volume and the resistance to forward motion. The FPSOs are not required to move forward; consequently resistance is not an issue. However, in a weathervaning mode the hull slenderness ratio (length to beam ratio) serves to present a low frontal area to the prevailing environment and assists in the natural weathervaning motion. A low slenderness ratio results in more favorable motions and mooring behaviour over a shorter and more bulky hull. A shorter hull would however offer savings in steel weight and possible cost reductions. Hull breadth to depth ratio comparisons are a different prospect since FPSOs, unlike tankers, are not constrained by maximum draughts, so the typical 2:1 ratio can be optimised. This helps structural design and seakeeping by increasing freeboard whilst permitting greater bow submergence to reduce slamming.

With regard to the issue of the double hull and the double bottom hull design, the former has become the norm for FPSOs, whilst there is little justification at present to impose a double bottom on a statically moored FPSO. It has become common practice to arrange ballast tanks outboard of the central cargo tanks. The ballast capacity depends on the range of operating draughts due to seakeeping requirements and offloading.

7.4.3 Example FPSO Design

Consider two different design options:

- New build Aframax tanker, modified to FPSO requirements (fig. 7.27)
- New build Barge, configured to FPSO requirements (fig. 7.28)

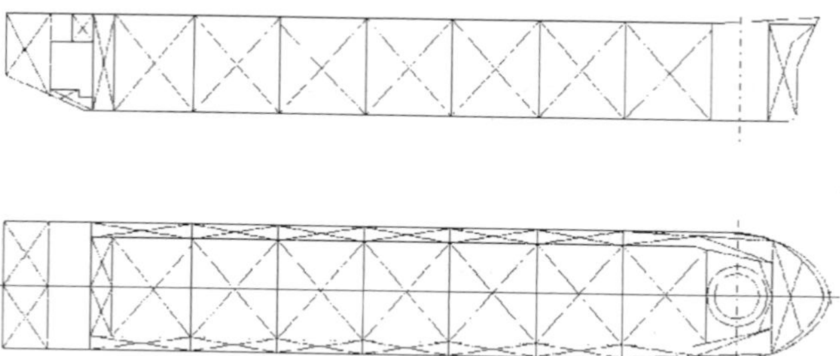

Figure 7.27 New build Aframax Tanker, 800,000 bbl storage

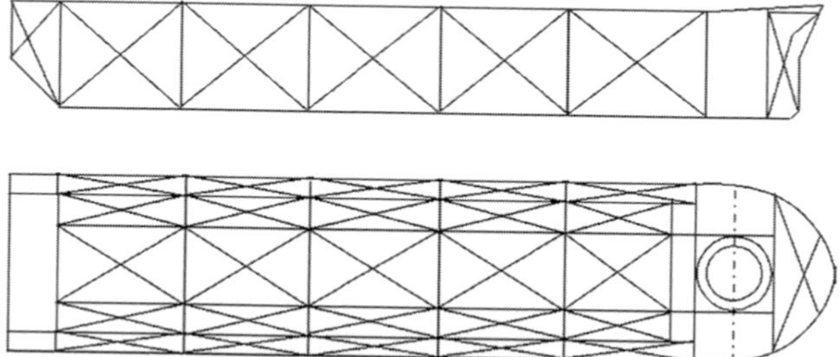

Figure 7.28 New build Aker Tentech Barge, 800,000 bbl storage

The tanker requirements include the following systems:

1. Cargo Oil Storage
2. Structural provision for accommodation, turret & cranes
3. Crude Oil Washing (COW) system
4. Inert Gas (IG) system
5. Slop tank system
6. Ballast water system
7. Fuel oil system
8. Safety systems

The standard approach when designing a new build FPSO is to start with a clean sheet and design an offshore structure as a permanently moored vessel according to regulatory requirements, e.g. Lloyds rules. This typically results in an optimised structure that bears little resemblance to a trading tanker, which is the source for a majority of converted FPSOs. The trading tanker business is a highly competitive market where tanker designs have evolved out of necessity into a "no frills" vessel that provide a minimum cost product that employ the maximum capability of the respective shipyards to manufacture a tanker within a tight fabrication schedule slot.

Accordingly, the tanker option would involve modification of a standard Aframax tanker at the design stage in order to introduce the specific requirements of the FPSO; whilst retaining as much of the trading tanker design as possible.

The modifications to the vessel would be similar to those employed for the converted FPSO, with the added benefit of new equipment instead of refurbishment of the old, and the optimisation of the marine systems that new fabrication allows. The design modifications to the hull are assumed to be minimal, hence no changes are made to the hull framing, plate thickness or welds on the basis that the proposed duty will not exceed that of a tanker working on worldwide duty. Secondly, the main changes in duty are configured through the turret and the topsides process loads, which are both accounted for via supplementary stiffening.

Hull Particulars: (800,000 bbls Storage Tanker)

Loa	243.20 m
Lbp	233.00 m
Breadth	41.80 m
Draught	13.60 m
DWT	97,046 mt
Displacement	110,884 mt

Hull Particulars: (800,000 bbl Storage Tentech Barge)

Loa	200.60 m
Breadth	42.00 m
Draught	19.50 m

7.4.4 Deck Structure

7.4.4.1 Accommodation

The position of the accommodation facilities can both dictate the position of the turret and potentially reduce the cargo carrying capacity of the tanker. The basic premise that accommodation should be placed forward on a weathervaning vessel is a compromise between safety issues and construction/design issues:

Bow Accommodation:

Advantages	Disadvantages
– Safety, process fires, smoke blows away from accommodation	– Accommodation displaces bow turret into cargo tanks, reducing capacity
– Reduced motions, higher crew comfort	– Lifeboats deploy into smoke in process fire
	– Loss of cargo tank capacity = longer vessel
	May need to move flare to stern, possible radiation problem with cargo offloading
	– Fire wall required to shield accommodation from turret.

Stern Accommodation:

Advantages	Disadvantages
Lifeboats deploy away from smoke in process fire.	Higher motions at stern, crew discomfort factor.
No fire wall required to segregate Process/turret and accommodation	Safety, process fires, smoke blows towards Accommodation.

Since no oil can be stored below the accommodation block and the propulsion machinery must be placed down-wind to aid the weathervaning of the vessel. This means that storage space is lost at both ends. If we look at say the Petrojarl 1 (Hewlett et al, 1987) or Gryphon A (Addy et al, 1994) as little as 31% of the cubic volume is made available for oil storage. In contrast the Alba FSU achieved 57% ratio using an internal turret and stern accommodation and the FPSO Challis Venture achieved 70% ratio with a similar set-up and optimised hull shape.

7.4.5 Turret Design and Selection

There is a wide range of FPSO turret designs available on the market to suit new build and/or tanker conversion with basically four main experienced designers providing the FPSO turret and the buoy systems worldwide:

1. SBM/Imodco (www.singlebuoy.com)
2. Bluewater (www.bluewater-offshore.com)

3. Sofec (www.fmcsofec.com)
4. APL (www.apl.com)

These main contractors offer a range of services from a straightforward turret delivery to a more comprehensive role with total FPSO delivery and operation. There are several other companies offering related turret designs on a niche basis.

7.4.5.1 Turret Designs

When designing turret mooring systems for ship-shaped vessels, one of the most important vessel design factors affecting the mooring line tensions is the location of the turret. The farther forward the turret is located away from the mid-ship, the easier it becomes for the vessel to weathervane into an equilibrium heading under non-collinear environments. However, the farther away the turret is placed, the more the vertical motions at the fairleads due to the vessel pitch will increase, which could have an adverse effect on the mooring line tensions in the line dynamic mooring analysis.

If the turret position is moved from a bow position to a position one-third the vessel length back from the bow, the intact minimum mooring line tension safety factor can improve by 15 to 18%.

In addition to considering the optimum position of the turret with respect to mooring line tension, it is equally important to evaluate the impact of the turret incursion into the vessel hull, and cost the impact. The following items can have a major design and cost impact on the vessel development:

- Loss of cargo tank volume
- Loss of longitudinal strength
- Introduction of additional stiffening steel to dissipate turret loads
- Diameter of the turret structure derived architecturally from the space requirements for the risers, the moorings and the turret equipment.
- Size of the turret bearing required to efficiently transmit mooring loads between the turret shaft and the FPSO hull.

When considering the total impact of the turret on the hull, the bow turret has proven more cost effective in both benign and harsh environments. The bow turret can be configured in two ways:

1. Integral bow turret (built within tanker bow) (fig. 7.29)
2. Cantilevered bow turret (fig. 7.30)

A fully weathervaning vessel has opex advantages over a controlled heading/limited rotation vessel, but the inherent requirements of a swivel joint for each flow path imposes practical limitations on the number of flow paths that can be provided for a fully weathervaning vessel.

Whilst all turret systems are disconnectable, the term is only used for turrets having the facility for quick connection and quick disconnection (QCDC). Most of the turret systems that have been designed for fairly benign weather and shallow water are disconnected when a typhoon is expected.

Figure 7.29 Internal bow turret (SBM–Imodco)

Figure 7.30 Cantilevered bow turret (SBM–Imodco)

A typical internal turret arrangement is shown in fig. 7.31. This is the SBM Top Mounted Internal Turret (TMIT) system. Similar systems can be supplied by any of the other turret contractors.

This turret design has been employed on several production FPSOs, e.g. Chevron's Alba FSU, Shell's CNS FPSO and Curlew FPSO.

The turret presents benefits from previous applications, and limits the number of critical mechanical components, such as the bearings, by using a single roller bearing located above the vessel main deck.

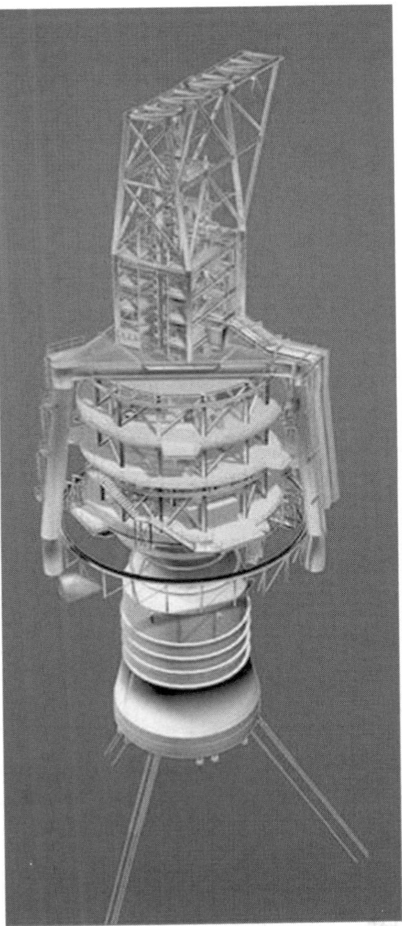

Figure 7.31 Top Mounted Internal Turret TMIT (SBM-Imodco)

The system comprises a chain-table structure anchored to the seabed anchor points via an array of nine (9) catenary mooring lines. The chain-table structure is attached to a turret cylinder structure which is suspended from a heavy duty roller bearing located above the vessel's main deck. The bearing is supported on the vessel's side and integrated into the forward part of the FPSO vessel's hull. The turret cylinder is mounted to the earth-fixed side of the main bearing. The turret cylinder houses the riser 'I-tubes' and supports the manifold deck (where the risers terminate), the manifold structure, the swivel deck together with the swivel stack and all the earth-fixed part of the rigid piping.

The roller bearing allows the turret to freely weathervane so that the tanker can take up the position of the least resistance to the prevailing weather, at all times.

Fluid transfer from the seabed flowlines is achieved by flexible risers which are suspended from, and routed through the turret mooring to the interface with rigid piping at the manifold deck level. After manifolding the product is routed through toroidal swivels to the vessel's main deck piping and process facilities. Lift gas is compressed on board the FPSO and transferred by a high-pressure swivel to the gas lift riser. Inject water is pressurized on board and transferred by a high-pressure water injection swivel to the water injection riser.

The turret's manifold area is protected from possible wave action (green water) by a wave breaker structure mounted on the vessel.

A pedestal crane serves the complete turret to handle swivels and perform other maintenance functions. A stair tower is integrated with the crane superstructure to provide access to the swivel stack area.

A hydraulic power unit provides power via the swivel to activate the different valves in the manifold. Swivels are also provided for well control fluid and signals.

Electric power is provided via a distribution system serving an exterior floodlighting and an interior accessway lighting of the complete turret.

The turret would be classified as the hazardous area, consisting of Zones 1 and 2.

The turret cylinder would be monitored for the presence of gas. Fire and gas detectors would be located in the manifold swivel area. All signals would be transmitted to the FPSO's central fire and gas control panel.

During inspection, temporary ventilation inside its annulus and cylinder would be provided by air fans. The manifold area and the swivel stack are situated above the vessel's main deck and are thus freely (naturally) ventilated.

Fire protection inside the turret cylinder area is effected using portable fire extinguishers, as the possibility of fire occurring in this area is considered remote.

7.4.5.2 Moorings Interface

The mooring forces applied to the vessel are introduced and accommodated by the turret components contained within the vessel's hull.

These components are:

1. The chain-table, including chainstoppers,
2. The turret cylinder,
3. The friction pad assemblies and support structure,
4. The main weathervaning bearing and its support structure,
5. The turret moonpool structure.

These components are described in more detail as follows:

Chain-table Structure

The chain-table forms the connection point for the anchor lines of the turret mooring system to the FPSO vessel. Mooring loads are transferred from the chainhawse connection points through the chain-table structure to the turret cylinder, and via the vessel friction pads and main weathervaning bearing into the ship's structure. The chain-table also provides the entry points for the product risers and houses the foundations for the bend restrictors (as required).

The chain-table deck, which is approximately 1.5 meters deep, is formed of welded steel plates. The deck consists of an external support ring box for the chainstoppers and a clear central opening which allows the entry of the risers and free passage of water flow in and out of the turret.

Chainstoppers and Chainhawses

Nine (9) chain hawse supports, each consisting of two cast steel pieces having a hook shape, are welded to the chain-table deck. The chain hawse assemblies complete with automatic chainstoppers are fitted into the hooks via a self lubricated bearing which allows the chain hawse to rotate and align with the chains. A tubular guide trumpet ensures proper alignment of the chain during tensioning and helps the articulation of the chainhawses.

This system allows the chain to be hauled in when the mooring line is pulled, and to be locked as soon as the tension is released.

The chain hawse and the automatic chain stoppers are of a design, which has successfully used on previous projects and reduces the time to hook-up the FPSO vessel to the pre-installed anchorlines.

The bearings of the chain hawse shafts are formed of low friction material, designed for full operational lifetime of the system, to allow rotation about a horizontal axis, which is perpendicular to the chain. This articulation allows the chain hawse assembly to pivot and thus minimise the interlink movements that are experienced by the top of anchor chain, virtually eliminating wear and chafing.

Turret Cylinder

This structure is the principal support element of the deck's turret section. It is supported from the main roller bearing and constitutes a circular stiffened shell member approximately five metres diameter, which extends from the chain-table up through the total hull depth. At the main deck level, a circular penetration allows the turret cylinder to continue until approximately four metres above the vessel's main deck, where it is supported by the main weathervaning bearing.

An annulus between the turret cylinder and the moonpool shell provides a flooded passage for the anchor chain connection lines. The chains pass to individual mooring line gypsy winches. The turret cylinder houses the riser and umbilical 'I-tubes', which provide protection to the flexible risers over their transit height inside the turret cylinder.

The turret cylinder extends from the chain-table up to the vessel's main deck and also supports the turret main deck, the manifold structure, the swivel stack and all earth bound parts of the fluid transfer system.

Bearing Support Structures

Weathervaning Bearing Support Structure – In order to effectively transit the vertical and the horizontal mooring loads from the turret chain-table and the turret cylinder into the FPSO vessel's structure, and to keep deformations at the main weathervaning bearing support within very tight tolerances, a conical piece of welded structure is fitted. The continuation of this cone forms the outside shell of the turret moonpool and interfaces with the vessel's structure. On the one hand, this conical structure must be flexible enough to decouple the bearing from the vessel's deformations, but on the other hand, the top surface of the insert piece must be formed of thick plate to rigidly support the rotating part of the weathervaning bearing.

Friction Pad Assembly Support Structure – The horizontal component of the mooring load is taken by an arrangement of friction pads. The support structure of the pads is integrated into the vessel keel and reinforced locally. The pads react on a heavy steel ring, which is an integral part of the turret cylinder.

Bearing Systems

Main Bearing – The link between the weathervaning vessel and the fixed turret column is a large diameter heavy duty three (3) race slewing bearing. The bearing is mounted to its support structure by high strength, hydraulically pre-tensioned bolts.

The slewing bearing provides the revolving connection of the mooring structure about the fixed (earth bound) structure of the TMIT. It therefore transmits forces and moments applied by the moored vessel under the combined action of wind, wave and current in either weathervaning or oscillatory rolling mode.

Friction Pads – Fourteen (14) friction pad assemblies are arranged to receive and distribute the horizontal component of the mooring force. They are disposed in two groups – eight (8) towards the bow and six (6) aft of the turret. Each pad is composed of a self-lubricating bearing surface which mates with nickel steel corrosion resistant overlay counterface. The pads are arranged as individual rams, which are supported by an hemispheric elastomeric bearing which reacts back into the vessel keel via a box structure.

Turret Moonpool Structure

The turret moonpool structure is a cylindrical bulkhead inserted through the hull from the FPSO main deck down to just below the keel. It is made of stiffened plate framed back into the vessel sections by radial stiffeners. At its lower end, the cylindrical bulkhead is equipped with inflatable seals, which can be pressurised against a support surface attached to the chain-table below. Once temporarily sealed, the water inside the turret annulus can be pumped out allowing dry inspection of the friction pad assemblies.

Deck Turret Structure

The turret extends between the top of the turret cylinder and the swivel stack, and includes two main areas:

1. The turret main deck and manifold area
2. The swivel deck

Turret Main Deck

This deck is supported by the turret cylinder and functions as a non-rotating base for the incoming fluid transfer system. This deck is fabricated in welded steel plate and beams, and has a circular plan form.

The deck is surrounded by a gutter to stop any waste products from leaking onto the FPSO deck. A drainage system, for piping and possible spills is integrated into the slops system.

The deck supports the manifold structure, which is hung off below, and the swivel deck. The manifold area serves as an access to crude oil production lines before their entry to the swivel stack.

Swivel Deck – this deck is located above the turret main deck and supports the swivel stack. A central opening is provided for the passage of the swivel entry piping and cabling.

Overhead Gantry Structure

The overhead gantry structure is mounted on the FPSO deck, over, but independent from the earth bound turret top part.

The gantry supports a traveling hoist crane suitable for maintenance activities in the turret area, and can handle the swivel stack modules, either in the maintenance mode or as part of a field expansion development.

Swivel Stack Assembly

The swivel stack assembly consists of an inner non-rotating ring and an outer rotating ring that encloses a toroidal shaped chamber. The individual swivel rings are stacked atop of each other to provide multiple independent flow paths. The inner and outer ring concentricity is maintained by triple race roller bearings. Seals around the periphery of the interface between the inner and the outer rings prevent leakage of the working fluid.

The swivel design is based upon the following guiding principles:

- Flow re-routing as back mode of operation
- Swivel redundancy
- Ease of maintenance with minimum intervention
- Maximum use of field proven designs

The driving arrangement for the swivel stack is an integral part of the overhead framework structure. An individual driving ring is supported at two opposite sides by a fork arrangement. Each swivel unit is therefore driven on both sides of the outer ring simultaneously resulting in a pure driving torque with no net shear being applied on the piping flanges.

The swivel sealing system incorporates double seal sets in each module. A leak detection system and recuperation system is built-in between the seals which collects any product fluid and returns it to the production flow.

The gas swivel sealing principle is based upon an over-pressurised back-up fluid that serves as a barrier between the gas to be sealed and the environment. This is achieved by using three seals on each side of the gas chamber. In order to assure the absolute gas tightness, the pressure on the back-up fluid is always maintained higher than the gas pressure. This is achieved through a simple accumulator.

7.4.6 Marine Systems

7.4.6.1 Inert Gas and Tank Venting System

To avoid the presence of an explosive mixture in the cargo tanks above the oil an inert gas is installed to provide a slight overpressure gas blanket at the top of the cargo tanks and slop tanks.

The inert gas system is designed to comply with the SOLAS 1974/1978 regulations for tankers (International Maritime Organization, 1974).

Inert gas generators are used to produce the inert gas with less than 5% oxygen would be provided. An inert gas main would be provided with branch connections to all cargo and slops tanks. A high velocity P/V valve would be provided on the inert gas main, to ensure the safe venting of the inert gas containing hydrocarbon gas without risk of ignition.

A common vent main is installed with branch connections to all cargo and slop tanks, to facilitate the gas freeing of the tanks for inspection or maintenance. The system has a high velocity type P/V valve to ensure that neither a dangerous overpressure nor under pressure can occur inside any tank whilst it is out of service for maintenance or inspection.

7.4.6.2 Tank Washing System

A Crude Oil Washing (COW) system serves the cargo and slop tanks. The COW is connected to the offloading line and branched off to all cargo and slop tanks. A booster pump installed in the system raises the pressure during offloading, should the loading pump pressure in the COW main be insufficient for crude oil washing.

7.4.6.3 Ballast Water System

A ballast water ring main system is used to fill and discharge the segregated wing ballast tanks. All ring main ballast piping runs over the deck so as to avoid any possible contamination with crude oil.

All valves used for routine operation in the ballast water system are hydraulically operated and remotely controlled from the central control room.

7.4.6.4 Fire Water System

The fire water system supplies:

Fire water to:

- Deck foam system
- Helideck foam system
- Accommodation sprinkler system

deluge/foam system to:

- heli-refuelling tanks
- process platform
- swivel on turret
- manually operated water jets and foam applicators

The capacity of the main fire water pump is estimated for the worst event, i.e. a cargo fire, requiring simultaneous operation of the following.

- foam system
- process deluge system
- turret deluge system
- two water jets

7.4.6.5 Fuel Oil System

A diesel oil system serves the various machinery on the vessel, comprising the following main components:

- Fuel oil storage tanks
- Piping from the filling station on the main deck to the storage tanks
- One settling tank with a capacity of approx. 40 m^3
- Two fuel oil transfer pumps (one as standby)
- One fuel oil purifier
- One pre-heater

7.4.6.6 Offloading

Cargo offloading can be arranged through several means, either direct transfer, i.e. FPSO to shuttle tanker through:

- Offloading reel
- Trailing hose
- Loading buoy

or indirectly through a permanently moored attending FSO.

A typical stern mounted offloading reel system is shown (fig. 7.32). The reel typically stores approximately 110 meters of 16 inch diameter bore flexible offloading hose string. The reel stores the hose between offloadings. Transfer rates are approximately 4000 m^3/hour (600,000 bbls in 24 hours) from the FPSO to a shuttle tanker moored approximately 80 meters astern.

The offloading reel is driven by multiple hydraulic motors which allows for satisfactory operation of the system even with one motor out of action.

Connection between the vessels is made without the need to transfer personnel or use of a support vessel.

This can be achieved via two methods, either by firing a line to the shuttle tanker and transferring the messenger line, chafe chain, main hawser and offloading hose from the FPSO to the shuttle. Alternatively, the hawser and offloading hose can be deployed prior to arrival of the shuttle tanker such that the shuttle can collect the messenger line and then haul in the hawser with the hose already attached to the chafe chain via a connecting wire.

The offloading hose is made-up of 10 meter lengths, joined so that any length can be easy replaced.

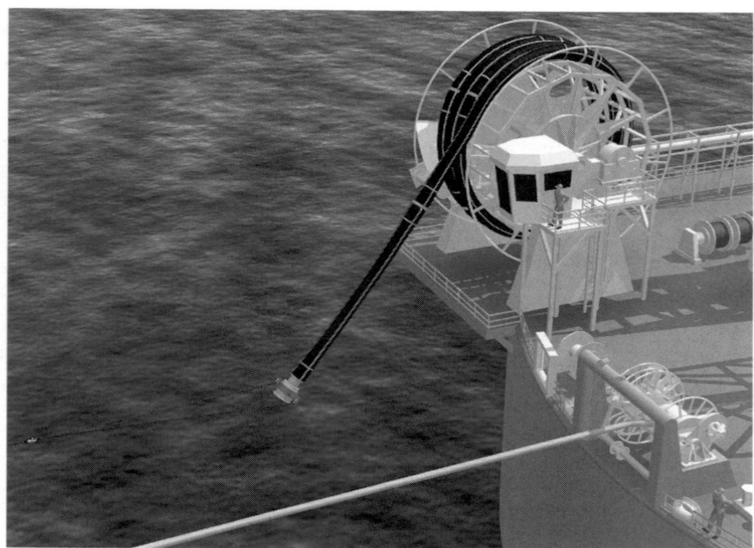

Figure 7.32 Stern offloading reel (Aker Kvaerner)

The inboard end of the hawser is attached to a bracket that forms an integral part of the drum, which is mounted on the aft deck next to the hose reel. This drum stores the main hawser in one layer, the chafe chain and the messenger line.

Each machine is hydraulically driven with the hydraulic motors powered by a power pack, located in a non-hazardous area.

The drives for the hose reel and hawser storage drum are controlled from a panel mounted between the two units, as is the spooling mechanism for the hawser storage.

Safety features in the export system include a breakaway coupling in the hose string that is designed to close automatically should hose tensions become excessive. A self closing coupling at the outboard end of the hose also closes when disconnected from the shuttle tanker.

Load sensing devices are incorporated to measure the tension in the hose and hawser line, and are continuously monitored during the offloading operation, with signals being fed into alarm and cargo offloading pump shut-down systems.

7.5 Semi-submersibles (John Filson, Consultant, Gig Harbor, Washington)

7.5.1 History of the Semi-Submersible

Semi-submersibles evolved from a drilling vessel type called a "submersible," which operated sitting on bottom in fairly shallow water and provided a working deck well above the highest expected waves (see Lim and Ronalds, 2000). These units transited afloat on pontoons and required "stability columns" to safely submerge to a bottom founded mode of operation. To operate in deeper water, the marine riser was developed and spread moorings were perfected allowing drilling afloat. This first application was with barges, however (e.g. "CUSS I"; see Burleson, 1999 for story of CUSS). To overcome the barges undesirable motions, the basic submersible design of the time was adapted to the floating drilling function. This was the "Bluewater." This was a Shell Oil sponsored development with Bruce Collip as the inventor of record.

Although highly evolved in size and configuration, these semi-submersibles remain fundamentally the same as they originated: a deck supported well above the sea by submerged pontoons, with a spread of large columns providing floatation stability. Both its parent, the submersible, and the semi-submersible are officially designated as "column stabilised units" (USCG, ABS, etc). The columns are "stability columns" and primarily provide flotation stability.

Drilling semis may be divided into four generations:

- 1st Generation: Before 1971
- 2nd Generation: 1971–1980
- 3rd Generation: 1981–1984
- 4th Generation: 1984–1998

Those most recently built, those since 1998, might be called "5th Generation," but a clear distinction has not emerged.

The first generation consists of a broad variety of configurations developed throughout the 1960s, beginning with the "Bluewater I" and including the notable SEDCO 135 designs and the variously configured ODECO designs (e.g. "Ocean Queen"). With the exception of the SEDCO 135, these designs all featured an array of multiple pontoons (replacing the bottom support mat of the submersible parent). Besides large diameter stability columns at the extremities, they also had many slender interior columns to support the working deck – but no diagonal trussing. The main element of global strength was the pontoon. Conversely, the SEDCO 135, a 3-column design, evolved from a Transworld submersible design and employed independent columns and footing tied together with a trussed space frame. The Forex Neptune "Pentagone 81," a 5-column, independent footing, trussed space frame design, is the culmination of this period. It is a true space frame, without trusses in a bent and also is an early user of hull-type superstructure.

The main elements of this period affecting design was a lack of technology exchange, a lack of consistent design philosophy, and the fact that no single design exhibited a complete understanding of the important principles of semi-submersible design. This period is more of historical interest that anything else. Nevertheless, some of the better 1st generation designs continued to be built well into the 1970s.

It is interesting to note that, as many of the 1st generation semi-submersibles became non-competitive as mobile drilling units, they found new lives as floating production systems (FPS). One of these, the "Transworld 58," was the first true FPS and deployed in the North Sea (1984). More recently (1995), another, the "Glomar Biscay II" (ODECO

"Ocean Victory Class"), was converted to an FPS for the Enserch Garden Banks 388 Development.

The 2nd generation produced the majority of the units built. In addition to better technology exchange, this generation was stimulated by competition from drillships. It became evident that a better mobility was required and that eliminating the transverse pontoons was important. Perhaps inspired by the aborted "Mohole Project" Semi-submersible (National Science Foundation, Deep Ocean Drilling Project of the late 1960s), it be came evident that two pontoons would be sufficient. However, this would require a different structural system. The obvious solution was evident in the SEDCO 135 and other similar designs and a truss system was employed which included transverse braces between the columns, and the various arrangements of diagonal bracing. An improvement in the hydrodynamic performance did not go unnoticed. Therefore, the principal feature of the second generation is the twin pontoon configuration and the high mobility it provides.

Although incomplete, due to budget cutting, the Mohole Project was a semi-submersible under construction (see Burleson, 1999). It was nevertheless a watershed of technology. The structural system was somewhat unique and maintained almost entirely by a hull-type superstructure, integrally built into the tops of the stability columns. It did have a shallow truss system with diagonal bracing, but did not have the low, horizontal transverse typical of later designs.

The "Zephyr/Marge Class," drilling units of the early 1970s, emulated the Mohole structural system but, upon entering service in the North Sea, the "Margie" suffered failures in the shallow truss. This event proved the necessity of the horizontal bracing and also the viability of the hull-type superstructure, without which, the "Margie" would have broken up. Another was the Offshore Company's "Chris Chenery." Without the horizontal transverses or the shallow truss, it had a hull-type superstructure, albeit heavily reinforced. It too demonstrated the strength of the hull-type super structure. The "Chris Chenery" and its single sister both have since been fitted with horizontal transverses to extend the fatigue life.

While no single design represented a complete understanding of the design principals, most 2nd generation semi-submersibles were relatively well designed from a performance point of view. An important aspect of this period was a higher level of dissemination of the design and performance knowledge. It is not a coincidence that the first USCG Regulations and the ABS Classing Rules were published in the late 1960s and that the OTC began in 1969. Most notable of this period, if only by its numbers, is the Aker H-3.0 design. Other notables include the Pacesetter (Friede & Goldman) and the SEDCO 700 classes.

The demarcation between the 2nd and the 3rd generation is rather sharp. In 1979, only two units were delivered and in 1980 there were none. By the time building resumed, the tragedies of the "Alexander Kjelland" and the "Ocean Ranger" had been reflected in new rules, performance requirements had increased, and costs had doubled. While the loss of the "Alexander Kjelland" revealed a structural redundancy flaw of a 1st generation design, ironically, the "Ocean Ranger," was a 3rd generation design, both configurationally and structurally, or at least a precursor.

Besides the increased size, payload, and higher standards of redundancy, the main features of the third generation are the continuation of the twin pontoon form, the use of hull-type superstructure, the well designed brace connections, and a generally thorough understanding of the design principles of semi-submersibles.

Notable 3rd generation designs are the "Bingo," "Ocean Odyssey" (an improved "Ocean Ranger"), "Scarabeo 5" and the "Zapata Arctic". It should be noted that a number of enhanced 2nd generation designs were built during this period, particularly the Sedco 700 and Pacesetter Classes and even a few 1st generation designs.

The 4th generation is rather difficult to define and is small. The most notable deigns are the GVA 4500 and its variants (two FPVs: "Balmoral" and "Petrobras XVIII"), the "Henry Goodrich" and the "Zane Barnes" (now "Jack Bates"). Typically, the 4th generation semi-submersibles are large, suitable for harsh environment operation, and deep water capable.

From a structural point of view, a marker of the 4th generation semi-submersible designs is that they rely fully on a hull-type superstructure with no bracing other than horizontals between the columns. One of the favorable aspects of this configuration is that, by the elimination of bracing, many inspection problems and the fatigue potential they represent are eliminated. The first use of this structural configuration was in the five Penrod semi-submersibles by Reineke in the early 1970s. However, due to immature connection detailing practice, some of these suffered from brace cracking. Some were fitted with diagonals and others were used in milder environments without any problem. One of these (with diagonals fitted), the "Penrod 72" is another obsolete drilling semi-submersible converted to an FPS. First serving as the Placid GC29 FPS in the Gulf of Mexico, it continues to serve as "Petrobras XXIV." The first new-built-FPS is an unbraced configuration with a hull-type superstructure designed by the Swedish shipbuilder GVA and serves as the "Balmoral" FPS in the North Sea.

The "Asgaard "B" FPS designed by Aker, shown in fig. 7.33, is the largest semi-submersible built to date (at 81,500 t displacement) and a good example of the unbraced hull with ring pontoon. This example also utilises an integrated truss deck which is installed by a floatover. Several FPS conversions have used cross pontoons to considerable benefit in this regard. While their towing speed is considerably slower, it is sufficient for the one-time deployment required. For reasons unknown, the two GVA designed FPSs are twin pontoon with transverse braces, but several proposed designs by GVA show the closed pontoon arrangement.

Figure 7.34 shows the 4th generation drilling semi-submersible, MDC "Marine 700." This unit was designed for a large variable deck load and large storage volumes for drilling and completion fluids (Moyer et al, 2001). It is designed for a maximum wave height of 32 m (105 ft) and a wave period of 11 to 15 s. Box-type elements and transitions provide a minimum of stress concentration at critical connections and an adequate fatigue life of 20 yr, based on North Sea criteria. The "Marine 700" has a transit draught of 9.75 m (32 ft), an operating draught of 23.75 m (77.9 ft) and a survival draught of 21 m (68.9 ft). It has an overall length of 119 m (390 ft) and an overall width of 71.4 m (234 ft).

The "Deepwater Nautilus," owned by Transocean (fig. 7.35) is an example of a 5th generation semi-submersible designed for ultra-deepwater drilling.

Figure 7.33 Hulls with ring pontoons: "Asgaard B" FPS (Øyvind Hagen, STATOIL), "Snorre B" FPS hull (NORSK HYDRO) and Brutus TLP (SHELL)

Figure 7.34 Marine 700 4th generation semi-submersible (Moyer, 2001)

Figure 7.35 "Deepwater Nautilus" – 5th generation drilling semi-submersible (Photo courtesy Transocean)

Other semi-submersibles have been used for accommodations, derrick vessels, pipelaying, and work vessels. While some are conversions, most are special purpose newbuilds, somewhat unique in their size or arrangements and generally include 3rd and 4th generation designs.

7.5.2 Distinctions between a MODU Semi-submersible and an FPS

Most semi-submersibles begin life as a "Mobile Offshore Drilling Unit" (MODU). The functions of a MODU and the operating patterns of a MODU are decidedly different from those of an FPS. Fortunately, the configuration and much of the equipment and systems are useful. A MODU has only the mission functions, drilling and well service, but is amply provided with support functions.

A particularly costly support function of a MODU is its mobility that is not required of an FPS. Mobility is the primary reason for the twin pontoon configuration. However, the twin hull configuration makes global strength more difficult to provide. There are a number of cases where mobility, bridge (and drydock) clearance has also dictated the vertical and lateral dimensions of the semi-submersible.

Mobility in general, and quick deployment in particular, is a major factor in the mooring system design. The MODU moorings are rarely suitable for long term applications. For deep water FPS applications moorings are a particularly significant system cost component, making minimisation of wind and current load a more important design priority than it is for a MODU.

Another important MODU distinction is its maintenance. Mobility provides considerably more favorable circumstances for maintenance. These and more subtle differences can be reasonably dealt with in a conversion and likewise need to be recognised in the design for new construction. For strength reasons discussed later, the production semi-submersibles tend to the closed array pontoon configuration.

7.5.3 Semi-submersible Design

Semi-submersibles consist of a deck, multiple columns and pontoons. They are "column stabilised", meaning that the centre of gravity is above the centre of buoyancy, and the stability is determined by the restoring moment of the columns. This contrasts with the spar platform, which achieves stability by placing the centre of gravity below the centre of buoyancy, and the TLP, whose stability is derived from the tendons.

The design of semi-submersibles depends on these principle considerations which are somewhat generic to floater concepts:

- Weights and CG's (cycle of steadily improving estimates)
- Hydrostatics; tank capacities
- Intact and Damaged Stability
- Wind Forces (stability and mooring loads)
- Current forces (mooring loads)
- Ballast System Performance
- Motions (seakeeping; drift and low frequency mooring loads)
- Global Strength
- Fatigue

Before initiating the design, there should be a definitive Functions List (e.g. production, drilling, quarters), a Systems Summary, and an Equipment List (mission and support). Trial equipment and systems layouts should be made and coordinated with any constraints needed in the initial design. The constraints might include, for example:

- Maximum lightship draft for quayside outfitting,
- Maximum beam for canal transit or dry transportation (see Section 7.9.2),
- Maximum lightship weight and VCG envelope for dry transport,
- Environmental criteria for operations, transit and survival,
- Maximum lateral eccentricity of the deck load which needs to be trimmed,
- Maximum allowable motions (angles, accelerations) for each given environmental and load condition,
- Applicable rules and standards.

There may be different constraints for various load cases: operations, transit, survival, and installation. These should be identified in order to be able to check the configurations for each case.

Weight estimates need to be made of all permanent payload and variable loads, including equipment and systems outfit for the functions (drilling, processing, utilities, quarters, flare, etc.). The equipment weights are to be supplanted by vendor equipment as it becomes available. The proforma arrangements and calculations should be developed to support the outfit estimates (piping, access, corrosion protection, etc.). In addition, variable load requirements in amount, distribution, and with respect to the operating state should be firmly established. And, if it matters, the installation weight-states for permanently sited platforms should be determined.

The variable weights should be identified for each load case as illustrated in the following table.

	Operating	Transit	Survival
Variable Ballast	x	minimal	x
Drilling Fluids	x		x
Process Fluids	x		
Drill Pipe	x		x
Casing	x		x

The recommended overall process is to execute initial design to conclusion, later addressing the specific shape (pontoon cross-section and ends, column section, flare, etc.) and the use and arrangement of bracing. More than one alternative might be kept for further consideration. Immediately following this, rigorous hydrostatic and intact stability analyses, revised weight estimates, and the motion analyses should be made for each retained alternative. From these results, a choice should be made and a final, definitive principal dimensions sketch should be produced. The outcome of this, particularly the column size and arrangement, should be more a matter of weight, buoyancy, and intact stability. The damaged stability is addressed later by subdivision. Adequate motions response should already have been achieved through refined cross sectional shape of the pontoons. Choices regarding station keeping loading would be resolved with a limited number of alternative initial design variants with addressed with rigorous analyses of the simple model forms. At this point, work should begin on the basic scantlings with the specific objective of establishing the framing system, initial scantlings, and refinement of the steel weight estimate. Concurrently, damaged stability analysis should also proceed in order to determine internal subdivision, internal weight distribution and piping requirements. Global strength analysis can only begin after the basic structural design (framing and scantlings) is largely complete. Fatigue analysis is reserved for detailed design.

7.5.4 Functions and Configurations of Semi-submersibles

General

In the design of a semi-submersible, and its configuration in particular, a clear idea of the functions it must perform should be in hand. These will strongly influence configurational choices. Besides drilling, these functions include production, heavy lift, accommodations, operational support (surface, subsea), and even space launch.

Apart from the mission and support functions, stated simply, there are two essential functions of a semi-submersible:

- To stably support a payload above the highest waves,
- To minimally respond to waves.

These are the principal factors that establish size. It is, however, the mission functions and associated support functions that most significantly contribute to configuration.

The four main configurational components are:

- Pontoons
- Stability columns
- Deck
- Space frame bracing

Figure 7.36 shows sectional views of four semi-submersible arrangements, identifying the above four components. Waterlines are shown at their typical operating state, "semi-submerged". While each has the noted components, each is distinctive. Case A is typical of 3rd generation semi-submersibles, whereas Case B is quite typical of the 2nd generation. Similarly Cases C and D are typical of the 3rd and 4th generations respectively.

Virtually, all semi-submersibles have at least two floatation states: semi-submerged (afloat on the columns) and afloat on the pontoons. The pontoons are the sole source of floatation of the semi when not semi-submerged. The stability columns are the principal elements of floatation and floatation stability while semi-submerged. Although they may function

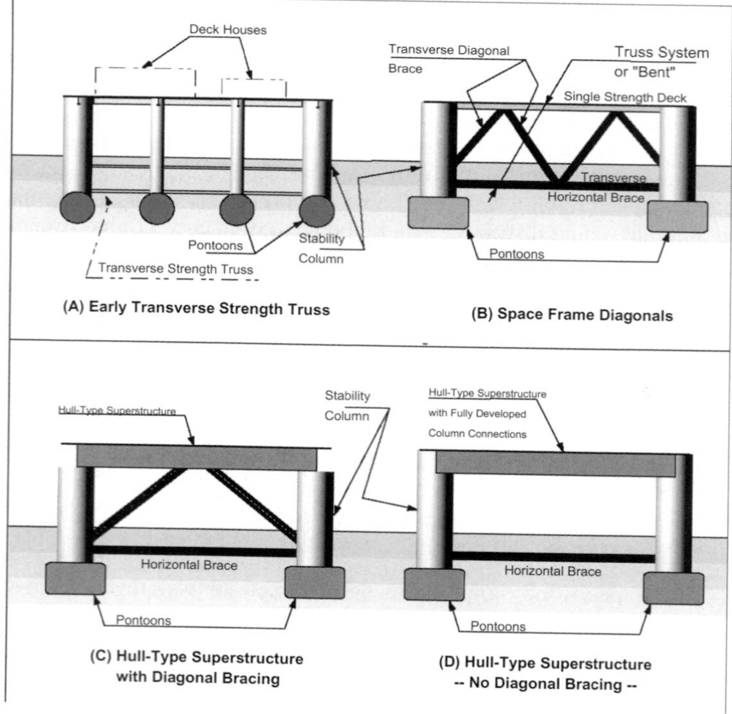

Figure 7.36 Semi-submersible sectional arrangements

structurally, structural strength is not the main function of the columns. It is notable that the pontoons are primarily filled with ballast when semi-submerged. Beyond this, the size, submergence, proportion and spacing of the columns and pontoons are major factors in the hydrodynamic performance of semi-submersibles.

Ostensibly, the deck provides the working surface for most of the semi-submersible's functions. It has the structural function to transfer the weight of the deck and its loading to the columns (and bracing). However, the deck is also a part of the overall global strength system, providing a structural connection between all of the columns.

The pontoons and columns are usually arranged and connected in a way that can provide considerable global strength. Generally the deck is likewise arranged and connected. Where this arrangement does not provide sufficient global strength, a space frame bracing system is employed (see fig. 7.36B and C). This has been very much the case in the earlier designs. However, bracing systems are problematic in that they are expensive to build, are vulnerable to fatigue, and are a costly maintenance item in regard to inspections and repairs.

Decks

The decks of the early semi designs were a single level structures with individual deckhouses arranged with no coherent interrelated structural function. This arrangement was often referred to as a "piece of toast with lumps of butter" (see fig. 7.36A). Support of a single deck requires a space frame bracing system and/or close column spacing. Single decks were favoured in earlier semi-design because of the then limited erection resources. What has since evolved is the hull-type superstructure (fig. 7.36C and D) with integral connection to the column tops (fig. 7.36D). Such a configuration can eliminate most, if not all space frame bracing. Among the advantages of the hull-type, integrally connected deck is superior strength, considerable usable interior space, and valuable floatation in damaged stability. If built with the rest of the hull in a modern shipyard, a hull-type deck is lighter, less costly, and of superior strength than other alternatives. A disadvantage, in some cases, is a necessity for mechanical ventilation and to fully outfit by a single builder.

A "cousin" to the hull-type deck is the "truss-deck." It is preferred in some production applications that favour open, natural ventilation as well as historical design and fabrication practices. Particularly where there is separate fabrication, outfitting, and joining of the deck ("split construction"), the truss-type decks can be preferred because most fabricators of production decks are not equipped to build plated-structures. Similarly, design organisations that specialise in "topsides" are not experienced in working with hull-type structures. The choice of the deck type is therefore of considerable importance in configuring a semi insofar as it determines whether a split or an integrated construction will be preferred.

For clarity of terminology, the pontoon, columns, and bracing (usually) are referred to as the "hull", the "deck" being distinguished separately. With hull-type decks, there is not such a distinction.

Columns/Pontoons

The number and arrangements of pontoons and columns distinguish many configurational variants employed in the evolution of the semi. This has included as few as three to as many as a dozen or more columns. It has likewise included a simple two parallel pontoon arrangement, up to six, and even a grillage of orthogonally intersecting pontoons. As noted in the historical discussion, a few major designs featured independent footing pontoons, one for each stability column. The SEDCO 135 design, for example, had three independent pontoons; the Pentagone design had five. Figures 7.37 and 7.38 show a number of typical column and pontoon arrangements.

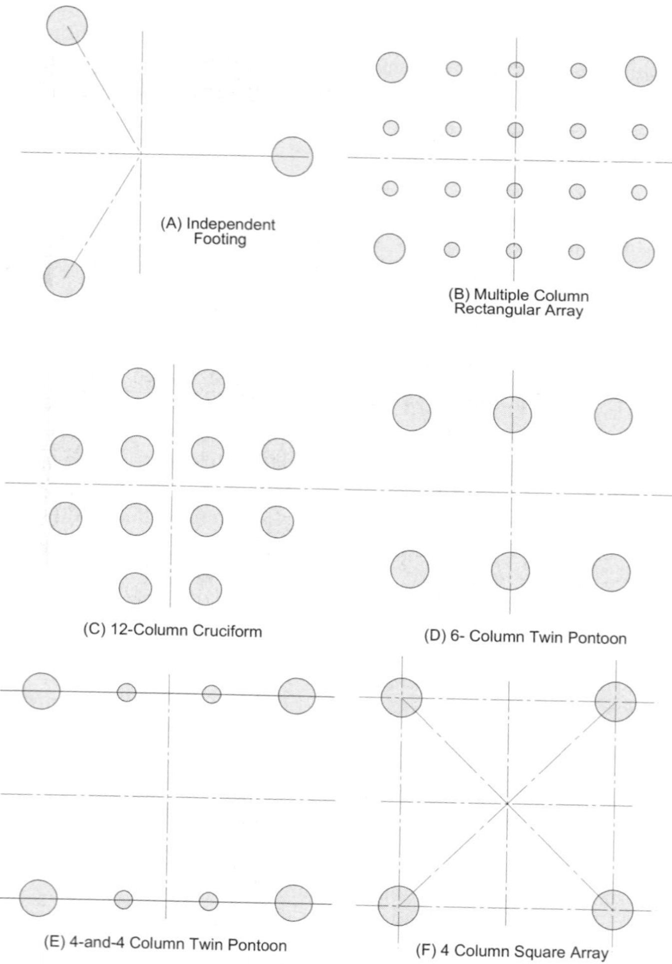

Figure 7.37 Semi-submersible column arrangements

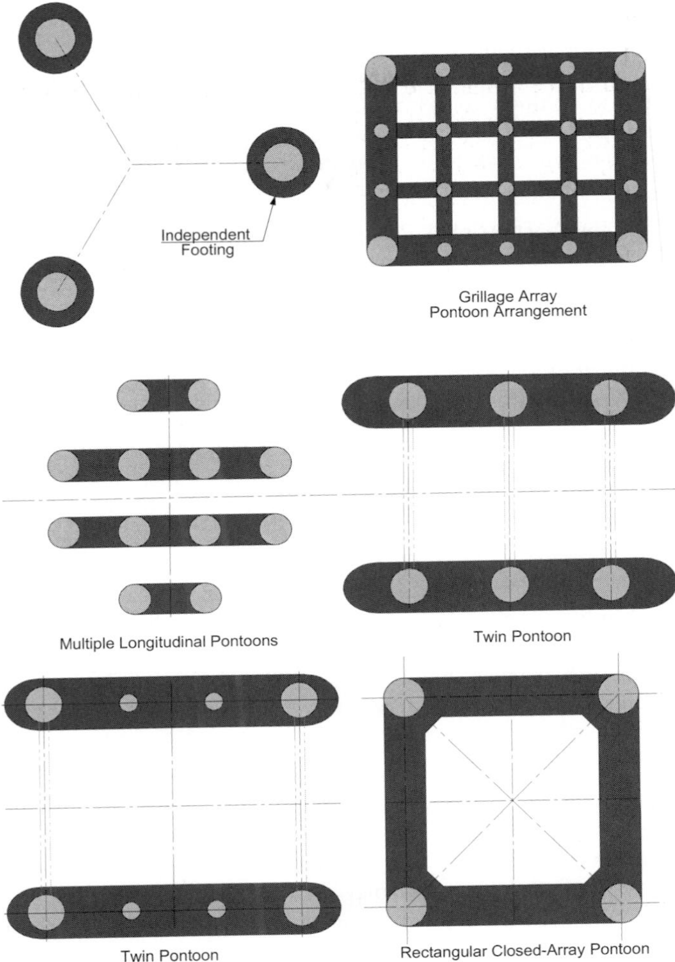

Figure 7.38 Semi-submersible pontoon arrangements

Only the 4-, 6-, and 8-column configurations continue in preference. Similarly only the twin pontoon and the closed array pontoon arrangements are currently used. A 3-column, closed array pontoon (triangular) arrangement has been proposed for both FPS semi-submersible and TLP applications, and offers a steel reduction opportunity, but these designs have not been successful, perhaps because of the more complex deck arrangements. The twin pontoon preference is principally because of its mobility. A preference for the 6- and 8-columns relates primarily to the twin pontoon option, and is influenced by the use of bracing systems.

A closed array, or "ring", pontoon arrangement is not very good for towing mobility, but is often preferred for a permanently sited system because it offers superior strength and an excellent potential for a braceless system. Transverse braces are not required and, with well designed column to pontoon connections, as well as special connection at the deck, the system can handle the racking loads. This is the basis of most TLP global strength systems. Fully developed hull-type deck to the column connections offer an even greater strength potential, and allows a wider column spacing.

As noted earlier, the function of the columns is to provide stability. A critical point of stability is when a semi is submerging, and when the flotation undergoes transitions from being afloat on the pontoons to being afloat on the columns. This operation is restricted to mild conditions and requires only that there be "positive GM". It limits the deck loading and otherwise discourages the particularly tall semis. For this reason, it is common to flare the columns at the pontoons to enhance stability through the critical range of drafts.

Deck area is sometimes considered a sizing factor. Usually, the spacing of the columns for stability provides adequate interior space, particularly if there are two decks. Moderate deck extensions outside the column are a practical option. Sometimes, the overall width is a limiting factor. A limited maximum width has had a role in selecting the 6- and 8-column arrangement.

Bracing

Bracing configurations vary considerably (see fig. 7.36). These principally include a transverse bracing, low on the columns, to resist squeeze/pry forces (discussed later) and, with these, a transverse diagonal bracing (fig. 7.36B and C only). The diagonal bracing is both to support the deck weight and, together with the horizontal transverse, provide the lateral racking strength. Often, a system of the horizontal diagonals is used to provide racking strength against quartering seas.

A bracing system commonly found on many of the 3rd generation drilling semis (and later) is shown in fig. 7.39, where transverse bracing is shown in heavy dark lines and horizontal diagonals are shown in heavy dashed lines. Where continuous, strong longitudinal pontoons are employed, the longitudinal diagonals are not particularly useful and are rarely used in contemporary designs. As a structural system, the strength of the space frame truss system is typically developed in parallel series of planes between columns, following civil engineering practice, called "bents." Each bent is a full truss, including the deck as a top chord and the horizontal, transverse brace as the bottom chord, all spanning between a pair of stability columns. Some use an "inverted-V" form of diagonals and some use an "inverted-W." Except for a deck girder, the members consist of large diameter, thin walled cylinders.

A well designed and well connected deck structure can eliminate the need for most bracing. Similarly a closed array pontoon can also eliminate the need for bracing (see fig. 7.37F), similar to a 4-column TLP. However, a twin pontoon structure will require horizontal transverses.

Station Keeping

Though not specifically the topic of this chapter, the station keeping function will need consideration. The principal options are spread mooring, dynamic positioning, and spread

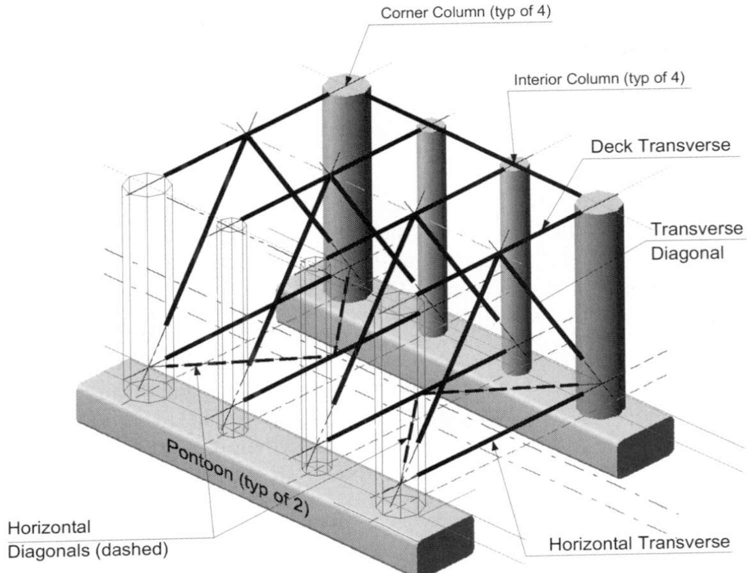

Figure 7.39 Typical 3rd generation semi-submersible bracing system

mooring with thruster assist. The key design issue of spread mooring is the onboard mooring equipment, particularly its arrangement, space requirements, and weight, not to mention significant foundations. Dynamic positioning, however, has particularly significant power requirements, substantial fuel storage capacity (and resupply), and the arrangement of a number of thruster units. Either choice presents design problems that can usually be resolved independent of the configurational choice. An exception relates to stability. The heeling moment (Section 7.3.4) is computed based on a couple between wind loads and the center of reaction (drag) for a free floating vessel. However, if the vessel is permanently moored, the heeling moment is computed for the reaction loads at the worst location: center of drag or fairlead locations. The burden on the payload (e.g. vertical component of mooring line tension) must be considered in the choice of configuration, however.

Risers

Risers and drilling is also not the subject of this section, however risers may set the constraints on allowable motions and have a significant impact on arrangements. Drilling semi-submersibles have considerable equipment for supporting and handling of the drilling riser and significant storage space is also required for riser storage. In the case of production semi-submersibles both the production and the export subsea risers will be required in addition to the likelihood that all or a part of the drilling function is also included. In an emerging class of production semi-submersibles, there will also be the top-tensioned risers as found on the TLPs. Proper motions at the riser hangoff point may make the difference as to whether a certain riser design will work or not. A preliminary assessment of the riser strength and fatigue should be performed before finalising a configuration using deepwater, top-tensioned and steel catenary risers.

7.5.5 Sizing of Semi-submersibles

General

The approach to be taken in the following discussion is to address overall design and not to focus on the analytical methods. Rigorous analytic methods are specifically addressed elsewhere in this handbook. "Design" is more a matter of making good decisions than it is about precision and rigor in determining the various aspects of performance. In the end, good design involves both these elements.

The main problem in semi-submersible design is to adopt the right configuration for the specific functions required and the construction program (Hung and Mangiavacchi, 1990). Beyond this, it is a problem of rational sizing. The immediate need in design is for an efficient means to evaluate the available choices more than it is for rigorous precision. To this end, a process is outlined, referred to as an *initial design*, whereby simple models of behavior are used to establish choices for key parameters of size.

One feature of the initial design process is that specialised subsystem design and analyses (mooring, riser, etc.) can begin on the basis of earliest models and proceed concurrently with independent design iterations.

The initial design forms a rationally based model for subsequent detail designs and rigorous analysis, from which a *preliminary design* is developed. One of the objectives of the process is to discourage multiple, trial-and-error, detailed analyses on what may be models of inappropriate configuration and size. It is hoped that, instead of precise answers for wrong designs, good answers for good designs prevail. Confirming the choices of the initial design process, and refining these choices, is achieved though subsequent detailed analysis.

As noted, rigorous hydrostatic, stability, hydrodynamic and structural analyses should be performed once the appropriate shape and size is determined for the *initial design*, this discussion does not address these analyses. Likewise the design and analysis of risers, mooring and dynamic positioning systems are singular subjects discussed elsewhere.

Initial Design Considerations

In sizing of a semi, it is informative to re-examine the most fundamental functions of the type:

- To stably support a payload above the highest waves
- To minimally respond to waves

Noting the words "*payload*" and "*above the most extreme waves*," two very important sizing issues are addressed:

- The number, size and spacing of the stability columns
- The height of the deck

The second basic function, "*minimum response to waves*" relates to the size, shape, and submergence of the pontoons relative to the column waterplane area, and the spacing of the pontoons and columns.

To initiate the initial design process, a configuration must be established. Some configurational issues have been introduced earlier. These decisions are closely interrelated with function and construction and are not always quantifiable in the usual sense. For a MODU, to favour mobility, a twin pontoon configuration is invariably the choice. Twin pontoon configurations will generally require horizontal transverse braces. For permanently moored semi-submersibles, an FPS for example, mobility is not a priority and the closed array pontoon is a more likely choice for its superior strength.

A choice must also be made regarding the use of an integral or an independent deck and whether bracing is to be used. Integral decks are used for most semi-submersible types, particularly when built by a single builder shipyard. An independent deck is preferred where split construction is to be considered, particularly for an FPS. The column connections of independent truss-decks usually involve highly designed, mechanical connectors (even if welded) and are generally designed to consider deflections and tolerance adjustment. Some connections are designed to pre-load the deck structure in a favourable manner. Because of fitup, tolerance, and connection design problems, bracing has yet to be a favoured option with independent decks. Because of their strength, the multi-cell, hull-type integral decks have been the prevalent choice with most currently designed mobile semi-submersibles, particularly if completed by a single builder and to reduce or eliminate space frame bracing. Single decks are now rare and require space frame bracing.

A four-column format (fig. 7.37F) has a minimum of surfaces and therefore steel area. It works well with the closed array pontoon and also with hull-type or deep truss decks. It also is used with the twin pontoon arrangement, but more often used are three equal columns on each pontoon (fig. 7.37D) or four (fig. 7.37E; the two interior columns smaller). With the closed array pontoon, the use of the additional, smaller columns mid way between the corner columns (to reduce spans) is also an option, but has fitup problems for independent decks.

The use of bracing and using more than four columns are largely a structural issue relating to the type of deck system, these are also germane to certain deck erection procedures. Besides reducing steel, fewer columns can reduce the cost of complexity. Each and every column has a top and a bottom connection that is necessarily complex and costly to fabricate and erect. Each column requires a setup, an erection, and an alignment process that is somewhat independent of size. Generally, the fewer the columns, the lower is the cost of the structure, even if the lesser number of columns must be more robust. For the twin pontoon configuration, each column pair at least requires a transverse brace between the columns to resist the squeeze-pry forces. These are usually associated with diagonal bracing. The issue of bracing and deck configuration can be avoided in the initial, parametric stages of design, but must be addressed before any efforts for serious, detail-based analyses begin.

The pontoon-to-column connection is especially important, particularly with regard to structural connectivity. For reasons noted before, the column may be flared at the pontoon, typically rectangularly. If rectangular, aligning the internal bulkheads within the pontoons as continuations of the column sides can significantly reduce stress. Generally there are at least two, often four, pump rooms in the pontoons under the corner columns. The mooring equipment arrangement is also a significant aspect of the column design, most notably chain lockers (or wire storage), hawsepipes, external fairleaders, and windlasses (or winches) at the top. Equipment, storage, and access are other considerations. While these aspects may not bear on the configurational choices, they often do affect the minimum column size.

Should thrusters be installed, space and special arrangements must be provided for the thrusters and their internal support systems, not to mention the fuel tanks. The importance of this is the overall space and size, particularly in that this is in or near the column base. Also, the internal space otherwise available for ballast is no longer, and column ballast may be required.

The column-to-pontoon connections for the closed array pontoons differ considerably from the twin-pontoon configuration. Two forms are typical. One is where the pontoons run between columns, much as with the TLPs. The second has the pontoon side faces flush with the column face and features pontoons much wider than the column. For strength reasons discussed later, a "haunch" (diagonal) connection between the pontoons is often employed.

The height of the deck and columns matter most to weight estimating and meeting stability requirements. As noted previously, the columns should be sufficiently tall to support the deck with sufficient wave clearance. With single deck semi-submersibles, the column tops are flush with the deck. With hull-type decks, particularly if the column is integrally connected, the column tops may be in level with the upper deck. The additional column height enhances stability and, if the associated deck structure is also watertight, it also does. Independently installed truss decks usually sit atop the columns. What matters, however, is the location of the bottom of the deck, and the vertical centre of gravity of the deck.

The ABS Rules and the API Codes require that there be 5 ft (1.5 m) clearance ("airgap") between the highest wave crest and the deck. The highest wave, or the crest level above still water, is usually specified with the design seastate data. There is some ambiguity as to whether the airgap requirement applies to the deck itself, or to the supporting girders on the underside. Examples of both interpretations exist. The fact is that most critical girders are usually quite robust, some very large, deep box girders. These can usually withstand wave impact. More vulnerable to damage are the smaller, local under-deck framing, but these are not critical to survival. Nevertheless, any girders extending below the top of the 5 ft line should ultimately be examined with respect to wave impact.

There can be too much airgap. Excessive airgap raises the centre of gravity and thereby impairs the payload performance. Determination of the effective airgap should consider the relative motions of the vessel. For large, long period waves, a semi will tend to rise and fall synchronously with the waves, possibly as much as 20% of the wave height (single amplitude). To recognise this, in initial design, it can be conservatively assumed that the semi rises 10% of the wave height. Then, setting the calm water deck height at 5 ft plus 90% of the extreme crest elevation should suffice. Part of finalising the preliminary design is to re-examine crest clearance by a more rigorous theory or by model testing.

It is noteworthy and discussed in the next section, that the airgap issue for TLPs is just the opposite. TLPs actually move downward ("set-down") with offset and are subject to the rising of tides as well as subsidence of the bottom foundation. Therefore, what might seem like too much airgap on a semi, when compared to a TLP, or even a fixed platform, can actually be just right.

A side issue is tolerable "wetness." "Wetness" is due to a run-up of waves along the columns. There are the many examples where wetness in extreme seas is tolerated. Wetness prediction is best discussed with model testing.

Having mentioned the height, the depth of column and pontoon below the surface remains to be determined. A variety of different functions drive this choice. The deeply submerged pontoons reduce the heave motion. With drilling units the, shorter columns are preferred for a lower centre of gravity for large deck loads. Drilling semi-submersibles achieve deep submergence by ballasting to a deeper draft for drilling, but otherwise deballast to a desirable airgap for severe storms. It is also desirable to minimise the ballasting time and the amount of ballast water to be handled. Consequently, mobile semi-submersibles are no taller than need be, with operating drafts no more than necessary.

For drilling, the maximum drafts would be in the 70–80 ft range, with a relatively small air gap. For the severe storm condition ("survival"), drafts in the 50–60 ft range would be used and a more generous air gap. As an example, with a typical hurricane survival condition with $H_s = 40$ ft, the extreme crest elevation would be about 45 ft above stillwater. Allowing a 5 ft of vessel heave at the crest, and 5 ft crest clearance, a 45 ft calm water airgap should be adequate. It is also considered undesirable that the pontoon tops be exposed in the trough of extreme waves. Under the same hurricane survival condition, the pontoon tops should be at least 40 ft below still water. With pontoon depth 25–30 ft deep, this would correspond to a 60–70 ft survival draft and 85 ft of column between the pontoon top and the deck. Correspondingly, the operating draft would be 80–90 ft with 25 ft stillwater airgap. The initial deign of a drilling semi-submersible would be based upon achieving the best drilling performance, and be based upon the shallower operating draft.

With other functional priorities, permanently sited semi-submersibles take advantage of reducing motion with deeper draft. While tall semi-submersibles have stability problems while submerging, a permanently sited unit would not require any deck variable load during installation and certain permanent deck payload may be absent. Also, because ballasting is only performed at installation (and de-commissioning), it need not be efficient or constrained. Additionally, operational management of the installation with respect to weather is much more favourable. Consequently, a deeper operating draft may be very feasible for a permanently sited unit, up to 100 ft being practical. A variety of very deep draft designs have been proposed reducing heave motions sufficiently to employ top-tensioned risers. These designs have drafts of 150 ft or greater. In these cases, the quay-side deck installation is not practical. Deck installation requires an offshore heavy lift, floatover, or a scheme for self-installation (LeGuennec et al, 2002; Halkyard et al 2002). When performing the initial design of a permanently sited semi, attention should be focused on the installation condition and procedure.

In the pontoons, between the corner columns, and also in the outer extensions (bow/stern), typically only liquids and mostly ballast and some consumables are carried. There may also be access trunks required for internal tank inspections. Most of the ballast is maintained in stasis for the duration of the operation, but part of the ballast, usually beneath or within the columns, is actively used to adjusting draft and trim. Locating such tanks ("trim tanks") in the corner columns maximise the trim effect for the minimum amount of ballast. It is noted that, since trim tanks are particularly vulnerable to corrosion, they should be kept small and given high performance corrosion protection.

As a part of the initial design process, definitive Functions List, a Systems Summary, and an Equipment List (all: mission and support) should be developed. Concurrent with the hull sizing, trial equipment and systems layouts should be made and coordinated with any

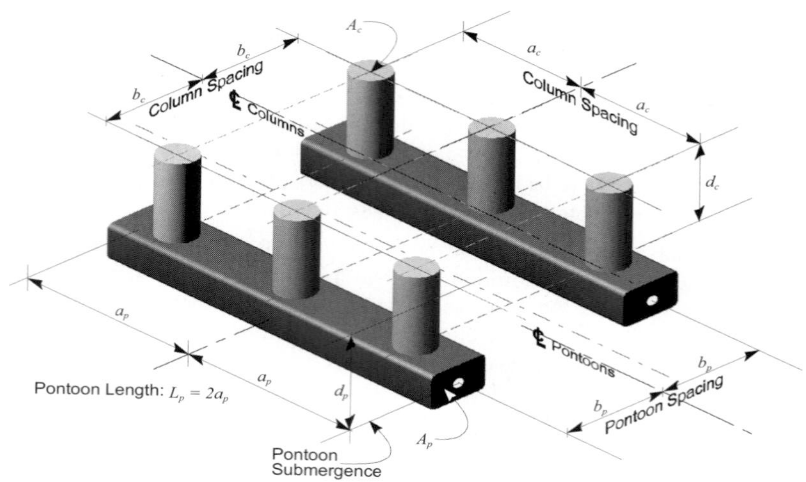

Figure 7.40 Initial design underwater model – twin pontoon, 6 columns

constraints needed in the initial design. Additionally, weight estimates need to be made of all equipment and outfit systems. Proforma arrangements and calculations should be developed to support the latter (piping, access, corrosion protection, etc.). In addition, variable load requirements in amount, distribution, and with respect to operating state should be firmly established. And, if it matters, the installation weight-states for permanently sited platforms should be determined.

Figure 7.40 shows, a twin pontoon, six-column semi, the underwater body model used for initial design computation of displacement, initial stability, and heave motions. The defining set of parameters for a six-column, twin pontoon semi-submersible is:

Pontoon:	Pontoon cross-section	A_p
	Pontoon length:	$L_p = 2a_p$
	Pontoon lateral spread:	b_p
	Pontoon center submergence:	d_p
	Pontoon freeboard afloat on pontoons	f_p
	Pontoon volume (total)	$\nabla_p = 2\ A_p\ L_c$
Column:	Column cross-section:	$A_c = \pi\ Dc/4$
	Column depth to pontoon top:	d_c
	Column longitudinal spread:	a_c
	Column lateral spread:	b_c
	Column height (wt top):	f_c
	Water plane area:	$\mathrm{A_{wp}} = 6\ A_c$
	Immersed Column Volume:	$\nabla_c = A_{wp}\ d_c = 6\ \mathrm{A}_c\ d_c$
Total Displaced Volume:		$\nabla_o = \nabla_p + \nabla_c$

The pontoons should extend beyond the corner columns.

Figure 4.41 has a similar illustration for a four-column, closed array pontoon semi of unequal spread. For this case, the defining set of parameters is:

Pontoon:	Side Pontoon cross-section:	$A_{p\text{-}side}$
	Side Pontoon length:	$L_{p\text{-}side} = 2a_{p\text{-}side}$
	Side Pontoon lateral spread:	$b_{p\text{-}side}$
	End Pontoon cross-section:	$A_{p\text{-}end}$
	End Pontoon length:	$L_{p\text{-}end} = 2b_{p\text{-}end}$
	End Pontoon lateral spread:	$a_{p\text{-}end}$
	Pontoon center submergence (all):	d_p
	Pontoon freeboard afloat on pontoons (all):	f_p
	Pontoon volume (total)	$\nabla_p = 2\ (A_{p\text{-}side} \times L_{p\text{-}side} + A_{p\text{-}end} \times L_{p\text{-}end})$
Column:	Column cross-section:	$A_c = w^2$
	Column depth	d_c
	Column longitudinal spread:	a_c
	Column lateral spread:	b_c
	Column height (wt top):	f_c
	Water plane area:	$A_{wp} = 4\ A_c$
	Immersed Column Volume:	$\nabla_c = A_{wp}\ d_c = 4\ A_c\ d_c$
Total Displaced Volume:		$\nabla_o = \nabla_p + \nabla_c$

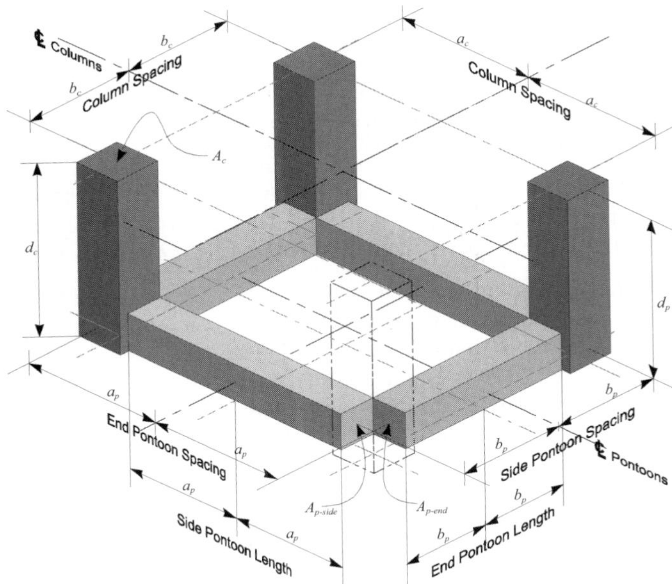

Figure 7.41 Initial Design Underwater Model – Closed Array Pontoon, 4 Columns

Deferring specific pontoon shape and dimensions to later, an effective column bottom must be estimated, particularly with regard to the column's contribution to total buoyancy. To start with, an approximate design draft, d_o should be established. Anticipating a rectangular pontoon cross-section (try $Depth_p = \sqrt{1/2}\ A_p$; $width_p = 2 \times depth_p$), set the pontoon center, $d_p = d_o - 1/2\ depth_p$. For a twin pontoon semi, the column bottom, d_c is taken at the pontoon top: $d_p = d_o - depth_p$. For the closed array pontoon, d_c is taken at pontoon bottom and therefore: $d_c = d_o$. Once the specific pontoon cross-sections are used, these should be adjusted accordingly and made compatible.

7.5.6 Initial Design Process

General

Given a configuration, the process is to formulate a simple, parametric model with which the buoyancy, a weight estimate, and the initial stability can be determined. Initial stability is discussed in another section of this chapter. Weight estimation for initial design is specifically discussed after this section. A major aspect of initial design of a semi-submersible is its motion performance. The initial design process outlined uses a simple, closed-form model to compute the most important motion attributes. An example for heave is also given. Other motion attributes could be similarly formulated. What is important is that it forms a tool to quickly cycle the parameters of size. Initial estimates of loads for mooring can be approximately determined from these models, with parametric computation from the underwater body. Wind and current forces may be estimated using appropriate code formulas (e.g. ABS MODU Rules, API RP2SK).

Each application will have unique constraints and freedoms to be considered in the design and sizing. The principal focus here will be on mobile drilling and production applications. However, the application of the principles to the design of semi-submersibles for other functions will be similar.

The recommended process is to execute the initial design to conclusion, later addressing specific shape (pontoon cross-section and ends, column section, flare, etc.) and the use and arrangement of bracing. More than one alternative might be kept for further consideration. Immediately following this, using just the simplified models, rigorous hydrostatic and intact stability analyses can be performed and revised weight estimates made.

At the beginning the focus of initial design, the pontoons are represented by their volume as prismatic, circular cylinders of length, L_p and sectional area, A_p, submerged at depth, d_p. The columns are addressed in terms of area and spacing, and. Once these are reasonably established, pontoon depth, width, length, and shape are addressed. Depth and width relate to the heave added mass needed to achieve the desired motion characteristics, while the depth as well as the freeboard relate to the transit displacement required. These aspects are discussed later.

Once the configuration, size, and key dimensional parameters have been established, refinement of the parameters and specific shape of the hull elements can proceed with specific, rigorous analytic models, specifically stability, motions, mooring, and preliminary structural design. From these results, a choice should be made and a final, definitive principal dimensions sketch should be produced, indicating all the important aspects of geometry. All other analytic and detailed design activities/analyses should follow this sketch. Finalising, rigorous hydrodynamic analysis should at this point proceed, particularly the main issues of specific shape. With the resultant revisions, the preliminary design is defined. At this point only, work should begin on the basic scantlings with the specific objective of establishing the framing system, initial scantlings, and refinement of the steel weight estimate. Concurrently, the damaged stability analysis should also proceed in order to determine internal subdivision, internal weight distribution, piping requirements. Global strength analysis can only begin after the basic structural design (framing and scantlings) is largely complete.

Parametric Model

The simple parametric model need only represent the pontoons and the columns as prismatic sections as shown in figs. 7.40 and 7.41. While the final design will no doubt be more complex than this, the objective is to determine the best fundamental dimensions for the design. This can be accomplished with a spreadsheet such that a minimum of parameters is specified and a minimum of determinants produced. It would be an iterative process.

The defining set of parameters for the six-column, the twin pontoon semi was given above. As few as a five-column and four-pontoon parameters are sufficient to describe the semi. From these parameters, the displacement, the centre of buoyancy, and the metacenter (KM) can be computed and, with an overall heave added mass coefficient for the pontoons, C_{azp}, the heave RAO also can be computed.

If the vertical centre of gravity (KG) of the semi is known, then its initial stability (GM) is known. Given a target GM, one can vary the column areas, A_c, and the spread, a_c and b_c, for their acceptable values. The target values of the initial GM might be in the 12–15 ft range for a drilling semi at the drilling draft and 18–22 ft at the survival draft. It would be easy to provide values for initial GM more specific than this, but this would be somewhat misleading. The best approach is to examine the operating manual of one or more existing semis of similar size and configuration (they will be different). It must be borne in mind that the objective at this stage is to produce design proportions that are likely to produce an adequate righting arm curve for the corresponding condition. Very little is initially known of the wind profile and down flood points to be more specific than this.

The GM for an FPS would be slightly higher, unless in a milder environment. For watertight, hull-type superstructures, a slightly lower GM (2–3 ft lower) will suffice. E.g. a production FPS in relatively severe environments might have a target GM of 18–22 ft with at truss deck, or 15–19 ft with a hull type superstructure. Meeting such target values of GM will generally produce design proportions that, with a more detailed stability analysis, will meet the necessary requirements for intact stability. The damaged stability requirements are met through internal subdivision and are not necessary at this point.

What is addressed above amounts to a constraint condition and needs the estimated KG for the corresponding design condition. This should include a vertical force balance that considers all, a weight, buoyancy and the external forces, as well as the vertical centre of each. Estimates of the weight and the vertical centre of gravity are discussed in Subsection 7.5.8. It includes a process whereby the hull steel weight and centre of gravity are computed as a function of the geometric parameters used here. Table 7.9 gives an outline structure of

Table 7.9 Semi-submersible hull weight and force breakdown groups

Pontoon Steel
Special Steel (e.g Riser Supports)
Pontoon Outfit and Equipment
Pontoon Subtotal
Column Steel
Special Steel (e.g. fairleader fdns, chain lkr, etc.)
Column Outfit and Equipment:
Column Subtotal
Deck Steel (basic structure)
Deck Steel – deck houses
Special Steel (e.g. substructure, crane fdns, etc.)
Deck Equipment and Outfit – Marine and Support (mooring, utilities, safety, accommodations)
Deck Equipment and Outfit – Mission Systems (drilling, production, tensioning, etc)
Deck Subtotal
Other Hull Weight
Deck Reserve/Margin
Pontoon Fixed Ballast
LIGHTSHIP
Mooring Tension
Drilling Riser Tensions**
Production Riser Tensions**
Export Riser Tensions**
Deck Variable Load
Column Variable Load
Water Ballast (pontoon/column)**
Subtotal External Load, Variable Load and Ballast
TOTAL SUPPORTED WEIGHT AND EXTERNAL FORCE

**Since risers may not be present throughout the life of a structure, the GM should be checked for risers being present or not. For a permanent facility, ballast may be added when risers are not present to maintain a constant draft.

the items to include. It includes all the vertical loading and weight that must be balanced by buoyancy.

With the dimensional parameters given above, the displacement, Δ_o at the relevant draft, d_o is directly computed. If water ballast is left as a dependent variable, a balanced system will result for any parameter set. Whether using a more sophisticated search algorithm, macros, spreadsheet, data modeling tools, or a simple trial and error, the values of A_p, a_c, and b_c can be found which will provide an adequate initial GM. This might be explored with two trial pontoon volumes before the parameter set is narrowed down (see next discussion). However, for a given pontoon displacement, what is produced is a range of suitable column diameters and spacing.

At this point the pontoon parameters need to be addressed. A vertical force balance should be produced for an afloat on the pontoon condition. This would reflect the lightship and an appropriately reduced variable load and should float on the pontoon at a draft that provides the requisite freeboard, f_p. As a practical matter, for initial design only, pending establishment of cross sectional dimensions, minimum freeboard, f_p is better expressed as a freeboard which results in a submerged cross section as a percentage of A_p; 92% should be a good starting value. Therefore, specifying the weight of this condition, W specifies the pontoon displacement: $2 \times 0.92 \times \Delta_p = W$. Then, specifying the practical length of the pontoon, $2a_p$, the cross-sectional area, can be specified. As a rule of thumb, for practical pontoon length, $a_p > (a_c + D_c)$. For a twin pontoon semi it is desirable for hydrodynamic performance to have as much pontoon volume as practical extending beyond the column. One design strategy (the "dog bone") is to narrow the pontoons between the columns and make them wider at and beyond the columns. Rectangular, closed array pontoons automatically achieve this effect. With very few trials, effective combinations of a_p and A_p can be found.

It is important to note that, until the ratio between water plane area, pontoon volume, and heave added mass is addressed, precision in the choice of pontoon volume is not important. However, good choices will produce heave periods between a 25–30 s for a MODU and somewhat higher for an FPS. What has been discussed up to this point will establish which choices of column area and column spacing will provide adequate initial stability.

7.5.7 Closed-form Heave RAO Calculation

The first order undamped heave RAO can be written in a simple, closed form equation as:

$$RAO_z(\omega) = \frac{1}{1-\beta^2} G(\omega) \tag{7.11}$$

Where β is the frequency ratio and T_z is the heave period of the semi, with

$$\beta \equiv \frac{T_z}{T_w} = \frac{\omega}{\omega_z} \tag{7.12}$$

and

$$T_z = 2\pi \sqrt{\frac{1}{g}\left[d_c + \frac{\nabla_p}{A_w}(1 + C_{az})\right]} \tag{7.13}$$

C_{az} is the heave added mass coefficient. It is 1.0 for a circular pontoon, higher for rectangular or flat pontoons (see Sarpkaya, 1981).

The dependence of the heave period on the ratio of the pontoon volume to the waterplane should be noted. For a drilling semi-submersible, the trial dimensional values should produce T_z between 25 and 30.

The undamped motions are suitable for optimisation on the basis of maximum drilling conditions (Hs < 15 ft), particularly if $T_z > 20$ s. However, if the design is based upon response to extreme seas, the equation can be rewritten for damped heave motions.

The effective linear damping factor for extreme seas can be obtained from model tests, but in any case should not produce a heave RAO peak of more than 2 at resonance, generally less. It is generally preferred in permanently sited systems to have high values of T_z (near 30 s or higher), in which case undamped response will suffice for current purposes.

$G(\omega)$ is a term addressing the specific geometry: size, spread, and submergence of the elements. Included is an assumed heave added mass coefficient for the pontoons, C_{az}. For a twin pontoon semi with head seas (longitudinal to the pontoons),

$$G(\omega) \equiv \left\{ \Phi_3(d_c)\Lambda(a_c) - \frac{k\nabla_p}{A_w}[1 + C_{az}]\Lambda(a_c)\Phi_2(d_c) \right\} \tag{7.14}$$

k is the wave number.

$$\begin{aligned} &\text{Columns :} \quad \Lambda(a_c) = \frac{1}{3}(1 + 2\cos\lambda_a) \quad (\text{note : } \lambda \equiv ka) \\ &\text{Pontoons (length } L = 2a_p) : \quad \Lambda(a_a) \equiv \frac{\sin\lambda_{a-pon}}{\lambda_{a-pon}} \end{aligned} \tag{7.15}$$

The spreading functions, head seas (longitudinal to pontoons), are given as:

$$\begin{aligned} &\text{Columns :} \quad \Lambda(a_c) = \frac{1}{3}(1 + 2\cos\lambda_a) \quad (\text{note : } \lambda \equiv ka) \\ &\text{Pontoons (length } L = 2a_p) : \quad \Lambda(a_a) \equiv \frac{\sin\lambda_{a-pon}}{\lambda_{a-pon}} \end{aligned} \tag{7.16}$$

For beam seas the spreading functions are:

$$\begin{aligned} &\text{Columns :} \quad \Lambda(b_b) \equiv \cos\lambda_{b-col} \quad (\text{note : } \lambda \equiv ka) \\ &\text{Pontoons :} \quad \Lambda(b_a) \equiv \cos\lambda_{b-pon} \end{aligned} \tag{7.17}$$

In either case, the depth attenuation functions:

$$\begin{aligned} &\text{Column :} \quad \Phi_3(d_c) \equiv \varepsilon^{-kd_c} \\ &\text{Pontoon :} \quad \Phi_2(d_p) \equiv \varepsilon^{-kd_p} \end{aligned} \tag{7.18}$$

For a four-column semi, with a rectangular, closed array pontoon, the RAO expression is the same but the heave period, Tz and the geometric function, $G(\omega)$ are written differently. These are as follows:

$$G(\omega) \equiv \left\{ \begin{array}{l} \Phi_3(d_c)\Lambda(a_c) \\ \quad - \dfrac{k\nabla_p}{A_w}\left[\dfrac{(1+C_{a-side})\nabla_{side}}{\nabla_p}\Lambda\left(a_{p-side}\right) + \dfrac{(1+C_{a-end})\nabla_{end}}{\nabla_p}\Lambda\left(a_{p-end}\right)\right]\Phi_2(d_p) \end{array} \right\} \tag{7.19}$$

This formulation is written for the head and beam seas where the longitudinal and transverse spread and pontoon cross-sections are different (areas: $A_{p\text{-}side}$ *and* $A_{p\text{-}end}$). However, both pontoons have the same submergence, d_p. ∇_p is the total pontoon volume, given as: $\nabla_p = \nabla_{p\text{-}side} + \nabla_{p\text{-}end} = 4(a_{p\text{-}side} \times A_{p\text{-}side} + b_{p\text{-}end} \times A_{p\text{-}end})$. $a_{p\text{-}side}$ and $b_{p\text{-}end}$

respectively are the half-length of the pontoon elements. (note: the lateral spread of the pontoon centres are given as: $b_{p\text{-}side}$ and $a_{p\text{-}end}$ respectively). Similarly $C_{az\text{-}side}$ and $C_{az\text{-}end}$ are the heave added mass coefficients of the pontoon segments.

The spreading functions for the head seas is written:

$$\begin{aligned}
&\text{Column}: && \Lambda(a_c) \equiv \cos\lambda_{a-col} \\
&\text{Side Pontoons}: L = 2a_{p-side}: && \Lambda(a_{a-side}) \equiv \frac{\sin\lambda_{a-side}}{\lambda_{a-side}} \\
&\text{End Pontoons}: L = 2b_{p-end}: && \Lambda(a_{a-end}) \equiv \cos\lambda_{a-end}
\end{aligned} \tag{7.20}$$

The spreading functions for the beam seas are similarly written:

$$\begin{aligned}
&\text{Column}: && \Lambda(b_c) \equiv \cos\lambda_{b-col} \\
&\text{Side Pontoons}: L = 2a_{p-side}: && \Lambda(b_{p-side}) \equiv \cos\lambda_{b-end} \\
&\text{End Pontoons}: L = 2b_{p-end}: && \Lambda(b_{p-end}) \equiv \frac{\sin\lambda_{b-eside}}{\lambda_{b-side}}
\end{aligned} \tag{7.21}$$

The depth attenuation functions:

$$\begin{aligned}
&\text{Columns}: && \Phi_3(d_c) \equiv \varepsilon^{-kd_c} \\
&\text{Pontoons}: && \Phi_2(d_p) \equiv \varepsilon^{-kd_p} \quad \text{(side and ends)}
\end{aligned} \tag{7.22}$$

These are simple comparative measures of the heave performance for sets of parameters for the simple models of the two given configurations. Similar expressions can be written for other motions as well as other configurations. These methods do not substitute for rigorous modeling (see Chapter 3). The purpose of the simple modelling is to be able to simply and quickly canvas a wide range of geometric parameters together with the aspects of design other than motions. Once the size and proportions are established, a specific shape is addressed and only then should rigorous modelling proceed.

All said, as interesting as it may be, the RAO is not completely useful. It needs to be further subjected to spectral analysis with the design sea state to produce a significant heave amplitude. Only with this, the response in the design sea state, is a truly useful comparison measure of heave motion produced.

Also, the RAO does not account for 2nd order drift which can load to very large responses, particularly in roll and pitch, if the natural periods are greater than 35 seconds. This is particularly the case for deep draft semi-submersibles.

This is not to say that the geometric parameters could not instead be used for more rigorous and general motion analysis of the simple models for similar use in evaluative software. However, without significant programming, what is outlined here, together with the spectral analysis, can be self contained in a spreadsheet and instantly cycled to a conclusion. The objective is to interactively determine the better choices of primary dimensions for later detailed analysis.

How this procedure would be used, having selected viable sets of pontoon parameters (A_p, L_p), the pontoon cross-section would be further represented by choices of heave added mass coefficients. Then, for various chosen column areas and spacings found to provide adequate initial stability, the best pontoon C_{azp} can be found for each set, each producing a maximum z_s for the design sea state. Heave being the principal performance measure here, the data set can be culled down to "good sets" of A_c, a_c, and C_{azp}. It is important to note that variation in pontoon C_{azp} in the end is a result of variation in depth/width ratios and corner radii. These should be represented in the parameter set when addressing pontoon steel weight.

Typically, the pontoon cross-sectional shape would be rectangular with radiused corners. For construction simplicity, pontoons are prismatic. As noted elsewhere, there are designs, which maximise the volume at the ends. Mobile units (twin-pontoon) will incorporate a bow and and a stern shape foreword and aft to facilitate towing and may incorporate propulsion and steering aft. Permanently sited, closed array pontoons are fairly simple and squared at the corners in this respect.

Specific to the cross-sectional shape and added mass coefficients, DnV Classification Notes, 31.6 Appendix B includes tables of coefficients for determining added mass coefficients, particularly for pontoons clear of the columns where the beam/draft ratio and the corner radius are addressed. In addition, there is guidance for considering end effects (pontoon extending beyond the columns) and the region in way of columns. The recommended process is, once the desired value of C_{az} is known, determine a composite heave added mass for the pontoon with two or more cross-sections and specific shape considerations. On the basis of this outcome, interpolate cross-sectional dimensions, which produce the necessary added mass.

If heave is the only measure of interest, the process stops here. Other possibilities are the total steel weight (or surface area) and current force, both computable on the simple geometric parameters. Here, the interactive tradeoffs play a key role. Taking hull steel as a cost measure, the total watertight surface area (or steel weight equivalent) generated by a parameter set will vary with the same choices. If a large difference in this measure corresponds to a small difference in heave performance, a good guide to compromise is at hand. Likewise, the mooring loads are quite sensitive to the column diameter as well as the pontoon cross-section and length. Again, a small compromise on one measure can produce material benefit on the other.

7.5.8 Weight and Buoyancy Estimates

General

Most obviously, there must be sufficient buoyancy to balance the weight of the semi and the external forces. The required buoyancy determines the underwater volume, or "displacement." This comprises the volume of the pontoons, the columns, and, sometimes, the bracing. Displacement is a primary determinant of size and proportions. Consequently, much of the initial design work is devoted to determining *all* the components of weight.

Although payload and its height above the most extreme of waves were specifically identified as the salient factors in design, payload is only a part of the total weight. It and all other weights as well as its centre of gravity is needed to proceed with a design, at least a first estimate. This estimate should be continuously refined throughout the designing process.

In as much as there are many components of weight, not all concurrent, and different operating states with different drafts and displacement, some definitions need to be made.

Weight is made up of two components, "Lightship" (W_o) and "Variable Load" (δW). The former comprises all the steel, equipment, and outfitting provided at completion and is usually defined and verified according to regulation. The latter comprises all weight beyond the light ship to be carried by the semi: the ballast, the consumable liquids, the operating liquids, bulk items, the personnel and effects, etc. and, as the name implies, varies according to the operating state of the vessel. In addition, there are a variety of external loads to consider (e.g. mooring tensions, riser tension, hook load, etc.).

Returning to the term *payload*, this comprises all of the *mission-related* equipment, variable load, and external load. The necessary support system weight that is needed, regardless of the mission function (e.g. mooring equipment and other "marine systems"), is *not* considered to be a part of the payload. Payload exclusive of deck structural steel, is referred to as *net payload*. However, if the deck structural steel is included, it is referred to as *gross payload*. The net and gross distinction is needed, particularly in comparing designs, because some mission functions can have a high impact on the amount of structural steel and is not an inherent property of the semi design. Such distinctions are particularly important when the same design is used for varied applications and also when conversion and upgrades are to be considered. This distinction is also needed in the evaluation of designs in as much as many designers are not consistent. Structure explicitly devoted to a mission function is, in fact, the payload.

As noted earlier, it is recommended that the initial design process be executed through a spreadsheet. This recommendation includes weights, centres of gravity, external loads and vertical force balance. Table 7.9 represents a summary table of a spreadsheet implementation where many parts are fixed givens and other parts are automatically computed within the spreadsheet from the given geometric parameters, particularly, steel weight. As with the rest of the initial design process, most of the weight (equipment, variable load, etc.) information is separately specified according to design requirements and is not a function of geometry of the semi. That which is dependent upon geometry can be automatically computed within the spreadsheet and is discussed here. These items are marked with an asterisk in the listing.

Particular attention should be given to the effective vertical centres of all items. In addition, intuitive adjustments to vertical centres expected non-even distribution will improve predictions. Mooring and the riser tensions must be estimated, pending concurrent initial design activities and refined in later iterations as appropriate. In deepwater, the riser tensions and mooring loads require special attention and need to be considered in the initial sizing. If there is a substantial "future" load to be applied, e.g. future satellite risers, the associated vertical load should be allocated to variable ballast that will be removed in the future. Note that the MODU stability rules state that the stability calculations shall take into account these external loads only if they worsen the stability of the vessel. Risers hung off at the topsides deck raise an effective KG and do indeed worsen the stability. On the other hand, risers or moorings that are hung off at the keel benefit the stability and should not be included in stability calculations.

If the draft is specified along with the other geometric parameters, the displacement is determined. If this is the case (recommended), the water ballast (double asterisk in table) should be left as an independent variable in the spreadsheet.

Lightship Weight Estimating

Weight estimating is addressed elsewhere in the handbook, but some factors unique to the semis should be noted. In general, lightship weight estimates should be broken down into three basic groups: steel, outfitting, and equipment. Each group has very different factors that determine weight and are generally developed by different technical disciplines. With semis, particularly if it is an FPS with a separately fabricated deck, this primary breakdown should be separately applied to both a "hull" and "deck" group. In the case of a separately fabricated deck, this follows the administrative breakdown of the development of engineering data. However, even with an integral deck, the fact that the deck is lifted (usually in blocks), this breakdown becomes useful.

In estimating the hull steel weight for a semi, it is important to note that about 85% of the steel relates to its hydrostatic design pressures. The balance of the hull steel is a result of a combination of global reinforcement and specific functional foundations (e.g. mooring fairleaders). The reasons for this is that hydrostatic loading is particularly high in semis, there is a significant amount of internal subdivision, and the fact that nearly half of the structural weight is framing and stiffening. In addition, the pontoons and the columns tend to have considerable global strength without reinforcement. Internal subdivision and framing in general have little role in global strength. However, shell stiffening is moderately increased (buckling strength). Providing global strength is generally a matter of small to moderate, local increases in shell plating thickness, particularly where columns and pontoon (or deck) connect. The significance of this fact is that, given the hydrostatic loading, the weight per square foot of all the watertight surfaces can be estimated and totaled. There are a variety of rational and empirical processes for weight estimates and initial design of hull structure. It is also noted that a minimum weight design does not represent the least cost to build, that cost effective design typically results in weights that are 10–15% higher than a least weight structure. This said, the unit weights range from about 30 lb/ft^2 for the upper columns to 50 lb/ft^2 or more for the pontoons and lower columns (depending on draft). This part is referred to as the "primary steel weight" and is a function of geometry and local loading only. The other steel weight can be added to this as a percentage (15–20%) or as specifically estimated allowances (e.g. fairleaders) and a smaller percentage. Other weight factors are included in the section on Hull Structure.

Equally important, particularly with regard to design, is the fact that the primary steel weight varies with the total area of *all* watertight surfaces (shell and internal subdivision). For weight estimating, the total area of all the watertight surfaces should be determined as a function of the dimensional choices to be considered, and unit weights for each surface type established. In this way, where the design process may parametrically alter key dimensions, the corresponding weight change can be reflected. It should be apparent that design choices that minimise the aggregate hull surface area will minimise steel weight. This surface area should be reported in the initial design spreadsheet as one of the several determined properties of the parameter set. Given the need for internal subdivision for damaged stability as well as consumable liquids, these factors need immediate consideration in the design process.

An alternative weight-estimating procedure used by some engineers is to use a steel density for the enclosed volume of the columns and pontoons. Values of between 10 lb/ft^3 in the upper sections to 15 lb/ft^3 in the pontoons are typical, and this can vary greatly depending on the subdivision used. This method has the advantage of not requiring a tabulation of internal bulkheads, which are not actually defined until a damaged stability assessment is performed. It has the disadvantage of inaccuracy, unless the density figures are well-benchmarked against similar designs. This method is not recommended for anything other than the most rudimentary conceptual sizing exercise. The area weight estimating procedure should be employed as early as possible in the design.

Some semi designs continue to employ major bracing systems, particularly transverses between the columns for the twin hull semis. These are often non-buoyant, but in any case are primarily sized according to axial loads, with end connections reinforced for bending and fatigue. The steel weight of such bracing ranges from 5% (transverses only) to 15% of the area-based weight. For a global strength system based principally on space frame bracing, a reduction of the other steel for global strength is reduced from 10% to 6%. There is very little actual hull steel weight difference between space frame braced and unbraced hull designs. There is some saving in deck steel, however, with space frame bracing. The big difference is higher cost to build and maintain for space frame bracing.

The deck steel estimates can follow a similar rationale, but, because of the functionality aspects, there is a less rigorous relationship. Unless it is very lightly loaded, the hull-type decks tend to build out at 20–25 lb/ft^2 (each deck). However, as a global strength element, a hull-type deck (minimum of two continuous decks with appropriate bulkheads), 20–30% additional steel is required for global strength and still additional steel is required for specific functional items, particularly drilling substructures, winch foundations, and deck houses. Truss-decks tend to be heavier than the hull-type decks and vary considerably. Weight estimates for truss decks usually require a structural frame model, properly loaded with equipment weights (including dynamic factors) and reaction loads at the connections with columns. It is typical to build a frame mode for the entire semi-submersible for this purpose. The primary steel for a truss deck typically ranges from 25 to 35% of the payload carried by the deck. It is important to select the method of deck installation when performing this frame analysis, e.g. float-over, module lifts or single lift, as this will affect the application of loads during the operation.

Outfitting weight includes piping, wiring, ducting, and corrosion protection (anodes/coatings). It should also include access items (ladders, walkways, gratings, rails, closures, etc). While the access outfit is usually made of steel, and often included in the steel "budget," it is considerably costlier per unit weight than structure, is truly distinct, and should be separately distinguished as an outfit. Other structural outfitting should be so treated. Specific estimating of outfit weight is left to other parts of this handbook, but suffice it to say that it typically is less than 10–12% of the hull weight and consists of mostly piping, access, and anodes. This number is higher for decks and more diverse, but very dependent upon functions.

Equipment weight will be entirely a product of functionality and cannot be addressed here. Estimates need to be provided by disciplines addressing functions of the semi. It should be a product of a comprehensive equipment list with no exclusions. It is typically the omissions that contribute to estimating errors. The estimate should include all piping and electrical interconnects as well as module steel used to support the equipment, independent of primary structural steel.

Finally, the lightship estimate should include explicit allowances and margins. The former is *expected items* (even if unidentified) and the latter is to address *uncertainty*. In the case of steel, items such as welding, brackets, plate thickness are known to exist, add to the steel weight, and have empirical allowances. Likewise certain equipment and outfit are known to exist, but will not be determinable. However, errors in design, quantification, item omission, etc. are probabilistic, tend to compensate to some degree, and have different uncertainties with different parts of the system. What is important in this regard is that explicit allowances and margins, ever so small, be addressed and available for review and recommendation.

Variable Load

Variable loads on semis is somewhat unique. At the operating draft, the variable load is mostly the pontoon ballast, but this may also include consumable liquids (e.g. fuel, drill water, etc.). This weight can be more than half of the operating displacement. Most of the remaining variable load is typically on the deck and related to the mission function. This is referred to as the variable deck load (VDL) and a particularly important measure of a semi's capability. More importantly, the amount and the height of the VDL is limited (through regulations) by stability or, more specifically, the vertical centre of gravity of the total weight, or the "allowable KG."

The specific limits on weight and centre of gravity were discussed in Section 7.3.4. However, stability is an extremely important issue in the initial design of a semi and some issues will be briefly addressed here.

Loading Conditions are the long-term operating states of the semi. There are also a variety of transitional conditions to be considered in design and operation. There are at least two particularly important and very different loading conditions: "transit" and "operating." Usually there is a third condition referred to as "severe storm" (sometimes "survival").

The transit condition varies considerably with the semi's function. Normally a transit condition is a state afloat on the pontoons with a modicum of pontoon freeboard. A drilling semi can have a field transit condition and an ocean transit condition. The former may require most of its variable load capacity and be weather limited. However, the latter condition will likely have more freeboard, reduced VDL, and be required to comply with unrestricted intact and damaged stability requirements. Long ocean transits are usually governed by conditions set by a Warranty Surveyor and based upon length of voyage, route, season, and other factors. This is so even for a "dry transit" on a barge. A production semi may have a short transit from base-port to the installation site, operational prerogative on the timing of transit relative to weather, and minimisation of variable load, including exclusion of certain lightship items. Such different transit conditions can produce very different designs. In the opposite extreme, semi-submersible

Derrick Vessels (SSCVs), section 7.9.3, can have considerable value by being able to transit with heavy variable loads.

The operating and the severe storm conditions are for a semi-submerged state. In a drilling semi, these states would be at quite different drafts. The drilling condition would provide the deepest possible draft (70–85 ft) and produce the best possible heave motions. Conversely, the survival condition is the opposite. In the extreme, the drilling riser is disconnected and the draft is reduced to provide clearance of extreme waves, but no less than that required for extreme conditions stability requirements. This includes a reduction in variable deck load, particularly pipe in the derrick ("setback") and possibly some liquids. For a drilling semi, good heave motions are less of a priority in survival than in the drilling state, although sometimes reduced roll/pitch is considered important. Between the full drilling and the survival, there are a number of intermediary states, depending on weather expectations and the time available. Although the practice varies, it is noted that the survival conditions may mean complete abandonment of the semi.

For a production semi, even with drilling, there may not be a change in draft distinction between "operating" and the "severe storm" conditions. There will no doubt be changes in weight and its distribution, but these would not be as extensive as with a mobile drilling semi. Typically, both heave and roll/pitch motions will have been optimised at an operating draft established primarily to clear the highest wave crests. If all risers are flexibles or SCRs, roll/pitch motions will be particularly important. However, if the semi is to have top-tensioned risers, minimising heave will be of atmost importance. In any case these factors lead to quite a different set of design parameters. This is particularly true where top tensioned risers are to be used and very deep draft is necessary to achieve the minimum heave motions.

7.5.9 Semi-submersible Hull Structure

General

While structural design is addressed more specifically in Section 7.8, several aspects of the structural design, specific to semi-submersibles, needs to be addressed in this section. It should be first noted that the prevalent design codes for semi-submersibles are the classification rules, particularly those of the American Bureau of Shipping (ABS) and det Norske Veritas (DnV). However, some oil companies may require aspects of the American Petroleum Institute (API) recommended practices to be followed. There may be inconsistencies between these and the classification rules. Also, some owners may have very specific additional requirements. These should be resolved on a case by case basis, but be no less than those indicated in the classification rules. It should be also noted that besides classing, there is a trend in the regulating authorities to rely on the classification societies for certification review.

A big issue in semi-submersible design criteria is longevity, inspectability, and repair. For permanently sited semi-submersibles, there are site-specific extreme environments and the fatigue requirements and the difficulties in structural maintenance, repair and inspections. Conversely, mobile units can be dry-docked and can also be inspected and repaired afloat on the pontoons. However, the MODU classification rules do represent unlimited, world class service and this is actually quite severe. Also, most semi-submersibles give 30 years or more in service life. Quite often, the extreme design loads for mobile units are more severe than those of the permanently sited units. The opposite is true for the mooring systems, whereby the permanent structure mooring is usually subject to more severe requirements than the mobile units. Mooring is covered in a separate chapter.

In the structural design of semi-submersibles it is important to consider the fact that most semi-submersibles have been built by shipbuilders, and their design construction reflects the fabrication methods and resources unique to this industry. Foremost among these considerations is that shipbuilders employ considerable automation in fabrication of the basic components, particularly the cross-stiffened plate panel, and have highly developed processes for the assembly of panels into structural blocks and the subsequent erection of these. To not fully consider these aspects in the structural design will compromise productivity and make it difficult to achieve cost levels demonstrated by past construction.

For both the semi-submersibles and the TLPs, the shape of the structural components can become a particularly lively issue, regarding the merits of round versus square columns. Hydrodynamicists prefer round columns to reduce current drag, important to mooring loads. However, rectangular columns with generously radiused corners ($R > D/5$) work nearly as well for current loading and produce little practical difference in motions and hydrodynamic loading. The real issue is fabrication and who is to build the columns and how. Shipyards can build both, but they are exceptionally efficient at fabricating flat panels as well as rounded corners (transversely framed and longitudinally stiffened). There are other compelling reasons which result in rectangular pontoons, but these typically have rounded corners.

The idea of large round columns with circumferential ring-frames only, is an illusion inasmuch as, to function as such, they must be perfectly circular and be free of discontinuities. Internal subdivision and a variety of hull appurtenances subvert the structural function of true ring frames. It seriously matters how the internal subdivision is arranged and framed into the shell. It could be arranged as intersecting vertical bulkheads with a minimum of flats, or, alternatively, a large number of flats without vertical bulkheads.

As an important aside, completely square corners are attractive to some builders. However, if not for hydrodynamic reasons, radiused corners of hull components are preferred for damage resistance. Under impact (likely at the waterline), square corners tend to open up, separating at the joining weld. Conversely, a rounded corner (e.g. 18 inch radius or more) will more likely simply dent and not compromise watertight integrity and may even remain otherwise "serviceable."

Local Strength of Semi-submersibles

As will be discussed more specifically in Section 7.8, the design of the hull structure is taken at two fundamental levels: Local Strength and Global Strength. Local strength is the consideration of whether the structure is sufficiently strong to resist the expected distributed load, particularly the hydrostatic pressures. This applies to the plating, the stiffening, and the framing of all watertight surfaces. It also applies to distributed deck loading as well as a variety of functional concentrated loads. In this connection it is significant that 80–85% of all hull steel is a consequence of local loading. In semi-submersibles, reinforcement for the global strength rarely accounts for more than 15% of the total hull steel, the reminder being for special local situations (e.g. fairleaders, foundations, etc.).

For the column and pontoon shell, and the internal subdivision surfaces, a variety of hydrostatic heads are to be considered as potential controlling design pressures. At a minimum, the shell plating must be designed to resist the static loading for the most extreme operating draft without consideration of internal pressure. However, the water-tight shell must be designed for no less than a 20 ft head. In addition, the increased external pressure from the extreme heel/trim must be considered as well as the combined static head and the dynamic pressure from waves. However, in both cases, these pressures are allowed to be 1/3 times greater than the static design pressure for the same scantlings.

Dynamic pressures are not usually a controlling factor for semi-submersibles. The combined static and environmental pressure generally does not exceed the dynamic capacity of the minimum static design head. Frequently, it is the design pressure for the damaged state that governs. In designing for extreme heel, the static design head for goes virtually to the column top (inclined).

For internal subdivision, and sometimes the external shell plating, internal pressures govern. There are special rules for these based upon the piping system. Dynamic pressures are rarely of consequence 70 ft or more below the surface and static criteria govern.

It is important to consider various potential governing pressures on both sides of any plate field and to determine which actually governs the design of that particular surface. Later, in Subsection 7.8.6, fig. 7.92 gives a summary of the design head requirements according to the ABS MODU Rules. For framing, there is an additional complexity in that different loadings of adjacent internal spaces may actually create a controlling load condition for the frame (see fig. 7.102 in Subsection 7.8.9).

Global Strength of Semi-submersibles

Global strength addresses the overall strength of the structure as a space frame, and of the main elements forming it. For a semi-submersible, the elements that form the space frame are the pontoons, columns, and deck and may include bracing. Given later in Section 7.8, fig. 7.73A shows a typical, single celled hull element as part of the global structure. Such elements can be modelled between connections according to the conventional beam theory. However, it should be noted that such shell structures do not behave strictly according to the engineering beam theory, especially if multi-cellular or if there is a sharp bending moment gradients, high shears, or torsion. Where questions of viability of the beam theory exists, the conventional hull girder theory can be applied, or local finite element models can be employed. The hull girder theory is beyond the scope of this handbook (see Taggart, 1980; Hughes, 1988).

The structural configuration of a semi-submersible is primarily distinguished by its transverse strength system. While many variants and hybrids exist, the four categories summarised in fig. 7.36, given earlier, show four basic structural configurations for drilling semi-submersibles. As discussed, these configurations follow an evolution. The earliest designs had structural configurations that dealt more with erection problems than strength. The global strength of the early semi-submersibles was based upon an array of interconnected pontoons (see fig. 7.38) and only later on space frame diagonal trussing. To a large extent, their designers did not fully appreciate the load patterns and the structural responses. Also noted, the proper analytic tools and adequate understanding of the relevant theory did not become prevalent until the late 1970s.

Global strength relates primarily to two types of loading systems: the gravity/buoyancy load and the environmental loading. The direct loading of waves and the inertial load from consequent response are the principal environmental loads. What is unique to the global strength of the semi-submersibles is the controlling load patterns. As previously discussed, two basic semi-submersible configurations have evolved, the twin-pontoon and the closed array pontoon arrangements. The latter, being prevalent with the TLPs, as well as many FPS, will be discussed with respect to TLPs in the next section. Due to their pontoon arrangement, each type has different controlling global design loading patterns.

To look first at the gravity/buoyancy load system, see fig. 7.42. A section and a profile view are shown. An idealised distribution of deck load and concentrations of buoyancy forces at

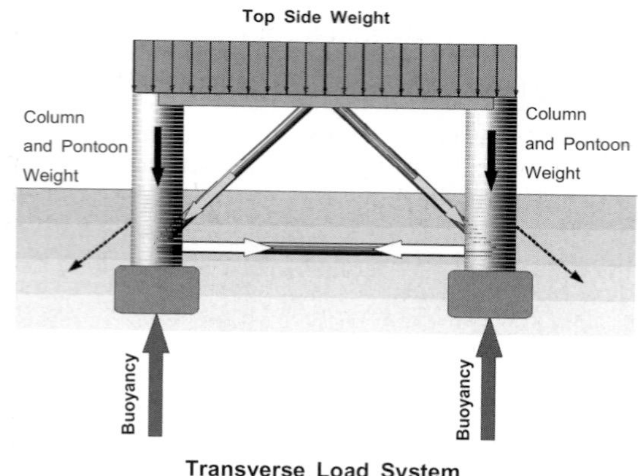

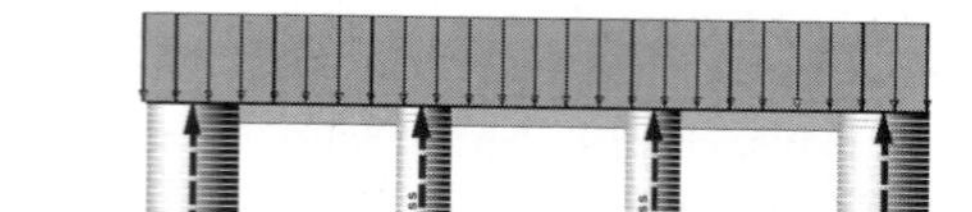

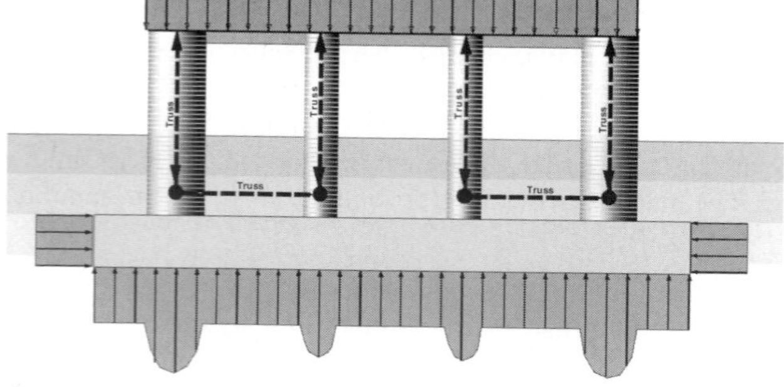

Figure 7.42 Semi-submersible Gravity/Buoyancy Load System

the pontoon and column lines is shown in the section view. An additional gravity load is included in the pontoons and the columns. A distribution of gravity loading on the superstructure must be supported by buoyancy concentrated at the extremities, causing a tendency to sag. This causes very large tensions in the horizontal brace to resist the sag. Additionally, the interior parts of the deck weight will transmit directly through the diagonals into the column. This exhibits one important function of the main bracing as primary structure. Particularly important are the end connections at the column, especially the efficiency of the load flow from the diagonal to the transverse. The loading of the mooring system is also shown. This load will transfer through the bracing also as tension.

Without the diagonal bracing, the deck can be supported only by shears and bending moments from the column tops. Still, however, the horizontal transverse is needed to resist these moments and thereby develops tensions. Alternatively, the transverse pontoon segment of a closed array pontoon system provides this tension.

Longitudinally, most semis have a series of 3 or 4, fairly close-spaced columns on each side, supported by a continuous pontoon. The lower part of fig. 7.42 shows the profile on one side of an 8-column configuration. Being comparatively short, with reasonable distributions of load, the pontoons and deck do not have significant longitudinal bending/shear problems and generally do not require longitudinal diagonals. As can be seen, the superstructure gravity loading is better supported, featuring shorter deck spans and continuity of longitudinal strength. The pontoons generally have considerable shear strength for their length. Secondary bending does occur at the column tops and the pontoon connections due to uneven deck loads and pontoon liquids, but stresses from these distributions rarely are significant (usually less than 5 ksi). Also noted on the figure is the end pressures on the pontoons. While small, they are not negligible (about 3 ksi).

The 8-column semi typically has large columns at the ends and smaller diameter interior columns. The interior columns are as much structural members as anything else and can have somewhat higher stress levels because they are sized for loading and not their waterplane contribution. Six-column semis generally have columns of equal diameter. While some have longitudinal diagonal bracing, the pontoons are sufficiently strong and the bracing is not truly necessary.

From an environmental load perspective, there are two wave load systems of importance. One is the "squeeze/pry" load. This is illustrated schematically in fig. 7.43. The squeeze/pry load system is primarily a lateral loading effect and vertical loading is not of serious consequence. There are two stages to the load system. With the "pry," the wave crest is centred and the collected wave forces (pressures) try to pull the semi apart. The second stage, the "squeeze" is with the wave trough is centred, where the effect is for the wave to push in from the sides. This effect applies at all wave headings, but most structural configurations are more vulnerable to beam seas. Length of the wave (period) relative to spacing of columns and pontoons is very important.

Oblique seas have a squeeze/pry effect, but vertical loading tends to twist, or "rack" the structure. This is illustrated 3-dimensionally in fig. 7.44. It is particularly severe for the widely spread, twin pontoon configuration. It is also a much more complicated situation. Each pontoon/column set can be viewed as a separate system. When the wave system is oblique, the heave force on one pontoon is concentrated at one end and, on the other

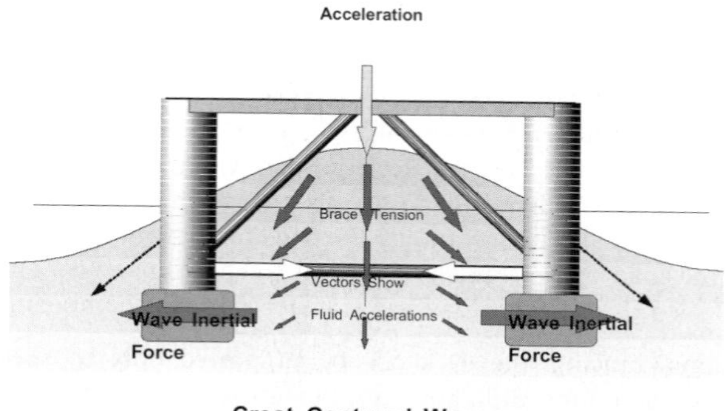

Crest Centered Wave

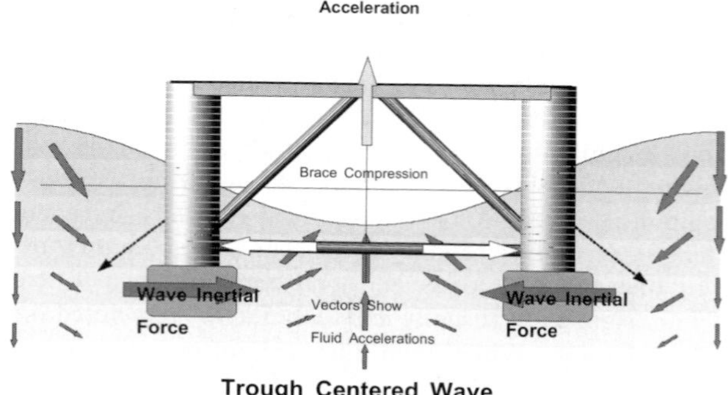

Trough Centered Wave

Figure 7.43 Twin pontoon semi-submersible squeeze/pry wave load system

pontoon, the force is concentrated at the other end. These combine to cause a global torque, or a "racking moment" on the hull. The vertical force distributions on the pontoons are shown on the figure along with their resultants. The racking moment is augmented by similar shifts in the lateral force systems. The distribution of these forces is also shown on the figure.

Like squeeze/pry, there are two stages to the racking load system also. One stage can be viewed as sagging on an oblique wave with wave crests centred at opposite *corners* of the semi-submersible. The other is the corresponding, crest centred hog. The sagging stage is illustrated. The effect is that the wave forces alternately twist the semi one way and reverses to the other. Racking occurs at any angle of oblique seas, with the critical heading and wave length slightly different for each location of critical stress. Conventionally, a single critical wave is used, usually one along the diagonal between the corner columns. Again, the length of the wave relative to spacing of columns and pontoons is very important, in this case a longer wave is critical to squeeze/pry.

The controlling load patterns for the closed array pontoon semi-submersibles are quite different from those given above for twin pontoon semi-submersibles. In as much as this configuration and its controlling load patterns are quite similar to those of the more typical

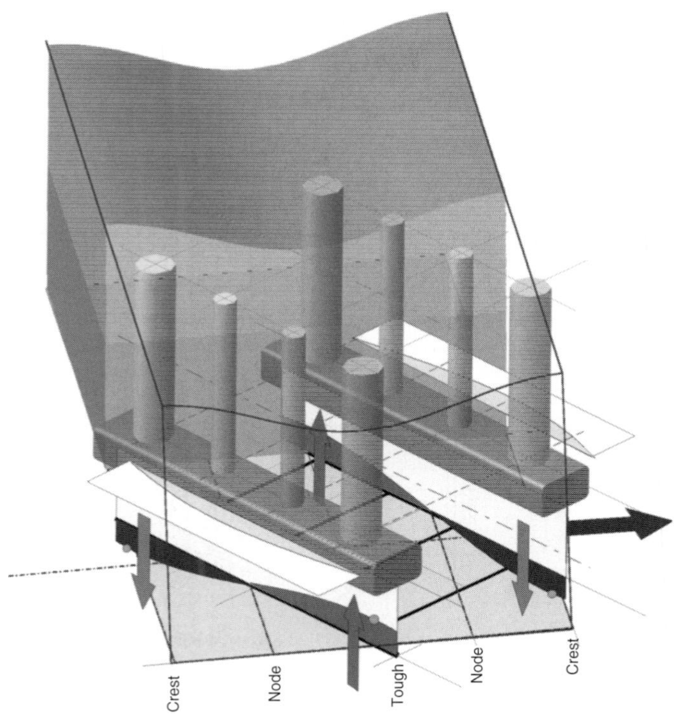

Figure 7.44 Twin pontoon semi-submersible wave racking load system

TLP configuration for TLP, the discussion of controlling global load patterns will be addressed in the following section, addressing TLPs. To some degree, the gravity/buoyancy load behaves like that loading illustrated in fig. 7.42, except without the interior pontoons. However, with the large transverse pontoon, the squeeze/pry loading as described is no longer critical. However, there is a diagonal equivalent to be discussed. Also critical to a TLP, due to their height, and also to a deep draft semi-submersible is a lateral shearing effect from accelerations applied to the deck mass.

7.6 Tension Leg Platforms

7.6.1 Introduction

Tension Leg Platforms have been used exclusively as production and drilling platforms, with the exception of the "East Spar" platform, which is a control buoy. Figure 1.18 shows the platforms installed as of 2002.

Like a semi-submersible, the TLPs consist of columns and pontoons. The unique feature is the mooring system, which consists of vertical tendons (sometimes called "tethers"),

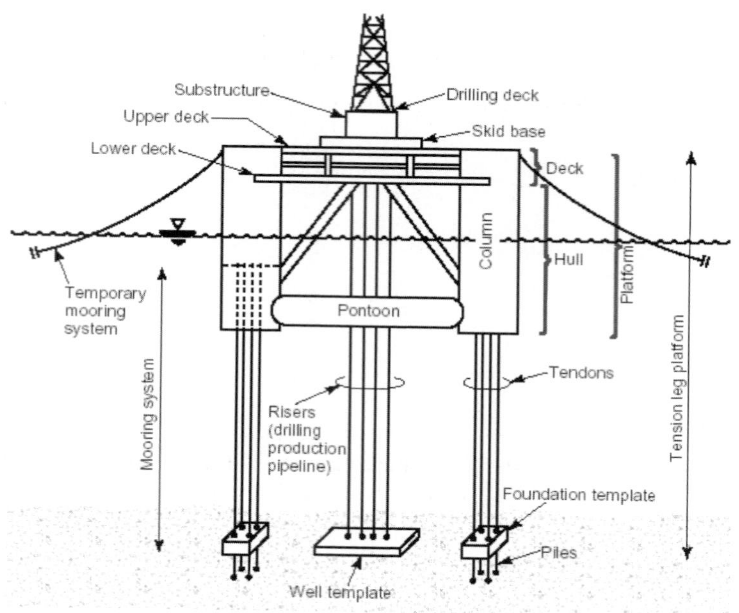

Figure 7.45 Tension leg platform terminology (API RP2T, Reproduced with permission of the American Petroleum Institute)

which restrain the heave motion. Figure 7.45 shows the configuration and terminology applied to the TLPs.

The previous section also applies to tension leg platforms (TLPs). The reader is therefore advised to have read Section 7.7 prior to continuing this discussion. The TLP is nevertheless fundamentally different in many respects, and this section will concentrate on these differences and note the contrast and the distinctions that make the difference.

The emphases in this section is on the design of the TLP as a platform, particularly with regard to its sizing, proportions, tendon arrangement, and the integration of many diverse, important design aspects. Moreover, this section addresses the issue of making the key fundamental decisions of overall design. Analysis is left to confirming those decisions and to refining the choices. It is hoped that, instead of precise answers for a wrong design, good answers for a good design prevails.

As with the semi-submersible, the main problem in TLP design is to address the specific functions and rational sizing. Even more so than semi-submersibles, the construction program consideration is a salient design issue. Here too the approach is to initiate the design with a straight forward process for sizing, and to discourage multiple trial-and-error analyses with inappropriately rigorous detail and methods.

The *initial design(s)* should be represented in a sketch indicating the chosen configuration and all important dimensions to be used in the *preliminary analysis* and design activity, including tendon arrangements. The internal hull subdivision need not be included, but a reasonably definitive conceptual arrangement of the deck should. *The initial design* should include the best working weight estimate, hull displacement, and hydrostatics available.

The key analytic areas for *preliminary analyses* of the *initial design* for a TLP include the following:

- Weights and CG's
- Wind Forces
- Current forces
- Global Performance Analyses
  - Motions
  - Drift force
  - Tendon tensions
- Global Strength

With a well executed *initial design*, a model can be quickly established for the above analyses to rigorously proceed in parallel without the need for additional major design iterations. Results for the *preliminary analyses*, based upon the *initial design*, can be used for a reasonably conclusive revision, this being the *preliminary design*. The *initial design* is also adequate for the beginning of specialised subsystem (topsides, tendons, riser, installation, etc.) design and analyses to proceed concurrently. The following is about how to develop such a model.

Unlike semi-submersibles, hydrostatics and stability are not salient design issues for a TLP, although they are important considerations for addressing transport and installation. Eventually, in the initial design process, internal subdivision does need to be addressed. For a TLP, internal subdivision primarily based upon avoiding excessive reduction (or increase) in tendon tension from internal flooding. However, an additional subdivision may be required for transit/installation damaged stability.

The design spiral for the TLP, represented in fig. 7.7, indicates that the parameters are continuously updated as more refined analysis or as criteria are changed. For a TLP, the preliminary design, as outlined above, should require much less refinement of the initial design than with a semi-submersible. The refinement should mainly be confined to minor changes in column diameter, pontoon dimensions, and initial tendon tension, and should probably not require a change in draft. With timely execution of the deck layout work and weight estimates, it is possible to conclusively fix the column spacing as well as vertical dimensions with the initial design.

Unlike the semi-submersible, the TLPs to date have been exclusively used for permanently sited production systems, most with drilling or workover functions. They have fewer functions to consider and therefore limited configurational variants.

Although it is implicit in any design, that the construction, transport and installation scenario is a particularly important aspect of TLP design with considerable impact on certain design choices. Usually, the hull and deck are separately fabricated, with either an inshore hull–deck mating or an offshore heavy lift. In either case transit, ballast to the operating draft, stability, and installation of the tendons are key issues in the design of the TLP. Many of the comments made for semi-submersible design are useful considerations addressing TLP transportation and installation issues. In particular, weight, vertical centre of gravity, and stability are design constraints. Otherwise, once installed, stability is not a design factor, although weight and weight distribution is very much so.

As with the semi-submersible, a definitive Functions List, Systems Summary, and Equipment List (all: mission and support) should be developed concurrently to the initial design work. This should likewise include a trial equipment and systems layout, with particular emphases on how and when the deck and hull are joined, identifying any design constraints needed in the initial design. Deck layout must address the number of levels, minimum spacing between columns, future addition/removal of equipment, and its installation. In addition, a weight estimate for all equipment, systems outfit, and the deck structure should be made, particularly as a necessary element of installation planning. Pending vendor equipment weight data, weights should be estimated from the historical data. The weight estimate should be based upon proforma arrangements and calculations and include outfitting (piping, access, corrosion protection, etc.). For the design operating states, variable load and its distribution should be firmly established. The conclusion of the initial design should reflect the preliminary topsides work, particularly with regard to column spacing, height of the deck, its weight, and the wind profile.

Apart from hull steel-weight estimates, development of the hull scantlings and framing system decisions should be deferred until the preliminary design is completed. However, a preliminary global strength analysis should be completed as a guide to the development of scantlings.

Before completion of the preliminary analyses and design, a re-evaluation of the installation process should be made, particularly with regard to its stability and ballasting. Any adjustments found necessary to facilitate construction and installation should become evident at this point and be incorporated.

7.6.2 Functions and Configurations of TLPs

Functions

Unlike the semi-submersible, the tension leg platform is a more recent development and is exclusively used as a permanently sited production platform. Mobility is not an issue. With a minimal range of functions, TLPs have very little configurational dependence upon function. Size does present certain operational logistics problems and permanent siting reduces the maintenance options. Also, TLPs inherently do not lend themselves to petroleum storage as an option.

Besides the mission and support functions, the essential function of a TLP is to the support a payload above the highest waves. More specifically, the hull is to provide buoyancy, both for the support of weight and to provide tendon tension. It should also be tall enough to give the deck wave clearance in all modes of operation. Tendon tension has as much influence on hull size as the payload. Functional requirements of the deck and well systems can also influence column spacing.

Unlike a semi-submersible, it is the columns that are the principal source of buoyancy, with the pontoons functioning more as structure. The pontoons do have a size relationship with the columns in regard to hydrodynamic force. Also, unlike the semi-submersible, in the operational state, the TLP employs very little ballast. Ballast is provided to even loading between tendons and also to offset unused payload capacity. Beyond this, like the semi-submersible, the size, submergence, proportion, and spacing of the columns and pontoons are major factors in the hydrodynamic performance.

While the heave motion of a semi-submersible is a salient design issue, vertical motion of a TLP is far less of an issue and entirely different. While the TLP does not heave, it will undergo set-down with offset. Like a semi-submersible, the TLP is laterally compliant and will surge, sway and yaw. In both platform types, there is relatively little design-wise that can be done to affect lateral motions, although steady offset can be minimised by increased tendon tension.

Configuration

The three main configurational components are:

- Pontoons
- Stability columns
- Deck

Figure 7.46 illustrates four TLP configurations. The Brutus Platform (fig 7.49C) is a typical example of a four column, "classic" TLP. Unlike semi-submersibles, TLPs do not employ space frame bracing. One reason is, given the closed pontoon array, there is less need of them.

Only recently have TLP configurations begun to significantly evolve. Except for the Hutton platform, in the late 70s, which was a rectangular, 6-column unit, all TLPs until the late 1990s were square 4-column units. Three-column designs have been proposed, but not built. Since the late 1990s, four single column designs (*SeaStar* fig. 7.46B) and two close-clustered, multi-column designs (*Moses*, fig. 7.46A) have been built. Both designs have external, radial pontoon arrangements. Another innovation is removable sponsons (external column additions) for installation. Yet another factor is the short, external, radial pontoons at the columns, connecting the tendons (fig. 7.46d). These extend the tendon spread and reduce maximum tendon tension, as well as reduction in deck spans.

Other configurational choices involve deck design and connection of the deck with the columns. The pontoon-to-column connection configuration is another, although this is driven primarily by structural design considerations. This is largely the case whether to adopt the circular or the square column and/or pontoon cross-sections and, in the case of square, whether to have radiused corners.

The mission functions and associated support functions generally do not significantly influence configuration. The well-system arrangement can have impact on column spacing and deck design. However, construction and installation are more likely to influence configuration. This is evident in some recent construction. Mating of the deck will have an important influence on the deck configuration, particularly its interface with the columns. Many of the innovations have addressed installation issues, tendons, and the deck, particularly risers and well systems.

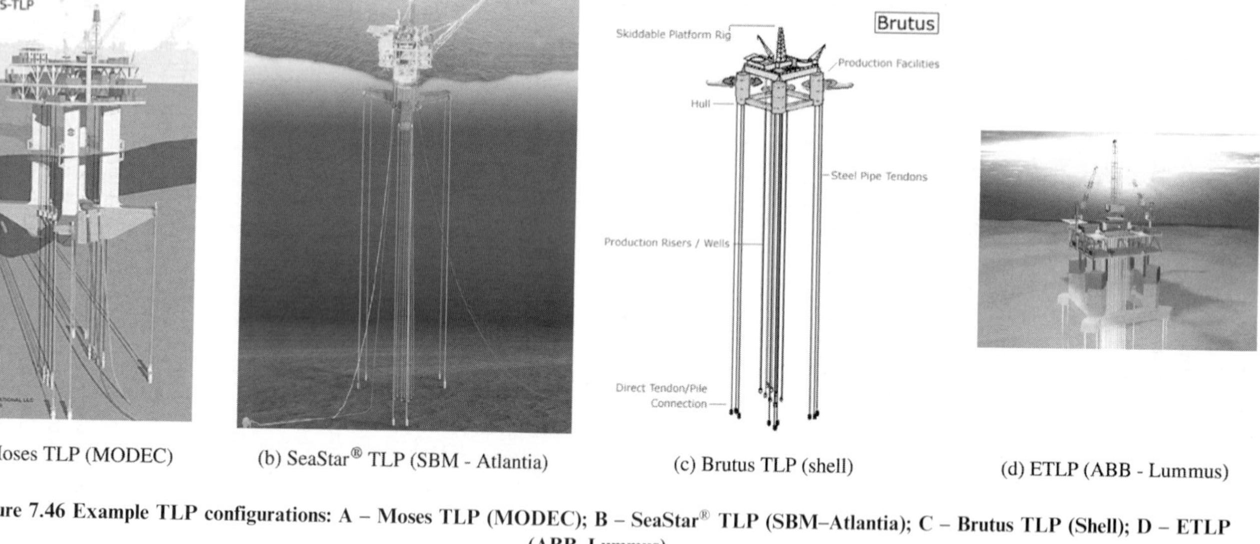

(a) Moses TLP (MODEC)

(b) SeaStar® TLP (SBM - Atlantia)

(c) Brutus TLP (shell)

(d) ETLP (ABB - Lummus)

Figure 7.46 Example TLP configurations: A – Moses TLP (MODEC); B – SeaStar® TLP (SBM–Atlantia); C – Brutus TLP (Shell); D – ETLP (ABB–Lummus).

Decks of TLPs (and some production semi-submersibles) are unique. Virtually all TLP decks are separately built from the hull (often on a different side of the world) and joined later, either at dockside, offshore, or in a separate, sheltered location. An underlying issue with this uniqueness is more administrative than technical. Except at the very early design stages, deck and hull are separately designed and managed. Because of their functional and design complexity, and an extensive design history, design of TLP decks is very different from those of other floating systems. This is also true of the FPSO production decks.

To date, classic TLP decks have always been long span, four-point corner support, open trusses. As such they are multilevel structures. The single column and close-clustered, multi-column designs have featured deck designs very much like those of fixed platforms, with a four-point support as if a 4-pile jacket supports it. A very similar arrangement has been used on the spars. The installation process is with an offshore heavy lift very much as it is with fixed platforms. Preference of the open truss is both to avoid ventilation requirements (as in a hull-type deck) and to facilitate existing fabrication and outfitting methodology. Outfitting makes extensive use of distinct, separately built modules. Besides the intrinsic structural function, to transfer the weight of the deck and its loading to the columns, the deck is also a part of the overall global strength system and must be considered so. Decks also have significant installation considerations to address.

With TLP hull installation benefitting from minimisation of topside weight, usually some parts of the deck system are placed after installation of the TLP. The deck-to-column interface is of considerable importance, with many options for load transfer and securing, and many being proprietary. These include the welded and mechanical connections, some shimmed to achieve a favorable pre-load. This connection could be an integral, full moment connection, as in most semis, but this has not been preferred.

As noted above, with the exceptions noted, the 4-column configuration has dominated the design choices until the late 90s. These all have the closed array pontoon arrangement. The reader is referred to the corresponding discussion for semi-submersibles for addition attributes of this arrangement. It should be noted, however, that the column to pontoon side relationship is very different in TLPs from semi-submersibles, not to mention important differences in load patterns.

For reasons discussed regarding decks, there has been an avoidance of any functionality (equipment, tanks, etc.) within the columns. There has been a small tendency away from this, especially the storage of liquids and the handling of saltwater. Otherwise, the columns and pontoons consist of vast empty spaces with much watertight subdivision.

7.6.3 TLP Mechanics

General

Despite some apparent similarities, TLPs and semi-submersibles are fundamentally different types of systems. Whereas a semi-submersible is a true, free floating structure, restrained with compliant spread moorings and/or dynamic positioning, a TLP is kept in place through lateral forces developed by the tendons when the TLP is moved off from centre. The lateral force is dependent upon the tendon tensions. Consequently, a major portion of the TLP buoyancy is devoted to development of tendon tension. Besides, while dynamic mooring loads of other floating structures are largely mitigated by platform

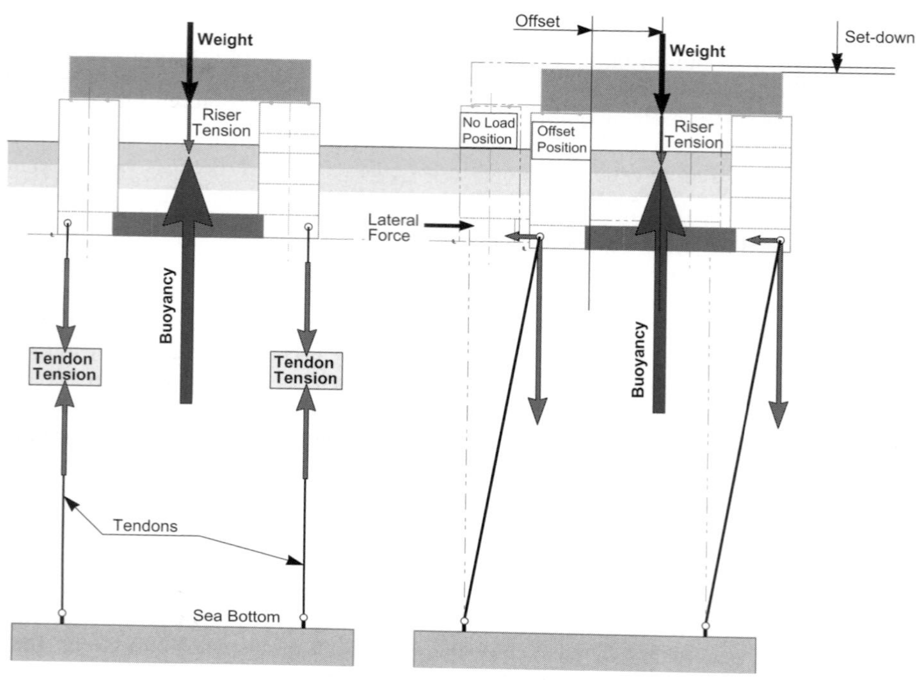

Figure 7.47 TLP tendon mechanics

inertia, the mooring loads of TLPs are directly linked to first order wave loads on the structure. In heave, the TLP is "fixed".

Pretension

The left of fig. 7.47 shows the forces acting on a TLP in still water without lateral loading. The total weight is given as:

$$W = W_o + \delta W \tag{7.23}$$

Where W_o is the "lightship weight" and δW is the variable load (the distinction is discussed later in Subsection 7.6.4). In addition there are riser loads, T_r. On a TLP with surface trees and top-tensioned risers, this would be a downward force amounting to a significant part of the payload. It may also include other risers (export, subsea, drilling, etc.), but, for this discussion, no separate distinction is made. The tendons add another downward force, T_t.

Where ∇ is the displaced volume of the hull, ρ is the mass density of the seawater, and g is the gravitational, the buoyant force is expressed as $\rho g \nabla$. This must support the downward load weight of the TLP, W, the riser tensions, T_r and develop the requisite tendon tension. The tendon tension is therefore given as:

$$n_t T_t = \rho g \nabla - W - T_r \tag{7.24}$$

where T_t is the individual tendon tension and n_t is the number of tendons.

It is important to note that the displacement, ∇ will depend upon the draft of the TLP and that this could change with tide. Therefore there must be a basis displacement, ∇_o, based upon a basis draft d_o. The basis draft will typically be the stillwater draft at the mean low water tide reference. The tide will be reference to this and can included storm tide as well as lunar tide. Any tide change, δd_t, is relative to d_o and would correspondingly change the displacement and tendon tension. This is as follows:

$$\delta\nabla_o = \rho g A_w \times \delta d \tag{7.25}$$

A_w is the water plane area of the columns and δd is the sum of the relevant draft corrections. Depending on design circumstances, there can be draft increases due to bottom subsidence (δd_o) as well as an allowance for error in the installation (δd_e). This may be recognised in a variety of ways, and may not be trivial, but is not further addressed. In addition, discussed below, when the TLP is offset from centre, there is also an increase in draft referred to as "set-down." What is important in this regard is that there will be a minimum and a maximum tendon tension to consider, and that the freeboard will be reduced in the offset position.

At the basis draft, d_o, the individual tendon tension is:

$$T_o = (\rho g \nabla_o - W - T_r)/n_t \tag{7.26}$$

This is referred to as the "pre-tension." Relative to this, due to tide or other draft change, the increment of tension is:

$$\delta T = \rho g A_w \times \delta d/n_t \tag{7.27}$$

In this discussion, it is assumed that the riser tension, T_r is constant. Except for drilling risers, this is normally true. However, top-tensioned, production risers are incrementally installed over a period of three years or longer, often with some deferred longer. Also, over a limited range, their tension is also adjustable. In the end, this issue requires various completion states to be considered. Generally, the basis riser tension is maintained by adjustment of ballast.

Offset and Set-down

The right of fig. 7.47 shows the change in forces acting on the TLP with lateral loading. This is the crux of a TLP. Where a TLP is offset by distance x, the tendons, maintaining length, L_t, cause the TLP to submerge, or set-down, δz given as follows:

$$\delta z = L_t\left[1 - \sqrt{1 - (x/L_t)^2}\right] \tag{7.28}$$

$$F_x = n_t T_t \tan\phi \qquad \text{and} \qquad \phi = \arcsin(x/L_t) \tag{7.29}$$

For completeness, a small tension increment, $\delta T = \rho g A_w \times \delta z / n_t$ may be considered. However, what is important is the lateral force developed. Where $n_t \times T_o = T_t$ is the total tendon tension, the following lateral force is developed:

This is near linear for small offsets, expressible as:

$$F_x \approx \left(\frac{n_t T_t}{L}\right) x \tag{7.30}$$

It is important to note that top tensioned riser tensions can be substantial and that these can act as virtual tendons. These can be included and, if so, they should be considered as centred and at a constant value. Also, the above assumes that the tendons are neutrally buoyant. For a uniform tendon weight in water of w, per unit length, the horizontal restoring force becomes, approximately [Demirbilek, 1989]:

$$F_x \approx \left(n_t T_t - \frac{wL}{2}\right) \frac{x}{L} \tag{7.31}$$

The limits on offset will, of course, depend upon the design specifics, but keeping a maximum offset in the 6–8% of water depth range is a good starting point. For 2000 ft or more water depth, this will produce extreme tendon angles just under 5°. However, considering tendon flexure, and other factors, mechanically the tendon connectors cannot exceed 10°.

An offset is made up of several components. The steady components are due to wind (F_w), current (F_c), and steady wave drift forces (F_d). Depending upon specific environmental parameters, these forces will account for half or more of the maximum offset. The remaining part of offset is dynamic. One part is due to wave frequency surge, $x_w(t)$ and the other part due to slow drift oscillation x_d (t). The former is a simple function of wave dynamics with a maximum amplitude of 10–12% of the extreme wave height, depending upon the draft and lateral spread of the hull. The latter is a small low-frequency response of maxima not concurrent with the wave frequency maxima. It is typically estimated from model test or time domain analyses. Lacking such data, taking the low frequency component at 30% the wave frequency maximum will suffice until such data is available. Specific analysis methods are discussed in Chapter Three and extensively in the literature (see, e.g. Demirbilek, 1989; API RP2T).

Tendon Tension

For design purposes, the initial value of tendon tension can be set to limit offset to 5% of water depth. Therefore, computing the sum of steady forces, $F_w + F_c + F_d$ will lead to a determination of T_t. Depending on the size of the TLP and the environmental specifics, the steady forces can be 1000 kips or larger. It is important to note that F_c can include significant current force from tendons and risers. As a first approximation, therefore, the tendon pretension can be taken as twenty times the mean horizontal environmental force to be resisted. This nominally results in 5% offset if the added tension due to setdown is ignored.

The increased tendon tension from the offset is due primarily to the overturning moment if the wind force. This does not affect the total tension, but results in an increase in the upwind tensions and a decrease in the downwind tensions. There is significant increases (and decreases) in tendon tension due to wave loading. Wave frequency response and forces are the dominant component of tendon tensions, and these are readily determinable from linear wave theory. There are three important exceptions, however.

One of the exceptions is the vertical drag force of extreme waves on the pontoons. While a much smaller component than other components of wave force, they are not trivial.

The second is a variety of second order effects directly from waves.

The third exception is dynamic response of the TLP mass on the elasticity of the tendons. This includes vertical response (heave) and rotational response (pitch/sway). Except for very deep water, these motions have periods in the 2–3 s. range, and respond with very little damping. As a result, a persistent absorption of energy from wave will excite these modes ("springing") and thereby increase the maximum tendon tension. A second form of mechanical response ("ringing") results from short duration, impact loads, typical being the drag force from a particularly large wave crest impacting a column.

While considerable research and theory has gone into the springing, ringing and second order effects, meticulous model testing and factors based upon earlier design have been the most reliable basis for addressing these (Engebretsen et al, 2002).

Wave forces increase tendon tensions in two principal ways. One is with the crest (or trough) is centred. In this case, the net wave vertical force is usually resisted equally by the tendons. This is summarised in the upper part of fig. 7.48. The second is summarised in the lower part of fig. 7.48. In this case the wave force system at the passing of the wave nodes ($L_w/4$ before or after the crest) causes an overturning moment (pitch). This moment is resisted in part by lateral inertial forces due to surge acceleration on the fixed and hydrodynamic masses of the system. Surge response is very subresonant, and it can be assumed that the surge wave force is entirely resisted by surge inertial forces. The lateral tendon reactions are typically 2–3% of the force.

The following is a summary of the tendon reaction for the two defined cases:

$$T_r = \frac{1}{n_t}\left\{\sum F_{zc} + \sum F_{zp}\right\} \tag{7.32}$$

$$T_i = \frac{1}{s_t n_t}\left\{F_{wx} z_w - (M_o \bar{z}_G + \delta M_o \bar{z}_a)x'' \quad + \quad \sum F_{wz-c} a_c + M_{w-p}\right\} \tag{7.33}$$

T_r is taken at the crest (T_r here refers to tendon forces for a crest centered wave, not to be confused with riser tension used earlier). F_{zc} is positive upward and F_{zp} is downward. Both are a function of wave frequency, ω and amplitude, H_w. T_i is taken at the trailing node, after the passing of the crest. F_{wx} is the surge wave force (total from columns and pontoons) and is centred at z_w above the baseline of the TLP. $F_{wz\text{-}c}$ is the heave force on the bottom of each column, centred at a_c off centreline, and $M_{w\text{-}p}$ is the collective moment of all pontoon heave forces. These are all resisted by a symmetric set of n_t tendons,

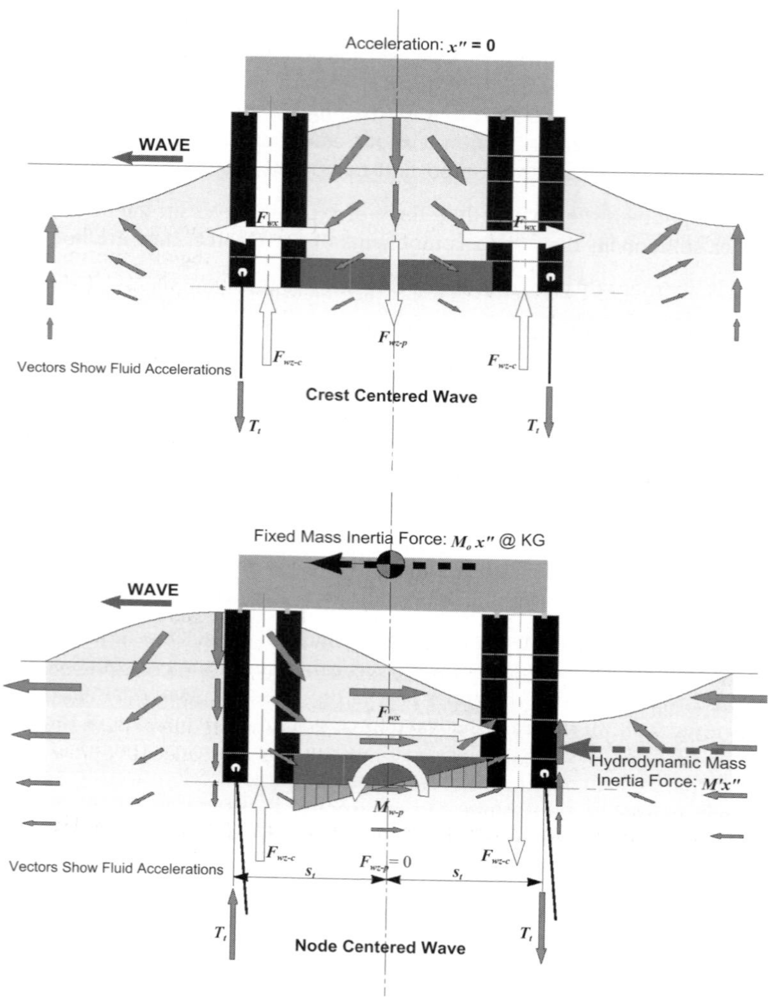

Figure 7.48 TLP wave load systems

all at distance s_t from centreline. Additionally, surge acceleration, x'' develops inertial forces in the fixed mass, M_o and the hydrodynamic mass, δM, centred respectively at z_G and z_a above baseline. Like those of T_r, the components of T_i are functions of wave frequency, ω and amplitude, H_w. These functions are all 90° out of phase with those of T_r.

There is much more rigor and detail to the writing of the above equations, but the intent is to demonstrate behavior, relying as much on fig. 7.48, and illustrate the two components of wave frequency tendon force. Then, if $T_r(\omega)$ is the tension in each tendon for a passing crest, and $T_i(\omega)$ is the tension for a particular tendon at the advancing node, at that wave frequency, ω, the maximum tension is given as:

$$\max T_x(\omega) = \sqrt{T_r^2(\omega) + T_i^2(\omega)} \tag{7.34}$$

This is addressed most specifically in later discussion. What is important is that each term is sensitive to different design parameters. T_r depends largely upon the distribution of volume between the pontoons and columns whereas T_i depends upon the tallness of the TLP and the spread distance of the tendons.

There are other components to tendon reaction. Some of these do not matter in the parametric analysis because they are small and cannot be directly related to controllable parameters. In this sense true maxima will not be given and appropriate allowances should be made where this matters. The reactions to steady forces (wind, current and drift) can be readily determined and should be used appropriately. Likewise, the change in tensions due to set-down should be incorporated.

Motions

A TLP is highly compliant to lateral forces and, at the same time, is highly resistant to vertical forces. Offset from lateral forces is not altogether different from surge/sway response of any compliantly restrained floater. However, what is truly different for a TLP is *set-down*. As discussed earlier (see fig. 7.47), this is a corresponding downward motion, geometrically coupled to offset. Offset includes a steady component from steady wind, current and drift forces. Dynamic offset includes wave frequency and low frequency responses.

In the usual sense of floating structures, there are no heave or pitch motions. However, as a mechanically elastic system, there are vertical and rotational responses to the heave forces and pitch moments. Since the natural period in heave and pitch is lower than the wave energy, global heave and pitch wave loads on the platform are carried directly by the tendons. Wave energy close to the natural period of the platform results in an amplified response at resonance known as "springing". While typically small, springing is usually not negligible and it is important for the estimation of fatigue of the tendons and supporting structure (API RP2T). A springing response of a TLP can also result in vertical accelerations, which affect the comfort of personnel. Similar, but different in origin, is a second high frequency response referred to as *ringing*. This results from an impulse load in an extreme sea state, which will excite a transient response. The ringing response is important for tendon extreme tension estimates.

The TLP dynamics, including both the high-frequency and low-frequency responses, are addressed in other discussions in this handbook. Nevertheless, it is noted that tendon response periods higher than 3 s should be avoided in the initial design. To achieve this, tendons need sufficient stiffness relative to the TLP mass. This precludes certain material for tendons and tends to pose limits on TLP water depth.

7.6.4 Sizing of TLP

General

As we have seen (fig. 7.46), there are a variety of TLP configurations currently in use. The following addresses only the 4-column arrangement, but does include short external pontoons (fig. 7.49).

When selecting critical dimensions, the function, construction and installation should all be kept in mind. Column/pontoon spacing may be constrained by the spread of the well system array and/or deck space requirements. Extreme spacing of the columns can introduce global strength problems, particularly for the deck (While interior support of the deck by bracing or small interior columns is possible, these have yet to prove viable choices). Also, dry transport of the hull and/or deck may impose constraints. While also important, the choices regarding deck-to-column connection are more a matter of structural/mechanical, with little bearing on TLP size.

As with semi-submersibles, the connection of the column to the pontoon is especially important. It is notable, however, that TLP columns are considerably larger than those found on semi-submersibles and for reasons which will become clear, the pontoons are considerably smaller in cross-section. Columns have generally been circular in cross-section, but square or rectangular columns are viable choices. It is more of a fabrication choice with some hydrodynamic implications. Large corner radii ($r \approx D/4$) on square

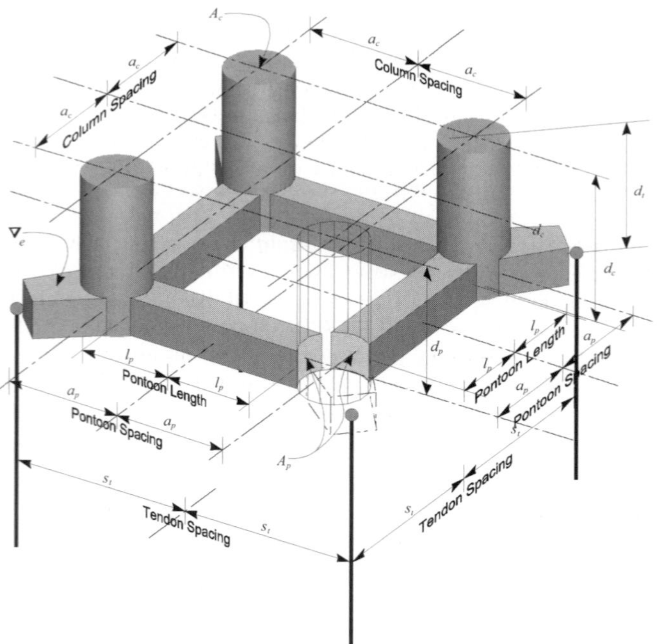

Figure 7.49 TLP Initial design underwater model

column achieves most of the hydrodynamic benefits of round columns. Although circular pontoons have been used, the trend is toward rectangular cross-sections. What will influence this issue the most is the structural design joining the pontoons with the columns (see semi-submersible discussion) as well as tendon connections.

There are important benefits to employing short, external pontoons, extending outward from the columns to make the tendon connections. The main benefit is the reduction of extreme tendon tensions from pitch moment. It also makes closer column spacing more viable, making for a more efficient deck structure.

Generally the internal outfitting within the TLP hulls is considerably less than found on semi-submersibles. Typically, there is no mooring equipment. In as much as there is very little active ballast, piping systems are much different. In service, most internal spaces are considered to be voids and may be piped differently than in a semi-submersible. The ballast is used for installation (and removal), but this may be through a temporary system deactivated after installation. Otherwise, any "operable" system must be maintained in working order. Often a less robust system is employed for damage control dewatering, but this depends upon the damage mediation strategy. There will be some piping requiring space for manifolds, but all fluid paths will likely be through the column top and not penetrate the shell. There is a possibility that consumable liquid storage will be provided internal to the hull. Inspection access must be included however.

Initial Design Considerations

Given the configuration, the sizing of the columns and pontoons, and determination of their spacing, is a relatively simple process. It should be kept in mind that the principal function of the hull is to provide the buoyancy to support the weight, most of which is in the deck load and not the hull, the weight of which is minor in comparison. However, 25–45% of the buoyancy can be devoted developing tendon tension. It should also be kept in mind that it is also the function of the hull to support the deck above the highest waves.

Apart from buoyancy, the principal function of the pontoons is to provide a vertical hydrodynamic force (heave) opposite in direction to the hydrodynamic force on the column bottoms. For multi-column TLPs, the pontoons also have a structural function part of a space frame consisting of the deck, columns and the pontoons. The upper part of fig. 7.48 shows, schematically, the heave force system for a crest-centred wave. As will be discussed further, it is a major design objective to minimise this force. Typically, the pontoon volume will be a third or less of the total displacement.

Besides a reversal in proportions in column-to-pontoon size, a major difference in TLPs from semi-submersibles is that the hull and contents contribute relatively little mass to the system. The centre of mass of the TLP is comparatively high. Conversely, particularly with a large mass of ballast in the pontoons, the centre of mass of a semi-submersible is much lower.

The tendons must be connected at a sufficient spread that pitch moment does not cause excessive reactions.

The height of the columns should be sufficient to support the deck with adequate clearance above the highest waves taking into account maximum set-down. The API RP2T allows for designs which experience wave impact with portions of the deck, as long as these impact loads are properly accounted for. Additionally, since the tendons are mechanical fixed elements, connected to the sea bottom, there must be allowance for tides, sea bottom subsidence, and installation error tolerance. These factors establish the height for the bottom of the deck. To set the column height relative to the deck, it will be important to have conceptually addressed the deck to column connection. Generally, a considerable effort is expended in determining the above data and it is represented in the design basis. The design seastate data will specify the crest level above still water.

While the height of columns on a TLP tend to make "wetness" above the columns less likely, very large columns can distort the wave form and cause "wave run-up." As with the semi-submersible, wetness prediction is best addressed with model testing. What tolerable wetness may be is subjective, however.

The past design tendency is for fairly deep draft columns and pontoons. There are hydrodynamic benefits for doing so. However, there are also structural penalties. Most obvious is the higher hydrostatic design heads of the lower column and the pontoons, not to mention the heavier column steel. Shorter, larger diameter columns have less surface area and therefore less steel. A taller structure also suffers higher racking moments generated by surge/sway accelerations on the high centre of mass. This is discussed more in the following pages.

Stability during installation is an important issue, although there need not be any deck variable load and certain permanent deck items can be absent. Also, because installation is a one-time operation, it need not be efficient and operational management can invoke options with regard to favourable weather.

Figure 7.49 shows the underwater body model used for initial design computation of displacement, motions, and forces of 4-column TLP. The defining set of parameters is:

Columns:	Column cross-section:	$A_c = \pi\ D_c/4$
	Column draft:	d_c
	Column spread:	a_c
	Column height (WT top):	f_c
	Waterplane area:	$A_{wp} = 4\ A_c$
	Immersed Column Volume:	$\nabla_c = A_{wp}\ d_c = 4\ \mathrm{A}_c\ d_c$
Interior Pontoons:	Pontoon cross-section:	A_p
	Pontoon length:	$L_p = 2b_p$
	Pontoon spread:	a_p
	Pontoon centre submergence:	d_p
	Pontoon volume (total)	$\nabla_p = 4\ A_p\ L_c$
Exterior Pontoons:	Cross-section:	A_e
	Pontoon length:	L_e
	Radial volume centre (from col centre)	r_e
	Pontoon spread:	$a_c + (D_c/2 + r_e)/2^{1/2}$
	Vertical centre submergence:	d_p (same as interior)
	Pontoon volume (total)	$\nabla_e = 4\ A_e\ L_e$
	Total Displaced Volume:	$\nabla_o = \nabla_p + \nabla_c + \nabla_{ext}$

Tendons:	Tendon spread:	s_t
	Depth of pivot point:	d_t

The simple parametric model need only represent the pontoons and columns as prismatic sections as shown in fig. 7.49. The interior and exterior pontoons are represented only by a cross-sectional area and length. Similarly the columns are represented by cross-sectional area and column draft. The cross-sectional shape of the columns and pontoons is represented by the added mass coefficients (see Section 7.7.4).

The columns would not necessarily be round and could possibly be square, or square with rounded corners (see discussion in Section 7.4 regarding square versus round column, considering structural versus hydrodynamic priorities). If square columns and external pontoons are used, columns oriented 45° to the pontoons will be preferred for structural reasons.

Pontoons will in all likelihood be rectangular in cross-section, with small corner radii. The main variable would be the depth/width ratio. A ratio of 2:3 is a good starting assumption. Since it affects heave force, sensitivity to depth/width ratio might be a good secondary issue to examine in the first stage of initial design.

It is assumed that there is a square, truss type deck with widths equal to the column spacing and depth, D_{deck}. The bottom of the deck is assumed to be s_{deck} above the column top.

Initial Design Process

Given the configuration, the process is to first determine the payload and weight of the hull and deck and a range of tendon pretensions to consider. Weight estimation for initial design is specifically discussed following this section. The tendon pretension will depend upon the water depth, the estimated steady lateral forces, and the allowable offset for the steady forces. It will also depend upon the range of tensions caused by environmental loading. The buoyancy, $\rho g \nabla$ is simply the sum of the total weight, tendon tensions, and riser tensions.

Using tendon tension as the principal performance measure, an outline is given below of a suggested parametric process to determine the best set of design parameters. Initially, the minimisation of heave force is used to determine the best distribution of volume between column and pontoons for given combinations of tendon pretension, T_o and column draft, d_c.

Subsequently, considering heave force and pitch moment from waves, the combination of draft and column spacing that minimises the increased tendon tension is determined. However, at this stage of initial design, the column spacing and pontoon lengths should be set at provisional values, these being more related to deck and system design and transport and installation issues. Also, if external pontoons are to be used, these should be set to a fixed portion of the total pontoon volume.

As indicated in Section 7.6.3, the net heave force is dependent on the apportionment of displacement between the columns and pontoons (i.e. ∇_c:(∇_p +∇_e)). A simple closed-form expression for the heave force at the crest-centred wave is given later. With this expression, the combination of column and pontoon sizes that minimise heave force can be established.

This should be considered for two or more column drafts and a range of tendon pretension for each. The result is a determination for each case the column sectional area, A_c and the pontoon volume, $\nabla_p + \nabla_e$. Column spacing is addressed later.

The variation of tendon tension is dependent upon pitch moment in addition to heave force. A second closed-form expression is given which gives the tendon tension variation from pitch moment at the node-centred wave. This, with the tensions for the crest-centred wave, provides a basis for evaluation of parameters of draft, column height and spacing and the tendon spread.

In these analyses, the tendon tension maximum is $T_o + \max T$, where the second term is the spectral maximum amplitude. While it is an objective to minimise this sum, it is also important that $T_o - \max T$ be not be less than zero. Therefore, with the pretension, T_o being initially set to only to control steady offset, T_o may need to be adjusted to avoid negative tension and the process repeated. If the tendon is not neutrally buoyant, the minimum tendon tension should be equal to the wet weight of the tendon in order that the effective tension at the sea floor remains positive. This criterion may be relaxed if analysis demonstrates that negative tendon tension at one corner does not result in tendon failure, however it is not recommended that the criteria be relaxed for initial sizing.

As indicated, the closed-form expressions are only a tool to quickly cycle the parameters of size and are not rigorous representations of tendon tensions. For reasons discussed elsewhere, $\max T$ is perhaps 10–20% less than its actual maximum. Therefore, when looking specifically at tendon tensions, these may be better considered with an empirical multiplier, fac, such that tendon tension can then be written as:

$$T_o + \text{fac} \times \max T \tag{7.35}$$

Owing to the initial estimates of the steady loading (wind, current, and drift force) being not especially accurate early in the initial design process, it is also appropriate to recompute these to verify maximum steady offset. It is for this reason, and the positive minimum tension requirement, that a range of pretensions should be considered.

Without elaborating, later in the initial design, $T_o + \text{fac} \times \max T$ can be computed for a less extreme seastate corresponding to tank flooding. In this way, the minimum tension can be used as a basis for considering internal subdivision. The maximum size of a tank to cause $T_{\text{min-damage}}$ to be zero can be computed.

The parametric models can be evaluated by using the defining set of parameters indicated above. An outline of the process is given below. There are four column parameters (A_c, a_c, d_c, f_c), four for the interior pontoons (A_p, a_p, b_p, d_p) and three for the exterior pontoons (A_e, l_e, r_e) to define the hull geometry. In addition, there are three parameters for the tendons (T_o, s_t, n_t). Added mass parameters should be provisionally fixed at representative values pending cross-sectional shape refinement, pontoons in particular. As discussed below, some parameters may be made dependent upon and computed from others.

Payload (including riser tensions) and other weight data would be given as input as well as centres of gravity relative to the still waterline. Any expected eccentricities of the given weight should be balanced by appropriate ballast at the pontoon level (d_p). The steel and hull outfit weight can be computed by formula as a function of geometric parameters. With the total weight and supported riser load determined, with given tendon pretension, T_o, the required displacement, Δ_o is established. It is recommended that the computation be constructed such that the column draft, d_c, the tendon pretension, T_o, and the column/pontoon displacement ratio, ∇_c:$(\nabla_p + \nabla_e)$ be independent variables, taking ∇_e/∇_p, a_c, l_e, and r_e as fixed. As such, A_c, A_p, A_e will be computed as a function of the input, (T_o, d_o, and ∇_c: $(\nabla_p + \nabla_e)$).

Initially, the pontoon draft, d_p can be based upon the column draft, d_c, minus some factor reflecting the pontoon sectional area and the anticipated shape. Similarly, the pontoon spread, a_p and length, $2b_p$ can be initially determined by a formula based upon the column spread and diameter. The tendon spread, s_t should be computed by a fixed formula based upon the column spacing and diameter (A_c, a_c), the external pontoon length (l_e), and a set relationship to the latter. It is suggested that the tendon pivot point be based upon the pontoon draft (d_p).

PROCESS OUTLINE

1. Fix provisional column spacing, a_c
   Set tendon count, n_t
   External pontoons: Set l_e and ratio ∇_e/∇_p
2. ***First Stage***:
   Input tendon pretensions, T_o and column drafts, d_c;
   For each, consider a series of 3 to 5 column/pontoon volume ratios: ∇_c:$(\nabla_p + \nabla_e)$
   Determine ratio with least spectral heave force maximum for chosen T_o and d_o;
   Result: displacement, ∇_o for each pretension, T_o and, for each T_o and d_c, the best apportionment of ∇_c:$(\nabla_p + \nabla_e)$; the corresponding column and pontoon areas, A_c, A_p, and A_e are given.
   Reconsideration: sensitivity to pontoon shape; vary C_{azp}
3. ***Second Stage***: (minimum tendon tension with heave force pitch moment)
   Input pretensions, T_o, and drafts, d_o; set ratio∇_c:$(\nabla_p + \nabla_e)$ as determined best in first stage; drafts may be different from first stage; interpolate column/pontoon volume ratios.
   For each, consider 3 to 4 column spacings, a_c; set ∇_e/∇_p and l_e for each as is geometrically appropriate. Suggest $\nabla_e/\nabla_p \approx 1$:5 and $l_e \approx 1.0$–1.5 times column diameter, D_c. Set tendon spacing by formula on the basis of column spacing, column diameter, and length of external pontoons.
   Final Varied Parameter Set: T_o, d_o and a_c (best ∇_c:$(\nabla_p + \nabla_e)$ for each are set; ∇_e/∇_p and l_e are prescribed for each);
   Result: Each case (parameters: T_o, d_o and a_c) will have a max$T$. That which has the least max$T$ will be the best parametric set for the initial design. Reconsider with respect to fac $\times$ max$T > T_o$ to avoid slack tendon.
   Reconsideration: Constrain T_o and/or d_o to vary a_c; make choice on a_c.

Reconsideration: Constrain T_o, d_o and a_c to vary pontoon shape (C_{azp}); make choice on pontoon cross-section.

Conclusion: best combination of T_o, d_o, and a_c.

4. ***Finalisation***:

Fix draft, d_o and column spacing; also make final shape determination for columns and pontoons; sets A_c, A_p (depth, width and corners), and A_e (length, width, depth and taper); set pontoon spread, a_p if different a_c; revise d_p and b_p to match pontoon and column dimensions. This input will determine displacement, Δ_o and pretension, T_o. Consider 2–4 distinct models.

Result: Each model will produce a T_o and a max$T$. Also each will have a determined steel weight. Choose.

Closed-form Heave Force and Tendon Tension Calculations

The main objective in initial design is to minimise the maximum tendon tension and, at the same time, to avoid zero tendon tension. The initial design will use a fixed volume, operating mass (weight) and tendon tension, although this may change to increase or reduce the initial tendon tension. With this, the first issue, before any further optimisation of proportions, is to ensure that a reasonable range of column areas and pontoon volumes are used. This is approached by simply minimising the heave force. Subsequently, however, a more complex approach is used where the computation of tendon tension is used including the pitch moment. The first order heave force can be written in the simple, closed form equation as:

$$F_z(\omega) = \frac{H}{2}\rho\omega^2\, G(\omega) \tag{7.36}$$

where $G(\omega)$ is a term addressing the specific geometry: size, spread and submergence of the elements.

This formulation is highly simplified. Apart from other components, there is a component of heave force 90° out of phase with the above due to drag ("heave drag"). This component is not trivial for extreme waves. It is an upward force at the wave node before the advancing crest. It will not appear in linear theory, but will appear in model testing and prototypical behaviour. It can add as much as 10% to the maximum tendon tension, and therefore it is recommended that final optimisation incorporate a linearised approximation for the heave drag component.

Included are the assumed heave added mass coefficients for the internal and the external pontoons, C_{azp} and C_{aze} respectively. For the 4-column, square TLP, with the external as well as the internal pontoons (orthogonal seas), the following is given:

$$G(\omega) \equiv \left\{ \begin{aligned} &\frac{A_c}{k}\Phi_3(d_c)\cos\lambda_c \\ &\qquad - \Phi_2(d_p)\left[(1 + C_{azp})\frac{\nabla_p}{2}\left(\cos\lambda_a + \frac{\sin\lambda_l}{\lambda_l}\right) + \nabla_e(1 + C_{aze})\cos\lambda_e\right] \end{aligned} \right\} \tag{7.37}$$

The first term is the upward force on the column bottoms for a crest-centred wave and the following terms represent the concurrent downward forces on the pontoons. The terms in-λ are spreading functions, representing the lateral distribution of the bodies in the wave train, and the Φ-terms are the depth attenuation terms, representing the submergence of the bodies.

The depth attenuation functions are:

$$\begin{aligned} \text{Column}: \quad & \Phi_3(d_c) \equiv \varepsilon^{-kd_c} \\ \text{Pontoon}: \quad & \Phi_2(d_p) \equiv \varepsilon^{-kd_p} \end{aligned} \tag{7.38}$$

In this, it is assumed that all pontoons, internal and external, are submerged at the same depth, d_p.

The sole objective of the expression for heave force is as a parametric model for determination of the best distribution of volume between column and pontoon. It will provide a means to narrow the parametric range for sizing the remaining structure to minimise the maximum tendon tension. As with the semi-submersible, the heave force response function, $F_z(\omega)$, needs to be subjected to spectral analysis for determination of its maximum value for the extreme design seastate.

As discussed, the hull parameters representing the distribution of displacement between column and pontoons, are to be varied, with all other parameters maintained as constant. The approach outlined above starts with a matrix of given tendon pre-tensions and pontoon drafts. Each pre-tension will have a different total displacement, ∇_o. Corresponding to each combination, several apportionments of ∇_o between columns and pontoons are considered. For the design seastate, pre-tension, draft and apportionment, the total heave force, F_z is determined. For each parameter set, that apportionment which minimises the spectral maximum of F_z represents the best division of displacement for that set.

These ratios should be associated with the pre-tension and draft combinations used to determine them and maintained when the other parameters are varied. For intermediate values of pre-tension and draft, interpolation is appropriate. To address the other parameters, the pitch moment needs to be considered.

Taking pre-tension, T_o as constant, the tendon tension due to heave force alone is given as $T_r(\omega) = (F_z/n_t)\cos(\omega t)$. However, the pitch moment, also causes tendon tension, $T_i(\omega) = M_\varphi(\omega)\sin(\omega t)/s^*n_t$. As discussed in Section 7.6.3, neither is the true maximum. The importance of this is that $T_i(\omega)$ is more a function of the vertical dimensions of the TLP and the spread of the tendons. The latter is the reason for external pontoons. Conversely, $T_r(\omega)$ is more a function of the distribution of displacement between columns and pontoons.

To consider the pitch moment in a closed form, we first consider the horizontal component of inertial wave forces:

$$\text{SurgeForce: } F_x(\omega) = F_{xc} + F_{xp} + F_{xe} \tag{7.39}$$

where

Column: $F_x(\omega) \equiv \rho\Phi_c A_o \nabla_c (1 + C_{axce}) \cos\lambda_c$

Interior Pontoon: $F_x(\omega) \equiv \rho\Phi_p A_o \frac{\nabla_p}{2}\left[(1 + C_{axp}) \cos\lambda_a + \frac{\sin\lambda_b}{\lambda_b}\right]$

Exterior Pontoon: $F_x(\omega) \equiv \rho\Phi_p A_o \nabla_e (1 + C_{axe}) \cos\lambda_e$

(7.40)

The column surge force due to waves is centred at z_c below the surface and the pontoon surge force at z_p below the surface. While the centre of the horizontal column force z_c is a function of wave frequency, it is sufficient to approximate it as a constant value. The position of z_c does not vary significantly over the range of relevant wave frequencies and, for all practical purposes, z_c can be taken at the value computed at spectral peak.

The vertical force components also generate a pitch moment. These are as follows:

Pitch Moment: $M_\phi(\omega) = M_{\phi c} + M_{\phi p} + M_{\phi e}$ (7.41)

where

Column: $M_{\phi c}(\omega) \equiv \frac{\rho A_o}{k} \Phi_c c A_{wp} \cos\lambda_c$

Interior Pontoon: $M_{\phi p}(\omega) \equiv \rho A_o \Phi_p \frac{\nabla_p}{2}(1 + C_{a_{zp}}) \times \left[a_p \cos\lambda_a + \frac{b_p}{\lambda_b^2}(\sin\lambda_b - \lambda_b \cos\lambda_b)\right]$

Exterior Pontoon: $M_{\phi e}(\omega) \equiv \rho A_o \Phi_p \nabla_e (1 + C_{axe}) \cos\lambda_e$

(7.42)

Combining the vertical and horizontal wave force components, the total pitch moment directly due to waves is given as:

Pitch Moment: $M_\phi(\omega) = M_{\phi c} + M_{\phi p} + M_{\phi e} + F_{xc} z_c + (F_{xp} + F_{xe}) z_p$

(7.43)

In addition to the pitch moment generated by wave forces, there is also a moment due to the horizontal inertial loads resisting surge accelerations. Horizontal force components are generated by both the fixed mass, M_o, centred at z_o, and the surge added mass δM, centred at z_a. Surge is highly subresonant such that the lateral component of tendon restoration force is negligible. Therefore, essentially the wave surge force F_x equates the resisting inertial force: i.e. $F_x = (M_o + \delta M)\, x''$. The centre of M_o is known. The centre of the hydrodynamic mass can be determined as follows:

Surge Added Mass : $\delta M_\phi = \rho\left(C_{axc}\nabla_c + 1/2\ C_{axp}\nabla_p + C_{axe}\nabla_{pe}\right)$

Centre of Added Mass : $\bar{z}_a = \frac{\rho\left[C_{axc}\nabla_c \bar{z}_c + \left(1/2\ C_{axp}\nabla_p + C_{axe}\nabla_{pe}\right)\bar{z}_p\right]}{\delta M_\phi}$

(7.44)

Therefore, assuming equality of the surge inertial forces and the surge wave forces,

$$F_{aw} = (M_o + \delta M)x'' \quad \text{or} \quad x'' = F_{aw}/(M_o + \delta M) \tag{7.45}$$

The inertial moment can be given as follows:

$$\text{Inertial Moment :} \quad x''[M_a\bar{z}_o + \delta M_a\bar{z}_a] = \frac{M_a\bar{z}_o + \delta M_a\bar{z}_a}{M_a + \delta M}$$
$$F_w = (f_o\bar{z}_o + f_a\bar{z}_a)F_w \tag{7.46}$$

where f_o and f_a are partition functions defined as follows:

$$f_w = \frac{M_a}{M_a + \delta M} \quad \text{and} \quad f_w = \frac{\delta M}{M_a + \delta M} \tag{7.47}$$

Combining the wave generated pitch moment to the inertial pitch moment, the following pitch moment is given:

$$\text{Total Pitch Moment :} \quad M_\phi(\omega) - (f_o\bar{z}_o + f_a\bar{z}_o)F_x(\omega) \tag{7.48}$$

The tendon tension response function is simply the total pitch moment, as given above, divided by the spread between tendons across the platform, s_t; for a given tendon, this is divided by the number of resisting tendons on that side of the platform (1/2 n_t). Referring to the earlier discussion associated with fig. 7.47, this is T_i a given as follows:

$$T_i(\omega) = \frac{M_\phi(\omega) - (f_o\bar{z}_o + f_a\bar{z}_o)F_x(\omega)}{1/2n_t s_t} \tag{7.49}$$

$T_i(\omega)$ corresponds to the node centred wave, following the passing of the crest. Similarly, $T_r(\omega)$ can be computed for the total heave force, $F_z(\omega)$ corresponding to a crest-centred wave as follows:

$$T_r(\omega) = \frac{F_{wz}(\omega)}{n_t s_t} \tag{7.50}$$

The following is a summary of the tendon reaction for the two defined cases:

$$\max T(\omega) = \sqrt{T_i^2(\omega) + T_r^2(\omega)} \tag{7.51}$$

$\max T(\omega)$ is only the maximum amplitude at a particular wave frequency, ω representing only the maximum for that frequency between crest and node events. As with the heave force response, $F_z(\omega)$, a crest event only, the function, $\max T(\omega)$ must be subjected to spectral analysis to determine its maximum with respect to the design extreme design seastate.

The given equations represent wave directions along the axis of symmetry. However, a 45° approach can actually produce a higher tendon tension amplitude. While the resisting arm increases (by $1.41s_t$), only half as many tendons participate and, for the same moment, the tendon tension increases by 1.41×. There is, however, a small decrease in pitch moment due to the greater hull spread. Ultimately, both directions need consideration. Being somewhat more complex, and possibly obscuring the intent of this discussion, the skew direction equations are not presented. The orthogonal relationship (given) will still enable discrimination between parameter and choices. Nevertheless, it is recommended that the skew direction equations be developed and used. Having made parametric choices for hull dimensions, more rigorous methods are more appropriate for determination of tendon tensions.

For small amplitudes, the linear, inviscid wave theory, the above is symmetric in that the maximum tension and maximum compression are the same and independent of positive wave direction. A variety of second order effects produces important differences. Significant among these is heave drag. This peaks at the wave node and, for extreme waves, not at all small. It is important in that it adds to the maximum tendon tension amplitude and introduces an asymmetry to the wave force system, in that it accounts for a difference in the maximum tension with respect to wave direction between an "up-wave tendon" and a "down-wave tendon". Such asymmetries are seen in model tests and time-domain simulations that include drag force. Another, smaller second order effect is due to the finite height of the crest and trough.

Clauss and Birh (1998) present an automated method for performing optimisation of TLPs, semis and spars using a parametric grid generator, together with a diffraction–radiation motions program (WAMIT). This provides a sophisticated and more rigorous alternative to the above procedure. It is recommended that several configurations be selected for more rigorous analysis before embarking on detailed design in order to verify the selection made using the simplified, interactive approach discussed here.

If tendon tension is the only measure of interest, the process stops here. The design spreadsheets could report the total steel weight (or surface area), current force, and surge/sway acceleration. Here is where interactive tradeoffs play a key role.

7.6.5 Weight Estimates of TLPs

General

As discussed above, the TLP is essentially a fixed-draft, constant buoyancy system and, once installed, does not rely on floatation stability. It was also noted that a large part of the buoyancy is provided to develop tendon tension. While small changes in the sea level (e.g. tide) and set-down do occur, these result in small changes in tendon tension.

While buoyancy and stability are important design issues during transport and installation, they are not so for in-place operation and therefore not specifically addressed here. In as much as the TLP is functionally a semi-submersible during the installation phase, the reader is referred to Section 7.5.5 where this subject is discussed as it relates to semi-submersibles. The remainder of this discussion is, therefore, focused on weight and other downward loading.

As with the semi-submersible discussion, the weight includes not only the payload, but also all other weight contributing to the total weight (W). Similarly, the centre of gravity is needed as an essential property of the system. While not a stability issue, the vertical centre of mass is important to dynamic behaviour, tendon tension and structural loading. The weight/mass properties are required throughout the design phase and must be continuously refined throughout the design process. In this connection weight represents the fixed mass (M_o). The gyradii of this mass is also required. Surge-sway dynamics should also include mass participation of tendons and risers. Attention should be given to the effective vertical and lateral centres of all items.

In addition to the gravitational force from the mass of the system (i.e. total weight, W), other downward forces, particularly those of risers, need to be considered. These tend to be significant parts of the TLP payload. As with the semi-submersible discussion, definitions for weight and external loading needs to be made.

The division of the total weight (W) into the two components, "Lightship" (W_o) and "Variable Load" (δW) is less obvious for a TLP, but it is still useful and is followed. *Lightship* includes all the steel, equipment, and outfitting provided at completion and is usually defined and verified according to regulation. In the case of a TLP, however, there is necessarily a complete dichotomy between the hull and the deck lightship.

Variable load (δW) comprises all weight to be carried other than lightship: operating items, bulk and liquids, ballast and consumable liquids in the hull as well as personnel and effects. In addition, often referred to as *variable load* are the massless external loads such as riser tensions and the drilling hook load (noted as T_r in the earlier discussion). Therefore, using the notation of Section 7.8.3, the following notation is suggested for the total downward forces:

$$W + \delta W + T_r \tag{7.52}$$

δW and T_r respectively represent the mass and massless parts of *variable load*.

Variable load varies according to the operating state of the TLP. Relative to a semi-submersible, there is very little weight change between the operating and the storm conditions for a TLP and, of course, no change in draft. There would be a change in weight distribution, particularly with the drilling payload and perhaps some liquids. Conversely, there can be big changes over the lifetime of a TLP if major components of equipment are added or removed. Particularly important is that a considerable part of the payload is riser tensions and that a full suit of risers may take three or more years to develop. Additionally, riser tension is adjustable.

Provision for ballast needs to be made for future payload increases, including riser tension, and for ballast to balance tendon tensions due to eccentric payload.

Payload comprises all of the *mission-related* equipment, variable load, and external load. The necessary support system weight that is needed, regardless of the mission function (e.g. "marine systems"), is *not* considered to be a part of the payload. Payload, exclusive of deck structural steel, is referred to as the *net payload*. If deck structural steel is included, it is referred to as the *gross payload*. The net and the gross distinction is needed, particularly in comparing designs, because some mission functions can have a high impact on the amount of structural steel required. Such structure explicitly devoted to a mission function is in fact *payload* and is not an inherent property of the TLP design.

Table 7.10 TLP weight and force input groups for initial design

Pontoon Steel
Special Steel (e.g riser supports):
Pontoon Outfit and Equipment:
Pontoon Subtotal
Column Steel
Special Steel (e.g. tendon porches, etc.)
Column Outfit and Equipment:
Column Subtotal
Deck Steel (basic structure)
Deck Steel – deck houses
Special Steel (e.g. substructure, crane fdns, etc.)
Deck Equipment and Outfit – Marine and Support (mooring, utilities, safety, accomodations)
Deck Equipment and Outfit – Mission Systems (drilling, production, tensioning, etc.)
Deck Subtotal
Other Hull Weight
Deck Reserve/Margin
Pontoon Fixed Ballast
LIGHTSHIP
Tendon Tension
Drilling Riser Tensions
Production Riser Tensions
Export Riser Tensions
Deck Variable Load
Column Variable Load
Water Ballast (pontoon/column)
Subtotal External Load, Variable Load and Ballast
TOTAL SUPPORTED WEIGHT AND EXTERNAL FORCE

As with the semi-submersible, it is recommended that the initial design process be executed through a spreadsheet that interactively includes weights, centres of gravity, external loads and vertical force balance. Similar to that given for semi-submersibles, table 7.10 represents a breakdown summary for a TLP initial design spreadsheet. Some parts are input parameters and other parts, particularly steel weight, are computed from the geometric parameters. However, most of the weight information (equipment, variable load, etc.) needs to be separately specified according to design requirements and is not a function of geometry. Weight items that are dependent upon geometry and computed within the spreadsheet are marked with an asterisk in the listing.

Lightship Weight Estimation

Weight estimation is addressed elsewhere in the handbook, along with a discussion of semi-submersibles. Much of the latter applies to TLPs, but some factors unique to TLPs should be noted. The breakdown of the lightship weight between the three basic groups, steel, outfitting, and equipment, should be maintained. With the deck is separately designed and fabricated, and separately managed, the hull and deck lightship weight estimates should be separately maintained.

As with semi-submersibles, about 85% of the steel is a function of local hydrostatic design pressures. Likewise, the balance of hull steel is from global reinforcement and specific functional foundations. Even more so than semi-submersibles, the hydrostatic loading is particularly high, with nearly half of the structural weight as framing, stiffening and internal subdivision.

Framing and internal subdivision, in general, do not contribute to global strength. Conversely, the design of the shell plating and stiffening for high hydrostatic loading produces pontoons and columns of considerable global strength with comparatively little need for reinforcement. Small to moderate local increases in shell plating thickness are typically required where columns and pontoon (or deck) connect. Fatigue of these joints has, in the past, required expensive cast nodes, but modern designs utilise more cost-effective fabricated nodes.

As with semi-submersibles, given the hydrostatic loading, the weight per square foot of all the watertight surfaces can be estimated and totalled. Unit weights of TLP structure ranges from about 30 lb/ft^2 for the upper columns to 60 lb/ft^2 or more for the pontoons and lower columns. Other steel weight can be added to this as a percentage for fabrication: 7–8% for welding, steel thickness overage and brackets; 6–10% for local reinforcement; plus specifically estimated allowances (e.g. tendon porches).

It is difficult to address deck steel estimating in as much as there are so many design possibilities and functionality aspects. For local loading the decks should build out at 15–20 lbs/ft^2 (each deck) and somewhere between 10–20 lbs/ft^2 for trusses. These numbers are very approximate and depend much on specific loading and structural design. Needless to say, the parameters should be used with caution and only for initiating the design. They should be replaced as early as possible by estimates based upon the conceptual design of the deck. For estimating future designs of a similar form, the deck steel should be broken down as indicated into "local" and "major truss systems. The former should be normalised on the basis of literal deck area provided and the latter normalised on the basis of overall format area (column centres) and the supported weight.

For a TLP, the outfitting weight of the hull will be $<10\%$ of hull steel, principally consisting of piping, corrosion protection and access (ladders, walkways, gratings, rails, closures, etc.). For the deck, outfitting is considerable and very dependent upon functions and is beyond the scope of this section. Approximately, it is on the order of 25% of the topsides equipment weight.

Equipment weight is entirely a product of functionality and must be provided by others who are addressing the functions of the TLP. Nevertheless, it should be a product of a comprehensive equipment list including the entire TLP, hull as well as deck, with no exclusions.

As with all platform types, the lightship estimate should include explicit allowances and margins. Allowances represent *expected items* (even if unidentified) and a margin is for *uncertainty*. Equipment and outfit allowances and margins should be prepared by those specialised in the systems. Margins should account for errors in design, quantification, item omission, etc., and should be appropriate to the specific uncertainty. They are probabilistic, tend to compensate to some degree, and have different uncertainties with different parts of the system. What is important in this regard is that explicit allowances and margins, ever so small, should be addressed and available for review and reconsideration.

Variable Load

Variable load on TLPs is mostly on the deck and includes operating liquids, supplies, drilling and other consumables as well as personnel and effects. Liquid consumables (e.g. fuel, drill water, etc.) may be on the deck or within the hull. Additionally, and significantly, is the tension of top tensioned risers. Hull variable load will include ballast and risers, export in particular, but sometimes subsea well riser systems. Hull variable load on a TLP is usually quite small.

While there may be very little difference between the normal operating loading condition and that for a severe storm, there can be significant differences related to the state of production riser deployment. There may, however, be special ballast distributions related to tendon tensions. There also may be special ballast distributions related to tendon failure or internal flooding. Without further detail, transport and installation will address a number of loading conditions, most expedient to the objectives.

7.6.6 TLP Hull Structure

General

While the structural design is more generally covered in Section 7.8, several aspects of structural design as they specifically relate to TLPs should be discussed in this section. Unlike semi-submersibles, the API code, API RP2T is specifically developed for TLP design and is more prevalently used than Classification Rules. However, apart from the more global aspects and specific TLP issues, the structural response is not altogether different from that of semi-submersibles. The classification rules are applicable, particularly as discussed in Section 7.8, and are being more increasingly recognised thus. While many a major oil company owners have very specific additional requirements, as noted in the semi-submersible discussion, the regulating authorities are increasingly reliant on the classification societies for certification review.

In the same way as semi-submersibles, a TLP hull structure design is taken at two levels: Local Strength and the Global Strength. There are design issue differences from semi-submersibles at both levels and those are discussed here. Perhaps more important for a TLP than for a semi-submersible is considerations of loading from construction and installation. Ocean transit of the hull from a remote builder to the installation site can pose structural design challenges (see fig. 7.33).

Due to permanent siting, the longevity, inspectability and repair issues noted in the semi-submersible discussion are highly significant factors for TLPs. Typically TLPs are designed and built to highly considered design bases developed for a specific application and do not suffer from the less than specific, but versatile requirements needed for a mobile unit. The design standards for mobile units reflect the anticipation of highly variable use for well in excess of 20 years.

While more prevalently built by fabricators in the past, there is a persistent shift to shipbuilders for TLP hull construction. Therefore, many of the comments made for semi-submersible fabricating practice are relevant to TLPs, particularly those regarding cross-stiffened panel construction and block fabrication and erection practices. Also, the discussion of column and pontoon configuration is relevant.

Local Strength of TLPs

For a TLP, local strength follows largely the same issues as discussed for semi-submersibles but there are some notable differences. For TLPs also, 80–85% of all hull steel is a consequence of local loading. This may be even more true of TLPs in that they tend to have significantly deeper draft, requiring higher pontoon and lower column design pressures. Also, since TLPs require relatively little ballast, the internal spaces tend to be empty. Most internal subdivision therefore need not be designed for tank service; i.e. they are "void spaces." Generally the watertight flats and bulkheads are designed for watertight integrity only, a less demanding criteria. Also, for the shell plating, external pressure is usually the controlling designed pressure where as for a semi-submersible the internal pressure of a tank with a particularly high overflow vent may govern.

For a TLP, dynamic wave pressure can be significant. Unlike semi-submersibles and floaters, TLPs cannot rise with the crest (*heave*) and thereby reduce pressures. A TLP will, in fact, go through set-down. The heave of floaters actually mitigates enough of the dynamic pressure from a passing wave crest that the 20 ft code-based minimum design head is sufficient for the upper columns of semi-submersibles. This is not sufficient for a TLP and dynamic pressures must be considered. The underwater hull also endures significantly higher pressures. Figure 7.50 shows the external pressures applicable to a particular TLP column. Superposed over the static design head curve is the dynamic capacity for that static design head. The constant upper part reflects that the 20 ft minimum static head has a dynamic capacity of 26.7 ft. However, also shown is a combined static and dynamic wave pressure curve for the indicated wave crest. From about 68 ft below the design waterline to about 14 ft above it, the combined static and dynamic pressures exceeds the capacity required for static criteria (elsewhere, the curve is shown with a heavy dash). It is notable that below about 70 ft (e.g. pontoons), wave pressures do not sufficiently exceed the static pressures capacity. (The code basis for this is discussed more specifically in Section 7.8) While this illustration is specific to TLPs, similar effects will occur for minimal heave structures (e.g. SPARs).

Although a consideration for transit and installation (criteria adjusted as appropriate), stability is not an in-place design issue for a TLP. Not only is there less internal subdivision, high damaged stability waterlines are not needed for the shell plating design. Rather than stability, the internal subdivision is dictated by a need to prevent slack tendons. This is less demanding than damage stability compartmentation.

TLP framing is subject to considerable compressing. Figure 7.51 shows two typical framing sections for a TLP. On each is shown the distributed hydrostatic load transferred to the frames from the intervening plate and stiffeners. Notably for TLPs and deep draft semi-submersibles, particularly high pressures are developed. Particularly for TLPs, as noted above, there generally is no compensating internal fluid pressures. In the case of the

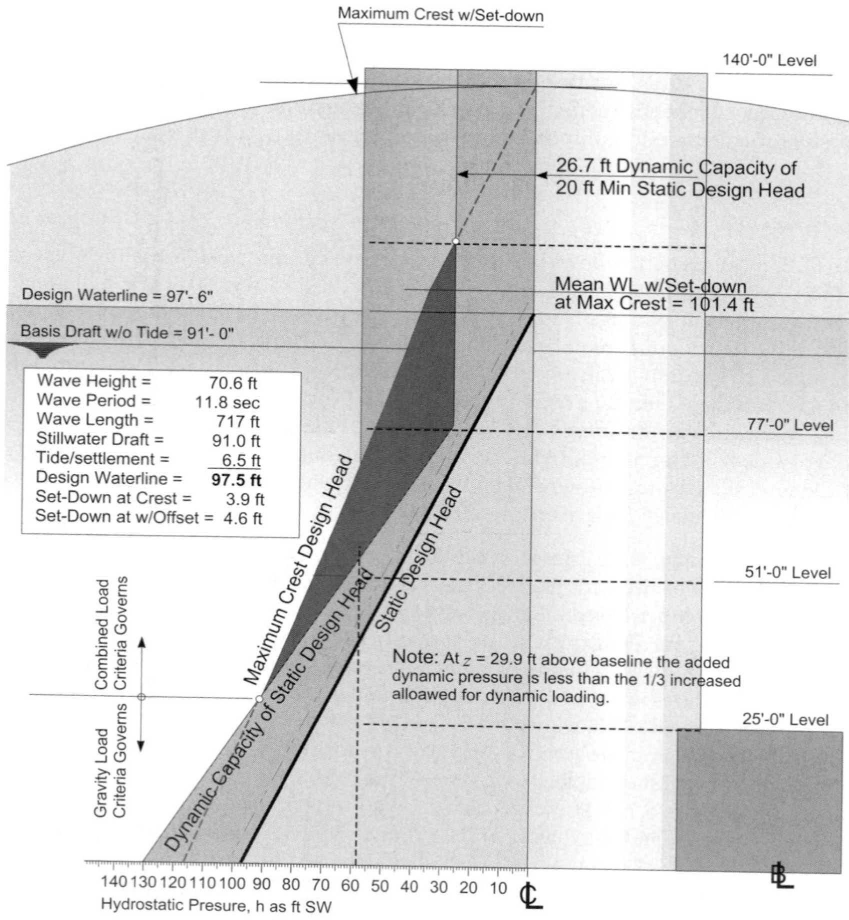

Figure 7.50 TLP column external pressure example

pontoon section, the frame is not only highly loaded in flexure, but each frame element is under considerable compression. In the case of the circular column, there is a circular, unsupported ring frame. The strength of this type of frame is governed by buckling and not flexure. However, if there are internal supports (e.g. bulkheads, struts, etc.), the frame is governed by flexure, as is the pontoon frame.

Global Strength of TLPs

All of the TLPs use an unbraced arrangement. Although the superstructure of TLPs is in a truss form, the behaviour is flexure and similar to the column and pontoon behaviour.

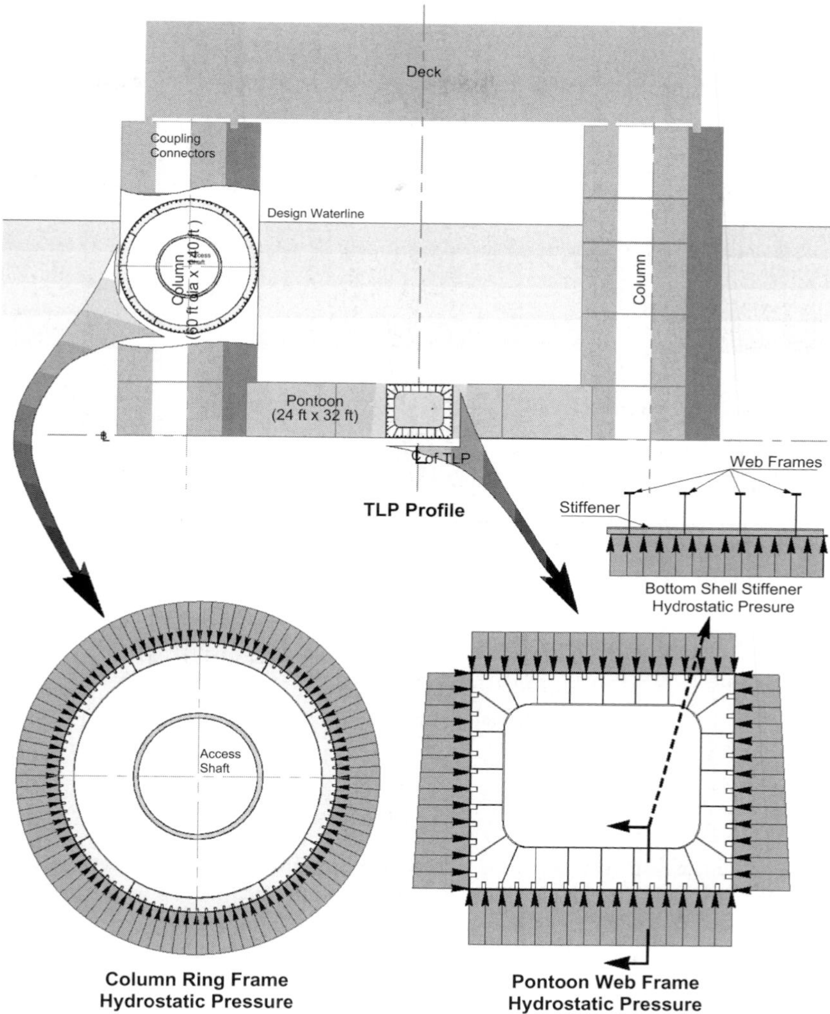

Figure 7.51 Typical hydrostatic frame loading for a TLP

In all cases of this discussion, the subject is a 4-column, closed array pontoon TLP. The deck moment is coupled to the column tops.

Figure 7.52 shows a profile of a side of a typical TLP with an elastic frame superposed over it. Also shown is the still water gravity/buoyancy load system corresponding to conditions shown in fig. 7.47 given previously. The hull loading includes the column weight, the upward bottom pressures, the net of pontoon buoyancy and weight, and the downward

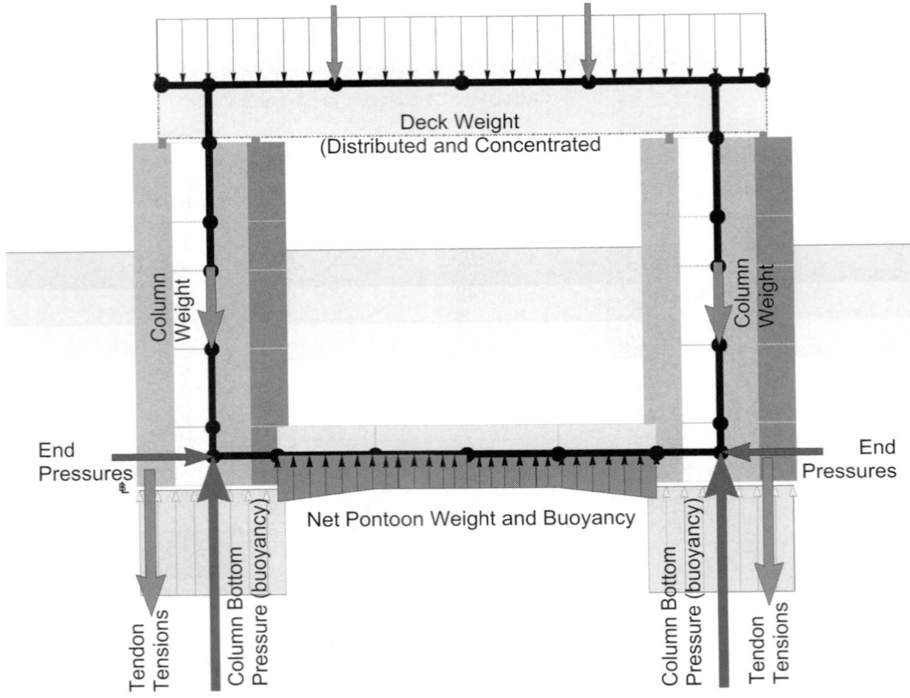

Figure 7.52 Global loading – gravity/buoyancy load

tendon loads. The deck loading included a distributed load and concentrated load corresponding to the well system (risers and/or drilling equipment).

This load system is shown in fig. 7.53. The corresponding distortion pattern is shown in fig. 7.54. What should be evident is the hog in the pontoons, the sag in the deck and the constant moment in the columns. Shear is small in the columns and there is no torsion in the pontoons and columns. By symmetry, this is the same in all four faces. Although the variable load can change, it is important to note that this is the background stress level under any additional environmental loading.

Figure 7.55 schematically shows the load system corresponding to the node centred environmental loading shown in the lower part of fig. 7.48, given previously. As with the gravity/buoyancy loading, fig. 7.56 shows the moments and shears and fig. 7.57 shows the distortion pattern. As can be seen, a very large part of the wave surge force is resisted by the deck inertia reactions. In the extreme, lateral deck accelerations can exceed 0.2 g. The frame deformation clearly is a *side-sway shear*. The proportions on a TLP are such that this loading can be particularly severe for the pontoons, which tend to be on the small side (relative to semi-submersibles).

This load system is sometimes called the "horizontal shear load" and is dominated by large horizontal shear loads between the deck and column tops. This is, of course, created by

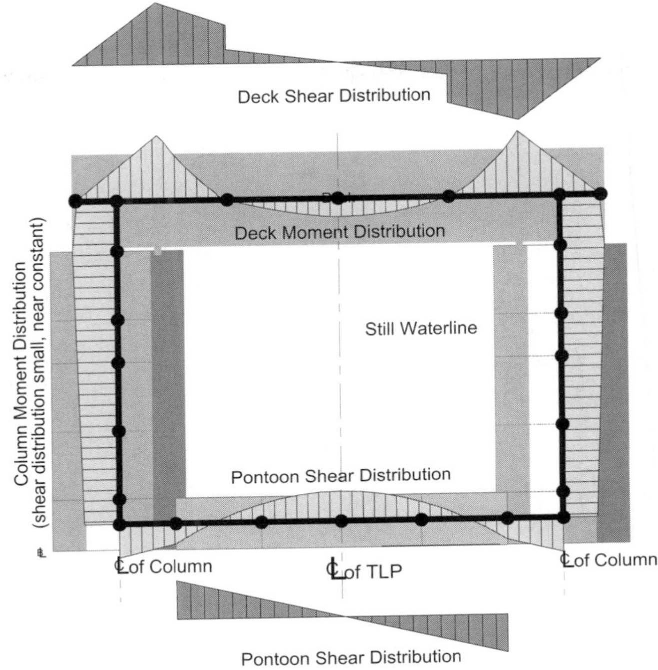

Figure 7.53 Global loading – gravity/buoyancy load, shear and moment

Figure 7.54 Global loading – gravity/buoyancy load deformation pattern

lateral acceleration on the mass of the deck. It is notable of TLPs that there is a high concentration of mass high in the structure. This is a result of the fact the TLPs are comparatively tall, are essentially empty within the hull, and are not stability limited. This can be the controlling load case for some parts of the structure.

The most severe loading for a closed array pontoon structural system, for TLPs, and semi-submersibles, is the crest (or trough) centred, oblique seas racking conditions. This is actually a form of squeeze/pry loading and corresponds to the upper part of fig. 7.48,

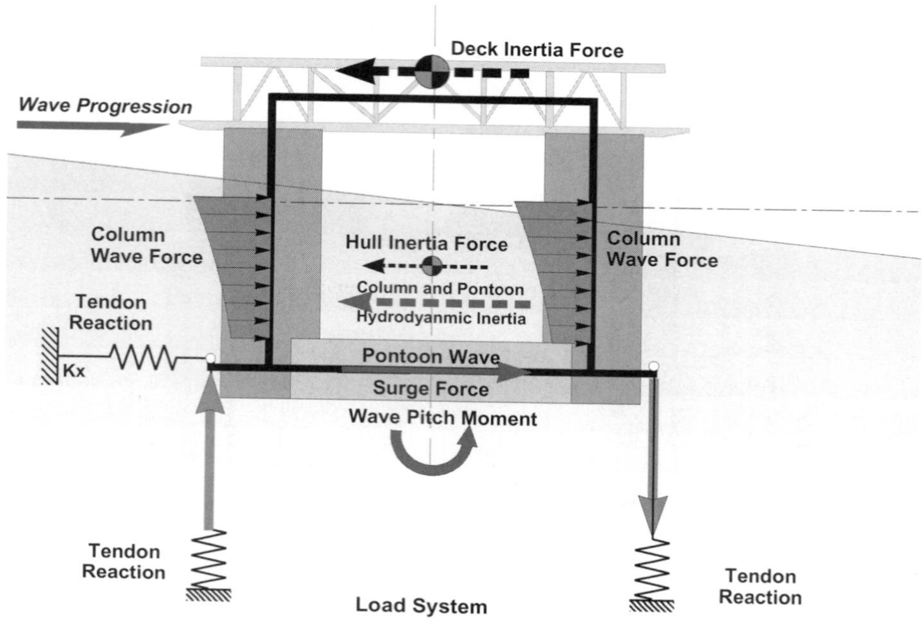

Figure 7.55 TLP global loading – node centred wave

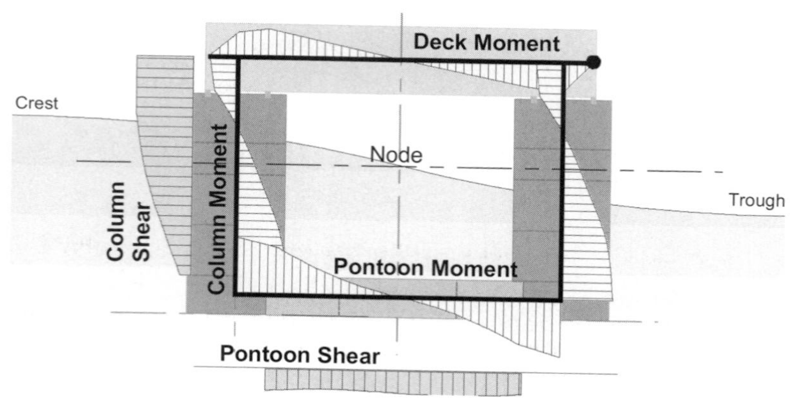

Figure 7.56 Node centred wave – shear and moment

given previously. The essence of this load pattern is seen in figs. 7.58 and 7.59. The former shows the pontoon array sitting atop a wave crest that is essentially trying to push the up- and down-wave corners outward. Forces on the pontoons indicated in fig. 7.58, are more specifically shown in profile in fig. 7.59. The consequent deformation pattern is shown in fig. 7.60.

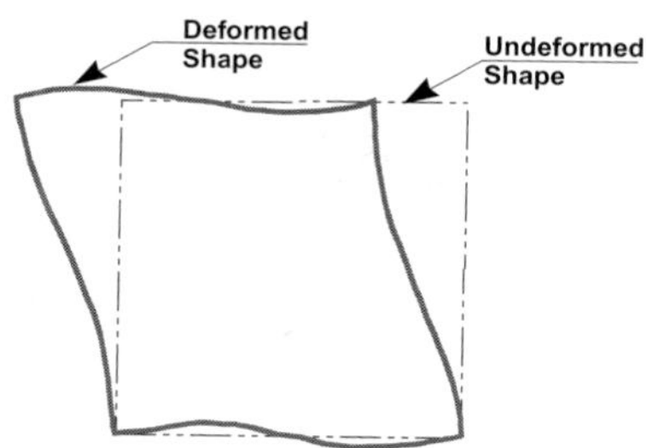

Figure 7.57 Node centred wave – deformation pattern

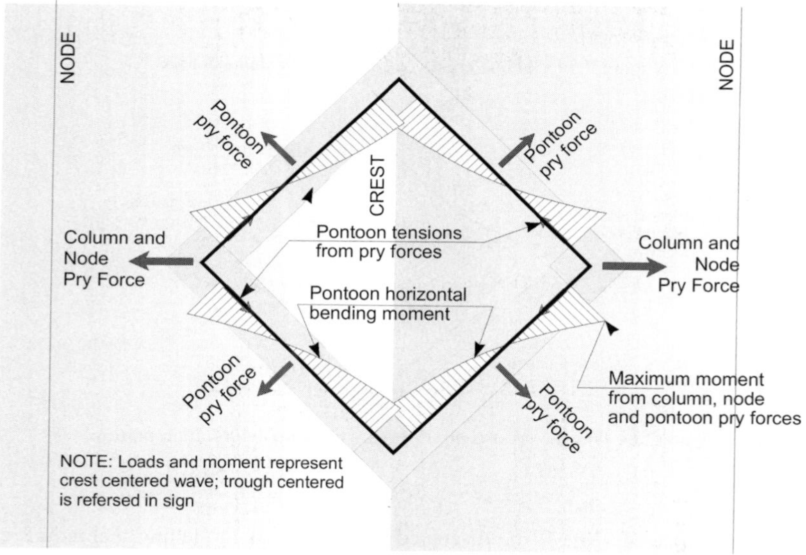

Figure 7.58 TLP Global loading – oblique crest centred wave ("Squeeze/Pry")

The horizontal plane moments are shown in fig. 7.58. There are also shears and significant axial force (not shown). However, it is not quite this simple and vertical plane bending and shear are considerable, not to mention torque. One part of the vertical plane action is an imbalance of vertical force on the column bottoms. Figure 6.60 summarises this action. As would be expected, the column force shown in fig. 7.59 imposes end moments at the pontoon connection. Figures 7.61 and 7.62 summarise the action of these moments. The column end moments create reacting moments and torques in the pontoons as shown.

The foregoing has been largely qualitative and is given as a guide to strength analyses as well as design. For preliminary sizing (prior to analysis), simple estimater can be made of various force magnitudes and some sizing estimates can be made. However, what is most important is that the analyses include the controlling load cases and that the results

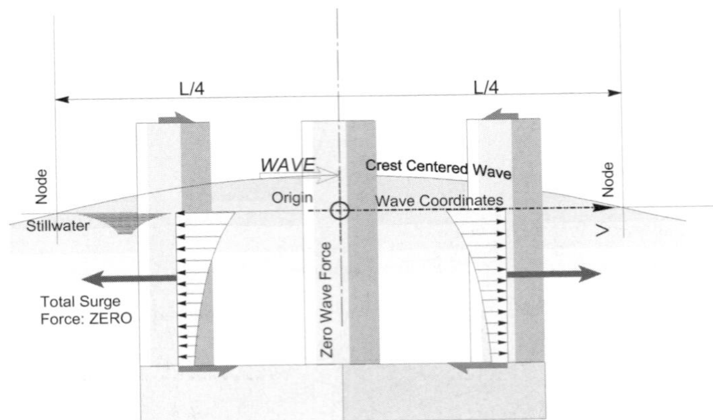

Figure 7.59 Oblique crest centred wave – column forces

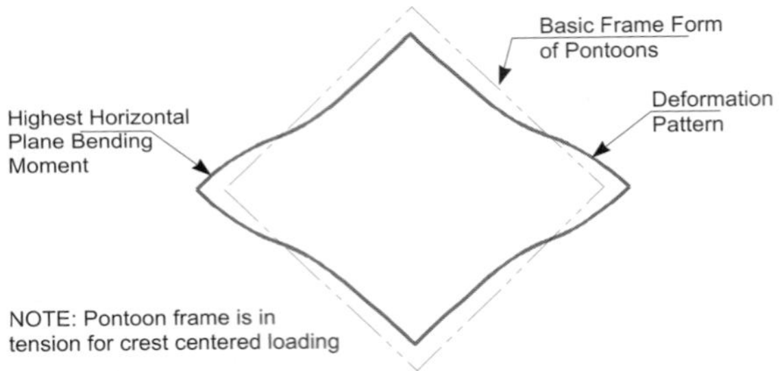

Figure 7.60 Oblique crest centred wave – pontoon deformation pattern

be correctly interpreted. As will be discussed later, various modelling techniques can be employed for strength and fatigue analysis. It should be noted in closing that the moments, shears, etc. (*stress resultants*) used in this discussion apply to the large-scale hull elements actually used for the hull. Figure 7.63 schematically shows the relationship. Given the end stress resultants, and load distribution between the ends, local global stress within the element can be determined, including shear flow from torsion as well as shear and axial stress from biaxial bending and axial load. While the hull girder theory, particularly shear lag considerations, will improve accuracy, the engineering theory of bending shear and torsion are generally adequate unless the segments are multi-cellular.

7.7 Spar Design

7.7.1 History of Spars

Spars have been used for decades as marker buoys and for gathering oceanographic data. The first significant Spar for our purposes is Flip, a structure owned by the

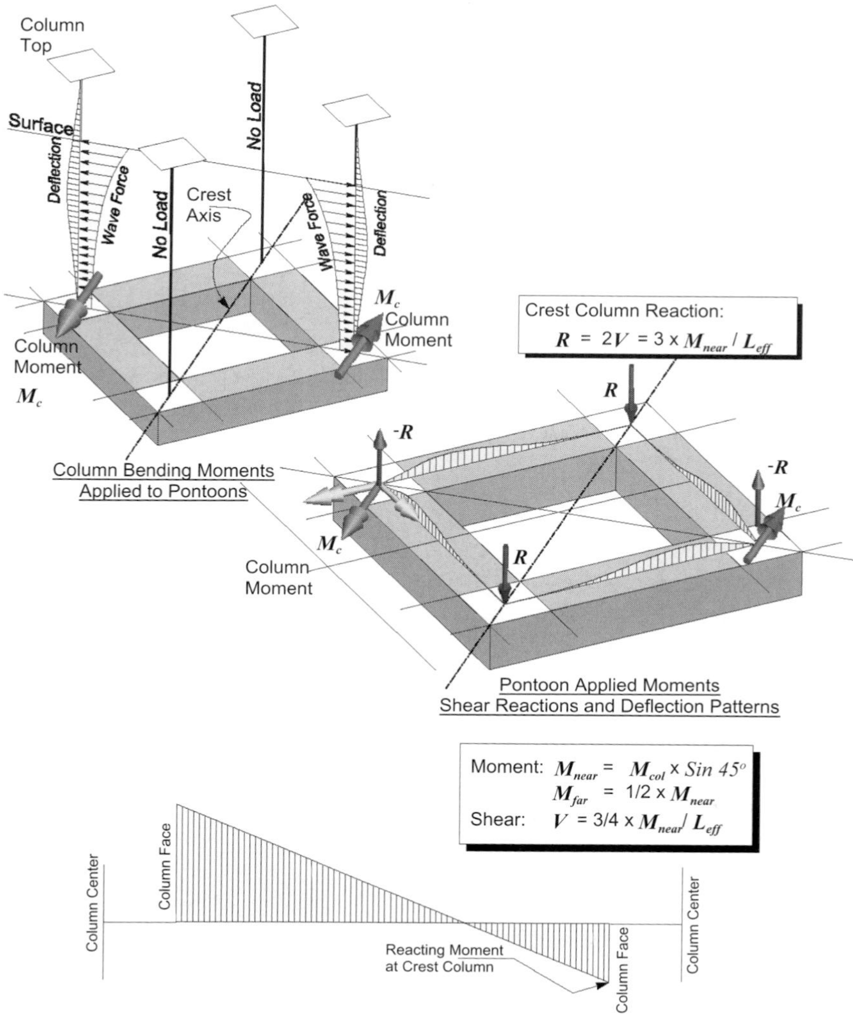

Figure 7.61 Pontoon vertical plane bending from crest centred oblique wave – pontoon component

U.S. Navy and operated by the Scripps Institution of Oceanography in California (Fisher, 1963). Flip was put into service in 1965 and was used primarily for ocean acoustic measurements. Operating draft is about 270 feet. Its diameter is 12 ft at the waterline tapering to 20 ft over most of the hull. Its heave natural period is 29 s. Although Flip occasionally has been tethered to the sea floor, it is more commonly allowed to drift in the ocean currents.

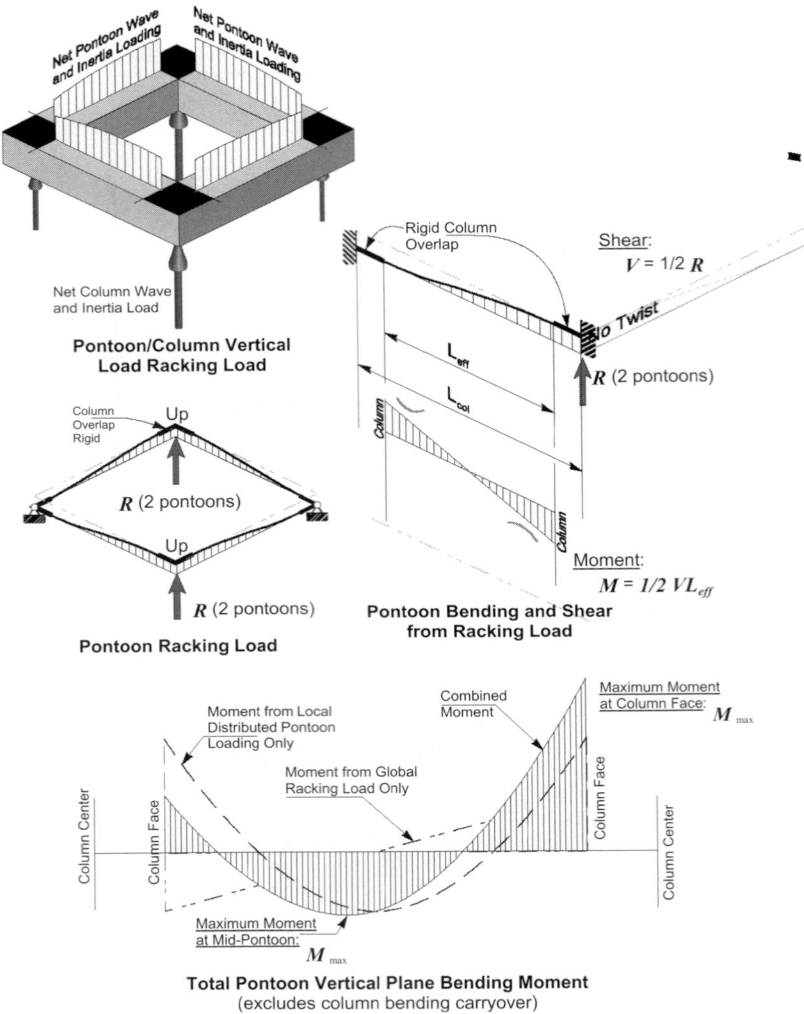

Figure 7.62 Pontoon vertical plane bending from crest oblique wave – column bending component

In the early 1960s, Nippon Telegraph installed a Spar off the coast of Japan to carry a microwave relay station. This Spar was 445 ft long with a stepped hull ranging in diameter from 10 to 20 ft. The topside structure is a cylinder, 50 ft in diameter by 33 ft high, with equipment, accommodations and a heliport on top. A four-point catenary mooring of 3 in.-chain connected to a 175 ton clump weights keeps the Spar in place. Operating draft is 330 ft. In the mid-seventies Shell installed an oil storage and offloading

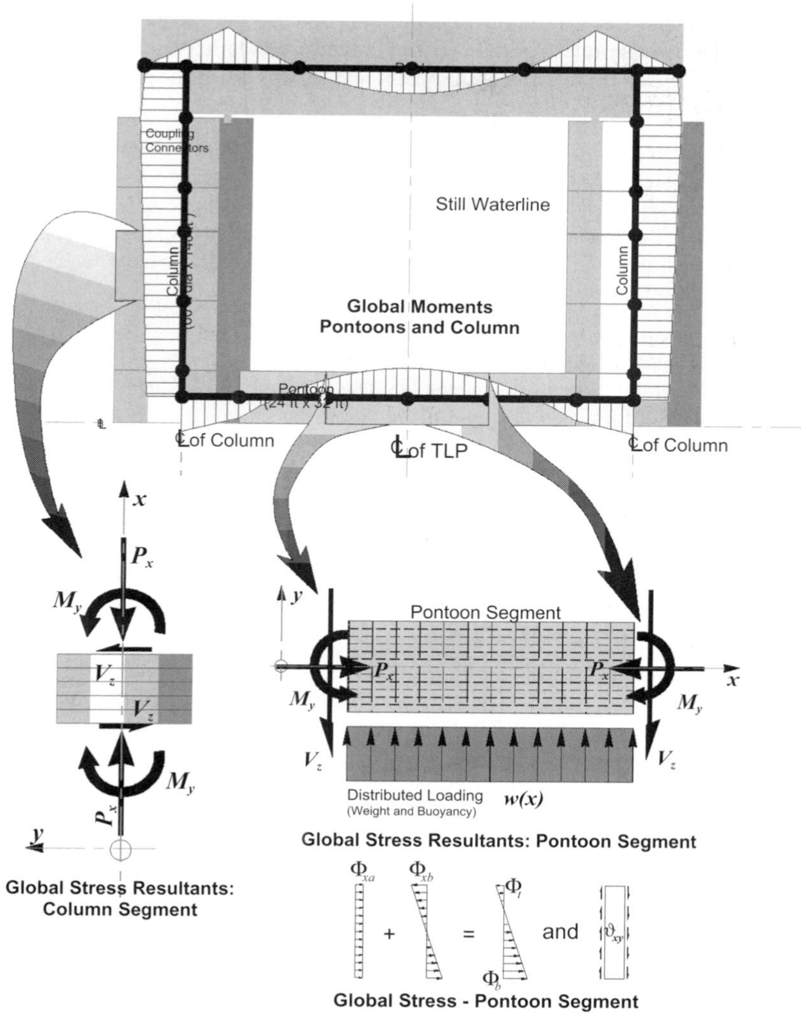

Figure 7.63 Hull elements as components of a TLP hull

Spar at Brent Field, in the North Sea. The hull is 95 ft in diameter, necks down to 55 ft at the water plane, and the operating draft is 357 ft. This Spar was designed to store 300,000 barrels of produced crude and to transfer it to bow loading tankers. The oil storage system used a water displacement principle which allowed the tanks to be designed for ambient pressures. The mooring system consists of six lines, each made up of a 1000 ton concrete gravity anchor, 2600 ft. of 3.5 in. wire rope, and 935 ft. of 4 in. chain. The top part of the superstructure rotates to allow the tanker to weathervane about the Spar. Accommodations, power plant, other equipment and pumps are in the superstructure and a heliport is fitted.

Agip installed a flare Spar off West Africa in 1992 designed to flare 100 million cubic feet of gas per day. The Spar is 233 ft long, with a diameter of 7.5 ft which tapers to 5.5 ft through the water plane. With a draft of 170 ft, it is held in place with a four leg chain catenary mooring. In 1993, Shell installed a loading Spar at Draugen. The Spar topsides include a rigid boom to which the flexible loading hose is attached. The hull diameter is 28 ft and the operating draft about 250 ft.

The world's first production spar was the Neptune Spar installed in 1996 by Oryx Energy Company (now Kerr McGee) and CNG (Vardeman et al, 1997). Design production rates were 25 mbod and 30 mmcfd respectively. The maximum topsides weight with the workover rig is 5500 tons. Wells would be predrilled with a semi-submersible and completed with a platform workover rig placed temporarily on the Spar. The Neptune spar has a hull 705 ft long with a 32 ft by 32 ft centrewell and a diameter of 72 ft. The six-point mooring system consisted of driven pile anchors, 4.75 in. spiral strand wire rope and chain for the section leading up to the fairleads and onto the hull (fig. 1.15).

As of this writing there are 14 spars in production or under construction. Figure 1.16 shows the progression of spars built by Technip Offshore, Inc. Three additional spars have been built by J. Ray McDermott.

The Spar, along with the Tension Leg Platform, is the only "floating" platform which up to 2003 has been used for dry trees. The reason for this is that these are the only platforms with small enough heave and pitch motions to allow the risers to be safely and economically supported by the floater.

7.7.2 Spar Description

Three types of production spars have been built to date: the "Classic" and "Truss" spars shown in fig. 1.17, and recently the third generation cell spar (Finn and Maher, 2003). The basic parts of the classic and truss spar include (refer to fig. 1.17):

1. Deck
2. Hard Tank
3. Midsection (steel shell or truss structure)
4. Soft Tank

The Topsides **Deck** is typically a multi-level structure in order to minimise the cantilever requirement. For decks up to about 18,000 tons, the deck weight is supported on four columns which join the hard tank at the intersection of a radial bulkhead with the outer shell. Additional columns are added for heavier decks. Figure 7.64 shows the arrangement of the top of a spar with the deck supports. Decks up to about 10,000 tons may be installed offshore with a single lift. Larger decks require multiple lifts.

The **Hard Tank** provides the buoyancy to support the deck, hull, ballast and vertical tensions (except the risers). The term "Hard Tank" means that its compartments are designed

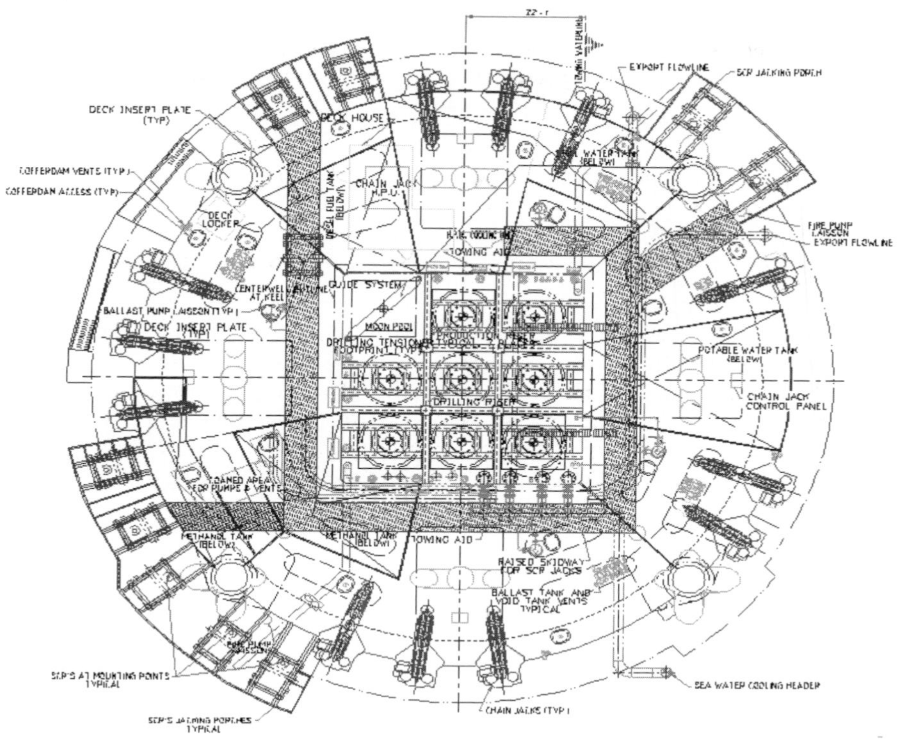

Figure 7.64 Spar deck arrangement (Technip Offshore)

to withstand the full hydrostatic pressure (see section of Hull Structure). Figure 7.65 illustrates a typical structural arrangement. The profile is shown in fig. 1.17. There are typically five to six tank levels between the spar deck and the bottom of the hard tank, each level separated by a watertight deck. Each level is further divided into four compartments by radial bulkheads emanating from the corner of the centerwell. The tank level at the waterline includes additional cofferdam tanks to reduce the flooded volume in the event of a penetration of the outer hull from a ship collision. Thus there are up to 28 separate compartments in the hard tank. Typically, only the bottom level is use for variable ballast, the other levels being void spaces.

The **Midsection** extends below the hard tank to give the spar its deep draft. The selection of the spar draft is discussed below. In the early "classic" spars the midsection was simply an extension of the outer shell of the hard tanks. There was no internal structure, except as required to provide support for the span of risers in the midsection. The scantlings for the midsection were determined by construction loads and bending moments during upending. Later spars replaced the midsection with a space frame truss structure. This "truss spar"

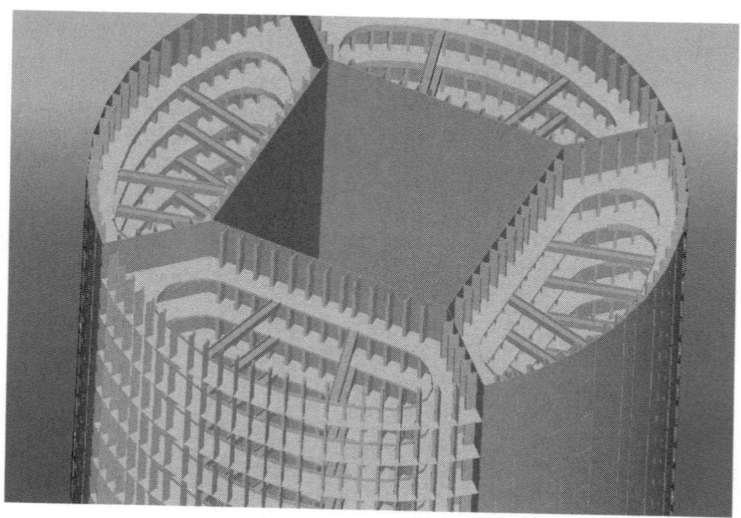

Figure 7.65 Hard tank structural arrangement (Technip Offshore)

arrangement resulted in a lower weight, less expensive hull structure. Also, the truss has less drag and reduces overall mooring loads in high current environments.

The **Soft Tank** at the bottom of the spar is designed to provide floatation during the installation stages when the spar is floating horizontally. It also provides compartments for the placement of fixed ballast once the spar is upended. The soft tank has a centrewell and a keel guide which centralises the risers at that point.

7.7.3 Spar Riser Systems

Top tensioned risers are placed in a centrewell which runs the length of the hard tank. The risers are typically supported on buoyancy cans which provide the top riser tension instead of hydraulic tensioners used on TLPs. The buoyancy cans are located in guides, which provide lateral constraint. As the spar heaves the buoyancy cans remain in place while the spar moves relative to them. As the spar offsets, the geometry of the risers causes draw down and again the guides allow the risers to slide relative to the hull. The buoyancy can/guide system allows stroke of the risers to be relatively large compared to that for typical hydraulic tensioners.

A schematic of the riser arrangement in the centrewell is shown in fig. 7.66. The buoyancy cans have an outer shell and an inner stem. The stem has an internal diameter sufficient to allow passage of the riser tie back connector. When in place, the riser itself is centralised within the stem. The stem extends above the buoyancy cans to the tree elevation at the production deck level. The risers and the production tree are supported by the stem there. Production fluids from the risers are carried to a production manifold by flexible jumpers, which accommodate the relative movement between the risers and the hull.

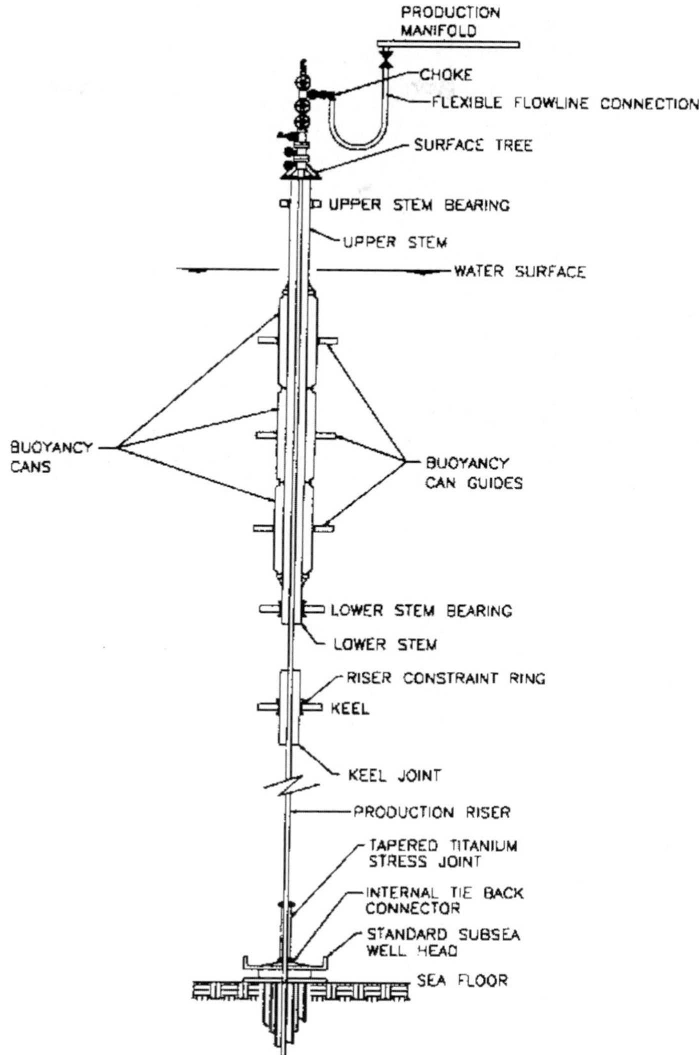

Figure 7.66 Schematic of riser and buoyancy cans (Glanville et al, 1997)

Figure 7.67 shows one of the buoyancy cans being installed. Figure 7.68 shows the upper stem, X-mas trees and flexible jumpers.

Drilling risers on spars have typically not used the buoyancy cans for tensioning. Instead, hydraulic tensioners have been used. Figure 7.69 shows a RAM type tensioner in which the

Figure 7.67 Buoyancy can being installed (Technip Offshore)

hydraulic piston rods are in compression. This type of tensioner requires that the riser remain aligned with the hull, as it is on a spar. More traditional tensioner cassettes with the rods in tension have also been used (Genesis Spar).

The use of tensioners for drilling eliminates the need for the drilling riser slot to be increased in diameter to accommodate larger buoyancy cans that might be required for drilling.

7.7.4 Spar Mooring

The mooring system for the spars built till date consists of a chain-wire-chain taut catenary system similar to the one shown in fig. 7.70 for the Neptune Spar. A "taut" mooring is defined here as one in which the anchor loads have an uplift component for all load conditions, i.e. the anchor chain or wire never lies on the seabed.

The spar motions are small enough, even in the 100-year hurricane, that the taut system can be used without synthetic mooring lines. The taut system saves a considerable length of wire and chain needed for a conventional catenary mooring.

Figure 7.68 Surface trees on a spar (Courtesy Kerr McGee)

The platform chain is tensioned using chain jacks or windlasses, which are installed on the periphery of the upper deck of the hull (the "spar deck"). Figure 7.64 shows an example of the arrangement of this deck.

The chain runs from the chain jack to a fairlead, which is located from up to 350 ft. below the mean waterline. The length of platform chain is determined by the amount chain, which needs to be pulled in or paid out to maneuver the spar.

The midsection of the mooring system consists of a spiral strand wire rope or polyester line. For long life the steel strand is typically sheathed with a urethane coating. The lower end is attached to a length of anchor chain. The length of chain and the mooring tension is selected so that the wire will not make contact with the sea bottom except under the most extreme conditions.

The anchor chain is connected to a piled anchor which can sustain uplift and lateral loads. The pile padeye is usually about 50 ft. below the mudline, so that the bending moment from the mooring forces is minimised.

The total scope of the taut moor from the fairlead to the anchor is typically about 1.5 times the water depth.

7.7.5 Spar Sizing

The main design criteria for sizing a spar are:

1. Maximum weight of topsides and risers supported by the spar that needs to be accommodated,

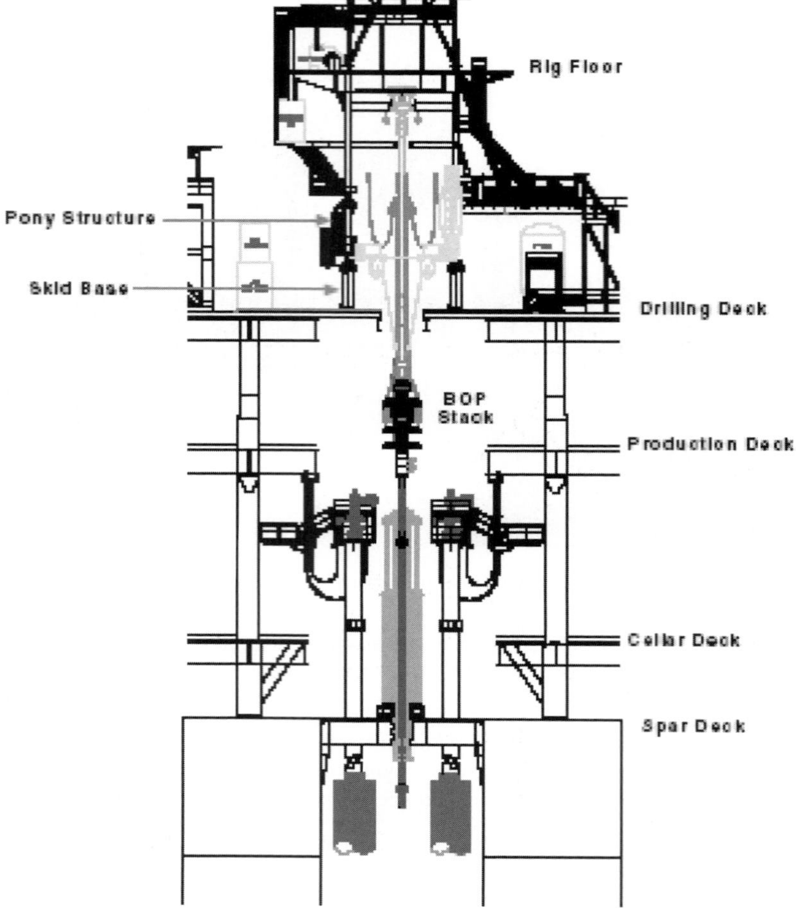

Figure 7.69 Tensioning of a spar drilling riser (Bates et al, 2000)

2. Eccentricity of the deck which needs to be trimmed by variable ballast,
3. Area required in the centrewell to enclose risers and buoyancy cans,
4. Maximum pitch motions in a 100-year event less than about 10°,
5. Centrewell sufficient to support the riser buoyancy cans and other riser requirements,
6. Maximum riser stroke less than about ± 15 ft,
7. Ability to be single piece transported on a heavy lift vessel,
8. Ability to float-off heavy lift vessel (maximum draft less than 10 m).

Topsides weight and other payload weights are obviously key drivers. Apart from this, the most critical first decision is the size of the centrewell, because this sets the minimum diameter for the hull.

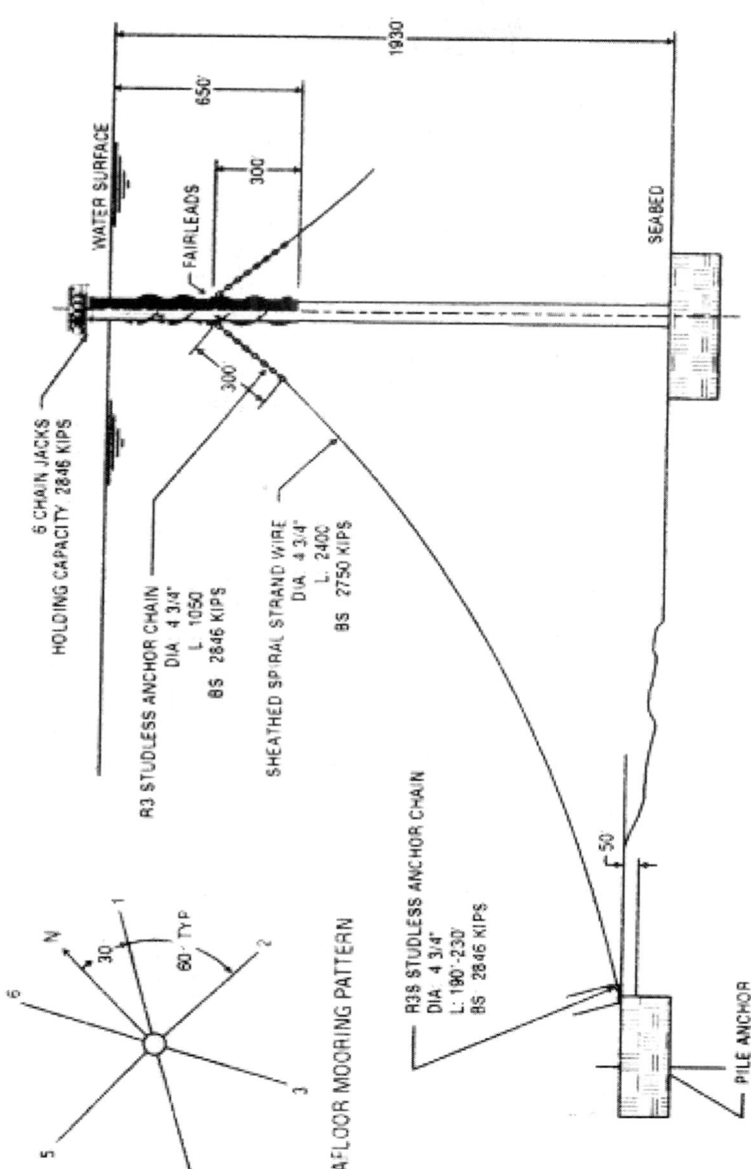

Figure 7.70 Neptune spar mooring system (Vardeman et al, 1997)

Centrewell Sizing

All the spar centrewells to date have been square, lending themselves to 4×4 (16), 5×5 (25) or 6×6 (36) well slots. Sometimes the four slots in the centre of the pattern are allocated to a drilling riser. Besides, four slots may be opened to provide a moonpool for running of tools, ROVs, etc. Space may also be allocated for SCR J-tubes, or even access trunks for personnel entry to lower tanks. Figure 7.71 shows a typical layout with two slots dedicated to running of flowlines, umbilicals and SCRs.

Slot spacing is dictated by the diameter of the buoyancy cans. It is best to perform sufficient riser analysis early in the design in order to determine the maximum top tensions required. Once this is known the buoyancy cans can be sized. Spacing for existing spar centrewell slots has ranged from 8 ft (for the Neptune Spar) to 14 ft. The size increases with water depth because of the higher tensions required. The recommended spacings would be:

Up to 3000 ft water depth	12 ft
3000–5000 ft	13 ft
Greater than 5000 ft	14 ft or more

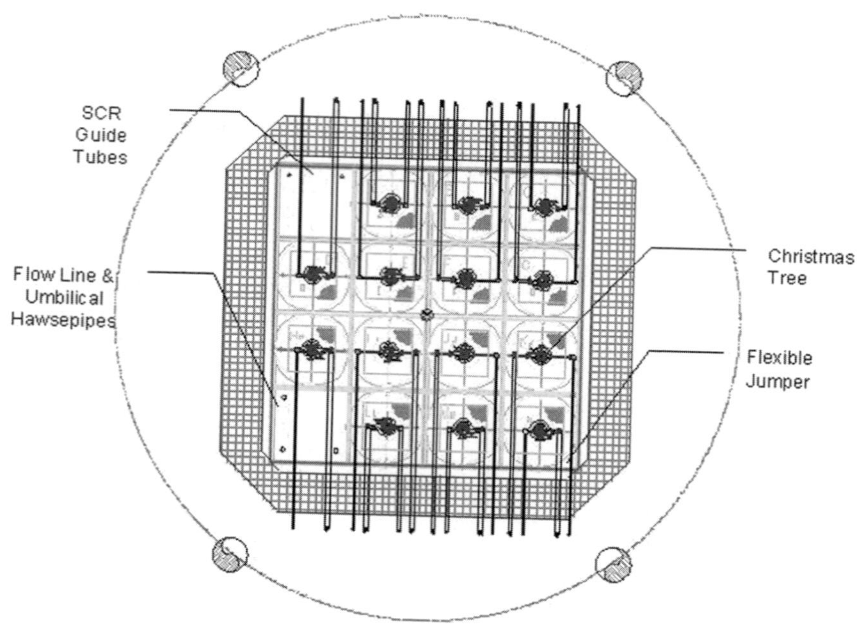

Figure 7.71 Typical centrewell layout (Technip Offshore)

Spar Hull Sizing Parameters
Apart from the centrewell size, the key sizing parameters for the spar hull include:

1. Diameter
2. Hard Tank Depth
3. Fixed Ballast
4. Draft
5. Fairlead Elevation

The initial classic spar designs were based on a draft of 650 ft. This was based on providing adequate separation between the wave energy and the natural heave period of the spar so that resonant heave would not be excited (Glanville et al, 1997). The heave response in the 100-year hurricane, maximum wave height of 74 ft, was predicted to be less than $\pm$ 4 ft. Experience with full-scale spars has proven that the heave response is less than predicted, and also that the spar heave was not controlling the amount of riser stroke, hence more recent spars have had shallower drafts, around 500 ft.

Spar sizing is primarily determined by its heel and pitch response. "Heel" is the static inclination in response to a ballast shift or steady environmental loads. "Pitch" is the total inclination including transient or dynamic responses. For the Gulf of Mexico environment, the maximum pitch response is achieved by designing for a **static** heel angle in the 100-year hurricane of less than 5°. Figure 7.72 shows a free body diagram of a truss spar. The "hull weight" includes lightship weight, topsides deck operating weight, trim ballast, fixed ballast and on board mooring weight. The static heel angle is determined by the resultant moment caused by the couple between the steady environmental forces (wind, current and wave drift loads) and the resisting force of the mooring lines. To a first order this moment is

$$M_{env} = F_{env}(KF_{env} - KF_{moor}) \tag{7.54}$$

where M_{env} = moment due to environmental forces, F_{env} = total wind, current and wave drift force, KF_{env} = distance from keel to centre of action of the environmental forces, and KF_{moor} = distance from keel to fairlead elevation. This moment is resisted by the restoring moment stiffness of the hull:

$$\begin{aligned} K_{Pitch} &= GM * \Delta \\ GM &= KB - KG + I/\forall \end{aligned} \tag{7.55}$$

where K_{pitch} = Initial restoring moment stiffness (N-m/rad), GM = Metacentric height (m), Δ = Hull displacement in force units (N), KB = Distance from keel to center of buoyancy (m), KG = Distance from keel to the centre of gravity (m), I = Moment of inertia of the waterplane (m^4), $\forall$ = Hull displacement in volumetric units.

The waterplane inertia, I, consists of the moment of inertia of the outer shell ($\pi D^4/64$) ***minus*** the inertia of the centrewell and the surface of any tanks which are not totally

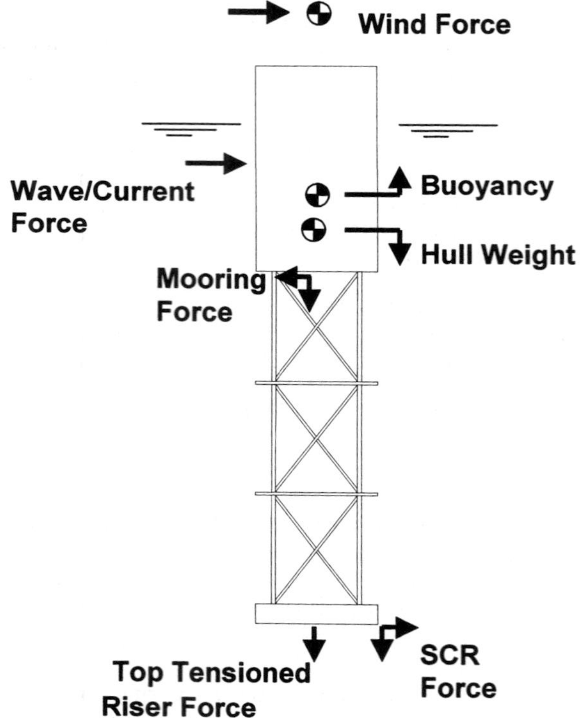

Figure 7.72 Free body diagram for spar hull sizing (Technip Offshore)

flooded. Unlike a "column stabilised" semi-submersible, the pitch stiffness of a spar hull is derived primarily by the fact that the centre of gravity is below the centre of buoyancy ($KG < KB$). The waterplane effect, I/∇, is usually insignificant.

Initial sizing of the hull is determined by the following steps:

1. Minimum variable ballast is selected for trimming the topsides,
2. Trial values for the diameter, hard tank depth and draft are selected,
3. An initial estimate of hull weight and VCG is obtained (see section on hull structure for estimating hull weight),
4. Total buoyancy (displacement) and the center of buoyancy is calculated and the amount of fixed ballast is computed as the difference between buoyancy and the total payload, ballast and vertical loads,
5. The equilibrium angle of heel is computed as

$$\theta_{env} = \frac{M_{env}}{K_{pitch}}. \tag{7.56}$$

The above process is repeated for a range of diameters, hard tank depths and drafts. For each configuration an estimate of heave responses is made. The hull weight (or cost) is tabulated for each configuration which meets a maximum allowable heel angle and heave response.

Internal to this process, the horizontal draft of the lightship spar needs to be checked to determine if its floating draft is below the requirement for transportation. The criteria could, of course, be waived if an alternate method of moving the spar from a fabrication yard skid way to the installation site is available. The method of constructing and transporting the spar should in any event be considered prior to the sizing exercise.

The configuration with the minimum weight (or cost) which meets all the criteria is selected as the baseline configuration. Certain constraints might be considered at this stage. In addition to the limitation on draft for transport, other criteria might include:

- Maximum hull diameter for easy construction at a particular yard,
- Minimum length of hard tank to enclose buoyancy cans,
- Optimum diameter for accommodating deck leg spacing

The sizing question usually becomes an issue of selecting a short, "fat" hard tank and perhaps a longer midsection, or a long "skinny" hard tank and perhaps a shorter mid-section. Optimisation on steel weight rarely shows a significant different in these options, so the designer needs to pick a compromise which satisfies all the criteria.

Table 7.11 shows the main dimensions for spars currently constructed.

A third generation spar, the "cell spar", was installed for in 2004 (Finn et al, 2003). This design embodies a new construction technique using ring stiffened tubulars assembled in a hexagonal formation to form a spar. Figure 7.73 shows an illustration of this concept. The first cell spar is designed for wet trees only.

7.7.6 Drilling from a Spar

The spar has been used for drilling and workover of wells (see Moyer et al, 2001, Bates et al, 2001, Glanville and Vardeman, 1999). The Genesis, Diana and Holstein spars are equipped for full drilling using platform rigs. Drilling operations on the Genesis spar are illustrated in fig. 7.74. This figure shows the arrangement of wells at the spar, and the corresponding well layout on the seabed. The drilling riser is first run through the central slot of the wellbay while the spar is positioned in the centre of the seafloor pattern (neutral position). When the drilling riser is close to the seabed, the mooring lines are adjusted on the spar to position the bottom of the drilling riser next to the seafloor wellhead (Drilling A-1 position). The active mooring on the spar allows the connection of the drilling riser to the wellhead without guidelines or special seafloor guidance. The connection process is controlled by the chain jack operator while monitoring a television camera on the riser and an ROV camera. Positioning is controlled by a computer controlled joy-stick type of maneuver. When the drilling riser connector is stabbed into the wellhead, it is latched in place by the ROV actuating a lockdown mechanism in the tieback connector. The well is now sealed and the drilling may proceed. If there is a large current, the connection may still

Table 7.11 Spar dimensions (Technip Offshore)

Design Topsides Weight	Draft	Centerwell	Diameter	Hard Tank Depth
Tons	ft	ft	ft	ft
5500	650	32	72	220
7700	493	40	90	188
8800	493	40	90	188
9300	505	52	106	176
10800	499	42	98	182
18500	505	60	128	195
19800	650	58	122	220
24000	650	44	122	295
28000	691	75	149	236

be made by maneuvering the spar up current. As long as the entry angle is less than around two degrees the riser can be stabbed and connected.

Once the well is drilled and ready for completion, the drilling riser is unlatched and moved to another well where it is "parked" while the well is completed. The drilling riser BOP is removed. The drilling riser is then hung off in the centre slot while the production riser is run and the well is completed. In order to complete the well, the drill rig is skidded to the appropriate well slot on the platform (tieback A-1 position). The production riser is run and the spar mooring system is again used to position the bottom of the riser over the wellhead where it is latched using the same methods as the drilling riser. Once latched, a surface tree and BOP is installed and the well is completed. Once the well is completed it may commence production. After this, the rig may be moved to the central drilling slot and drilling of another well may commence.

Most of the spars in service do not have full drilling; instead they have workover and completion rigs. These operations are conducted through the production risers and do not require a special drilling riser. Nevertheless, the spar allows for simultaneous drilling and production even with this reduced capability. This is accomplished by using "pullover drilling", a procedure pioneered on the Neptune spar (Glanville and Vardeman, 1999). The procedure is illustrated in fig. 7.75.

Pullover drilling involves using the mooring system of the spar to move far enough from the location of the wells that a MODU may be moored alongside to drill the wells. In the case of the Neptune spar this involved moving about 250 ft from its neutral position. This large offset in 1930 ft of water necessitated using a Titanium stress joint at the base of the production risers (Berner et al, 1997). Once this maneuver is done, a MODU is moored in place over the wells alongside the spar. This requires a preinstalled mooring with buoys to

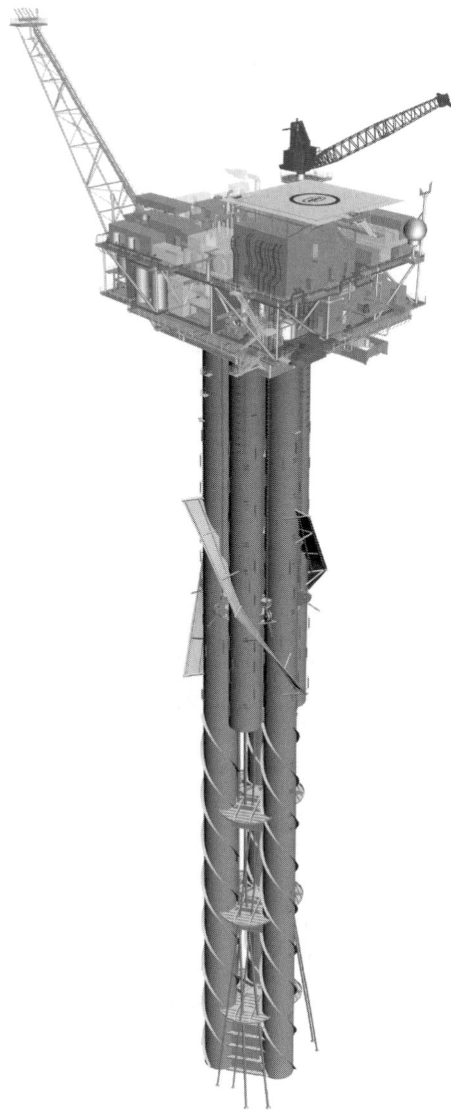

Figure 7.73 Cell spar (Technip Offshore)

prevent interference between the spar and the MODU moorings. The MODU may spud in new wells, or drill wells already batch set to an intermediate casing point. Figure 7.76 shows the seafloor well pattern for Neptune as of 1999. The original wells and new wells are shown. New wells were drilled with the spar offset to the southwest. Future wells will be drilled with the spar moved to the northeast.

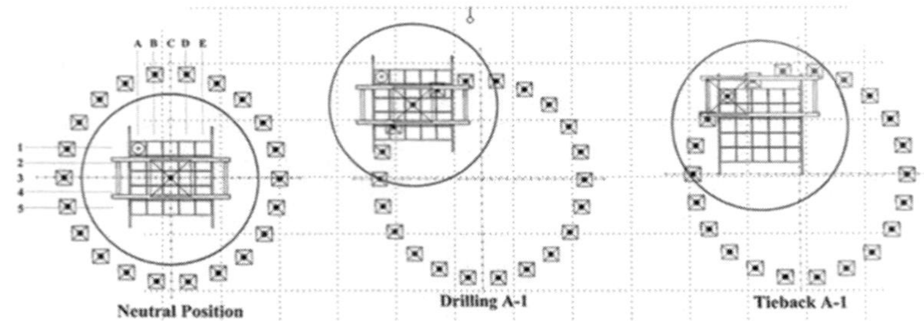

Figure 7.74 Drilling operations from a spar (Technip Offshore)

Figure 7.75 Pullover drilling (Technip Offshore)

Once the MODU drills the wells, it may be removed and the spar can be moved back over the wells where the production riser is run and the wells are completed and/or tied back for production. In this design, the offset mooring system with the MODU is designed for a 10-year hurricane survival condition.

Kerr McGee found that pullover drilling helped improve their project in two ways. First, by delaying drilling some of the wells, the production of the first wells could be examined to determine reservoir characteristics prior to committing to a specific drilling program. Secondly, the costs of drilling could be deferred until revenue was being produced.

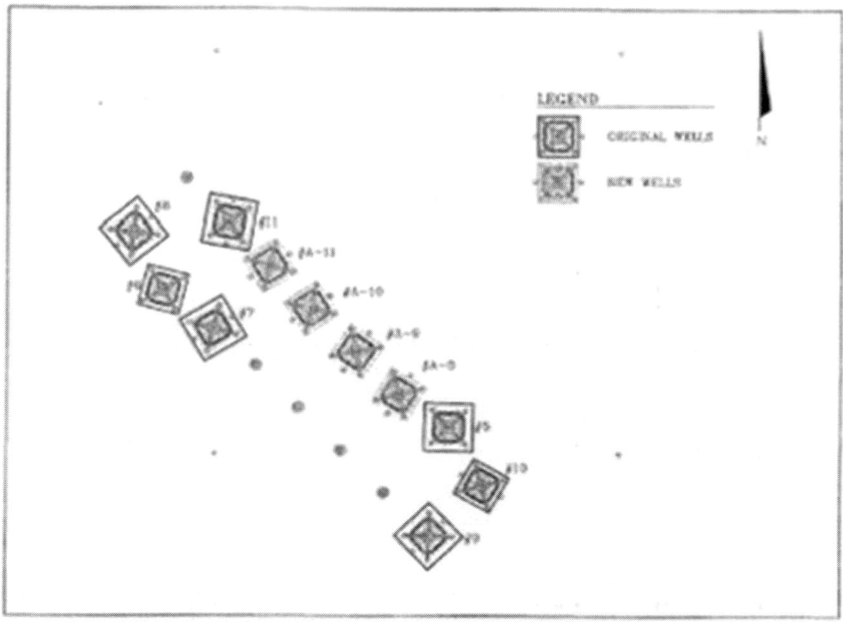

Figure 7.76 Seafloor well pattern for neptune (Glanville and Vardeman, 1999)

7.7.7 Spar Construction and Installation

Construction

Spars are traditionally built and transported horizontally. They are upended while afloat near the installation site. This facilitates use of the conventional shipyard and fabrication yard equipment to fabricate and assemble the hull.

The particular method of fabrication and assembly will be unique to each yard. It depends upon the shop space and the equipment available. Over 70% of the spars built till date have been constructed at the Mantiluoto, Finland yard owned by Technip Offshore. Figures 7.77–7.82 show the sequence of construction used there.

Subassemblies (fig. 7.77) are fabricated in a shop equipped with automatic plate cutting and welding equipment. The size of the subassemblies is limited by the lifting and transporting capacity available. The subassemblies are joined to form half round ring sections of the spar (fig. 7.78). If the spar is very large, the half section may be too large for the fabrication shop and instead smaller sections may be constructed. The upper partial ring section is transported to a lifting tower and raised. The lower section is transported below this section and the two sections are welded together (figs 7.79–7.80). As each ring section is assembled, the entire section is moved to a skidway and connected to other sections to make the entire spar. This procedure would apply to the hard tank of a truss spar. The truss section would be fabricated separately and joined to the hard tank on the skid ways.

After construction, the spar is loaded out onto a heavy lift vessel for transportation using winchs and skid beams to move the hull. The hull may then be transported to a staging area in protected waters near the installation site (fig. 7.82).

Figure 7.77 Fabrication of subassemblies (Technip Offshore)

Figure 7.78 Rollout of Half-section (Technip Offshore)

Figure 7.79 Lifting of upper half-section (Technip Offshore)

Figure 7.80 Connection of two half-sections (Technip Offshore)

Mooring and Hull Installation

The installation of a permanent mooring requires that the anchor piles and mooring lines be preinstalled several months before the spar is installed. Anchors usually consist of steel piles, which are either driven into the soil with hydraulic hammers, or steel piles

Figure 7.81 Construction of the spar (Technip Offshore)

Figure 7.82 Transportation of the spar (Technip Offshore)

which are forced into the soil using suction (Suction Piles). Figure 7.83 shows a pile driving installation using a derrick barge. A wire to support the hammer and pile is run through the main block of the derrick. Separate windlasses and winches are used to pay out the main anchor chain, spiral strand and an umbilical for the hammer.

Once the pile is driven to its design penetration, the chain and spiral strand are laid on the sea floor in a pattern that will facilitate retrieval and connection to the spar upon its arrival (fig. 7.84).

Once the spar is offloaded from the transportation vessel and rigged for installation, it is wet towed to the installation site. The spar is potentially vulnerable to large bending moments and high loads on heave plates (in a truss spar), so this wet tow is limited to the smaller sea states, typically an annual storm event. When the spar arrives at the installation location, it is upended by flooding the soft tank and midsection (fig. 7.85).

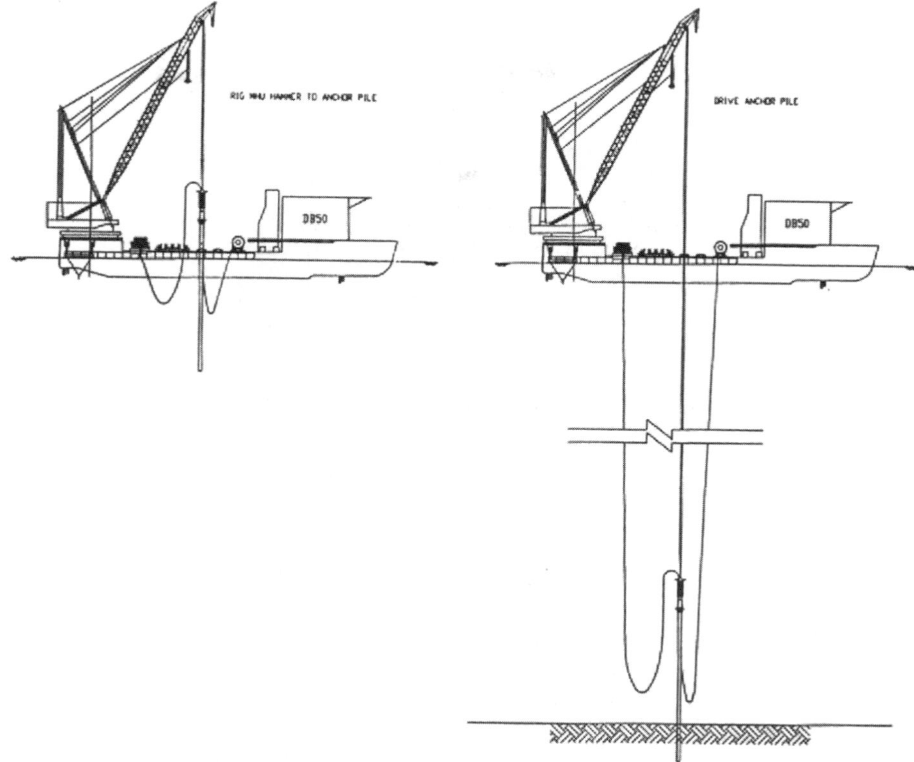

Figure 7.83 Running pile from derrick barge (Kocaman et al, 1997)

Once the hull is upended it is made ready for attachment of the mooring lines. This has typically involved these steps:

1. Ballasting of the spar for an even trim and acceptable freeboard,
2. Lifting of a temporary deck containing power supply, messenger wire and winches,
3. Paying out of the messenger wire to the derrick barge,
4. Retrieving the bitter end of the pre-laid wire rope and chain using an ROV (fig. 7.86),
5. Connecting the platform chain to the messenger wire,
6. Pull-in of the chain.

This procedure is repeated for each mooring line. When two opposite lines are installed, chain jacks are activated to pretension the lines. Once all the lines are attached and the mooring is secure, the temporary deck is removed.

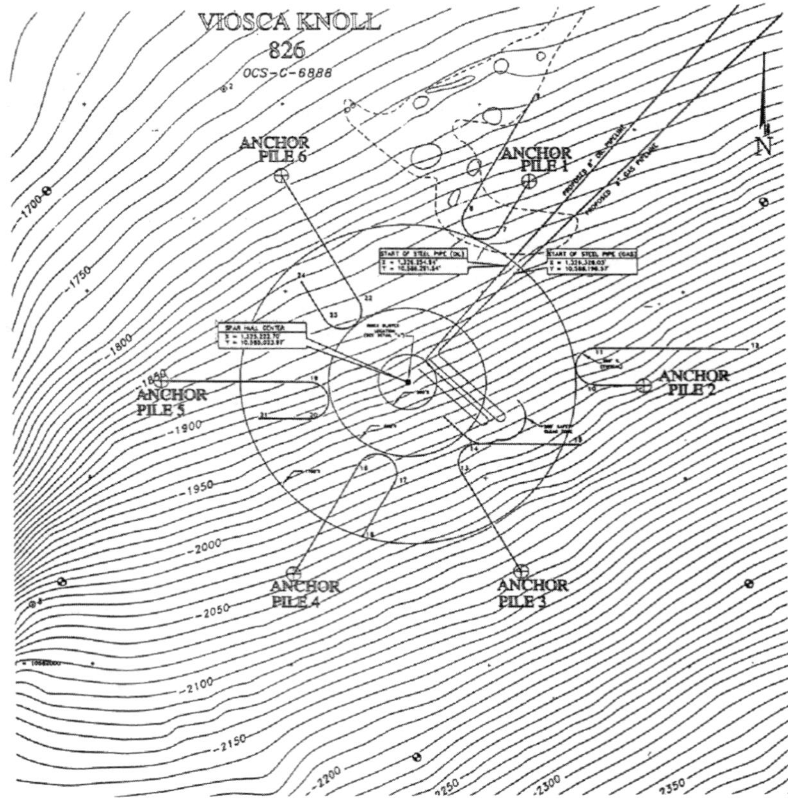

Figure 7.84 Lay down pattern for pre-Installed mooring (Kocaman, et al 1997)

Deck Installation

Once the mooring system is installed, the spar is ready to receive the topsides deck. This requires a derrick vessel to lift the deck. Decks weighing from 3000 to more than 20,000 tons have been installed on spars. Figure 7.87 shows the 3300 ton lift of the Neptune spar deck using the McDermott DB 50 derrick barge.

Larger decks require multiple lifts. This entails a protracted period for offshore hookup and commissioning of the deck. An alternative to deck lifting is the floatover deck method, which has been used extensively for fixed platforms (O'Neill et al, 2000). A spar floatover requires a catamaran transfer similar to that used on the Hibernia Gravity Based Structure. This method allows for the installation of very large, fully integrated decks on the spar (Maher et al, 2001). The sea state limits for the float over installation are similar to those for a derrick barge installation.

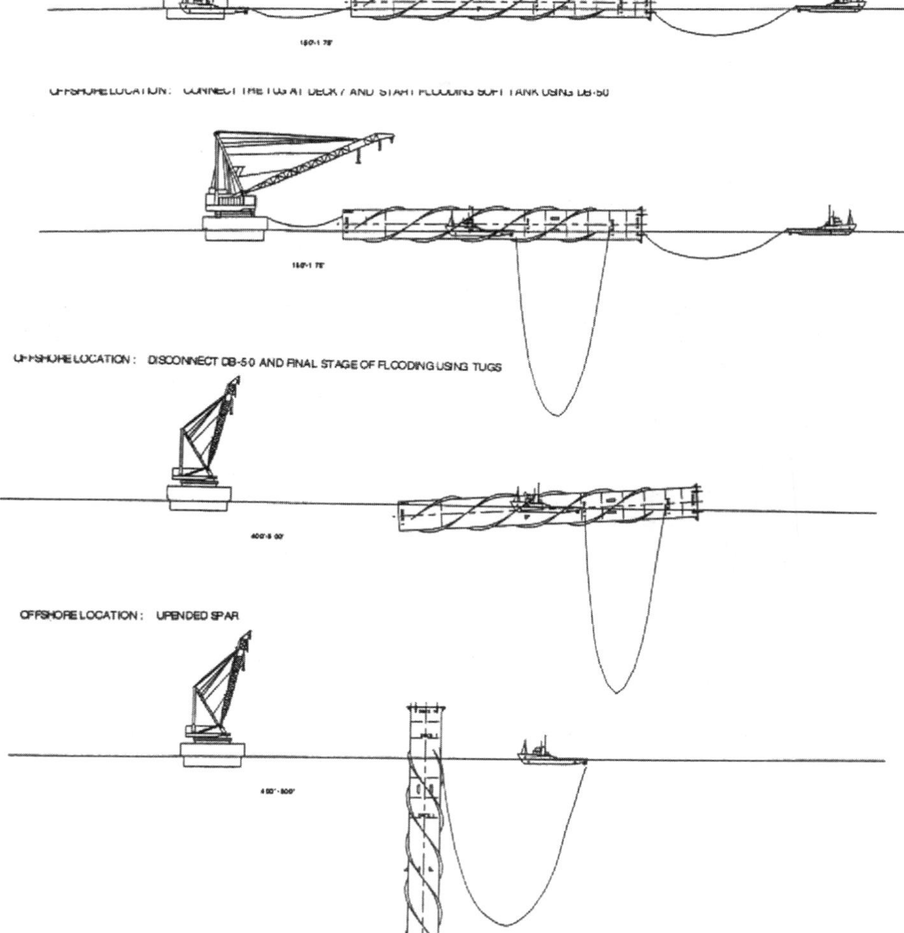

Figure 7.85 Spar upending sequence (Kocaman et al, 1997)

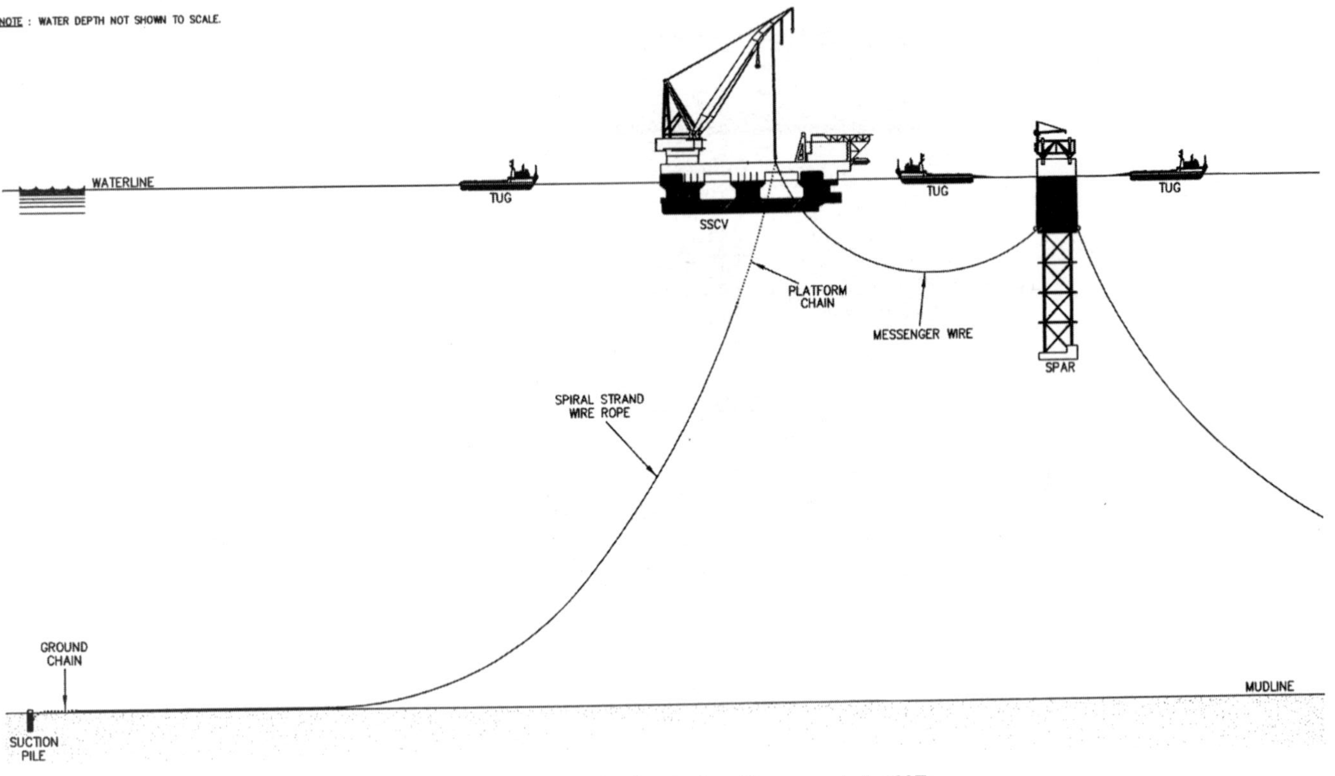

Figure 7.86 Spar mooring line hookup (Kocaman et al, 1997)

Figure 7.87 Derrick barge setting deck on spar platform (Technip Offshore)

7.8 Hull Structure (John Halkyard and John Filson)

7.8.1 Hull and Deck Definition

The term "hull" generally refers to that section of a platform that provides buoyancy. In the offshore industry, the term "deck" generally refers to the supported functional superstructure. However, for a hull, the term is also used to refer to the horizontal flats separating compartments in the hull. In the discussion on hull structures, to make a distinction, deck structure above and between the columns will be referred to as the "superstructure." The term "deck" will be reserved to refer to the horizontal, local parts of the superstructure. With most production structures, spars, TLPs and most semi-submersibles, the superstructure deck is a physically separate structure. For semi-submersibles and TLPs, the pontoons, columns, deck and bracing system compose a single, integral structural system.

7.8.2 Applicable Code

Generally, the most applicable codes for an FPS design are the classification rules for the MODUs. Virtually all semi-submersibles and spar production platforms built till date follow the maritime practice of "classing" and are designed and built according to the classification rules. Rooted in shipbuilding practice, but highly developed for MODUs, the most prevalent code is the American Bureau of Shipping "Rules for Building and Classing Mobile Offshore Drilling Units". This code has been applied to production platforms as well. Other classification societies have similar rules, Det Norske Veritas (DnV) being a major factor in the MODU classing also. These and other codes are listed in the references at the end of this chapter.

The MODU classification rules represent several decades of service experience. Adjustments are evolving where MODU service is different from a FPS, particularly with regard to permanent siting. The differences relate primarily to site specific environmental criteria, inspections, and longevity issues, particularly fatigue and corrosion. Beyond these issues, the service is considered to be generally the same as for a MODU, particularly with regard to the local and global strengths.

API RP2A is specifically for fixed structures, but may influence some aspects of an FPS design, particularly the environmental criteria and the design of certain functional structures. However, it fails to address many design issues important to semi-submersibles, spars and other floating platforms. Closer, but with significant differences, is API RP 2T and the associated Bulletins 2V (Buckling of Cylindrical Shells) and 2U (Stiffened Flat Plate). API RP 2T specifically applies to TLP design. 2U and 2V are relatively generic and can be applied to spar and semi-submersible structures, but were developed primarily to support TLP structural design and reflect that.

The responsibility for "seaworthiness" and marine safety is usually placed with an agency responsible for maritime regulation. Such agencies tend to look to classification societies in the area of structural integrity. This is the case with the Floating Production Systems (FPS). In US waters, USCG is the responsible agency for certification of FPSs with respect to "seaworthiness." In the UK, the DEn recognises and delegates certain classification societies as "certifying authorities."

Newly built FPSOs in the offshore industry follow similar design practices to the other FPS types, but their ship-like hull configuration leads to extensive use of tanker design practice, but with special differences. This is especially true of FPSO conversions that are ships to begin with and are designed and built to commercial shipbuilding practice.

The following is a description of the practice of design of the commercial ships extracted from SSC-419 (Ship Structures Committee, 2002):

"The primary basis for designing the structure of commercial ships is contained in the rules of various classification societies, of which about 80 exist worldwide. The most significant are those who belong to the International Association of Classification Societies (IACS), namely:

- American Bureau of Shipping (USA)
- Bureau Veritas (France)
- China Classification Society (China)
- Det norske Veritas (Norway)
- Germanischer Lloyd (Germany)

- Korean Register of Shipping (South Korea)
- Lloyds Register of Shipping (UK)
- Nippon Kaiji Kyokai (Japan)
- Registro Italiano Navale (Italy)
- Russian Maritime Register of Shipping (Russia)

The IACS also includes the following Associate Members:

- Hrvatski Registar Brodova – Croatian Register of Shipping
- Indian Register of Shipping
- Polish Register of Shipping

The technical base of the IACS is provided in the IACS Bluebooks, which represent a set of standards that have been developed through cooperation among all the member societies. The standards for ship structure deal principally with the strength of the hull girder. The book contains unified requirements, recommendations, and interpretations for material, hull girder strength, superstructure and deckhouses, equipment (anchors and chain), and rudders. Each member society in the IACS is expected to adopt the unified requirements into their rules. By basing their rules on the IACS standards, the member societies compete on the basis of factors such as the services that they will give to owners and not on the basis of permitting lower structural standards than competing societies.

Ship owners cannot go from one IACS member to another looking for lower requirements in critical areas, because they are all the same. However, these IACS unified requirements do not cover local criteria for plate, frames or support structure. Therefore, the statement that a ship design will not differ between societies is the ideal but not the fact."

7.8.3 Structural Design Considerations

General

Structural design of floating platforms can be broken down in two fundamental levels:

1. Local Strength
2. Global Strength

In addition, the buckling strength needs to be checked and this may be on either level or, more likely, involves stresses from both global and local effects. Besides buckling, fatigue needs and endurance needs are to be checked. Although this is usually related to global environmental loading, there are also possible fatigue situations related to local loading, both environmental and functional loads.

The design of most structural steel on floating platforms is based upon local loading and most of this is based upon gravity/buoyancy loading. Buoyancy loading, while distinct, is a gravitational load. The demands of global loading are primarily met with reinforcement (increased thickness) of plating otherwise in-place for local load. With braces on semi-submersibles, however, given their purpose, the global load governs their design. As with the rest of the hull, most superstructure design is based upon local load, however, many of the main girders are part of the space frame bracing and subject to considerable global loading, particularly with hull-type superstructures.

Local Strength

Design for local loading is mostly based upon empirical, classification rules and gravity/buoyancy loading, although stress-based checks are made for some components. Steel sized according to these rules is generally referred to as "basic scantlings." The empirical equations given in the scantling rules have been long established and are the product of considerable service experience and research. They reflect the intended service, including effects of environmental loading, dynamics, corrosion and maintenance. Loading on the pontoons and the columns is generally expressed as an equivalent hydrostatic head. Local design in the superstructure is primarily based upon rated, distributed deck load. Regardless of load, a number of minimums are specified, the superstructure bulkheads being an example.

The loaded structure functions also in a global context. Pontoons and the column shell plating, while locally designed for hydrostatic loading, assembled as a hull element, are the primary strength elements too. Figure 7.88 shows the assembled hull element as part of the hull global strength system of several typical floating offshore structure forms. Hull elements, given their large cross-sectional size, have sufficient sectional properties to keep global stresses very low. Only the column-to-pontoon and the superstructure-to-column connection have high stresses, and these are usually concentrations due to internal structure that are moderated with reinforcements. The most important global stress related area interacting with the local design is the connections of the braces with the columns, with the superstructure, and with each other.

Global Strength

Consideration of global strength uses stress-based, rational analysis to examine the entire structure as a space frame or, in the case of a ship-type hull or a spar, as a single slender beam. Both the distributed gravity/buoyancy loading and the combination with environmental loading are applied. In addition to the effects of wind, current, and wave load, environmental loading includes the inertial forces due to vessel motions and also the reactions of mooring lines or tendons. The wave loading and the corresponding inertia load are the most important environmental load on floating structures. In some cases, transportation and/or installation loading can form controlling design loads. This is particularly the case for spar platforms, in that they are built and transported horizontally, and upended offshore. In this case, the wave loads during transportation and gravitational and buoyancy loads during upending may be the most critical global loads.

The Classification Rules (ABS, DnV) lay out a rational, allowable stress-based design basis which is, in essence, an adoption of the 1967 AISC Code (both since evolved). Particularly notable are the separate factors of safety for gravity load only and the combination of extreme environmental loading with gravity load, the latter allowing an elevated allowable stress.

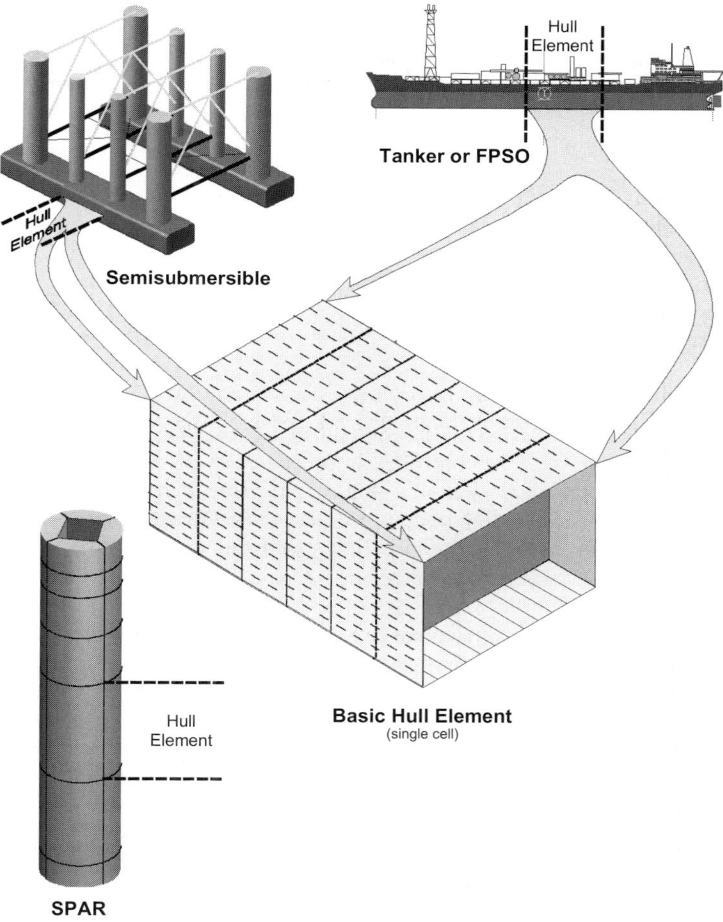

Figure 7.88 Hull elements in offshore structure

Exclusive of buckling, the allowable stress is given as $F_{allowable} = F_y/FS$ where the factor of safety (*FS*) is:

Gravity Load Only

1.67 for axial and bending stress

2.50 for shear stress

Combined Environmental and Gravity

1.25 for axial and bending stress

1.88 for shear stress

Where buckling is a consideration, the allowable stress is given as $F_{allowable} = F_{cr}/FS$ where F_{cr} is the critical buckling or shear stress and the FS is:

Gravity Load Only

1.67 for axial and bending stress

Combined Environmental and Gravity

1.25 for axial and bending stress

Under certain codes, structural redundancy must be considered, but allowing elevated limit states (allowable stresses). Also considered are the damaged conditions through loss of watertight integrity. Besides lateral load due to severe heel, this increases hydrostatic loading on deeply submerged elements.

Buckling Strength

Most floating platform designs are controlled by the local and global strength considerations. Buckling may be controlled in special cases of slender chords and braces in truss structures and the upper deck of a ship shaped vessel that is subjected to a hogging condition under still water loading conditions. The DnV Classification Note 30.1 (or its most recent designation) is widely used for the buckling analysis of stiffened plate.

Fatigue

Except for designs that do not rely on space frame bracing, fatigue is almost invariably in the bracing connections to the columns or to other braces, invariably at a weld, and usually associated with a known form of stress raiser. Connections of the superstructure at the column tops can be critical to fatigue loading. Brace connections to the superstructure sometimes experience a form of low cycle fatigue, but this is not amenable to the conventional fatigue analysis, only the highest waves cause the problem and the relationship to waves is highly non-linear.

Most of the 4th generation drilling semis and almost all the production semis are not braced and resist global loading mostly by frame bending, with moment/shear transfer at the column ends. This is more prevalent in the longitudinal direction, where it is found that the pontoons provide adequate shear strength. Where such high moment connections are employed, stress concentration and local analyses with respect to global strength as well as fatigue become more problematic. However, section geometry tends to provide considerable strength with moderate plating thickness.

Fatigue design is important for ships and spars in the framing close to the waterline. Also, the transition between the truss and "hard tank" of a truss spar is a critical area for fatigue. This is identical to the connection of the superstructure to the hull structure in many respects.

7.8.4 Hull Structure Design

General

There are two very important distinctions for "hull structures" which distinguish them from the typical space frame structure familiar to civil engineers and designers of fixed

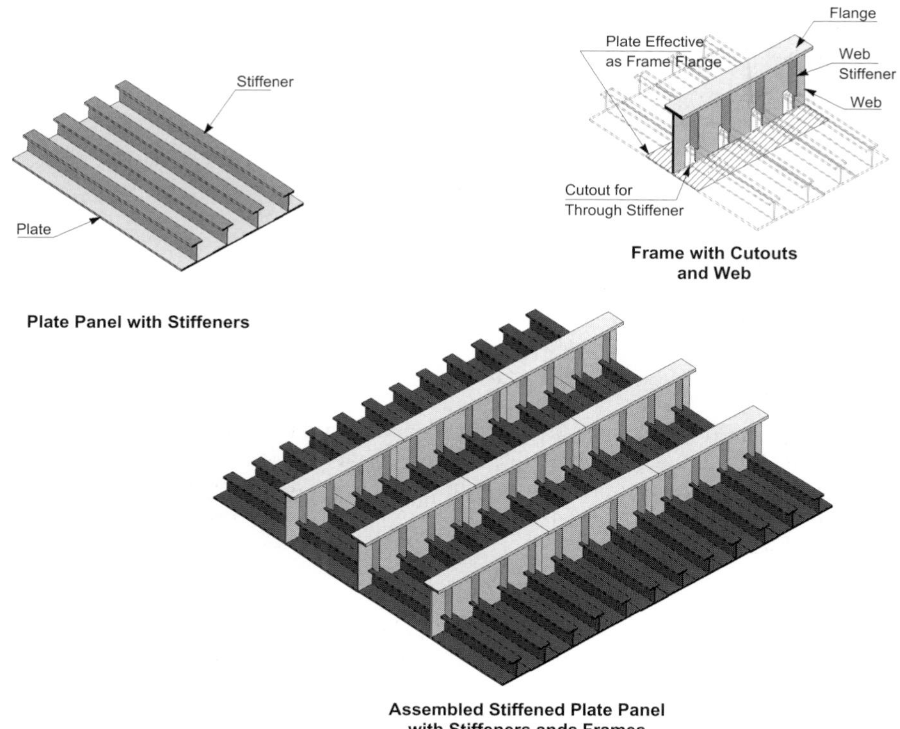

Figure 7.89 Typical cross-stiffened plated panel

platforms. One is the formation into box forms with considerable strength as girders. The second is cross-stiffened plate panels.

With regard to the structural design, some comments on the structural systems typically used in shipbuilding and hull structures should be noted. These structures are primarily composed of panels of cross-stiffened plate as illustrated in fig. 7.89. The plate provides watertightness and resists hydrostatic loading. Assembled into a box form, as shown in fig. 7.90, the assemblage forms a box girder of considerable global strength in bending, shear and torsional strength. Schematically indicated are the global loading of the element.

Cross-Stiffened Plate Systems

As shown in fig. 7.89, the cross-stiffened plate panel consists of the *plate field*, the *stiffeners* in one direction, and the *frames* supporting these in the other. Where necessary, frames are supported by *girders*, if not by bulkheads. The stiffeners are either angles, tees, or specially rolled "bulb flats," integrally welded to the plate. The stiffener and plate are structurally considered to be a single unit. The same is true of the frames and girders. Taken together, they are referred to as a "panel." The cross-stiffened panels are integral structural

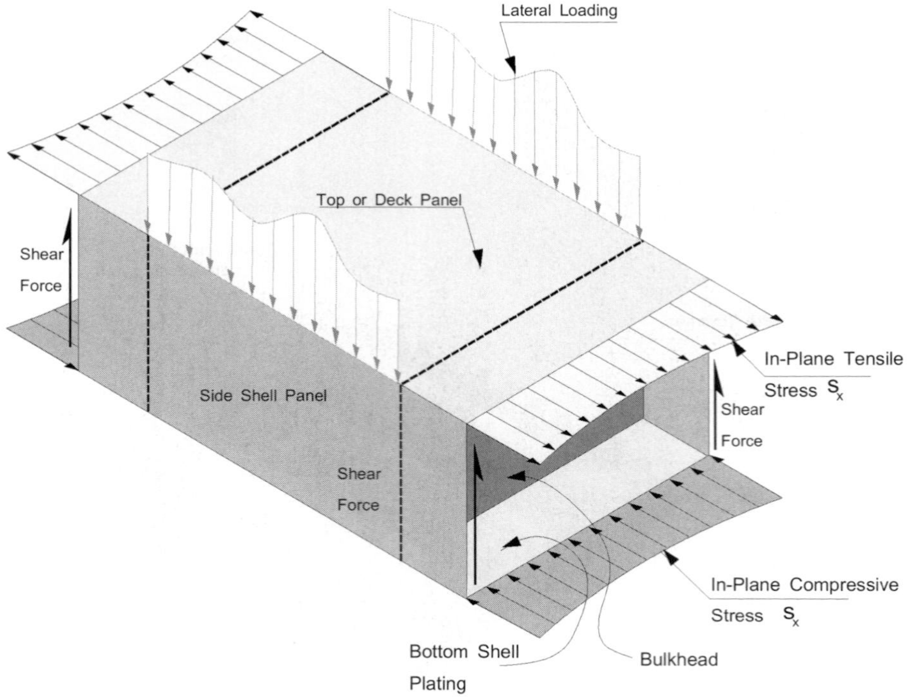

Figure 7.90 Hull type box girder and loading

components and fabricated by shipbuilders as a unit though highly automated processes. The plating in this configuration works as a flange to both the stiffener *and* the frame (and girder) in both directions.

Panels are not necessarily flat. The shell plating on columns of semi-submersibles and spars are formed of curved panels. If cross-stiffened, lateral load is resisted by stiffener flexure and are not true "shells", which fail by buckling, especially ring stiffened. For large diameters, as found in semi-submersibles, TLPs, and spars, the stiffeners run parallel to the axis of the cylinder, supported by perpendicular ring frames. The shell plating is considered as virtually flat and is analyzed as flat panels spanning the distance between frames bulkheads. Unless totally without internal support (struts), the frames are a connected series of curved beams and not designed as rings solely governed by buckling criteria. Figure 7.91 shows such a structural form as built for a spar. In both the flat and the cylindrical cases, the stiffeners run *through* the cutouts in the frames.

It is noteworthy that, as integral structures, stressed in both directions, including the plating as flange to the stiffeners, etc., such systems provide considerable strength for their weight and cost. Considerable effort is spent optimising, plate thickness, stiffener and frame spacing, etc. with regard to weight and fabrication cost. With stiffener spacing

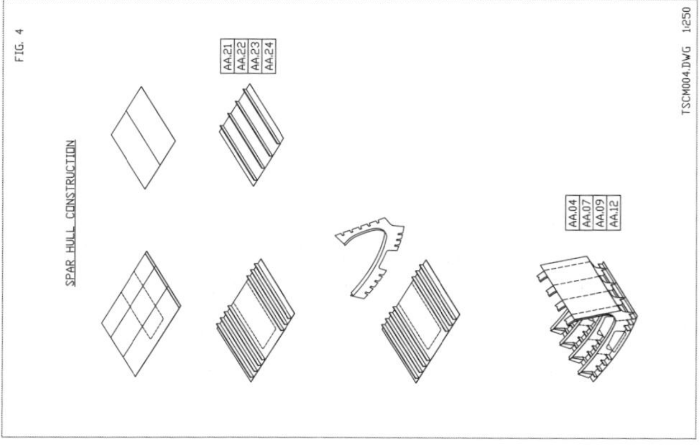

Figure 7.91 Forming the outer shell of spar platform, plating, stiffeners and bulkhead shown (Technip Offshore)

typically 24–30 in., plating runs from typically 3/8 in. at the column tops to 3/4 in. or more on the pontoon bottoms. For very deep draft structures heavier scantlings are used. The framing system generally adds 65–90% to the shell plating weight of pontoons and columns. It is somewhat less for internal bulkheads and flats.

Hull Girders

In the box section hull element shown previously in fig. 7.90, a number of flat plate panels are assembled into a prototypical "hull girder." The box girder form is relatively straightforward and is common for pontoons and effectively equivalent for columns. The pontoons and the columns locally perform the same primary strength function as a ship hull in this respect. The column-to-pontoon connection is equivalent to a concentrated load and not unlike a derrick on a crane ship.

The same strength principles are used in some types of superstructure design. It is actually a simple matter to consciously arrange superstructure deckhouses into well-supported box girders providing considerable hull strength to the superstructure. A fully developed system, running in both directions, between the columns is referred to as a "hull-type superstructure" and found in many of the newer semis and is capable of reducing and even eliminating bracing. Multi-cellular construction is a very common form in aircraft and other transportation structures and is also called "stressed skin" construction.

Figure 7.90 also shows an idealised loading and stresses on the hull girder element. Notable is the shear transferred by the side plating and in-plane stresses in the top and bottom. These stresses are typically not uniform across the width of the plate field. This is an effect referred to a "shear lag." It is an important concept in hull structures and can be addressed rather easily. This will be discussed further below for local structure. The discussion also applies to other structural forms.

7.8.5 Local Strength Design

General

Local strength design addresses primarily plate thickness and the spacing, span and cross-sectional design of framing systems, all to resist lateral distributed load. This part of the local strength design is usually based upon the gravity/buoyancy loading and uses empirical, classification society, rule-based, minimum scantlings. The girders supporting frames, and sometimes also the frames, are designed with rational, allowable stress analysis. Highly repetitive frames are often optimised with rational allowable stress analysis.

The plate is integrally welded to, and considered to be part of the flexural section of the stiffeners, frames and girders. Bulkhead, deck and shell plating thickness so determined are the minimums for local load purposes and sometimes increased for primary or secondary strength purposes. Generally, the resulting scantlings produce a hull girder section suitable for both the local, distributed design load as well as the global loading. Otherwise, as noted elsewhere, shell plating is increased in thickness to accommodate global strength requirements.

Other local loads are from functional components (typically foundations) such as derrick/substructure foundations, riser supports, and foundations for mooring equipment and tendon connectors. Design and analysis of these "special structures" are local in nature and follow allowable stress design practice and often employ AISC and/or API RP 2A codes.

The following is a brief introduction to the "local scantling design," and is given to characterise the design process. It can and has been taken much further than given here, however. The expressions and citations apply to tanks, voids, and shell plating, and are primarily applicable to all watertight hull structures, internal as well as external. This includes the pontoons and columns on the TLPs and semi-submersibles as well as SPARs and ships. The ship rules are somewhat different in that they are locally specialised to specific parts of the hull. What is given herein is primarily from the ABS MODU Rules. The DnV Rules are noted for comparison. They are similar and largely equivalent, but are different. Designers should refer to the latest editions of the various rule documents for updates and clarifications.

The ABS *MODU Rules* for minimum scantlings (plating, beams, and frames, and girders) are given in Part 3 (*Hull Construction and Equipment*), Chapter 2 (*Hull Structures and Arrangements*), Section 2 (*Common Structures*). (The ABS MODU Rules will be cited as follows: 3-2-2/*Subsection*.) All service applications separately refer back to one of two, fundamentally different criteria sets in this section for the design of scantlings: *Watertight Bulkheads and Flats* (Section 7) and *Tank Bulkheads and Flats* (Section 9). The scantling rules in 3-2-2/9 are commonly known as the *deep tank rules* and have been in the MODU Rules from their inception (1969) and the ship rules before that. The former has reduced criteria used for internal, watertight subdivision that is only intended to prevent progressive flooding from damage. All other design requirements relate to normal service and refer to the scantling rules given for tanks (Section 9).

While similar in formulations for scantlings, the DnV Rules do not make the distinction between the normal service and the watertight subdivision (i.e. damage) in the same way. Instead, the same formulation for scantlings is used in both situations, but a higher usage factor (stress) is used for damage conditions.

7.8.6 Hydrostatic Loading

General

The predominant form of local loading is hydrostatic pressure, both static and hydrodynamic. This can be the static external pressure of the sea as well as the dynamic augment of waves. In addition, the pressures on internal surfaces from internal liquids are to be considered both from service as tankage and from internal flooding. Internal flooding can result from a breech in the shell plating or from internal piping system failure. All plating surfaces of the watertight envelope and internal watertight subdivision should be considered to be subject to hydrostatic loading.

Where local loading is not a fluid, particularly on flats, decks, and platforms, including the superstructure, a pressure-based loading is used, and usually a *rated loading* not to be exceeded (e.g. 250 psf). Certain equipment and functional loading is treated on a case by case basis. In practice, deck loading is random and usually employs measures to spread the load (e.g. dunnage). Concentrated loading of plating is generally taken on a case by case basis, often requiring only headers between framing. However, where necessary, more complex foundations are developed to diffuse the load. There are also special cases, such as wheel loading or tie-down pads, which have special criteria.

Although distributed loading can be expressed in various units (ksf, psi, etc.), the traditional form, still in persistent use, is to express hydrostatic loading as the equivalent head of *seawater*. The density of seawater is conventionally taken at 64 lb/ft^3. Therefore a 1 ft head = 64 psf = 0.064 ksf = 0.444 psi. A 100 ft head is 6.40 ksf = 44.4 psi. For DnV equivalency, density of seawater is 1026 kg/m^3. A 100 ft (30.48 m) head is 0.307 MPa (307000 kN/m^2) pressure. ABS Rules specifies local loading in terms of the *design head, h*, either feet or metres. The DnV Rules use pressure, *p*, kN/m^2 (Pa). Using the equivalent head of seawater has a tangible value in that a geometric relation to water levels provides an intuitive "reality check" which helps avoid errors.

Local loading can be "static," referring to the Gravity Load (only) condition, and "dynamic," referring to environmental loading, as a component of Combined Loading.

The classification rules specify minimum design loads for various plated surfaces. These loadings apply to the *minimum* scantling requirements as mentioned above (e.g. "ABS 3-2-2") and extend to plate thickness, stiffener spacing and size, and frame size and spacing. While, in some aspects, these rules may seem to be prescriptive, they are companions to the scantling rules and serve as part of their empirical nature. They are specific to a variety of structural forms used in hull construction (framed shells) and consider the function of the local structure, the type and magnitude of pressures, and the intended service. The following presents local loading according to the 2001 ABS MODU Rules, Part 3, Chapter 2, as mentioned above. Section 4 (Column-Stabilised Drilling Units) is used as being representative of service requirements for most applications to be considered (i.e. columns and pontoons). Section 3 is *Self-elevating Drilling Units* and Section 5 is *Surface-type Drilling Units* (drillships). The reader is referred to the ABS Ship Rules for FPSOs and similar units.

Finally, it should be emphasised that the design hydrostatic pressures determine most of the hull steel (> 80%) and have far more impact on the steel required than global strength (5–15%). Furthermore, there are many subtle complexities regarding which of several criteria truly govern.

Design Heads

Subsection 5 (of 3-2-4) addresses minimum scantling requirements for *Columns*, *Lower Hulls*, and *Footings*. While specific to semi-submersibles, they are applicable to nearly all watertight parts of offshore structures. There are, however, special cases for specific unit types with particular known problems locations (e.g. forefoot, etc.). Paragraph 5.3 of this subsection gives required design heads, *h* for determination of *Scantlings of Framed Shells*. The design heads are applied in formulae for scantlings given in Section 2 (*Common Structures*). These are given in four service categories:

(a) Tank Space
(b) Void Compartment Space
(c) Areas Subject to Wave Immersion
(d) Minimum Scantlings

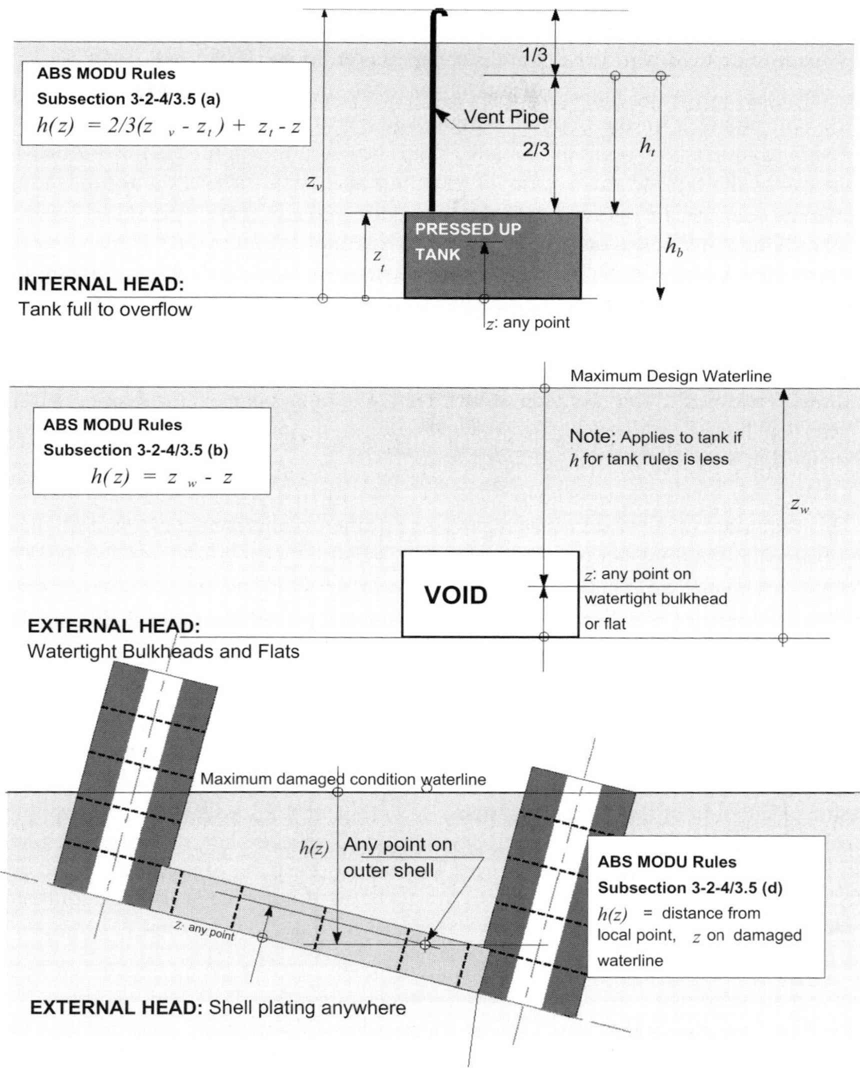

Figure 7.92 Summary – ABS design head requirements

According to the service defined above, the design heads are specifically defined as follows (interpreted for clarity). Figure 7.92 graphically summarises these requirements

(a) 5.3.1: *Tank Space* (*numbering is from the ABS MODU Rules*)
A *tank* is an internal space used for fluids such as ballast water or liquid consumables and is usually connected to a pump driven piping system. It is required that there be a

vent (or overflow) to relieve the inlet pressure or suction. A closure may be used to prevent ingress of water (for stability), but this must be of the ball check variety to insure adequate water ingress. The tank space requirements relate to pressure relief via the vent piping, with the required design head based upon the height of the vent. The design head, h at any point in the tank, z is to be taken to a point 2/3 the distance from the *top* of the tank to the overflow (vent), but at least 3 ft above the top of the tank. Where the specific gravity of the contents exceeds 1.05, the head is to be correspondingly increased.

(b) 5.3.2: *Void Space*
Any internal space, which is *not a tank*, is considered to be a *void*. Additionally, any tank is to be considered as a void when slack or empty. Watertight void spaces are considered to be floodable if it is adjacent to shell plating or there is piping running through the space. The design head of a void, h is to be taken to the maximum service waterline (*Design Waterline* – see later). This is essentially a minimum that applies to all shell plating as well as internal subdivision.

(c) 5.3.3: *Areas subject to wave immersion*
This requirement stipulates that, regardless of other requirements, any watertight shell plating shall be designed to a minimum 20 ft head.

(d) 5.3.4: *Minimum Scantlings*
This requirement relates to all watertight boundaries, including the shell. The design head, h, is to be taken to the most extreme one-compartment damaged waterline for any point on the shell (static). It applies only to freely floating structure.

Requirement (a), addresses the internal pressures of tank spaces, giving particular attention to excessive pressures from pumping. Such tanks are required to have vent pipes of certain sizes to obviate excessive pressures. The DnV Rules actually include a dynamic pressure calculation of the pumping system. Requirement (b) is essentially a minimum that applies to all shell plating and internal subdivision. It reflects that all watertight surfaces below the design waterline should at least be able to resist the pressure from the design waterline. Both requirements (a) and (b) are to be used with the scantling requirements for tanks (ABS 3-3-2/9, discussed later).

The design waterline, as mentioned in requirement (b), is conceptually rooted in floating structures and represents the waterline of the deepest operating draft. This waterline is used for Gravity Loading (static). Any submergence below this level, be either the downward movement of the vessel (heave) or the upward movement of the wave surface, is considered to be environmental and relevant to Combined Loading criteria. Some hull-type structures (e.g. TLPs) are connected in some way to the sea bottom, such that tides and settlement can raise the water line. In such cases, a tide and settlement allowance should be added to the design waterline.

Requirement (c), the minimum 20 ft design head, essentially applies to the upper columns and the watertight superstructure of semi-submersibles. It is prescriptive and is generally considered to address dynamic wave pressure on the above water parts of the watertight hull, but is non-specific in this regard. However, this may be inadequate in some cases and dynamic pressures should be specifically determined.

Requirement (d) applies to all watertight surfaces, both internal as well as external. It is a damaged criterion with separate scantling requirements (ABS 3-2-2/7, discussed later). These requirements produce slightly lighter scantlings than the static requirements (tanks), but at a materially higher loading. They are intended to avoid progressive damage from flooding. DnV has a similar requirement, but it is treated as an "extraordinary load" with a lower factor of safety. In the case of a TLP, flooding damage usually causes reduced tendon tension and not a change in waterline. However unlikely, should flooding result in a viable damaged waterline, requirement (d) should be considered.

The ABS MODU Rules are less than explicit about dynamic pressures, although the DnV is quite so. Paragraph 5.9 simply indicates that "provision for wave and current" meet the general allowable stress requirements. In so far as local scantlings are concerned this is generally applied by using the pressure of combined static and environmental loading (dynamic pressure), reduced by one-fourth for use with static scantling criteria for tanks. (This is consistent with DnV Rules; these allow a 0.8 usage factor instead of 0.6.).

Dynamic pressure for point-z_j (above BL), for a mean draft of d_s, can be determined as follows:

$$p(z) = {}^1\!/_2 \rho g H e^{-k(ds-zj)}$$

This is simply the "wave form pressure" of deep water, the linear wave theory (small amplitude) and follows the ABS *MODU Rules* Part 3, Appendix 3A. It does not account for pressures developed by the presence of the body (hull). For bottom-connected structures, d_s should include set-down. Above the mean waterline ($z > d_s$), the theory breaks down. A useful approximation is to take the total pressure as linear from the $z = d_s$ value to zero at the wave surface. Above the still waterline, it should be noted that that pressure computation from diffraction–radiation wave theory (linear) is based upon infinitesimal wave height and provides little useful information.

In addressing loading from wave dynamics, particularly in conjunction with platform motions, it is important to consider also the phase relationships of different components of load.

7.8.7 Plate Thickness

General

Plating under a lateral load does not behave according to simple rules of stress suitable for use in stress-based design. At best, this would require large deflection plate theory and specific knowledge of edge boundary conditions, not to mention an assumption of the out-of-plane deformation. This is in part due to the fact that welded steel plating is not sufficiently flat for a simple theory, and is subject to indeterminable in-plane stresses. However, this limitation has been overcome through empirical formulae based upon considerable theoretical and experimental work as well as service experience. On such bases, each classification society has developed rules for minimum plating thickness in comparatively simple form. Generally these formulations account for other service factors such as thickness loss due to corrosion.

Scantling rules for plating are generally based upon a uniform pressure loaded plate in a continuous field bounded by stiffening on the long edges (a) and frames supporting the stiffeners on the short edges (b). (These lengths in notation used by ABS are respectively l and s). The upper part of fig. 7.93 illustrates the context and notation for an individual plate panel. Generally speaking (as indicated in the figure), plate bending due to lateral loading is most severe at the middle of the long edge. As illustrated, normal to the long side stiffener, bending tensile stress (σ_b) develops in the top surface of the plate and a

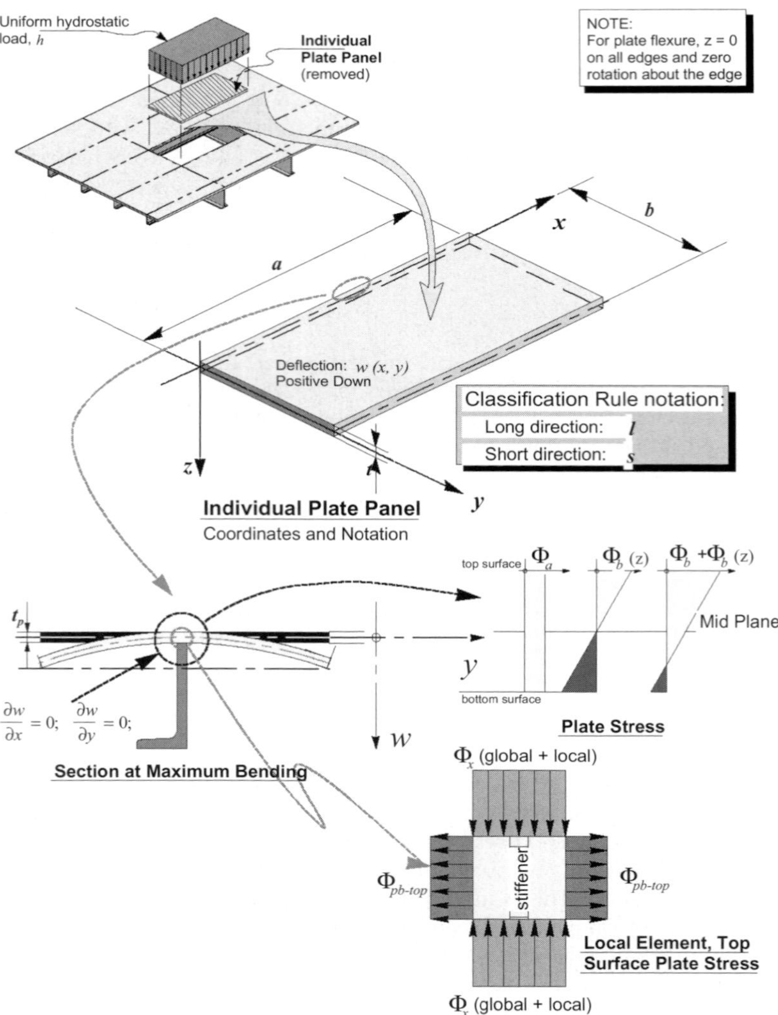

Figure 7.93 Plate bending geometry, notations, and conventions

corresponding compressive stress develops at the bottom surface. Typically, the plate is constrained and sufficiently flexible such that a catenary-like tensile stress (σ_a) also develops. The combination of these stresses ($\sigma_a + \sigma_b$) is indicated on the right of the figure. In addition, as indicated in the lower right of the figure, there is frequently a global in-plane stress, σ_x, normal to these stresses.

Plate bending is further illustrated in fig. 7.94 showing the typical continuity between the adjacent panels, particularly noting the edge boundary condition of end fixity. While the

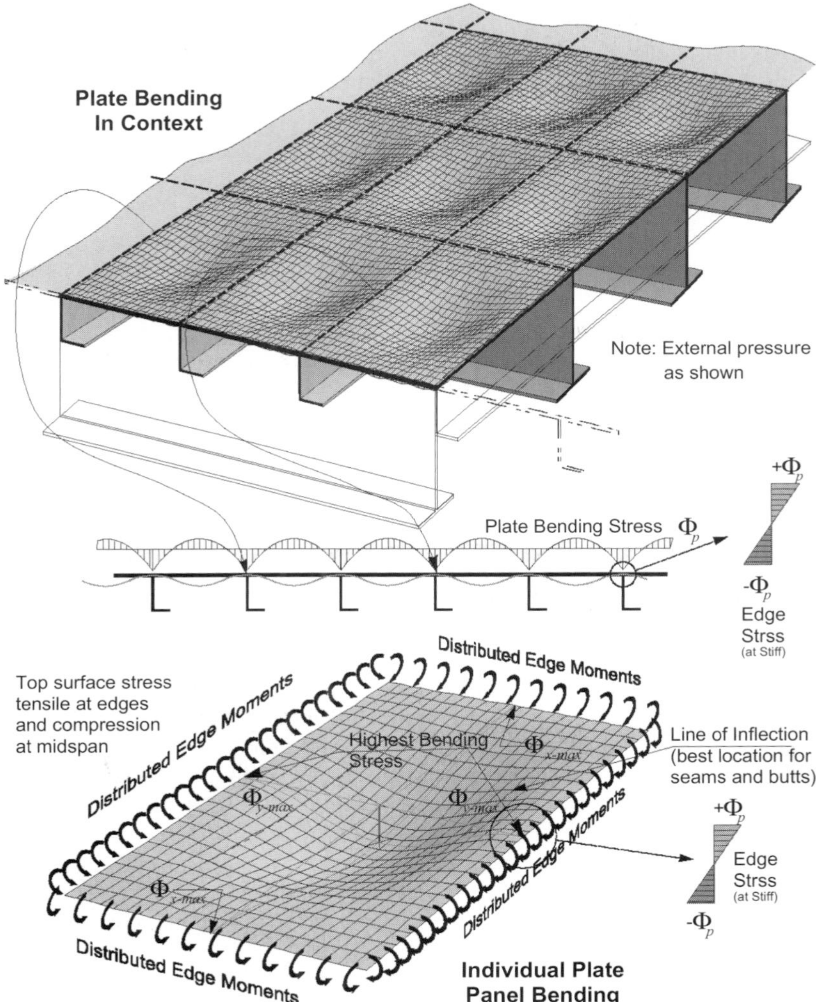

Figure 7.94 Bending of plating between stiffeners – uniform pressure loading

illustrated deflections may seem large, and exaggerated, plate deflections are indeed large relative to simple theory and large deflection theory must be used to actually capture behaviour. To complicate theoretical representation further, once welded, dishing will inevitably occur. Apart from the theoretical difficulties, such plating is very well understood and represented rather simply in the classification rules, albeit with empirical help. Both the ABS and the DnV rules for the thickness of plating under lateral load is discussed.

ABS Criteria

It is noted here that, for stiffener sizing and framing design as well as plating thickness, the ABS rules fall into basically two general cases: *Watertight Bulkheads and Watertight Flats* and *Tank Bulkheads and Tank Flats*. The rules are given in Section 2 (*Common Structures*), respectively in Sections 2.7 and 2.9. The latter is the most general case while the former is a special case specifically addressing watertight integrity after damage. There are also a variety of special cases given in Section 2.

It is also noted that in shell plating, the plate thickness as used are generally more than that required for local loading in order to achieve global strength. These are in-plane stress strength issues. Often buckling strength is an issue, however, and discussed elsewhere. Therefore, in the code, the criteria is expressed a "minimum thickness," and is the minimum for hydrostatic loading, provided that other factors do not require more thickness.

For the most general case (*Tanks*), scantlings for plating thickness is given in Paragraph 9.3 as follows (metric, English units):

$$t = \frac{sk\sqrt{qh}}{254} + 2.5\,\text{mm} \qquad t = \frac{sk\sqrt{qh}}{460} + 0.10\,\text{in.} \tag{7.57}$$

For the special cases, governed by watertight integrity, *Watertight Bulkheads and Watertight Flats*, scantlings for plating is given in Paragraph 7.3 as follows (As noted above this also applies to shell plating submerged due to flooding damage.). In both cases

$$t = \frac{sk\sqrt{qh}}{290} + 1.5\,\text{mm} \qquad t = \frac{sk\sqrt{qh}}{525} + 0.06\,\text{in.} \tag{7.58}$$

where t = thickness; mm or in., s = stiffener spacing: mm or in., h = specified design head to lower edge of plate; m or ft; q is a factor reflecting the ratio of the yield strength of the plate to the nominal yield of mild steel; metric: $235/F_y[\text{MN/mm}^2] = 24/F_y$ [kgf/mm^2]; English: $q = 34/F_y$ [ksi]. k however is a special factor considering additional strength of plates with short aspect ratios. $k = 1$ for $\alpha > 2$, but for $\alpha < 2$, the following is used:

$$k = \frac{3.075\sqrt{\alpha}}{(\alpha + 0.272)} \tag{7.59}$$

It should be noted that $\alpha > 3$ is more typical and setting $k = 1$ will suffice in most cases. α is the ratio of beam length to width.

Both the expressions are no more than the simple beam formula, albeit "adjusted to fit the facts" (service and research). The "+0.10" (and "+0.06") is often considered to be a corrosion allowance and, to some extent, it is. However, the origins of this and the divisor "460" (and "525") is largely a curve fit to a basic beam formulation, with the factor k addressing short plates. It therefore includes service experience, testing and other research, and some conservatism.

As a final note on plate thickness, to repeat, the rules for thickness given above are the minimums for local loading only. The plating still must also be fit for service in global strength, including buckling.

DnV Criteria

It would be useful here to briefly look at the DnV plating thickness rule. In Subsection 6, *Plating and Stiffening* (DnV MODU Rules Part 3, Ch. 1), the rule for plating thickness (B 100: *Plating*) thickness is given as:

$$t = \frac{15.8\ K_a K_r s\sqrt{p}}{\sqrt{\sigma_p k_p}} + t_k \text{ mm} \tag{7.60}$$

While seemingly more complex than the ABS rule, this expression is in fact in the same form. We will examine it term by term and address the metric units for a comparison. Looking first at the denominator, σ_p is the nominal, yield-based allowable stress (ABS: $f_a = F_y/FS$). k_p represents the long side (stiffener) boundary condition ($k_p = 1.0$ for clamped and $k_p = 0.5$ for simply supported edges). In as much as virtually all plates in a watertight panel are continuous, $k_p = 1.0$ can be assumed.

In the numerator, k_a is an aspect ratio factor similar to the k in the ABS formulation. In this case, $k_a = (1.1 - 0.25\,s/l)^2 \leq 1.0$. For a square plate, $k_a = 0.72$. In the ABS formulation, $k = 0.78$. However, for $l/s > 2$, which is nearly always the case, $k = 1$ in both cases. (The difference is most likely due to a different form of curve fit.).

Also in the numerator, k_r is a curvature factor. This is not used by ABS. Considering curvature of radius, r (m), perpendicular to stiffeners (only), $k_r = (1 - 0.5\,s/r)$. For flat plate, $k_r = 1$. Taking a 2 ft stiffener spacing example, for any radius above 15 ft (e.g. a 30 ft dia column), $k_r > 0.97$. Therefore, for our purposes $k_r = 1$ unless we are dealing with a fairly small radius. Where column radii are 20 ft or less, it is likely to be ring stiffened and the formulation is inapplicable. However, for radiused corners of rectangular sections it is very useful. For r = 5 ft and s = 2 ft k = 0.8. The origin of the factor is from bilge radii in ships. Taking k_p, k_a, and k_r, as unity, the DnV plate thickness equation becomes,

$$t = \frac{15.8\ s\sqrt{p}}{\sqrt{\sigma_p}} + t_k \quad \text{mm} \tag{7.61}$$

The expression is now more recognisable. s is the stiffener spacing in meters. Inside the radical, the pressure p, in kN/m^2 can be expressed as head, h in feet by the relation: $h = 0.3413\ p$. The equivalency fails in part here because the computation of p from DnV rules include a round-off factor ($p = 10\ h$, where h is in meters).

The DnV design heads/pressures are prescriptive and specific for jackup and semi-submersible units. They are necessarily quite different and direct comparison is problematic. These are given in Section 4, *Design Loads*, Subsection D, *Local Pressures and Forces*.

While *Sea Pressures* (Paragraph D 200) is totally prescriptive, *Tank Pressures* (Paragraph D 300) is similar to the ABS, yet different.

The factor, $\sigma_p^{1/2}$ in the denominator is an allowable stress. This is where DnV varies considerably from ABS. In Section 6, B100, the DnV give a variety of different values for allowable stress depending upon load condition, in-plane global stresses, application, member location, and material. For comparison, consider the basic allowable stress for gravity load. ABS mild steel (Grade A) is specified at 34 ksi yield and the corresponding gravity load allowable is 20.36 ksi. DnV mild steel is nominally 240 kN/mm^2 yield, which converts to 34.8 ksi and an equivalent allowable of 144 kN/mm^2 (20.8 ksi). These are used for comparison.

The second term, t_k, is an explicitly computed corrosion allowance, determined according to Subsection 9, B 500. *Corrosion Addition*, t_k, is specified as 1.5 mm (0.059 in.), if one side only is coated, and 2.0 mm (0.079 in.), if both sides of the plate are coated. This is slightly less than the 0.1 in. in the ABS formulation. Putting all these factors together, the DnV formulation (in US units) becomes:

$$t = \frac{s\sqrt{h}}{434} \quad + \quad 0.08 \quad \text{in.} \tag{7.62}$$

This is not altogether different from the ABS formulation in which the denominator is 460 (equivalent to 5.6% less thickness) and added 0.10 in. The difference is in part due to the DnV relationship between pressure and head ($p = 10\ h$; metric) and also because the basis steel is slightly different. However, this must also be taken in the light that DnV includes an application specific set of allowable stresses.

7.8.8 Stiffener Sizing

General

Stiffeners are employed to resist lateral loading of the plate and are usually made from the rolled shapes integrally welded to the plate. Such stiffeners are distinct from the other stiffeners used to prevent plate buckling. Typically, stiffeners run continuously through the supporting frames. Otherwise they are referred to as "intercostal" and require special connections at each end at each frame.

A summary of stiffener bending is given in fig. 7.95, the upper part showing context in the stiffened panel, between frames and uniform loading. The lower part of the figure shows the typical bending patterns and bending stresses in the stiffener/plate combination.

Figure 7.96 shows some of the typical stiffeners sections used. The US Unequal angles sections are the prevalently available rolled shape in the US, but do not offer a sufficiently wide selection, particularly in the larger shapes. The T-section, while a particularly efficient shape material wise, is problematic and expensive in relation to connections and frame penetrations and prevalent only in naval construction or where special builders tooling is used. International hull construction uses both the *profile section* (Europe primarily) and the *JIS ship angle*. Both are manufactured primarily for ship construction, particularly efficient, and have distinct advantages, especially large size. The latter are distinctive in that the tall leg is thin, with the short leg thick, making an excellent flange.

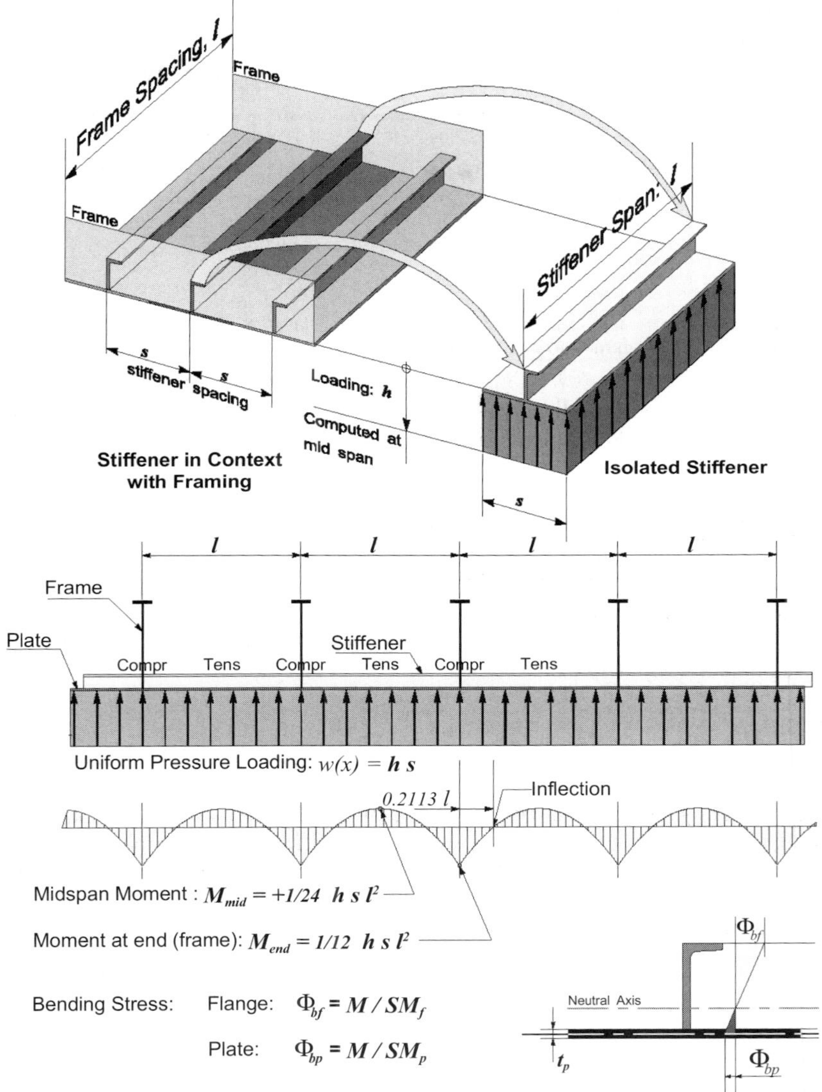

Figure 7.95 Summary of stiffener bending

Tables 7.12, 7.13 and 7.14 respectively give dimensions and section properties of the US unequal angles, profiles, and JIS ship angles. The latter two include combined stiffener/plate properties as attached to a 20 in. wide, 1/2 in. plate. Figure 7.97 illustrates the notational conventions for the US unequal angle. The others are similar.

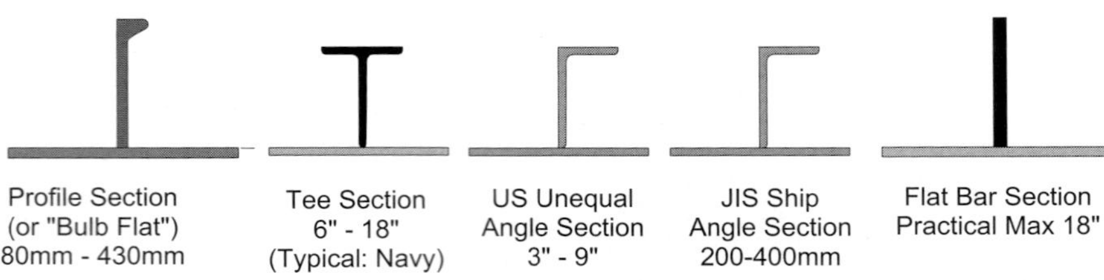

Figure 7.96 Alternative stiffener sections

Table 7.12 US Standard Unequal Leg Angles – with 1/2 in. plate at 20 in. width attached

Stiffener					Section Properties w/20″ × 1/2″ Plate				
Size	ds	As	ys	Is	1st mom	NA	Ixx	SMp	SMs
5″×3″×1/4″	5	1.94	1.66	5.10	4.0	0.33	26.04	31.2	5.6
5″×3″×5/16″	5	2.40	1.68	6.30	5.5	0.44	30.97	32.9	6.8
5″×3″×3/8″	5	2.86	1.70	7.40	6.9	0.54	35.43	34.1	7.9
6″×4″×5/16″	6	3.03	1.92	11.40	9.9	0.76	55.00	43.8	10.5
6″×4″×3/8″	6	3.61	1.94	13.50	12.2	0.89	62.77	45.1	12.3
6″×4″×7/16″	6	4.18	1.96	15.50	14.4	1.01	69.75	46.1	14.0
6″×4″×1/2″	6	4.75	1.99	17.20	16.5	1.12	75.64	46.6	15.5
7″×4″×3/8″	7	3.98	2.37	20.60	15.9	1.14	88.40	53.9	15.1
7″×4″×7/16″	7	4.62	2.39	23.70	18.8	1.29	98.34	55.1	17.2
7″×4″×1/2″	7	5.25	2.42	26.70	21.5	1.41	107.01	55.9	19.2
7″×4″×9/16″	7	5.87	2.44	29.60	24.3	1.53	115.18	56.8	21.1
7″×4″×5/8″	7	6.48	2.46	32.40	26.9	1.63	122.62	57.5	22.8
8″×4″×1/2″	8	5.75	2.86	38.50	27.1	1.72	144.56	65.2	23.0
8″×4″×9/16″	8	6.43	2.88	42.80	30.4	1.85	155.66	66.2	25.3
8″×4″×5/8″	8	7.11	2.91	46.90	33.7	1.97	165.40	67.0	27.4
8″×4″×3/4″	8	8.44	2.95	54.90	40.1	2.18	183.47	68.6	31.5
8″×4″×1″	8	11.00	3.05	69.60	52.0	2.47	211.24	71.0	38.2

Before proceeding, a few comments on shear-lag and the effective breadth of the plating should be made. As a flange attached to the stiffener, the bending stress in the plate is not laterally uniform (maximum at the stiffener) and can be considered as not fully effective. Figure 7.98 illustrates bending stress distribution in the plate, a schematic of an equivalent uniformly stressed plate (effective breadth, $2b_e$), and provides notational reference.

Table 7.13 Profile sections (HP) – with 1/2 in. plate at 20 in. width attached

Profile Size; millimeters				Area	Prof-NA	Prof-Inert	US Units				Combined Sect w/ 1/2 in. Plate @ 20 in. width Attached					
Depth	thick-ness	b + t	b	Wx, cm^2	ex, cm	lx, cm^4	ds, in.	As, in.	ys, in.	ls, $in.^4$	1st Morr	NA	lxx	SMp	SMs	Ax
160	7	29.0	22.0	14.6	9.66	373	6.3	2.26	2.50	8.96	6.1	0.50	39.28	39.4	6.8	12.3
160	8	30.0	22.0	16.2	9.49	411	6.3	2.51	2.56	9.87	6.9	0.55	41.77	39.8	7.3	12.5
160	9	31.0	22.0	17.8	9.36	448	6.3	2.76	2.61	10.76	7.7	0.60	44.25	40.2	7.8	12.8
180	8	33.0	25.0	18.9	10.90	609	7.1	2.93	2.80	14.63	10.1	0.78	61.36	48.0	9.7	12.9
180	9	34.0	25.0	20.7	10.70	663	7.1	3.21	2.87	15.93	11.0	0.83	64.30	48.2	10.3	13.2
180	10	35.0	25.0	22.5	10.60	717	7.1	3.49	2.91	17.23	12.1	0.89	67.82	48.7	11.0	13.5
180	11	36.0	25.0	24.3	10.50	770	7.1	3.77	2.95	18.50	13.1	0.95	71.08	49.0	11.6	13.8
200	9	37.0	28.0	23.5	12.10	941	7.9	3.64	3.11	22.61	14.9	1.09	89.73	56.5	13.2	13.6
200	10	38.0	28.0	25.6	11.90	1020	7.9	3.97	3.19	24.51	16.1	1.15	93.69	56.7	13.9	14.0
200	11	39.0	28.0	27.6	11.80	1090	7.9	4.28	3.23	26.19	17.4	1.22	98.00	57.1	14.7	14.3
200	12	40.0	28.0	29.6	11.70	1164	7.9	4.59	3.27	27.97	18.6	1.28	102.14	57.5	15.5	14.6
220	10	41.0	31.0	29.0	13.40	1400	8.7	4.50	3.39	33.64	21.2	1.46	128.32	65.4	17.8	14.5
220	11	42.0	31.0	31.2	13.20	1500	8.7	4.84	3.46	36.04	22.6	1.53	132.75	65.5	18.6	14.8
220	12	43.0	31.0	33.4	13.00	1590	8.7	5.18	3.54	38.20	24.0	1.58	136.50	65.6	19.3	15.2
240	10	44.0	34.0	32.4	14.70	1865	9.4	5.02	3.66	44.81	26.6	1.77	166.66	73.5	21.7	15.0
240	11	45.0	34.0	34.9	14.60	2000	9.4	5.41	3.70	48.05	28.6	1.86	174.35	74.0	23.0	15.4
240	12	46.0	34.0	37.3	14.40	2130	9.4	5.78	3.78	51.17	30.3	1.92	179.53	74.2	23.8	15.8

(Continued)

Table 7.13 Continued

Profile Size; millimeters				Area	Prof-NA	Prof-Inert	US Units				Combined Sect w/ 1/2 in. Plate @ 20 in. width Attached					
Depth	thick-ness	b + t	b	Wx, cm^2	ex, cm	lx, cm^4	ds, in.	As, in.	ys, in.	ls, $in.^4$	1st Morr	NA	lxx	SMp	SMs	Ax
260	11	48.0	37.0	38.7	16.00	2606	10.2	6.00	3.94	62.61	35.3	2.21	223.43	82.6	27.8	16.0
260	12	49.0	37.0	41.3	15.80	2770	10.2	6.40	4.02	66.55	37.3	2.28	229.96	82.9	28.9	16.4
260	13	50.0	37.0	43.9	15.60	2940	10.2	6.80	4.09	70.63	39.3	2.34	236.06	83.2	29.9	16.8
280	11	51.0	40.0	42.6	17.40	3334	11.0	6.60	4.17	80.10	42.7	2.57	280.60	91.3	33.2	16.6
280	12	52.0	40.0	45.5	17.20	3550	11.0	7.05	4.25	85.29	45.3	2.65	289.20	91.7	34.6	17.1
280	13	53.0	40.0	48.3	17.00	3760	11.0	7.49	4.33	90.33	47.6	2.72	296.71	92.1	35.7	17.5
300	11	54.0	43.0	45.7	18.90	4194	11.8	7.08	4.37	100.76	50.2	2.94	346.02	100.6	39.0	17.1
300	12	55.0	43.0	49.7	18.70	4461	11.8	7.70	4.45	107.18	54.2	3.06	359.32	100.9	41.1	17.7
300	13	56.0	43.0	52.8	18.50	4720	11.8	8.18	4.53	113.40	57.1	3.14	368.83	101.3	42.5	18.2
300	14	57.0	43.0	55.8	18.30	4980	11.8	8.65	4.61	119.64	59.8	3.21	377.38	101.8	43.9	18.6
320	12	58.0	46.0	54.2	20.10	5528	12.6	8.40	4.69	132.81	64.0	3.48	437.06	109.9	47.9	18.4
320	13	59.0	46.0	57.4	19.90	5852	12.6	8.90	4.76	140.59	67.2	3.56	448.33	110.5	49.6	18.9
320	14	60.0	46.0	60.7	19.70	6170	12.6	9.41	4.84	148.23	70.5	3.63	458.94	111.1	51.2	19.4
320	15	61.0	46.0	63.9	19.50	6480	12.6	9.90	4.92	155.68	73.5	3.69	468.38	111.7	52.6	19.9
340	12	61.0	49.0	58.8	21.50	6761	13.4	9.11	4.92	162.43	74.6	3.91	524.55	119.1	55.3	19.1
340	13	62.0	49.0	62.2	21.30	7156	13.4	9.64	5.00	171.92	78.3	3.99	538.00	119.8	57.3	19.6
340	14	63.0	49.0	65.5	21.10	7539	13.4	10.15	5.08	181.13	81.8	4.06	550.01	120.6	59.0	20.2
340	15	64.0	49.0	68.9	20.90	7919	13.4	10.68	5.16	190.25	85.4	4.13	561.48	121.3	60.7	20.7

Table 7.14 JIS (Japanese Industrial Standard) Ship Angles – (leg and flange unequal)

Section	Dimensions, mm		Thickness, mm		Area	Prof-NA	Prof-Inert	US Units				Combined Section w/ 1/2 in. Plate @ 20 in. width Attached					
millimeters	Depth	Flange	Long leg	Short leg	Wx, cm^2	ex, cm	lx, cm^4	ds, in.	As, in.	ys, in.	ls, $in.^4$	1st Morr	NA	lxx	SMp	SMs	Ax
Angles, Constant Thickness																	
200×90×8/14	200	90.0	8	14	27.80	13.98	1120	7.87	4.31	2.37	26.91	20.7	1.27	131.88	74.5	20.0	16.3
200×90×9/14	200	90.0	9	14	29.66	13.64	1210	7.87	4.60	2.50	29.07	21.7	1.31	134.06	74.2	20.4	16.6
250×90×9/14	250	90.0	9	14	34.31	16.54	2240	9.84	5.32	3.33	53.82	31.6	1.83	222.30	95.6	27.7	17.3
250×90×9/15	250	90.0	9	15	35.12	16.64	2270	9.84	5.44	3.29	54.54	32.7	1.87	227.76	96.0	28.6	17.4
250×90×10/15	250	90.0	10	15	37.47	16.39	2440	9.84	5.81	3.39	58.62	34.5	1.94	234.45	96.2	29.7	17.8
250×90×11/16	250	90.0	11	16	40.61	16.26	2640	9.84	6.29	3.44	63.43	37.3	2.04	246.10	96.9	31.5	18.3
250×90×12/16	250	90.0	12	16	42.95	16.01	2790	9.84	6.66	3.54	67.03	39.0	2.09	250.91	96.9	32.4	18.7
300×90×10/16	300	90.0	10	16	43.38	19.40	4100	11.81	6.72	4.17	98.50	48.4	2.58	366.62	118.9	39.7	18.7
300×90×11/16	300	90.0	11	16	46.22	19.00	4470	11.81	7.16	4.33	107.39	50.6	2.64	375.46	119.6	40.9	19.2
300×90×12/17	300	90.0	12	17	49.84	18.90	4590	11.81	7.73	4.37	110.28	54.5	2.76	388.27	119.0	42.9	19.7
300×90×13/17	300	90.0	13	17	52.67	18.70	4940	11.81	8.16	4.45	118.68	57.1	2.83	400.21	120.1	44.6	20.2
350×100×11/17	350	100.0	11	17	54.4	22.30	7030	13.8	8.43	5.00	168.90	71.0	3.48	572.71	144.0	55.6	20.4
350×100×12/17	350	100.0	12	17	57.7	22.00	7440	13.8	8.95	5.12	178.75	74.5	3.56	585.85	144.4	57.3	20.9
400×100×11.5/18	400	100.0	11.5	16	61.1	24.70	10300	15.7	9.47	6.02	247.46	89.1	4.15	774.02	166.5	66.7	21.5
400×100×12/18	400	100.0	12	18	64.8	24.90	10900	15.75	10.04	5.94	261.87	95.4	4.33	814.32	168.6	71.3	22.0
400×100×13/18	400	100.0	13	18	68.6	24.60	11500	15.7	10.63	6.06	276.29	100.0	4.42	832.71	169.3	73.5	22.6

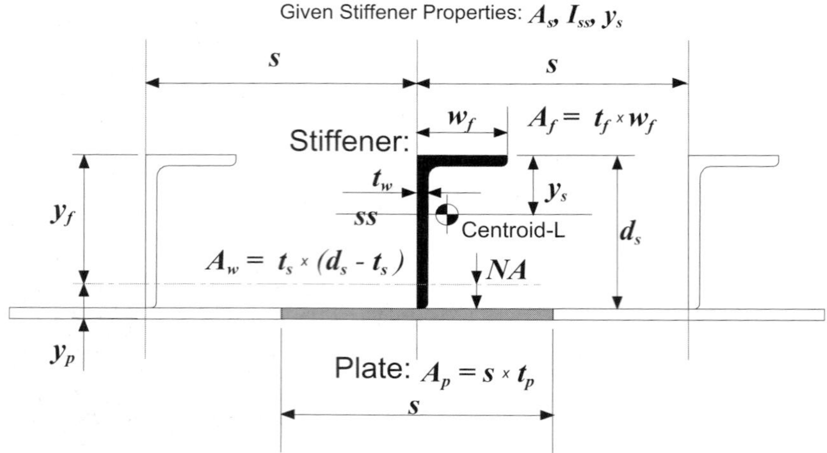

Figure 7.97 Notation for stiffener properties (US Unequal Leg Angles)

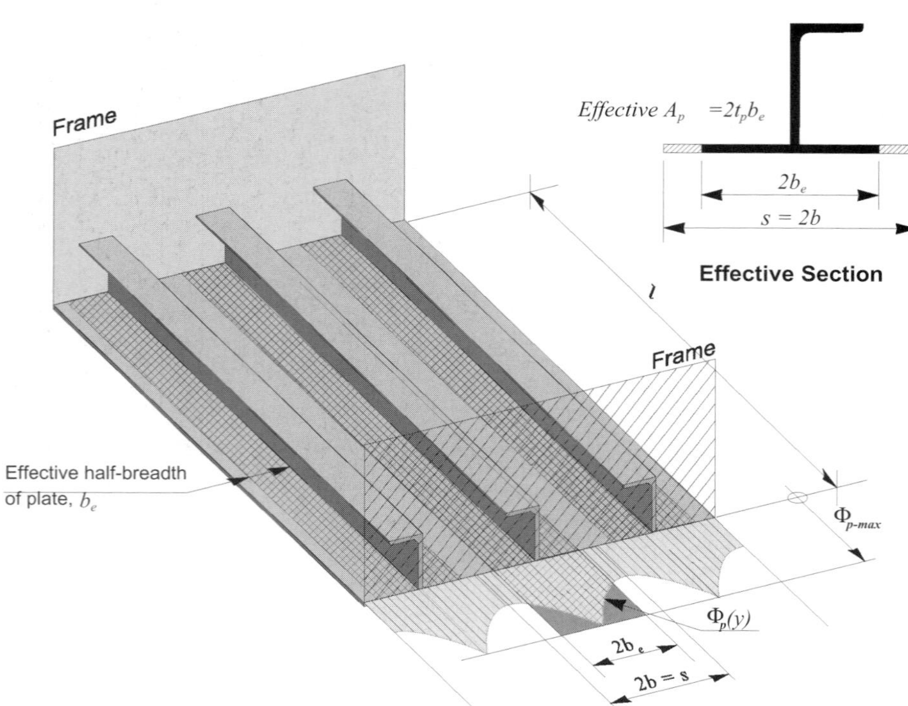

Figure 7.98 Effective breadth of plating with stiffeners

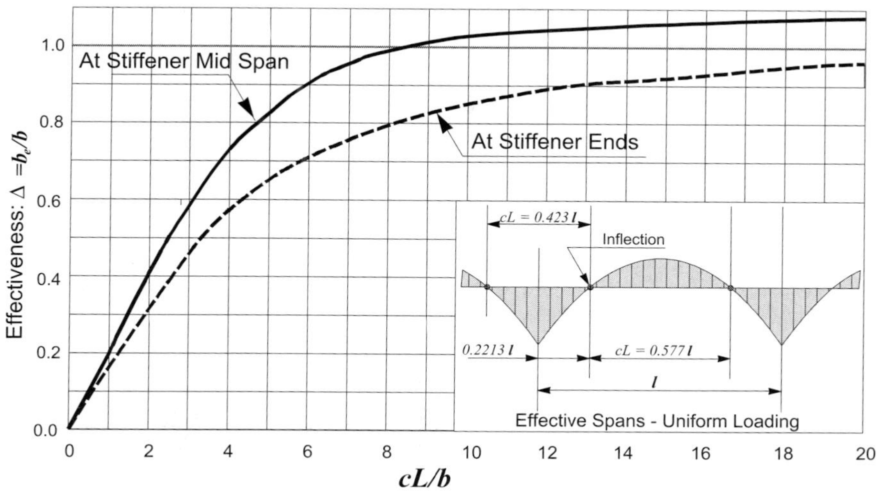

Figure 7.99 Effectiveness ratio, ρ, of attached plating for stiffeners (Mid-span and Ends)

Figure 7.99 is a plot of the effectiveness of the plate ($\rho = b_e/b$) as a function of the ratio of the effective span (cL) to the breadth of the panel ($b = s/2$). Pending a more complete discussion of shear-lag, $c = 0.577$ for mid-span locations and $c = 0.423$ at the stiffener ends. What is important is that in most instances l > 4 s and most of the plate is effective.

Finally, it should be noted that, even if the effective breadth is small, the neutral axis of the combined section is very close to the plate. This makes section modulus at the flange relatively insensitive to the accuracy of the effective plate area. The upper part of fig. 7.100 shows the neutral axis as a function of effective breadth for a typical case. The lower plot shows the resulting section modulus, SM_f as a function of effectiveness, ρ from 0.4 to 1.0. SM_f is no less than 80% of its maximum value. The point is that it is of relatively little consequence for stiffener design that shear lag is ignored and the properties of the combined section are calculated on the basis of the full width of the plating, s. For marginal cases, this should be re-examined, however.

For design, the classification rule-based stiffener sizing criteria are given as formulae for the section modulus of the stiffener and plate combination, SM_f without consideration of shear lag effects. The formulae are essentially based upon beam flexure as outlined above. The DnV formulation uses a specific allowable stress and the ABS formulation does not and is empirical. These are for minimum scantlings only. The shell plating stiffeners tend to be larger in areas of high global stress both to increase section (in lieu of thicker plating) and to improve stiffener buckling resistance.

ABS Criteria

As noted with plating, the ABS rules give two general cases. For the most general case (*Tanks*), stiffener scantlings are given in Paragraph 9.5 (*Stiffeners and Beams*). For the

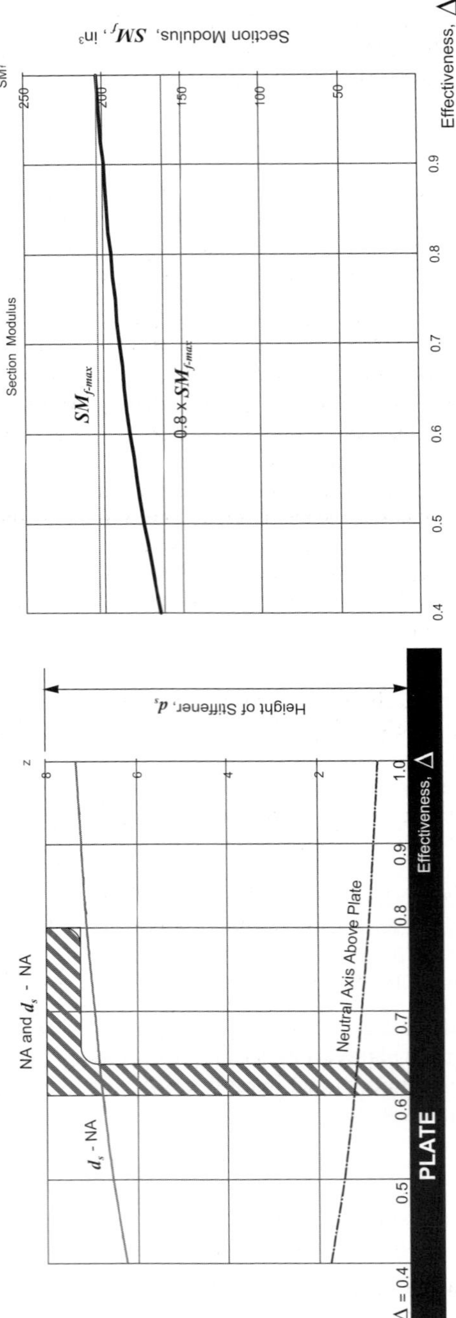

Figure 7.100 Stiffener neutral axes and section modulus variation with effective breadth of plate

special cases governed by watertight integrity, in Paragraph 7.5 (*Stiffeners and Beams*), the same expression is given, but the coefficient values are different. The required, minimum section modulus is given as follows (metric, US units):

$$SM = f\,c\,h\,s\,l^2\,Q; \qquad (\text{cm}^3;\ \text{in.}^3) \qquad (7.63)$$

where [metric (US)] $f = 7.8$ (0.0041), s = stiffener spacing (m, ft), l = effective stiffener span (m, ft), h = the specified design head (discussed earlier) taken at stiffener mid span (m, ft). The constant c is specifically discussed below. The factor Q is material constant adapted from the Ship Rules and is not given in the MODU Rules. It is also discussed below.

The section modulus, SM, taken to the stiffener flange, is computed on the basis of an *integral section*, taking both the stiffener and the full width (s) of the attached plating. However, the rules note that, for short panels ($l/s < 4$), a lesser *effective breadth* of the plate, s_e should be determined according to shear lag theory (at $l/s < 4$, $se/s < 0.80$). Because the stiffener-plate combination typically produces a quite unbalanced section, accurate effective plate area has minimal effect on the section modulus. Typically the stiffener flange is 3-times or more distant from the neutral axis as is the plate, making bending stress in the plate of little consequence. Even at $l/s = 3$, the effective breadth is still 0.65 and using the full plate width would still be of little serious consequence. In any case, the combination of global in-plane stresses from that of local bending is still required to be checked when the former is high. Often, in such cases, there is sufficient global compression that buckling will govern. When the buckling controls, the plate effectiveness is also reduced.

The factor c represents stiffener end fixity and the stress pattern of the specific application. Although the *Ship Rules* have a number of specific cases represented, the *MODU Rules* give primarily two sets of factors:

- Tank boundaries and shell (9.5) : $c = 1.00$, both ends fixed or continuous
  $c = 0.90$, terminated with clip or bracket
- Waterlight subdivision (7.5) : $c = 0.60$, no end attachments
  $c = 0.56$, ends attached to clips or brackets

In the most general situation, shell plating, and tank boundaries (3-2-2/9.5), stiffeners run continuously through the frames and are considered to function as a series of fixed-fixed beams ($c = 1.0$). However, where this pattern ends (at a deck, flat or bulkhead), the stiffener is terminated by a bracket or clip and is considered to have less bending moment ($c = 0.9$). In the case of internal, watertight subdivision (3-2-2/7.5), the factors reflect that strength being a damage issue, i.e. internal flooding.

Further to the above, non-watertight flats and decks are addressed in Subsection 3, Paragraph 7.3 (*Upper Structure*), where $c = 0.6$, clear of tanks, and $c = 1.0$ in way of tanks. The latter is simply a reiteration of tank boundary requirements. Otherwise, $c = 0.6$ is taken in association with a non-tank loading, h specific to the flat or deck.

The rules also give a number of comments regarding the determination of effective span, l. Unless effective brackets or clips are used, the effective span is the centre-to-centre distance between supports. However, the effective span may be reduced according to the effectiveness of the brackets meeting criteria specified in the Rules. With brackets meeting requirements of table 3/5.2 (ABS MODU Rules, 2001), the effective span may be reduced by 75% of the bracket length.

Of the various specific application in the *Ship Rules* that uniquely give a value for c, the one for bottom plating of a longitudinally stiffened ship gives $c = 1.30$. Where otherwise 1.0, this factor empirically recognises that bottom stiffeners are part of the overall hull section modulus and will sustain considerable global axial stress over and above that of local bending. A non-documented, but very useful practice, for preliminary design for offshore structures is to use a value of c from 1 to 1.3 for shell stiffeners of primary hull elements where they are known to have a high global strength role. This is the case for parts of pontoons and columns. These have a global loading pattern quite similar to that found in ship hulls: hydrostatic load of stiffeners, participating in longitudinal strength. Using a higher value of c will anticipate the added stiffener section area needed to sustain global hull bending stress and buckling and improve the prospects that sizes chosen in preliminary design will indeed be adequate.

The *Ship Rules* employ a multiplier "Q" to reflect the benefits of high tensile steels. This can be used for offshore structures as well but has not been widely incorporated in to the *MODU Rules*. Taken as unity for mild steel, various values are specified for high tensile steels. Taking HT36 grade steels (Fy = 51 ksi), for example, $Q = 0.72$. High tensile steels are prevalently used for the more highly loaded portions of the hull.

As a rule of thumb, for hull plating, where mild steel thickness would exceed 5/8 in., high tensile steel should be used. If buckling controls, the threshold might better be 3/4 in. In any case, for thickness in excess of 3/4 in., high tensile steels should be used. The minor additional cost of the higher grade steel is more than offset by reduced welding of thinner sections, not to mention larger, fewer modules for a given lift limit. Conversely, high tensile steels should generally not be used for thickness less than 1/2 in. as much as buckling will control and little benefit of the higher yield strength will be achieved.

With regard to the *ABS Rules* and considerations of corrosion protection, Paragraph 3-2-2/1.4, *Scantlings and Corrosion Control* indicates that rule based scantlings *include* a corrosion allowance in as much as it addresses reduction of scantlings. It indicates that scantlings may be reduced should suitable corrosion protection be provided. Conversely, this paragraph also indicates, that, should corrosion protection not be provided, stress-based calculation should consider the *net section* with a suitable reduction. This would apply only to the stiffener's role in global strength.

DnV Criteria

As per the plating thickness criteria, the DnV stiffener design criteria is given in the DnV *MODU Rules* Part 3, Chapter 1, Subsection 6, *Plating and Stiffening*. This criteria is much more specific that the one given by the ABS. In Paragraphs B 200 (*Stiffeners*), in the DnV notation, the following section modulus rule is given:

$$Z = \frac{1000\, l^2 sp}{m\sigma_p k_s} + Z_k \quad (\mathrm{cm}^3); \quad \text{minimum} : 15\ \mathrm{cm}^3 \tag{7.64}$$

Like the plate thickness rule given previously, the DnV stiffener rule is essentially the same form as that given by ABS, again with some differences. The terms, l and s have the same meaning, if not units, and p is the pressure in kN/m^2. The m in the denominator is simply the denominator of the applicable beam equation. DnV provides table B2 in Section 7, *Girders and Girder Systems* which is simply a complete set of bending coefficients, m taken in the form: $M = hsl^2/m$. For the fixed-fixed case, $m = 12$ for the beam end and 24 for mid span, the same as used above under the discussion of *Loading and Stress*. Again, as was done for plates, σ_p is the allowable stress. $k_s = 1$ unless both ends of the stiffeners are pinned end boundary conditions (in which case it is 0.9).

The additional term, Z_k is added modulus allowing for corrosion. However, if an effective corrosion protection system is provided, Z_k may be taken as zero. The provisions for Z_k are in Paragraph B206. Being somewhat in detail, they have not been presented here. Generally speaking, 2 mm is added to all surface of the stiffener.

Simplifying, as was done with plating, the following can be written for the stiffener ends:

$$Z = \frac{83.33\, l^2 sp}{\sigma_p} \qquad (\text{cm}^3) \qquad (7.65)$$

Continuing to simplify, and converting from the metric, p (kN/m^2) is expressed as head, h in feet by the relation: $h = 0.3413p$. To take the l^2s term into ft^2-in. units, divide by $(3.2808^3 \times 12)$. Then, using the basic mild steel allowable stress for gravity load (20.36 ksi) the following results:

$$Z = 0.0033\, hsl^2 \qquad (\text{in.}^3) \qquad (7.66)$$

This is essentially the same form as ABS, except that the ABS coefficient is 0.0041. While the DnV coefficient is 20% less, it does not have the corrosion allowance that is implicit in the ABS rule.

What is more notably different in the DnV Rules is the use of allowable stresses in the formulation. For stiffeners, these are as follows (N/mm^2):

	Loading Condition	
	Gravity Only	Combined Loading
Non-Primary Structure:	$145\ f_1$	$190\ f_1$
Primary Structure:	$145\ f_1 - \sigma_{af}$	$190\ f_1 - (\sigma_{af} + \sigma_{ac})$
Watertight Subdivision:	$240\ f_1 - (\sigma_{af} + \sigma_{ae})$	

The "non-primary structure" category actually expresses local loading criteria alone, without global stress adjustments. f_1 is the material factor, which, for mild steel, is 1. As with allowable stress for plating, the global stress is deducted from the local allowable for combined stress criteria. The definition of these deductions is as follows:

σ_{af} : Global axial stress from gravity loading only;

σ_{ae} : Global axial stress from environmental loading only;

For preliminary design, it is suggested that $\sigma_{af} = 50\ f_1$ and $\sigma_{ae} = 80\ f_1$.

To make the comparison more direct, the above is converted to ksi for mild steel ($F_y = 34$ ksi) as follows:

	Loading Condition	
	Gravity Only	Combined Loading
Non-Primary Structure:	21.0	27.6
Primary Structure:	$21.0 - \sigma_{af}$	$27.6 - (\sigma_{af} + \sigma_{ac})$
Watertight Subdivision:	$F_y - (\sigma_{af} + \sigma_{ac})$	

The mild steel equivalent values, for preliminary design, global stress are $\sigma_{af} = 7.2$ ksi and $\sigma_{ae} = 8.2$ ksi.

One might say, rather than equivalent, the DnV stiffener rule is consistent with the ABS rule. Once the corrosion addition is factored in, it is found to be reasonably close. While the DnV rules are much more detailed with the allowable stress formulation, the same is accomplished in the ABS rules with the general requirement for combined stresses not exceeding the allowable stress. In the end they are the same except for the corrosion allowance.

Another important factor in the stiffener design appears in the DnV Classification Notes 31.4 (*Column Stabilised Units*), Paragraph 6.4.2: *Columns; Shell Plating and Stiffeners*. Therein is a discussion of curved plating that recognises that, a longitudinally stiffened shell, with sufficient curvature, will share the stiffener load in the form of circumferential stress in the shell. There is given a curvature reduction factor as follows (DnV notation; consistent units):

$$\frac{P_{stiffener}}{P_{shell}} = \frac{cr^2 I}{L^4 ts} \tag{7.67}$$

where r = Radius of shell mid surface, I = Inertia of the stiffener and plate combination, L = Distance between frames (same as l), t = Shell thickness (same as t_p), and s = Stiffener Spacing. The quantity c is a factor reflecting the boundary conditions of the stiffener as a beam. For fixed ends, $c = 384$; for pinned ends $c = 384/5$.

For large column diameters (> 60 ft), this factor does not provide much benefit (< 5% stiffener load born by the shell). However, for smaller columns, reduction of the stiffener load by recognition of the shells participation is worth considering. For columns and pontoons with radiused corners, along with plating reductions previously discussed, one stiffener can be eliminated at each corner.

Shear Strength

While the shear strength of rolled shape stiffeners are rarely a problem, and are not explicitly addressed in the rules, there can be difficulty at their connections, particularly to frames. This is particularly the case where the stiffener are highly loaded and high strength steels are used for plating and stiffeners and the frames are of mild steel. Normally there is a double-fillet weld connection between the stiffener and the frame web. This may at first seem to be a welding problem, but is actually a problem in the frame web size in that the depth of the stiffener times the thickness of the frame web cannot provide the shear strength required. This is a stiffener problem in that it may be the basis choosing deeper or ever over size stiffeners. Thicker frame webs are not always an attractive solution and over size stiffeners can benefit global strength. Also, there are standard details to address this problem (clips and *collars*), but these are costly.

7.8.9 Framing

General

Typically, a *frame* is a system of connected beams that support stiffeners. Usually a frame is an interacting, closed unit. The functions of framing are purely that of load distribution, both in the support of stiffening and also in the transfer of local load to the connecting shell structure. Generally, frames are not participants in global strength.

The upper left of fig. 7.101 shows three-dimensionally, in-context, a frame and its various components. The individual elements of a frame are usually associated with adjoining

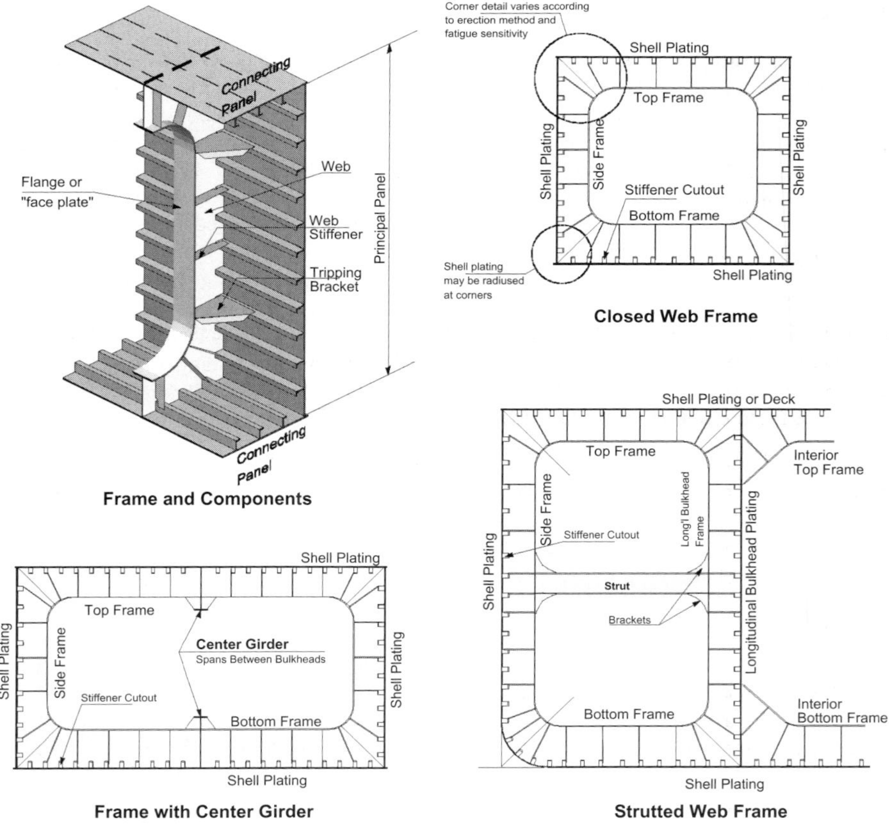

Figure 7.101 Typical frame configurations

panels. While frequently referred to as "beams" or "girders," relating more to form than function, here they will be referred to as *frame elements*. When built-up, particularly with a deep web, they are referred to as *web frames*. The flange is often referred to as a *face plate*. The single-celled, closed web frame shown at the upper right is typical of columns and pontoon section of semi-submersibles and TLPs. They could be circular, partly curved, or rectangular as shown in fig. 7.101.

The lower part of fig. 7.101 shows other typical frame arrangements. In the lower left the arrangement is such that the top and bottom frame span is excessive and is therefore supported at mid-span by an orthogonal girder. Sometimes mid-span support is achieved with a longitudinal bulkhead (tight or non-tight), itself with a frame element. The term *girder* usually refers to an element perpendicular to a frame in a more complex framing system. In the lower right is a similar long span situation (but vertical) where the span is broken by a *strut*. This is a prevalent form from tankers found in FPSOs. In addition, the term *stringer* is used, usually referring to an isolated frame-like element, with similar elements of adjoining panels.

Framing behaviour is primarily that of flexure, but in offshore structures, axial loading can be quite significant. Also, because frames and girders are often short and deep, shears, shear stress, and shear deflection are important. Buckling is another important issue for framing and girder design. Shown previously in the upper parts of fig. 7.101, *Web stiffeners* are used to control web buckling and *tripping brackets*, to provide stability against lateral buckling of the flange.

Although scantling rules are given for framing systems, these are largely based on the *Ship Rules* and, in the end, framing design requires a stress-based, analytic approach. The principal reason for this is that the given rules are largely flexure-based, whereas, in offshore structures, the connectivity of interacting framing members (moment redistribution) and the transfer of high axial load from adjacent, perpendicular panels are just as important. Figure 7.102 shows some of the load patterns typical for a pontoon. In addition to the high external head for a deeply submerged, empty tank, there can alternatively be significant internal heads in a tank when not submerged. Even more complex is with one tank filled and the other empty. Nevertheless, the scantling rules are useful for initial sizing. The suggested design approach is to use the scantling rules for initial sizing, but to subsequently use allowable stress based, frame analysis for the final design.

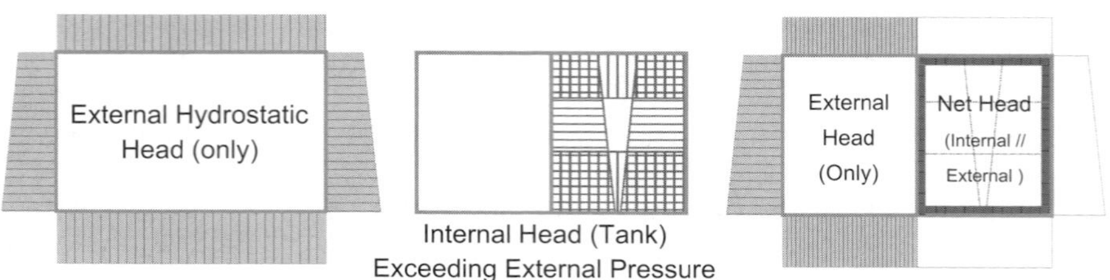

Figure 7.102 Common frame loading patterns in offshore hull structures

Although the scantling rules for framing are about as straightforward as the stiffener rules, they do not consider the axial load and moment redistribution. Frames, usually being two-dimensional, simple two-dimensional analysis can be used for refinement with respect to the latter. However, for more complex frames, a three-dimensional modeling is required, particularly if there are intersecting girders or the frames are not straight. In offshore structures, most frames are simple and orthogonal to the global section. In these cases stress is entirely from local loading because the global stress is orthogonal to the frame stress and virtually does not interact. There is interaction in the sense that the state of stress in the shell plating includes global which, in combination, may establish the limit state for plate buckling. Consequently the framing design can proceed reasonably well early in the design process without addressing other aspects of the structure design. This is not true of girders oriented in the direction of global stress, however.

While this discussion focuses almost entirely on the strength aspects, some comments are in order regarding the practical aspects of frame design. Although, as a component, the frames themselves are a small contribution to the total hull steel, they are nevertheless quite labour intensive. Much attention should be given to simplicity of the framing system with regard to spacing so that joining pieces easily match. Assembly is a major cost factor. In addition, repetition in design is very beneficial to the reduction of fabrication labour. Because of the importance of efficiency in the assembly process, frame design is driven by fabrication related choices more than material efficiency.

The Design Process

Various design processes can be pursued to design framing. A three stage process is recommended as outlined below. The objective is to, in the *1st Stage*, quickly establish the essential geometric character of the overall hull structural design, this being the principal dimensions and formulation of the framing system. The objective is also to determine the initial, basic minimum scantlings. The methods of the *1st Stage* are simple and allow an interactive approach where multiple priorities may be harmonised to achieve a balanced structure design.

The objective of the *2nd Stage* is to address complexities, refinement, and validation of the primary framing decisions of the *1st Stage*. It is the sort of activity, when complete, supports the development of scantling drawings and initiation of detailed FEM modeling.

The 3rd Stage is the FEM modeling. Its objective is to verify the design and to identify deficiencies that may need correction. This stage depends upon how evident adequacy may be demonstrated in the *2nd Stage* and may not be necessary.

1st Stage

1. Determine the controlling hydrostatic loads on a panel-by-panel basis.
2. Consider the implications of internal subdivision and identify viable framing strategies.
3. Determine for each frame and girder, one or more loading diagrams for the frame for potentially controlling loads.
4. Determine preliminary configuration of the frames with particular attention to connections, repeated frame element designs, and to a practical range of element depths.
5. Dismember the frame and loads to examine loading of individual members; estimate shears and axial loads.

6. Establish the range of section depth needed for shear strength and total area required for axial load.
7. Determine the section modulus requirement for individual frame elements according to the rules.
8. Size section considering the minimum modulus, shear area, and axial requirements.
9. Complete simple, stress-based verification.
10. Adjust scantlings and revise connection geometry accordingly.

2nd Stage: (The 1st stage has established the essential geometry)

1. Develop a plane frame analysis model (beam elements); include shear deflections.
2. Develop section properties considering shear lag for effective plating.
3. Adapt loading system to model.
4. Prepare moment and shear diagrams for frame;
5. Shears will be accurate; make final determination of web depth and thickness according to allowable stress criteria, including buckling checks.
6. Mid-span moments will be accurate; finalise flange; verify adequacy of web-depth for flexure.
7. Refine corner connections to adequately transfer loads between the connected frame elements.

3rd Stage: An FEM analysis of selected frames. The 2nd stage design may require verification. The 3rd stage is only for critical frames or prototype of highly repeated frames.

1. Replicate scantlings and geometry from mid frame to mid frame;
   - Include web stiffeners;
   - Higher detail at end connections;
   - Presentation quality model;
2. Include shell plating; stiffeners can be ignored or simplified.
3. BCs: should be appropriate and realistic.
4. Loading: can ignore self-weight and global stress.

ABS Criteria

The minimum requirements for *Girders and Webs* (includes frames) are given in the ABS MODU Rules, Paragraph 3/2.7.9 for *Watertight Bulkheads and Watertight Flats* and Paragraph 3/2.9.9 for *Tank Bulkheads and Tank Flats*. The latter also applies for shell plating by reference from Subsection 4/5.3 (*Scantlings of Framed Shells*). Strength requirements for girders, individual frames, and the elements forming web frames are given in terms as a minimum required section modulus, *SM* as follows:

$$SM = fchsl^2Q; \quad in.^3 \quad f = 0.0025 \qquad (\mathrm{cm}^3;\ \mathrm{in.}^3) \tag{7.68}$$

where [metric (English)] $f = 4.74$ (0.0025), s = frame spacing or mean breadth of space supported (m, ft), l = effective span (m, ft); (see discussion below), h = the specified design head (discussed earlier) taken at stiffener mid span (m, ft), and Q = material constant as discussed previously.

For shell plating and tank boundaries, the factor $c = 1.5$. For watertight subdivision only, $c = 1.0$. As with plating and stiffeners, for a combined loading criteria including the dynamic pressure, 0.75 h_d can be used for h.

The effective span, depends upon the end connections. Without brackets, l is taken as the flange-to-flange span (or *clear span*), but there must be effective moment resisting end connections. However, if the frames employ substantial brackets that effectively reduce the span, the rules provide a span reduction. As with stiffening, l may be reduced by 75% of the bracket length provided that the bracket is at least 45°. In Section 2, table 2, the rules specify bracket sizes for medium to small frames (plain: without flanges and flanged brackets). In the *ABS Ship Rules*, however, effective span is considered in much more specificity and detail with attention to actual detailing practice, particularly regarding radiused brackets. These guidelines are applicable to similar framing forms in offshore structures.

Figure 7.103 illustrates the effective spans of frames and girders with respect to effectiveness of brackets and different bracket forms. At the top is the *radiused corner* connection, a type of *fully continuous face plate* end connection. The terms "face plate" and "flange" refer to the same thing and are interchangeable. Flange is more commonly used in the flexural function sense. Face plate is more related to physical form and fabrication and more often applies to brackets. The span of the girder is taken where the horizontal, at 1.25× the depth of the frames/girder web (d_w), intersects the *web* on the radius. The radiuses should be checked.

In the middle of the figure are two variations of straight brackets where the face plates of the bracket functionally continue the flange. This is the *fully continuous, bracketed face plate*. Provided that the slope of the brackets is 45° or steeper, the span may be taken as the clear span plus 1/4 of the bracket lengths. If less steep, it is treated and a non-prismatic section and the face plate must be terminated by and effective moment resisting connection. For the bracket shown at the left, the girder flange runs level and through and is cut short at the end (*sniped*). If sufficiently steep, the face plate may also be sniped at the end. The bracket face plate must be at least 50% of the area of the girder flange and connected to the flange and chocked with full penetration welds. The alternative is to simply continue the girder flange, through a small radius, as a face plate to the bracket, as shown at the right. The radius must be chocked.

The bottom of the figure shows a less robust detail, *brackets with non-continuous face plates*. These are more common in ships, but are suitable for offshore structures where there is minimal cyclic loading. In this arrangement, the frame/girder and flange are straight and continuous, bulkhead to bulkhead, with the flanges sniped at the ends. The brackets are fitted *atop* the flange and *do not have connected face plates*, these being sniped at both ends. The face plate function here is mainly buckling control. The inclination must be at least 45°.

Additionally, the *Rules* include various requirements for proportions, tripping brackets, and detailing (Paragraphs 2/9.9.2 and 2/9.9.3 for shell plating and tank boundaries and

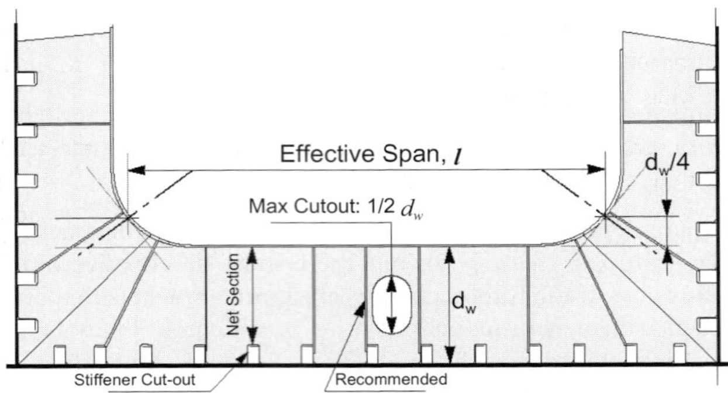

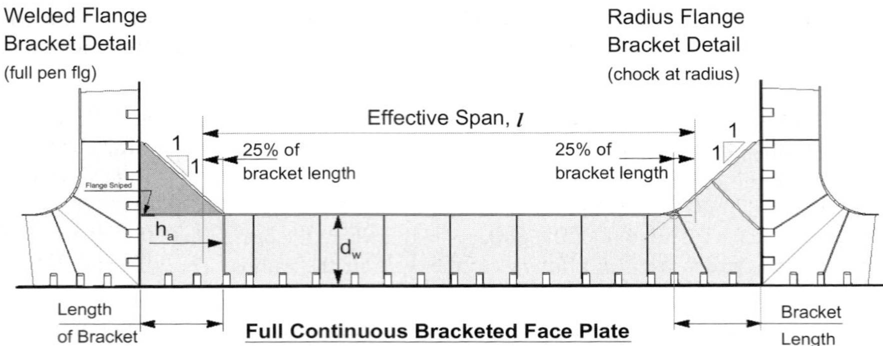

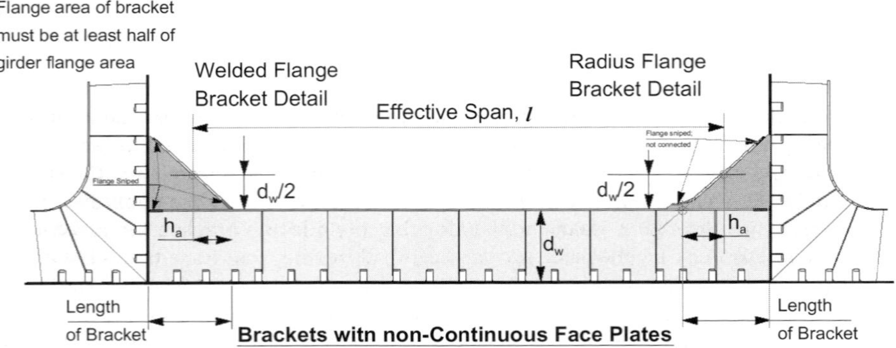

Figure 7.103 Effective spans for frames and girders

Paragraphs 2/7.7.2 and 2/7.7.3 for watertight subdivision). For depth of web, $d_w < l/8$ (watertight subdivision: $d_w > l/12$). For web thickness, $t_w > 0.001\ d_w + 0.12$ in., but not more than 0.44 in., unless required for shear strength or buckling. In addition, if there are to be cutouts (e.g. openings for piping or accesses), d_w must be twice the depth of the cutout. This is not the same as cutouts for through stiffening. These may not be more than 20% of d_w. Noting that cutouts exceed the stiffener depth, d_s, by 1 in., using $d_w > 3 \times d_s$ is also a generally used design practice.

Tripping brackets are perpendicular, triangular attachments between the frame/girder web and the stiffeners. They provide lateral stability to the frame or girder. They are required to have intervals of about 10 ft, but do need to match the spacing of stiffeners. An attachment to the frame/girder flange is required, if more than 8 in. of the flange are unsupported.

DnV Criteria

Unlike the previous discussions, comparing the DnV Rules to the ABS Rules, this one will be brief in as much as the ultimate disposition is largely the same: A stress analysis is pursued and essentially the same allowable stress criteria are used. However, the DnV Rules do have some very useful ancillary criteria related to detailing which are different and more extensive than those given by the ABS.

The DnV Criteria for frames and girders is given in the *MOU Rules*, Section 7, *Girders and Girders Systems*, Subsection B, *Bending and Shear*. The rules make it clear that they apply to *simple, regular framing system* only and, in Paragraph B 102, state: "*When boundary conditions for individual girders are not predictable due to dependence on adjacent structures, direct calculations according to procedures given in Subsection D will be required.*" Quite apparently, DnV requires an allowable stress approach such as outline in the *2nd Stage* of the design process discussed above.

In Paragraph B 400, the following section modulus rule is given (the DnV notation):

$$Z = \frac{1000\, S^2 bp}{m\sigma_p} + Z_k \qquad (\text{cm}^3); \quad \text{minimum}: \ 15\ \text{cm}^3 \tag{7.69}$$

Like the stiffener modulus rule given previously, this rule is essentially the same form as that given by the ABS, but again there are some differences. The terms, S and b are, respectively, the effective span and frame (or girder) spacing, in metres. p is the design pressure in kN/m^2. The m, in the denominator, is the denominator of the applicable beam equation given in table B2 of Section 7. This table provides a complete set of bending coefficients, m taken in the form: $M = hsl^2/m$ several addressing load pattern and boundary conditions. This approach is especially useful for frames and girders because variation in load pattern and boundary conditions are much more likely. Noting that the boundary conditions include *pinned connections*, this accommodates assumed inflection points. The insightful placement of inflection points is an effective and realistic approach to break down complex frames.

Again, as is the case for stiffeners and plates, σ_p is the allowable stress and Z_k is the added modulus allowing for corrosion. As with other structure, provided an effective corrosion protection system is provided, Z_k may be taken as zero. Otherwise, a thickness of $t_k = 2\,\text{mm} =$ is added to the web and flange (or deducted from actual thicknesses for *net section*).

Noting that frames and girders are *not* primary structure, the mild steel allowable stresses (N/mm^2) are given as follows:

	N/mm^2	ksi
Gravity Only	125	17.7
Combined Loading	170	24.1

As stresses alone, without the other considerations, the gravity load case is exactly the same as determined above for the equivalent stress in the ABS rule. Is noted also that the allowable Combined Loading is 36% more than Gravity Loading only case. The combined load allowable stress is 36% more than the gravity load allowable, which is slightly more that the nominal 1/3 increase allowed for environmental loading. It is consistent, however.

Without going into the detail as done for plating and stiffening, simplifying and converting to ABS units and format, using $\Phi_p = 17.7$ ksi, this expression, for a fixed–fixed span of l becomes (see stiffener discussion),

$$Z = 0.0038\, hsl^2 \qquad (in.^3) \tag{7.70}$$

If one takes $c = 1.5$ in the ABS rule, the ABS multiplier becomes $1.5 \times 0.0025 = 0.00375$. This is the same. The ABS rule-based scantlings do include a corrosion allowance and this would make the DnV requirement slightly more demanding. It is still reasonably close and certainly consistent. This may be mute in as much as, for all but the simplest frames, the stress-based design would be pursued, in which case, the rules are quite the same.

Shear Lag

As with stiffening, the required section modulus, *SM* is to be taken to the flange and to is include the *effective* attached plating. Unlike stiffeners, the effective breadth of plating (as flange to the frame section) is often much less than the full width. This must be considered in as much as the frame spacing (s) and the spans between inflections (cL) can produce low values of effective breadth, ρ. Also, because frames and girders produce a more balanced section, accurate representation of the plate is more important than for stiffeners.

The previous discussion of shear lag and effective breadth of plating with stiffeners, particularly fig. 7.99, is relevant here. Corresponding to fig. 7.98 for stiffeners, fig. 7.104 is given for frame elements, showing stress patterns in the plate. However, only the solid, upper curve of fig. 7.99 (mid-*span*) is truly relevant. As with stiffeners, application of this curve relies on determination (or estimation) of the inflection points. Unlike stiffener ends, the ends of frames and girders are complicated by end connection (brackets, radiuses, etc.) and interaction with connecting structure. Some simple frames will fit this model, however. Figure 7.105 is given showing a typical frame/girder section and notation.

Buckling

Where the stiffeners are orthogonal, typical of girder, and the loading is external (plate side compression), the buckling strength of the plating is reduced and post buckling considerations may be required. This reduction is similar in effect to shear lag, but not in nature. In both cases, however, effectiveness of the plate, as a flange, is reduced.

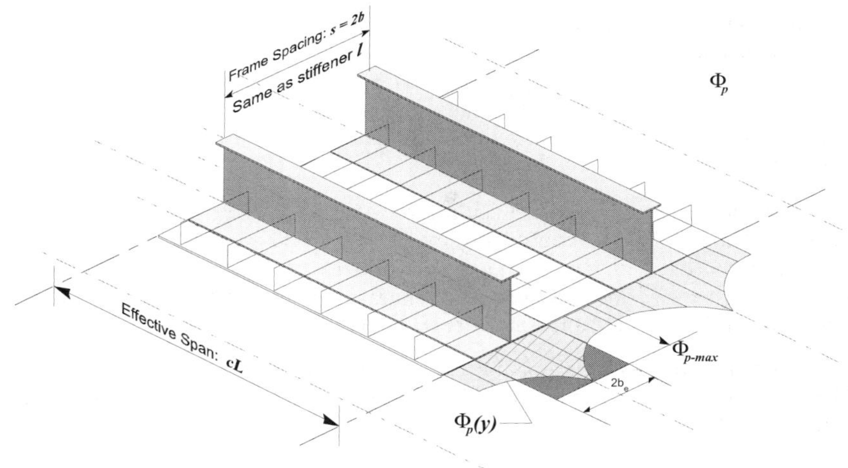

Figure 7.104 Effective breadth of plate for frames and girders

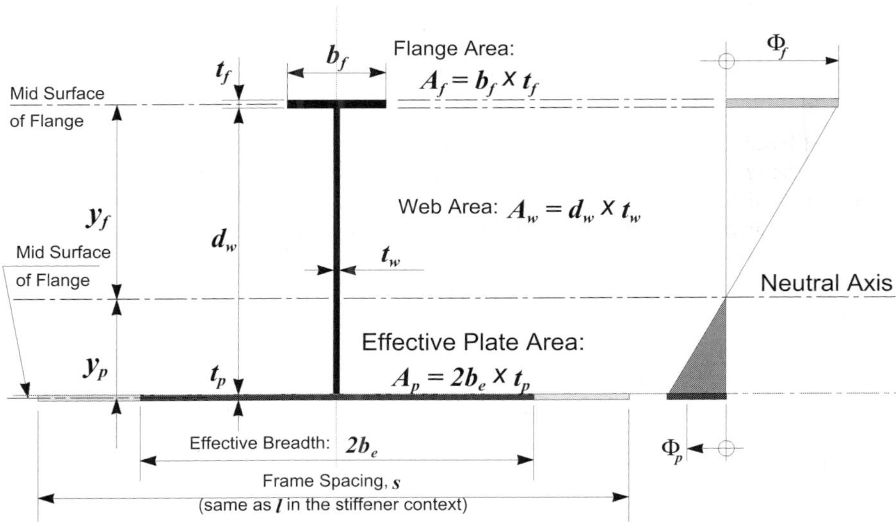

Figure 7.105 Notation for frame/girder properties

Additionally, girder webs are particularly deep and thin and subject to buckling. As noted in the beginning of the discussion of framing, web stiffeners are employed to alleviate this problem. Similarly, the flanges are narrow and can buckle laterally. A particularly thorough guide on girder-buckling is contained within the DnV Class Notes 30.1.

As a final brief note on framing, some columns use *ring frames* to support vertical stiffening. A circular frame is only a ring frame if it is totally without radial support and, in such a case it is subject to buckling. DnV Class Notes 30.1 also provide guidance in this case. However, if the frame is at all radially supported, it is a collection of curved beams subject to flexure, as well as compression, and treated as any straight frame.

7.8.10 Global Strength

Structural Configurations

The approach to considering global strength depends greatly on its structural configuration. Fundamentally there are two, that of a long, slender beam and that of a space frame. Typical of the former is the SPAR paltform and various ship-type hulls, e.g. the FPSO, the drillships, and certain forms of heavy lift vessels. The latter case includes the semi-submersibles and the TLPs. The structural configuration of each particular type has been discussed to some extent in the sections that address the various platform types.

The structural configuration of a semi-submersible is primarily distinguished by its transverse strength system. Figure 7.36, given previously, shows the four basic structural configurations for drilling semi-submersibles. As discussion therein, they follow an evolution. While many variants and hybrids exist, these four categories summarise the basic differences.

Global Load Systems

Global strength relates primarily to two types of loading systems: the gravity/buoyancy load and the environmental loading. Wave load inertial loading is the principal environmental load.

To look at the gravity/buoyancy load system on a semi, see fig. 7.106. Shown is a section and a profile view. An idealised distribution of deck load and concentrations of buoyancy forces at the pontoon and the column lines is shown in the section view. Additional gravity load is included in the pontoons and columns. A reasonably well spread gravity loading on the superstructure must be supported by buoyancy concentrated at the extremities, causing a the tenancy to sag. This causes very large tensions in the horizontal brace to resist the sag. Additionally, the interior parts of the deck weight will transmit directly through the diagonals into the column. This exhibits one important function of the main bracing as primary structure. Particularly important are the end connections at the column, especially the efficiency of the load flow from the diagonal to the transverse. Shown also in the loading of the mooring system. This load will transfer through the bracing also as tension.

Longitudinally, most semis have a series of 3 or 4, fairly close-spaced columns on each side, supported by a continuous pontoon. A profile of an 8-column semi is also shown in fig. 7.42. Drilling semis, being comparatively short, with reasonable distributions of load, do not have significant longitudinal bending/shear problems and generally do not require longitudinal diagonals. As can be seen, the superstructure gravity loading is better supported, featuring shorter deck spans and continuity of longitudinal strength. The pontoons generally have considerable shear strength for their length. Secondary bending does occur at the column tops and pontoon connections due to uneven deck loads and pontoon liquids, but stresses from these distributions rarely are significant (usually less than 5 ksi). Noted on the figure is also the end pressures on the pontoons. While small, they are not negligible (about 3 ksi).

The 8-column semi typically has large columns at the ends and smaller diameter interior columns. The interior columns are as much structural members as anything else and can have somewhat higher stress levels because they are sized for loading and not their waterplane contribution. The 6-column semis generally have six columns of equal diameter and, despite the longer spans, less severe longitudinal stresses. Nevertheless, the Pacesetter Class does have longitudinal bracing.

Environmental Loading

From an environmental load perspective, there are two systems of wave load of importance. One is the "squeeze/pry" load. Squeeze/pry is a lateral loading effect. The vertical loading is not of serious consequence except for oblique seas, particularly with the widely spread, twin pontoon configuration. This results in the second important wave load system: "racking" load.

Figure 7.43 schematically illustrates the squeeze/pry load system applied to a semi. There are two stages to the load system. With "pry," the wave crest is centred and the collected wave forces (pressures) are trying to pull the semi apart. The second stage, "squeeze" is with the wave trough is centred, where the effect is for the wave to push in from the sides. This effect applies at all wave headings, but most structural configurations are more vulnerable to beam seas. The length of the wave (period) relative to spacing of columns and pontoons is very important.

A much more complicated situation, shown in fig. 7.44, schematically illustrates the racking load system. Each pontoon/column set can be viewed as a separate system. When the wave system is oblique, the heave force on one pontoon is concentrated at one end and the force on the other pontoon at the other end. These combine to cause a global torque, or "racking moment" on the hull. Vertical force distributions on the pontoons are shown on the figure along with their resultants. The racking moment is augmented by similar location shifts in the lateral force systems. The distribution of these forces is also shown in the figure.

Like the squeeze/pry, there are two stages to the racking load system also. One can be viewed as sagging on an oblique wave with the crests centred at opposite *corners* of the FPS. The other is the corresponding, crest centred hog. The sagging stage is illustrated. The effect is that the wave forces alternately twist the semi one way and reverses to the other. Racking occurs at any angle of oblique seas, with the critical heading and the wave length slightly different for each location of critical stress. Conventionally, a single critical wave is used, usually one along the diagonal between the corner columns. Again, the length of the wave relative to spacing of columns and pontoons is a very important, in this case a longer wave than critical to squeeze/pry.

Spar Hull Structural Configuration

Spar hulls and other single column hulls such as the SeaStar® Mini TLP are similar to semi-submersible columns. A typical spar construction is shown in fig. 7.65. The outer shell, inner shell (centrewell), decks and bulkheads are all plate and frame structures designed to local scantling rules as discussed above. Notice the radial columns between the frames of the centrewell and the outer shell. These have the same function as the strut in fig. 7.101 (lower right).

Even though the spar diameter is typically greater than a semi-submersible column, the presence of a large centrewell minimises the in-plane hoop stresses on the outer shell. Local scantling rules usually result in a robust design from a global strength and buckling perspective.

An exception to this is the global bending arising from towing and upending, which can control the strength of the hull midsection or truss to hard tank connection in the case of a truss spar.

Basis of Global Analysis

General

Global analysis of an FPS or MODU requires solution of the equations of motion for the platform under combined wind, wave and current loading. The resultant gravitational, inertial and environmental loads must be applied to the structure to determine the internal structural loads and ultimately stresses.

Hydrodynamic loading is covered in Chapter 3 and physical modeling in Chapter 13. We will not repeat the discussions there, but will summarise the practical implementation for Global Strength analysis.

Hydrodynamics of Loading

As shown in Chapter 3, diffraction effects may be neglected if:

$$D \leq \frac{\lambda}{2\pi} \tag{7.71}$$

In practical terms this means that diffraction (and radiation) may be ignored for member diameters less than about 40 ft. In these cases slender body theory may be used with a modified Morison equation to account for the relative motions between the floating body and the wave. Figure 7.106 and Equation 7.72 show the application of the modified Morison equation to a member of a floating structure.

$$\vec{f} = \iint p\vec{n}ds + \rho\frac{\pi D^2}{4}C_a\vec{a}_n + \frac{1}{2}\rho D C_d\vec{u}_n|u_n| \tag{7.72}$$

where $\vec{f}$= force per unit length normal to centerline of the cylindrical element, p = fluid pressure on the surface of the cylinder computed as though the member does not disturb the flow, ρ = density of water, C_a = added mass coefficient, C_d = drag coefficient, $\vec{u}_n$ = resultant velocity normal to the element centreline, $\vec{a}_n$ = resultant acceleration normal to the centreline.

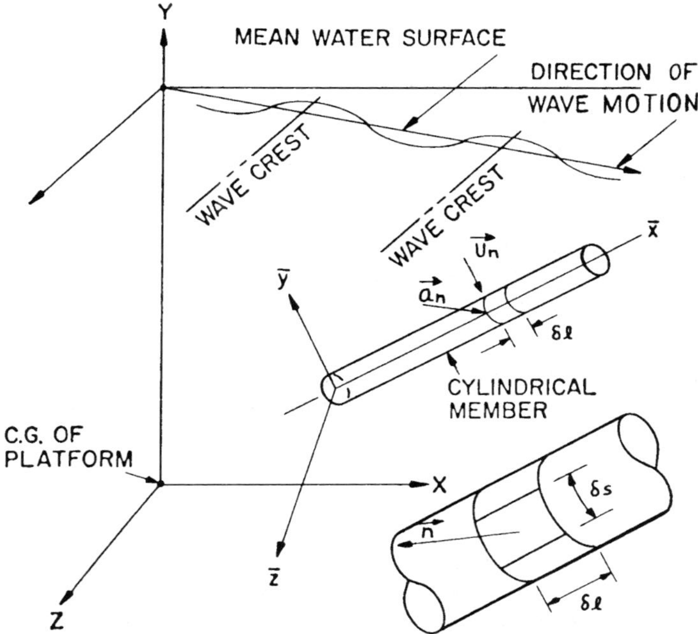

Figure 7.106 Nomenclature for fluid forces on a cylinder (Paulling, 1995)

End forces on the columns and pontoons are important to model. Applying (7.72) to the cylinder ends results in:

$$f_e(t) = \iint p dA + \rho C_{ma} \frac{D^3}{6} [\ddot{v} - \ddot{y}] + \frac{\rho}{2} C_{da} \frac{\pi D^2}{4} (v - \dot{y}) |v - \dot{y}| \tag{7.73}$$

where p = incident wave pressure (without diffraction/radiation effects), C_{ma} = added mass coefficient of a thin disk normal to flow, C_{da} = drag coefficient for flow normal to the column end, $\ddot{v}$ = wave particle velocity normal to end face of column/pontoon, $\dot{y}$ = motion of column/pontoon end perpendicular to its surface.

The first term is the Froude–Krylov force, i.e. the force associated with the pressure field of the incoming "undisturbed" wave. The second term is one half the added mass of a thin disk and the third term is the drag force. C_{ma} and C_{da} are about equal to 1.0.

Global strength analysis requires mapping of the hydrodynamic, the gravitational and the inertial forces onto a structural model consisting of plate, shell, beam and/or truss elements. This may be done in the frequency or time domain. Frequency domain is efficient and fast but may not capture some important non-linearities such as gravitational loads from large heel angles. Several commercial computer codes incorporate hydrodynamic and structural modeling in a single package, with load mapping done automatically. Some commercial codes commonly used in the offshore industry which do this include:

Shell and Beam loadings

WAMIT (www.wamit.com)
SESAM Suite (www.dnvsoftware.com)
ASAS/AQWA (http://century-dynamics.co.uk)
DIODORE (http://www.principia.fr)
MOSES (http://www.ultramarine.com)

Mapping of shell loads is currently limited to linear theory.

Beam Loadings

STRUCAD (http://www.zentech-usa.com)
SACS (http://www.sacs-edi.com)
ABAQUS (http://www.hks.com)

Beam loadings, based on a modified Morison equation, are non-linear.

Diffraction/Radiation solutions are important for operational seas or very large structures, which include the FPSOs. Mapping of the panel loads from waves onto an FEA of the hull is a specialised problem. Some of the commercial codes identified above have this capability, however it is common among offshore contractors to tailor a radiation–diffraction program (e.g. WAMIT: www.wamit.com) with a favoured FEA code for a specific fit-for-purpose program for a particular floater. These codes may run more efficiently than the general purpose codes listed above, but they do require effort for software development and support within the contractor's business.

FPSO Global Strength

Loads

FPSO loads include the deadload, cargo and ballast weight, mooring, riser load and wave load. Morison's equation may be applied to derive wave loads on column stabilised platforms, spars and TLPs. FPSOs fall under the category of ships, and represent a slightly different class of vessel than the semi-submersible, TLP or Spar, which were largely covered above. Local scantling design still dominates the primary steel structure and may be applied as above. Longitudinal strength considerations could control the required main girder section modulus and other properties of the hull. Cargo management under normal operations is usually designed so that these conditions do not control the design from a bending moment and shear point of view. Wave induced loads at the operational site should be considered: these are usually less severe than those required by the rule formulas given below. The most critical conditions from a bending moment and shear perspective are typically associated with the cargo and ballast tank inspection conditions. The FPSOs undergo continuous tank inspections following a five-year cycle. Contrary to the trading tanker practice, inspections are not carried out in a drydock. The inspections are performed on site, while producing. This means that the selected tanks may be empty while others are either empty or full. This loading can result in large "hogging" or "sagging" moments exceeding those normally associated with waves.

An FPSO design is usually based on an allowable stress method using direct stress calculation. This is particularly necessary to design the turret sections of an FPSO. The Classification Rules, however, may form the basis for an initial design. The following discussion presents some highlights of the ABS Rules for Building and Classing Steel Vessel as modified by the "Guide for Building and Classing Floating Production Installations".

The Steel Vessel rules are based on the experiences in ship design based upon unrestricted service, typically North Atlantic environments. The Offshore Industry traditionally relies more heavily on direct calculation methods employing hydrodynamic models coupled to the detailed structural FEA models to perform a rational stress based design. Studies comparing the two methods have shown that they both result in hull designs of similar reliability (Hambden et al, OMAE 2002). An FEA analysis is required for turret moored vessels in order to design the turret to appropriate loadings. The ABS Class Rules also require the FEA analysis for any FPSO over 150 m in length.

The ABS *Guide for Building and Classing Floating Production Installations* (ABS FPS Guide) allows the ABS *Steel Vessel Rules* criteria to be modified to suit a site specific application by use of an Environmental Severity Factor (ESF) which takes into account site specific criteria. ABS uses α and β to represent ESF values for fatigue and strength, respectively.

While the ESF values are intended to allow designers of FPSs some relief from the requirements of unrestricted service, ABS has set minimums on loads and strength that need to be observed. The Rules state that:

- the required value of wave bending moment, M_w, shall not be less than 50% of the value given in Section 3.5.1 of the Steel Vessel Rules,
- nor will the value used for vessel properties (e.g. section modulus) be less than 85% of the corresponding value in Steel Vessel Rules.

The formulas for minimum wave induced bending moments from Steel Vessel Rules, incorporating the ESF, from the ABS Guide, 2000 edition, Appendix 2, Section 8) are reproduced here for reference. Specific ESF values require analysis of several factors and are not presented here (refer to the Guide and to other class rules). Steel Vessel Rules assume an ESF of 1.0, hence the actual value for a specific design will be between the minimums mentioned above and 1.0.

Wave Bending Moment Amidship

Sign convention for bending moments is shown in fig. 107. Wave maximum bending moments are given here in units of kN-m.

$$M_{ws} = -k_1 \beta_{VBM} C_1 L^2 B(C_b + 0.7) \times 10^{-3} \quad \text{Sagging Moment} \quad (7.74)$$

$$M_{wh} = +k_2 \beta_{VBM} C_1 L^2 B C_b \times 10^{-3} \quad \text{Hogging Moment} \quad (7.75)$$

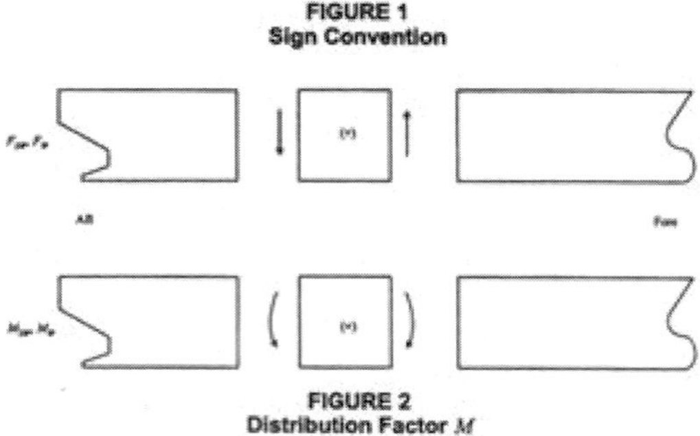

Figure 7.107 Sign convention for ABS bending moments and shear (ABS FPS Guide)

where $k_1 = 110$, $k_2 = 190$, $\beta_{VBM} =$ ESF for vertical bending moment, L = length of vessel, $mB =$ breadth of vessel, $mC_b =$ block coefficient of vessel, and

$$C1 = 10.75 - \left(\frac{300 - L}{100}\right)^{1.5} \qquad 90 \leq L \leq 300\,\text{m} \tag{7.76}$$

$$= 10.75 \qquad 300 < L \leq 350\,\text{m} \tag{7.77}$$

$$= 10.75 - \left(\frac{L - 350}{150}\right)^{1.5} \qquad 350 \leq L \leq 500\,\text{m} \tag{7.78}$$

Distribution of Wave Bending Moment

Figure 7.108 shows the distribution factor for the bending moment computed above along the length of the vessel.

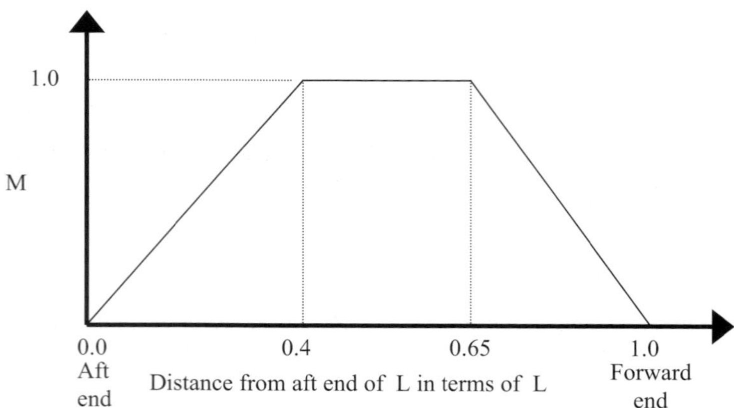

Figure 7.108 Distribution of ship bending moment (ABS FPS Guide)

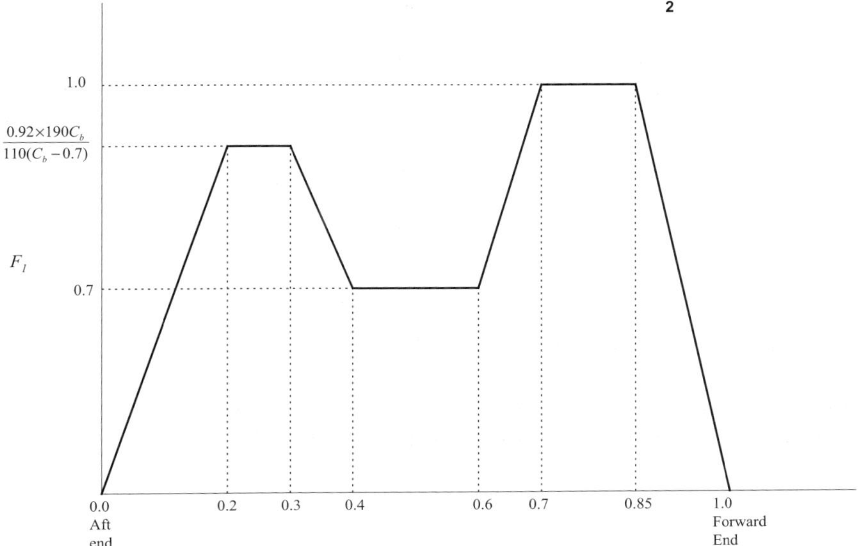

Figure 7.109 Distribution factor F_1 (ABS FPS Guide)

Wave Shear Force

Maximum shearing forces are given as:

$$F_{wp} = +k\beta_{VSF}F_1C_1LB(C_b + 0.7) \times 10^{-2} \quad \text{for positive shear force, in kN} \tag{7.79}$$

$$F_{wn} = +k\beta_{VSF}F_2C_1LB(C_b + 0.7) \times 10^{-2} \quad \text{for negative shear force, in kN} \tag{7.80}$$

where $k = 30$. F_1 and F_2 are distributed as shown in figs. 7.109 and 7.110, respectively.

The total load on the vessel requires the addition of these loads with the appropriate dead load which consists of the lightship (with equipment), cargo, mooring, riser and ballast weight. The Total Bending Moment is given by:

$$M_t = M_{sw} + M_w \tag{7.81}$$

Where M_t = total bending moment, kN-m, M_{sw} = still water bending moment for an appropriate loading condition, kN-m, M_w = maximum wave induced bending moment in accordance with the above formulas, kN-m.

The still water bending moment should consider a bending loading condition associated with the extreme environmental event, e.g. tank inspection suspended, production

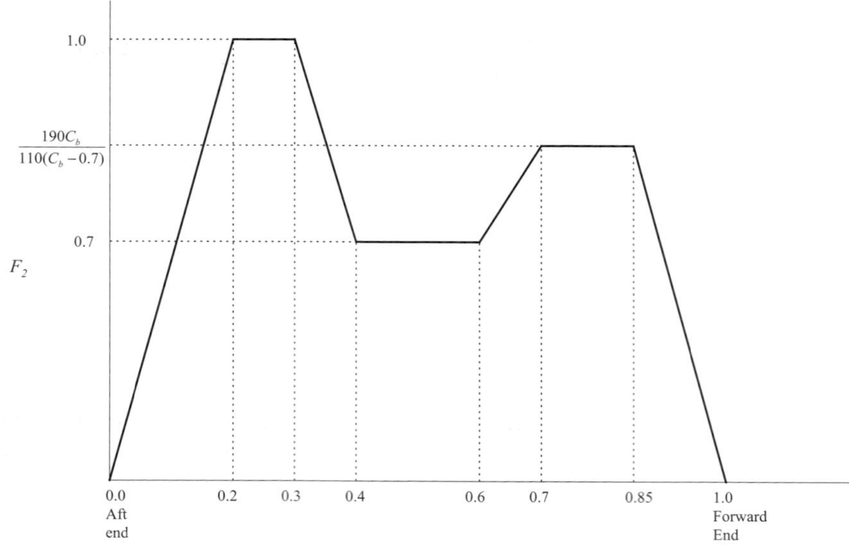

Figure 7.110 Distribution factor F_2 (ABS FPS Guide)

Table 7.15 Comparison of Still Water Bending Moment for Tanker and FPSO (Neto, 2000)

Condition	Fully Loaded	Minimum Load
Tanker	−722,000 t-m	522,000 t-m
FPSO (Turret)	−554,000 t-m	554,000 t-m
FPSO (Spread Mooring)	−695,000 t-m	520,000 t-m

suspended, etc. In some cases a loading condition such as tank inspection in a lower environment may dominate.

Table 7.15 shows a comparison of still water bending moments for a trading tanker and two FPSOs, one turret moored and one spread moored. The FPSOs are designed for the Brazilian environment. All vessels are designed for 2,000,000 BBL storage. The internal turret system has 40 risers. The spread moored FPSO has 70 risers distributed longitudinally on both sides amidships.

Section Modulus

$$SM = \frac{M_t}{f_p} \tag{7.82}$$

$$SM = C_1 C_2 L^2 B(C_b + 0.7) \tag{7.83}$$

where SM = section modulus, cm^2-m, f_p = nominal permissible bending stress, 17.5 kN/ cm^2, C_1 is given in Eqns. 7.76–7.78. $C_2 = 0.01$, L = length of vessel as defined in *Rules*, m, B = greatest molded breadth of the vessel, m, C_b = block coefficient, defined as

$$C_b = \frac{\Delta}{1.025LBd} \tag{7.84}$$

Δ = molded displacement, metric tons, and d = draft at summer load line, m.

The allowable stress, $f_p = 17.5$ kN/cm^2 or 25.5 ksi, assumes ASTM A36 grade steel. Proportionally lower section moduli are allowed for higher strength steels.

The above section modulus is required over the midship for a length equal to 0.4 L.

Shear Strength

The Steel Vessel Rules specify that the nominal total shear stress shall be less than 11.0 kN/ cm^2 (16 ksi) assuming the ASTM A36 construction, or 15.3 kN/ cm^2 (22 ksi) for higher strength H36. The shear stress is given as:

$$fs = (F_{sw} + F_w)\frac{m}{2t_s I} \tag{7.85}$$

Where I = moment of inertia of hull girder at the section under consideration, cm^4, m = first moment, in cm^3, about the neutral axis, of the area of the effective longitudinal material between the horizontal level at which the shear stress is being determined and the vertical extremity of effective longitudinal material, taken at the section under consideration., t_s = thickness of the side shell plating at the section under consideration, cm, F_{sw} = hull girder shearing force in still water, hat, $F_w = F_{wp}$ or F_{wn} from Eqns. 7.79, 7.80.

The Rules contain additional formulas for the shearing stress with longitudinal bulkheads. There is also a reduction in Fs used for calculating shear stress in the side shell for ships with a double bottom.

Global Modelling

FPSOs

Direct calculation of wave loads may be performed either by a two-dimensional strip theory, or a three dimensional panel calculation. The strip theory has been the method of choice for ship designers, particularly for ships with a forward speed. The Canadian Defense Research Establishment Atlantic developed the frequency domain strip theory program SHIPMO in the 1970s and it has become a standard in ship design. It is a public domain code, and some vendors have adapted it to PC use and added pre- and post-processing capabilities. The latest enhanced version is distributed by BMT Fleet Technology (http://www.fleetech.com). This program may be used to compute beam loadings on FPSOs. There are several three dimensional panel codes, some of which are linked to finite element programs (See Section 7.8.12). In general, the three dimensional codes are limited to zero speed, which is suitable for FPSO applications. The ABS SafeHull® (www.eagle.org) program incorporates a range of frequency and time domain programs for hull loading (Shin, 1997). The U.S. Navy has sponsored the development of a fully non-linear (but ideal fluid) Large Amplitude Motion Program (LAMP) for time domain simulation of ship motions and loads (see http://www.ship.saic.com/overview_lamp.htm; Weems et al, 2000; Lin et al, 1999).

A great deal of research on ship motions and loads has been consolidated by the Ship Structures Committee (SSC), a non-profit interagency organisation which has membership from the U.S. and the Canadian navies, classification and professional societies. More information can be found at www.shipstructure.org.

The Classification Rules require a Finite Element Model (FEM) of the entire hull for hulls over 150 m in length, and it is recommended for shorter hulls of double hull or "unusual construction". FPSOs with turret moorings qualify as having unusual construction, and in any event turret design usually involves a sophisticated FEM model of the hull and the turret components.

Global modelling requires the use of a separate hydrodynamic code with load mapping to an FEA. Some commercial codes incorporate this capability. Many designers choose to develop their own mapping tools to accommodate their hydrodynamic and FEM codes of choice. This type of global modelling can be very computationally intensive and time consuming, particularly if the results are to be used for fatigue analysis.

Column Stabilised Platforms

Unlike a fixed structure, the forces shown on the semi are not equilibrated by earth based boundary conditions. Vertically, the forces on the semi are equilibrated by changes in buoyancy due to vertical movement (heave) and by the inertial force developed by vertical accelerations. Changes in the vertical component of mooring force is very small compared to water plane forces. Horizontal forces are equilibrated primarily by the force developed by horizontal accelerations. Mooring load is still a much smaller component, but not unimportant. For rotational motions, roll and pitch, the importance of forces is a mixture of the preceding.

Most structural analysis programs will not accept a structure freely suspended in space and require some stable form of boundary elements. The physical equivalent of this is the "waterplane" buoyancy for vertical displacements and the mooring system. In some form or another, such restraints must be included, even if only to satisfy imperfections in a "balanced force system."

The mooring system represents a "spring-like" device in a boundary element scheme. However, spread and taut moorings are quite non-linear. A practical solution is to apply spring constants for the extended position of the mooring system and to *separately* apply mooring reactions as loads appropriate to any *steady* loads on the structure and the wave load (for the wave position of the load case). Should there be any unbalanced mean horizontal loading, it will appear in the "spring reactions" and should not be large or important.

The simplest way to represent the buoyancy related restoration forces, at least for a frame-type, global model, is to employ vertical spring elements under each column bottom node *in proportion* to the water plane area of the column. Also, include the vertical stiffness of the mooring restoration system. Many structural programmes have difficulties with large global displacements mixed with elastic displacements of the structure model. It may be necessary to "scale" the boundary elements (springs).

The force system is a dynamic balance of direct hydrodynamic force, elastic restraint, and inertial reaction. It could be written in the more generic, simplified dynamical form as

$$[M] \cdot \ddot{\vec{z}} + [K] \cdot \vec{z} = \vec{F}(t) \quad \text{Dynamic equation} \tag{7.86}$$

Damping may be ignored as long as the resonant frequency of the structure is well separated from the forcing frequency. Since the natural period of most floating platforms is below the period of wave energy, the inertial loads may be applied as quasistatic d'Alembert forces. The solution at each time step may be represented as

$$\vec{z} = [K]^{-1} \cdot \left\{ \vec{F} - [M] \cdot \ddot{\vec{z}} \right\} \quad \text{Quasi-static solution} \tag{7.87}$$

$\vec{F}$ is the (nodal) force vector, $[K]$, $[M]$ are the stiffness and mass matrices, respectively.

The nodal force and acceleration terms may be derived from rigid body motion calculations were the displacements due to structural deflection are ignored. Either frequency domain or time domain methods may be used. Frequency domain methods are easier and faster to implement because the responses can be characterised by the linear transfer functions (Response Amplitude Operators) and the spectral analysis techniques may be applied to assess both extremal values and fatigue. Formally, the frequency domain solution is found by mapping the real and imaginary parts of the rigid body forcing and inertial terms to the structural model and tabulating the resultant responses.

Linear analysis does not capture the effects of heeling responses typically associated with wind moments or, in the case of deep draft floaters like spars, the slowly varying drift forces associates with waves. These effects may be included in a linear analysis as static buoyancy and gravitational forces which are superimposed upon the dynamic wave loads.

Alternately, time domain motions analysis can be used to capture the non-linearities in both the wave and the wind loadings, and the motions. In this case, instantaneous forces and accelerations need to be mapped to the structural model. For random sea responses a large number of force and acceleration data sets are required to ensure that extreme values for the stresses and deflections are captured.

Structural Modelling

Global structural modelling is best taken at two levels. The overall global model with hydrodynamic and mass loading is best handled as a line element, space frame model. Detail stress distribution is best determined using an FEM. Attempts to model the full structure with a FEM model rarely provide enough detail to do an adequate job in the hot spot locale and unnecessarily burden the global model with complication.

A variation of global modelling, which is quite useful should be mentioned. It is actually a form of local design. The variation follows from the fact that nearly all deck steel is based upon gravity loading, the exceptions being local connections immediately at the column tops, brace connections, and special structures (substructure foundation, crane foundations, etc.). The variation is to use two variants of the model. For the global strength design, it uses a relatively coarse model of the superstructure, perhaps only six nodes across. This will have virtually no impact on the results in the columns, pontoons, and braces. Similarly, use a coarse hull, column and brace model. These only serve as elastic and load boundary conditions for the super structure model. This allows a more direct pursuit of design in the superstructure, and provided that the models are coordinated and of the same "parent," provides a helpful division of the work.

Structural analysis and design can be divided into five phases:

Phase A: Local Load and Rule Scantlings

Phase B: Global – Gravity/Buoyancy Load

1. Operating Conditions
2. Severe Storm Loading Conditions
3. Afloat on Pontoons
4. Severe Flooding Damage
5. Global – Gravity/Buoyancy + Linear Environmental Load
6. Three headings and range of wave periods
7. Global – Gravity/Buoyancy + Large Wave Environmental Load
8. Selected waves, headings, and crest positions

Phase C: Local Reinforcement

1. Global Strength – Gravity Load
2. Global Strength – Combined Loads
3. Local Strength – Global/Local stress checks
4. Local Strength – Extreme List/Trim

Phase D: Fatigue Analysis

1. Local FEM (if required)

Phase E: Transit: Identify limiting conditions

Phase A has been described earlier. Phase B requires a global space frame model. The global model would be examined for three types of load systems. One is simply the static gravity loading and is examined for several loading conditions, including one or more operational loadings, at least one representing alteration of load for severe storms (possibly draft change), and two cases afloat on the columns (on with maximum deck load and minimal ballast and the other minimum deck load with maximum ballast). In some situations, a case with extreme trim and heel needs to be examined, representing flooding damage (depends of applicable code).

Environmental loading can be treated as additive to the appropriate gravity load case. Phase B considers environmental loading in two ways. First a large range of wave parameters can be analyzed with linear analysis. This can also be used for fatigue analysis. Second, selected extreme amplitude waves cane examined using non-linear approach. This would be important to capture effects, which are not well represented in the linear analysis. Specifically, this includes drag dominated wave actions and loading significantly affected by wave elevation or wind forces.

Phase C involves a variety of secondary analyses considering extremes of global stress and local loading. It is necessarily selective and intended as a basis to determine reinforcement, generally highly localised increases in shell plating and increases in side shell and bulkhead thicknesses subject to high shear stresses. It is important to consider torsion of the element as well as bending shear in then latter.

Phase D proceeds on the basis of the results in the linear environmental load analysis in Phase C and produces suitable local models, perhaps the FEM models which quantify local intensities of cyclic stress. The same FEMs or other models may be required to address local strength issues from Phase C not satisfied through more conventional means.

Finally, where stress during transit to the builder is exposed to severe environmental conditions, Phase E, it will be necessary to determine strength limitations. The alternatives are to change route (dry transport) or to "ballast down" (wet transport). Where deployment is in a harsh environment, the same would be required to determine limitation for floating on the pontoons.

7.8.11 Buckling

General

The Classification Rules require that structural stability for the structure as a whole and for each part of the structure. This topic is covered extensively in several industry publications, notably the DNV Classification Note 30.1, API RP2A and Bulletin 2U (cylindrical shells) and the ABS Steel Vessel Rules (for ship shaped hulls). Stability analysis tends to be more complicated than strength analysis. Thankfully, most floaters designed for local and global strength as discussed above have adequate margins on buckling, with some notable exceptions listed below. While a great amount of mathematical theory is available to draw upon, actual stability limit states depend upon subtle design details, loading conditions and imperfections in fabrication which need to be considered empirically. The best buckling data has been derived from very large scale models designed and built to industrial standards. It is not unusual for the critical buckling stress to be 1/2 of the theoretical value.

The Recommended Practices, Classification Notes and Rules referenced herein provide guidance on suitable design values and serve as the best reference. This section includes numerous examples from these references, especially DNV CN 30.1. However these should be considered for reference only. Always refer to the latest edition of these references for design guidance.

Buckling is of concern for "slender" members in compression and shear. Some examples where this is the case include:

- FPSOs and ships which experience hogging stresses under still water conditions: the upper deck plating and framing may be controlled by buckling,
- Deep draft semi-submersibles or spars: the column scantlings may be controlled by buckling.
- Deep frames and girders on plated structures: the web may require stiffening

- Cylinders under hydrostatic loading, including braces, chords, columns and TLP tendons, may require ring stiffening or wall thicknesses governed by buckling,
- Large diameter hull sections under construction may require ring stiffening to resist buckling (see fig. 7.81).

Buckling checks are typically performed in conjunction with global strength analysis. Compressive loads, in-plane plate loadings, etc. for particular details are derived from the FEM and compared to the various rule formulas to check for adequacy. Global buckling, plate buckling are rarely issues, however it is common to have to stiffen deep girder webs, at tripping brackets and or increase stiffener dimensions to meet buckling criteria, especially for panels designed fro high hydrostatic heads. Pressure vessels are a special case. TLP tendons and risers which are designed to be operated with air on the interior are buckling candidates.

Buckling may be either *global* or *local*. Global buckling involves more than one member and represents a complete collapse of a structure. This is usually a catastrophic event. Local buckling involves the yielding of a single member, for example the buckling of a plate section between frames. This type of buckle may not jeopardise the safety of the structure.

Buckling behaviour is usually characterised by a reference stress, σ, which may either be a single stress component or an "equivalent" stress. The buckling strength of a structure is defined by a critical value of this reference stress, σ_{cr}. σ_{cr} is typically represented non-dimensionally as the ratio of critical buckling stress to material yield stress, $\sigma_{\mathrm{cr}}/\sigma_{\mathrm{y}}$. Buckling strength is functionally related to a property of the structure generally known as structural slenderness, represented by the general reduced slenderness:

$$\lambda = \sqrt{\frac{\sigma_F}{\sigma_E}} \tag{7.88}$$

where σ_F = material yield stress, and σ_E = elastic buckling stress.

Figure 7.111 illustrates a typical non-dimensional buckling curve. For zero "slenderness" buckling is not a consideration and yield strength governs the design. For slenderness greater than a quantity λ_0 buckling governs.

Column Buckling

Column buckling is a classic problem in fixed offshore structures, which are constructed of tubular members with large slenderness and axial loads. Global column buckling strength for a tubular under nominal axial compressive loading is given by (API RP2A):

$$\frac{\sigma_{cr}}{\sigma_F} = 1.0 - .25\lambda^2 \qquad \text{for } \lambda \leq \sqrt{2} \tag{7.89}$$

$$\frac{\sigma_{cr}}{\sigma_F} = \frac{1}{\lambda^2} \qquad \text{for } \lambda \geq \sqrt{2} \tag{7.90}$$

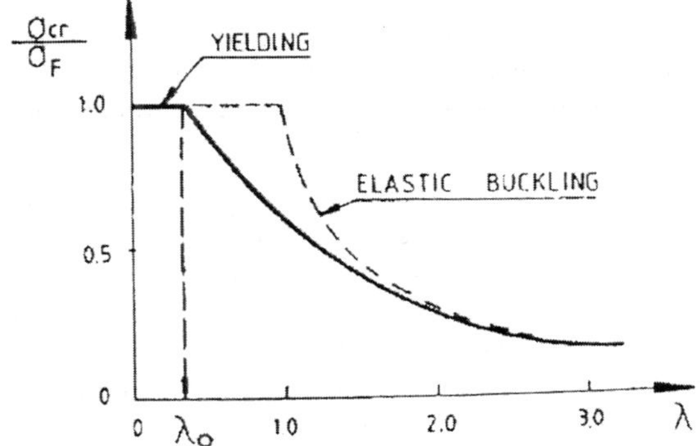

Figure 7.111 Typical non-dimensional buckling curve (DNV CN 30.1)

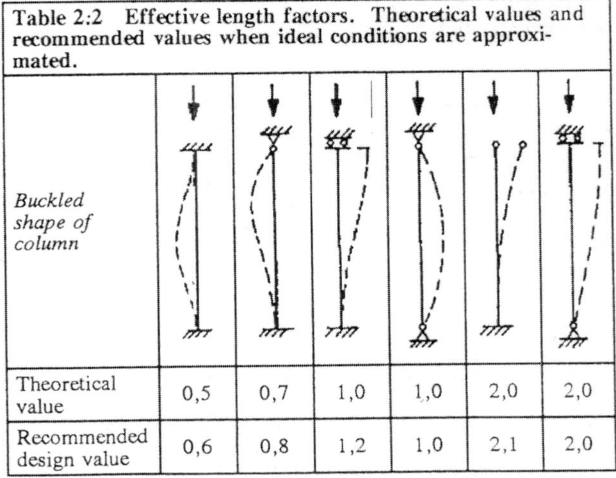

Table 2:2 Effective length factors. Theoretical values and recommended values when ideal conditions are approximated.

Buckled shape of column						
Theoretical value	0,5	0,7	1,0	1,0	2,0	2,0
Recommended design value	0,6	0,8	1,2	1,0	2,1	2,0

Figure 7.112 Recommended effective length factor, K (DNV CN 30.1)

where:

$$\lambda = \frac{KL}{\pi r}\left[\frac{\sigma_F}{E}\right]^{0.5} \tag{7.91}$$

E = Young's modulus, K = Effective length factor, L = Unbraced length of column, and r = radius of gyration.

The effective length factor is a function of end conditions. Recommended values from DNV 30.1 are given in fig. 7.112. API RP2A presents values from 0.7–1.0 suitable for fixed jacket type structures.

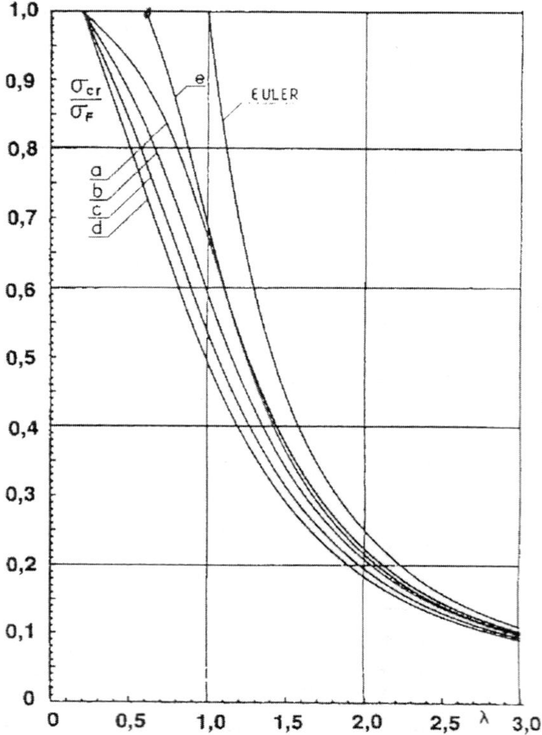

Figure 7.113 Non-dimensional buckling curves for columns (DNV CN 30.1)

DNV CN 30.1 gives a more general formulation suitable for a variety of structural members.

$$\frac{\sigma_{cr}}{\sigma_F} = \frac{1 - \mu + \lambda^2 - \sqrt{(1 + \mu + \lambda^2)^2 - 4\lambda^2}}{2\lambda^2} \qquad \lambda > \lambda_0 \tag{7.92}$$

where $\mu = \alpha(\lambda - \lambda_0)$. The values λ_0 and α depend upon the section geometry as shown in figs. 7.113 and 7.114, and table 7.16.

The formulation for λ depends on the mode of buckling. Equation (7.91) applies to flexural buckling. Torsional buckling strength is (DNV CN 30.1):

$$\lambda = \sqrt{\frac{\sigma_F}{\sigma_{ET}}} \tag{7.93}$$

$$\sigma_{ET} = \frac{\pi^2 E I_f}{A_e (KL)^2} \tag{7.94}$$

Shape of section		Buckling about axis	Column curve
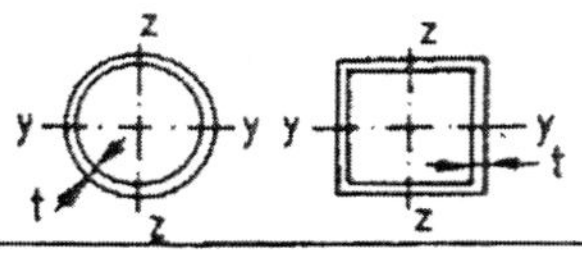	Rolled tubes	y-y z-z	a
	Welded tubes (hot finished)	y-y z-z	
y-y: b,t; z-z: h,d 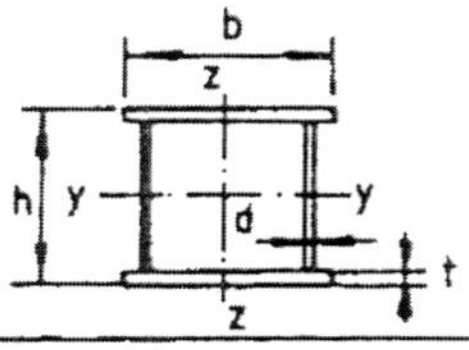	Welded box sections	y-y z-z	b
	Heavy welds (full penetration) and b/t < 30	y-y z-z	c
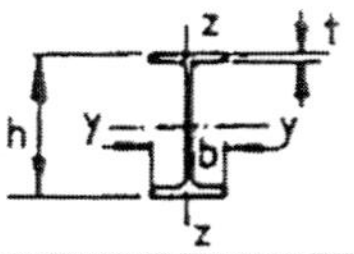 I and H rolled sections	h/b > 1,2	y-y z-z	a b
	h/b ≤ 1,2	y-y z-z	b c
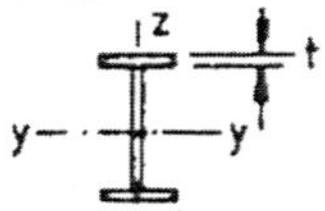 I and H welded sections	Flame cut flanges	y-y z-z	b
	Rolled flanges	y-y z-z	b c
I and H sections with welded flange cover plates $t = t_{1max}$	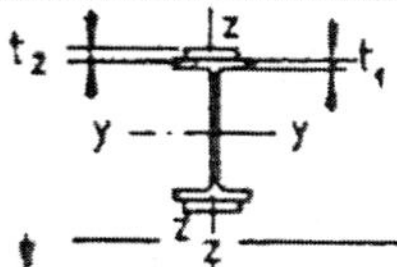	y-y z-z	b a
Box sections, stress relieved by heat treatment	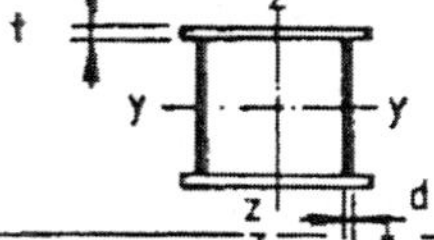	y-y z-z	a
I and H sections, stress relieved by heat treatment	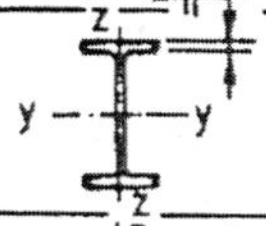	y-y z-z	a b
T and L sections y-y: d z-z: t	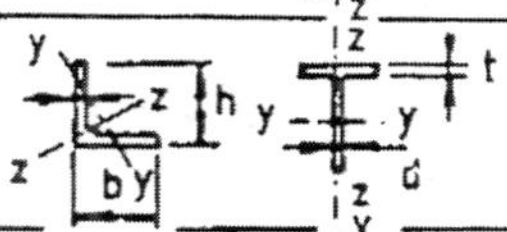	y-y z-z	c
Channels	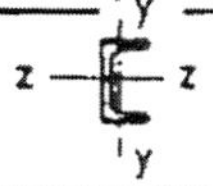	y-y z-z	c

Figure 7.114 Column selection chart (DNV CN 30.1)

Table 7.16 Numerical Values of λ_0 and α

Curve	λ_o	α
a	0.2	0.20
b	0.2	0.35
c	0.2	0.50
d	0.2	0.65
e	0.2	0.35

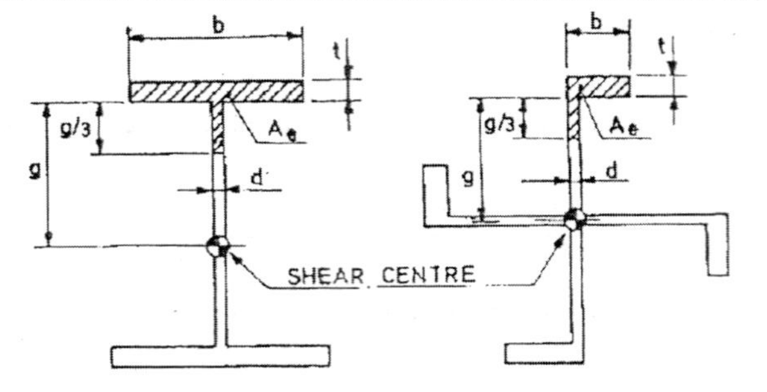

Fig. 2.5
Cross-sectional properties to be used for simplified evaluation of the torsional buckling strength.

a) $I_f = \frac{1}{12}\, t\, b^3$

b) $I_f = \frac{1}{12}\, t\, b^3 \left(\frac{1 + 4\left(g \frac{d}{b} t\right)}{1 + \left(g \frac{d}{b} t\right)} \right)$

Figure 7.115 Cross-sectional properties used for simplified evaluation of torsional buckling strength (DNV CN 30.1)

where I_f = moment of inertia of flange, A_e = effective cross-section (fig. 7.115). The buckling strength from pure bending is (DNV CN 30.1):

$$\sigma_{EV} = \frac{\pi^2 E I_{zc} h}{Z_{yc} l_e^2} \tag{7.95}$$

where Z_{yc} = section modulus with respect to compression flange, I_{zc} = moment of inertia of the compression flange, h = web height.

Buckling of beam-columns (combined axial and bending) is governed by a usage factor, η, for example given in DNV 30.1 as:

$$\eta = \frac{\sigma_a}{\sigma_{acr}} + \frac{\alpha\sigma_b}{(1 - \sigma_a/\sigma_E)\sigma_{bcr}} \tag{7.96}$$

where σ_a = axial stress due to compression, σ_b = axial stress due to bending (DNV 30.1), σ_{acr} = characteristic buckling stress for axial compression, σ_E = Euler buckling stress about weak axis, σ_{bcr} = characteristic buckling stress due to pure bending (Equation 7.96), α = coefficient depending upon type of structure and reduced slenderness.

Stiffened Plate Buckling

Most floating structures are constructed from stiffened plates which may experience one of the following buckling modes:

Plate buckling	Local buckling of plate between stiffeners
Panel buckling	Buckling of stiffeners and attached plating between stiffeners.
Girder buckling	Overall buckling involving bending of stiffeners and girders with attached plating.
Local buckling of stiffeners and girders	Webs of flanged stiffeners and flanges of box girders act as plates and are subject to the general issue of plate buckling mentioned above. Rules provide guidance on the proportions for these members to relieve the designer of the need for a separate buckling analysis.

The elastic buckling resistance of plate panels is given as (DNV CN 30.1):

$$\sigma_E = C\frac{\pi^2 E}{12(1 - \nu^2)}\left(\frac{t}{s}\right)^2 \tag{7.97}$$

where C = Buckling coefficient dependent upon loading condition, aspect ratio, boundary conditions, s = span between stiffeners (direction of in-plane stress), t = plate thickness, ν = plate thickness. The coefficient, C, is presented graphically in DNV 30.1. The ABS Steel Vessel Rules give the following formulas (C is substituted here for m, the nomenclature used in ABS Rules).

For plating with longitudinal stiffeners (parallel to compressive stress):

$$C = \frac{8.4}{\Psi + 1.1} \qquad (\text{for } 0 \leq \Psi \leq 1) \tag{7.98}$$

For plating with transverse stiffeners (perpendicular to compressive stress):

$$C = c\left[1 + \left(\frac{s}{l}\right)^2\right]^2 \frac{2.1}{\Psi + 1.1} \qquad (\text{for } 0 \leq \Psi \leq 1) \tag{7.99}$$

where l = longer side of plate panel, $c = 1.3$ when plating stiffened by floors or deep girders, = 1.21 when stiffeners are angles or T-sections, = 1.10 when stiffeners are bulb flats, =1.05 when stiffeners are flat bars; and Ψ = ratio of smallest to largest compressive stress varying linearly across panel. For plates subjected to lateral pressure, the stresses may be checked by the following formula (DNV CN 30.1):

$$P_d \leq 4\eta_p \sigma_F \left(\frac{t}{s}\right)^2 \left[\psi_y + \left(\frac{s}{l}\right)^2 \psi_x\right] \tag{7.100}$$

where p_d = design lateral pressure, η_p = maximum usage factor allowed in the Rules.

$$\Psi_y = \frac{1 - (\sigma_e/\sigma_F)^2}{\sqrt{1 - 3/4(\sigma_x/\sigma_F)^2 - 3(\tau/\sigma_F)^2}} \tag{7.101}$$

$$\Psi_x = \frac{1 - (\sigma_e/\sigma_F)^2}{\sqrt{1 - 3/4(\sigma_y/\sigma_F)^2 - 3(\tau/\sigma_F)^2}} \tag{7.102}$$

σ_e = von Mises equivalent stress, σ_x and σ_y are in-plane stresses along and perpendicular to the stiffener axes, respectively (fig. 7.116):

The buckling of stiffeners is treated as a column buckling problem with an equivalent axial load given by (DNV CN 30.1):

$$N_x = \sigma_x(A + st) + B\alpha\sigma_y st + C\tau st \tag{7.103}$$

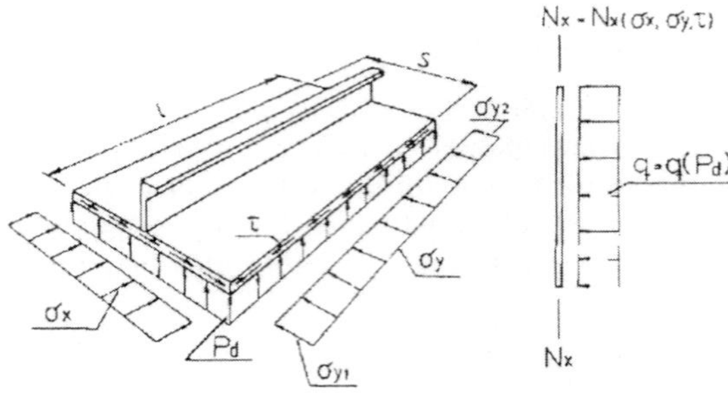

Figure 7.116 Plate stiffener treated like an equivalent beam-column (DNV CN 30.1)

where A = cross-sectional area of stiffener, $B = 0$ if σ_y is less than the elastic buckling stress in the transverse direction in the absence of the stiffener, otherwise B is a function of the stress level (see DNV CN 30.1), $\alpha = 1$ for uniform σ_y, <1 otherwise, and C = function of critical shear stress, simplified, $= l/s$.

Girder Buckling

Girders are connected to the plate and may fail by *plate induced failure* or *flange induced failure*. Plate induced failure results in the girder deflecting away from the plate with yielding in compression at the connection between the girder and the plate. Flange induced failure is caused by torsional buckling (see above) and is typically mitigated by the addition of tripping brackets so that the unsupported length does not exceed the value S_{TO} given by (DNV CN 30.1):

$$S_{TO} = C\sqrt{\frac{EA_f}{\sigma_f(A_f + A_w/3)}} \tag{7.104}$$

where b = flange width, $C = 0.55$ for symmetric flanges, $= 1.10$ for one sided flanges.

Buckling Strength of Longitudinal Hull Girders

Specific buckling criteria for FPSOs are included in the ship design rules, e.g. the ABS Steel Vessel Rules. The critical area is the main deck and longitudinals when subjected to a hogging moment. The plate buckling was covered above (see Equations 7.97–7.103). The following is the critical elastic buckling stress for the longitudinal frames for buckling perpendicular to the plane of the plating.

$$\sigma_E = \frac{EI_a}{c_1 A l^2}\,\mathrm{N/mm^2} \tag{7.105}$$

where I_a = moment of inertia, in cm^4, of longitudinal including plate flange, A = Cross-sectional area of longitudinal including plate flange, l = span, in m, of longitudinal, $C_l = 1000$.

Other buckling modes are covered in the *Rules* (ABS Steel Vessel Rules). The effective flange width is equivalent to that determined for strength considerations. Other failure modes, especially the torsional buckling mode, are covered in the *Rules*.

The ABS Rules call for the design buckling stress, σ_c, of plate panels and longitudinals to be greater than $1.0\sigma_a$ for plate panels and $1.1\sigma_a$ for stiffeners. σ_a is the design stress computed including the deadload and wave loads, i.e. the same stress used for strength design. This should incorporate adjustments as discussed under the strength section for site specific environments. σ_c is given as:

$$\sigma_c = \sigma_E \qquad \text{when } \sigma_E \le \frac{\sigma_F}{2} \tag{7.106}$$

$$\sigma_c = \sigma_F\left(1 - \frac{\sigma_F}{4\sigma_E}\right) \qquad \text{when} \quad \sigma_E \ge \frac{\sigma_F}{2} \tag{7.107}$$

Cylindrical Shells

Large diameter cylindrical shells common to column and pontoon design for many semi-submersibles and TLPs, and typical production spars, are treated as stiffened flat plate structures with the plates representing the shell structure between internal bulkheads. Stiffened plate buckling using documents such as DNV CN 30.1, API Bull 2V or ABS Steel Vessel Rules are typically applied to these plates. Cylindrical shell formulations are important for columns which are ring stiffened, for pressure vessels such as the buoyancy cans supporting spar risers, or the cylinders comprising a cell spar (fig. 7.73).

Cylindrical shells may be ring stiffened, longitudinally stiffened or orthogonally stiffened. The buckling modes (fig. 7.117) consist of the following (DNV CN 30.1).

1. Shell buckling: buckling of plates between the rings and the longitudinal stiffeners,
2. Panel stiffener buckling: buckling of shell plating including longitudinal stiffeners. Rings are nodal lines.
3. Panel ring buckling: buckling of shell plating including rings. Longitudinal stiffeners are nodal lines.
4. General buckling: buckling of shell plating including longitudinal stiffeners and rings.
5. Column buckling: buckling of the cylinder as a column
6. Local buckling of longitudinal stiffeners and rings.

Figure 9 from DNV CN 30.1 shows the buckling modes associated with these configurations.

As in any cases of shell buckling, the first step is to compute the state of stress for appropriate load cases for the assumed unbuckled geometry. These stresses are applied to the appropriate rule formula to determine the buckling usage factor.

For simple geometry the stresses may be calculated from the beam or shell theory solutions, which are readily, available in the references cited here. *Roark's Formulas for Stress and Strain* (Roark et al, 1989) is also useful reference for this purpose.

The most common cylindrical shell load in floaters consists of external pressure accompanied by axial compression or bending. Membrane stresses in the shell plating are completely defined by three stress components:

σ_x = normal stress in the longitudinal direction
σ_θ = normal stress in the circumferential direction
τ = shear stress tangential to the shell

The axial stress arises from the overall axial force, N, and the bending moments about the Principal axes, M_1 and M_2.

$$\sigma_x = \sigma_a + \sigma_b \tag{7.108}$$

$$\sigma_a = \frac{N}{2\pi r t} \quad \text{axial stress} \tag{7.109}$$

$$\sigma_b = \frac{M_1}{\pi r^2 t}\sin\theta + \frac{M_2}{\pi r^2 t}\cos\theta \quad \text{bending stress} \tag{7.110}$$

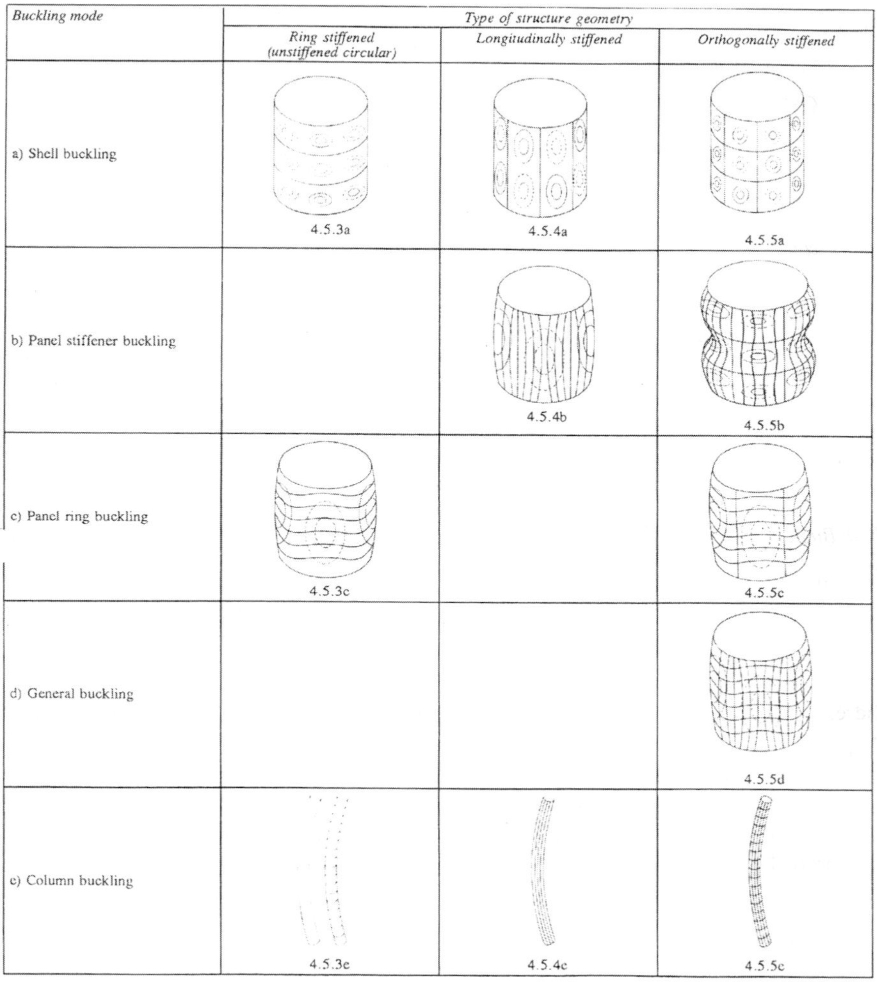

Buckling mode	*Type of structure geometry*		
	Ring stiffened (unstiffened circular)	*Longitudinally stiffened*	*Orthogonally stiffened*
a) Shell buckling	4.5.3a	4.5.4a	4.5.5a
b) Panel stiffener buckling		4.5.4b	4.5.5b
c) Panel ring buckling	4.5.3c		4.5.5c
d) General buckling			4.5.5d
e) Column buckling	4.5.3e	4.5.4e	4.5.5e

Figure 7.117 Buckling modes for different types of stiffeners (DNV CN 30.1)

For a shell with longitudinal stiffeners it is permissible to replace t with an equivalent t_e, which accounts for the area of the stiffeners. Shear stress for a simple cylinder arises from torque, T, and shearing forces, Q. Similar formulas define the shear forces.

The membrane hoop stress due to an external pressure p, midway between to ring stiffeners, is given as:

$$\sigma_\theta = \frac{pr}{t} - \frac{\alpha\zeta}{\alpha + 1}\left(\frac{pr}{t} - \nu\sigma_x\right) \quad (7.111)$$

where,

$$\zeta = 2\frac{\sinh\beta\cos\beta + \cosh\beta\sin\beta}{\sinh\beta + \sin 2\beta} \tag{7.112}$$

$$\beta = \frac{l}{1.56\sqrt{rt}} \tag{7.113}$$

$$\alpha = \frac{A_R}{l_{eo}t} \tag{7.114}$$

$$l_{eo} = \frac{l}{\beta}\frac{\cosh 2\beta - \cos 2\beta}{\sinh 2\beta + \sin 2\beta} \tag{7.115}$$

(effective shell plating flange width, see fig. 7.98)

For multi-axial stress states a von-Mises equivalent stress may be used (see DNV CN 30.1). The formulas for critical buckling stresses may be found in DNV CVN 30.1, API Bulletin 2U and other references. Some of the basic load cases are presented here.

Shell Buckling

The critical elastic buckling strength of an unstiffened circular cylinder is:

$$\sigma_E = C\frac{\pi^2 E}{12(1-\nu^2)}\left(\frac{t}{l}\right)^2 \tag{7.116}$$

where,

$$C = \psi\sqrt{1 + \left(\frac{\rho\xi}{\psi}\right)^2} \tag{7.117}$$

The coefficients ψ, ξ and ρ are given in table 7.17. Z is defined as:

$$Z = \frac{l}{rt}\sqrt{1-\nu^2} \tag{7.118}$$

Table 7.17 Buckling Coefficient for Unstiffened Cylinder (DNV CN 30.1)

	ψ	ξ	ρ
Axial stress	1	0.702 Z	$0.5\left(1+\frac{r}{150t}\right)^{-0.5}$
Bending	1	0.702 Z	$0.5\left(1+\frac{r}{300t}\right)^{-0.5}$
Torsion and shear force	5.34	$0.856\ Z^{3/4}$	0.6
Lateral pressure	4	$1.04\sqrt{Z}$	0.6
Hydrostatic pressure	2	$1.04\sqrt{Z}$	0.6

Ring Stiffened Cylinder

Shell buckling for a ring stiffened cylinder is as above with l being the ring spacing.

The moment of inertia for the ring frames is required to meet certain minimum standards for each loading condition. For a uniform external pressure the moment of inertia, I_θ shall be less than (DNV CN 30.1):

$$I_\theta = \frac{prr_o^2 l}{3E}\left[2 + \frac{3Ez_t\delta_o}{r_o^2(\sigma_k/2 - \sigma_{\theta R})}\right] \quad (7.119)$$

where $\sigma_k \cong 0.9$ to $1.0\ \sigma_F$ provided certain conditions are met (see DNV CN 30.1), z_t and r_o are defined in fig. 7.118, δ_o is the initial out of roundness assumed to equal 0.005r.

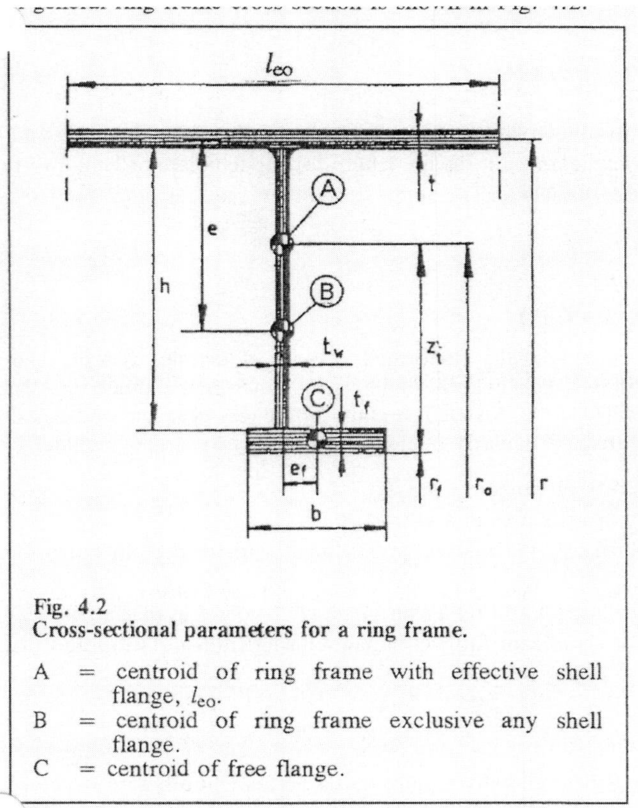

Figure 7.118 Cross-sectional parameters for a ring frame (DNV CN 30.1)

Tolerances

Buckling of cylindrical shells is particularly sensitive to fabrication tolerances. Buckling formulas in the literature cited here are generally applicable to cylinders built within the following tolerances:

Difference between the maximum and the minimum diameter at any cross-section should not exceed 1% of the nominal diameter for that cross-section (Bull 2U),

For cylinders subject to axial compression, the deviation from a straight line measured along a meridian over a gage length, L_x, should not exceed $0.01L_x$ (Bull 2U).

The out-of-straightness of longitudinal stiffeners shall not exceed. $0015L_x$, where L_x is the unsupported length of the stiffener (DNV CN 30.1),

The misalignment of longitudinal stiffeners shall not exceed .02s, where s is the stiffener spacing (DNV CN 30.1),

The local out-of-roundness and local out of straightness of the shell are to be measured from a circular template or a straight rod held anywhere on the shell. The local shell imperfection shall not exceed the value:

$$\delta = \frac{.01g}{1 + \frac{g}{r}} \tag{7.120}$$

where g = length of template or rod (see CN 30.1 for details). Bull. 2U and DNV CN 30.1 both provide guidance on the arc length for the template or length of the rod used for the above measurements. The out of straightness requirements apply to stiffeners on flat plate as well.

7.8.12 Fatigue

Fatigue analysis is typically performed as part of detailed design. The procedure for floating structures in principle the same as for other civil engineering structures (see, for example AASHTO, 1990, SAE). The main differences between land based structures and offshore structures, particularly floaters, with respect to fatigue include:

- The complexity of loads in a seaway,
- Complexity of responses in a seaway,
- The complexity of the marine environment with respect to corrosion and cathodic protection,
- Difficulty to inspect and repair an offshore structure in place,
- Potential for significant fatigue damage during transportation and installation of the platform

Fatigue Life

Offshore Class Rules generally require that a permanent offshore structure meet a fatigue life which is at least three times the intended service life of the platform. Areas of a platform which are not accessible for inspection must meet a fatigue life of ten times the service life.

Also, critical components such as TLP tendons and risers from floaters typically are designed for a fatigue life of ten times the service life.

The minimum fatigue life for MODUs is 20 years or the design life of the vessel, whichever is greater. MODUs are a special case because they may be drydocked at regular intervals.

For comparison, fixed offshore structures typically require a design fatigue life of two times the intended service life (API RP2A).

The requirement for longer design lives for floaters is a result of perceived uncertainty in the dynamic responses of these platforms. Shorter fatigue lives might be acceptable if responses and structural loads are validated by model test, coupled analysis, etc., and if a monitoring program is incorporated into the operations.

Conservative design practices have generally paid off. There are few documented failures of offshore platforms due to fatigue failure. A notable exception is the semi-submersible Alexander Kielland in 1984 which failed due to critical fatigue crack growth (Thomas, 1984). This example illustrates a general issue with offshore platform design. The Kielland failed because an attachment was welded to a column of the semi after it was commissioned. This attachment introduced a stress concentration which resulted in a fatigue crack. The basic design of the Alexander Kielland was safe, but a modification added after delivery was not subjected to the same engineering rigor as the basic structure.

Other marine disasters, notably Piper Alpha (U. of California, 1993), Ocean Ranger (Natl. Research Council of Canada, 1987) and P-36 (Filho, 2002) have illustrated the fact that the biggest problems arise from operational decisions, not basic design. The response of the offshore community to these disasters has generally been to implement standards which make the consequences of human error less disastrous.

Fatigue Loads

Fatigue loads are generally computed using the same methods used for Global Strength analysis discussed above. For column stabilized platforms, spars and TLPs the method usually employed is a space frame model employing a modified Morison's equation for wave loading. The "modification" amounts to inclusion of the relative wave and body motions into Morison's equation. Since a large part of fatigue damage can come from relatively smaller sea states, it may be important to modify the application of Morison's equation to account for wave diffraction and radiation effects. Generally, however, Morison's equation tends to overpredict the loads from shorter waves (Datta, 2000) and might be conservative.

For ships and FPSOs it is recommended to use a three dimensional radiation-diffraction program to compute responses (ABS Guidance Notes on Spectral Fatigue Analysis). Loads and accelerations from a three dimensional radiation-diffraction solution are mapped onto a structural model of a full vessel for this analysis. The hydrodynamic grid is generally coarser than even a coarse structural grid, so this pressure mapping involves interpolation of the pressures. A relatively coarse grid structural model may be used, with fine grids or sub-structuring used for the fatigue prone areas. Part models of a vessel may be analyzed as long as the imposed loads satisfy equilibrium conditions for that part of the ship, however shear lag effects are difficult to capture with this approach. The linear wave radiation-diffraction solutions have the deficiency of failing to represent the non-linear wave loadings at the waterline. Recent research on this subject has lead to some empirical representations for the wave pressure fields for use in spectral fatigue analysis (Kaminski et al, 2004).

Spectral Fatigue Analysis

Spectral fatigue analysis procedures specified by ABS follow the development of spectral methods for fixed platform design (see, e.g., Sherf and Tuestad, 1987) and can be applied to ship shaped hulls and other offshore floaters equally, with a few caveats discussed below. The first step is to derive a transfer function (RAO), $H_\sigma(\omega/\theta)$, for the stress range of interest. This function is in general a function of vessel heading.

The stress energy spectrum becomes,

$$S_\sigma(\omega|H_S, T_Z, \theta) = ||H_\sigma(\omega|\theta)|^2 \cdot S_\eta(\omega|H_S, T_Z) \quad (7.121)$$

And the spectral moments are,

$$m_n = \int_0^\infty \omega^n S_\sigma(\omega|H_S, T_Z, \theta) d\omega \quad (7.122)$$

ABS Guidelines suggest applying a $\cos(\theta)^2$ spreading function to the wave spectra to represent a directional sea spectrum. It is common in offshore engineering to assume collinear fatigue sea states as a conservative assumption (Liu and Wang, 2002).

Spectral damage calculations are derived from the assumption that the short term probability distribution of the stress cycles is Rayleigh distributed, the probability distribution given by

$$g(S) = \frac{S}{\sigma^2} \exp\left[-\frac{S}{2\sigma^2}\right] \quad (7.123)$$

The zero-crossing period and bandwidth parameter are:

$$f = \frac{1}{2\pi}\sqrt{\frac{m_2}{m_0}} \qquad \varepsilon = \sqrt{1 - \frac{m_2^2}{m_0 m_4}} \quad (7.124)$$

Assuming a Palmgren-Miner superposition of fatigue damage, and assuming the S–N curve coefficients are constant over the range of stresses, the fatigue damage from an individual sea state can be computed in closed form. The damage from a number of sea states is given by

$$D = \frac{T}{A} 2^{m/2} \Gamma(m/2 + 1) \sum_i \lambda(m, \varepsilon_i) f_{0i} p_i \cdot \sigma_i^m \quad (7.125)$$

where

σ_i = standard deviation of S for the ith sea state (m_0)

A and m are parameters of the S–N curve (see below)

T = the duration for the damage calculation

$\lambda(m, \varepsilon_i)$ = rainflow correction factor to account for the bandwidth of the spectrum (Wirshing et al, 1981)

$$\lambda(m, \varepsilon_i) = a(m) + [1 - a(m)][1 - \varepsilon_i]^{b(m)} \quad (7.126)$$

where

$a(m) = 0.926 - 0.033m$
$b(m) = 1.587m - 2.323$
ε_i = Spectral Bandwidth

ABS Guidance notes provide the formulation for an S–N curve with non-constant coefficients.

Other spectral fatigue formulations have been proposed for broad banded processes (see, e.g. Karadeniz, Bishopp, Jha).

This approach appears satisfactory for most floating structures, however it is becoming more common to perform fatigue analysis using explicit rainflow counting methods with time domain simulation of loads (see, e.g., Wang et al, 2001). This method allows inclusion of effects that might not be captured in linear wave analysis, e.g. pitch motions of a spar due to wind gusts, gravitational loads due to pitch motions. Figure 7.119 shows an example of the response spectrum for a spar in an extreme sea. A large part of the slowly varying pitch moment is due to the gravitational load due to pitch (g*sin(θ)).

Tension Leg Platforms have an important natural period which is below the range of wave energy. Higher order (non-linear) wave forces cause responses at the natural period

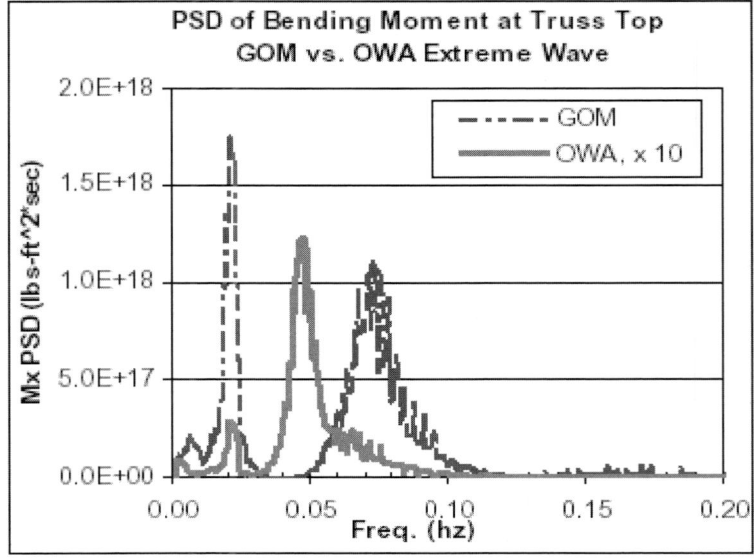

Figure 7.119 Bending moment spectrum for truss spar truss-hard tank Connection (Wang et al, 2001)

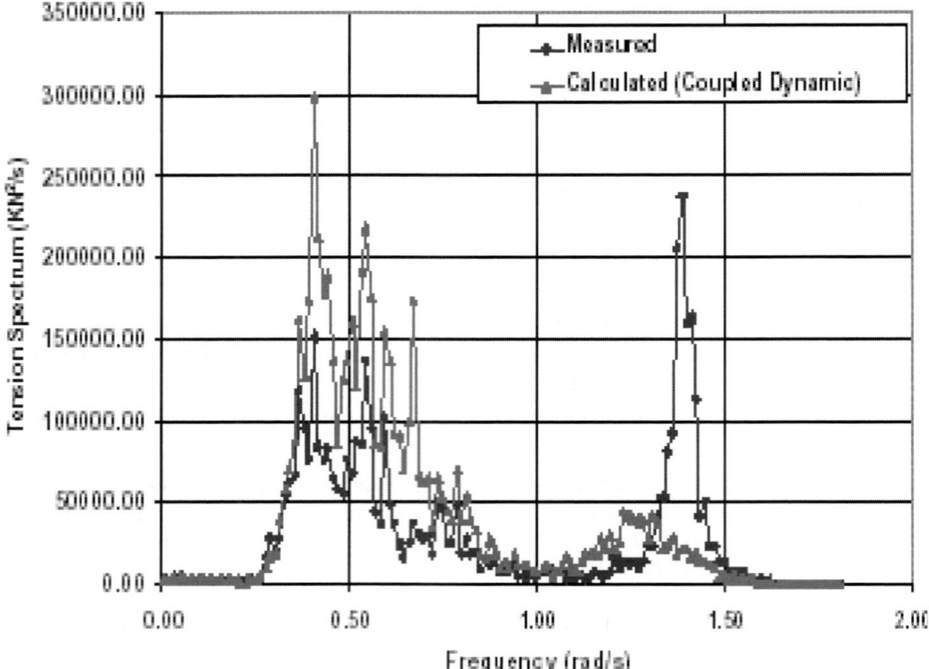

Figure 7.120 TLP tendon tension (Chen, 2002)

("springing"), which are important contributors to the fatigue of the tendons and the hull. Figure 7.120 illustrates this response. The peak at 1.40 radians per second shows the springing response which is not captured in the analysis.

Neither of the illustrated responses described here would be included in a linear analysis. The low frequency responses are best captured by a non-linear time domain solution (Glanville et al, 1991). The high frequency springing responses for TLPs have been found to be more of a factor in operational (i.e. fatigue) seas than they are in extreme seas (Sircar, 1993).

The best method for capturing these effects is to use a validated time domain analysis tool which captures these non-linear effects. Time domain analysis allows standard rainflow counting procedures (ASTM E 1045) to be employed.

Rainflow Counting Method

The rainflow method was first introduced by (Matsuishi and Endo, 1968). It is based on the theory that fatigue damage is related to the number of closed stress loops in a time series. That is, a closed loop exists when the stress starts from a local minimum or maximum and completes a cycle back to the same value. Formally, the method is applied by representing the time series as a series of straight lines connecting local maxima and minima. To eliminate counting half-cycles the series begins with a maximum value. A raindrop is initiated at each local maxima and minima, and it continues until:

1. The rain passes a local maxima (minima) point equal to or greater (lesser) than it's starting point, or
2. It encounters a previous rainflow

The principle is illustrated in fig. 7.121. There are 8 local maxima/minima here and 8 raindrops. The are 4 closed loops which can be characterized by a maximum, minimum and a mean value. Fatigue damage is computed by tabulating the damage from each of these closed loops.

Numerous rainflow counting algorithms exist, many can be found by searching the web. An example of a simple algorithm is given in fig. 7.122 (University of Iowa). The input consists of an order tabulation of the local maxima and minima in the time series. Time series data must be processed to derive the local peaks and valleys in order to apply this algorithm.

The "rainflow factor" used above in the spectral method has been shown to yield reasonably consistent results with time domain simulations using the actual rainflow method (Liu et al, 2003).

Fatigue Stresses

Application of either the spectral or the rainflow method requires selection of a fatigue S–N curve and a load-stress transfer function. The stress used for fatigue calculations should be the principle stress perpendicular to the direction of crack formation.

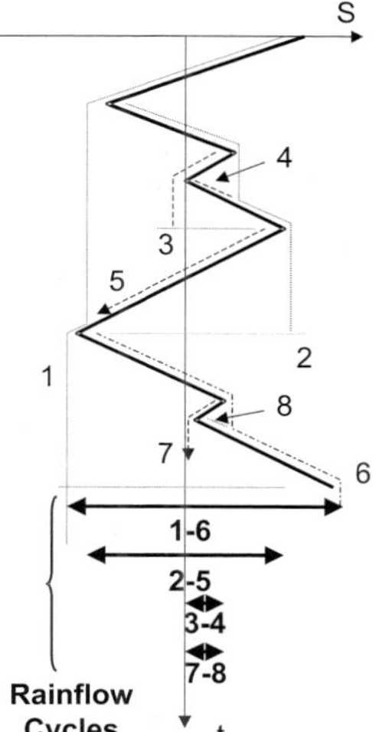

1. passes an equally large maximum
2. Passes a larger minimum
3. Passes a larger maximum
4. Reaches the run of drop 2
5. Reaches the run of drop 1
6. Falls out
7. Falls out
8. Reaches the run of drop 6

Figure 7.121 Example of rainflow counting

```
        Dimension E(50)
        N = 0
1       N = N + 1
        Call Data(E(N),K)
2       If (N.LT.3) Go To 1
        X = ABS(E(N)–E(N-1)
        Y = ABS(E(N-1)–E(N-2))
3       If (X.LT.Y) Go To 1
4       Range = Y
        Xmean = (E(N-1) + E(N-2))
        E(N) = E(N + 2)
        Go To 2
        End
```

Figure 7.122 Example of rainflow counting algorithm

There are two basic methods for specifying stress:

1. Nominal stress
2. "hot spot" stress

The nominal stress is the stress in a plate or beam away from the discontinuity of interest without taking into account the stress concentration at the discontinuity. This method requires selection of S–N curves which pertain to a particular structural detail. We will include examples of details and appropriate S–N curves based on industry standards later in this section. The hot-spot stress is the surface stress at the toe of a weld where fatigue life is desired. This includes stress concentrations due to local geometry, but not due to the weld itself. The hot-spot stress method has evolved since the 1970s for the analysis of tubular joints in fixed structures (see Efthymiou, 1988), but it has not been widely used in plated structures until recently (Bergan and Lotsberg, 2004). Hot-spot stress calculations for plated structures require extrapolation of finite element analysis results from a fine mesh to the position of the weld toe. An example of this is illustrated in fig. 7.123 where the stresses at a distance of t/2 and 3t/2 are linearly extrapolated to the position of the weld toe to derive a hot-spot stress. Classification rules and recommended practices set out guidance on the exact application of finite element analysis methods to derived these stresses (see DNV RP-C203, 2000 and American Bureau of Shipping, 2003). It is still recommended that, for a particular finite element procedure used for hot-spot stress analysis, the results of the hot-spot analysis be compared with nominal stress S–N curve results for a standard detail before application of the method to a general plated structure (DNV RP-C203).

The hot-spot stress approach suffers from a wide scatter in the derived stress based on the mesh size and type used in the analysis (Healy, 2004, Dong, 2004). A new method, termed

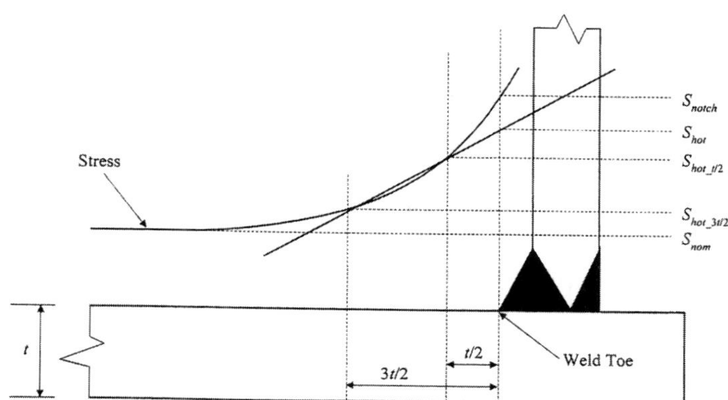

Figure 7.123 Definition of hot spot stress (ABS guidance notes on spectral fatigue)

the "structural stress" approach, has recently been introduced (Dong, 2004) which appears to be mesh insensitive. This method determines an equivalent linear through-thickness stress distribution at the weld toe based on energy considerations. The surface value of this equivalent stress has been shown in some examples to be mesh insensitive. So far this method has not been incorporated in classification rules or other industry standards, but research is continuing under industry sponsorship.

S–N Curves

The S–N curves developed by the UK Department of Energy (UK HSE, BS 7608) are commonly used for fatigue analysis of floating platforms. These curves have been adopted by the IACS Classification Societies. The curves are represented in the form

$$N = AS^{-m} \tag{7.127}$$

N is the cycles to failure, S is the stress range (either hot-spot or nominal stress range depending on the application), A and m are constants, however there is an inflection point where these constants change for high cycle fatigue.

The coefficients A, and m for the "Basic" HSE design curves are given in table 7.18 (based on stresses in MPa). Also included for reference are S–N parameters for the API X and X′ curves. These latter curves are not used for floating structures, however they are sometimes used for riser joints. The curves presented here are derived from experimental results on joints of particular geometries. The design curves represent a 2-sigma upper bound to the experimental data. A1, m1 and A2, m2 are values of A and m in eqn. 7.127 for low cycle and high cycle fatigue, respectively. All of the HSE curves have an inflection point at $N = 10^7$ cycles. The curves are plotted in fig. 7.124. Curve B represents the base metal curve,

Table 7.18 Parameters for S–N Curves

Class	A1	m1	A2	m2	Stress	Example Applications
X	1.15E + 15	4.38	(35 Mpa Endurence Limit)		hot spot	Tubular joints with profile control, no undercut (see API RP2A)
X′	2.50E + 13	3.74	(23 Mpa Endurance Limit)		hot spot	Tubular joints without profile control (see API RP2A)
B	1.01E + 15	4	1.01E + 17	6	nominal	Plain steel in as rolled condition (UK DOE)
C	4.23E + 13	3.5	2.93E + 16	5.5	nominal	Welds parallel to the direction of stress (UK DOE)
D	1.52E + 12	3	4.24E + 15	5	nominal or hot spot	Full penetration welds with smooth profile (UK DOE), or DNV RP-C203 recommends using this with hot spot stress.
E	1.04E + 12	3	2.30E + 15	5	nominal or hot spot	Full penetration welds without profile control (UK DOE), or ABS recommends using this hot spot stress (ABS Fatigue Guide)
F	6.32E + 11	3	9.97E + 14	5	nominal	Parent metal attached to the end of discontinuous welds with cope holes (UK DOE)
F2	4.30E + 11	3	5.28E + 14	5	nominal	Fillet welded lap joints (UKDOE)
G	2.48E + 11	3	2.14E + 14	5	nominal	Parent metal at ends of load carrying fillet welds which are essentially parallel to the direction of stress (UK DOE)
W	1.57E + 11	3	1.02E + 14	5	nominal	Any lap joint based on stress at weld throat area (UK DOE)

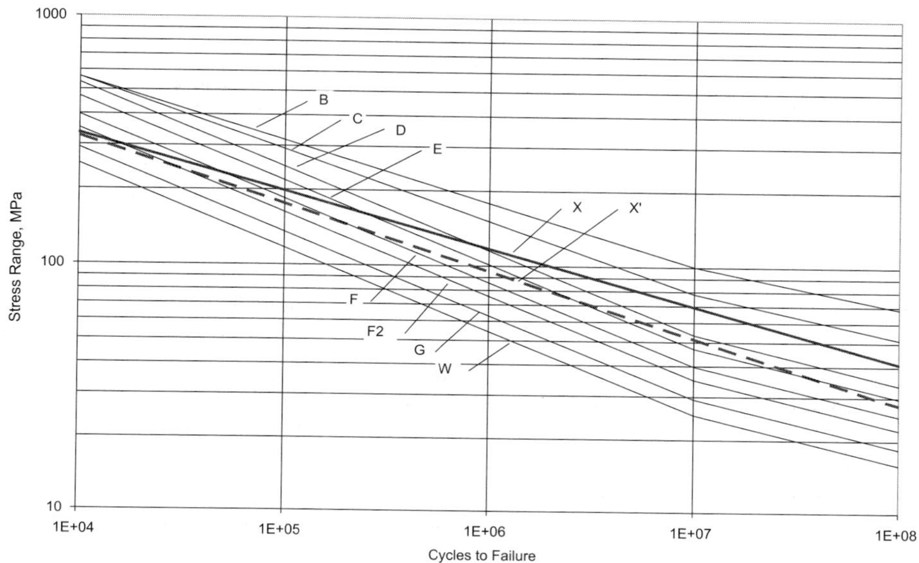

Figure 7.124 S–N curves

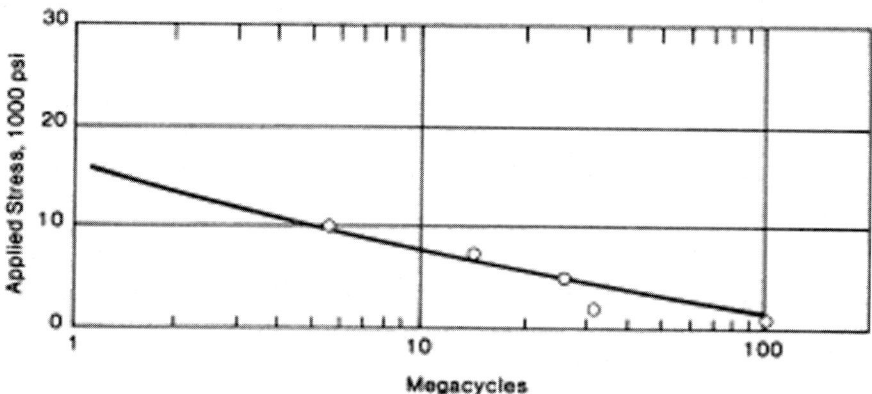

Figure 7.125 Corrosion fatigue of mild steel rotated at 1400 rpm in seawater (Kirk, 1968)

all others are for welded joints of various types. Figures 7.126–7.131 (from BS 7608) show examples of the classification of geometries for application of these curves. Table 7.18 indicates the most common applications for each class of S–N curve.

ABS Guidelines recommend using the "E" curve in conjunction with a hot spot stress. DNV (RP-C203) recommends use of the "D" curve with hot spot stress.

Table 1. Classification of details: plain material free from welding

Product form	Location of potential crack initiation	Dimensional requirements	Manufacturing requirements	Special inspection requirements	Design stress area	Type number	Class	Notes	Sketch
Rolled steel structural plates and sections	Away from all welds or structural connections	Member of constant or smoothly varying cross section with no holes or re-entrant corners	All surfaces fully machined and polished. No flame cutting			1.1	A	See **B.2.2**	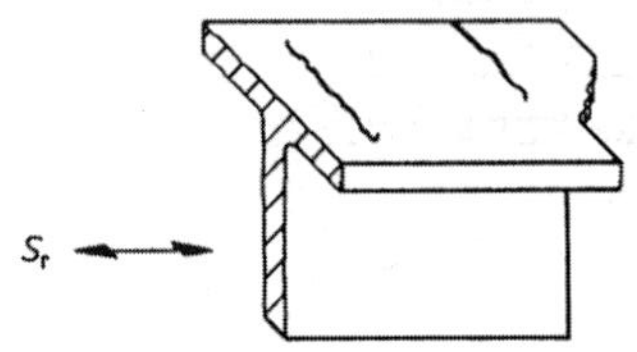 S_r
			Edges as rolled or machined smooth. No flame cutting			1.2	B		
	At any external or internal edge	Member with apertures, re-entrant corners or other discontinuities	Any flame cut edges subsequently machined or ground smooth		Net cross section	1.3	B	All visible signs of drag lines should be removed from the flame cut edge by grinding or machining. See also note for type 1.4.	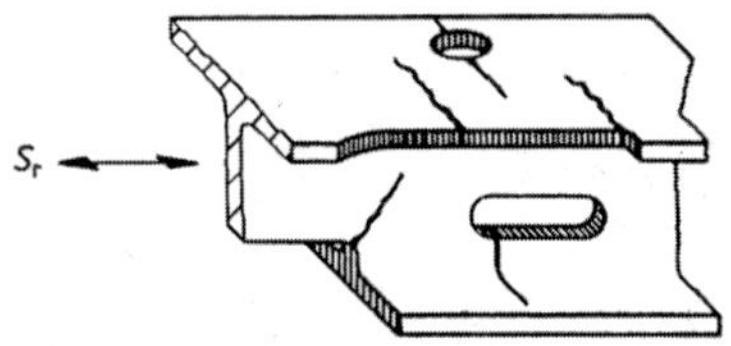 S_r
			Any cutting of edges by planing or machine flame cutting with controlled procedure		Net cross section	1.4	C	The controlled flame cutting procedure should ensure that the resulting surface hardness is not sufficient to cause cracking. Types 1.3 and 1.4. The presence of an aperture, re-entrant corner or other discontinuity implies the existence of a stress concentration and the design stress should be the stress on the net section multiplied by the relevant stress concentration factor (see annex C).	
	At a small hole (may contain bolt for minor fixtures)		Hole drilled or reamed		Net cross section	1.5	D	This type may be deemed to include bolt holes for attaching light bracing members where there is negligible transference of stress from the main member in the direction S_r. The classification includes allowance for the stress concentration created by the hole.	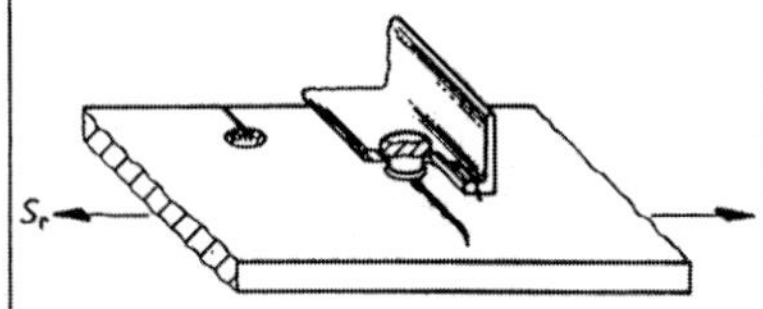 S_r

Figure 7.126 Plane material free from welding (BS 7608)

Table 4. Classification of details: continuous welded attachments essentially parallel to the direction of applied stress

Product form	Location of potential crack initiation	Dimensional requirements	Manufacturing requirements	Special inspection requirements	Design stress area	Type number	Class	Notes	Sketch
Rolled steel structural plates, sections and built-up members	At a long welded attachment (in the direction of S_r), away from the weld end	Butt weld with full penetration and no backing strip	Weld reinforcement dressed flush	Proved free of all flaws which are likely to degrade the joint below its stated classification (see 2.4.3)	Minimum transverse cross section of member at location of potential crack initiation	4.1	B	Finish machining should be in the direction of S_r. The significance of flaws should be determined with the aid of specialist advice and/or by the use of a fracture mechanics analysis. The non-destructive testing (NDT) technique should be selected with a view to ensuring the detection of such significant flaws. This type is only recommended for use in exceptional circumstances.	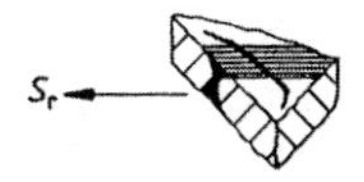 S_r
			Automatic weld with no stop/starts			4.2	C	Accidental stop/starts are not uncommon in automatic processes. Repair to the standard of a C classification should be the subject of specialist advice and inspection and as a result, the use of this type is not recommended.	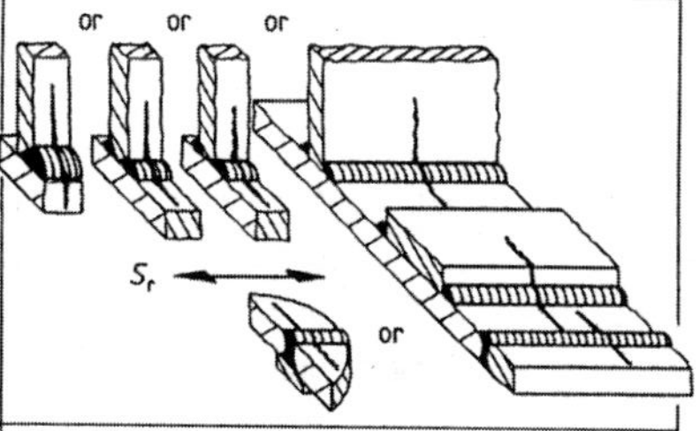 or or or
			Welds with stop/starts			4.3	D	For situation at the ends of flange cover plates see joint type 5.4. Backing strips, if used, need to be continuous and either not attached or attached by continuous fillet welds. If the backing strip is attached by discontinuous fillet welds see type 4.6.	S_r or
	At an intermediate gap in a longitudinal weld	Intermittent fillet weld with $\frac{g}{h} \leq 2.5$				4.4	E	The limiting gap ratio g/h applies even though adjacent welds may be on opposite sides of a narrow attachment (as in the case of a longitudinal stiffener with staggered fillet welds). Long gaps between intermittent fillet welds are not recommended as they increase the risk of corrosion and, in the case of compression members, may cause local buckling. If intermediate gaps longer than 2.5h are required the class should be reduced to F.	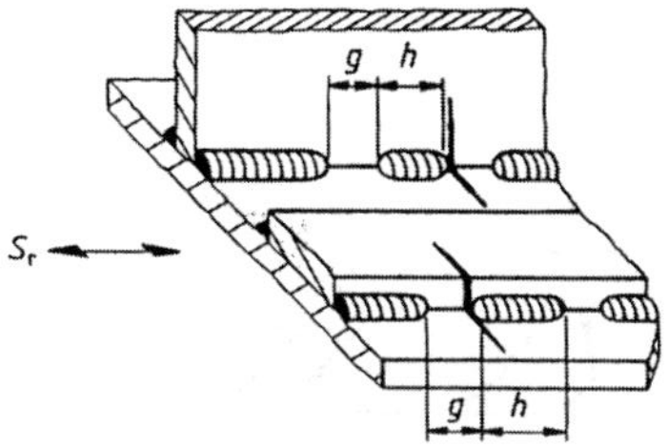 g h S_r g h

Figure 7.127 Continous welded attachments essentially parallel to the direction of applied stress (BS 7608)

Table 5. Classification of details: welded attachments on the surface or edge of a stressed member

Product form	Location of potential crack initiation	Dimensional requirements	Manufacturing requirements	Special inspection requirements	Design stress area	Type number	Class	Notes	Sketch
Rolled steel structural plates, sections, tubes, and built-up members	At weld toe of stud shear connector				Minimum transverse cross section of member at location of potential crack initiation	5.1	F		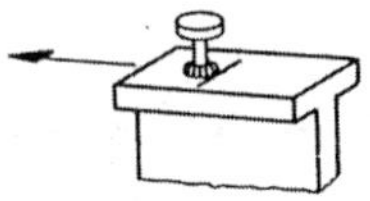
	At weld toe or end of a short attachment (in direction of S_r)	Weld length (parallel to S_r) $l \leq 150$ mm				5.2	F	Butt welded joints should be made with an additional reinforcing fillet so as to provide a similar toe profile to that which would exist in a fillet welded joint.	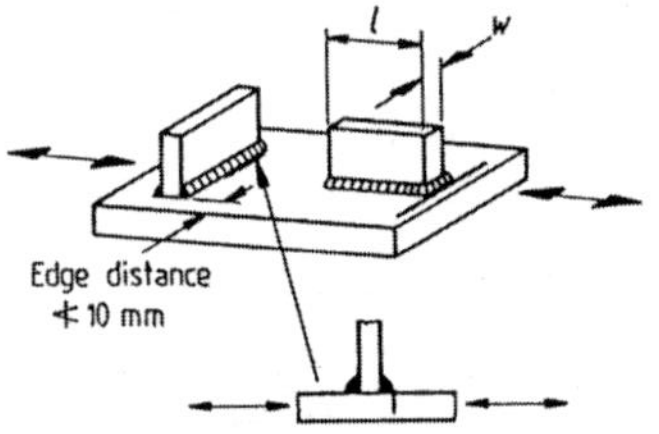 Edge distance ≮ 10 mm
	At weld toe or end of a long, narrow attachment	$l > 150$ mm $w \leq 50$ mm Weld not within 10 mm of member edge				5.3	F2	The decrease in fatigue strength with increasing attachment length is because more load is transferred into the longer gusset, giving an increase in stress concentration.	
	At weld toe or end of a long, wide attachment	$l > 150$ mm $w > 50$ mm Weld not within 10 mm of member edge				5.4	G	The classification may be deemed to include stress concentrations arising from normal eccentricities in the thickness direction. This type includes parent metal adjacent to the ends of flange cover plates regardless of the shape of the ends.	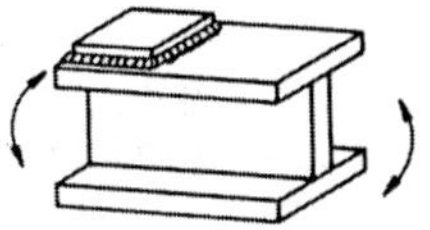
	At weld toe or end of any attachment close to edge of stressed member	Weld within 10 mm of member edge	Grind out undercut			5.5	G	This type applies regardless of the shape of the end of the attachment. In all cases, care should be taken to avoid undercut on element corners or to grind it out to a smooth profile should it occur. In particular, weld returns across a corner should be avoided and the use of cover plates wider than the flange, to which they are attached, is not recommended. The classification applies to all sizes of attachment. It would therefore include, for example, the junction of two flanges at right angles. In such situations a low fatigue classification can often be avoided by the use of a transition plate (see also joint type 6.5).	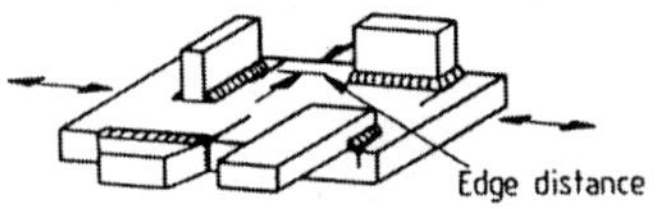 Edge distance

Figure 7.128 Welded attachments on the surface or edge of a stressed member

Table 6. Classification of details: full penetration butt welds between co-planar plates[1)]

Product form	Location of potential crack initiation	Dimensional requirements	Manufacturing requirements	Special inspection requirements	Design stress area	Type number	Class	Notes	Sketch
Rolled steel plates only	At transverse butt weld joining two single plates end to end	Plates of equal width and thickness Longitudinal axes in line	Misalignment slope ≤ 1 in 4 Dress flush reinforcement	Proved free of all flaws which are likely to degrade the joint below its stated classification (see **4.3**)	Minimum transverse cross section of member at location of crack initiation	6.1	C	The significance of flaws should be determined with the aid of specialist advice and/or by the use of a fracture mechanics analysis. The NDT technique should be selected with a view to ensuring the detection of such significant flaws. This class should not normally be used in structural work. (See **B.2.2** and **B.6.2.6**.)	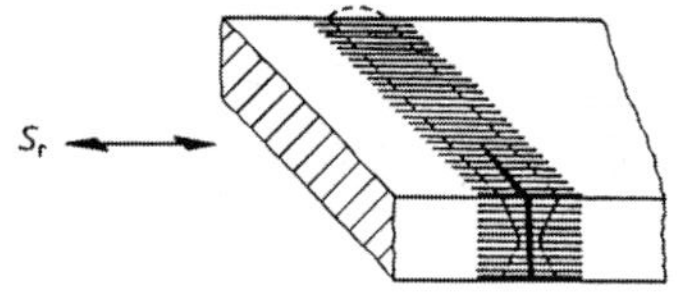 S_r
		Longitudinal axes in line Any width or thickness change ≤ 1 in 4 slope	Flat position shop welds, not submerged arc. Grind smooth any undercut			6.2	D	Shop welds made entirely in the flat position, either manually or by an automatic process other than submerged arc, tend to have a better reinforcement shape from the point of view of fatigue than positional, site or submerged arc welds, i.e. larger re-entrant angles at the toes.	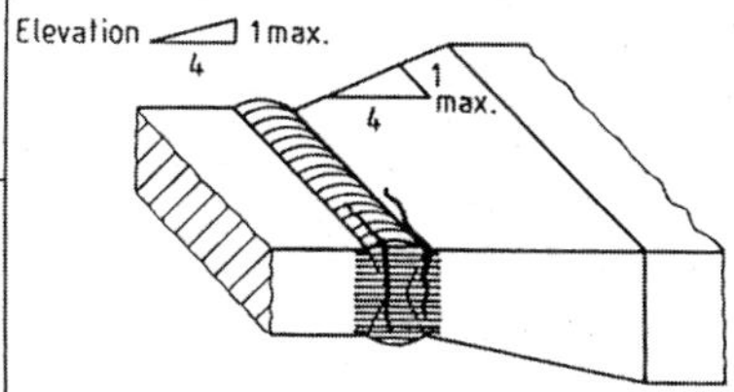 Elevation 1 max. 4; 1 max. 4
			Other weld processes and positions than those in type 6.2. Grind smooth any undercut			6.3	E	This category can be raised to class D provided that the reinforcement is ground smooth until flush with the plate surface on both sides, in addition to grinding smooth any undercut. Any NDT should be carried out after grinding.	
			On permanent backing strip No permanent tack welds within 10 mm of edge. Grind smooth any undercut			6.4	F	If the backing strip is fillet or tack welded to the plate (type 5.2) the detail class will not be reduced below class F unless permanently tacked within 10 mm of the member edge, in which case it will be class G (type 5.5).	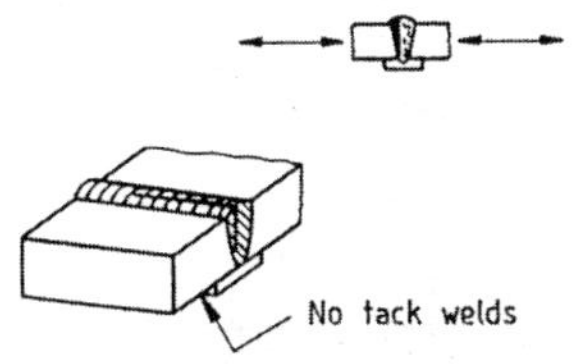 No tack welds

[1)] The joints covered by this table may also fail from internal weld flaws if they are more severe than the external geometrical discontinuity. Weld quality is therefore pertinent to the various classifications (see **2.4.3**).

Figure 7.129 Full penetration butt welds between co-planar plates (BS 7608)

Table 7. Classification of details: transverse butt welds in sections and tubes

Product form	Location of potential crack initiation	Dimensional requirements	Manufacturing requirements	Special inspection requirements	Design stress area	Type number	Class	Notes	Sketch
Rolled steel sections and built-up members						7.1	F2	Butt welds between rolled sections or between built-up sections are prone to weld flaws, which are difficult to detect, in the region of the web/flange junction. Special preparations, procedures and inspection may be undertaken in exceptional circumstances and type 6.3 may then be applied unless the weld is made on a permanent backing (type 6.4). Dressing of the weld reinforcement is advised to overcome poor reinforcement shape resulting from the greater misalignments which may occur in the jointing of sections. This joint is frequently made using a cope hole. This gives improved access to the flange butt welds when webs or longitudinal stiffeners have already been attached. The end of the web butt weld at the cope hole can be considered to be equivalent to class D with the appropriate stress concentration factor provided that the end of the butt weld and the reinforcement within a distance equal to the radius r are ground flush (see figure B.8). If there is no grinding class E with the appropriate stress concentration factor should be used. Cope holes of triangular shape are not recommended.	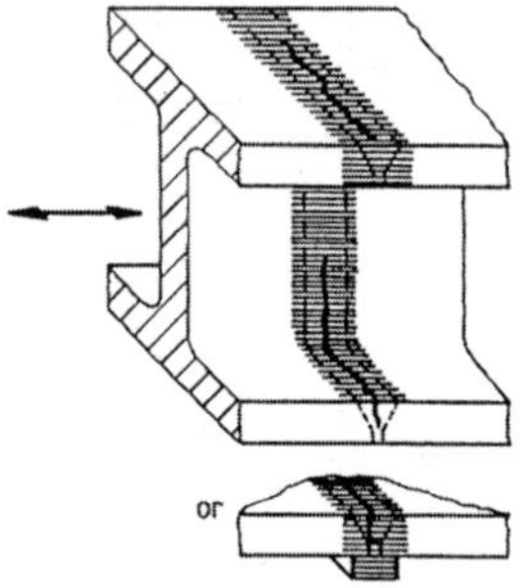 or
Steel tubes	At the toes of circumferential butt welds in tubes		Weld made from both sides with overfill dressed flush	Proved free of all flaws which are likely to degrade the joint below its stated classification (see **2.4.3**)		7.2	C	The significance of flaws should be determined with the aid of specialist advice and/or by the use of a fracture mechanics analysis. The NDT technique should be selected with a view to ensuring the detection of such significant flaws. Use of this type is not recommended.	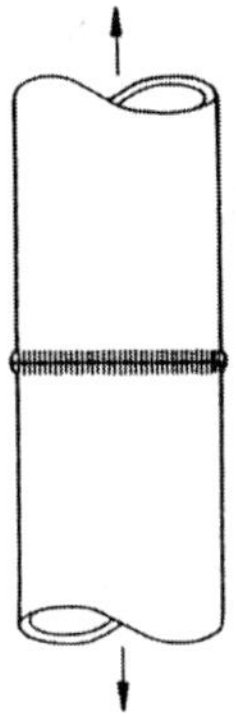
			Weld made from both sides			7.3	E		
			Weld made from one side on permanent backing strip			7.4	F		
			Weld made from one side with no backing strip	Weld to be checked for full penetration		7.5	F2		

Figure 7.130 Transverse butt welds in sections and tubes

Table 8. Classification of details: load carrying fillet and T-butt joints between plates in different planes

Product form	Location of potential crack initiation	Dimensional requirements	Manufacturing requirements	Special inspection requirements	Design stress area	Type number	Class	Notes	Sketch
Rolled steel struc-tural plates, sections and built-up members	At toe of weld joining two members end to end with third member transverse through joint	Full penetration butt weld with longitudinal axes in line	Any undercut should be ground smooth particularly on the corners of member X Weld ends should be ground flush with edge of member X		Cross section of member X	8.1	F	Weld metal failure will not govern with full penetration welds. Member Y can be regarded as one with a non-load-carrying weld (see joint types 5.2 and 5.5). In this instance the edge distance limitation applies.	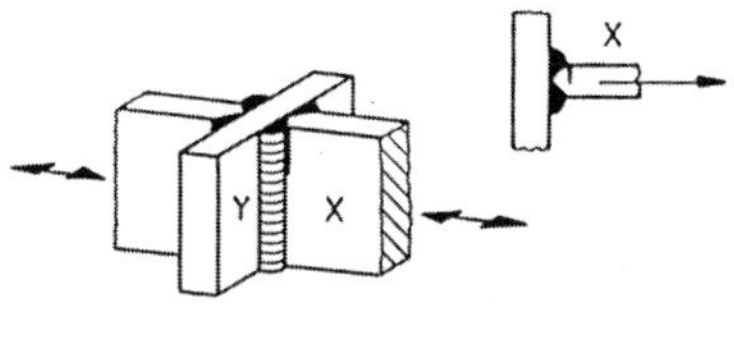 X Y X
		Partial penetration butt or fillet weld				8.2	F2	In this type of joint failure is likely to occur in the weld throat unless the weld is made sufficiently large. (See joint type 8.5.)	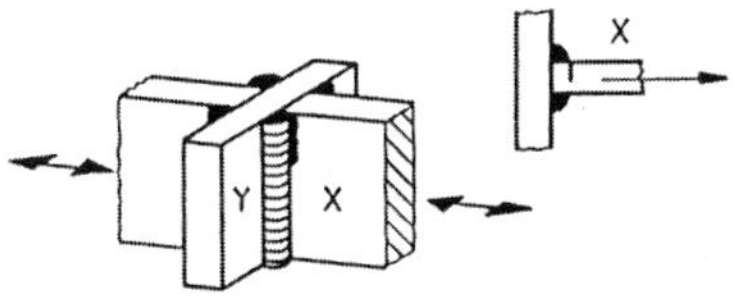 X Y X
	At toe of weld joining the end of one member to the surface of another	Full penetration butt weld				8.3	F	See joint type 8.1.	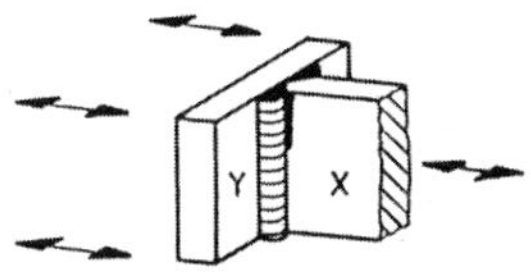 Y X
		Partial penetration butt or fillet weld				8.4	F2	See joint type 8.2.	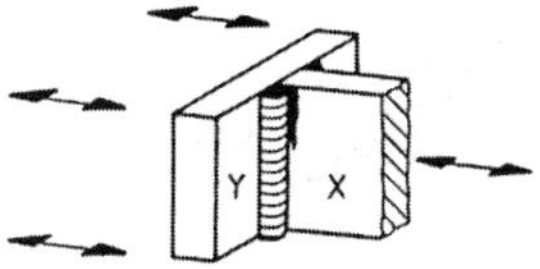 Y X

NOTE. Butt welded joints should be made with an additional reinforcing fillet so as to provide a similar toe profile to that which would exist in a fillet welded joint (see B.5.2.2).

Figure 7.131 Load carrying fillet and T-butt joints between plates in different planes

Corrosion Fatigue

S–N curves presented in fig. 7.124 and table 7.18 assume that the structure is in air. Earlier industry standards (HSE, BS 7608) recommended using these curves for joints in seawater if "adequately protected" from corrosion. If the joint was exposed to seawater and freely corroding, a reduction in life by a factor of two was recommended (BS 7608). There appears to be no endurance limit in a corrosive environment (Cozens and Kovach, 1984). The reduction factor depends on many variables including the fluid properties, temperature, frequency of loading, etc. (Kirk, 1968, Fink, 1960). The US Minerals Management Service has sponsored research on corrosion fatigue of offshore structural materials (Hartt, 1994, www.mms.gov/tarprojects/144.htm).

Figure 7.125 shows results of corrosion fatigue tests on mild steel. When compared with the data curve for base metal (curve B) a reduction factor of about 2.5 on life is indicated, and there appears to be no endurance limit. Industry standards are currently recommending a reduction on life of 2.5 for cathodically protected joints in seawater, and 3.0 for freely corroding joint in seawater (in addition to a modification to the high cycle curve to eliminate an endurance limit).

Although more applicable for risers than for floating structures, a reduction on life of a factor of 20 has been recommended for welded joints in a mildly sour (H_2S) environment (Buitrago and Zettlemeyer, 2002).

7.9 Construction and Installation

Construction and installation of FPSs may be as critical to their success as their functional design. Before picking a concept, the availability of yards to build it and means fort transporting it and installing it should be thought through carefully.

7.9.1 Fabrication

Fabrication falls roughly into two classes: the hull type and the truss type fabrication.

A Hull type fabrication is similar to shipyard construction with flat plate and orthogonal stiffeners and girders (framing) applied to the plate. This method is most efficiently employed when automatic cutting welding can be applied to apply the plates and framing. Spar platforms, FPSOs, TLP and semi-submersible hulls are typically constructed this way. Figure 7.132 shows a modern profile welding machine use for fabricating the Prince TLP in Brownsville, Texas. Figure 7.91 illustrates the fabrication of a subassembly for a spar hull.

Figure 7.133 shows fabrication yard assembly of the truss for a truss spar. Here there is a combination of stiffened plate construction and conventional tubular frame construction in the same yard. Fabrication yards are not typically efficient at the construction of stiffened plates, and visa versa shipyards are not efficient at tubular structures. Construction of this type often involves significant sub-contracting of the various pieces.

Figure 7.134 shows in a dry dock a mini-TLP under fabrication.

Figure 7.132 An automatic profile welder for stiffened plate fabrication (Keppel Fels)

Figure 7.133 Spar truss fabrication (Technip Offshore)

Figure 7.134 Seastar® Matterhorn platform in dry dock (SBM Atlantia)

While many offshore structures may appear to use shipyard construction practices, the specifications for offshore structure construction differ greatly from standard shipyard practice. The reason for this is that offshore structures are not subject to routine drydocking and inspection as ships are. Offshore structures are designed and built for long service away from maintenance facilities. Specifications for welding and fit up are designed to yield prolonged fatigue life. This often requires a 100% weld inspection and non-destructive evaluation (NDE). Standard shipyard practice is not sufficient. Shipyards experienced in offshore construction should always be surveyed and consulted early in the design phase to make sure that the FPS will be compatible with their facilities and schedule. Construction specialists should be consulted on the selection of a configuration.

7.9.2 Transportation

Transportation begins at the fabrication yard. A means for moving the structure away from the yard often is one of the most difficult engineering problems faced by project engineers. An offshore structure may weigh 10,000 tons or more. If constructed in a fabrication yard it must be moved, typically by skidding, into the water or onto a barge or heavy lift vessel that can transport it to its destination. Key technical constraints for offloading structures include quay load capacity, water depth at the key and exposure to waves. The fabrication contractor should be consulted early in the design phase of a project to ensure constructability and transportability of the design.

Transportation usually requires a barge or heavy lift vessel (HLV). Barges are commonly used to transport offshore jackets and topsides short distances offshore. Transocean transportation is more commonly performed with self propelled, heavy lift ships or open deck cargo ships. Tables 7.19 and 7.20 show particulars for typical cargo barges and launch barges operated by Heerema. Launch barges are specifically designed to launch jacket type structures, but they may also be used for general transportation.

Heavy Lift Vessels (HLVs) (table 7.21) are specially designed for transporting large floating structures. Figure 7.135 shows a spar hull being loaded by the Mighty Servant 1. HLVs are capable of submerging their deck up to 10 m below the water surface, so a floating structure may be offloaded. Barges are not typically submersible, so the offloading of a structure presents a challenge. In either case, the structure may be loaded onto the barge or HLV by skidding from the fabrication yard, provided the skidway and quay have adequate load

Table 7.19 Heerema cargo barges (www.heerema.com)

	Length	Breadth	Depth	Max. Loading Capacity
H-302	91.4 m	27.4 m	5.5 m	8,600 Mt
	300 ft	90 ft	18 ft	
H-404	122.0 m	36.6 m	7.6 m	21,700 Mt
	400 ft	120 ft	25 ft	

Table 7.20 Heerema launch/cargo barges (www.heerema.com)

	Length	Breadth	Depth	Max. loading Capacity	Max. launching Capacity
H-122	122.0 m	30.5 m	7.6 m	17,294 Mt	6,000 Mt
	(400 ft)	(100 ft)	(25 ft)		
H-401	122.0 m	36.6 m	8.0 m	22,441 Mt	20,000 Mt
(until late 2003)	(400 ft)	(120 ft)	(26 ft)		
MWB-403	122.0 m	31.9 m	7.6 m	16,322 Mt	6,800 Mt
	(400 ft)	(105 ft)	(25 ft)		
H-114	160.0 m	42.0 m	10.7 m	40,596 Mt	20,000 Mt
	(525 ft)	(138 ft)	(35 ft)		
H-541	165.0 m	42.0 m	10.7 m	41,718 Mt	20,500 Mt
	(541 ft)	(138 ft)	(35 ft)		
H-627	176.8 m	48.8 m	11.0 m	52,481 Mt	26,000 Mt
	(580 ft)	(160 ft)	(36 ft)		
H-851	260.0 m	63.0 m	15.0 m	130,514 Mt	60,000 Mt
	(853 ft)	(207 ft)	(49 ft)		

Table 7.21 Heavy transport vessel specifications

	Mighty Servant 1	Mighty Servant 3	Transhelf	Black Marlin	Blue Marlin	mv Tai An Kou/mv Kang Sheng Kou (sister vessels)
	Dockwise	Dockwise	Dockwise	Dockwise	Dockwise	NMA
Length o.a.	190.03 m	181.23 m	173.00 m	217.80 m	224.50 m	156.0 m
Length b.p.	174.70 m	168.93 m	162.00 m	206.50 m		145.0 m
Breadth	50.00 m	40.00 m	40.00 m	* 49.00 m	63.00 m	32.2 m
Depth	12.00 m	12.00 m	12.00 m	13.30 m	13.30 m	10.0 m
Draft sailing	8.77 m	9.06 m	8.80 m	10.11 m	10.08 m	7.5 m
Draft submerged	26.00 m	22.00 m	21.00 m	23.33 m	29.30 m	
Deadweight	40,190 t	27,720 t	34,030 t	57,021 tons	76,061 tons	18,000 tons
Deck space	50 × 150 m	40 × 140 m	40 × 132 m	42 × 178.20 m	63 × 178.2 m	32.2 × 126 m
Deck load	19–40 t/sq.m	19–25 t/sq.m	19–25 t/sq.m	27.5 t/sq.m		18 t/sq.m
Cargo hold	50 × 16 × 7.5 m	100 × 16 × 7.5 m	2 × 10 × 4 × 6 m			
Hatch	31 × 14.6 m	31 × 14.6 m				
Speed						
Service	14 kn	14 kn	14 kn	13 knots		
Maximum	15 kn	15 kn	15 kn	14.5 knots		15 kn
Range	59 days	44 days	44 days	65 days		

Figure 7.135 Offloading of spar hull for transport (Dockwise Shipping B.V., www.dockwise.com)

bearing capacity. The loading and transportation of a large offshore structure may take a considerable amount of engineering in its own right. During loading, the ballast of the vessel needs to be constantly adjusted to maintain the draft and trim compatible with the loading operation. The stiffness of the vessel and the cargo may be quite different resulting in deflections, which can lead to high stresses in either structure. A detailed stress analysis is typically required of both the structure and the vessel for the loadout and transport phases. For a long ocean transport, seafastenings must be able to withstand extreme storms. The loads on the structure during transport may be the controlling loads for the design and should be checked in the early design phases.

Transport of structures in US waters poses a special challenge because of The Merchant Marine Act of 1920 ("Jones Act", *46 USC 861*). This law requires, among other things, that passengers and cargo transported from one US port to another be carried on US owned and US flag vessels. An offshore platform located in the US Exclusive Economic Zone is considered a US port. Hence, transport of a deck to an existing platform, for example, could require a US flag vessel. Since most of the high capacity barges are non-US flag, this requirement places a burden on transportation of structures from US fabrication yards and in some cases may make foreign construction more attractive. Some foreign barge operators have been able to obtain waivers on a case by case basis, however this situation needs to be carefully considered when planning the execution of a project in a US facility. The list of web sites below includes one link, which lists US flag cargo barges.

Figure 7.136 Semi-submersible crane vessel (Heerema Offshore)

7.9.3 Derrick Barges

Offshore crane or derrick vessels are the workhorses for heavy lift. They are used for a variety of tasks from pipelaying, jacket installation and removal, mooring installation and deck installation. The following shows the largest crane vessels currently in operation. "DP" refers to Dynamically Positioned vessels which can operate in deep water. Moored vessels are typically limited to less than around 1000 ft., but they can typically operate in deep water with a pre-installed mooring spread.

1. McDermott DB-50 – 3300 MT Lift – DP
2. McDermott DB 101 (SSCV) – 3200 MT – Moored
3. McDermott DB 30 – 2800 MT – Moored (S.E. Asia)
4. McDermott Shear Leg SL5000 – 4500 MT – Moored
5. Heerema Balder (SSCV) – 7700 MT Dual Lift – DP
6. Heerema Heremod (SSCV) – 7000 MT Dual Lift – Moored
7. Heerema Thialf (SSCV) – 14,000 MT Dual Lift – DP
8. Saipem S-7000 (SSCV) – 14,000 ST Dual Lift – DP

Recently, the 5500 short ton Kerr–McGee Nansen spar deck was installed using the DB-50 and a smaller crane vessel, the SL5000, in tandem. Figure 7.136 shows a Semi-Submersible Crane Vessel (SSCV) setting a deck on a spar platform.

The following web sites provide additional information on marine transportation and construction vessels.

http://www.offshore-technology.com/ — Links to offshore suppliers and information on projects

http://www.offshore-vessels.net	Private web site devoted to Derrick Barges
http://www.jraymcdermott.com/	Barges, Derrick Barges, Pipelay
http://www.heerema.com/	Barges, Derrick Barges and Dockwise HLVs
http://www.saipem.it/	Derrick Barges, Pipelay Vessels
http://www.stoltoffshore.com/	Derrick Barges, Pipelay Vessels
http://www.nmamaritime.com/	HLVs
http://www.coltoncompany.com/	US Flag Barges, Drilling Rigs, Shipyards, etc.
http://www.texbrick.com/offshore/	Crane Vessels

References

AASHTO (1990). *Guide Specification for Fatigue Evaluation of Existing Steel Bridges*, American Association of State Highway and Transportation Officials.

Addy, P. D., Dickerson, J, Doble, P. A., Smith, G., and Young, R. J. (1994). "Gryphon A: the first purpose-built permanently moored FPSO in the North Sea", OTC 7179.

American Bureau of Shipping (1999). SafeHull Load Criteria for Tanker Structures, Commentary on Load Criteria.

American Bureau of Shipping (2003). "Guide for the Fatigue Assessment of Offshore Structures".

American Bureau of Shipping, Guidance Notes on Spectral-Based Fatigue Analysis for Floating Production, Storage and Offloading (FPSO) Systems, (latest edition).

American Bureau of Shipping, Guidance Notes on the Application of Synthetic Ropes for Offshore Mooring, (latest edition).

American Bureau of Shipping, Guide for Building and Classing Facilities on Offshore Installations, (latest edition).

American Bureau of Shipping, Guide for Building and Classing Floating Production Installations, (latest edition).

American Bureau of Shipping, Guide for Finite Element Analysis for Tanker Structures, (latest edition).

American Bureau of Shipping, Guide for Offshore Mooring Chain, (latest edition).

American Bureau of Shipping, Rules for Building and Classing Mobile Offshore Drilling Units, (latest edition).

American Bureau of Shipping, Rules for Building and Classing Single Point Mooring Systems, (latest edition).

American Bureau of Shipping, Rules for Building and Classing Steel Vessels, (latest edition).

American Institute of Steel Construction, AISC Manual of Steel Construction (latest edition).

American Petroleum Institute, API Recommended Practice 2I, Recommended Practice for In-Service Inspection of Mooring Hardware for Floating Drilling Units (latest edition).

American Petroleum Institute, API RP2A WSD, Recommend Practice for Planning, Designing and Constructing Fixed Offshore Platforms, Working Stress Design (latest edition).

American Petroleum Institute, API RP2RD, Recommended Practice for Design of Risers for Floating Production Systems and Tension Leg Platforms (latest edition).

American Petroleum Institute, API RP2SK, Recommended Practice for the Design and Analysis of Stationkeeping Systems for Floating Structures (latest edition).

American Petroleum Institute, API RP2T, Recommended Practice for Planning, Designing and Constructing Tension Leg Platforms (latest edition).

American Petroleum Institute, Bulletin 2U, Bulletin on Stability Design of Cylindrical Shells, (latest edition).

American Petroleum Institute, Bulletin 2V, Bulletin on Design of Flat Plate Structures, (latest edition).

American Petroleum Institute, Spec 2B, Specification for Fabricated Structural Steel Pipe, (latest edition).

American Petroleum Institute, Spec 2F, Specification for Mooring Chain, (latest edition).

American Petroleum Institute, Spec 9A, Specification for Wire Rope, (latest edition).

American Society of Mechanical Engineers, ASME Boiler and Pressure Vessel Code, Section 8, Div. II, (latest edition).

Anonymous, (2002). "Taking the EPIC Challenge", UPSTREAM magazine, Aug. 15.

ANSI/AWS D1.1.92, Structural Welding Code – Steel, American Welding Society Specification.

ASTM (1996). E 1045 – Standard Practices for Cycle, Counting in Fatigue Analysis.

Bangs, A. S., Miettinen, J. A., Mikkola, T. P. J., Silvola, I., and Beattie, S. M. (2002). "Design of the Truss Spars for the Nansen/Boomvang Field Development", Proceedings, Offshore Technology Conference, Paper 14090, Houston.

Barusco, P. (2002). "The Sinking of the P-36", Proceedings, 21st International Conference on Offshore Mechanics and Arctic Engineering, OMAE2002-28606, Oslo, Norway.

Bates, John B., Kan, Wan C., Allegra Allen P., and Yu, C. Alan (2001). "Dry Tree and Drilling Riser System for Hoover DDCV", Proceedings, Offshore Technology Conference, OTC 13084.

Bea, R., and Schulte-Strathaus, R. (1997). "Fatigue Classification of Critical Structural Details in Tankers", Ship Structure Committee Report SSC-395.

Bergan, Pal G. and Lotsberg, Inge (2004). "Advances in Fatigue Assessment of FPSOs", Proceedings, OMAE-FPSO'02-0012, Houston, Texas.

Bergan, Pål G., Lotsberg, Inge, Fricke, Wolfgang, Francois, Michel and Pisarski, Henryk (2002). "Overview of the FPSO – Fatigue Capacity JIP", Proceedings, 21st International Conference on Offshore Mechanics and Arctic Engineering, OMAE2002–28538, June 23–28, Oslo, Norway.

Berner, P. C., Gendron, K., Baugus, V. E., and Young, R. (1997). "Neptune Project: Production Riser System Design and Installation," Proceedings, Offshore Technology Conference, OTC 8386.

Bindingsbø, Arne Ulrik, and Bjørset, Arve (2002). "Deep Draft Semi-submersible", Proceedings of OMAE.02, Oslo, Norway, OMAE2002–28369.

Bishop, N. W. M., and Sherratt, F. (1989). "Fatigue Life Prediction from Power Spectral Density Data", Part 1, Traditional Approaches and Part 2, Recent Developments, Environmental Engineering, Vol 2, Nos. 1 and 2.

Blevins, R. (2003). *Applied Fluid Dynamics Handbook*. Krieger Publishing Company, Melbourne, Florida.

British Standard Institute, 1993. "Code of Practice for Fatigue Assessment of Steel Structures", BS 7608.

Buitrago, Jaime and Zettlemoyer, N. (2002). "Experimental Fatigue of Deepwater Risers in Mild Sour Service", Deep Offshore Technology Conference. Paper 25-01, November, New Orleans.

Burleson, C.W (1999), *Deep Challenge! The true epic story of our quest for energy beneath the sea.* Gulf Publishing Company, Houston, TX.

Chedzoy, Claire and Lim, Frank (2003). "Design Challenges of Deepwater Dry Tree Riser Systems for Different Vessel Types", Paper 2003-JSC-335, 2H Offshore Engineering Ltd., http://www.2hoffshore.com/papers/docs/2003.pdf.

Chen, Xiaohong, Zhang, Jun, Liagre, Pierre-Yves F., Niedzwecki, John M., and Teigen, Per (2002). "Coupled Dynamic Analysis of a Mini TLP: Comparison with Measurements", Proceedings 21st International Conference on Offshore Mechanics and Arctic Engineering, Oslo, Norway OMAE2002–28536.

Clauss, Günther F. and Birk, Lothar (1998). "Design of Optimum Offshore Structures Based on Long Term Wave Statistics", Proceedings of OMAE-98, Lisbon, Portugal OMAE 98-0521.

Congress, Trondheim, Norway, (1997). (Ship Motions and Loads Computer Programs).

Comstock, J. P. (editor), 1979. *Principles of Naval Architecture*. Society of Naval Architecture. New York.

Cozens, K. O., and Kovach, P. J. (1984). "Seawater corrosion fatigue of 2 1/4 Cr-1 Mo and 4130 steels for marine riser application", OTC 5208.

Datta, Indranath, Prislin, Igor, Halkyard, John E., Greiner, William L., Bhat, Shankar, Perryman, Steve and Beynet, Pierre A. (1999). "Comparison of Truss Spar Model Test Results with Numerical Predictions", Proceedings of OMAE '99: 18th Int. Conf. on Offshore Mechanics and Arctic Engineering, St. John's, Newfoundland, Canada OMAE99/OFT-4231.

Davison, C. (1997). "Development of a Gas Field Offshore Thailand Using an FPSO", Proceedings, Offshore Technology Conference, OTC 8377 (example of FPSO Hull Conversion).

Demirbilek, Z. (ed.) (1989). Tension Leg Platforms: A State of the Art Review, ASCE Task Group on Compliant Offshore Structures, 89–324.

Department of Energy (U.K.) (1990). Offshore Installation: Guidance on Design, Construction, and Certification, Fourth Edition, London, HMSO (fatigue curves).

Det Norske Veritas (1993). "WAVESHIP, 6.1. Wave Loads on Slender Vessels – Users Manual", SESAM, Hovik.

Det Norske Veritas (2000). "Fatigue Strength Analysis of Offshore Steel Structures".

Det Norske Veritas, Classification Notes 30.1, Buckling Strength Analysis (latest edition).

Det Norske Veritas, OS C102 Structural Design of Offshore Ships (latest edition).

Det Norske Veritas, OS-C101 Design of Offshore Steel Structures (latest edition).

Det Norske Veritas, OS-C103 Structural Design of Column Stabilised Units (latest edition).

Det Norske Veritas, OS-C104 Structural Design of Self Elevating Units (latest edition).

Det Norske Veritas, OS-C105 Structural Design of TLPs by the LRFD Method (latest edition).

Det Norske Veritas, OS-C106 tructural Design of Deep Draught Floating Units (latest edition).

Det Norske Veritas, OS-C201 Structural Design of Offshore Units by Working Stress Method (latest edition).

Det Norske Veritas, OS-C301 Stability and Watertight Integrity (latest edition).

Det Norske Veritas, OS-C401 Fabrication and Testing of Offshore Structures (latest edition).

Det Norske Veritas, Rules for Mobile Offshore Units, Part 3, Chapter 1 (latest edition).

DNV RP-C203.

Dong, P. (2004). "The Mesh-Insensitive Structural Stress and Master S–N Curve Method for Ship Structures". Proceedings, OMAE-FPSO'02-0021, Houston, Texas.

Donley, M. G., and Spanos, P. D. (1991). "Stochastic response of a tension leg platform to viscous drift force", Journal of Offshore Mechanics and Arctic Engineering, ASME, Vol. 113, pp. 148–155.

Efthymiou, M. (1988). "Development of SCF Formulae and Generalized Influence Functions for use in Fatigue Analysis". Recent Developments in Tubular Joint.

Engebretsen, Knut, Hannus, Henrik, and Husvik, Jorgen (2002). "TLP Operational Experience Enhancing Margins In-field as well as New Design", Proceedings, Offshore Technology Conference, Paper 14179, Houston.

Faltinsen, O. M., (1990). Sea Loads on Ships and Offshore Structures, Cambridge University Press, Cambridge.

Filho, Pedro Jose Barusco (2002). "The Sinking of P-36", Proceedings, 21st International Conference on Offshore Mechanics and Arctic Engineering, Oslo, Norway.

Fink, F. W. (1960). "Corrosion of Metals in Seawater," U.S. Dept. Interior Office of Saline Water, Report No. 46.

Finn, Lyle and Maher, J. (2003). "The Cell Spar for Marginal Field Development.". Deep Offshore Technology Conference, Marseille, France.

Finn, L. D., Maher, J. V. and Gupta, H. (2003). "The Cell Spar and Vortex Induced Vibrations". Proceedings, Offshore Technology Conference, OTC 15244.

Fricke, W. (2001). "Recommended Hot Spot Analysis Procedure for Structural Details of FPSO's and Ships Based on Round-Robin FE Analyses", Proceedings, ISOPE, Stavanger, June.

Glanville, Roger S. and Vardeman, R. Don (1999). "The Neptune Spar – Performance Over the First Two Years of Production", Proceedings, Offshore Technology Conference, OTC 11073.

Glanville, R. S., Halkyard J. E., Davies, R. L., Steen, A., and Frimm, F. (1997). " Neptune Project: Spar History and Design Considerations", Proceedings, Offshore Technology Conference, OTC 8382.

Glanville, R., Paulling, J. R. Jr., Halkyard, J., and Lehtinen, T. (1991). "Analysis of the Spar Floating Drilling, Production and Storage Structure", Proceedings, Offshore Technology Conference, OTC 6701.

Goodwin, P., and Ahilan, R. V. (2000). "Integrated Mooring and Riser Design: Target Reliabilities and Safety Facors", Proceedings of ETCE/OMAE2000 Joint Conference Energy for the New Millenium, OMAE 2000–6131, February 14–17, 2000, New Orleans, LA.

Gurney, T. R. (1976). "Fatigue Design Rules for Welded Steel Joints". The Welding Institute Research Bulletin. Volume 17, Number 5.

Halkyard, J. (2001). "Evolution of the Spar Floating Drilling, Production and Storage System", Proceedings, SNAME (TEXAS) Annual Meeting, Offshore Technology: Evolution and Innovation, Houston, Texas.

Halkyard, J., Chao, J., Abbott, P., Dagleish, J., Banon, H., and Thiagarajan, K. (2002). "A Deep Draft Semi-submersible with a Retractable Heave Plate", Proceedings, Offshore Technology Conference, Paper 14304, Houston.

Hamdan, F. H., Bassam, A. B., and White, R. (2002). "Margins of Safety in FPSO Hull Strength", Proceedings, 21st International Conference on Offshore Mechanics and Artic Engineering, OMAE2002–28012, June 23–28, Oslo, Norway.

Hansen, P. F., and Winterstein, S. R. (1995). "Fatigue damage in the side shells of ships", *Marine Structures*, 8, 631655.

Haslum, H. A., and Faltinsen, O. M. (1999). "Alternative Shape of Spar Platforms for Use in Hostile Areas", OTC, OTC 10953.

Hartt, W. H. (1994). "Growth Rate of Short Fatique Cracks as Relevant to Higher Strength Steels for Offshore Structures", Final Report, Florida Atlantic University College of Engineering, Department of Ocean Engineering, Boca Raton, Florida.

Health and Safety Executive (UK) (1995). "Offshore Installations: Guidance on design, construction, and certification".

Hewlett, C., Shaw, G., Glorstad, T., and Tennoy, O. (1987). "Petrojarl I – Subsea systems installation experience", Offshore Europe'87: SPE of AIME conference; Aberdeen, UK; Technical Paper SPE 16569/1.

Heyburn, R., and Riker, D. (1994). "Effect of High Strength Steels on Strength Considerations of Design and Construction Details of Ships", Ship Structure Committee Report SSC-374.

Hughes, O. F. (1988). Ship Structural Design, SNAME.

Hung, S., and Mangiavacchi, A. (1990). "Semi-submersible Sizing and Modification Strategies", OTC, OTC 6275.

ISSC (2003), Proceedings of the 15th International Ship and Offshore Structures Congress, Two Volumes, San Diego, CA, Elsevier, UK.

Jha, A. K. (1997c). "Nonlinear stochastic models for loads and responses of offshore structures and vessels", Ph.D. thesis, Stanford University.

Kaminski, M.L., van der Cammen, J. and Francois, M. (2004). "Is Uncertainty of Wave Induced Fatigue Loading on FPSO's Uncertain?" Proceedings, OMAE-FPSO 04-0011, Houston, Texas.

Karadeniz H., (1990). "Fatigue analysis of offshore structures under non-narrow banded stress processes", Proceedings of the First European Offshore Mechanics Symposium, pp. 221–228, Trondheim, Norway.

Karadeniz, H., Vrouwenvelder, A. C. W. M., and Shi, C. (1998). "Spectral Fatigue Analysis of a Tensioned Riser Compliant Tower", Proceedings, 17th International Conference on Offshore Mechanics and Arctic Engineering, Lisbon, OMAE98–1203.

Kaster, F., Falkenberg, E., and Waclawek, I. (1997). "DICAS – A New Mooring Concept for FPSO's", OTC 8439.

Kibbee, Stephen E., and Snell, David C. (2002). "New Directions in TLP Technology", OTC, Paper 14175.

Kihl, D. P. (1999). "Fatigue Strength and Behavior of Ship Structural Details", NSWCCD-65-TR-1998/23.

Kirk, W. W., Covert, R. A., and May, T. P. (1968). *Metals Engineering Quarterly*, November.

Kocaman, A., Verdin, E., Toups, J. (1997). "Neptune Project: Spar Hull, Mooring and Topsides Installation," Paper OTC 8385, Houston.

Koon, J. R., Heijermans, Bart, and Wybro, P. G. (2002). "Development of the Prince Field", OTC 14173.

Kuuri, J., Lehtinen, T. J., and Miettinen, J. (1997). "Neptune Project: Spar Hull and Mooring System Design and Fabrication," Proceedings, Offshore Technology Conference, OTC 8384.

Landau, L. D., and Lifschitz, E. M. (1976). *Mechanics*, 3rd ed., Pergamon Press, Oxford, England.

Le Guennec, S., Hough, C., Chao, J., Abbott, P., Banon, H., Beynet, P., and Dagleish, J. (2002). "A Self-Installing Deep Draft Dry Tree Floater", Proceedings, Deep Offshore Technology Conference, New Orleans.

Leite, A. J. P., Aranha, J. A. P., Umeda, C., and de Conti, M. B. (1998). "Current forces in tankers and bifurcation of equilibrium of turret systems: hydrodynamic model and experiments", *Applied Ocean Research*, 20:145–156.

Leverette, S., Rijken, O., Spillane, M., and Williams, N., (2004) "Integrated Global Performance of the Matterhorn SeaStar TLP", Proceedings, Offshore Technology Conference, OTC 16609.

Lim, E. F. H., and Ronalds, B. F. (2000). "Evolution of the Production Semi-submersible", SPE Annual Technical Conference, Dallas SPE 63036.

Lin, W. M., and Yue, D. K. P. (1990). Numerical Solutions for Large-Amplitude Ship Motions in the Time Domain. In: 18th Symposium on Naval Hydrodynamics. University of Michigan, Ann Arbor, MI, USA (LAMPS Program – Ship Loads).

Lin, W. M., Zhang, S., Weems K., and Yue, D. K. P. (1999). "A Mixed Source Formulation for Nonlinear Ship-Motion and Wave-Load Simulations", Presented at the Seventh International Conference on Numerical Ship Hydrodynamics, Nantes, France.

Liu, Michael Y. H., Wang, Jim J. and Lu, Roger R. (2003). "Spar Topsides-to-Hull Connection Fatigue – Time Domain vs. Frequency Domain". Proceedings, ISOPE-03, Paper 2003-YSC-03, Honolulu, Hawaii.

Lloyd's Register of Shipping, Rules and Regulations for the Classification of a Floating Offshore Installation at a Fixed Location, (latest edition).

Maddox, S. (2001) "Recommended Design S-N Curves for Fatigue Assessment of FPSOs", Proceedings, ISOPE, Stavanger, June.

Maher, J. V., Prislin, I., Chao, J. C., Halkyard, J. E., and Finn, L. D. (2001). "Floatover Deck Installation for Spars", Proceedings, Offshore Technology Conference, OTC 12971.

Martindale, S. G., and Wirsching, P. H. (1983). "Reliability-based progressive fatigue collapse", ASCE Journal of Structural Engineering, Vol. 109, No. 8.

Matsuishi, M., and Endo, T. (1974) "Damage Evaluation of Metals for Random or Varying Loading," Proc. 1974 Symp. on Mech. Behavior of Materials, Vol. 1, The Soc. of Materials Sci., Japan, pp. 371–380.

Matsuishi, M., and Endo, T. Mar., (1968) "Fatigue of Metals Subjected to Varying Stress," Proceedings Japan Society of Mechanical Engineers, Fukuoka, Japan.

Matten, R. B., Provost, M. J., Pastor, S., and Young, W. S. (2002). "Typhoon SeaStar TLP", OTC, Paper 14123, Houston.

Minerals Management Service (2002). Deepwater Gulf of Mexico 2002: America's Expanding Frontier, OCS Report MMS 2002–021.

Moyer, M. C., Barry, M. D., and Tears, N. C. (2001). "Hoover-Diana Deepwater Drilling and Completions", Proceedings, Offshore Technology Conference, OTC 13081.

Natl. Research Council of Canada (1987). "A Wind Simulation System for the Ocean Ranger Hydrodynamic Model Study", OTC 5361.

Nestegýrd, A., and Krokstad, J. R. (1999). "JIP-DEEPER: Deepwater Analysis Tools", OTC, OTC010811, Houston.

Neto, T. G. (2000). "FPSO Conversion Design", ABS FPSO Seminar, Sept. 13–14, Houston, Texas.

Newman, J. N. (1963). "The Motions of a Spar Buoy in Regular Waves", U. S. Dept. of the Navy, David Taylor Model Basin, Report 1499.

Nordstrom, C. D., Grant, B., Lacey, P. B., and Hee, D. D. (2002). "Impact of FPSO Heading on Fatigue Design in Non-Collinear Environments", Proceedings, OMAE Conference, Oslo.

Norwegian Petroleum Directorate (1996). "Acts, Regulations and Provisions for Petroleum Activities".

OCIMF (1994): Prediction of Wind and current loads on VLCCs, Oil Companies International Marine Forum, Witherby and Co, Ltd, London.

O'Neill, L. A., Fakas, E., Ronalds, B. F., and Christiansen, P. E. (2000). "History, Trends and Evolution of Float-Over Deck Installation in Open Waters", Proceedings, SPE Annual Technical Conference and Exhibition, SPE 63037, Dallas, Texas.

Parker, W. J., Ghosh, S., and Praught, M. W. (1997). "Neptune Project: Regulatory and Classification Issues and Requirements", Proceedings, Offshore Technology Conference, OTC 8383.

Paulling, J. R. (1995). "MULTISIM Time Domain Platform Motion Simulation For Floating Platform Consisting of Multiple Interconnected Modules", Theoretical Manual, Third Edition, J. Randolph Paulling, Inc., Geyserville, California, Nov. 2, 1995.

Paulling, J. R., and Shin, Y. S. (1985). "On The Simulation of Large-Amplitude Motions of Floating Ocean Structures", Proceedings of International Conference on Ocean Space Utilization, Nihon University, Tokyo (Pub. By Springer Verlag, Tokyo 1985) Vol. I, pp. 235–243.

Portella, Ricardo B., Mendes, Cristiane (2003). "DICAS Mooring System: Practical Design Experience to Dismystify the Concept", OTC 14309, Houston.

Prislin, I., Halkyard, J., DeBord, F. Jr., Collins, J. I., and Lewis, J. M. (1999). "Full-Scale Measurements of the Oryx Neptune Production Spar Platform Performance", Proceedings, Offshore Technology Conference, OTC 10951.

Rijkens, Frederick, Allen, Marcus, Hassold, Thomas (2003). "Overview of the Canyon Express project, business challenges and 'industry firsts'", Proceedings, Offshore Technology Conference, Paper 15093, Houston.

Roark, Raymond J., Budynas, Richard G., and Young, Warren C. (1989). Roark's Formulas for Stress and Strain, sixth edition, McGraw Hill Book Company, NY.

Rogers, W. S., Bloomfield, J. S. C. (1995). "Engineering Requirements for the Classification and Certification of Floating Production, Storage and Offloading Units", OTC, OTC 7836.

Ronalds, B. F., and Lim, E. F. H. (2001). "Deepwater Production with Surface Trees: Trends in Facilities and Risers", SPE Asia Pacific Oil and Gas Conference, Jakarta, SPE 68761.

Ronalds, B. F., and Lim, E. F. H. (1999). "FPSO Trends", SPE Annual Technical Conference, Dallas, SPE 56708.

SAE AE-10 – "Fatigue Design Handbook,".

Salvesen, N, Tuck, E. O., and Faltinsen, O. (1970). "Ship Motions and Sea Loads", Transactions of the Society of Naval Architects and Marine Engineers, Vol. 78, 1970.

Sarpkaya, T. (1981). *Mechanics of Wave Force on Offshore Platforms*. Van Nostrand Reinhold Co., New York, New York.

Sherf, I. and Tuestad, T. (1987). "Fatigue Design of the Oseberg Jacket Structure". Proceedings, OMAE, Houston, Texas.

Shin, Y. S., Chung, J. S., Lin, W. M., Zhang, S., and Engle, A. (1997). "Dynamic Loadings for Structural Analysis of Fine Form Container Ship Based of a Nonlinear Large Amplitude Motions and Loads Program", Transactions, Society of Naval Architects and Marine Engineers.

Ship Structures Committee (1997). "Assessment of Reliability of Ship Structures", SSC-398.

Ship Structures Committee (2002). "Supplemental Commercial Design Guidance for Fatigue", SSC-419.

Sircar, S., Kleinhans J. W., and Prasad, J. (1993). "Impact of Coupled Analysis on Global Performance of Deep Water TLPs", Proceedings, Offshore Technology Conference, Paper 7145.

Skeels, H. B., Johnson, B., Koleilat, B., and Frishmuth, R. E. (1997). "Development of a Novel Tieback Connector with High Mechanical Strength Capacity Spar Floating Production Facilities and Their Production Risers", Proceedings, Offshore Technology Conference, OTC 8574.

Society of Naval Architects and Marine Engineers, (1988). Principles of Naval Architecture, Volume 1, Stability and Strength.

Standing, R. G., Eichaker, R., Lawea, B. and Corr, R. B. (2002). "Benefits of Applying Response-Based Design Methods to Deepwater FPSOs". Proceedings, Offshore Technology Conference, OTC 14232.

Stein, W. R., and Gardiner, N. H. (1997). "Multidisciplined Team Solutions Address Deepwater Challenges in the Gulf of Mexico", Proceedings, Offshore Technology Conference, OTC 8379.

Stiff, J. J., Smith, D. W., Casey, N. F., 1996, "Fatigue of Mooring Chain in Air and Water – Results and Analysis," OTC Paper No. 8147, Proceedings of the 25th Offshore Technology Conference, Houston, Texas, USA, May.

Taggart, R. ed. (1980). Ship Design and Construction, SNAME.

Taylor, R. and Palo, P. (2000). "U.S. Mobile Base Technology Report", 23rd Annual United States Japan Natural Resource (UJNR) Marine Facilities Panel Meeting, Tokyo, Japan.

Technology, OTJ'88, London.

Teigen, P., Karunakaran, D. (1998). "A Consistent Approach to Riser Fatigue Analysis including Diffraction", Proceedings, 17th International Conference on Offshore Mechanics and Arctic Engineering, Lisbon, 1998, OMAE98–1394.

Thomas, J. M. (1984). "Analysis of the Environmental Conditions and the Structural Failure Which Led to the Alexander L. Kielland Accident," Failure Analysis Associates, Inc., Report FaAA-84-3-4.

University of California (1993). "Human and Organizational Error in Operations of Marine Systems: Occidental Piper Alpha and High-Pressure Gas Systems on Offshore Platforms", OTC Paper 7121.

University of Iowa, Class Notes, ME515, www.me.iastate/me515_commer/Lecture/lecture25.pdf.

Vardeman, Don, Richardson, Steve, McCandless, C. R. (1997). "Neptune Project: Overview and Project Management", Proceedings, Offshore Technology Conference, OTC 8381.

Vughts, J. H., and Kinra, R. K. (1976). "Probabilistic Fatigue Analysis of Fixed Offshore Structures", OTC 2608.

Wang, J., Berg, S., Luo, Y. H., Sablok, A., and Finn, L. (2001). "Structural Design of the Truss Spar – An Overview", Proc. Of 11th ISOPE Conference, Stavanger, Norway.

Wang, J., Luo, Y. H., and Lu, R. (2002). "Truss Spar Structural Design for West Africa Environment", Proceedings 21st International Conference on Offshore Mechanics and Arctic Engineering, Oslo, Norway OMAE2002–28245.

Wanvik, Leiv and Koos, John (2000). "Two Tier Well Riser Top Tensioning System". Proceedings, Offshore Technology Conference, OTC-11904, Houston.

Watanabe, I., and Soares, C. Guedes (1999). "Comparative Study on the Time-Domain Analysis of Nonlinear Ship Motions and Loads", Marine Structures, Vol. 12, pp. 153–170.

Weems, K., Lin, W. M., Zhang, S., and Treakle, T. (2000). "Time Domain Predictions for Motions and Loads of Ships and Marine Structures in Large Seas Using a Mixed-Singularity Formulation", Presented at the Fourth Osaka Colloqium on Seakeaking Performance of Ships (OC2000), Osaka, Japan.

Westwood, John (2003). "The Prospects for the Deepwater Market: The New 4th edition of The World Deepwater Report 2003–2007", Proceedings, Deep Ocean Technology Conference.

Wijngaarden, A. M. van and Heemskerk, R. J. (2000). "New Stability Approach for Offshore Construction Vessels", OTC 11952, Houston.

Wirsching, P. H., et al, (1981). "Probability-based fatigue design criteria for offshore structures", Final Report, American Petroleum Institute, PRAC Project 81-15.